To the Student

As authors our highest goal has been to write a solid mathematics textbook that will truly help you succeed in your mathematics course. After years of refining our textbooks, we have now developed a feature that we think will help ensure your success in mathematics. It is called the *How Am I Doing? Guide to Math Success.*

The *How Am I Doing? Guide to Math Success* shows how you can effectively use this textbook to succeed in your mathematics course. This clear path for you to follow is based upon how our successful students have utilized the textbook in the past. Here is how it works:

EXAMPLES and PRACTICE PROBLEMS: When you study an Example, you should immediately do the Practice Problem that follows to make sure you understand each step in solving a particular problem. The worked-out solution to every Practice Problem can be found in the back of the text starting at page SP-1 so you can check your work and receive immediate guidance in case you need to review.

EXERCISE SETS—Practice, Practice, Practice: You learn math by *doing* math. The best way to learn math is to *practice, practice, practice.* The exercise sets provide the opportunity for this practice. Be sure that you complete every exercise your instructor assigns as homework. In addition, check your answers to the odd-numbered exercises in the back of the text to see whether you have correctly solved each problem.

HOW AM I DOING? MIDCHAPTER REVIEW: This feature allows you to check if you understand the important concepts covered to that point in a particular chapter. Many students find that halfway through a chapter is the point of greatest need because so many different types of problems have been covered. This review covers each of the types of problems from the first half of the chapter. Do these problems and check your answers at the back of the text. If you need to review any of these problems, simply refer back to the section and objective indicated next to the answer. Before you go further into the chapter, it is important to understand what has been learned so far.

HOW AM I DOING? CHAPTER TEST: This test (found at the end of every chapter) provides you with an excellent opportunity to both practice and review for any test you will take in class. Take this test to see how much of the chapter you have mastered. By checking your answers, you can once again refer back to the section and the objective of any exercise you want to review further. This allows you to see at once what has been learned and what still needs more study as you prepare for your test or exam.

HOW AM I DOING? CHAPTER TEST PREP VIDEO CD: If you need to review any of the exercises from the *How Am I Doing? Chapter Test,* this video CD found at the back of the text provides worked-out solutions to each of the exercises. Simply insert the CD into a computer and watch a math instructor solve each of the Chapter Test exercises in detail. By reviewing these problems, you can study through any points of difficulty and better prepare yourself for your upcoming test or exam.

These steps provide a clear path you can follow in order to successfully complete your math course. More important, the *How Am I Doing? Guide to Math Success* is a tool to help you achieve an understanding of mathematics. We encourage you to take advantage of this new feature.

John Tobey
Jeffrey Slater
North Shore Community College

MORE TOOLS FOR SUCCESS

In addition to the *How Am I Doing? Guide to Math Success,* your Tobey/Slater textbook is filled with other tools and features to help you succeed in your mathematics course. These include:

Chapter Organizers

The key concepts and mathematical procedures covered in each chapter are reviewed at the end of the chapter in a unique Chapter Organizer. This device not only lists the key concepts and methods but provides a completely worked-out example for each type of problem. The Chapter Organizer should be used in conjunction with the *How Am I Doing? Chapter Test* and the *How Am I Doing? Chapter Test Prep Video CD* as a study aid to help you prepare for tests.

Developing Your Study Skills

These notes appear throughout the text to provide you with suggestions and techniques for improving your study skills and succeeding in your math course.

Blueprint for Problem Solving

The Mathematics Blueprint for Problem Solving provides you with a consistent outline to organize your approach to problem solving. You will not need to use the blueprint to solve every problem, but it is available when you are faced with a problem with which you are not familiar, or when you are trying to figure out where to begin solving a problem.

RESOURCES FOR SUCCESS

In addition to the textbook, Prentice Hall offers a wide range of materials to help you succeed in your mathematics course. These include:

Student Study Pack

Includes the *Student Solutions Manual* (fully worked-out solutions to odd-numbered exercises), access to the *Prentice Hall Tutor Center,* and the CD *Lecture Series Videos* that accompany the text. The *Student Study Pack* is available at no charge when packaged with a new textbook.

MyMathLab

MyMathLab offers the entire textbook online with links to video clips and practice exercises in addition to tutorial exercises, homework, and tests. *MyMathLab* also offers a personalized Study Plan for each student based on student test results. The Study Plan links directly to unlimited tutorial exercises for the areas you need to study and retest so you can practice until you have mastered the skills and concepts. *MyMathLab* is available at no charge when packaged with a new textbook.

This book is dedicated to John Tobey III, Marcia Tobey Salzman, and Melissa Tobey LaBelle. They are three college graduates who would make any parent proud. They have each worked hard to achieve a master of arts degree and have proved that they all can do two things at once and do well at both.

Contents

Intermediate
Algebra

FIFTH EDITION

John Tobey

North Shore Community College
Danvers, Massachusetts

Jeffrey Slater

North Shore Community College
Danvers, Massachusetts

PEARSON
Prentice
Hall

Upper Saddle River, NJ 07458

Library of Congress Cataloging-in-Publication Data

Tobey, John

Intermediate algebra / John Tobey, Jeffrey Slater. —5th ed.
p. cm.
Includes indexes.
ISBN 0-13-149078-8
  1.   Algebra–Textbooks. I. Slater, Jeffrey, II. Title
QA154.3.T64 2006
512.9—dc22   2004058659

Senior Acquisitions Editor: *Paul Murphy*
Project Manager: *Dawn Nuttall*
Editor in Chief: *Christine Hoag*
Production Editor: *Lynn Savino Wendel*
Senior Managing Editor: *Linda Mihatov Behrens*
Executive Managing Editor: *Kathleen Schiaparelli*
Assistant Manufacturing Manager/Buyer: *Michael Bell*
Executive Marketing Manager: *Eilish Collins Main*
Development Editor: *Tony Palermino*
Marketing Assistant: *Rebecca Alimena*
Editor in Chief, Development: *Carol Trueheart*
Media Project Manager, Developmental Math: *Audra J. Walsh*
Managing Editor, Digital Supplements: *Nicole M. Jackson*
Media Production Editor: *Zachary Hubert*
Print Supplements Editor: *Christina Simoneau*
Art Director: *Jonathan Boylan*
Interior and Cover Designer: *Susan Anderson-Smith*
Editorial Assistant: *Mary Burket*
Art Editor: *Thomas Benfatti*
Director of Creative Services: *Paul Belfanti*
Director, Image Resource Center: *Melinda Reo*
Manager, Rights and Permissions: *Zina Arabia*
Manager, Visual Research: *Beth Brenzel*
Manager, Cover Visual Research & Permissions: *Karen Sanatar*
Image Permission Coordinator: *Joanne Dipple*
Photo Researcher: *Melinda Alexander*
Art Studio: *Scientific Illustrators*
               Laserwords
Compositor: *Interactive Composition Corporation*
Cover Photo Credits: © Alamy Images

Photo credits appear on page P-1, which constitutes a continuation of the copyright page.

Pearson Education Ltd., London
Pearson Education Australia Pty. Limited,
Sydney Pearson Education Singapore Pte. Ltd.
Pearson Education North Asia, Ltd, Hong Kong

Pearson Education Canada, Ltd., Toronto
Pearson Educación de Mexico, S.A., de C.V.
Pearson Education, Japan, Tokyo
Pearson Education Malaysia, Pte. Ltd.

CHAPTER 8

Quadratic Equations and Inequalities   421

CHAPTER 9

The Conic Sections   485

CHAPTER 10

Additional Properties of Functions   539

EXAMPLE 6   Solve $\dfrac{x}{5} + \dfrac{1}{2} = \dfrac{4}{5} + \dfrac{x}{2}$.

Solution $10\left(\dfrac{x}{5} + \dfrac{1}{2}\right) = 10\left(\dfrac{4}{5} + \dfrac{x}{2}\right)$ Multiply each term by the LCD 10.

$10\left(\dfrac{x}{5}\right) + 10\left(\dfrac{1}{2}\right) = 10\left(\dfrac{4}{5}\right) + 10\left(\dfrac{x}{2}\right)$ Use the distributive property.

$2x + 5 = 8 + 5x$ Simplify.

$2x - 2x + 5 = 8 + 5x - 2x$ Subtract $2x$ from each side.

$5 = 8 + 3x$

$5 - 8 = 8 - 8 + 3x$ Subtract 8 from each side.

$-3 = 3x$

$-\dfrac{3}{3} = \dfrac{3x}{3}$ Divide each side by 3.

$-1 = x$

Check: See if you can verify this solution.

Practice Problem 6   Solve and check $\dfrac{y}{3} + \dfrac{1}{2} = 5 + \dfrac{y - 9}{4}$.

$\left(\text{\textit{Hint:} You can write } \dfrac{y - 9}{4} \text{ as } \dfrac{y}{4} - \dfrac{9}{4}.\right)$

An equation that contains many decimals can be multiplied by an appropriate power of 10 to clear it of decimals.

EXAMPLE 7   Solve and check $0.9 + 0.2(x + 4) = -3(0.1x - 0.4)$.

Solution

$0.9 + 0.2x + 0.8 = -0.3x + 1.2$ Remove the parentheses.

$10(0.9) + 10(0.2x) + 10(0.8) = 10(-0.3x) + 10(1.2)$ Multiply each side of the equation by 10 and use the distributive property.

$9 + 2x + 8 = -3x + 12$ Simplify.

$2x + 17 = -3x + 12$ Collect like terms.

$2x + 3x + 17 = -3x + 3x + 12$ Add $3x$ to each side.

$5x + 17 = 12$

$5x + 17 - 17 = 12 - 17$ Add -17 to each side.

$5x = -5$

$x = -1$ Divide each side by 5.

Check: $0.9 + 0.2(-1 + 4) \overset{?}{=} -3[(0.1)(-1) - 0.4]$

$0.9 + 0.2(3) \overset{?}{=} -3[-0.1 - 0.4]$

$0.9 + 0.6 \overset{?}{=} -3(-0.5)$

$1.5 = 1.5$ ✓

Thus, -1 is the solution.

Practice Problem 7   Solve and check $4(0.01x + 0.09) - 0.07(x - 8) = 0.83$.

Not every equation has a solution. Some equations have no solution at all.

EXAMPLE 8   Solve $7x + 3 - 9x = 14 - 2x + 5$.

Solution

$$-2x + 3 = -2x + 19 \qquad \text{Collect like terms.}$$
$$-2x + 3 - 3 = -2x + 19 - 3 \qquad \text{Subtract 3 from each side of the equation.}$$
$$-2x = -2x + 16 \qquad \text{Simplify.}$$
$$-2x + 2x = -2x + 2x + 16 \qquad \text{Add } 2x \text{ to each side.}$$
$$0 = 16 \qquad \text{We obtain a false equation.}$$

No matter what value we use for x in this equation, we get a false sentence. This equation has **no solution.**

NOTE TO STUDENT: Fully worked-out solutions to all of the Practice Problems can be found at the back of the text starting at page SP-1

Practice Problem 8   Solve $7 + 14x - 3 = 2(x - 4) + 12x$.

Now we examine a totally different situation. There are some equations for which any real number is a solution.

EXAMPLE 9   Solve $5(x - 2) + 3x = 10x - 2(x + 5)$.

Solution

$$5x - 10 + 3x = 10x - 2x - 10 \qquad \text{Remove parentheses.}$$
$$8x - 10 = 8x - 10 \qquad \text{Collect like terms.}$$
$$8x - 10 + 10 = 8x - 10 + 10 \qquad \text{Add 10 to each side of the equation.}$$
$$8x = 8x \qquad \text{Simplify.}$$
$$8x - 8x = 8x - 8x \qquad \text{Subtract } 8x \text{ from each side of the equation.}$$
$$0 = 0 \qquad \text{This is an equation that is always true.}$$

Replacing x in this equation by any real number will always result in a true sentence. This is an equation for which **any real number is a solution.**

Practice Problem 9   Solve $13x - 7(x + 5) = 4x - 35 + 2x$.

Developing Your Study Skills

Class Participation

People learn mathematics through active participation, not through observation from the sidelines. If you want to do well in this course, be involved in classroom activities. Sit near the front where you can see and hear well, and where your focus is on the instruction process and not on the students around you. Ask questions, be ready to contribute toward solutions, and take part in all classroom activities. Your contributions are valuable to the class and to yourself. Class participation requires an investment of yourself in the learning process, which you will find pays huge dividends.

2.1 EXERCISES

| Student Solutions Manual | CD/ Video | PH Math Tutor Center | MathXL®Tutorials on CD | MathXL® | MyMathLab® | Interactmath.com |

Verbal and Writing Skills

1. Is -20 a root of the equation $3x - 15 = 45$? Why or why not?

2. Is 21 a root of the equation $2x + 12 = -30$? Why or why not?

3. Is $\frac{2}{7}$ a solution to the equation $7x - 8 = -6$? Why or why not?

4. Is $\frac{3}{5}$ a solution to the equation $5y + 9 = 12$? Why or why not?

5. What is the first step in solving the equation $\frac{x}{3} + \frac{3}{4} = 2 - \frac{x}{2}$? Why?

6. What is the first step in solving the equation $0.7 + 0.03x = 4$? Why?

7. When solving the equation $x + 3.6 = 8$, would you clear it of decimals by multiplying each term by 10? Why or why not?

8. When solving the equation $x - \frac{1}{4} = 3$, would you clear it of fractions by multiplying each term by 4? Why or why not?

Solve Exercises 9–44. Check your solutions.

9. $-11 + x = -3$

10. $26 + x = -35$

11. $-9x = 45$

12. $-15x = -75$

13. $-14x = -70$

14. $-12x = 72$

15. $8x - 1 = 11$

16. $10x + 3 = 15$

17. $9x + 3 = 5x - 9$

18. $15x + 4 = 12x + 1$

19. $16 - 2x = 5x - 5$

20. $-12x - 8 = 10 - 3x$

21. $3a - 5 - 2a = 2a - 3$

22. $5a - 2 + 4a = 2a + 12$

23. $4(y - 1) = -2(3 + y)$

24. $5(2 - y) = 3(y - 2)$

25. $2 - y = 5 - 2(y - 1)$

Solve the following exercises. You may leave answers in either decimal or fraction form.

26. $3y + 3 = 7(y + 2) - 3y$

27. $\frac{2}{3}x = 8$

28. $-\frac{5}{6}x = 5$

29. $\frac{y}{2} + 4 = \frac{1}{6}$

30. $\frac{y}{3} + 2 = \frac{4}{5}$

31. $\frac{2}{3} - \frac{x}{6} = 1$

32. $\dfrac{4x}{5} + \dfrac{3}{2} = 2x$

33. $\dfrac{1}{2}(x + 3) - 2 = 1$

34. $5 - \dfrac{2}{3}(x + 2) = 3$

35. $5 - \dfrac{2x}{7} = 1 - (x - 4)$

36. $6 + 2(x - 1) = \dfrac{3x}{5} + 4$

37. $0.3x + 0.4 = 0.5x - 0.8$

38. $0.7x - 0.2 = 0.5x + 0.8$

39. $0.6 - 0.02x = 0.4x - 0.03$

40. $0.1x - 0.12 = 0.04x + 0.3$

41. $0.2(x - 4) = 3$

42. $0.6(2x + 1) = 1$

43. $0.05x - 2 = 0.3(x - 5)$

44. $0.3(x + 2) - 2 = 0.05x$

Mixed Practice

Solve each of the following.

45. $2x + 12 = 3 + 4x - 7$

46. $6y - 15 - 8y = 24 - 5$

47. $\dfrac{x - 2}{7} + \dfrac{5}{2} = \dfrac{9}{2}$

48. $\dfrac{1}{6} - \dfrac{x}{2} = \dfrac{x - 5}{3}$

49. $\dfrac{1}{3} - \dfrac{x + 1}{5} = \dfrac{x}{3}$

50. $\dfrac{y + 5}{7} = \dfrac{5}{14} - \dfrac{y - 3}{4}$

51. $2 + 0.1(5 - x) = 1.3x - (0.4x - 2.5)$

52. $3(0.4 - x) + 2 = x + 0.4(x + 8)$

53. $\frac{1}{2}(x + 2) = \frac{2}{3}(x - 1) - \frac{3}{4}$

54. $x - \frac{5}{3}(x - 2) = \frac{1}{9}(x + 2)$

To Think About

Some of the equations in this section have no solution. For some of the equations, any real number is a solution. Other equations have one real solution. Work carefully and solve each of the following equations.

55. $7x + 4 = 3x - 8 - 7x + 1$

56. $8x - 1 = 6x + 11 - 8x + 8$

57. $2x - 4(x + 1) = -2x + 14$

58. $3x - 17 = 8x - 5(x - 2)$

59. $7(x + 1) - 4 = 10x + 3(1 - x)$

60. $9(x + 1) - 4 = 4x + 5(x + 1)$

61. $6 + 8(x - 2) = 10x - 2(x + 4)$

62. $2x + 4(x - 5) = -x + 7(x - 1) + 3$

63. $x - 2 + \frac{2x}{5} = -2 + \frac{7x}{5}$

64. $x + \frac{2x + 8}{3} = \frac{5x + 5}{3} + 1$

Cumulative Review

Simplify. Do not leave negative exponents in your answers.

65. $5 - (4 - 2)^2 + 3(-2)$

66. $\left(\frac{3xy^2}{2x^2y}\right)^3$

67. $(-2)^4 - 12 - 6(-2)$

68. $(2x^{-2}y^{-3})^2(4xy^{-2})^{-2}$

69. *Circus Attendance* Last year, the Ringling Brothers and Barnum and Bailey Circus played to twenty-seven million people. At three shows in Memphis that year, the circus played to 6200 people, 8420 people, and 12,065 people. The circus put on four thousand shows during the year.

(a) What was the average attendance per show?

(b) What was the average attendance at the three shows in Memphis?

Student Learning Objective

After studying this section, you will be able to:

 1 Solve literal equations for the desired unknown.

1 Solving Literal Equations for the Desired Unknown

A first-degree **literal equation** is an equation that contains variables other than the variable that we are solving for. When you solve for an unknown in a literal equation, the final expression will contain these other variables. We use this procedure to deal with formulas in applied problems.

EXAMPLE 1   Solve for x: $5x + 3y = 2$

Solution

$$5x = 2 - 3y \qquad \text{Subtract } 3y \text{ from each side.}$$

$$\frac{5x}{5} = \frac{2 - 3y}{5} \qquad \text{Divide each side by 5.}$$

$$x = \frac{2 - 3y}{5} \qquad \text{The solution is a fractional expression.}$$

Practice Problem 1   Solve for W: $P = 2L + 2W$

Where possible, collect like terms as you solve the equation.

When solving more complicated first-degree literal equations, use the following procedure.

PROCEDURE FOR SOLVING FIRST-DEGREE (OR LINEAR) LITERAL EQUATIONS

1. Remove grouping symbols in the proper order.
2. If fractions exist, multiply all terms on both sides by the LCD.
3. Collect like terms if possible.
4. Add or subtract a term with the desired unknown on both sides of the equation to obtain all terms with the desired unknown on one side of the equation.
5. Add or subtract appropriate terms on both sides of the equation to obtain all other terms on the other side of the equation.
6. Divide each side of the equation by the coefficient of the desired unknown.
7. Simplify the solution (if possible).

Some equations appear difficult to solve because they contain fractions and parentheses. Immediately remove the parentheses. Then multiply each term by the LCD. The equation will then appear less threatening.

EXAMPLE 2   Solve for b: $A = \frac{2}{3}(a + b + 3)$

Solution

$$A = \frac{2}{3}a + \frac{2}{3}b + 2 \qquad \text{Remove parentheses.}$$

$$3A = 3\left(\frac{2}{3}a\right) + 3\left(\frac{2}{3}b\right) + 3(2) \qquad \text{Multiply all terms by the LCD 3.}$$

$$3A = 2a + 2b + 6 \qquad \text{Simplify.}$$

$$3A - 2a - 6 = 2b \qquad \begin{array}{l}\text{Subtract } 2a \text{ from each side.}\\ \text{Subtract 6 from each side.}\end{array}$$

$$\frac{3A - 2a - 6}{2} = \frac{2b}{2}$$ Divide each side of the equation by the coefficient of b.

$$\frac{3A - 2a - 6}{2} = b$$ Simplify.

Practice Problem 2 Solve for a: $H = \frac{3}{4}(a + 2b - 4)$

NOTE TO STUDENT: *Fully worked-out solutions to all of the Practice Problems can be found at the back of the text starting at page SP-1*

Be sure to collect like terms after removing the parentheses. This will simplify the equation and make it much easier to solve.

EXAMPLE 3 Solve for x: $5(2ax + 3y) - 4ax = 2(ax - 5)$

Solution

$10ax + 15y - 4ax = 2ax - 10$ Remove parentheses.

$6ax + 15y = 2ax - 10$ Collect like terms.

$6ax - 2ax + 15y = -10$ Subtract $2ax$ from each side to obtain terms containing x on one side.

$4ax = -10 - 15y$ Simplify and subtract $15y$ from each side.

$\dfrac{4ax}{4a} = \dfrac{-10 - 15y}{4a}$ Divide each side by the coefficient of x.

$x = \dfrac{-10 - 15y}{4a}$

Practice Problem 3 Solve for b: $-2(ab - 3x) + 2(8 - ab) = 5x + 4ab$

The Boston Marathon is a grueling 26-mile endurance race over city streets from Hopkinton to Boston, Massachusetts. The race has been held annually for over one hundred years and has been watched by millions of people. The winning time on April 17, 2000, was 2 hours, 9 minutes, 47 seconds. If we round to the nearest minute, this time is approximately equal to 130 minutes. The bar graph to the right shows the approximate winning time in minutes for this race for selected years from 1900 to 2000.

Winning Times in the Boston Marathon

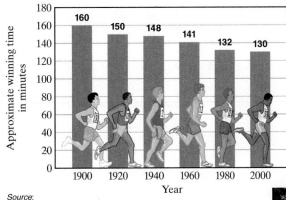

Source:
The Boston Globe, April 18, 2000

EXAMPLE 4 The winning times in minutes for the Boston Marathon are approximated by the equation $t = -0.3x + 159$, where x is the number of years since 1900. Solve this equation for x. Determine approximately what year it will be when the winning time for the Boston Marathon is 126 minutes.

Solution

$t = -0.3x + 159$

$0.3x = 159 - t$ Add $0.3x - t$ to each side.

$3x = 1590 - 10t$ Multiply each side by 10 to clear decimals.

$x = \dfrac{1590 - 10t}{3}$ Divide each side by 3.

Now we will use this equation, which has been solved for x (the number of years since 1900), to find when the winning time is 126 minutes.

$$x = \frac{1590 - 10(126)}{3} \qquad \text{We substitute in 126 for } t.$$

$$x = \frac{1590 - 1260}{3}$$

$$x = \frac{330}{3} = 110$$

It will be 110 years from 1900. Thus, we estimate that this winning time will occur in the year 2010.

NOTE TO STUDENT: Fully worked-out solutions to all of the Practice Problems can be found at the back of the text starting at page SP-1

Practice Problem 4   The winning time in minutes for a 15-mile race to benefit breast cancer research is given by the equation $t = -0.4x + 81$, where x is the number of years since 1990. Solve this equation for x. Determine approximately what year it will be when the winning time for this 15-mile race is 75 minutes.

▲ | **EXAMPLE 5**

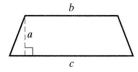

(a) Solve the formula for the area of a trapezoid, $A = \frac{1}{2}a(b + c)$, for c.

(b) Find c when $A = 20$ square inches, $a = 3$ inches, and $b = 4$ inches.

Solution

(a)
$$A = \frac{1}{2}ab + \frac{1}{2}ac \qquad \text{Remove the parentheses.}$$

$$2A = 2\left(\frac{1}{2}ab\right) + 2\left(\frac{1}{2}ac\right) \qquad \text{Multiply each term by 2.}$$

$$2A = ab + ac \qquad \text{Simplify.}$$

$$2A - ab = ac \qquad \text{Subtract } ab \text{ from each side to isolate the term containing } c.$$

$$\frac{2A - ab}{a} = c \qquad \text{Divide each side by } a.$$

(b) We use the equation we derived in part **(a)** to find c for the given values.

$$c = \frac{2A - ab}{a}$$

$$= \frac{2(20) - (3)(4)}{3} \qquad \text{Substitute the given values of } A, a, \text{ and } b \text{ to find } c.$$

$$= \frac{40 - 12}{3} = \frac{28}{3} \qquad \text{Simplify.}$$

Thus, side $c = \frac{28}{3}$ inches or $9\frac{1}{3}$ inches.

▲ | **Practice Problem 5**

(a) Solve for h: $A = 2\pi rh + 2\pi r^2$

(b) Find h when $A = 100$, $\pi \approx 3.14$, and $r = 2.0$. Round your answer to the nearest hundredth.

Solve for x.

1. $6x + 5y = 3$

2. $7x + 2y = 5$

3. $2x + 3y = 24 - 6x$

4. $10x - 5 = 10y + 2x$

5. $y = \dfrac{2}{3}x - 4$

6. $y = -\dfrac{1}{3}x + 2$

Solve for the variable specified.

7. $x = -\dfrac{3}{4}y + \dfrac{2}{3}$; for y

8. $x = \dfrac{5}{2}y - \dfrac{1}{5}$; for y

▲ **9.** $A = lw$; for l

▲ **10.** $V = lwh$; for w

▲ **11.** $A = \dfrac{h}{2}(B + b)$; for B

12. $C = \dfrac{5}{9}(F - 32)$; for F

▲ **13.** $A = 2\pi rh$; for r

▲ **14.** $V = \pi r^2 h$; for h

15. Solve for b. $H = \dfrac{2}{3}(a + 2b)$

16. $H = \dfrac{3}{4}(5a + b)$; for a

17. $2(2ax + y) = 3ax - 4y$; for x

18. $3(4ax + y) = 2ax - 3y$; for x

Follow the directions given.

▲ **19.** **(a)** Solve for a: $A = \dfrac{1}{2}ab$

20. **(a)** Solve for C: $F = \dfrac{9}{5}C + 32$

(b) Evaluate when $A = 20$ and $b = \dfrac{5}{2}$.

(b) Evaluate when $F = 23°$.

21. (a) Solve for n: $A = a + d(n - 1)$

(b) Evaluate when $A = 28$, $a = 3$, and $d = 15$.

▲ **22. (a)** Solve for h: $V = \frac{1}{3}\pi r^2 h$

(b) Evaluate when $V = 6.28$, $r = 3$, and $\pi \approx 3.14$.

Applications

23. *Boston Marathon* The winning times in minutes for women runners in the Boston Marathon are approximated by the equation $t = -0.7x + 160$, where x is the number of years since 1975. Solve this equation for x. Use this new equation to determine approximately what year it will be when the winning time for a woman in the Boston Marathon is 139 minutes.

24. *Boston Marathon* The winning times in minutes for wheelchair participants in the Boston Marathon are approximated by the equation $t = -1.3x + 117$, where x is the number of years since 1975. Solve this equation for x. Use this new equation to determine approximately what year it will be when the winning time for a wheelchair participant in the Boston Marathon is 78 minutes.

25. *Mariner's Formula* The mariner's formula can be written in the form $\frac{m}{1.15} = k$, where m is the speed of a ship in miles per hour and k is the speed of the ship in knots (nautical miles per hour).
(a) Solve the formula for m.
(b) Use this result to find the number of miles per hour a ship is traveling if its speed is 29 knots.

26. *Patient Appointments* Some doctors use the formula $ND = 1.08T$ to relate the variables N (the number of patient appointments the doctor schedules in one day), D (the duration of each patient appointment), and T (the total number of minutes the doctor can use to see patients in one day).
(a) Solve the formula for N.

(b) Use this result to find the number of patient appointments N a doctor should make if she has 6 hours available for patients and each appointment is 15 minutes long. (Round your answer to the nearest whole number.)

In Exercises 27 and 28, the variable C represents the consumption of products in the United States in billions of dollars, and D represents disposable income in the United States in billions of dollars.

 27. *An Economic Model* Suppose economists use as a model of the country's economy the equation $C = 0.6547D + 5.8263$.
(a) Solve the equation for D.
(b) Use this result to determine the disposable income D if the consumption C is $9.56 billion. Round your answer to the nearest tenth of a billion.

28. *An Economic Model* Suppose economists use as a model of the country's economy the equation $C = 0.7649D + 6.1275$.
(a) Solve the equation for D.
(b) Use this result to determine the disposable income D if the consumption C is $12.48 billion. Round your answer to the nearest tenth of a billion.

Cumulative Review

Write with positive exponents in simplest form.

29. $(2x^{-3}y)^{-2}$

30. $\left(\dfrac{5x^2y^{-3}}{x^{-4}y^2}\right)^{-3}$

31. Simplify. $7 + 6 \div 2 - (5 - 2)^2$

32. Simplify. $6x + 3[x - (y + 5)]$

33. *Education Fund* Sharon and James want to begin an education fund for their two daughters. They invest $5000 in a certificate of deposit for one year, with an annual return of 5%. $4000 is invested in a more risky venture that they hope will have an annual return of 9%. How much money will they have at the end of 1 year if their risky investment does well?

34. *Automobile Costs* Drew wants to go to college in Pennsylvania. He and his parents take a long weekend to drive there from Kansas City, Missouri. His odometer read 45,711.3 when he left the college in Pennsylvania and 46,622.1 when he arrived back home. He started and ended his trip on a full tank of gas. He made gas purchases of 9.9 gallons, 11.7 gallons, 10.6 gallons, 5.8 gallons, and 8 gallons during the trip. How many miles per gallon did the car get on the trip?

Developing Your Study Skills

Taking Notes in Class

An important part of learning mathematics is taking notes. To take meaningful notes, you must be an active listener. Keep your mind on what the instructor is saying, and be ready with questions whenever you do not understand something.

If you have previewed the lesson material, you will be prepared to take good notes. The important concepts will seem somewhat familiar. You will have a better idea of what needs to be written down. If you frantically try to write all that the instructor says or copy all the examples done in class, you may find your notes to be nearly worthless when you are home alone. You may find that you are unable to make sense of what you have written.

Write down *important* ideas and examples as the instructor lectures, making sure that you are listening and following the logic. Include any helpful hints or suggestions that your instructor gives you or refers to in your text. You will be amazed at how easily these are forgotten if they are not written down.

Successful note taking requires active listening and processing. Stay alert in class. You will realize the advantages of taking your own notes over copying those of someone else.

Student Learning Objectives

After studying this section, you will be able to:

1 Solve absolute value equations of the form $|ax + b| = c$.

2 Solve absolute value equations of the form $|ax + b| + c = d$.

3 Solve absolute value equations of the form $|ax + b| = |cx + d|$.

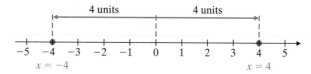

Solving Absolute Value Equations of the Form $|ax + b| = c$

From Section 1.2, you know that the absolute value of a number x can be pictured as the distance between 0 and x on the number line. Let's look at a simple absolute value equation, $|x| = 4$, and draw a picture.

Thus, the equation $|x| = 4$ has two solutions, $x = 4$ and $x = -4$. Let's look at another example.

$$\text{If} \quad |x| = \frac{2}{3},$$

$$\text{then} \quad x = \frac{2}{3} \quad \text{or} \quad x = -\frac{2}{3},$$

$$\text{because} \quad \left|\frac{2}{3}\right| = \frac{2}{3} \quad \text{and} \quad \left|-\frac{2}{3}\right| = \frac{2}{3}.$$

We can solve these relatively simple absolute value equations by recalling the definition of absolute value.

$$|x| = \begin{cases} x, & \text{if } x \geq 0 \\ -x, & \text{if } x < 0 \end{cases}$$

Now let's take a look at a more complicated absolute value equation: $|ax + b| = c$.

> The solutions of an equation of the form $|ax + b| = c$, where $a \neq 0$ and c is a positive number, are those values that satisfy
>
> $$ax + b = c \quad \text{or} \quad ax + b = -c.$$

EXAMPLE 1 Solve $|2x + 5| = 11$ and check your solutions.

Solution Using the rule established in the box, we have the following:

$$\begin{array}{lcl} 2x + 5 = 11 & \text{or} & 2x + 5 = -11 \\ 2x = 6 & & 2x = -16 \\ x = 3 & & x = -8 \end{array}$$

The two solutions are 3 and -8.

Check: **if $x = 3$**   **if $x = -8$**

$$\begin{array}{ll} |2x + 5| = 11 & |2x + 5| = 11 \\ |2(3) + 5| \overset{?}{=} 11 & |2(-8) + 5| \overset{?}{=} 11 \\ |6 + 5| \overset{?}{=} 11 & |-16 + 5| \overset{?}{=} 11 \\ |11| \overset{?}{=} 11 & |-11| \overset{?}{=} 11 \\ 11 = 11 \ \checkmark & 11 = 11 \ \checkmark \end{array}$$

NOTE TO STUDENT: Fully worked-out solutions to all of the Practice Problems can be found at the back of the text starting at page SP-1

Practice Problem 1   Solve $|3x - 4| = 23$ and check your solutions.

EXAMPLE 2   Solve $\left|\dfrac{1}{2}x - 1\right| = 5$ and check your solutions.

Solution   The solutions of the given absolute value equation must satisfy

$$\frac{1}{2}x - 1 = 5 \quad \text{or} \quad \frac{1}{2}x - 1 = -5.$$

If we multiply each term of both equations by 2, we obtain the following:

$$x - 2 = 10 \quad \text{or} \quad x - 2 = -10$$
$$x = 12 \qquad\qquad x = -8$$

Check: **if $x = 12$**   **if $x = -8$**

$$\left|\frac{1}{2}(12) - 1\right| \stackrel{?}{=} 5 \qquad\qquad \left|\frac{1}{2}(-8) - 1\right| \stackrel{?}{=} 5$$
$$|6 - 1| \stackrel{?}{=} 5 \qquad\qquad |-4 - 1| \stackrel{?}{=} 5$$
$$|5| \stackrel{?}{=} 5 \qquad\qquad |-5| \stackrel{?}{=} 5$$
$$5 = 5 \;\checkmark \qquad\qquad 5 = 5 \;\checkmark$$

Practice Problem 2   Solve and check your solutions.

$$\left|\frac{2}{3}x + 4\right| = 2$$

Solving Absolute Value Equations of the Form $|ax + b| + c = d$

Notice that in each of the previous examples the absolute value expression is on one side of the equation and a positive real number is on the other side of the equation. What happens when we encounter an equation of the form $|ax + b| + c = d$?

EXAMPLE 3   Solve $|3x - 1| + 2 = 5$ and check your solutions.

Solution   First we will rewrite the equation so that the absolute value expression is alone on one side of the equation.

$$|3x - 1| + 2 - 2 = 5 - 2$$
$$|3x - 1| = 3$$

Now we solve $|3x - 1| = 3$.

$$3x - 1 = 3 \quad \text{or} \quad 3x - 1 = -3$$
$$3x = 4 \qquad\qquad 3x = -2$$
$$x = \frac{4}{3} \qquad\qquad x = -\frac{2}{3}$$

Check: **if $x = \dfrac{4}{3}$**   **if $x = -\dfrac{2}{3}$**

$$\left|3\left(\frac{4}{3}\right) - 1\right| + 2 \stackrel{?}{=} 5 \qquad \left|3\left(-\frac{2}{3}\right) - 1\right| + 2 \stackrel{?}{=} 5$$
$$|4 - 1| + 2 \stackrel{?}{=} 5 \qquad\qquad |-2 - 1| + 2 \stackrel{?}{=} 5$$
$$|3| + 2 \stackrel{?}{=} 5 \qquad\qquad |-3| + 2 \stackrel{?}{=} 5$$
$$3 + 2 \stackrel{?}{=} 5 \qquad\qquad 3 + 2 \stackrel{?}{=} 5$$
$$5 = 5 \;\checkmark \qquad\qquad 5 = 5 \;\checkmark$$

Practice Problem 3   Solve $|2x + 1| + 3 = 8$ and check your solutions.

3 **Solving Absolute Value Equations of the Form** $|ax + b| = |cx + d|$

Let us now consider the possibilities for a and b if $|a| = |b|$.

Suppose $a = 5$; then $b = 5$ or -5.
If $a = -5$, then $b = 5$ or -5.

To generalize, if $|a| = |b|$, then $a = b$ or $a = -b$.

We now apply this property to solve more complex equations.

EXAMPLE 4   Solve and check: $|3x - 4| = |x + 6|$

Solution   The solutions of the given equation must satisfy

$$3x - 4 = x + 6 \quad \text{or} \quad 3x - 4 = -(x + 6).$$

Now we solve each equation in the normal fashion.

$$3x - 4 = x + 6 \quad \text{or} \quad 3x - 4 = -x - 6$$
$$3x - x = 4 + 6 \qquad\qquad 3x + x = 4 - 6$$
$$2x = 10 \qquad\qquad\qquad 4x = -2$$
$$x = 5 \qquad\qquad\qquad x = -\frac{1}{2}$$

We will check each solution by substituting it into the *original equation*.

Check: **if** $x = 5$ **if** $x = -\frac{1}{2}$

$$|3(5) - 4| \stackrel{?}{=} |5 + 6| \qquad \left|3\left(-\frac{1}{2}\right) - 4\right| \stackrel{?}{=} \left|-\frac{1}{2} + 6\right|$$

$$|15 - 4| \stackrel{?}{=} |11| \qquad\qquad \left|-\frac{3}{2} - 4\right| \stackrel{?}{=} \left|-\frac{1}{2} + 6\right|$$

$$|11| \stackrel{?}{=} |11| \qquad\qquad \left|-\frac{3}{2} - \frac{8}{2}\right| \stackrel{?}{=} \left|-\frac{1}{2} + \frac{12}{2}\right|$$

$$11 = 11 \;\checkmark \qquad\qquad \left|-\frac{11}{2}\right| \stackrel{?}{=} \left|\frac{11}{2}\right|$$

$$\frac{11}{2} = \frac{11}{2} \;\checkmark$$

NOTE TO STUDENT: Fully worked-out solutions to all of the Practice Problems can be found at the back of the text starting at page SP-1

Practice Problem 4   Solve and check: $|x - 6| = |5x + 8|$

TO THINK ABOUT: Two Other Absolute Value Equations   Explain how you would solve an absolute value equation of the form $|ax + b| = 0$. Give an example. Does $|3x + 2| = -4$ have a solution? Why or why not?

Verbal and Writing Skills

1. The equation $|x| = b$, where b is a positive number, will always have how many solutions? Why?

2. The equation $|x| = b$ might have only one solution. How could that happen?

3. To solve an equation like $|x + 7| - 2 = 8$, what is the first step that must be done? What will be the result?

4. To solve an equation like $|3x - 1| + 5 = 14$, what is the first step that must be done? What will be the result?

Solve each absolute value equation. Check your solutions for Exercises 5–24.

5. $|x| = 30$

6. $|x| = 14$

7. $|x - 6| = 16$

8. $|x - 3| = 13$

9. $|2x - 5| = 13$

10. $|7x - 3| = 11$

11. $|5 - 4x| = 11$

12. $|2 - 3x| = 13$

13. $\left|\dfrac{1}{2}x - 3\right| = 2$

14. $\left|\dfrac{1}{4}x + 5\right| = 3$

15. $|2.3 - 0.3x| = 1$

16. $|0.7 - 0.3x| = 2$

17. $|x + 2| - 1 = 7$

18. $|x + 3| - 4 = 8$

19. $\left|\dfrac{1}{2} - \dfrac{3}{4}x\right| + 1 = 3$

20. $\left|\dfrac{2}{3} - \dfrac{1}{2}x\right| - 2 = -1$

21. $\left|1 - \dfrac{3}{4}x\right| + 4 = 7$

22. $\left|4 - \dfrac{5}{2}x\right| + 3 = 15$

23. $\left|\dfrac{1 - 3x}{2}\right| = \dfrac{4}{5}$

24. $\left|\dfrac{3x - 2}{3}\right| = \dfrac{1}{2}$

Solve each absolute value equation.

25. $|x + 6| = |2x - 3|$

26. $|x - 4| = |2x + 5|$

27. $\left|\dfrac{x - 1}{2}\right| = |2x + 3|$

28. $\left|\dfrac{2x + 7}{3}\right| = |x + 2|$

29. $|1.5x - 2| = |x - 0.5|$

30. $|2.2x + 2| = |1 - 2.8x|$

31. $|3 - x| = \left|\dfrac{x}{2} + 3\right|$

32. $\left|\dfrac{2x}{5} + 1\right| = |1 - x|$

Solve for x. Round to the nearest hundredth.

 33. $|1.62x + 3.14| = 2.19$

34. $|-0.74x - 8.26| = 5.36$

Mixed Practice

Solve each equation, if possible. Check your solutions.

35. $|3(x + 4)| + 2 = 14$

36. $|4(x - 2)| + 1 = 19$

37. $\left|\dfrac{5x}{3} - 1\right| = 0$

38. $\left|\dfrac{3}{4}x + 9\right| = 0$

39. $\left|\dfrac{4}{3}x - \dfrac{1}{8}\right| = -5$

40. $\left|\dfrac{3}{4}x - \dfrac{2}{3}\right| = -8$

41. $\left|\dfrac{2x - 1}{3}\right| = \dfrac{5}{6}$

42. $\left|\dfrac{3x + 1}{2}\right| = \dfrac{3}{4}$

43. $|1.5x - 2.5| = |x + 3|$

44. $|6x - 0.3| = |5.6x + 11.9|$

Cumulative Review

45. Simplify. $(3x^{-3}yz^0)\left(\dfrac{5}{3}x^4y^2\right)$

46. Evaluate. $\dfrac{\sqrt{3 - 2 \cdot 1^2} + 5}{4^2 - 2 \cdot 3}$

47. *Chemistry Laboratory Supplies* A scientist bought three new Bunsen burners and twenty-five new beakers last month for $975. This month she bought three Bunsen burners and twenty beakers for $825. How much did each beaker cost? How much did each Bunsen burner cost?

48. *Cost of Potato Chips* In Pennsylvania, a bag of barbecue-flavored potato chips costs $3.29. The same-sized bag of chips costs $2.69 in South Carolina. If the Pennsylvania stores are offering $1.50 off each bag of potato chips, and the South Carolina stores are offering three bags for the price of two, the stores in which state are offering the better deal? How much would it cost to buy three bags in each place?

Developing Your Study Skills

How to Do Homework

Set aside time each day for your homework assignments. Do not attempt to do a whole week's worth on the weekend. Two hours spent studying outside of class for each hour in class is usual for college courses, and you may need more than that for mathematics.

Before beginning your homework exercises, read your textbook very carefully. Expect to spend much more time reading a few pages of a mathematics textbook than several pages of another text. Read for complete understanding, not just for the general idea.

As you begin your homework assignments, read the directions carefully. You need to understand what is being asked for. Concentrate on each exercise, taking time to solve it accurately. Rushing through your work will cause you to make errors. Check your answers with those given in the back of the textbook. If your answer is incorrect, check to see that you are doing the right exercise. Redo the exercise, watching for little errors. If it is still wrong, check

with a friend. Perhaps the two of you can figure out where you made an error.

Also, check the examples in the textbook or in your notes for a similar exercise. Can this one be solved in the same way? Give it some thought. You may want to leave it for a while and take a break or do a different exercise. But come back later and try again. If you are still unable to figure it out, ask your instructor for help during office hours or in class.

Work on assignments every day, and do as many exercises as it takes for you to know what you are doing. Begin by doing all the exercises that have been assigned. If there are more available in the section in your text, then do more. When you think you have done enough exercises to understand fully the kind at hand, do a few more to be sure. This may mean that you do many more exercises than the instructor assigns, but you can never practice too much. Practice improves your skills and increases your accuracy, speed, competence, and confidence.

Student Learning Objective

After studying this section, you will be able to:

 Solve applied problems by using equations.

1 Solving Applied Problems by Using Equations

The skills you have developed in solving equations will allow you to solve a variety of applied problems. The following steps may help you to organize your thoughts and provide you with a procedure to solve such problems.

1. **Understand the problem.**
   (a) Read the word problem carefully to get an overview.
   (b) Determine what information you will need to solve the problem.
   (c) Draw a sketch. Label it with the known information. Determine what needs to be found.
   (d) Choose a variable to represent one unknown quantity.
   (e) If necessary, represent other unknown quantities in terms of the same variable.

2. **Write an equation.**
   (a) Look for key words to help you translate the words into algebraic symbols.
   (b) Use a relationship given in the problem or an appropriate formula in order to write an equation.

3. **Solve the equation and state the answer.**

4. **Check.**
   (a) Check the solution in the original equation.
   (b) Be sure the solution to the equation answers the question in the word problem. You may need to do some additional calculations if it does not.

Symbolic Equivalents of English Phrases

English Phrase	Mathematical Symbol
and, added to, increased by, greater than, plus, more than, sum of	$+$
decreased by, subtracted from, less than, diminished by, minus, difference between	$-$
product of, multiplied by, of, times	$\cdot$ or $(\ )(\ )$ or $\times$
divided by, quotient of, ratio of	$\div$ or fraction bar
equals, are, is, will be, yields, gives, makes, is the same as, has a value of	$=$

Although we often use x to represent the unknown quantity when we write equations, any letter can be used. It is a good idea to use a letter that helps us remember what the variable represents. (For example, we might use s for speed or h for hours.) We now look at some translations of English sentences into algebraic equations.

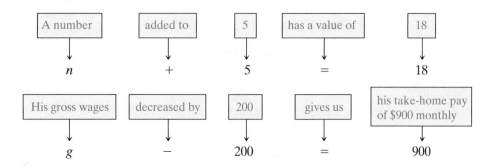

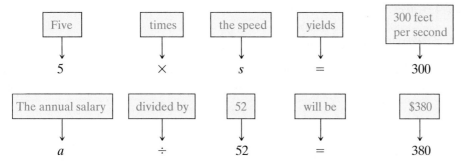

Be careful when translating the expressions "more than" and "less than." The order in which you write the symbols does not follow the order found in the English sentence. For example, "5 less than a number will be 40" is written as follows:

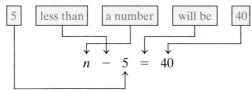

"Two more than a number is −5" is written as shown next.

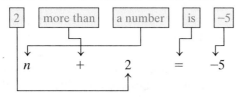

Since addition is commutative, we can also write $2 + n = -5$ for "2 more than a number is −5." Be careful, however, with subtraction. "5 less than a number will be 40" must be written as $n - 5 = 40$.

We will now illustrate the use of the four-step procedure in solving a number of word problems.

EXAMPLE 1   Nancy works for an educational services company that provides computers and software for local public schools. She went to a local truck rental company to rent a truck to deliver some computers to the Newbury Elementary School. The truck rental company has a fixed rate of $40 per day plus 20¢ per mile. Nancy rented a truck for 3 days and was billed $177. How many miles did she drive?

Solution

1. ***Understand the problem.*** Let n = the number of miles driven.
   Since each mile driven costs 20¢, we multiply the 20¢- (or $0.20) per-mile cost by the number of miles n.
   Thus, $0.20n$ = the cost of driving n miles at 20¢ per mile.

2. ***Write an equation.***

Fixed costs for 3 days	plus	mileage charge	equals	$177
$(40)(3)$	$+$	$(0.20)(n)$	$=$	177

3. ***Solve the equation and state the answer.***

   $$120 + 0.20n = 177 \quad \text{Multiply } (40)(3).$$
   $$0.20n = 57 \quad \text{Subtract 120 from each side.}$$
   $$\frac{0.20n}{0.20} = \frac{57}{0.20} \quad \text{Divide each side by 0.20.}$$
   $$n = 285 \quad \text{Simplify.}$$

   Nancy drove the truck for 285 miles.

4. *Check.* Does a truck rental for 3 days at $40 per day plus 20¢ per mile for 285 miles come to a total of $177?
We will check our values in the original equation.

$$(40)(3) + (0.20)(n) = 177$$
$$120 + (0.20)(285) \stackrel{?}{=} 177$$
$$120 + 57 \stackrel{?}{=} 177$$
$$177 = 177 \quad ✓$$

It checks. Our answer is correct.

NOTE TO STUDENT: Fully worked-out solutions to all of the Practice Problems can be found at the back of the text starting at page SP-1

Practice Problem 1 Western Laboratories rents a computer terminal for $400 per month plus $8 per hour of computer use time. The bill for 1 year's computer use was $7680. How many hours did Western Laboratories actually use the computer?

EXAMPLE 2 The Acetones are a barbershop quartet. They travel across the country in a special bus and usually give six concerts a week. This popular group always sings to a sell-out crowd. The concert halls have an average seating capacity of three thousand people each. Concert tickets average $12 per person. The onetime expenses for each concert are $15,000. The cost of meals, motels, security and sound people, bus drivers, and other expenses totals $100,000 per week. How many weeks per year will the Acetones need to be on tour if each wants to earn $71,500 per year?

Solution

1. *Understand the problem.* The item we are trying to find is the number of weeks the quartet needs to be on tour. So we let

$$x = \text{the number of weeks on tour.}$$

Now we need to find an expression that describes the quartet's income. Each week there are six concerts with three thousand people costing $12 per person. Thus,

$$\text{weekly income} = (6)(3000)(12) = \$216{,}000, \text{ and}$$
$$\text{income for } x \text{ weeks} = \$216{,}000x.$$

Now we need to find an expression for the total expenses. The expenses for each concert are $15,000, and there are six concerts per week. Thus, concert expenses will be $(6)(15{,}000) = \$90{,}000$ per week. Now the cost of meals, motels, security and sound people, bus drivers, and other expenses total $100,000 per week. Thus,

$$\text{total weekly expenses} = \$90{,}000 + \$100{,}000 = \$190{,}000, \text{ and}$$
$$\text{total expenses for } x \text{ weeks} = \$190{,}000x.$$

The four Acetones each want to earn $71,500 per year, so the group will need $4 \times \$71{,}500 = \$286{,}000$ for the year.

2. *Now we can write an equation.*

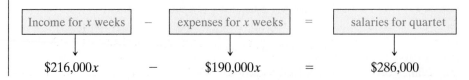

Income for x weeks	−	expenses for x weeks	=	salaries for quartet
$216,000x$	−	$190,000x$	=	$286,000$

3. *Solve the equation.*

$$26{,}000x = 286{,}000 \quad \text{Collect like terms.}$$

$$\frac{26{,}000x}{26{,}000} = \frac{286{,}000}{26{,}000} \quad \text{Divide each side of the equation by 26,000.}$$

$$x = 11$$

Thus, the Acetones will need to be on tour for 11 weeks to meet their salary goal.

4. *Check.* The check is left to the student.

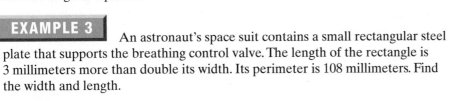

Practice Problem 2   A group of five women in a rock band plans to go on tour. They are scheduled to give five concerts a week with an average audience of six thousand people at each concert. The average ticket price for a concert is $14. The onetime expenses for each concert are $48,000. The additional costs per week are $150,000. How many weeks per year will this group need to be on tour if each wants to earn $60,000 per year?

Sometimes we need a simple formula from geometry or some other science to write the original equation.

▲ **EXAMPLE 3**   An astronaut's space suit contains a small rectangular steel plate that supports the breathing control valve. The length of the rectangle is 3 millimeters more than double its width. Its perimeter is 108 millimeters. Find the width and length.

Solution

1. *Understand the problem.* We draw a sketch to assist us. The formula for the perimeter of a rectangle is $P = 2w + 2l$, where w = the width and l = the length. Since the length is compared to the width, we will begin with the width.
Let w = the width. then $2w + 3$ = the length.

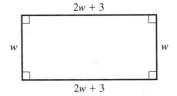

2. *Write an equation.*

$$P = 2w + 2l$$
$$108 = 2w + 2(2w + 3)$$

3. *Solve the equation and state the answer.*

$$108 = 2w + 4w + 6$$
$$108 = 6w + 6$$
$$102 = 6w$$
$$17 = w$$

Because $w = 17$, we have $2w + 3 = 2(17) + 3 = 37$. Thus, the rectangle is 17 millimeters wide and 37 millimeters long.

4. *Check.*

$$P = 2w + 2l$$
$$108 \overset{?}{=} 2(17) + 2(37)$$
$$108 \overset{?}{=} 34 + 74$$
$$108 = 108 \ \checkmark$$

▲ **Practice Problem 3**   The perimeter of a triangular lawn is 162 meters. The length of the first side is twice the length of the second side. The length of the third side is 6 meters shorter than three times the length of the second side. Find the dimensions of the triangle.

Student Solutions Manual · CD/ Video · PH Math Tutor Center · MathXL®Tutorials on CD · MathXL® · MyMathLab® · Interactmath.com

Write an algebraic equation and use it to solve each problem.

1. Three-fifths of a number is -54. What is the number?

2. Five-sixths of a number is -60. What is the number?

Applications

3. *Gym Membership Costs* Allyson can pay for her gym membership on a monthly basis, but if she pays for an entire year's membership in advance, she'll receive a $50 discount. Her discounted bill for the year would then be $526. What is the monthly membership fee at her gym?

4. *Parking Garage Fees* The monthly parking fee at the Safety First Parking Garage is one and one-half times last year's monthly fee, but compact cars will now receive a $10 per month discount. If the owner of a compact car pays $98 per month this year, what was his monthly fee last year?

5. *Computer Shipments* Barbara Stormer's computer company ships its packages through Reliable Shipping Express. Reliable charges $3 plus $0.80 per pound to ship each package. If the charge for shipping a package to Idaho was $17.40, how much did the package weigh?

6. *Parking Garage Fees* The Springfield City Garage charges $6 for the first hour and then $3.50 for each additional hour. How long has a car been parked at this garage if the parking charge is $69?

7. *Laundry Expenses* Roberto and Maria Santanos spend approximately $11.75 per week to wash and dry their family's clothes at a local coin laundry. A new washer and dryer would cost them a total of $846. How many weeks will it take for the laundromat cost to equal the cost of a new washer and dryer?

8. *Personal Banking* Manchester Community Bank charges its customers $8 per month plus 10¢ per check for the use of a standard checking account. Sonja had a checking account there for 4 months and was charged $39.70 in service charges. How many checks did she write during that period?

9. *Guest Speaker Costs* Mr. Ziglar is a famous motivational speaker. He travels across the country by jet and gives an average of eight presentations a week. He gives his speeches in auditoriums that are always full and hold an average of 2000 people each. He charges $15 per person at each of these speeches. His average expenses for each presentation, including renting the auditorium, are about $14,000. The cost of meals, motels, jet travel, and support personnel total $85,000 per week. How many weeks will Mr. Ziglar need to travel if he wants to earn $129,000 per year?

10. *Music Tour Costs* The Three Tenors are planning a nationwide tour. They will travel across the country by jet and give four concerts a week. They will sing in concert halls that hold an average of 5000 people each. Concert tickets average $18 per person, and most concerts are totally sold out. The advance expenses for each concert are $55,000. The cost of meals, motels, jet travel, and support personnel total $110,000 per week. How many weeks will the Three Tenors need to be on tour if each of them wants to earn $120,000 per year?

11. *Commuting to Work* While commuting to work in Orlando, Florida, Marcia drives one-half the distance that Melissa drives each day. John drives 17 miles more than Melissa each day. In one day, these three people drive 112 miles commuting to work. How far does each drive each day?

12. *Helping a Local Charity* Alex, Ramon, and Charley volunteer a total of 17 hours per week at My Brother's Table, an organization that feeds needy people. Last week, Alex worked $\frac{1}{3}$ as many hours as Ramon worked. Charley worked $1\frac{1}{2}$ as many hours as Ramon. How many hours did each of them work?

▲ **13.** *Geometry* A new Youth Opportunity Center is being built in Roxbury. The perimeter of the rectangular playing field is 340 yards. The length of the field is 6 yards less than triple the width. What are the dimensions of the playing field?

▲ **14.** *Geometry* Dave and Jane Wells have a new rectangular driveway. The perimeter of the driveway is 164 feet. The length is 12 feet longer than four times the width. What are the dimensions of the driveway?

▲ **15.** *Geometry* A vacant city lot is being turned into a neighborhood garden. The neighbors want to fence in a triangular section of the lot and plant flowers there. The longest side of the triangular section is 7 feet shorter than twice the shortest side. The third side is 6 feet longer than the shortest side. The perimeter is 59 feet. How long is each side?

▲ **16.** *Geometry* A leather coin purse has the shape of a triangle. Two sides are equal in length and the third side is 3 centimeters shorter than one and one-half the length of the equal sides. The perimeter is 28.5 centimeters. Find the lengths of the sides.

17. *Telephone Expenses* Minh Tran owns two appliance repair shops. This year the telephone bill for his Salem shop was $610 less than twice the telephone bill for his Saugus shop. The total telephone expense for both shops was $2054. What was the cost of each shop's telephone service for the year?

18. *Pharmacist Salary* Huyen Doan begins her job as a pharmacist this week. Her weekly salary will be $110 less than triple the weekly salary at her previous job as a pharmacy clerk. The difference between the weekly salaries is $590. What is Huyen's weekly salary as a pharmacist? What was her salary as a pharmacy clerk?

19. *Cell Phone Costs* Juanita Perez is choosing a cell phone plan. Three companies offer a different number of free minutes of phone calls per month. Reliable Cell Phone offers 50 fewer minutes per month than Clear Call Cell Phone. Nationwide Phone Company offers 300 more than one-third the number that Clear Call offers. The sum of the free minutes offered by Clear Call and Nationwide is exactly equal to twice the number of free minutes offered by Reliable.
(a) How many free minutes does each company offer?
(b) If Reliable charges $71.50 per month, Clear Call charges $90, and Nationwide charges $70, which company offers the lowest cost per minute?

20. *Cleaning Supplies* Mr. and Mrs. Wong purchased three boxes of Cascade dishwashing detergent. The largest box contains triple the number of ounces in the smallest box. The medium box contains 12 ounces less than double the number of ounces in the smallest box. The three boxes together contain 228 ounces of dishwashing detergent.
(a) How many ounces are in each size of box?
(b) If the largest box sells for $4.90, the medium box for $2.70, and the small box for $1.95, which box is the best buy?

Cumulative Review

Name the property that justifies each statement.

21. $57 + 0 = 57$

22. $(2 \cdot 3) \cdot 9 = 2 \cdot (3 \cdot 9)$

23. Evaluate. $7(-2) \div 7(-3) - 3$

24. Evaluate. $(7 - 12)^3 - (-4) + 3^3$

1. _____

2. _____

3. _____

4. _____

5. _____

6. _____

7. _____

8. _____

9. _____

10. _____

11. _____

12. _____

13. _____

14. _____

15. _____

16. _____

How are you doing with your homework assignments in Sections 2.1 to 2.4? Do you feel you have mastered the material so far? Do you understand the concepts you have covered? Before you go further in the textbook, take some time to do each of the following problems.

2.1

Solve for x.

1. $2x - 1 = 12x + 36$

2. $\dfrac{x - 2}{4} = \dfrac{1}{2}x + 4$

3. $4(x - 3) = x + 2(5x - 1)$

4. $0.6x + 3 = 0.5x - 7$

2.2

5. Solve for y: $5x - 8y = 15$

6. Solve for a: $5ab - 2b = 16ab - 3(8 + b)$

7. Solve for r: $A = P + Prt$

8. Use your results from Problem 7 to find r when $P = 100$, $t = 3$, and $A = 118$.

2.3

Solve for x.

9. $|3x - 2| = 7$

10. $|8 - x| - 3 = 1$

11. $\left|\dfrac{2x + 3}{4}\right| = 2$

12. $|5x - 8| = |3x + 2|$

2.4

Use an algebraic equation to find a solution for each exercise.

▲ **13.** The perimeter of a rectangle is 64 centimeters. Its length is 4 centimeters less than three times its width. Find the rectangle's dimensions.

14. Eastern Bank charges its customers a flat fee of $6 per month for a checking account plus 12¢ for each check. The bank charged Jose $9.12 for his checking account last month. How many checks did he use?

15. Alan and Cindi pick up food donations for their local food bank from grocery stores. Alan picked up 80 more than half as many pounds of food as Cindi did. Together they picked up 455 pounds of food donations. How many pounds did each pick up?

▲ **16.** In Freeport, Maine, the north end of L. L. Bean has a triangular parking lot for bicycles. The longest side is 5 feet shorter than twice the shortest side. The third side is 9 feet longer than the shortest side. The perimeter is 62 feet. How long is each side?

Now turn to page SA-3 for the answers to each of these problems. Each answer also includes a reference to the objective in which the problem is first taught. If you missed any of these problems, you should stop and review the Examples and Practice Problems in the referenced objective. A little review now will help you master the material in the upcoming sections of the text.

2.5 SOLVING MORE-INVOLVED WORD PROBLEMS

① Solving More-Involved Word Problems by Using an Equation

To solve some word problems, we might need to understand percents, simple interest, mixtures, or some other concept before we can use an algebraic equation as a model. We review one of these concepts in this section.

From arithmetic you know that to find a percent of a number we write the percent as a decimal and multiply the decimal by the number.

Thus, to find 36% of 85, we calculate $(0.36)(85) = 30.6$. If the number is not known, we can represent it by a variable.

EXAMPLE 1 The Wildlife Refuge Rangers tagged 144 deer. They estimate that they have tagged 36% of the deer in the refuge. If they are correct, approximately how many deer are in the refuge?

Solution

1. **Understand the problem.** Let n = the number of the deer in the refuge.
   Let $0.36n$ = 36% of the deer in the refuge.

2. **Write an equation.**

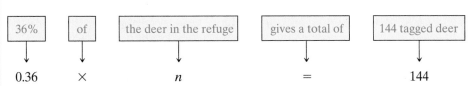

36%	of	the deer in the refuge	gives a total of	144 tagged deer
0.36	×	n	=	144

3. **Solve the equation and state the answer.**

$$0.36n = 144$$
$$\frac{0.36n}{0.36} = \frac{144}{0.36} \quad \text{Divide each side by 0.36.}$$
$$n = 400 \quad \text{Simplify.}$$

There are approximately 400 deer in the refuge.

4. **Check.** Is it true that 36% of 400 is 144?

$$(0.36)(400) \stackrel{?}{=} 144$$
$$144 = 144 \quad \checkmark$$

It checks. Our answer is correct.

Practice Problem 1 Technology Resources, Inc., sold 6900 computer workstations, a 15% increase in sales over the previous year. How many computer workstations were sold last year? (*Hint:* Let x = the amount of sales last year, and let $0.15x$ = the increase in sales over last year.)

Adding two numbers yields a total. We can call one of the numbers x and the other number (total $- x$). We will use this concept in Example 2.

Student Learning Objective

After studying this section, you will be able to:

① Solve more-involved word problems by using an equation.

NOTE TO STUDENT: Fully worked-out solutions to all of the Practice Problems can be found at the back of the text starting at page SP-1

EXAMPLE 2 Bob's and Marcia's weekly salaries total $265. If they both went from part-time to full-time employment, their combined weekly income would be $655. Bob's salary would double, while Marcia's would triple. How much do they each make now?

Solution

1. ***Understand the problem.*** Let b = Bob's part-time salary.
   Since the total of the two part-time weekly salaries is $265, we can let
   $265 - b$ = Marcia's part-time salary.

2. ***Write an equation.***

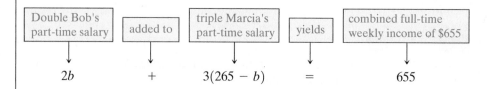

$$2b \qquad + \qquad 3(265 - b) \qquad = \qquad 655$$

3. ***Solve the equation and state the answer.***

   $$2b + 3(265 - b) = 655$$
   $$2b + 795 - 3b = 655 \qquad \text{Remove parentheses.}$$
   $$795 - b = 655 \qquad \text{Simplify.}$$
   $$-b = -140 \qquad \text{Subtract 795 from each side.}$$
   $$b = 140 \qquad \text{Multiply each side by } -1.$$

   If $b = 140$, then $265 - b = 265 - 140 = 125$. Thus, Bob's present part-time weekly salary is $140, and Marcia's present part-time weekly salary is $125.

4. ***Check.*** Do their present weekly salaries total $265? Yes: $140 + 125 = 265$.
   If Bob's income is doubled and Marcia's is tripled, will their new weekly salaries total $655? Yes: $2(140) + 3(125) = 280 + 375 = 655$. ✓

NOTE TO STUDENT: Fully worked-out solutions to all of the Practice Problems can be found at the back of the text starting at page SP-1

Practice Problem 2 Together Alicia and Heather sold forty-three cars at Prestige Motors last month. If Alicia doubles her sales and Heather triples her sales next month, they will sell 108 cars. How many cars did each person sell this month?

Simple interest is an income from investing money or a charge for borrowing money. It is computed by multiplying the amount of money borrowed or invested (called the *principal*) by the rate of interest and by the period of time it is borrowed or invested (usually measured in years unless otherwise stated). Hence

$$\text{interest} = \text{principal} \times \text{rate} \times \text{time}.$$
$$I = prt$$

All interest problems in this chapter involve simple interest.

EXAMPLE 3   Maria has a job as a financial advisor in a bank. She advised a customer to invest part of his money in a money market fund earning 12% simple interest and the rest in an investment fund earning 14% simple interest. The customer had $6000 to invest. If he earned $772 in interest in 1 year, how much did he invest in each fund?

Solution

1. *Understand the problem.* Let x = the amount of money invested at 12% interest.

   The other amount of money is $(\text{total} - x)$.

   $6000 - x$ = the amount of money invested at 14% interest.

2. *Write an equation.*

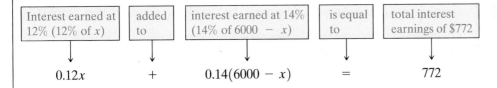

Interest earned at 12% (12% of x)	added to	interest earned at 14% (14% of 6000 $- x$)	is equal to	total interest earnings of $772
$0.12x$	$+$	$0.14(6000 - x)$	$=$	772

3. *Solve the equation and state the answer.*

$$0.12x + 0.14(6000 - x) = 772$$

$$0.12x + 840 - 0.14x = 772 \qquad \text{Remove parentheses.}$$

$$840 - 0.02x = 772 \qquad \text{Add } 0.12x \text{ to } -0.14x.$$

$$-0.02x = -68 \qquad \text{Subtract 840 from each side.}$$

$$\frac{-0.02x}{-0.02} = \frac{-68}{-0.02} \qquad \text{Divide each side by } -0.02.$$

$$x = 3400 \qquad \text{Simplify.}$$

   If $x = 3400$, then $6000 - x = 6000 - 3400 = 2600$. Thus, $3400 was invested in the money market fund earning 12% interest, and $2600 was invested in the investment fund earning 14% interest.

4. *Check.* Do the two amounts of money total $6000?
   Yes: $3400 + $2600 = $6000.
   Does the total interest amount to $772?

$$(0.12)(3400) + (0.14)(2600) \stackrel{?}{=} 772$$

$$408 + 364 \stackrel{?}{=} 772$$

$$772 = 772 \quad \checkmark$$

Our answers are correct.

Practice Problem 3   Tricia received an inheritance of $5500. She invested part of it at 8% simple interest and the remainder at 12% simple interest. At the end of the year she had earned $540. How much did Tricia invest at each interest amount?

Sometimes we encounter a situation in which two or more items are combined to form a mixture or solution. These types of problems are called **mixture problems.**

EXAMPLE 4 A small truck has a radiator that holds 20 liters. A mechanic needs to fill the radiator with a solution that is 60% antifreeze. He has 70% and 30% antifreeze solutions. How many liters of each should he use to achieve the desired mix?

Solution

1. **Understand the problem.** Let x = the number of liters of 70% antifreeze to be used.
   Since the total solution must be 20 liters, we can use $20 - x$ for the other part. So $20 - x$ = the number of liters of 30% antifreeze to be used.

   In this problem a chart or table is very helpful. We will multiply the entry in column (A) by the entry in column (B) to obtain the entry in column (C).

	(A) Number of Liters of the Solution	(B) Percent Pure Antifreeze	(C) Number of Liters of Pure Antifreeze
70% antifreeze solution	x	70%	$0.70x$
30% antifreeze solution	$20 - x$	30%	$0.30(20 - x)$
Final 60% solution	20	60%	$0.60(20)$

2. **Write an equation.** Now we form an equation from the entries in column (C).

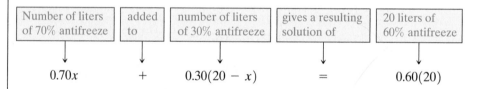

Number of liters of 70% antifreeze	added to	number of liters of 30% antifreeze	gives a resulting solution of	20 liters of 60% antifreeze
$0.70x$	$+$	$0.30(20 - x)$	$=$	$0.60(20)$

3. **Solve the equation and state the answer.**

$$0.70x + 0.30(20 - x) = 0.60(20)$$
$$0.70x + 6 - 0.30x = 12$$
$$6 + 0.40x = 12$$
$$0.40x = 6$$
$$\frac{0.40x}{0.40} = \frac{6}{0.40}$$
$$x = 15$$

If $x = 15$, then $20 - x = 20 - 15 = 5$. Thus, the mechanic needs 15 liters of 70% antifreeze solution and 5 liters of 30% antifreeze solution.

4. **Check.** Can you verify that this answer is correct?

NOTE TO STUDENT: Fully worked-out solutions to all of the Practice Problems can be found at the back of the text starting at page SP-1

Practice Problem 4 A jeweler wishes to prepare 200 grams of 80% pure gold from sources that are 68% pure gold and 83% pure gold. How many grams of each should he use?

Some problems involve the relationship distance = rate × time or $d = rt$.

EXAMPLE 5   Frank drove at a steady speed for 3 hours on the turnpike. He then slowed his traveling speed by 15 miles per hour on the secondary roads. The entire trip took 5 hours and covered 245 miles. What was his speed on the turnpike?

Solution

1. *Understand the problem.* Let x = speed on the turnpike in miles per hour. So $x - 15$ = speed on the secondary roads in miles per hour.

   Again, as in Example 4, a chart or table is very helpful. We will use a chart that records values of $(r)(t) = d$.

	Rate (miles per hour) r	Time (hours) t	Distance (miles) $(r)(t) = d$
Turnpike	x	3	$3x$
Secondary roads	$x - 15$	2	$2(x - 15)$
Entire trip	Not appropriate	5	245

2. *Write an equation.* Using the distance entries (third column of the table), we have

Distance on turnpike	+	Distance on secondary roads	=	Total distance
$3x$	+	$2(x - 15)$	=	245

3. *Solve the equation and state the answer.*

$$3x + 2(x - 15) = 245$$
$$3x + 2x - 30 = 245$$
$$5x - 30 = 245$$
$$5x = 275$$
$$x = 55$$

   Thus, Frank traveled at an average speed of 55 miles per hour on the turnpike.

4. *Check.* Can you verify that this answer is correct?

Practice Problem 5   Wally drove for 4 hours at a steady speed. He slowed his speed by 10 miles per hour for the last part of the trip. The entire trip took 6 hours and covered 352 miles. How fast did he drive on each portion of the trip?

Student Solutions Manual   CD/Video   PH Math Tutor Center   MathXL®Tutorials on CD   MathXL®   MyMathLab®   Interactmath.com

Write an algebraic equation for each problem and solve it.

1. **U.S. Population** The population of the United States in 1999 was estimated to be 273.1 million people. This was an increase of 34% from the population in 1969. What was the population of the United States in 1969? *Source*: U.S. Census Bureau.

2. **National Debt** The U.S. national debt on November 14, 2000, was approximately $5.676 trillion. This was an increase of 29% from the national debt on November 14, 1993. What was the national debt on November 14, 1993? *Source*: U.S. Treasury Department.

3. **Consumer Purchases** Eastwing dormitory just purchased a new Sony color television on sale for $340. The sale price was 80% of the original price. What was the original price of the television set?

4. **West Nile Virus** This year Larchmont County has reported finding 27 mosquitoes that are infected with West Nile Virus. This is an increase of 12.5% from the number reported last year. How many were reported last year?

5. **Unemployment Rates** Allentown's employment statistics show that 969 of its residents were unemployed last month. This is a decrease of 15% from the previous month. How many residents were unemployed in the previous month?

6. **Lyme Tick Disease** A wildlife expert at the Crane Wildlife Reserve has found fifteen deer with ticks that are carrying Lyme disease. She estimates that this number is about 60% of the total number of deer carrying infected ticks. Approximately how many such deer are there on the Reserve?

7. **Reforestation Program** In a reforestation program, 1400 seedling trees were planted. Twice as many spruce as hemlocks were planted. The number of balsams planted was twenty more than triple the number of hemlocks. How many of each type of seedling was planted?

8. **Summer Rental** Lynn and Judy are pooling their savings to rent a cottage in Maine for a week this summer. The rental cost is $950. Lynn's family is joining them, so she is paying a larger part of the cost. Her share of the cost in $250 less than twice Judy's. How much of the rental fee is each of them paying?

9. *Salary Increases* When Angela and Walker first started working for the supermarket, their weekly salaries totaled $500. Now during the last 25 years Walker has seen his weekly salary triple. Angela has seen her weekly salary become four times larger. Together their weekly salaries now total $1740. How much did they each make 25 years ago?

10. *Salary Increases* When Grace and Tony started as junior engineers at a manufacturing company, their weekly salaries totaled $1300. Now ten years later they have both become senior engineers. Grace had had her salary double. Tony has had his salary triple. Together their weekly salaries now total $3200. How much did they each make 10 years ago?

11. *Medicine Dosage* A hospital received a shipment of 8-milligram doses of a medicine. Each 8-milligram package was repacked into two smaller doses of unequal size and labeled packet A and packet B. The hospital then used 17 doses of packet A and 14 doses of packet B in one week. The hospital used a total of 127 milligrams of the medicine during that week. How many milligrams of the medicine are contained in each A packet? In each B packet?

12. *Cookie Sales* This year, two Girl Scout Troops together sold 460 boxes of cookies. Half of the Rockland troop's sales were Thin Mints and $\frac{2}{5}$ of the Harrisville troop's sales were Thin Mints. Together they sold 205 boxes of Thin Mints. How many boxes of cookies did each troop sell?

13. *Interest on a Loan* Alice and Samuel borrowed $600 at a simple interest rate of 12% for a period of 3 years. What was the interest?

14. *Interest on a Loan* Juanita and Carlos borrowed $4800 at a simple interest rate of 11% for a period of 2 years. What was the interest?

15. *College Fund Investment* The Vegas want to invest $5000 of their daughter's college fund for 18 months in a certificate of deposit that pays a simple interest rate of 3.1%. How much interest will they earn?

16. *Earned Interest* David had $4000 invested for one-fourth of a year at a simple interest rate of 6.1%. How much interest did he earn?

17. *Investment Income* Cynthia is a hardworking single mother on a very limited budget. She invests her savings only in safe investments because she is not in a position to take risks and lose any of her hard-earned money. She invested $6400 in two types of accounts for one year. The first type earned 5% simple interest, and the second type earned 8% simple interest. At the end of the year, Cynthia had earned $395 in interest. How much did she invest at each rate?

18. *Investment Income* The Johnson's family business did well *this* year due to their investments *last* year. The business earned $6570 on an investment of $45,000 in mutual funds. There were two types of funds that Jenna Johnson invested in. The first was a pharmaceutical fund, which paid out simple interest of 13%. The second was a genetech fund, which paid out simple interest of 16%. How much did Jenna Johnson invest in each fund last year for her family's business?

19. *Retirement Fund Investment* Jim Jacobs decided to invest $18,000 of his retirement fund conservatively. He invested part of this money in a certificate of deposit that pays 3.5% simple interest and part in a fixed interest account that pays 2.2% simple interest. Last year he earned $552 in interest. How much did he invest in each type of account?

20. *Investment Income* Julio invested $7000 in money market funds. Part was invested at 8% simple interest, and the rest was invested at 6% simple interest. At the end of 1 year, Julio had earned $532 in interest. How much did he invest in each fund?

21. *Fat Content of Food* A chef has one cheese that contains 45% fat and another cheese that contains 20% fat. How many grams of each cheese should she use in order to obtain 30 grams of a cheese mixture that is 30% fat?

22. *Chemical Mixtures* Becky was awarded a college internship at a local pharmaceutical research corporation. She has been given her own corner of the laboratory to try her hand at pharmacology. One of her assignments is to mix two solutions, one that is 16% strength and the other that is 9% strength. How many milliliters of each should she use in order to obtain 350 milliliters of a 12% strength solution?

23. *Fat Content of Food* The meat department manager at a large food store wishes to mix some hamburger with 30% fat content and some hamburger that has 10% fat content in order to obtain 100 pounds of hamburger with 25% fat content. How much of each type of hamburger should she use?

24. *Cost of Tea* A grocer at a specialty store is mixing tea worth $6 per pound with tea worth $8 per pound. He wants to obtain 144 pounds of tea worth $7.50 per pound. How much of each tea should he use?

25. *Fertilizer Mix* A landscaping company needs 150 gallons of 18% fertilizer to fertilize the shrubs in an office park. They have in stock 25% fertilizer and 15% fertilizer. How much of each type should they mix together?

26. *Insecticide Manufacturing* Most mosquito repellents contain the ingredient deet. A manufacturer needs to produce 10-ounce spray cans of 40% deet. How much 100% deet and how much 25% deet should be mixed to produce each of these spray cans?

27. *Auto Travel* Susan drove for 4 hours on secondary roads at a steady speed. She completed her trip by driving 2 hours on an interstate highway. Her total trip was 250 miles. Her speed on the interstate highway portion of the trip was 20 miles per hour faster than her speed on the secondary roads. How fast did she travel on the secondary roads?

28. *Airplane Flight* Alice and Wendy flew a small plane for 930 miles. For the first 3 hours they flew at maximum speed. After refueling they finished the trip at a cruising speed that was 60 miles per hour slower than maximum speed. The entire trip took 5 hours of flying time. What is the maximum flying speed of the plane?

29. **Walking on a Treadmill** Yissania and Charlotte walked on treadmills at the gym for the same amount of time. Yissania walks at 5 miles per hour, and Charlotte walks at 4.2 miles per hour. If Yissania walked 0.6 miles farther than Charlotte, how long did they use the treadmills?

30. **Boating on a Lake** The Clarke family went sailing on a lake. Their boat averaged 6 kilometers per hour. The Rourke family took their outboard runabout for a trip on the lake for the same amount of time. Their boat averaged 14 kilometers per hour. The Rourke family traveled 20 kilometers farther than the Clarke family. How many hours did each family spend on their boat trip?

To Think About

31. **Business Profits** A bottled iced tea manufacturer saw profits increase 65% last year. This year, due to so much new competition, profits fell 40%. Profits this year were $17,820,000. What was the realized profit for the iced tea company 2 years ago?

32. **Food on an Airline** Northwest Airlines gives out peanuts to all passengers who want them. On several recent flights to Boston, 72% of the passengers who were asked accepted the peanuts, and 23% rejected the peanuts. A total of thirty-six passengers were asleep and were not asked. How many passengers were on these flights?

▲ 33. **Geometry** The area of the shaded triangle is 6 cm². Find the area of parallelogram *ACDF*.

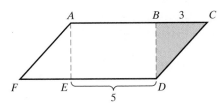

▲ 34. **Geometry** A sphere is constructed inside a cylinder. Assume that the sphere and the cylinder have the same radius. The volume of the sphere is 38,808 cm³. Find the volume of the cylinder. (Use $\pi \approx \frac{22}{7}$.)

Cumulative Review

Evaluate.

35. $5a - 2b + c$ when $a = 1, b = -3, c = -4$

36. $2x^2 - 3x + 1$ when $x = -2$

37. $\dfrac{2 - 6(-1) + 5^2}{|4 - 7|}$

38. $\dfrac{\sqrt{6^2 - 20}}{5(8) - 2^2(9)}$

2.6 LINEAR INEQUALITIES

Student Learning Objectives

After studying this section, you will be able to:

1 Determine whether one number is less than or greater than another number.

2 Graph a linear inequality in one variable.

3 Solve a linear inequality in one variable.

① Determining Whether One Number Is Less Than or Greater Than Another Number

We briefly introduced inequalities in Chapter 1. Now we will discuss this concept more completely.

A **linear inequality** is a statement that describes how two numbers or linear expressions are related to one another. We can use the number line to visualize the concept of inequality.

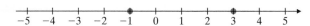

It is a mathematical property that -1 is less than 3. We write this in the following way:

$$-1 < 3.$$

Notice the position of these numbers on the number line. -1 is to the *left* of 3 on the number line. We say that a first number *is less than* a second number if the first number is *to the left* of the second number on the number line.

We could also say that 3 is greater than -1 (because it is *to the right* of -1 on the number line). We then write this in the following way:

$$3 > -1.$$

An inequality symbol can face either right or left, but its opening must face the larger number.

In general, if $a < b$, then it is also true that $b > a$.

EXAMPLE 1   Insert the proper symbol between the numbers.

(a) 8 _____ 6   **(b)** -2 _____ 3   **(c)** -4 _____ -2

(d) $\dfrac{1}{2}$ _____ $\dfrac{1}{3}$ **(e)** -0.033 _____ -0.0329

Solution

(a) $8 > 6$ because 8 is to the right of 6 on the number line.

(b) $-2 < 3$ because -2 is to the left of 3 on the number line.

(c) $-4 < -2$ because -4 is to the left of -2 on the number line.

(d) When comparing two fractions, remember to rewrite them with a common denominator.

$\dfrac{1}{2}$ ____ $\dfrac{1}{3}$ Rewrite fractions with the same denominator.

$\dfrac{3}{6}$ ____ $\dfrac{2}{6}$ Compare the numerators: $3 > 2$.

$\dfrac{3}{6} > \dfrac{2}{6}$

Thus, $\dfrac{1}{2} > \dfrac{1}{3}$ because $\dfrac{3}{6} > \dfrac{2}{6}$.

(e) $-0.033 < -0.0329$ because $-0.0330 < -0.0329$.

Notice that since both numbers are negative, -0.0330 is to the left of -0.0329 on the number line.

Practice Problem 1   Insert the symbol $<$ or $>$ between the two numbers.

(a) -1 _____ -2 **(b)** $\dfrac{2}{3}$ _____ $\dfrac{3}{4}$ **(c)** -0.561 _____ 0.5555

NOTE TO STUDENT: Fully worked-out solutions to all of the Practice Problems can be found at the back of the text starting at page SP-1

Numerical or algebraic expressions as well as numbers can be compared.

EXAMPLE 2   Insert the proper symbol between the expressions.

(a) $(5 - 8)$ _____ $(2 - 3)$ **(b)** $|1 - 7|$ _____ $|-4 - 12|$

Solution

(a) $(5 - 8)$ _____ $(2 - 3)$ Evaluate each expression and compare.
$$-3 < -1$$
Thus, $(5 - 8) < (2 - 3)$ because $-3 < -1$.

(b) $|1 - 7|$ _____ $|-4 - 12|$ Evaluate each expression and compare.
$$6 < 16$$
Thus, $|1 - 7| < |-4 - 12|$ because $6 < 16$.

Practice Problem 2   Insert $<$ or $>$ between the expressions.

(a) $(-8 - 2)$ _____ $(-3 - 12)$ **(b)** $|-15 + 8|$ _____ $|7 - 13|$

In addition to the $<$ and $>$ symbols, we will also encounter two other notations when we deal with inequalities. We briefly mentioned these in Chapter 1. If we want to say that a number x is greater than or equal to 5, we would write this using the following notation:

$$x \geq 5.$$

Likewise, if we want to say that a number x is less than or equal to 8, we would write it using the following notation:

$$x \leq 8.$$

The symbol $\leq$ means **less than or equal to,** and the symbol $\geq$ means **greater than or equal to.**

2 Graphing Linear Inequalities in One Variable

Inequality symbols are often used with variables. For example,

$$x > 4$$

means that x can be any number that is greater than 4. The variable x cannot equal 4. We can use the number line to graph this inequality.

On the number line, we shade the portion that is to the right of 4. Any point in the shaded portion will satisfy the inequality since all points to the right of 4 are greater than 4. The open circle at 4 means that x cannot be 4.

Let's look at the inequality $x < -1$. This means that x can be any number that is less than -1. To graph the inequality, we will shade all points to the left of -1 on the number line.

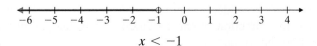

$$x < -1$$

Inequality symbols used with variables can include the equal sign. The inequality $x \geq -2$ means all numbers greater than or equal to -2. We graph this inequality as follows.

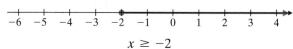

$$x \geq -2$$

The shaded circle at -2 means that the graph includes the point -2. -2 is a solution to the inequality.

Similarly, $x \leq 1$ is graphed as follows.

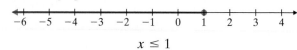

$$x \leq 1$$

EXAMPLE 3   Graph each inequality.

(a) $x < 0$ **(b)** $x \leq 0$ **(c)** $x > -5$ **(d)** $x \geq -5$ **(e)** $2 < x$

Solution

(a) $x < 0$
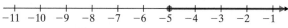

(b) $x \leq 0$
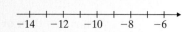

(c) $x > -5$
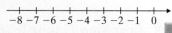

(d) $x \geq -5$

(e) We read an inequality starting with the variable. Thus, we read $2 < x$ as "x is greater than 2" and graph the expression accordingly.

NOTE TO STUDENT: Fully worked-out solutions to all of the Practice Problems can be found at the back of the text starting at page SP-1

Practice Problem 3   Graph each inequality.

(a) $x > 3.5$

(b) $x \leq -10$

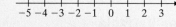

(c) $x \geq -2$

(d) $-4 > x$

③ Solving Inequalities in One Variable

Inequalities that have the same solution are said to be **equivalent.** Solving a first-degree inequality is similar to solving first-degree equations. We use various properties of real numbers.

> ### ADDITION AND SUBTRACTION PROPERTY FOR INEQUALITIES
>
> For all real numbers a, b, and c, if $a < b$, then
>
> $$a + c < b + c \quad \text{and} \quad a - c < b - c.$$
>
> If the same number is added to or subtracted from both sides of an inequality, the result is an equivalent inequality.

The same number can be added to or subtracted from both sides of an inequality without affecting the direction of the inequality. (Any inequality symbol can be used. We used $<$ for convenience.)

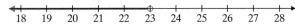

EXAMPLE 4 Solve the inequality. Graph and check the solution. $x - 8 < 15$

Solution
$$x - 8 < 15$$
$$x - 8 + 8 < 15 + 8 \quad \text{Add } +8 \text{ to each side.}$$
$$x < 23 \qquad\qquad \text{Simplify.}$$

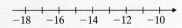

To check, choose any numerical value that lies on the number line where the red arrow is indicated. See whether a true statement results when you substitute it into the inequality. We will choose 22.5.

$$x - 8 < 15 \qquad \text{Substitute 22.5 for } x \text{ in the original inequality.}$$
$$22.5 - 8 \overset{?}{<} 15$$
$$14.5 < 15 \checkmark$$

Practice Problem 4 Solve the inequality. Graph and check the solution.
$x + 2 > -12$

MULTIPLICATION OR DIVISION BY A POSITIVE NUMBER

For all real numbers a, b, and c when $c > 0$, if $a < b$, then

$$ac < bc \quad \text{and} \quad \frac{a}{c} < \frac{b}{c}.$$

If both sides of an inequality are multiplied or divided by the same *positive* number, the result is an equivalent inequality.

When we multiply or divide both sides of an inequality by a positive number, the direction of the inequality is not changed. That is, if $5x < -15$ and we divide both sides by 5, we obtain $x < -3$.

To check this solution, we choose the value -4. Is $5(-4) < -15$ true? Yes.

EXAMPLE 5 Solve $6x + 3 \le 2x - 5$. Graph your solution.

Solution

$$6x + 3 - 3 \le 2x - 5 - 3 \quad \text{Subtract 3 from each side.}$$
$$6x \le 2x - 8 \qquad\qquad \text{Simplify.}$$
$$6x - 2x \le 2x - 2x - 8 \quad \text{Subtract } 2x \text{ from each side.}$$
$$4x \le -8 \qquad\qquad \text{Simplify.}$$
$$\frac{4x}{4} \le \frac{-8}{4} \qquad\qquad \text{Divide each side by 4. The inequality is not reversed.}$$
$$x \le -2$$

To check, we choose -2 and -3.

$$6x + 3 \le 2x - 5 \qquad\qquad 6x + 3 \le 2x - 5$$
$$6(-2) + 3 \overset{?}{\le} 2(-2) - 5 \qquad 6(-3) + 3 \overset{?}{\le} 2(-3) - 5$$
$$-9 \le -9 \checkmark \qquad\qquad\qquad -15 \le -11 \checkmark$$

NOTE TO STUDENT: Fully worked-out solutions to all of the Practice Problems can be found at the back of the text starting at page SP-1

Practice Problem 5 Solve $8x - 8 \geq 5x + 1$. Graph your solution.

$$\begin{array}{ccccccccc} \leftarrow\!\!\!&+&+&+&+&+&+&+&+&\!\!\!\rightarrow \\ -1 & 0 & 1 & 2 & 3 & 4 & 5 & 6 & 7 \end{array}$$

When we multiply or divide both sides of an inequality by a *negative number,* the direction of the inequality is *reversed.* That is, if we divide both sides of $-3x < 21$ by -3, we obtain $x > -7$. If we divide both sides of $-4x \geq -16$ by -4, we obtain $x \leq 4$.

To check the solution of the first inequality, we choose the value 1. Is $-3(1) < 21$? Yes. To reverse the inequality symbol might seem to be an unusual move. Let's see what would happen if we did not reverse the symbol. If we did not, the solution would be $x < -7$. To check this solution, we choose -8. Is $-3(-8) < 21$ true? No, 24 is not less than 21.

MULTIPLICATION OR DIVISION BY A NEGATIVE NUMBER

For all real numbers a, b, and c when $c < 0$, if $a < b$, then

$$ac > bc \quad \text{and} \quad \frac{a}{c} > \frac{b}{c}.$$

If both sides of an inequality are multiplied or divided by the same *negative* number *and the inequality symbol is reversed,* the result is an equivalent inequality.

TO THINK ABOUT: Why Reverse the Inequality Sign? Show why you must reverse the inequality symbol when you multiply or divide by a negative number. Provide several examples.

EXAMPLE 6 Solve. $-8x - 12 < -4(x - 4) + 8$

Solution

$-8x - 12 < -4x + 16 + 8$	Use the distributive property to remove the parentheses.
$-8x - 12 < -4x + 24$	Simplify.
$-8x + 4x - 12 < -4x + 4x + 24$	Add $4x$ to each side.
$-4x - 12 < 24$	Simplify.
$-4x - 12 + 12 < 24 + 12$	Add 12 to each side.
$-4x < 36$	Simplify.
$\dfrac{-4x}{-4} > \dfrac{36}{-4}$	Divide each side by -4 and reverse the inequality.
$x > -9$	Simplify.

Practice Problem 6 Solve. $2 - 12x > 7(1 - x)$

An inequality that contains all decimal terms is best handled by first multiplying both sides of the inequality by the appropriate power of 10.

EXAMPLE 7   Solve. $-0.3x + 1.0 \leq 1.2x - 3.5$

Solution $10(-0.3x + 1.0) \leq 10(1.2x - 3.5)$ Multiply each side by 10.

$$-3x + 10 \leq 12x - 35$$

$$-3x - 12x + 10 \leq 12x - 12x - 35 \quad \text{Add } -12x \text{ to each side.}$$

$$-15x + 10 \leq -35 \qquad\qquad \text{Simplify.}$$

$$-15x + 10 - 10 \leq -35 - 10 \qquad \text{Add } -10 \text{ to each side.}$$

$$-15x \leq -45 \qquad\qquad \text{Simplify.}$$

$$\frac{-15x}{-15} \geq \frac{-45}{-15} \qquad\qquad \begin{array}{l}\text{Divide each side by } -15 \text{ and} \\ \text{reverse the direction of the} \\ \text{inequality.}\end{array}$$

$$x \geq 3$$

Practice Problem 7   Solve. $0.5x - 0.4 \leq -0.8x + 0.9$

To solve an inequality that contains fractions, multiply both sides of the inequality by the LCD to clear the fractions.

EXAMPLE 8   Solve. $\dfrac{1}{7}(x + 5) > \dfrac{1}{5}(x + 1)$

Solution

$$\frac{x}{7} + \frac{5}{7} > \frac{x}{5} + \frac{1}{5} \qquad\qquad \begin{array}{l}\text{Using the distributive property, remove} \\ \text{the parentheses.}\end{array}$$

$$35\left(\frac{x}{7}\right) + 35\left(\frac{5}{7}\right) > 35\left(\frac{x}{5}\right) + 35\left(\frac{1}{5}\right) \quad \text{Multiply each term by the LCD.}$$

$$5x + 25 > 7x + 7 \qquad\qquad \text{Simplify.}$$

$$5x + 25 - 25 > 7x + 7 - 25 \qquad \text{Subtract 25 from each side.}$$

$$5x > 7x - 18 \qquad\qquad \text{Simplify.}$$

$$5x - 7x > 7x - 7x - 18 \qquad \text{Subtract } 7x \text{ from each side.}$$

$$-2x > -18 \qquad\qquad \text{Simplify.}$$

$$\frac{-2x}{-2} < \frac{-18}{-2} \qquad\qquad \begin{array}{l}\text{Divide each side by } -2 \text{ and reverse the} \\ \text{direction of the inequality.}\end{array}$$

$$x < 9$$

Practice Problem 8   Solve. $\dfrac{1}{5}(x - 6) < \dfrac{1}{3}(x - 2)$

TO THINK ABOUT: Multiplying by the LCD   Another approach to solving the inequality in Example 8 would be to multiply each side of the inequality by the LCD before removing the parentheses. Try it. Think about the pros and cons of this approach. Choose the method you like best.

EXAMPLE 9   Lexi and her mother are making a long distance phone call from Honolulu to West Chicago, IL. The charge is $4.50 for the first minute and 85¢ for each additional minute. Any fractional part of a minute will be rounded up to the nearest whole minute. What is the maximum time that Lexi and her mother can talk if they have $15.55 in change to make the call?

Let $x =$ the number of minutes after the first minute that they place the call. The cost must be less than or equal to $15.55. So we write the inequality

$$4.50 + 0.85x \le 15.55$$
$$0.85x \le 11.05 \quad \text{We subtract 4.50 from each side}$$
$$x \le 13 \quad \text{We divide each side by 0.85}$$

Now we add the 13 minutes to the one minute that cost $4.50. This gives us 14 minutes. Thus the maximum amount of time they can talk is 14 minutes.

Practice Problem 9   Snowflake and her mother are making a long distance phone call from Anchorage to Darien, CT. The charge is $3.50 for the first minute and 65¢ for each additional minute. Any fractional part of a minute will be rounded up to the nearest whole minute. What is the maximum time that Snowflake and her mother can talk if they have $13.90 in change to make the call?

Verbal and Writing Skills

True or false?

1. The statement $6 < 8$ conveys the same information as $8 > 6$.

2. Adding $-5x$ to each side of an inequality reverses the direction of the inequality.

3. Dividing each side of an inequality by -4 reverses the direction of the inequality.

4. The graph of $x > -2$ is the set of all points to the right of -2 on the number line.

5. The graph of $x \leq 6$ does not include the point at 6 on the number line.

6. To solve the inequality $\frac{2}{3}x + \frac{3}{4} \geq \frac{1}{2}x - 4$, multiply each fraction by the LCD.

Insert the symbol $<$ or $>$ between each pair of numbers.

7. 6 _____ -3

8. -15 _____ 4

9. -7 _____ -2

10. -5 _____ -9

11. $\frac{3}{4}$ _____ $\frac{2}{3}$

12. $\frac{5}{6}$ _____ $\frac{5}{7}$

13. $-\frac{2}{9}$ _____ $-\frac{3}{14}$

14. $-\frac{7}{16}$ _____ $-\frac{6}{13}$

15. -3.4 _____ -3.41

16. -2.69 _____ -2.7

17. $|3 - 7|$ _____ $|9 - 2|$

18. $|-8 + 2|$ _____ $|6 - 13|$

Graph each inequality.

19. $x \geq -2$

20. $x \geq -4$

21. $x < 15$

22. $x < 80$

Solve for x and graph your solution.

23. $2x - 7 \leq -5$

24. $3 + 5x \geq 18$

25. $3x - 7 > 9x + 5$

26. $2x + 5 > 4x - 5$

27. $0.5x + 0.1 < 1.1x + 0.7$

28. $1.7 - 0.6x \leq x + 0.1$

Solve for x.

29. $4x - 1 > 15$

30. $5x - 1 > 29$

31. $7x + 2 \leq 2x - 8$

32. $8x + 9 \leq 3x - 16$

33. $2x + \frac{5}{3} > \frac{2}{5}x - 1$

34. $2x + \frac{5}{2} > \frac{3}{2}x - 2$

35. $3x - 11 + 4(x + 8) < 0$

36. $2x - 11 + 3(x + 2) < 0$

37. $\frac{2}{3}x - (x - 2) \geq 4$

38. $-2(x - 3) + \frac{x}{2} - \frac{5}{2} < 0$

Mixed Practice

Solve for x.

39. $0.6x + 1 \leq 2.8$

40. $-0.7x + 1.1 \geq 2 - x$

41. $0.1(x - 2) \geq 0.5x - 0.2$

42. $1.2 - 0.8x \leq 0.3(4 - x)$

43. $2 - \frac{1}{5}(x - 1) \geq \frac{2}{3}(2x + 1)$

44. $\frac{3}{4} + \frac{1}{2}(x - 7) \leq 1 - \frac{x}{4}$

45. $\frac{2x - 3}{5} + 1 \geq \frac{1}{2}x + 3$

46. $4 - \frac{3x - 1}{3} > \frac{x}{6} + \frac{7}{2}$

Applications

For Exercises 47–56, describe the situation with a linear inequality and then solve the inequality.

47. *Tip Income* A waitress earns $3 per hour plus an average tip of $4 for every table served. How many tables must she serve to earn more than $52 for a 4-hour shift?

48. *Telemarketing* A phone solicitor selling long-distance services earns $7.75 per hour plus $25 for every new customer she signs up. How many customers must she sign up to earn more than $401.50 during the next 26 working hours?

49. *Telephone Rates* Rusty Slater is making a long-distance phone call to Orlando, Florida, from a pay phone. The operator informs him that the charge will be $3.95 for the first minute and 55¢ for each additional minute. Any fractional part of a minute used will be rounded up to the nearest whole minute. What is the maximum time Rusty can talk if he has $13.30 in change in his pocket?

50. *Aircraft Cargo Capacity* A small plane takes off with packages from Beverly Airport. Each package weighs 68.5 pounds. The plane has a carrying capacity for people and packages of 2395 pounds. The plane is carrying a pilot who weighs 180 pounds and a passenger who weighs 160 pounds. How many packages can be safely carried?

51. *Elevator Capacity* Molly and Denton from the computer services department are delivering several new computers to faculty offices using the college elevator. The elevator has a maximum capacity of 1100 pounds. Molly weighs 130 pounds, and Denton weighs 155 pounds. Each computer weighs 59 pounds. How many computers can Molly and Denton place on the elevator and then safely ride with the computers up to the next floor?

52. *Mailing Costs* Before the most recent price increase at the U.S. Postal Service, DeWolf Associates sent out several boxes of literature. They planned for a mailing budget of $4.51 per box. The post office charged $0.33 for the first ounce and $0.22 for each additional ounce. What was the most that a box could weigh and still be mailed at a cost that did not exceed the budget?

53. *Business Profit* A time-share salesperson sells a lifetime right to 1 week's vacation per year in San Diego, California, for $12,000. The corporation's overall fixed costs are $500,000 per month. The cost to the corporation to obtain a lifetime right to 1 week's vacation is $6000. The board of directors wants the salespeople to sell enough time-shares to realize a profit of $100,000 per month. How many time-shares must be sold each month?

54. *Commission Sales* A shoe salesperson earns $10 per hour plus 10% commission on all of her shoe sales. How many dollars of shoe sales must she have in order to earn at least $450 in a 40-hour week?

55. *Museum of Science Admission Costs* Admission to the Museum of Science costs the Hernandez family $36. A family membership costs $115 per year. How many times in a year must the family visit the museum in order to make a family membership less expensive than paying admission for each visit?

56. *Landscaping Costs* Morticia and Gomez want to put in a new garden. The landscaper will charge them either $250 to dig up the old garden plus $6 per hour to put in the new plants or just $8 per hour for the entire job. How many hours must be spent digging up the garden before the second payment option becomes the more cost-efficient one?

Cumulative Review

Simplify.

57. $3xy(x + 2) - 4x^2(y - 1)$ **58.** $\frac{2}{3}ab(6a - 2b + 9)$ **59.** $\left(\frac{3x}{2y^2w^{-4}}\right)^3$ **60.** $(-4x^{-2}y^4z^{-6})^{-2}$

2.7 COMPOUND INEQUALITIES

1 Graphing Compound Inequalities Using the Connective *and*

Some inequalities consist of two inequalities connected by the word *and* or the word *or*. They are called **compound inequalities.** The solution of a compound inequality using the connective *and* includes all the numbers that make both parts true at the same time.

EXAMPLE 1 Graph the values of x where $7 < x$ *and* $x < 12$.

Solution We read the inequality starting with the variable. Thus, we graph all values of x, where x is greater than 7 and where x is less than 12. All such values must be between 7 and 12. Numbers that are greater than 7 and less than 12 can be written as $7 < x < 12$.

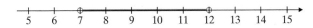

Practice Problem 1 Graph the values of x where $-8 < x$ *and* $x < -2$.

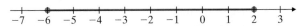

EXAMPLE 2 Graph the values of x where $-6 \le x \le 2$.

Solution Here we have that x is greater than or equal to -6 and that x is less than or equal to 2. We remember to include the points -6 and 2 since the inequality symbols contain the equal sign.

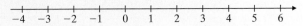

Practice Problem 2 Graph the values of x where $-1 \le x \le 5$.

EXAMPLE 3 Graph the values of x where $-8.5 \le x < -1$.

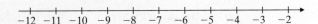

Solution Note the shaded circle at -8.5 and the open circle at -1.

Practice Problem 3 Graph the values of x where $-10 \le x \le -5.5$.

NOTE TO STUDENT: *Fully worked-out solutions to all of the Practice Problems can be found at the back of the text starting at page SP-1*

Student Learning Objectives

After studying this section, you will be able to:

1 Graph a compound inequality using the connective *and*.

2 Graph a compound inequality using the connective *or*.

3 Solve a compound inequality and graph the solution.

EXAMPLE 4 Graph the salary range (*s*) of the full-time employees of Tentron Corporation. Each person earns at least $190 weekly, but not more than $800 weekly.

Solution "At least $190" means that the weekly salary of each person is greater than or equal to $190 weekly. We write $s \geq 190$. "Not more than" means that the weekly salary of each person is less than or equal to $800. We write $s \leq \$800$. Thus, *s* may be between 190 and 800 and may include those end points.

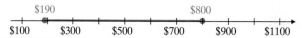

NOTE TO STUDENT: Fully worked-out solutions to all of the Practice Problems can be found at the back of the text starting at page SP-1

Practice Problem 4 Graph the weekly salary range of a person who earns at least $200 per week, but never more than $950 per week.

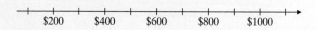

2 Graphing Compound Inequalities Using the Connective *or*

The solution of a compound inequality using the connective *or* includes all the numbers that belong to either of the two inequalities.

EXAMPLE 5 Graph the region where $x < 3 \; or \; x > 6$.

Solution Notice that a solution to this inequality need not be in both regions at the same time.

Read the inequality as "*x* is less than 3 or *x* is greater than 6." Thus, *x* can be less than 3 or *x* can be greater than 6. This includes all values to the left of 3 as well as all values to the right of 6 on the number line. We shade these regions.

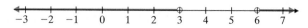

Practice Problem 5 Graph the region where $x < 8 \; or \; x > 12$.

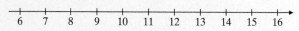

EXAMPLE 6 Graph the region where $x > -2 \; or \; x \leq -5$.

Solution Note the shaded circle at -5 and the open circle at -2.

Practice Problem 6 Graph the region where $x \leq -6 \; or \; x > 3$.

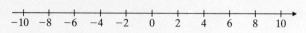

EXAMPLE 7   Male applicants for the state police force in Fred's home state are ineligible for the force if they are shorter than 60 inches or taller than 76 inches. Graph the range of rejected applicants' heights.

Solution   Each rejected applicant's height h will be less than 60 inches ($h < 60$) or will be greater than 76 inches ($h > 76$).

50 inches   60 inches   70 inches   80 inches

Practice Problem 7   Female applicants are ineligible if they are shorter than 56 inches or taller than 70 inches. Graph the range of rejected applicant's heights.

50 in.   60 in.   70 in.   80 in.   90 in.

---

③ Solving Compound Inequalities

When asked to solve a compound inequality for x, we normally solve each individual inequality separately.

EXAMPLE 8   Solve for x and graph the solution of $3x + 2 > 14 \text{ or } 2x - 1 < -7$.

Solution   We solve each inequality separately.

$$
\begin{array}{ccc}
3x + 2 > 14 & or & 2x - 1 < -7 \\
3x > 12 & & 2x < -7 + 1 \\
x > 4 & & 2x < -6 \\
& & x < -3
\end{array}
$$

The solution is $x < -3 \text{ or } x > 4$.

−5 −4 −3 −2 −1 0 1 2 3 4 5

Practice Problem 8   Solve for x and graph the solution of $3x - 4 < -1 \text{ or } 2x + 3 > 13$.

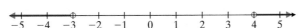

−2 −1 0 1 2 3 4 5 6 7 8

EXAMPLE 9   Solve for x and graph the solution of $5x - 1 > -2 \text{ and } 3x - 4 < 8$.

Solution   We solve each inequality separately.

$$
\begin{array}{ccc}
5x - 1 > -2 & and & 3x - 4 < 8 \\
5x > -2 + 1 & & 3x < 8 + 4 \\
5x > -1 & & 3x < 12 \\
x > -\dfrac{1}{5} & & x < 4
\end{array}
$$

The solution is the set of numbers between $-\frac{1}{5}$ and 4, not including the end points.

$$-\frac{1}{5} < x < 4$$

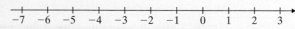

NOTE TO STUDENT: Fully worked-out
solutions to all of the Practice Problems
can be found at the back of the text
starting at page SP-1

Practice Problem 9   Solve for x and graph the solution of
$3x + 6 > -6$ *and* $4x + 5 < 1$.

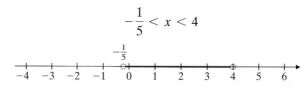

EXAMPLE 10   Solve and graph $2x + 5 \le 11$ *and* $-3x > 18$.

Solution   We solve each inequality separately.

$$\begin{array}{ccc}
2x + 5 \le 11 & \text{and} & -3x > 18 \\
2x \le 6 & & x < -\dfrac{18}{3} \\
x \le 3 & & x < -6
\end{array}$$

The solution is $x < -6$ *and* at the same time $x \le 3$.

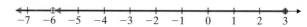

The only numbers that satisfy the statements $x \le 3$ *and* $x < -6$ at the same time are $x < -6$. Thus, $x < -6$ is the solution to the compound inequality.

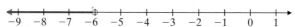

Practice Problem 10   Solve and graph $-2x + 3 < -7$ *and* $7x - 1 > -15$.

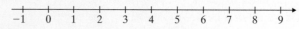

EXAMPLE 11   Solve $-3x - 2 < -5$ *and* $4x + 6 < -12$.

Solution   We solve each inequality separately.

$$\begin{array}{ccc}
-3x - 2 < -5 & \text{and} & 4x + 6 < -12 \\
-3x - 2 + 2 < -5 + 2 & & 4x + 6 - 6 < -12 - 6 \\
-3x < -3 & & 4x < -18 \\
\dfrac{-3x}{-3} > \dfrac{-3}{-3} & & \dfrac{4x}{4} < \dfrac{-18}{4} \\
x > 1 & & x < -4\dfrac{1}{2}
\end{array}$$

Now, clearly it is impossible for one number to be greater than 1 *and* at the same time less than $-4\frac{1}{2}$.

Thus, there is *no solution*. We can express this by the notation $\varnothing$, which is the **empty set**. Or we can just state, "There is no solution."

Practice Problem 11   Solve $-3x - 11 < -26$ *and* $5x + 4 < 14$.

Graph the values of x that satisfy the conditions given.

1. $3 < x$ and $x < 8$

2. $5 < x$ and $x < 10$

3. $-4 < x$ and $x < 2$

4. $-7 < x$ and $x < 1$

5. $7 < x < 9$

6. $3 < x < 5$

7. $-2 < x \leq \dfrac{1}{2}$

8. $-\dfrac{3}{2} \leq x \leq 4$

9. $x > 8$ or $x < 2$

10. $x \geq 2$ or $x \leq 1$

11. $x \leq -\dfrac{5}{2}$ or $x > 4$

12. $x < 3$ or $x > \dfrac{11}{2}$

13. $x \leq -10$ or $x \geq 40$

14. $x \leq -6$ or $x \geq 2$

Solve for x and graph your results.

15. $2x + 3 \leq 5$ and $x + 1 \geq -2$

16. $4x - 1 < 7$ and $x \geq -1$

17. $2x - 3 > 0$ or $x - 2 < -7$

18. $x + 1 \geq 5$ or $x + 5 < 2.5$

19. $x < 8$ and $x > 10$

20. $x < 6$ and $x > 9$

Applications

Express as an inequality.

21. **Toothpaste** A tube of toothpaste is not properly filled if the amount of toothpaste t in the tube is more than 11.2 ounces or fewer than 10.9 ounces.

22. **Clothing Standards** The width of a seam on a pair of blue jeans is unacceptable if it is narrower than 10 millimeters or wider than 12 millimeters.

23. **Interstate Highway Travel** The number of cars c driving over Interstate 91 during the evening hours in January was always at least 5000, but never more than 12,000.

24. **Campsite Capacity** The number of campers c at a campsite during the Independence Day weekend was always at least 490, but never more than 2000.

Temperature Conversion *Solve the following application problems by using the formula $C = \dfrac{5}{9}(F - 32)$. Round to the nearest tenth.*

25. When visiting Montreal this spring, Marcos had been advised that the temperature could range from $-20°C$ to $11°C$. Find an inequality that represents the range in Fahrenheit temperatures.

26. The temperature in Mexico City during February can range from $8°C$ to $23°C$. Find an inequality that represents the range in Fahrenheit temperatures.

Exchange Rates *At one point the exchange equation for converting American dollars into Japanese yen was $Y = 129(d - 4)$. In this equation, d is the number of American dollars, Y is the number of yen, and 4 represents a one-time fee that banks sometimes charged for currency conversion. Use this equation to solve the following problems. (Round answers to the nearest cent.)*

27. Betty is traveling to Japan for 2 weeks and has been advised to have between 18,000 yen and 33,000 yen for spending money for each week she is there. Write an inequality that represents the number of American dollars she will need to bring to the bank to exchange money for this 2-week period.

28. Paul is traveling to Japan for 2 weeks, and has been advised to have between 17,000 yen and 29,000 yen for spending money for each week he is there. Write an inequality that represents the number of American dollars he will need to bring to the bank to exchange money for this 2-week period.

Mixed Practice

Solve the compound inequality.

29. $x - 3 > -5$ *and* $2x + 4 < 8$

30. $x + 3 < 7$ *and* $x - 2 < -3$

31. $-3x + 2 \geq -1$ *and* $4 - x \leq 6$

32. $7 - x \geq 4$ *and* $9x - 8 \geq -17$

33. $2x - 5 < -11$ *or* $5x + 1 \geq 6$

34. $3x + 2 < 5$ *or* $5x - 7 > 8$

35. $-0.3x + 1 \geq 0.2x$ *or* $-0.2x + 0.5 > 0.7$

36. $-0.3x - 0.4 \geq 0.1x$ *or* $0.2x + 0.3 \leq -0.4x$

37. $\dfrac{5x}{2} + 1 \geq 3$ *and* $x - \dfrac{2}{3} \geq \dfrac{4}{3}$

38. $\dfrac{5x}{3} - 2 < \dfrac{14}{3}$ *and* $3x + \dfrac{5}{2} < -\dfrac{1}{2}$

39. $2x + 5 < 3$ *and* $3x - 1 > -1$

40. $6x - 10 < 8$ *and* $2x + 1 > 9$

41. $3x - 1 \geq -10$ *and* $4x + 7 \leq 2x + 1$

42. $8x - 5 \geq 4x + 3$ *and* $x - 1 \leq 1$

43. $\dfrac{2 + 3x}{4} \leq -2$ *or* $\dfrac{2x - 7}{3} < 1$

44. $\dfrac{x + 1}{6} > -1$ *or* $\dfrac{2 - x}{5} > 2$

To Think About

45. $2(x + 1) - 5 > -7$ *or* $4 - (x + 3) < 5 - 2x$

46. $11 - 3(x + 2) > 5x - 1$ *or* $6x - 5 < 7$

Solve the compound inequality.

47. $\dfrac{1}{4}(x + 2) + \dfrac{1}{8}(x - 3) \leq 1$ *and* $\dfrac{3}{4}(x - 1) > -\dfrac{1}{4}$

48. $\dfrac{x - 4}{6} - \dfrac{x - 2}{9} \leq \dfrac{5}{18}$ *or* $-\dfrac{2}{5}(x + 3) < -\dfrac{6}{5}$

Cumulative Review

Solve for the specified variable.

49. Solve for x: $3y - 5x = 8$

50. Solve for y: $7x + 6y = -12$

51. Evaluate $|x + 4| - x^2 + 3x$ for $x = -1$.

52. Evaluate $\sqrt{7x + 4} + 2x^3 - x^2$ for $x = 3$.

53. *Online Auction* A table and four chairs are being auctioned online. The starting price is $40. The first bid is $50, the second bid is $120, the third bid is $190, and the fourth bid is $230. If the same people bid again, using the same pattern of increase, what will the next four bids be?

54. *Biology* Jeremiah has a beautiful plant in his office. The plant has 7 main branches. The first branch has 12 fewer leaves than the second branch. The third branch has 22 more leaves than the second branch. The fourth and fifth branches each have 2 fewer leaves than the sixth and seventh branches, respectively. The sixth and seventh branches have 42 leaves each. The second branch has double the number of leaves on the fourth branch. How many leaves are on each branch?

2.8 ABSOLUTE VALUE INEQUALITIES

1. Solving Absolute Value Inequalities of the Form $|ax + b| < c$

We begin by looking at $|x| < 3$. What does this mean? The inequality $|x| < 3$ means that x is less than 3 units from 0 on the number line. We draw a picture.

This picture shows all possible values of x such that $|x| < 3$. We see that this occurs when $-3 < x < 3$. We conclude that $|x| < 3$ and $-3 < x < 3$ are equivalent statements.

DEFINITION

If a is a positive real number and $|x| < a$, then $-a < x < a$.

EXAMPLE 1   Solve $|x| \le 4.5$.

Solution   The inequality $|x| \le 4.5$ means that x is less than or equal to 4.5 units from 0 on the number line. We draw a picture.

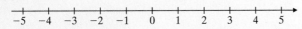

Thus, the solution is $-4.5 \le x \le 4.5$.

Practice Problem 1   Solve $|x| < 2$.

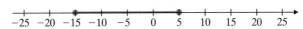

This same technique can be used to solve more complicated inequalities.

EXAMPLE 2   Solve and graph the solution of $|x + 5| \le 10$.

Solution   We want to find the values of x that make $-10 \le x + 5 \le 10$ a true statement. We need to solve the compound inequality.
To solve this inequality, we add -5 to each part.

$$-10 - 5 \le x + 5 - 5 \le 10 - 5$$
$$-15 \le x \le 5$$

Thus, the solution is $-15 \le x \le 5$. We graph this solution.

Practice Problem 2   Solve and graph the solution of $|x - 6| < 15$.
(*Hint:* Choose a convenient scale.)

Student Learning Objectives

After studying this section, you will be able to:

1 Solve absolute value inequalities of the form $|ax + b| < c$.

2 Solve absolute value inequalities of the form $|ax + b| > c$.

NOTE TO STUDENT: Fully worked-out solutions to all of the Practice Problems can be found at the back of the text starting at page SP-1

EXAMPLE 3 Solve and graph the solution of $\left| x - \dfrac{2}{3} \right| \le \dfrac{5}{2}$.

Solution

$$-\dfrac{5}{2} \le x - \dfrac{2}{3} \le \dfrac{5}{2}$$ If $|x| < a$, then $-a < x < a$.

$$6\left(-\dfrac{5}{2}\right) \le 6(x) - 6\left(\dfrac{2}{3}\right) \le 6\left(\dfrac{5}{2}\right)$$ Multiply each part of the inequality by 6.

$$-15 \le 6x - 4 \le 15$$ Simplify.

$$-15 + 4 \le 6x - 4 + 4 \le 15 + 4$$ Add 4 to each part.

$$-11 \le 6x \le 19$$ Simplify.

$$-\dfrac{11}{6} \le \dfrac{6x}{6} \le \dfrac{19}{6}$$ Divide each part by 6.

$$-1\dfrac{5}{6} \le x \le 3\dfrac{1}{6}$$ Change to mixed numbers to facilitate graphing.

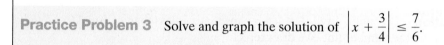

Practice Problem 3   Solve and graph the solution of $\left| x + \dfrac{3}{4} \right| \le \dfrac{7}{6}$.

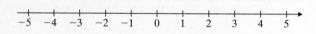

NOTE TO STUDENT: Fully worked-out solutions to all of the Practice Problems can be found at the back of the text starting at page SP-1

EXAMPLE 4   Solve and graph the solution of $|2(x - 1) + 4| < 8$.

Solution   First we simplify the expression within the absolute value symbol.

$$|2x - 2 + 4| < 8$$

$$|2x + 2| < 8$$

$$-8 < 2x + 2 < 8$$ If $|x| < a$, then $-a < x < a$.

$$-8 - 2 < 2x + 2 - 2 < 8 - 2$$ Add -2 to each part.

$$-10 < 2x < 6$$ Simplify.

$$\dfrac{-10}{2} < \dfrac{2x}{2} < \dfrac{6}{2}$$ Divide each part by 2.

$$-5 < x < 3$$

Practice Problem 4   Solve and graph the solution of $|2 + 3(x - 1)| < 20$.

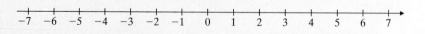

② Solving Absolute Value Inequalities of the Form $|ax + b| > c$

Now consider $|x| > 3$. What does this mean? This inequality $|x| > 3$ means that x is greater than 3 units from 0 on the number line. We draw a picture.

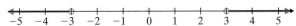

This picture shows all possible values of x such that $|x| > 3$. This occurs when $x < -3$ or when $x > 3$. (Note that a solution can be either in the region to the left of -3 on the number line or in the region to the right of 3 on the number line.) We conclude that $|x| > 3$ and $x < -3$ or $x > 3$ are equivalent statements.

> **DEFINITION**
>
> If a is a positive real number and $|x| > a$, then $x < -a$ or $x > a$.

EXAMPLE 5   Solve and graph the solution of $|x| \geq 5\frac{1}{4}$.

Solution   The inequality $|x| \geq 5\frac{1}{4}$ means that x is more than $5\frac{1}{4}$ units from 0 on the number line. We draw a picture.

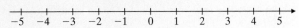

Thus, the solution is $x \leq -5\frac{1}{4}$ or $x \geq 5\frac{1}{4}$.

Practice Problem 5   Solve and graph the solution of $|x| > 2.5$.

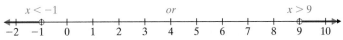

This same technique can be used to solve more complicated inequalities.

EXAMPLE 6   Solve and graph the solution of $|x - 4| > 5$.

Solution   We want to find the values of x that make $x - 4 < -5$ or $x - 4 > 5$ a true statement. We need to solve the compound inequality.

We will solve each inequality separately.

$$x - 4 < -5 \qquad or \qquad x - 4 > 5$$
$$x - 4 + 4 < -5 + 4 \qquad\qquad x - 4 + 4 > 5 + 4$$
$$x < -1 \qquad\qquad\qquad x > 9$$

Thus, the solution is $x < -1$ *or* $x > 9$. We graph the solution on the number line.

Practice Problem 6   Solve and graph the solution of $|x + 6| > 2$.

EXAMPLE 7   Solve and graph the solution of $|-3x + 6| > 18$.

Solution   By definition, we have the following compound inequality.

$$-3x + 6 > 18 \qquad or \qquad -3x + 6 < -18$$
$$-3x > 12 \qquad\qquad\qquad -3x < -24$$
$$\frac{-3x}{-3} < \frac{12}{-3} \longleftarrow \text{Division by a negative} \longrightarrow \frac{-3x}{-3} > \frac{-24}{-3}$$
$$\text{number reverses the}$$
$$\text{inequality sign.}$$
$$x < -4 \qquad\qquad\qquad x > 8$$

The solution is $x < -4$ *or* $x > 8$.

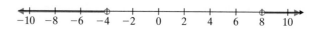

NOTE TO STUDENT: *Fully worked-out solutions to all of the Practice Problems can be found at the back of the text starting at page SP-1*

Practice Problem 7   Solve and graph the solution of $|-5x - 2| > 13$.

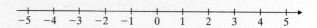

EXAMPLE 8   Solve and graph the solution of $\left|3 - \frac{2}{3}x\right| \geq 5$.

Solution   By definition, we have the following compound inequality.

$$3 - \frac{2}{3}x \geq 5 \qquad or \qquad 3 - \frac{2}{3}x \leq -5$$
$$3(3) - 3\left(\frac{2}{3}x\right) \geq 3(5) \qquad 3(3) - 3\left(\frac{2}{3}x\right) \leq 3(-5)$$
$$9 - 2x \geq 15 \qquad\qquad 9 - 2x \leq -15$$
$$-2x \geq 6 \qquad\qquad\quad -2x \leq -24$$
$$\frac{-2x}{-2} \leq \frac{6}{-2} \qquad\qquad \frac{-2x}{-2} \geq \frac{-24}{-2}$$
$$x \leq -3 \qquad\qquad\quad x \geq 12$$

The solution is $x \leq -3$ *or* $x \geq 12$.

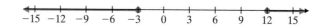

Practice Problem 8   Solve and graph the solution of $\left|4 - \frac{3}{4}x\right| \geq 5$.

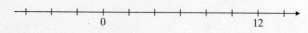

EXAMPLE 9   When a new car transmission is built, the diameter d of the transmission must not differ from the specified standard s by more than 0.37 millimeter. The engineers express this requirement as $|d - s| \leq 0.37$. If the standard s is 216.82 millimeters for a particular car, find the limits of d.

Solution

$$|d - s| \leq 0.37$$

$$|d - 216.82| \leq 0.37 \qquad \text{Substitute the known value of } s.$$

$$-0.37 \leq d - 216.82 \leq 0.37 \qquad \text{If } |x| \leq a, \text{ then } -a \leq x \leq a.$$

$$-0.37 + 216.82 \leq d - 216.82 + 216.82 \leq 0.37 + 216.82$$

$$216.45 \leq d \leq 217.19$$

Thus, the diameter of the transmission must be at least 216.45 millimeters, but not greater than 217.19 millimeters.

Practice Problem 9 The diameter d of a transmission must not differ from the specified standard s by more than 0.37 millimeter. Solve to find the allowed limits of d for a truck transmission for which the standard s is 276.53 millimeters:

SUMMARY OF ABSOLUTE VALUE EQUATIONS AND INEQUALITIES

It may be helpful to review the key concepts of absolute value equations and inequalities that we have covered in Sections 2.3 and 2.8. For real numbers a, b, and c, where $a \neq 0$ and $c > 0$, we have the following:

Absolute value form of the equation or inequality	Equivalent form without the absolute value	Type of solution obtained	Graphed form of the solution on a number line
$\|ax + b\| = c$	$ax + b = c$ or $ax + b = -c$	Two distinct numbers: m and n	
$\|ax + b\| < c$	$-c < ax + b < c$	The set of numbers between the two numbers m and n: $m < x < n$	
$\|ax + b\| > c$	$ax + b < -c$ or $ax + b > c$	The set of numbers less than m or the set of numbers greater than n: $x < m$ or $x > n$	

Developing Your Study Skills

Problems with Accuracy

Strive for accuracy. The mistakes students make are often simple ones and not the result of a lack of understanding. Such mistakes are frustrating. A simple arithmetic or copying error can lead to an incorrect answer.

These five steps will help you to cut down on errors.

1. Work carefully and take your time. Do not rush through a problem just to get it done.

2. Concentrate on one problem at a time. Sometimes problems become mechanical, and your mind begins to wander. You can become careless and make a mistake.

3. Check your problem. Be sure that you copied it correctly from the book.

4. Check your computations from step to step. Check the solution in the problem. Does it work? Does it make sense?

5. Keep practicing new skills. Remember the old saying "Practice makes perfect." An increase in practice will result in an increase in accuracy. Many errors are due simply to lack of practice.

There is no magic formula for eliminating all errors, but these five steps will be a tremendous help in reducing them.

2.8 EXERCISES

| Student Solutions Manual | CD/ Video | PH Math Tutor Center | MathXL®Tutorials on CD | MathXL® | MyMathLab® | Interactmath.com |

Solve and graph the solutions.

1. $|x| \le 8$

2. $|x| < 6$

3. $|x + 4.5| < 5$

4. $|x + 6| < 3.5$

Solve for x.

5. $|x - 3| \le 5$

6. $|x - 7| \le 10$

7. $|2x - 5| \le 7$

8. $|3x + 2| \le 12$

9. $|5x - 2| \le 4$

10. $|2x - 3| \le 1$

11. $|0.5 - 0.1x| < 1$

12. $|0.9 - 0.2x| < 2$

13. $\left|\frac{1}{4}x + 2\right| < 6$

14. $\left|\frac{1}{5}x + 1\right| < 5$

15. $\left|\frac{3}{4}(x - 1)\right| < 6$

16. $\left|\frac{4}{5}(x - 1)\right| < 8$

17. $\left|\frac{3x - 2}{4}\right| < 3$

18. $\left|\frac{5x - 3}{2}\right| < 4$

Solve for x.

19. $|x| > 5$

20. $|x| \ge 7$

21. $|x + 2| > 5$

22. $|x + 4| > 7$

23. $|x - 1| \ge 2$

24. $|x - 2| \ge 3$

25. $|3x - 8| \ge 7$

26. $|5x - 2| \ge 13$

27. $|6 - 0.1x| > 5$

28. $|0.4 - 0.2x| > 3$

29. $\left|\frac{1}{5}x - \frac{1}{10}\right| > 2$

30. $\left|\frac{1}{4}x - \frac{3}{8}\right| > 1$

31. $\left|\frac{1}{3}(x - 2)\right| < 5$

32. $\left|\frac{2}{5}(x - 2)\right| \le 4$

Mixed Practice

33. $|3x + 5| < 17$

34. $|2x + 3| < 5$

35. $|2 - 9x| > 20$

36. $|2 - 5x| > 2$

Applications

Manufacturing Standards *In a certain company, the measured thickness m of a helicopter blade must not differ from the standard s by more than 0.12 millimeter. The manufacturing engineer expresses this as* $|m - s| \le 0.12$.

37. Find the limits of *m* if the standard *s* is 18.65 millimeters.

38. Find the limits of *m* if the standard *s* is 17.48 millimeters.

Computer Chip Standards *A small computer microchip has dimension requirements. The manufacturing engineer has written the specification that the new length n of the chip can differ from the previous length p by only 0.05 centimeter or less. The equation is* $|n - p| \le 0.05$.

39. Find the limits of the new length if the previous length was 9.68 centimeters.

40. Find the limits of the new length if the previous length was 7.84 centimeters.

To Think About

41. A student tried to solve the inequality $|4x - 8| > 12$. Instead of writing $4x - 8 > 12$ or $4x - 8 < -12$ as he should have done, he wrote $12 < 4x - 8 < -12$. What was wrong with his approach?

42. A student tried to write the solution to the compound inequality $6 < 4 - 3x < 19$. He used the following steps.

Step 1 $6 - 4 < 4 - 4 - 3x < 19 - 4$
Step 2 $2 < -3x < 15$
Step 3 $\dfrac{2}{-3} < \dfrac{-3x}{-3} < \dfrac{15}{-3}$
Step 4 $-\dfrac{2}{3} < x < -5$

What error did he make?

Cumulative Review

Perform the correct order of operations to simplify.

43. $(6 - 4)^3 \div (-4) + 2^2$

44. $12 \div (-2)(3) - (-5) + 2$

In Problems 41 and 42 use $\pi \approx 3.14$. Round answers to the nearest hundredth.

▲ **45.** *Geometry* The Outward Bound program in the United States is famous for teaching self-esteem and personal achievement to young people. One of its physical challenges is for a student to hang on to a rope 19 meters long and swing from one shore to another and then back. The rope swings through a circular arc, measuring $\frac{1}{8}$ of the circumference of a circle. How many seconds does it take for one full *round-trip* swing if the student moves at 3 meters per second?

▲ **46.** *Geometry* The rigging on a sailboat comes loose from the mast. The end of the wire rigging that is hanging down is 30 feet from the top of the mast. This end swings through a circular arc, measuring $\frac{1}{6}$ of the circumference of a circle. How many seconds does it take the rigging wire, swinging like a pendulum, to make one *round-trip* swing, if the wind is blowing the end at 8 feet per second?

47. *CD Costs* Three compact disc racks are for sale online. One rack holds 160 CDs and costs $39.95 plus $6.50 for shipping and handling. The second rack holds 120 CDs and costs $24.95 plus $5.95 for shipping and handling. The third rack holds 75 CDs and costs $18.95 plus $4.75 for shipping and handling. Based on the above information, what is the cost per CD space on each rack? Don't forget the shipping and handling charges. Round your answer to the nearest cent. Which rack is the least expensive in terms of cost per CD space?

Projected Fuel Savings with Hybrid Vehicles

Hybrid vehicles are becoming more common in the United States. The Honda Civic Hybrid, Honda Insight, and Toyota Prius are three hybrid vehicles available to consumers in 2004. By the year 2007, hybrid pickup trucks and SUVs will also be available. Facts about the 2004 Toyota Prius and the conventional Toyota Camry LE (both automatic transmission) are given in the following table.

	Miles per Gallon (city)	Miles per Gallon (highway)	Manufacturer's Retail Price	Size of Gas Tank
Prius	51	60	$20,510	11.9 gal
Camry	23	32	$19,875	18.5 gal

(*Source:* www.Toyota.com)

Problems for Individual Investigation

1. **(a)** How many city miles could you drive in the Camry on a full tank of gas?

   (b) How many city miles could you drive in the Prius on a full tank of gas?

2. On average, George commutes to work 20 days each month. The distance from home to work is 35 highway miles. How many miles does he drive between home and work each month? If George drove a Camry, how many gallons of gas would he use in one month for his commute? If he drove a Prius, how many gallons would he use? Round your answers to the nearest tenth.

Problems for Group Investigation and Cooperative Study

3. Write an equation that finds the cost to travel *x* miles in a Camry at highway speeds if gasoline costs $2.10 per gallon.

4. Write an equation that finds the cost to travel *x* miles in a Prius at highway speeds if gasoline costs $2.10 per gallon.

5. Use the equations you obtained in problems 3 and 4 to answer the following question: If George continues to work on average 20 days per month, how many months would George need to commute to work so that it is worth buying the Prius over the Camry?

6. Use the equations you obtained in problems 3 and 4 to answer the following question: If George continues to work an average 20 days per month, how many months would George need to commute to work so that it is $1000 less expensive to have bought the Prius rather than the Camry?

To graph the equation, we could graph all its solutions. However, this would be impossible. Since there are an infinite number of solutions. It is a mathematical property that an equation of the form $Ax + By = C$, where A, B, and C are constants, is a straight line. Hence, to graph the equation we graph three ordered pair solutions and connect them with a straight line. (The third ordered pair solution is used to check the line.)

EXAMPLE 1 Graph the equation $y = -3x + 2$.

Solution We choose three values of x and then substitute them into the equation to find the corresponding values of y. Let's choose $x = -1$, $x = 1$ and $x = 2$.

For $x = -1$, $y = -3(-1) + 2 = 5$, so the first point, or solution, is $(-1, 5)$.

For $x = 1$, $y = -3(1) + 2 = -1$, and for $x = 2$, $y = -3(2) + 2 = -4$.

We can condense this procedure by using a table.

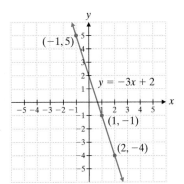

x	y
-1	5
1	-1
2	-4

Practice Problem 1 Graph the equation $y = -4x + 2$.

PRACTICE PROBLEM 1

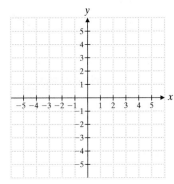

NOTE TO STUDENT: Fully worked-out solutions to all of the Practice Problems can be found at the back of the text starting at page SP-1

② Using *x*- and *y*-intercepts to Graph a Linear Equation

We can usually graph a straight line by using the x- and y-intercepts. A straight line that is not vertical or horizontal has these two intercepts.

> The **x-intercept** of a line is the point where the line crosses the x-axis (that is, where $y = 0$). It is described by an ordered pair of the form $(a, 0)$. The **y-intercept** of a line is the point where the line crosses the y-axis (that is, where $x = 0$). It is described by an ordered pair of the form $(0, b)$.

EXAMPLE 2 Find the x-intercept, the y-intercept, and one additional ordered pair. Then graph the equation: $4x - 3y = -12$.

Solution

Find the x-intercept by using $y = 0$.

$$4x - 3(0) = -12$$
$$4x = -12$$
$$x = -3$$

The x-intercept is $(-3, 0)$.

Find the y-intercept by using $x = 0$.

$$4(0) - 3y = -12$$
$$-3y = -12$$
$$y = 4$$

The y-intercept is $(0, 4)$.

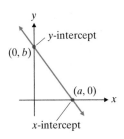

We can now pick any value of x or y to find our third point. Let's pick $y = 2$.

$$4x - 3(2) = -12$$
$$4x - 6 = -12$$
$$4x = -12 + 6$$
$$4x = -6$$
$$x = -\frac{6}{4} = -\frac{3}{2}$$

PRACTICE PROBLEM 2

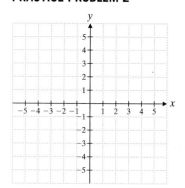

Graphing Calculator

 Graphing a Line

You can graph a line given in the form $y = mx + b$ using a graphing calculator. For example, to graph the equation in Example 2, first rewrite the equation by solving for y.

$$4x - 3y = -12$$
$$-3y = -4x - 12$$
$$y = \tfrac{4}{3}x + 4$$

Enter the right-hand side of the resulting equation in the Y = editor of your calculator and graph. Choose an appropriate window to show all the intercepts. The following window is $[-10, 10]$ by $[-10, 10]$.
　Display:

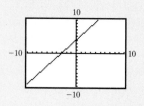

Hence, the third point on the line is $\left(-\dfrac{3}{2}, 2\right)$. The graph of the equation is shown below.

x	y
-3	0
0	4
$\dfrac{3}{2}$	2

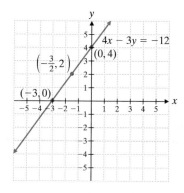

Practice Problem 2　Find the x-intercept, the y-intercept, and one additional ordered pair. Then graph the equation: $3x - 2y = -6$.

③ Graphing Horizontal and Vertical Lines

Let's look at the standard form of a linear equation, $Ax + By = C$, when $B = 0$.

$$Ax + (0)y = C$$
$$Ax = C$$
$$x = \frac{C}{A}$$

Notice that when we solve for x we get $x = \dfrac{C}{A}$, which is a constant. For convenience we will rename it a. The equation then becomes

$$x = a.$$

What does this mean? The equation $x = a$ means that for any value of y, x is a. The graph is a vertical line.

What happens to $Ax + By = C$ when $A = 0$?

$$(0)x + By = C$$
$$By = C$$
$$y = \frac{C}{B}$$

We will rename the constant $\dfrac{C}{B}$ as b. The equation then becomes

$$y = b.$$

What does this mean? The equation $y = b$ means that for any value of x, y is b. The graph is a horizontal line.

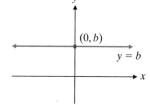

The graph of the equation $x = a$, where a is any real number, is a **vertical line** through the point $(a, 0)$.

The graph of the equation $y = b$, where b is any real number, is a **horizontal line** through the point $(0, b)$.

EXAMPLE 3 Graph each equation: **(a)** $x = -3$ **(b)** $2y - 4 = 0$

Solution

(a) The equation $x = -3$ means that for any value of y, x is -3. The graph of $x = -3$ is a vertical line 3 units to the left of the origin.

(b) The equation $2y - 4 = 0$ can be simplified.

$$2y - 4 = 0$$
$$2y = 4$$
$$y = 2$$

The equation $y = 2$ means that, for any value of x, y is 2. The graph of $y = 2$ is a horizontal line 2 units above the x-axis.

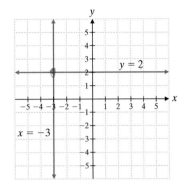

Practice Problem 3 Graph each equation. **(a)** $x = 4$ **(b)** $3y + 12 = 0$

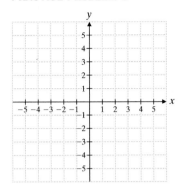
4 Graphing a Linear Equation Using Different Scales for the Axes

By common convention, each tick mark on a graph's axis indicates 1 unit, so we don't need to use a marked scale on each axis. But sometimes a different scale is more appropriate. This new scale must then be clearly labeled on each axis.

EXAMPLE 4 A company's finance officer has determined that the monthly cost in dollars for leasing a photocopier is $C = 100 + 0.002n$, where n is the number of copies produced in a month in excess of a specified number. Graph the equation using $n = 0$, $n = 30,000$, and $n = 60,000$. Let the n-axis be the horizontal axis.

Solution For each value of n we obtain C.

When $n = 0$,
then $C = 100 + 0.002(0) = 100 + 0 = 100$.
When $n = 30,000$,
then $C = 100 + 0.002(30,000)$
$= 100 + 60 = 160$.
When $n = 60,000$,
then $C = 100 + 0.002(60,000)$
$= 100 + 120 = 220$.

The table of values and graph are shown next.

n	C
0	100
30,000	160
60,000	220

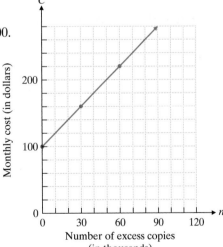

Since n varies from 0 to 60,000 and C varies from 100 to 220, we need different scales on the axes. We let each square on the horizontal scale represent 10,000 excess copies and each square on the vertical scale represent \$20.

Graphing Calculator

Changing Windows

To do Example 4, your window values would be
$X_{min} = 0$
$X_{max} = 120,000$
$Y_{min} = 0$
$Y_{max} = 300$

PRACTICE PROBLEM 4

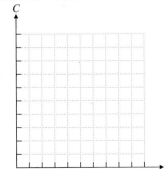

Practice Problem 4 The cost of a product is given by $C = 300 + 0.15n$, where n is the number of products produced. Graph the equation using an appropriate scale. Use $n = 0$, $n = 1000$, and $n = 2000$.

3.1 EXERCISES

Student Solutions Manual | CD/ Video | PH Math Tutor Center | MathXL®Tutorials on CD | MathXL® | MyMathLab® | Interactmath.com

Verbal and Writing Skills

1. Graphs are used to show the relationships among the _____ in an equation.

2. The *x*-axis and the *y*-axis intersect at the _____.

3. Explain in your own words why the point (a, b) in a rectangular coordinate system is an *ordered* pair. In other words, what is the importance of the word *ordered* when we say it is an ordered pair? Give an example.

4. $(5, 1)$ is a solution to the equation $2x - 3y = 7$. What does this mean?

Find the missing coordinate.

5. $(-2, \underline{\hspace{1cm}})$ is a solution of $y = 3x - 7$.

6. $(-3, \underline{\hspace{1cm}})$ is a solution of $y = 4 - 3x$.

7. $\left(\underline{\hspace{1cm}}, \frac{1}{2}\right)$ is a solution of $7x + 14y = -21$.

8. $\left(\underline{\hspace{1cm}}, \frac{1}{3}\right)$ is a solution to $-x + 6y = -3$.

Graph each equation.

9. $y = 2x - 3$

10. $y = 3x + 2$

11. $y = 4 - 2x$

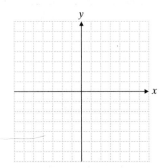
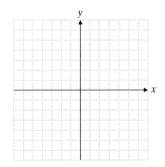
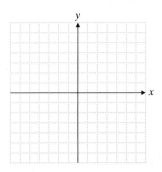

12. $y = -5x - 2$

13. $y = \frac{2}{3}x - 4$

14. $y = \frac{5}{2}x + 1$

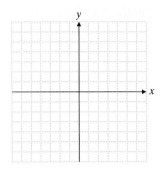
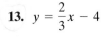
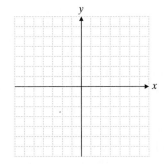
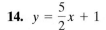
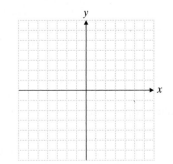

Simplify the equation if possible. Find the x-intercept, the y-intercept, and one or two additional ordered pairs that are solutions to the equation. Then graph the equation.

15. $2y - 3x = 6$

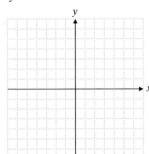

16. $2y + 5x = 10$

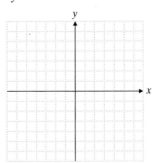

17. $2x - y = 6$

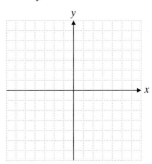

18. $4x - y = -4$

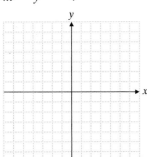

19. $-3x - 4y = 8$

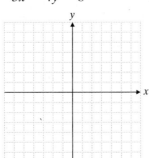

20. $2x - 3y = -9$

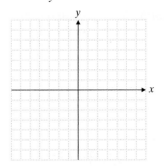

21. $5y - 4 = 3x - 4$

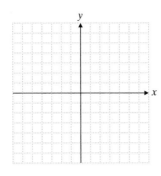

22. $4x + 6y + 2 = 2$

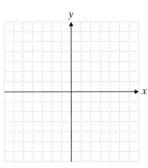

Simplify the equation if possible. State whether the equation represents a horizontal or a vertical line. Then graph the equation.

23. $x = -5$

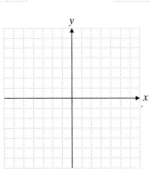

24. $x = 2$

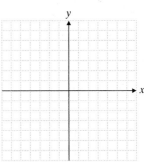

25. $3x - 18 = 0$

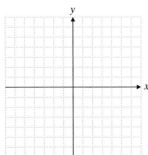

26. $2x - 3 = 3x$

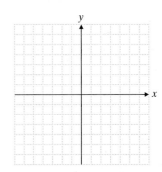

27. $2y + 8 = 0$

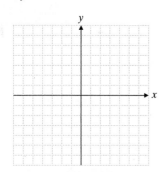

28. $5y + 6 = 2y$

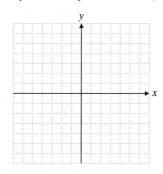

Mixed Practice

Simplify each equation if possible. Then graph the equation by any appropriate method.

29. $y = -1.5x + 2$

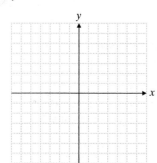

30. $y = 0.5x + 4$

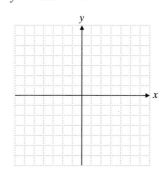

31. $2x + 5y = -5$

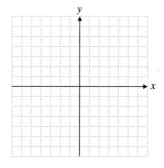

32. $4x - 3y = 6$

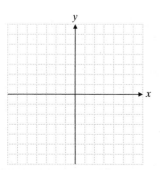

33. $5x + y + 4 = 8x$

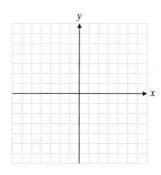

34. $5x - 4y - 4 = 4x$

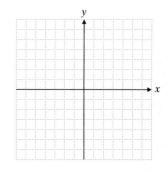

Graph each equation. Use appropriate scales on each axis.

35. $y = 82x + 150$

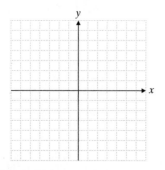

36. $y = 0.06x - 0.04$

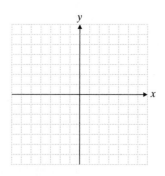

To Think About

Income of Men Versus Women   *The following graph shows the median weekly earnings of men and women in the United States during the period 1980 to 2005. Use the graph to answer Exercises 37–42.*

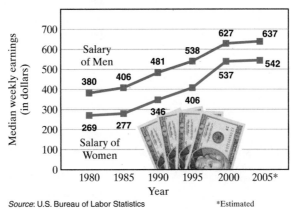

Median Weekly Earnings of Full-Time Workers in the U.S.

Source: U.S. Bureau of Labor Statistics   *Estimated

37. During what 5-year period did the greatest increase in the median weekly earnings of men occur?

38. During what 5-year period did the greatest increase in the median weekly earnings of women occur?

39. In what year did the median weekly earnings of men and women have the largest difference?

40. In what year did the median weekly earnings of men and women have the smallest difference?

41. What was the percent of increase in earnings for women during the period 1980 to 2005? Round your answer to the nearest tenth of a percent.

42. What was the percent of increase in earnings for men during the period 1980 to 2000?

Applications

43. ***Baseball*** If a baseball is thrown vertically upward by Paul Frydrych when he is standing on the ground, the velocity of the baseball V (in feet per second) after T seconds is $V = 120 - 32T$.
  (a) Find V for $T = 0, 1, 2, 3,$ and 4.

  (b) Graph the equation, using T as the horizontal axis.
  (c) What is the significance of the negative value of V when $T = 4$?

44. ***Gasoline Storage Tank*** A full storage tank on the Robinson family farm contains 900 gallons of gasoline. Gasoline is then pumped from the tank at a rate of 15 gallons per minute. The equation $G = 900 - 15m$ describes the number of gallons of gasoline G in the tank m minutes after the pumping began.
  (a) Find G for $m = 0, 10, 20, 30,$ and 60.

  (b) Graph the equation, using m as the horizontal axis.
  (c) What happens when $m = 61$?

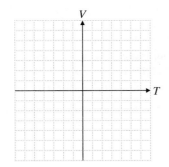

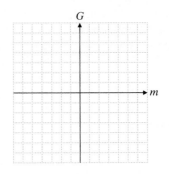

45. *Atmospheric Pressure* Graph the equation $P = 14.7 - 0.0005d$ for $d = 0$, 1000, 2000, 3000, 9000, and 15,000. (This equation predicts the atmospheric pressure in pounds per square inch that a person would experience at a height of d feet above sea level.)

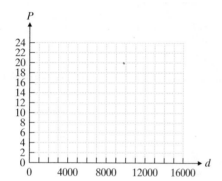

46. *Parachuting* Bob and Cheryl Finkelstein belong to a Sky Diving Club. When jumping out of a plane, they try to open their sport parachutes at 5600 feet above the ground. If they accomplish that goal, then their height above the ground measured in feet is given by the equation $h = 5600 - 190t$, where t is the number of seconds since the parachute was opened. Find h for $t = 0$, 5, 10, 20, and 29. Graph the equation using t as the horizontal axis.

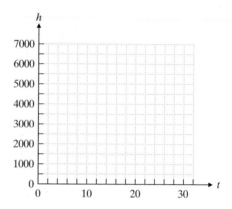

Optional Graphing Calculator Problems

 If you have a graphing calculator, use it to graph the following equations. Choose an appropriate window that will allow you to see both the x-intercept and the y-intercept. (The appearance of your graph will depend on the window selected.)

47. $y = -2.15x + 2.73$ **48.** $y = 1.36x - 1.83$ **49.** $y = 0.713x + 25.82$ **50.** $y = -0.819x - 43.82$

Cumulative Review

51. Evaluate. $36 \div (8 - 6)^2 + 3(-4)$

52. Solve for x. $3(x - 6) + 2 \le 4(x + 2) - 21$

53. *Balloon Giveaway* A novelty company is giving away balloons in a shopping mall. There are twice as many red balloons as green balloons. There are three times as many blue balloons as red balloons. There are half as many white balloons as there are yellow balloons. There are half as many yellow balloons as there are red balloons. There are 130 white balloons. How many balloons of each color are being given away?

54. *Commission Sales* At Greenland Realty, a salesperson receives a commission of 7% on the first $100,000 of the selling price of a house and 3% on the amount that exceeds $100,000. Ray Peterson received a commission of $9100 for selling a house. What was the sale price of the house?

3.2 SLOPE OF A LINE

1 Finding the Slope If Two Points Are Known

The concept of slope is one of the most useful in mathematics and has many practical applications. For example, a carpenter needs to determine the slope (or pitch) of a roof. (You may have heard someone say that a roof has a 5:12 pitch.) Road engineers must determine the proper slope (or grade) of a roadbed. If the slope is steep, you feel as if you're driving almost straight up. Simply put, slope is a measure of steepness. That is, slope measures the ratio of the vertical change (*rise*) to the horizontal change (*run*).

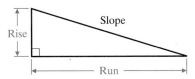

Mathematically, we define slope of a line as follows:

> The **slope of a straight line** with points (x_1, y_1) and (x_2, y_2) is
>
> $$\text{slope} = m = \frac{y_2 - y_1}{x_2 - x_1} \qquad x_2 \neq x_1.$$

In the sketch, we see that the rise is $y_2 - y_1$ and the run is $x_2 - x_1$.

EXAMPLE 1 Find the slope of the line passing through $(-2, -3)$ and $(1, -4)$.

Solution Identify the y-coordinates and the x-coordinates for the points $(-2, -3)$ and $(1, -4)$.

$$\overbrace{(-2, -3) \qquad (1, -4)}^{y_2 - y_1}$$
$$\underbrace{}_{x_2 - x_1}$$

Use the formula.

$$\text{slope} = m = \frac{y_2 - y_1}{x_2 - x_1}$$
$$= \frac{-4 - (-3)}{1 - (-2)}$$
$$= -\frac{1}{3}$$

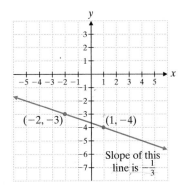

Practice Problem 1 Find the slope of the line passing through $(-6, 1)$ and $(-5, -2)$.

NOTE TO STUDENT: Fully worked-out solutions to all of the Practice Problems can be found at the back of the text starting at page SP-1

Notice that it does not matter which ordered pair we label (x_1, y_1) and which we label (x_2, y_2) as long as we subtract the x-coordinates in the same order that we subtract the y-coordinates. Let's redo Example 1.

$$\begin{array}{cc} (x_1, \ y_1) & (x_2, \ y_2) \\ \downarrow \ \ \downarrow & \downarrow \ \ \downarrow \\ (1, \ -4) & (-2, \ -3) \end{array} \qquad m = \frac{-3 - (-4)}{-2 - 1} = \frac{-3 + 4}{-3} = -\frac{1}{3}$$

Student Learning Objectives

After studying this section, you will be able to:

1 Find the slope of any nonvertical straight line if two points are known.

2 Determine whether two lines are parallel or perpendicular by comparing their slopes.

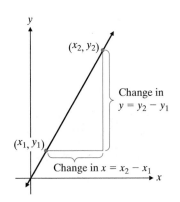

Graphing Calculator

Slopes

Using a graphing calculator, graph

$$y_1 = 3x + 1,$$

$$y_2 = \frac{1}{3}x + 1,$$

$$y_3 = -3x + 1, \text{ and}$$

$$y_4 = -\frac{1}{3}x + 1$$

on the same set of axes. How is the coefficient of x in each equation related to the slope of the line? Will the graph of the line $y = -2x + 3$ slope upward or downward? How do you know? Verify using your calculator.

1. Lines sloping upward to the right have positive slopes.
2. Lines sloping downward to the right have negative slopes.

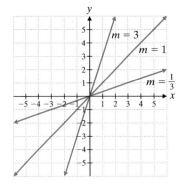

What is the slope of a horizontal line? Let's look at the equation $y = 4$. This equation $y = 4$ is the horizontal line 4 units above the x-axis. It means that for any x, y is 4. The slope is

$$m = \frac{y_2 - y_1}{x_2 - x_1} = \frac{4 - 4}{x_2 - x_1} = \frac{0}{x_2 - x_1} = 0.$$

In general, for all horizontal lines, $y_2 - y_1 = 0$. Hence, the slope is 0.

What is the slope of a vertical line? The equation of the vertical line 4 units to the right of the y-axis is $x = 4$. It means that for any y, x is 4. The slope is

$$m = \frac{y_2 - y_1}{x_2 - x_1} = \frac{y_2 - y_1}{4 - 4} = \frac{y_2 - y_1}{0}$$

Because division by zero is not defined, we say that a vertical line has no slope or the slope is undefined.

> The slope of a horizontal line is 0. The slope of a vertical line is undefined.

EXAMPLE 2 Find the slope if possible of the line passing through each pair of points.

(a) $(1.6, 2.3)$ and $(-6.4, 1.8)$ **(b)** $\left(\frac{5}{3}, -\frac{1}{2}\right)$ and $\left(\frac{2}{3}, -\frac{1}{4}\right)$

Solution

(a) $m = \dfrac{1.8 - 2.3}{-6.4 - 1.6} = \dfrac{-0.5}{-8.0} = 0.0625$

(b) $m = \dfrac{-\dfrac{1}{4} - \left(-\dfrac{1}{2}\right)}{\dfrac{2}{3} - \dfrac{5}{3}} = \dfrac{-\dfrac{1}{4} + \dfrac{2}{4}}{-\dfrac{3}{3}} = \dfrac{\dfrac{1}{4}}{-1} = -\dfrac{1}{4}$

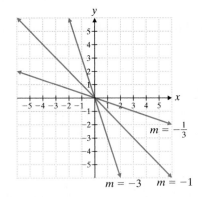

The Mount Washington Cog Railway in New Hampshire has a train track that in some places rises 37 feet for every 100 feet horizontally. This is a slope of $\frac{37}{100}$.

NOTE TO STUDENT: Fully worked-out solutions to all of the Practice Problems can be found at the back of the text starting at page SP-1

Practice Problem 2 Find the slope if possible of the line through each pair of points. **(a)** $(1.8, -6.2)$ and $(-2.2, -3.4)$ **(b)** $\left(\frac{1}{5}, -\frac{1}{2}\right)$ and $\left(\frac{4}{15}, -\frac{3}{4}\right)$

When dealing with practical situations, such as the grade of a road or the pitch of a roof, we can find the slope by using the formula

$$\text{slope} = \frac{\text{rise}}{\text{run}}.$$

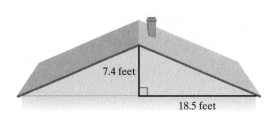

> **EXAMPLE 3** Find the pitch of a roof as shown in the sketch.
>
> $$\text{Slope} = \frac{\text{rise}}{\text{run}} = \frac{7.4}{18.5} = 0.4.$$
>
> 7.4 feet
>
> 18.5 feet
>
> **Solution** This could also be expressed as the fraction $\frac{2}{5}$. A builder might refer to this as a *pitch* (slope) of $2:5$.
>
> **Practice Problem 3** Find the slope of a river that drops 25.92 feet vertically over a horizontal distance of 1296 feet. (*Hint:* Use only positive numbers. In everyday use, the slope of a river or road is always considered to be a positive value.)

2 Determining Whether Two Lines Are Parallel or Perpendicular

We can tell a lot about a line by looking at its slope. A positive slope tells us that the line rises from left to right. A negative slope tells us that the line falls from left to right. What might be true of the slopes of parallel lines? Determine the slope of each line in the following graphs, and compare the slopes of the parallel lines.

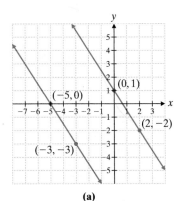

(a)

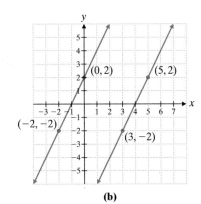

(b)

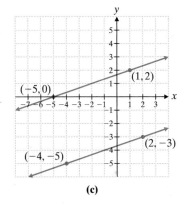

(c)

Since parallel lines are lines that never intersect, their slopes must be equal.

PARALLEL LINES

Two different lines with slopes m_1 and m_2 are *parallel* if $m_1 = m_2$.

Now take a look at perpendicular lines. Determine the slope of each line in the following graphs, and compare the slopes of the perpendicular lines. What do you notice?

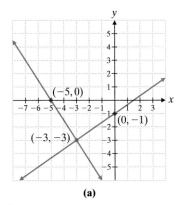

(a)

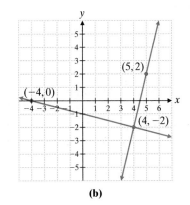

(b)

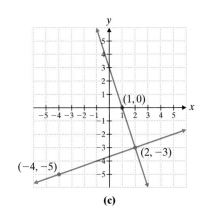

(c)

By definition, two lines are perpendicular if they intersect at right angles. (A right angle is an angle of 90°.) We would expect that one slope would be positive and the other slope negative.

In fact, it is true that if the slope of a line is 5, then the slope of a line that is perpendicular to it is $-\frac{1}{5}$.

> **PERPENDICULAR LINES**
>
> Two lines with slopes m_1 and m_2 are *perpendicular* if $m_1 = -\dfrac{1}{m_2}$ ($m_1, m_2 \neq 0$).

EXAMPLE 4   Find the slope of a line that is perpendicular to the line l that passes through $(4, -6)$ and $(-3, -5)$.

Solution   The slope of line l is

$$m_l = \frac{-5 - (-6)}{-3 - 4} = \frac{-5 + 6}{-7} = \frac{1}{-7} = -\frac{1}{7}.$$

The slope of a line perpendicular to line l must have a slope of 7.

Practice Problem 4   If a line l passes through $(5, 0)$ and $(6, -2)$, what is the slope of a line h that is perpendicular to l?

It is helpful to remember that if two lines are perpendicular, their slopes are negative reciprocals. One slope will be positive and the other negative. If one line had a slope of $-\frac{3}{7}$, the slope of a line perpendicular to it would have a slope of $\frac{7}{3}$.

Now we will use our knowledge of slopes to determine if three points lie on the same line.

EXAMPLE 5   Without plotting any points, show that the points $A(-5, -1)$, $B(-1, 2)$, and $C(3, 5)$ lie on the same line.

Solution   First we find the slope of the line segment from A to B and the slope of the line segment from B to C.

$$m_{AB} = \frac{2 - (-1)}{-1 - (-5)} = \frac{2 + 1}{-1 + 5} = \frac{3}{4}$$

$$m_{BC} = \frac{5 - 2}{3 - (-1)} = \frac{3}{3 + 1} = \frac{3}{4}$$

Since the slopes are equal, we must have one line or two parallel lines. But the line segments have a point (B) in common, so all three points lie on the same line.

Practice Problem 5   Without plotting the points, show that $A(1, 5)$, $B(-1, 1)$, and $C(-2, -1)$ lie on the same line.

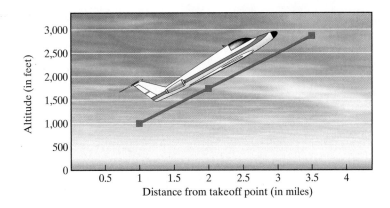

At takeoff an aircraft climbs into the sky at a certain rate of speed. This measurement of a slope is often called the *rate of climb*.

EXAMPLE 6   A gulfstream jet takes off from Orange County Airport in California. At 1 mile from the takeoff point, the jet is 1000 feet above the ground and begins a specified rate of climb. At 2 miles from the takeoff point, it is 1750 feet above the ground. At 3.5 miles from the takeoff point, it is 2865 feet above the ground. Look at the graph at the top of the page. Does it appear that the jet traveled in a straight line from the 1-mile point to the 3.5-mile point?

Solution   From the 1-mile point to the 2-mile point, the jet traveled 750 feet upward over 1 horizontal mile. From the 2-mile point to the 3.5-mile point, the jet traveled 1115 feet upward over 1.5 horizontal miles. Thus, the first slope is 750 feet per mile, and the second slope is $743\frac{1}{3}$ feet per mile. These slopes are not the same, so the jet has not traveled in a straight line.

Practice Problem 6   On the return flight the jet is at an altitude of 4850 feet when it is 4.5 miles from the airport. Its altitude is 3650 feet when it is 3.5 miles from the airport and 1010 feet when it is 1.3 miles from the airport. Does it appear that the jet is descending in a straight line?

NOTE TO STUDENT: Fully worked-out solutions to all of the Practice Problems can be found at the back of the text starting at page SP-1

Developing Your Study Skills

Getting Organized for an Exam

Studying adequately for an exam requires careful preparation. Begin early so that you will be able to spread your review over several days. Even though you may still be learning new material at this time, you can be reviewing concepts previously learned in the chapter. Giving yourself plenty of time for review will take the pressure off. You need this time to process what you have learned and to tie concepts together.

Adequate preparation enables you to feel confident and to think clearly with less anxiety.

Student Solutions Manual CD/ Video PH Math Tutor Center MathXL®Tutorials on CD MathXL® MyMathLab® Interactmath.com

Verbal and Writing Skills

1. Slope measures _____ change (rise) versus _____ change (run).

2. A positive slope indicates that the line slopes _____ to the right.

3. The slope of a horizontal line is ____.

4. Two different lines are parallel if their slopes are _____.

5. Does the line passing through $(-3, -7)$ and $(-3, 5)$ have a slope? Give a reason for your answer.

6. Let $(x_1, y_1) = (-6, -3)$ and $(x_2, y_2) = (-4, 5)$. Find $\dfrac{y_2 - y_1}{x_2 - x_1}$ and $\dfrac{y_1 - y_2}{x_1 - x_2}$. Are the results the same? Why or why not?

Find the slope, if possible, of the line passing through each pair of points.

7. $(2, 2)$ and $(6, -6)$

8. $(2, -1)$ and $(6, 3)$

9. $\left(\dfrac{3}{2}, 4\right)$ and $(-2, 0)$

10. $(6, 1)$ and $\left(0, \dfrac{1}{3}\right)$

11. $(6.8, -1.5)$ and $(5.6, -2.3)$

12. $(-2, 5.2)$ and $(4.8, -1.6)$

13. $\left(\dfrac{3}{2}, -2\right)$ and $\left(\dfrac{3}{2}, \dfrac{1}{4}\right)$

14. $\left(\dfrac{7}{3}, -6\right)$ and $\left(\dfrac{7}{3}, \dfrac{1}{6}\right)$

15. $(-7, -3)$ and $(10, -3)$

16. $(4, 12)$ and $(-5, 12)$

17. $\left(6, \dfrac{3}{2}\right)$ and $(2, 1)$

18. $\left(-5, -\dfrac{2}{3}\right)$ and $(3, -1)$

Applications

19. *Snowboarding* Find the slope (grade) of a snowboard "half-pipe" recreation hill that rises 48 feet vertically over a horizontal distance of 80 feet.

20. *Grade of a Driveway* Find the grade of a driveway that rises 4.5 feet vertically over a horizontal distance of 90 feet.

21. *Rock Formation* Find the slope (pitch) of a perfectly smooth rock formation that rises 35.7 feet vertically over a horizontal distance of 142.8 feet.

22. *Pitch of a Roof* Find the slope (pitch) of a roof that rises 3.15 feet vertically over a horizontal distance of 10.50 feet.

23. *River Flow* A river has a slope of 0.16. How many feet does it fall vertically over a horizontal distance of 500 feet?

24. *Slope of a Road* A Maine mountain road used by tractor-trailer trucks has a slope of 0.18. How many feet does it fall vertically over a horizontal distance of 700 feet?

Find the slope of a line parallel to the line that passes through the following points.

25. $(6, 7)$ and $(24, 3)$

26. $(35, -3)$ and $(5, 9)$

27. $(7, 1)$ and $(6.5, 2)$

28. $(3, 5)$ and $(2.8, 6)$

29. $\left(-9, \dfrac{1}{2}\right)$ and $(-6, 5)$

30. $\left(1, \dfrac{5}{2}\right)$ and $\left(\dfrac{1}{3}, 2\right)$

Find the slope of a line perpendicular *to the line that passes through the following points.*

31. $(8, 12)$ and $(3, 9)$

32. $(3, 9)$ and $(7, 15)$

33. $\left(2, -\dfrac{1}{2}\right)$ and $\left(1, \dfrac{5}{2}\right)$

34. $\left(-\dfrac{2}{3}, -2\right)$ and $\left(-3, \dfrac{1}{3}\right)$

35. $(-8.4, 0)$ and $(0, 4.2)$

36. $(0, -5)$ and $(-2, 0)$

To Think About

37. A line k passes through the points $(-3, -9)$ and $(1, 11)$. A second line h passes through the points $(-2, -13)$ and $(2, 7)$. Is line k parallel to line h? Why or why not?

38. A line k passes through the points $(4, 2)$ and $(-4, 4)$. A second line h passes through the points $(-8, 1)$ and $(8, -3)$. Is line k parallel to line h? Why or why not?

39. Show that $ABCD$ is a parallelogram if the four vertices are $A(2, 1)$, $B(-1, -2)$, $C(-7, -1)$, and $D(-4, 2)$. (*Hint:* A parallelogram is a four-sided figure with opposite sides parallel.)

40. Do the points $A(-1, -2)$, $B(2, -1)$, and $C(8, 1)$ lie on a straight line? Explain.

41. ***Handicapped Ramp*** Most new buildings are required to have a ramp for the handicapped that has a maximum vertical rise of 5 feet for every 60 feet of horizontal distance.

(a) What is the value of the slope of a ramp for the handicapped?

(b) If the builder constructs a new building in which the ramp has a horizontal distance of 24 feet, what is the maximum height of the doorway above the level of the parking lot where the ramp begins?

(c) What is the shortest possible length of the ramp if the architect redesigns the building so that the doorway is 1.7 feet above the parking lot?

42. ***Rate of Climb of Aircraft*** A small Cessna plane takes off from Hyannis Airport. When the plane is 1 mile from the airport, it is flying at an altitude of 3000 feet. When the plane is 2 miles from the airport, it is flying at an altitude of 4300 feet. Round your answers to the following questions to the nearest tenth.

(a) If the plane continues flying at the same slope (the same rate of climb), what will its altitude be when it is 4.8 miles from the airport?

(b) If the plane continues flying at the same rate of climb, how many miles from the airport will it be when it reaches an altitude of 6000 feet?

(c) A Lear jet leaves the airport at the same time and has the same altitude (3000 feet) as the Cessna when each plane is 1 mile from the airport. When the jet is 1.8 miles from the airport, it is flying at an altitude of 4040 feet. Is the plane being flown at the same rate of climb as the Cessna?

Cumulative Review

Evaluate.

43. $\dfrac{5 + 3\sqrt{9}}{|2 - 9|}$

44. $2(3 - 6)^3 + 20 \div (-10)$

Simplify.

45. $\dfrac{-15x^6y^3}{-3x^{-4}y^6}$

46. $8x(x - 1) - 2(x + y)$

1 Using the Slope–Intercept Form of the Equation of a Line

Recall that the standard form of the equation of a line is $Ax + By = C$. Although the standard form tells us that the graph is a straight line, it reveals little about the line. A more useful form of the equation is the **slope–intercept form.** The slope–intercept form immediately reveals the slope of the line and where it intersects the y-axis. This is important information that will help us graph the line.

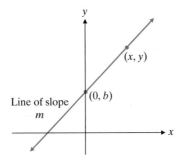

We will use the definition of slope to derive the equation. Let a nonvertical line with slope m cross the y-axis at some point $(0, b)$. Choose any other point on the line and label it (x, y). By the definition of slope, $\dfrac{y_2 - y_1}{x_2 - x_1} = m$. But here, $x_1 = 0$, $y_1 = b$, $x_2 = x$, and $y_2 = y$. So

$$\frac{y - b}{x} = m.$$

Now we solve this equation for y.

$$y - b = mx$$
$$y = mx + b$$

SLOPE–INTERCEPT FORM

The **slope–intercept form** of the equation of a line is $y = mx + b$, where m is the slope and $(0, b)$ is the y-intercept.

Write the Equation of a Line Given Its Slope and y-intercept.

EXAMPLE 1 Write an equation of the line with slope $-\frac{2}{3}$ and y-intercept $(0, 5)$.

Solution

$$y = mx + b$$
$$y = \left(-\frac{2}{3}\right)x + (5) \quad \text{Substitute } -\frac{2}{3} \text{ for } m \text{ and } 5 \text{ for } b.$$
$$y = -\frac{2}{3}x + 5$$

Practice Problem 1 Write an equation of the line with slope 4 and y-intercept $\left(0, -\frac{3}{2}\right)$.

Write the Equation of a Line Given the Graph. We can write an equation of a line if we are given its graph since we can determine the y-intercept and the slope from the graph.

Student Learning Objectives

After studying this section, you will be able to:

1 Use the slope–intercept form of the equation of a line.

2 Use the point–slope form of the equation of a line.

3 Write the equation of the line passing through a given point that is parallel or perpendicular to a given line.

Graphing Calculator

 Exploring y-intercepts

Using a graphing calculator, graph

$$y_1 = 2x,$$
$$y_2 = 2x + 1,$$
$$y_3 = 2x - 1, \text{ and}$$
$$y_4 = 2x + 2$$

on the same set of axes. Where does each graph cross the y-axis? What effect does b have on the graph of $y = mx + b$? What would the graph of the line $y = 2x - 5$ look like? Use your graphing calculator to verify your conclusion.

NOTE TO STUDENT: Fully worked-out solutions to all of the Practice Problems can be found at the back of the text starting at page SP-1

PRACTICE PROBLEM 2

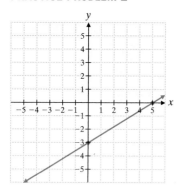

EXAMPLE 2   Find the slope and y-intercept of the line. Then use these to write an equation of the line whose graph is shown.

Solution   Looking at the graph, we can see that the y-intercept is at $(0, 5)$. That is, $b = 5$. If we can identify the coordinates of another point on the line, we will have two points and we can determine the slope. Another point on the line is $(3, 3)$.

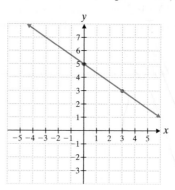

Thus,

$$(x_1, y_1) = (0, 5),$$
$$(x_2, y_2) = (3, 3), \text{ and}$$
$$m = \frac{y_2 - y_1}{x_2 - x_1} = \frac{3 - 5}{3 - 0} = -\frac{2}{3}.$$

We can now write the equation of the line in slope–intercept form.

$$y = mx + b$$
$$y = -\frac{2}{3}x + 5 \quad \text{Substitute } -\frac{2}{3} \text{ for } m \text{ and } 5 \text{ for } b.$$

Practice Problem 2   Find the slope and y-intercept of the line whose graph is shown in the margin on the left. Then use these to write an equation of this line.

Use the Slope–Intercept Form to Graph an Equation.   We have just seen that given the graph of a line, we can determine its equation by identifying the y-intercept and finding the slope. We can also draw the graph of an equation without plotting points if we can write the equation in slope–intercept form and then locate the y-intercept on the graph.

EXAMPLE 3   Find the slope and the y-intercept. Then sketch the graph of the equation $28x - 7y = 21$.

Solution   First we will change the standard form of the equation into slope–intercept form. This is a very important procedure. Be sure that you understand each step.

$$28x - 7y = 21$$
$$-7y = -28x + 21$$
$$\frac{-7y}{-7} = \frac{-28x}{-7} + \frac{21}{-7}$$

$$\overset{\text{slope}}{y = 4x + (-3)} \quad \underset{\text{gives } y\text{-intercept}}{y = mx + b}$$

Thus, the slope is 4, and the y-intercept is $(0, -3)$.

To sketch the graph, begin by plotting the point where the graph crosses the y-axis, $(0, -3)$. Plot the point. Now look at the slope. The slope, m, is 4 or $\frac{4}{1}$. This means there is a rise of 4 for every run of 1. From the point $(0, -3)$ go up 4 units and to the right 1 unit to locate a second point on the line. Draw a straight line that contains these two points, and you have the graph of the equation $28x - 7y = 21$.

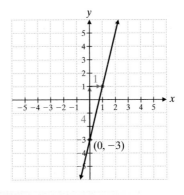

PRACTICE PROBLEM 3

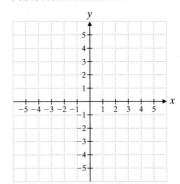

Practice Problem 3   Find the slope and the y-intercept. Then sketch the graph of the equation $3x - 4y = -8$.

NOTE TO STUDENT: Fully worked-out solutions to all of the Practice Problems can be found at the back of the text starting at page SP-1

② Using the Point–Slope Form of the Equation of a Line

What happens if we know the slope of a line and a point on the line that is not the y-intercept? Can we write the equation of the line? By the definition of slope, we have the following:

$$m = \frac{y - y_1}{x - x_1}$$

$$m(x - x_1) = y - y_1$$

That is, $y - y_1 = m(x - x_1)$.

This is the point–slope form of the equation of a line.

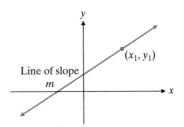

POINT–SLOPE FORM

The **point–slope form** of the equation of a line is $y - y_1 = m(x - x_1)$, where m is the slope and (x_1, y_1) are the coordinates of a known point on the line.

Write the Equation of a Line Given Its Slope and One Point on the Line.

EXAMPLE 4   Find an equation of the line that has slope $-\frac{3}{4}$ and passes through the point $(-6, 1)$. Express your answer in standard form.

Solution   Since we don't know the y-intercept, we can't use the slope–intercept form easily. Therefore, we use the point–slope form.

$$y - y_1 = m(x - x_1)$$

$$y - 1 = -\frac{3}{4}[x - (-6)] \qquad \text{Substitute the given values.}$$

$$y - 1 = -\frac{3}{4}x - \frac{9}{2} \qquad \text{Simplify. (Do you see how we did this?)}$$

$$4y - 4(1) = 4\left(-\frac{3}{4}x\right) - 4\left(\frac{9}{2}\right) \quad \text{Multiply each term by the LCD 4.}$$

$$4y - 4 = -3x - 18 \qquad \text{Simplify.}$$

$$3x + 4y = -18 + 4 \qquad \text{Add } 3x + 4 \text{ to each side.}$$

$$3x + 4y = -14 \qquad \text{Add like terms.}$$

The equation in standard form is $3x + 4y = -14$.

Practice Problem 4   Find an equation of the line that passes through $(5, -2)$ and has a slope of $\frac{3}{4}$. Express your answer in standard form.

Write the Equation of a Line Given Two Points on the Line. We can use the point–slope form to find the equation of a line if we are given two points. Carefully study the following example. Be sure you understand each step. You will encounter this type of problem frequently.

EXAMPLE 5   Find the equation of a line that passes through $(3, -2)$ and $(5, 1)$. Express your answer in slope–intercept form.

Solution   First we find the slope.

$$m = \frac{y_2 - y_1}{x_2 - x_1} = \frac{1 - (-2)}{5 - 3} = \frac{1 + 2}{2} = \frac{3}{2}$$

Now we substitute the value of the slope and the coordinates of either point into the point–slope equation. Let's use $(5, 1)$.

$$y - y_1 = m(x - x_1)$$

$$y - 1 = \frac{3}{2}(x - 5) \qquad \text{Substitute } m = \frac{3}{2} \text{ and } (x_1, y_1) = (5, 1).$$

$$y - 1 = \frac{3}{2}x - \frac{15}{2} \qquad \text{Remove parentheses.}$$

$$y = \frac{3}{2}x - \frac{15}{2} + 1 \qquad \text{Add 1 to each side of the equation.}$$

$$y = \frac{3}{2}x - \frac{15}{2} + \frac{2}{2} \qquad \text{Add the two fractions.}$$

$$y = \frac{3}{2}x - \frac{13}{2} \qquad \text{Simplify.}$$

Practice Problem 5   Find an equation of the line that passes through $(-4, 1)$ and $(-2, -3)$. Express your answer in slope–intercept form.

Before we go further, we want to point out that these various forms of the equation of a straight line are just that—*forms* for convenience. We are *not* using different equations each time, nor should you simply try to memorize the different variations without understanding when to use them. They can easily be derived from the definition of slope, as we have seen. And remember, you can *always* use the definition of slope to find the equation of a line. You will find it helpful to review Example 4 and Example 5 for a few minutes before going ahead to Example 6. It is important to see how each example is different.

Graphing Calculator

Using Linear Regression to Find an Equation

Many graphing calculators, such as the TI-83, will find the equation of a line in slope–intercept form if you enter the points as a collection of data and use the Regression feature. We would enter the data from Example 5 as follows:

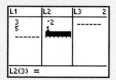

The output of the calculator uses the notation $y = ax + b$ instead of $y = mx + b$.

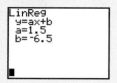

Thus, our answer to Example 5 using the graphing calculator would be $y = 1.5x - 6.5$.

NOTE TO STUDENT: *Fully worked-out solutions to all of the Practice Problems can be found at the back of the text starting at page SP-1*

3   Writing the Equation of Parallel and Perpendicular Lines

Let us now look at parallel and perpendicular lines. If we are given the equation of a line and a point not on the line, we can find the equation of a second line that passes through the given point and is parallel or perpendicular to the first line. We can do this because we know that the slopes of parallel lines are equal and that the slopes of perpendicular lines are negative reciprocals of each other.

We begin by finding the slope of the given line. Then we use the point–slope form to find the equation of the second line. Study carefully each step of the following example.

EXAMPLE 6   Find the equation of a line passing through the point $(-2, -4)$ and parallel to the line $2x + 5y = 8$. Express the answer in standard form.

Solution   First we need to find the slope of the line $2x + 5y = 8$. We do this by writing the equation in slope–intercept form.

$$5y = -2x + 8$$

$$y = -\frac{2}{5}x + \frac{8}{5}$$

The slope of the given line is $-\frac{2}{5}$. Since parallel lines have the same slope, the slope of the unknown line is also $-\frac{2}{5}$. Now we substitute $m = -\frac{2}{5}$ and the coordinates of the point $(-2, -4)$ into the point–slope form of the equation of a line.

$$y - y_1 = m(x - x_1)$$

$$y - (-4) = -\frac{2}{5}[x - (-2)] \qquad \text{Substitute.}$$

$$y + 4 = -\frac{2}{5}(x + 2) \qquad \text{Simplify.}$$

$$y + 4 = -\frac{2}{5}x - \frac{4}{5} \qquad \text{Remove parentheses.}$$

$$5y + 5(4) = 5\left(-\frac{2}{5}x\right) - 5\left(\frac{4}{5}\right) \qquad \text{Multiply each term by the LCD 5.}$$

$$5y + 20 = -2x - 4 \qquad \text{Simplify.}$$

$$2x + 5y = -4 - 20 \qquad \text{Add } 2x - 20 \text{ to each side.}$$

$$2x + 5y = -24 \qquad \text{Simplify.}$$

$2x + 5y = -24$ is the equation of the line passing through the point $(-2, -4)$ and parallel to the line $2x + 5y = 8$.

Practice Problem 6   Find the equation of a line passing through $(4, -5)$ and parallel to the line $5x - 3y = 10$. Express the answer in standard form. ∎

Some extra steps are needed if the desired line is to be perpendicular to the given line. Note carefully the approach in Example 7.

EXAMPLE 7   Find the equation of a line that passes through the point $(2, -3)$ and is perpendicular to the line $3x - y = -12$. Express the answer in standard form.

Solution   To find the slope of the line $3x - y = -12$, we rewrite it in slope–intercept form.

$$-y = -3x - 12$$
$$y = 3x + 12$$

This line has a slope of 3. Therefore, the slope of a line perpendicular to this line is the negative reciprocal $-\frac{1}{3}$.

Now substitute the slope $m = -\frac{1}{3}$ and the coordinates of the point $(2, -3)$ into the point–slope form of the equation.

$$y - y_1 = m(x - x_1)$$

$$y - (-3) = -\frac{1}{3}(x - 2) \qquad \text{Substitute.}$$

$$y + 3 = -\frac{1}{3}(x - 2) \qquad \text{Simplify.}$$

$$y + 3 = -\frac{1}{3}x + \frac{2}{3} \qquad \text{Remove parentheses.}$$

$$3y + 3(3) = 3\left(-\frac{1}{3}x\right) + 3\left(\frac{2}{3}\right) \qquad \text{Multiply each term by the LCD 3.}$$

$$3y + 9 = -x + 2 \qquad \text{Simplify.}$$
$$x + 3y = 2 - 9 \qquad \text{Add } x - 9 \text{ to each side.}$$
$$x + 3y = -7 \qquad \text{Simplify.}$$

$x + 3y = -7$ is the equation of a line that passes through the point $(2, -3)$ and is perpendicular to the line $3x - y = -12$.

Practice Problem 7   Find the equation of a line that passes through $(-4, 3)$ and is perpendicular to the line $6x + 3y = 7$. Express the answer in standard form.

NOTE TO STUDENT: Fully worked-out solutions to all of the Practice Problems can be found at the back of the text starting at page SP-1

Developing Your Study Skills

Making a Friend in the Class

Try to make a friend in your class. You may find that you enjoy sitting together and drawing support and encouragement from one another. Exchange phone numbers so you can call each other whenever you get stuck while doing your homework. Set up convenient times to study together on a regular basis, to do homework, and to review for exams.

You must not depend on a friend or fellow student to tutor you, do your work for you, or in any way be responsible for your learning. However, you will learn from one another as you seek to master the course. Studying with a friend and comparing notes, methods, and solutions can be very helpful. And it can make learning mathematics a lot more fun!

Verbal and Writing Skills

1. You are given two points that lie on a line. Explain how you would find the equation of the line.

2. Suppose $y = -\frac{2}{7}x + 5$. What can you tell about the graph by looking at the equation?

Write the equation of a line with the given slope and the given y-intercept. Leave the answer in slope–intercept form.

3. Slope $\frac{3}{4}$, y-intercept $(0, -9)$

4. Slope $-\frac{2}{3}$, y-intercept $(0, 5)$

Write the equation of a line with the given slope and the given y-intercept. Express the answer in standard form.

5. Slope $\frac{3}{4}$, y-intercept $\left(0, \frac{1}{2}\right)$

6. Slope $\frac{5}{6}$, y-intercept $\left(0, \frac{1}{3}\right)$

Find the slope and y-intercept of each of the following lines. Then use these to write the equation of the line.

7.

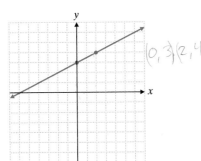

8.

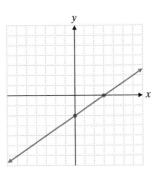

9.

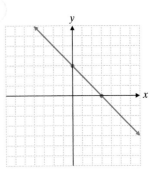

10.

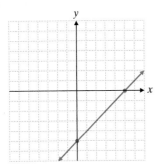

11.

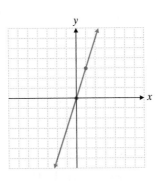

12.

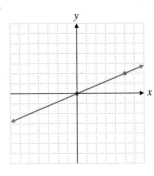

13.

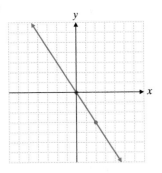

14.

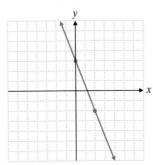

15.

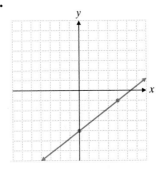

Write each equation in slope–intercept form. Then identify the slope and the y-intercept for each line.

16. $2x - y = 12$

17. $x - y = 5$

18. $2x - 3y = -8$

19. $5x - 4y = -20$

20. $\frac{1}{2}x + 4y = 5$

21. $3x + \frac{2}{3}y = -2$

For each equation find the slope and the y-intercept. Use these to graph the equation.

22. $y = 3x + 4$

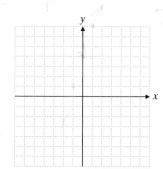

23. $y = \frac{1}{2}x - 3$

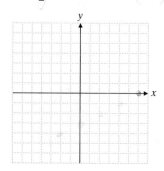

24. $5x - 4y = -20$

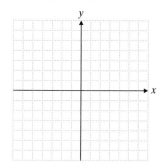

25. $5x + 3y = 18$

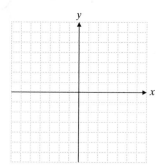

Find the equation of the line that passes through the given point and has the given slope. Express your answer in slope–intercept form.

26. $(4, 6), m = -\frac{1}{2}$

27. $(6, 4), m = -\frac{2}{3}$

28. $(8, 0), m = -3$

29. $(-7, -2), m = 5$

30. $(0, -1), m = -\frac{5}{3}$

31. $(6, 0), m = -\frac{1}{5}$

Find the equation of a line passing through the pair of points. Write the equation in the slope–intercept form.

32. $(7, -2)$ and $(-1, -3)$

33. $(-4, -1)$ and $(3, 4)$

34. $\left(\frac{7}{6}, 1\right)$ and $\left(-\frac{1}{3}, 0\right)$

35. $\left(\frac{1}{2}, -3\right)$ and $\left(\frac{7}{2}, -5\right)$

36. $(4, 8)$ and $(-3, 8)$

37. $(12, -3)$ and $(7, -3)$

Find the equation of the line satisfying the conditions given. Express your answer in standard form.

38. Parallel to $3x - y = -5$ and passing through $(-1, 0)$

39. Parallel to $5x - y = 4$ and passing through $(-2, 0)$

40. Parallel to $2y + x = 7$ and passing through $(-5, -4)$

41. Parallel to $x = 3y - 8$ and passing through $(5, -1)$

42. Perpendicular to $y = 5x$ and passing through $(4, -2)$

43. Perpendicular to $2y = -3x$ and passing through $(6, -1)$

44. Perpendicular to $x - 4y = 2$ and passing through $(3, -1)$

45. Perpendicular to $x + 7y = -12$ and passing through $(-4, -1)$

To Think About

Without graphing determine whether the following pairs of lines are (a) parallel, (b) perpendicular, or (c) neither parallel nor perpendicular.

46. $5x - 6y = 19$
$6x + 5y = -30$

47. $-3x + 5y = 40$
$5y + 3x = 17$

48. $y = \dfrac{2}{3}x + 6$
$-2x - 3y = -12$

49. $y = -\dfrac{3}{4}x - 2$
$6x + 8y = -5$

50. $y = \dfrac{3}{7}x - \dfrac{1}{14}$
$14y + 6x = 3$

51. $y = \dfrac{5}{6}x - \dfrac{1}{3}$
$6x + 5y = -12$

Optional Graphing Calculator Problems

If you have a graphing calculator, use it to graph each pair of equations. Do the graphs appear to be parallel?

52. $y = -2.39x + 2.04$ and $y = -2.39x - 0.87$

53. $y = 1.43x - 2.17$ and $y = 1.43x + 0.39$

Applications

Cost of Homes *The median price of homes in the United States has been increasing steadily. The increase can be approximated by a linear equation of the form $y = mx + b$. The U.S. Census Bureau reported that in 1980 the median cost of a home in the United States was $68,700. In 1998 the median cost of a home was $167,900. We can record the data as follows:*

Number of Years Since 1980	Cost of Home in Thousands of Dollars
0	68.7
18	167.9

Source: U.S. Census Bureau

Use the table of values for Exercises 54–57.

54. Using these two ordered pairs, find the equation $y = mx + b$ where x is the number of years since 1980 and y is the median cost of a home in thousands of dollars. Round your values of m and b to the nearest hundredth.

55. Use the equation obtained in Exercise 54 to find the expected median cost of a home in the year 2016 (36 years after 1980).

56. Graph the equation using the data for 1980, 1998, and 2016.

57. Use your graph to estimate the median cost of a home in 2007 (27 years after 1980).

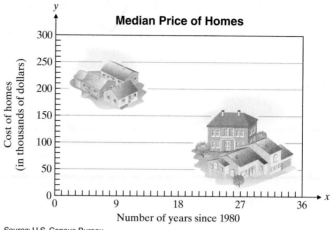

Median Price of Homes

Source: U.S. Census Bureau

Number of Housing Units *The number of housing units in the United States has been increasing steadily. The increase can be approximated by a linear equation of the form $y = mx + b$. The U.S. Census Bureau reported that in 1980 there were 87,700,000 housing units in the United States. In 1998 the figure was 117,300,000 housing units. We can record the data as follows:*

Number of Years Since 1980	Number of Housing Units in Millions
0	87.7
18	117.3

Source: U.S. Census Bureau

Use the table of values for Exercises 49–52.

58. Using these two ordered pairs, find the equation $y = mx + b$, where x is the number of years since 1980 and y is the number of housing units in millions. Round your values of m and b to the nearest hundredth.

59. Use the equation obtained in Exercise 58 to find the expected number of housing units in millions in the year 2016 (36 years after 1980).

60. Graph the equation using the data for 1980, 1998, and 2016.

61. Use your graph to estimate the number of housing units in 2007 (27 years after 1980).

Total Housing Inventory

Source: U.S. Census Bureau

Cumulative Review

Solve for x.

62. $11 - (x + 2) = 7(3x + 6)$

63. $0.3x + 0.1 = 0.27x - 0.02$

64. $70 + 70(0.01x) + 3 = 82.10$

65. $\dfrac{5}{4} - \dfrac{3}{4}(2x + 1) = x - 2$

How are you doing with your homework assignments in Sections 3.1 to 3.3? Do you feel you have mastered the material so far? Do you understand the concepts you have covered? Before you go further in the textbook, take some time to do each of the following problems.

3.1

1. Find the value of a if $(a, 6)$ is a solution to $5x + 2y = -12$.

Graph each equation.

2. $y = -\dfrac{1}{2}x + 5$

3. $5x + 3y = -15$

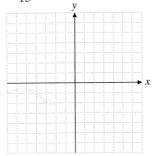

4. $4y + 6x = -8 + 9x$

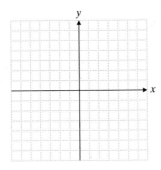

3.2

5. Find the slope of the line passing through the points $(-2, 3)$ and $(-1, -6)$.

6. Find the slope of a line parallel to the line that passes through $\left(\dfrac{2}{3}, 4\right)$ and $\left(\dfrac{5}{6}, -2\right)$.

7. Find the slope of the line *perpendicular* to the line passing through the points $(5, 0)$ and $(-13, -4)$.

8. A Colorado road has a slope of 0.13. How many feet does it rise vertically over a horizontal distance of 500 feet?

3.3

9. Find the slope and the y-intercept of $6x + 7y = 14$

10. Write an equation of the line of slope -2 that passes through $(7, -3)$.

11. Write an equation of the line passing through $(-1, -2)$ and perpendicular to $3x - 5y = 10$.

12. Find the equation of a line that passes through $(-1, -8)$ and $(2, 7)$.

Now turn to page SA-8 for the answers to each of these problems. Each answer also includes a reference to the objective in which the problem is first taught. If you missed any of these problems, you should stop and review the Examples and Practice Problems in the referenced objective. A little review now will help you master the material in the upcoming sections of the text.

1. _____

2. _____

3. _____

4. _____

5. _____

6. _____

7. _____

8. _____

9. _____

10. _____

11. _____

12. _____

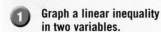
1 Graphing a Linear Inequality in Two Variables

A linear inequality in two variables is similar to a linear equation in two variables. However, in place of the = sign, there appears instead one of the following four inequality symbols: $<$, $>$, $\leq$, $\geq$.

LINEAR INEQUALITY IN TWO VARIABLES

A **linear inequality in two variables** is an inequality that can be written

$$as \quad Ax + By > C \quad or \quad Ax + By < C$$
$$or \quad Ax + By \geq C \quad or \quad Ax + By \leq C$$

where A, B, and C are real numbers and A and B are not both zero.

The graph of this type of linear inequality is a half-plane that lies on one side of a straight line. It will also include the boundary line if the inequality contains the $\leq$ or the $\geq$ symbols.

PROCEDURE FOR GRAPHING LINEAR INEQUALITIES

1. Replace the inequality symbol by an equal sign. This equation will be a boundary for the desired region.

2. Graph the line obtained in step 1. Use a dashed line if the original inequality contains a $<$ or $>$ symbol. Use a solid line if the original inequality contains a $\leq$ or $\geq$ symbol.

3. Choose a test point that does not lie on the boundary line. Substitute the coordinates into the original inequality. If you obtain an inequality that is true, shade the region on the side of the line containing the test point. If you obtain a false inequality, shade the region on the side of the line opposite the test point.

If the boundary line does not pass through $(0, 0)$, that is usually a good test point to use.

EXAMPLE 1   Graph $y < 2x + 3$.

Solution

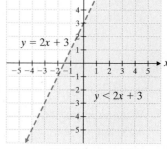

1. The boundary line is $y = 2x + 3$.

2. We graph $y = 2x + 3$ using a dashed line because the inequality contains $<$.

3. Since the line does not pass through $(0,0)$, we can use it as a test point. Substituting $(0,0)$ into $y < 2x + 3$, we have the following:

$$0 < 2(0) + 3$$
$$0 < 3$$

This inequality is true. We therefore shade the region on the same side of the line as $(0, 0)$. See the sketch. The solution is the shaded region *not including* the dashed line.

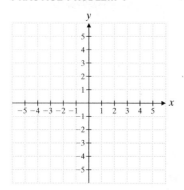
Practice Problem 1   Graph $y > 3x + 1$.

EXAMPLE 2 Graph $3x + 2y \geq 4$.

Solution

1. The boundary line is $3x + 2y = 4$.
2. We graph the boundary line with a solid line because the inequality contains $\geq$.
3. Since the line does not pass through $(0, 0)$, we can use it as a test point. Substituting $(0, 0)$ into $3x + 2y \geq 4$ gives the following:

$$3(0) + 2(0) \geq 4$$
$$0 + 0 \geq 4$$
$$0 \geq 4$$

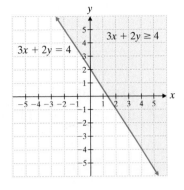

This inequality is false. We, therefore, shade the region on the side of the line opposite $(0, 0)$. See the sketch. The solution is the shaded region *including* the boundary line.

Practice Problem 2 Graph $-4x + 5y \leq -10$.

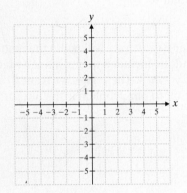

EXAMPLE 3 Graph $4x - y < 0$.

Solution

1. The boundary line is $4x - y = 0$.
2. We graph the boundary line with a dashed line.
3. Since $(0, 0)$ lies on the line, we cannot use it as a test point. We must choose another point not on the line. We try to pick some point that is *not* close to the dashed boundary line. Let's pick $(-1, 5)$. Substituting $(-1, 5)$ into $4x - y < 0$ gives the following:

$$4(-1) - 5 < 0$$
$$-4 - 5 < 0$$
$$-9 < 0$$

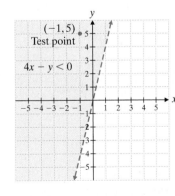

Graphing Calculator

Graphing Linear Inequalities

You can graph a linear inequality like $y \leq 3x + 1$ on a graphing calculator. On some calculators you can enter the expression for the boundary directly into the $Y =$ editor of your graphing calculator and then select the appropriate direction for shading. Other calculators may require using a Shade command to shade the region. In general, most calculator displays do not distinguish between dashed and solid boundaries.

Display:

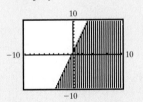

Graph the following:

1. $y < 3.45x - 1.232$
2. $y > -5.346x - 3.678$
3. $3.45y + 4.782x > 6.0238$

PRACTICE PROBLEM 3

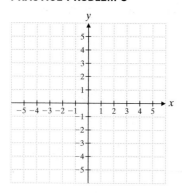

This is true. Therefore, we shade the side of the boundary line that contains $(-1, 5)$. See the sketch. The solution is the shaded region above the dashed line but *not including* the dashed line.

Practice Problem 3   Graph $3y + x < 0$.

2 Graphing a Linear Inequality for Which the Coefficient of One Variable Is Zero

EXAMPLE 4   Graph $3y \geq -12$.

Solution   First we will need to simplify the original inequality by dividing each side by 3.

$$3y \geq -12$$
$$\frac{3y}{3} \geq \frac{-12}{3}$$
$$y \geq -4$$

This inequality is equivalent to

$$0x + y \geq -4.$$

We find that if $y \geq -4$, any value of x will make the inequality true. Thus, x can be any value at all and still be included in our shaded region. Therefore, we draw a solid horizontal line at $y = -4$. The region we want to shade is the region above the line. The solution to our problem is the line and the shaded region above the line.

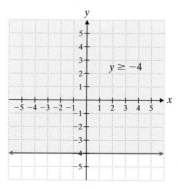

NOTE TO STUDENT: *Fully worked-out solutions to all of the Practice Problems can be found at the back of the text starting at page SP-1*

Practice Problem 4   Graph $6y \leq 18$.

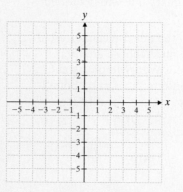

Verbal and Writing Skills

1. Explain when to use a dashed line as a boundary when graphing a linear inequality.

2. Explain when to use a solid line as a boundary when graphing a linear inequality.

3. When graphing $x > 5$, should the region to the left of the line $x = 5$ be shaded or the region to the right of the line $x = 5$?

4. When graphing $y < -6$ should the region below the line $y = -6$ be shaded or should the region above the line $y = -6$ be shaded?

5. When we graph the inequality $3x - 2y \geq 0$, why can't we use $(0, 0)$ as a test point? If we test the point $(-4, 2)$, do we obtain a false statement or a true one?

6. When we graph the inequality $4x - 3y \geq 0$, why can't we use $(0, 0)$ as a test point? If we test the point $(6, -5)$, do we obtain a false statement or a true one?

Graph each region.

7. $y > -2x + 4$

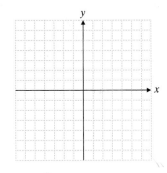

8. $y > -3x + 2$

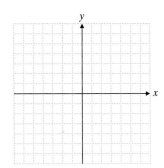

9. $y < \dfrac{2}{3}x - 2$

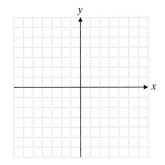

10. $y < \dfrac{3}{4}x - 3$

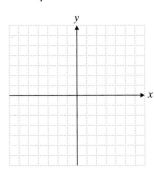

11. $y \geq -\dfrac{5}{3}x + 3$

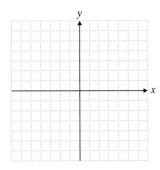

12. $y \geq -\dfrac{1}{5}x + 2$

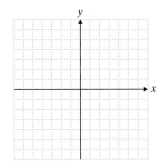

13. $5y - x \leq 15$

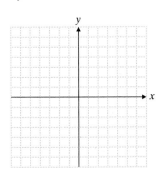

14. $-2x - y > 1$

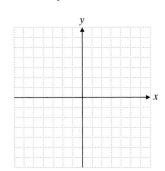

15. $2x + y > 0$

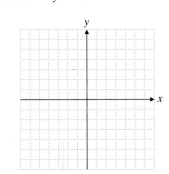

16. $x + y < 0$

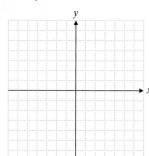

17. $5x - 2y \geq 0$

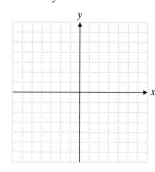

18. $x - 3y \geq 0$

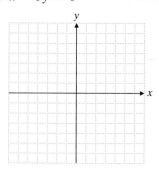

Graph each inequality in a rectangular coordinate system.

19. $x > -4$

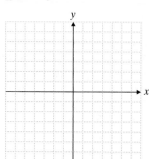

20. $x < 3$

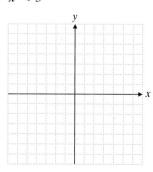

21. $y \leq -1$

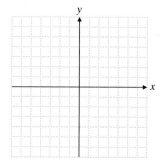

22. $y \geq -1$

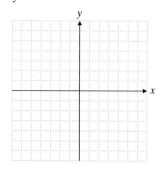

23. $-8x \leq -12$

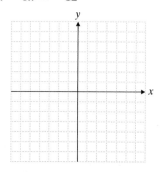

24. $-5x \leq -10$

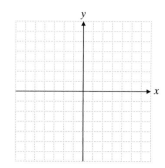

25. $4y \geq 2$

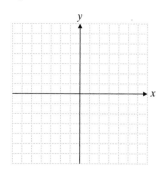

26. $3y + 2 > 0$

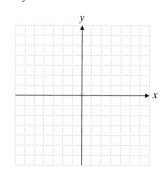

To Think About

27. What *one* region should be shaded to satisfy the inequalities $x + y \leq 3, x \geq 0,$ and $y \geq 0$ at the same time?

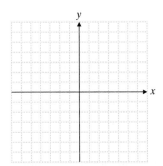

28. What *one* region should be shaded to satisfy the inequalities $x \geq 1, y \geq 2, x \leq 4,$ and $y \leq 5$ at the same time?

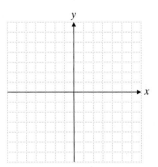

Applications

29. *Elevator Capacity* An elevator in an office building has a weight limit of 2100 pounds. Suppose x children each weighing 75 pounds and y adults each weighing 175 pounds board the elevator. Write a linear inequality using x and y that describes the weight limit of the children and adults. Graph the inequality.

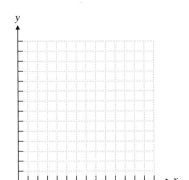

30. *SUV Capacity* The passenger and luggage weight capacity for many popular sport-utility vehicles (such as the Ford Explorer, Jeep Grand Cherokee, and Nissan Pathfinder) is around 1200 pounds. Suppose x adults each weighing 200 pounds and y pieces of luggage each weighing 60 pounds are put inside one of these SUVs. Write a linear inequality using x and y that describes the weight limitation for a load of passengers and luggage on an SUV. Graph the inequality.

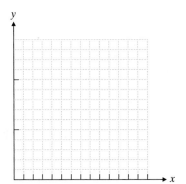

Cumulative Review

Evaluate each of the following.

31. $3x^2 + 6x - 2$ when $x = -2$

32. $A = \dfrac{1}{2}a(b + c)$ when $a = 6.0, b = 2.5,$ and $c = 5.5$

▲ **33.** *Floor Tiles in Living Room* A living room in a summer camp measures 12 feet by 18 feet. Dan and Marsha Perkins plan to cover the room with floor tiles. A package of 1-foot-square tiles contains twenty-four tiles. How many packages of tiles will they need?

Student Learning Objectives

After studying this section, you will be able to:

1. Describe a relation and determine its domain and range.

2. Determine whether a relation is a function.

3. Evaluate a function using function notation.

═ PIZZA PLUS ═

Size of Pizza	Diameter of Pizza	Price of Pizza
Small	5"	$4.75
Medium	10"	$7.00
Large	15"	$9.25
Party Size	20"	$11.50

1 Describing a Relation and Determining Its Domain and Range

Whenever we collect and study data, we look for relationships. If we can determine a relationship between two sets of data, we can make predictions about the data. Let's begin with a simple finite example.

A local pizza parlor offers four different sizes of pizza and prices them as shown in the table on the right. By looking at the table, we can see that the price depends on the diameter of the pizza. The larger the size, the more expensive it will be. This appears to be an increasing relation. We can use a set of ordered pairs to show the correspondence between the diameter of the pizza and the price.

$$\{(5 \text{ in.}, \$4.75), \quad (10 \text{ in.}, \$7.00), \quad (15 \text{ in.}, \$9.25), \quad (20 \text{ in.}, \$11.50)\}$$

We can graph the ordered pairs to get a better picture of the relationship.

Following convention, we will assign the **independent variable** to the horizontal axis and the **dependent variable** to the vertical axis. Note that the dependent variable is price because price *depends* on size.

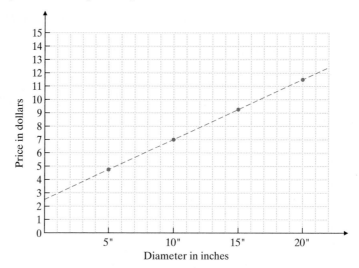

Just as we suspected, the relation is increasing. That is, the graph goes up as we move from left to right. We draw a line that approximately fits the data. This allows us to analyze the data more easily. Notice that if Pizza Plus decides to come out with a size between the small pizza and the medium pizza, it will probably be priced around $6. Thus, we can see that there is a relation between the diameter of a pizza and its price and that we can express this relation as a table of values, as a set of ordered pairs, or in a graph.

Mathematicians have found it most useful to define a relation in terms of ordered pairs.

> **RELATION**
>
> A **relation** is any set of ordered pairs.

All the first items of each ordered pair in a relation can be grouped as a set called the **domain** of the relation. The domain of the pizza example is the set of sizes (diameters measured in inches) {5, 10, 15, 20}. These are all the possible values of the independent variable *diameter*. The domain is the input. It is the starting value.

All the second items of each ordered pair in a relation can be grouped as a set called the **range** of the relation. The range of the pizza example is the set of prices {$4.75, $7.00, $9.25, $11.50}. These are the corresponding values of the dependent variable *price*. The range is the output. It is the result that comes out once you pick a diameter.

EXAMPLE 1   The information in the following table can be found in most almanacs.

Look at the data for the men's 100-meter run. Is there a relation between any two sets of data in this table? If so, describe the relation as a table of values, a set of ordered pairs, and a graph.

Olympic Games: 100-Meter Run for Men		
Year	**Winning runner, Country**	**Time**
1900	Francis W. Jarvis, USA	11.0 s
1912	Ralph Craig, USA	10.8 s
1924	Harold Abrahams, Great Britain	10.6 s
1936	Jesse Owens, USA	10.3 s
1948	Harrison Dillard, USA	10.3 s
1960	Armin Harg, Germany	10.2 s
1972	Valery Borzov, USSR	10.14 s
1984	Carl Lewis, USA	9.99 s
1996	Donovan Bailey, Canada	9.84 s
2000	Maurice Greene, USA	9.87 s

Source: The World Almanac.

Solution   A useful relation might be the correspondence between the year the event occurred and the time in which the race was won. Let's see how this looks in a table, as a set of ordered pairs, and on a graph. Because we will choose the year as the independent variable, we will list it first in the table as is customary.

Table.

Year	1900	1912	1924	1936	1948	1960	1972	1984	1996	2000
Time in Seconds	11.0	10.8	10.6	10.3	10.3	10.2	10.14	9.99	9.84	9.87

Ordered pairs.

$\{(1900, 11.0), (1912, 10.8), (1924, 10.6), (1936, 10.3), (1948, 10.3), (1960, 10.2),$

$(1972, 10.14), (1984, 9.99), (1996, 9.84), (2000, 9.87)\}$

Graph. To save space, we draw the graph so that the time values on the vertical axis range from 9 seconds to 11 seconds rather than from 0 seconds to 11 seconds. We indicate a break like this in the scale on a vertical axis with the symbol $\lessgtr$ on the axis. On a horizontal axis we use the symbol ⌐∨⌐. By looking at the graph, we can see that the winning time usually decreases each year. This is a decreasing relation most of the time. What might we expect the winning time to be in 2008? Can we expect the time to decrease indefinitely? Why or why not?

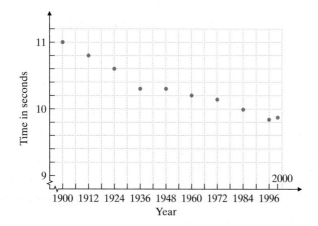

NOTE TO STUDENT: Fully worked-out solutions to all of the Practice Problems can be found at the back of the text starting at page SP-1

Practice Problem 1
The data for the women's 100-meter run in the Olympic Games for selected years are given in the table on the right. Describe the relation between the year and the winning time in a table, as a set of ordered pairs, and as a graph.

Olympic Games: 100-Meter Run for Women		
Year	**Winning runner, Country**	**Time**
1936	Helen Stephens, USA	11.5 s
1948	Francina Blankers-Koen, Netherlands	11.9 s
1960	Wilma Rudolph, USA	11.0 s
1972	Renate Stecher, East Germany	11.07 s
1984	Evelyn Ashford, USA	10.97 s
1996	Gail Devers, USA	10.94 s
2000	Marion Jones, USA	10.75 s

Source: The World Almanac.

Year							
Time in Seconds							

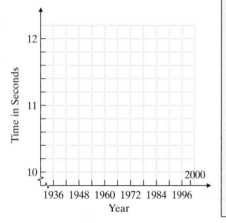

② Determining Whether a Relation Is a Function

Two of the relations we have just discussed have a special characteristic. Each value in the domain is matched with exactly one value in the range. This is true of our pizza example. One size of pizza does not have several different prices. This is also true of the relation between the year and winning time in the 100-meter run for men. No year has several different winning times.

Looking at this issue in terms of ordered pairs, we say that no two different ordered pairs have the same first coordinate. We will list each set of ordered pairs so that you can verify this characteristic.

{(Small, $4.75), (Medium, $7.00), (Large, $9.25), (Party Size, $11.50)}

{(1900, 11.0), (1912, 10.8), (1924, 10.6), (1936, 10.3), (1948, 10.3), (1960, 10.2), (1972, 10.14), (1984, 9.99), (1996, 9.84), (2000, 9.87)}

All such relations with this special property are called *functions*. Each input has only one output.

> **FUNCTION**
>
> A **function** is a relation in which no two different ordered pairs have the same first coordinate.

Notice that if we reverse the order of coordinates in the ordered pairs in Example 1, the resulting relation is not a function. We can see this readily if we list the ordered pairs.

{(11.0, 1900), (10.8, 1912), (10.6, 1924), (10.3, 1936), (10.3, 1948), (10.2, 1960), (10.14, 1972), (9.99, 1984), (9.84, 1996), (9.87, 2000)}

Two pairs, (10.3, 1936) and (10.3, 1948), have the same first coordinate. This relation is not a function.

EXAMPLE 2 Give the domain and range of each relation. Indicate whether the relation is a function.

(a) $g = \{(2, 8), (2, 3), (3, 7), (5, 12)\}$

(b) Individuals' Incomes and Taxes

Income in Dollars	14,000	18,000	24,500	33,000	50,000	50,000
Income Tax in Dollars	2350	2800	2900	3750	1350	7980

(c) Women's Tibia Bone Lengths and Heights

Length of Tibia Bone in Centimeters	33	34	35	36
Height of the Woman in Centimeters	151	154	156	159

Source: National Center for Health Statistics.

Solution Recall that the domain of a function consists of all the possible values of the independent variable or input. In a set of ordered pairs, this is the first item in each ordered pair.

The range of a function consists of the corresponding values of the dependent variable. In a set of ordered pairs, this is the second item or output in each ordered pair.

(a) Domain $= \{2, 3, 5\}$

Range $= \{8, 3, 7, 12\}$

$(2, 8)$ and $(2, 3)$ have the same first coordinate. Thus, g is not a function.

(b) Domain $= \{14,000, 18,000, 24,500, 33,000, 50,000\}$

Range $= \{2350, 2800, 2900, 3750, 1350, 7980\}$

$(50,000, 1350)$ and $(50,000, 7980)$ have the same first coordinate. This relation is not a function.

(c) Domain $= \{33, 34, 35, 36\}$

Range $= \{151, 154, 156, 159\}$

No two different ordered pairs have the same first coordinate. This relation is a function.

Practice Problem 2 Give the domain and range of the given relation. Indicate whether the relation is a function.

Car Performance

Horsepower	158	161	163	160	161
Top Speed (Mph)	98.6	89.2	101.4	102.3	94.9

If we are looking at the graph of a function, it will never have two different ordered pairs with the same x-value. We will examine some graphs in Exercises 17–25.

Often we can tell a lot about the graph of a relation. Any graph that is not a function will have at least one region in which a vertical line will cross the graph more than once.

> **VERTICAL LINE TEST**
>
> If a vertical line can intersect the graph of a relation more than once, the relation is not a function. If no such line can be drawn, then the relation is a function.

EXAMPLE 3   Determine whether each of the following is a graph of a function.

(a)

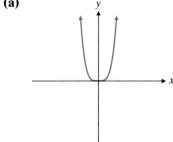

(b)

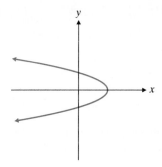

Solution

(a) This is a function. No vertical line will pass through more than one ordered pair on the curve.

(b) This is not a function. A vertical line could pass through $(2, 1)$ and $(2, -1)$. Locate those two points on the graph of **(b).** Do you see how a vertical line could pass through those two points?

NOTE TO STUDENT: *Fully worked-out solutions to all of the Practice Problems can be found at the back of the text starting at page SP-1*

Practice Problem 3   Determine whether each of following is a graph of a function.

(a)

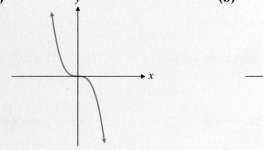

(b)

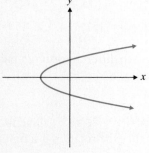

③ Evaluating a Function Using Function Notation

We can use **function notation** to indicate the relationship between the ordered pairs in a function. Looking back at our pizza example, we see that the diameter of a pizza determines its price. Since price is a function of diameter and a 5-inch diameter pizza costs $4.50 we can write

$$f(5) = 4.50.$$

Some equations describe functions. You are already familiar with linear equations. Let's look at one such equation.

$$y = 3x - 2$$

In this equation, y is a function of x. That is, for each value of x in the domain, there is a unique value of y in the range. We can use function notation when we determine the function value y for specific values of x. Let $x = 0$. Then we can write the following.

$$f(x) = 3x - 2$$
$$f(0) = 3(0) - 2 \quad \text{Substitute 0 for } x.$$
$$f(0) = -2 \quad\quad \text{Evaluate.}$$

We see that when x is 0, $f(x)$ is -2. That is, the value of the function is -2 when x is 0. We can write this ordered pair solution as $(x, f(x)) = (0, f(0)) = (0, -2)$.

Note: The notation $f(x)$ does not mean that f is multiplied by x. It means that for any specific value of x, there is only one value for y. $f(x)$ is read as "f of x." Although we commonly use f as the function name, we can also use other variables, like g and h.

It may help you to understand the idea of a function by imagining a "function machine." An item from the domain enters as input into the machine—the function—and a member of the range results as output.

Function, f

EXAMPLE 4   If $f(x) = 4 - 3x$, find

(a) $f(2)$,   **(b)** $f(-1)$, and   **(c)** $f\left(\dfrac{1}{3}\right)$.

Solution   In each case we replace x by the specific value in the domain.

(a) $f(2) = 4 - 3(2)$
$\qquad\quad = 4 - 6$
$\quad f(2) = -2$

(b) $f(-1) = 4 - 3(-1)$
$\qquad\quad\; = 4 + 3$
$\quad f(-1) = 7$

(c) $f\left(\dfrac{1}{3}\right) = 4 - 3\left(\dfrac{1}{3}\right)$
$\qquad\quad = 4 - 1$
$\quad f\left(\dfrac{1}{3}\right) = 3$

Practice Problem 4   If $f(x) = 2x^2 - 8$, find each of the following.

(a) $f(-3)$ **(b)** $f(4)$ **(c)** $f\left(\dfrac{1}{2}\right)$

3.5 EXERCISES

| Student Solutions Manual | CD/ Video | PH Math Tutor Center | MathXL®Tutorials on CD | MathXL® | MyMathLab® | Interactmath.com |

Verbal and Writing Skills

1. Explain the difference between a relation and a function.

2. Explain the difference between the domain and the range of a relation.

3. What are the three ways you can describe a function?

4. Write the ordered pair for $f(-5) = 8$ and identify the x- and y-values.

What are the domain and range of each relation? Is the relation a function?

5. $D = \{(0, 0), (5, 13), (7, 11), (5, 0)\}$

6. $C = \left\{\left(\frac{1}{2}, 5\right), (-3, 7), \left(\frac{3}{2}, 5\right), \left(\frac{1}{2}, -1\right)\right\}$

7. $F = \{(85, -12), (16, 4)(-102, 4), (62, 48)\}$

8. $E = \{(40, 10), (-18, 27), (38, 10), (57, -15)\}$

9. Women's Dress Sizes

USA	6	8	10	12	14
France	38	40	42	44	46

10. Women's Dress Sizes

France	40	42	44	46	48
Britain	14	15	16	17	18

11. Average Monthly Fahrenheit Temperature: Pago Pago, Samoa

Month	Jan.	Feb.	Mar.	Apr.	May	June	July	Aug.	Sept.	Oct.	Nov.	Dec.
Temperature	81	81	81	81	80	80	79	79	79	80	80	81

Source: United Nations Statistics Division.

12. Some of the World's Longest Rivers

River	Nile	Amazon	Chang Jiang	Ob-Irtysh	Huang He	Congo
Approximate Length in Miles	4160	4000	3964	3362	2903	2900

Source: United Nations Statistics Division.

13. Tallest Buildings, USA

City	Chicago	New York	New York	Chicago	Chicago	New York
Height in Feet	1454	1350	1250	1136	1127	1046

Source: World Almanac.

14. Metric Conversion

Miles	1	2	3	4	5
Kilometers	1.61	3.22	4.83	6.44	8.05

15. Mariner's Speed Conversion

Knots	10	20	30	40	50
Miles per Hour	11.51	23.02	34.53	46.04	57.55

16. Temperature Scales

Fahrenheit	32	41	50	59	68	95
Celsius	0	5	10	15	20	35

Determine whether the graph represents a function.

17.

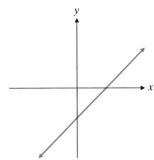

18.

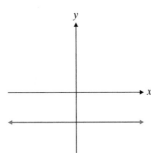

19.

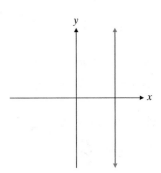

20.

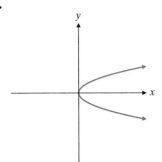

21.

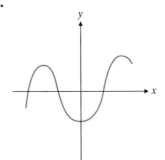

22.

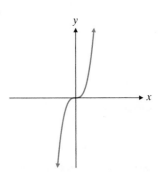

23.

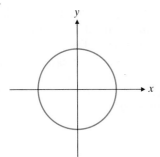

24.

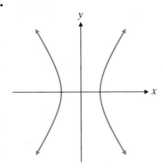

25.

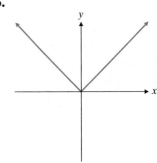

Given the function defined by $g(x) = 2x - 5$, *find the following:*

26. $g(-3)$ **27.** $g(-2.4)$ **28.** $g\left(\dfrac{1}{3}\right)$ **29.** $g\left(\dfrac{1}{2}\right)$

Given the function defined by $h(x) = \dfrac{2}{3}x + 2$, *find the following:*

30. $h(-3)$ **31.** $h(-6)$ **32.** $h\left(\dfrac{1}{2}\right)$ **33.** $h(2)$

Given the function defined by $r(x) = 2x^2 - 4x + 1$, *find the following:*

34. $r(1)$ **35.** $r(-1)$ **36.** $r(0.1)$ **37.** $r(-0.1)$

Given the function defined by $b(x) = -x^2 + 2x - 1$, *find the following.*

38. $b(1)$ **39.** $b(0)$ **40.** $b(-1)$ **41.** $b(5)$

Given the function defined by $t(x) = x^3 - 3x^2 + 2x - 3$, *find the following:*

42. $t(10)$ **43.** $t(4)$ **44.** $t(-1)$ **45.** $t(-2)$

46. If $f(x) = \sqrt{x + 6}$, find $f(-5)$.   **47.** If $f(x) = \sqrt{2 - x}$, find $f(-2)$.

48. If $g(x) = |x^2 - 8|$, find $g(-3)$.   **49.** If $g(x) = |6 - x^2|$, find $g(2)$.

To Think About

Find the range of the function for the given domain.

50. $f(x) = x + 3$ Domain $= \{-2, -1, 0, 1, 2\}$ **51.** $g(x) = x^2 + 3$ Domain $= \{-2, -1, 0, 1, 2\}$

Find the domain of the function for the given range. Be sure to find all possible values.

52. $h(x) = \dfrac{2}{3}x - 4$ Range $= \left\{0, 2, \dfrac{10}{3}, 4\right\}$ **53.** $d(x) = 3 - \dfrac{1}{4}x$ Range $= \left\{0, \dfrac{1}{4}, \dfrac{3}{4}, 4\right\}$

Cumulative Review

54. Solve: $|2x - 1| = 7$ **55.** Solve: $|x + 3| \leq 2$

56. *Personal Finance* Laurie has $763.21 in her checking account. She writes checks for $280.55, $78.91, $116.01, and $196.69. She needs to pay a bill of $424.98 next week, and she wants to keep a minimum deposit of $100 in her account at all times (so that she is not required to pay a penalty). How much should she deposit into her account?

57. *Environmental Science* To provide the average modern person living in a developed county with the necessities and luxuries of his or her life, at least twenty tons of raw material must be dug from the Earth every year. If the average life span of a U.S. female is 81 years of age and the average life span of a U.S. male is 76 years of age, how much raw material would have to be dug for the lifetime of an average couple in the United States?

1 Graphing a Function from an Equation

Frequently, we are given a function in the form of an equation and are asked to graph it. Each value of the function $f(x)$, often labeled y, corresponds to a value in the domain, often labeled x. This correspondence is the ordered pair (x, y) or $(x, f(x))$. The graph of the function is the graph of the ordered pairs.

If a function can be written in the form $f(x) = mx + b$, it is called a **linear function.** The graph of a linear function is a straight line.

If we can describe a real-life relationship with a function in the form of an equation, a table of values, and/or a graph, we can determine characteristics of the relationship and make predictions.

EXAMPLE 1 A salesperson earns \$15,000 a year plus a 20% commission on her total sales. Express her annual income in dollars as a function of her total sales. Determine values of the function for total sales of \$0, \$25,000, and \$50,000. Graph the function and determine whether the function is increasing or decreasing.

Solution Since we want to express income as a function of sales, let's see how income is determined.

$$\text{income} = \$15{,}000 + 20\% \text{ of total sales}$$

Income depends on total sales. Thus, total sales is the independent variable, and income is the dependent variable.

Let x = the amount of total sales in dollars and
$i(x)$ = income.
$$i(x) = 15{,}000 + 0.20x \quad \text{Change 20\% to 0.20.}$$
$$i(0) = 15{,}000 + 0.20(0) = 15{,}000$$
$$i(25{,}000) = 15{,}000 + 0.20(25{,}000) = 20{,}000$$
$$i(50{,}000) = 15{,}000 + 0.20(50{,}000) = 25{,}000$$

We can put this information in a table of values.

x (dollars)	$i(x)$ dollars
0	15,000
25,000	20,000
50,000	25,000

To facilitate the graphing, we will use a scale in thousands. Thus, we modify the table to make our task of graphing easier. Thus our equation is now $i(x) = 15 + 0.20x$ and not $i(x) = 15{,}000 + 0.20x$.

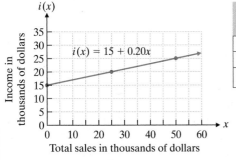

x (thousands of dollars)	$i(x)$ (thousands of dollars)
0	15
25	20
50	25

Practice Problem 1 A salesperson earns \$10,000 a year plus a 15% commission on her total sales. Express her annual income in dollars as a function d of her total sales x. Use the values $d(0)$, $d(40{,}000)$, and $d(80{,}000)$ to graph her salary.

Student Learning Objectives

After studying this section, you will be able to:

 Graph a function from an equation.

 Graph a function from a table of data.

Graphing Calculator

 Table of Values

You can use the Table feature of your graphing calculator to create a table of values. To create the table of values for Example 1, enter the income function in $y = 15 + 0.2x$ into the $Y =$ editor of your calculator and then set the table to compute values for $x = 0, 25, 50$. This will require you to use Table Setup:
Tb1 Start = 0
ΔTb1 = 25.
Display:

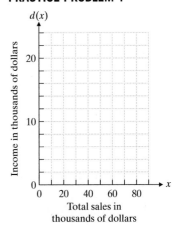

PRACTICE PROBLEM 1

NOTE TO STUDENT: Fully worked-out solutions to all of the Practice Problems can be found at the back of the text starting at page SP-1

Thus far, most of our work has been with linear functions. Let's look at some other functions and their graphs. We will begin with the **absolute value function** $f(x) = |x|$.

EXAMPLE 2 Graph $f(x) = |x|$.

Solution Let's find the function values for five values of the independent variable x. Because of the nature of the function, we will choose both negative and positive values for x. The table of values and the resulting graph are as follows:

x	f(x)
−2	2
−1	1
0	0
1	1
2	2

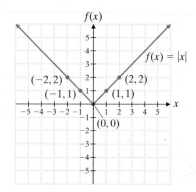

Notice that the graph is **symmetric about the y-axis.** That is, the graph on one side of the y-axis is a mirror image of the graph on the other side of the y-axis. This means that if the point (a, b) is on the graph, then the point $(-a, b)$ is also on the graph.

PRACTICE PROBLEM 2

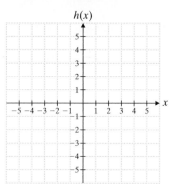

NOTE TO STUDENT: *Fully worked-out solutions to all of the Practice Problems can be found at the back of the text starting at page SP-1*

Practice Problem 2 Graph the function $h(x) = |x + 2|$. Use values for x from −4 to 0.

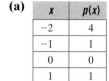

TO THINK ABOUT: Example 2 Follow-up How are the graphs of $f(x) = |x|$ and $h(x) = |x + 2|$ the same? How are they different? What would the graph of $k(x) = |x - 2|$ look like?

Now let's take a look at another function, $p(x) = x^2$. What happens when $x = 2$? When $x = -2$? Notice that when x is 2 or −2, the value of the function is 4. The ordered pairs are $(-2, 4)$ and $(2, 4)$. This graph is also symmetric about the y-axis as we shall see in the next example.

EXAMPLE 3 Graph: **(a)** $p(x) = x^2$ **(b)** $q(x) = (x + 2)^2$

Solution For each function we will choose both negative and positive values of x. Since these are *not* linear functions, we use a curved line to connect the points. The tables of values and the resulting graphs are as follows.

(a)

x	p(x)
−2	4
−1	1
0	0
1	1
2	4

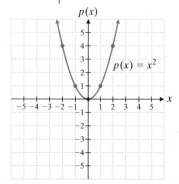

(b)

x	q(x)
−4	4
−3	1
−2	0
−1	1
0	4
1	9

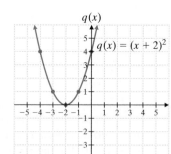

Practice Problem 3   Graph: **(a)** $r(x) = (x - 2)^2$ **(b)** $s(x) = x^2 + 2$

(a)

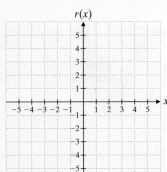

(b)

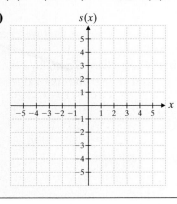

TO THINK ABOUT: Example 3 Follow-up How are the graphs of each of the previous functions the same? How are they different? Describe how changing $p(x) = x^2$ to $q(x) = (x + 5)^2$ and to $s(x) = x^2 + 5$ would affect the graph of the function.

Another interesting graph is the graph of the function $g(x) = x^3$. A quick look at $g(x)$ when x is 2 and -2 reveals that $g(x)$ is *not* symmetric about the y-axis. That is, there are different function values for 2 and -2. Let's see.

$$g(x) = x^3$$

$$g(2) = 2^3 \qquad\qquad g(-2) = (-2)^3$$
$$= 8 \qquad\qquad\qquad = -8$$

EXAMPLE 4   Graph: **(a)** $g(x) = x^3$ **(b)** $h(x) = x^3 + 1$

Solution   We will pick five values for x, find the corresponding function values, and plot the five points to assist us in sketching the graph.

(a)

x	$g(x)$
-2	-8
-1	-1
0	0
1	1
2	8

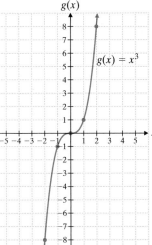

(b)

x	$h(x)$
-2	-7
-1	0
0	1
1	2
2	9

$h(x) = x^3 + 1$

Practice Problem 4   Look at the function $f(x) = x^3 - 2$. What do you think the graph of $f(x) = x^3 - 2$ will look like? Make a table of values for the function and draw the graph.

PRACTICE PROBLEM 4

$f(x)$

TO THINK ABOUT: Example 4 Follow-up Describe how the graph of $f(x) = x^3 - 2$ is related to the graph of $g(x) = x^3$.

In Examples 1–4 each function has had a domain of all real numbers. We were free to choose any real number for x. However, if a function is written in the form of a fraction and it contains a variable in the denominator, the domain of that function will not include any value for which the denominator becomes zero.

EXAMPLE 5 Graph $p(x) = \dfrac{4}{x}$.

Solution First we observe that we cannot choose x to be 0 because $\frac{4}{0}$ is not defined. (We can never divide by zero.) Therefore, the domain of this function is all real numbers except zero. To make a table of values, we will choose five values for x that are greater than 0 and five values for x that are less than 0.

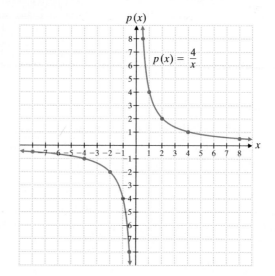

x	$p(x)$
8	$\frac{1}{2}$
4	1
2	2
1	4
$\frac{1}{2}$	8
$-\frac{1}{2}$	-8
-1	-4
-2	-2
-4	-1
-8	$-\frac{1}{2}$

The study of this type of function, called a **rational function,** will be continued in a more advanced course, such as college algebra or precalculus.

Practice Problem 5 Graph $q(x) = -\dfrac{4}{x}$.

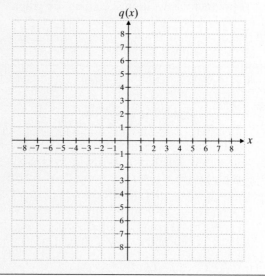

NOTE TO STUDENT: Fully worked-out solutions to all of the Practice Problems can be found at the back of the text starting at page SP-1

TO THINK ABOUT: Example 5 Follow-up Describe how the graph of $q(x) = -\dfrac{4}{x}$ is related to the graph of $p(x) = \dfrac{4}{x}$.

2 Graphing a Function from a Table of Data

Sometimes when we study an event in daily life, we record data to better understand the event. In many cases we will not have an equation to work with, but rather a table of values that gives a general indication of some functional relationship. A graph of the values may help us to understand this underlying function.

In the following example, the number of items sold is the domain and the profit obtained by the sale is the range.

EXAMPLE 6 The marketing manager of a shoe company compiled the data in the table.

(a) Plot the data values and connect the points to see the graph of the underlying function.

(b) From the graph, determine the profit from selling 4000 pairs of shoes in one month.

(c) What kind of profit would you expect to make from selling 0 pairs of shoes in a month? What value do you obtain on the graph for $x = 0$? What does this mean?

x Pairs of Shoes Sold in a Month (In Thousands of Pairs)	p(x) Monthly Profit from the Sales of Shoes (In Thousands of Dollars)
3	5
5	9
7	13

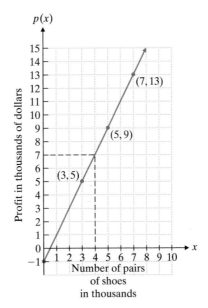

Solution

(a) We plot the three ordered pairs and connect the points by a straight line. This means the function is a linear function.

(b) We find that the value $x = 4$ on the graph corresponds to the value $y = 7$. Thus, we would expect that if we sold 4000 pairs of shoes, we would have a profit of $7000 for the month.

(c) In any business we would not expect to make a profit at all if we sold no items. Looking at our graph, we find that when $x = 0$, $y = -1$. Thus, we would predict a loss of $1000 in a month of zero sales of shoes. The loss would most likely be due to fixed expenses such as rent and supplies.

TO THINK ABOUT: Example 6 Follow-up Can x in Example 6 be any real number? Well, not really. Obviously the number of pairs of shoes sold monthly must be a whole number. You cannot sell 345.859 pairs of shoes. However, when we connect the points on the graph we use a *continuous* line (i.e., a line with no breaks in it). When we study the function for all real numbers greater than or equal to zero, it helps us better understand the relationship between sales and profit. Using a continuous line helps us make better predictions for the future. Now, of course, we know as well that the graphed line does not really extend forever. There are physical limitations on the number of shoes that can be made and sold by any one shoe manufacturer.

Practice Problem 6 An accountant reviewed the profitability of a doctor's practice by comparing the number of patients the doctor saw with the profit. The data collected are in the table.

x Average Number of Patients Per Hour	g(x) Weekly Profit
6	$2000
9	$4000
12	$6000

PRACTICE PROBLEM 6

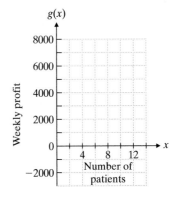

(a) Plot the data values and connect the points to see the graph of the underlying function.

(b) Describe the function in as much detail as possible.

(c) What would the profit be if the doctor saw three patients per hour?

(d) What would the weekly loss be if the doctor saw zero patients per hour?

(e) Can we expect the function to increase indefinitely? Why or why not?

Given a set of data, we do not always know what the graph of the underlying function will look like. Not all functions are linear. When points do not appear to lie on a line, we connect them with a smooth curve.

Let's look at windchill, the effect wind has on the outside temperature to the body. For example, if the temperature outside were 10° Fahrenheit (°F) and the wind were blowing at 20 miles per hour (mph), the effect would be the same as if the temperature were 24° below zero! A windchill table that describes this relationship is used in the next example.

EXAMPLE 7   The windchill values for various wind speeds when the temperature is 10° Fahrenheit are given in the table.

(a) Plot the points and connect them to see the graph of the underlying function.

(b) Estimate the windchill when the wind is blowing at 23 mph.

(c) Estimate the windchill when the wind is blowing at 40 mph.

(d) If the windchill is −15°F, at what speed is the wind blowing?

(e) Based on your analysis of the graph, does an increase in wind speed result in a greater change in the windchill at lower wind speeds or at higher wind speeds?

Wind Speed (mph)	Windchill (°F)
0	10
5	7
10	−9
15	−18
20	−24
25	−29
30	−33
35	−35

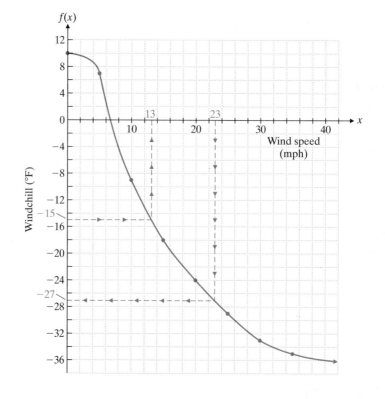

Solution

(a) First we need to determine which value is the independent variable and which value is the dependent variable. Since windchill depends on wind speed, wind speed is the independent variable and windchill is the dependent variable. Thus, the windchill values are the function values. Label the vertical axis "Windchill" and the horizontal axis "Wind Speed." Plot the points. Use a smooth curve to connect the points.

(b) Find 23 along the horizontal axis. Move down to the curve. Then move right until you intersect the vertical axis. The function value on the vertical axis for 23 mph is about −27°F.

(c) Since the last windchill number in the table is −35, we need to extend the curve. This is called *extrapolation*, and the function values obtained by this technique may be less accurate. At 40 mph the windchill is about −36°F.

(d) Find −15 along the vertical scale. Move to the right until you intersect the curve. Then move up until you intersect the horizontal axis. This is the wind speed when the windchill is −15°F. This is about 13 mph.

(e) The curve goes downward more quickly at lower wind speeds. Thus, lower wind speeds have a greater effect on windchill. For example, when the wind speed goes from 0 to 15 mph, the windchill goes from 10°F to −18°F. This is a change of 28°F. A change in wind speed from 20 to 35 mph produces a change in windchill of only 11°F.

Practice Problem 7   A table of windchill values at 0°F is given below.

Wind Speed (mph)	0	5	10	15	20	25	30	35
Windchill °F	0	−5	−22	−31	−39	−44	−48	−52

(a) Plot the points and connect them to graph the underlying function.
(b) Estimate the windchill when the wind is blowing at 32 mph.
(c) Estimate the windchill when the wind is blowing at 40 mph.
(d) If the windchill is −24°F, at what speed is the wind blowing?
(e) What is the domain of the function? What is the range?

PRACTICE PROBLEM 7

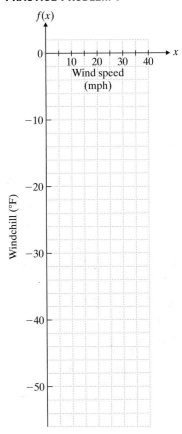

NOTE TO STUDENT: *Fully worked-out solutions to all of the Practice Problems can be found at the back of the text starting at page SP-1*

Scientists continue to study the present windchill table to see if it is truly accurate. Many scientists feel the figures need to be revised. Further research is needed to determine how the revision should be carried out. In addition to windchill tables, calculations can also be made using the Internet. There are several sites that feature a "windchill calculator."

An interesting example of a place where windchill is critical is on top of Mount Washington in New Hampshire. In 1934 the observatory recorded a wind gust of 231 miles per hour. You can learn more at www.mountwashington.org.

EXAMPLE 8   Graph the function $f(x)$ suggested by the data given in the following table. The domain is $x \geq 2$, and the range is $f(x) \geq 0$.

(a) Determine an approximate value for $f(x)$ when $x = 8$.

(b) Determine an approximate value for x when $f(x) = 1.5$.

(c) What do you notice about the graph as the values of x get larger?

x	2	3	6	11
f(x)	0	1	2	3

Solution   Assign the independent variable x to the horizontal axis. Assign the function values, the dependent variable, to the vertical axis. Plot the points and connect them with a smooth curve.

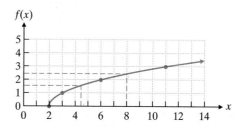

(a) Find $x = 8$ along the horizontal axis. Move up until you intersect the graph. Then move to the left until you intersect the vertical axis. This value is $f(x)$ when x is 8. Read the scale. Thus, $f(x)$ is about 2.5.

(b) Find $f(x) = 1.5$ along the vertical axis. Move to the right until you intersect the graph. Then move down until you intersect the horizontal axis. This value is x when $f(x)$ is 1.5. Read the scale. Thus, x is about 4.5.

(c) As the values of x get larger, the function values are increasing more slowly. In other words the *rate of change* decreases for larger values of x. Since the rate of change is the slope, we say that the slope is decreasing as x increases.

NOTE TO STUDENT: Fully worked-out solutions to all of the Practice Problems can be found at the back of the text starting at page SP-1

Practice Problem 8   Graph the function $g(x)$ suggested by the data given in the following table. The domain is $x \leq 4$, and the range is $g(x) \geq 0$.

(a) Determine an approximate value for $g(x)$ when $x = -3$.

(b) Determine an approximate value for x when $g(x) = 1.5$.

(c) As x goes from 3 to 0 to -5, what do you observe about the curve?

x	4	3	0	-5
g(x)	0	1	2	3

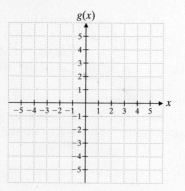

Graph each function.

1. $f(x) = \dfrac{3}{4}x + 2$

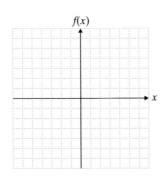

2. $f(x) = \dfrac{2}{3}x - 4$

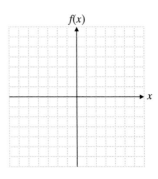

3. $f(x) = -3x - 1$

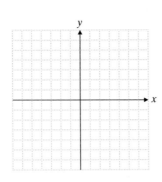

4. $f(x) = -2x + 3$

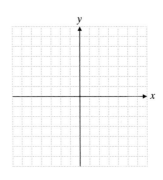

Applications

5. *Cost of a Newsletter* The cost of producing a newsletter is an initial charge of $25 for formatting and editing plus $0.15 for printing each copy. Express the cost $C(x)$ of printing the newsletter as a function of the number of copies printed x. Obtain values for the function when $x = 0$, $x = 100$, $x = 200$, and $x = 300$. Graph the cost function.

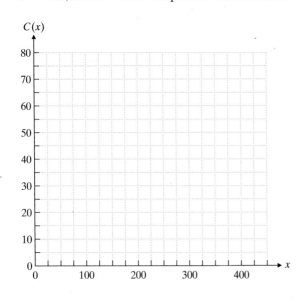

6. *Population Growth* In 1990 the population of Sommerville was 20,000. The population has increased 400 people each year since then and is expected to continue to do so. Express the population as a function $P(x)$, where x is the number of years since 1990. Obtain values for the function when $x = 0$, $x = 8$, $x = 12$, and $x = 16$ years. Graph the population function.

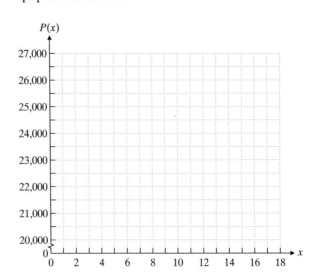

7. ***Water Pollution*** A wildlife biologist has determined that a certain stream in New Hampshire can support 45,000 fish if it is free of pollution. She has estimated that for every ton of pollutants in the stream, 1500 fewer fish can be supported. Express the fish population as a function $P(x)$, where x is the number of tons of pollutants found in the stream. Obtain values of the function when $x = 0$, $x = 10$, and $x = 30$. Graph the population function. What is the significance of $P(x)$ when $x = 30$?

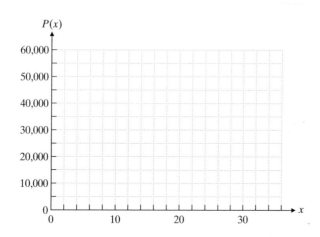

8. ***Insulation in a House*** The R value of insulation in a house is a measure of its ability to resist the loss of heat from the house. The R value of fiberglass insulation is a linear function of its thickness in inches. One type of fiberglass insulation that is 6 inches thick has an R value of 19. The R value in general of this type of insulation is obtained by multiplying 3.2 by the thickness x measured in inches and then adding the result to -0.2. Express the R value of this insulation as a function of x, the thickness in inches. Obtain values of the function when $x = 0$, $x = 1$, $x = 3.5$, and $x = 6$. Graph the function.

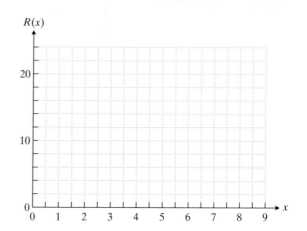

Graph each function.

9. $f(x) = |x - 1|$

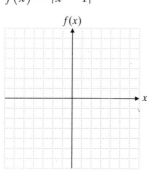

10. $g(x) = |x - 3|$

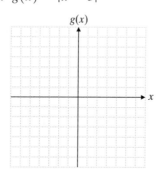

11. $g(x) = |x| - 5$

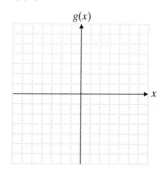

12. $f(x) = |x| + 2$

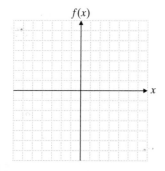

13. $g(x) = x^2 - 4$

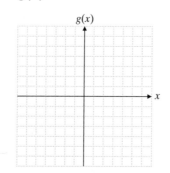

14. $f(x) = x^2 + 1$

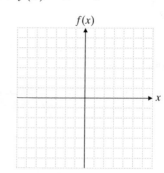

15. $g(x) = (x + 1)^2$

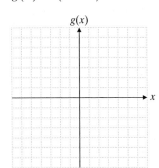

16. $f(x) = (x - 3)^2$

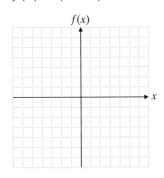

17. $g(x) = x^3 - 3$

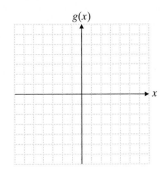

18. $f(x) = x^3 + 2$

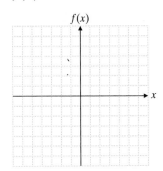

19. $p(x) = -x^3$

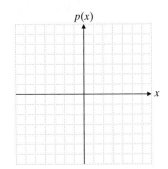

20. $s(x) = -x^3 + 2$

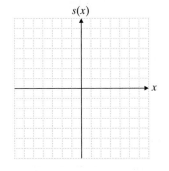

21. $f(x) = \dfrac{2}{x}$

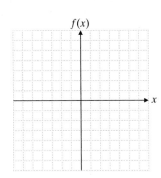

22. $g(x) = -\dfrac{3}{x}$

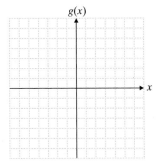

23. $h(x) = -\dfrac{6}{x}$

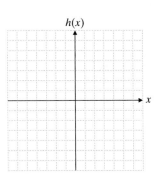

24. $t(x) = \dfrac{10}{x}$

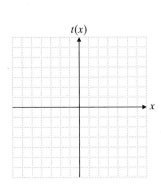

Optional Graphing Calculator Problem

25. On your graphing calculator, graph on the same set of axes: $y_1 = x^2$, $y_2 = 0.4x^2$, and $y_3 = 2.6x^2$. What effect does the coefficient have on the graph?

In each of the following problems, a table of values is given. Graph the function based on the table of values. Assume that both domain and range are all real numbers. After you have finished your graph in each case, estimate the value of $f(x)$ when $x = 2$.

26.

x	f(x)
−1	2
1	−2
3	−6

27.

x	f(x)
−5	−2
3	2
5	3

28.

x	f(x)
−2	0.25
−1	0.5
0	1
2.5	5.7
3	8

29.

x	f(x)
0	3
1	2.5
3	2.2
−1	4
−2	6

30.

x	f(x)
−3	−1
−2	2
−1	3
0	2
1	−1

31.

x	f(x)
−2	−8
−1	−3
0	0
1	2
3	−3

Applications

32. *Carbon Dioxide in the Atmosphere* The following table records the buildup of carbon dioxide in the atmosphere in eastern Canada for selected years.

Year, x	1960	1965	1970	1975	1980	1985	1990	1995	2000
Carbon Dioxide in Parts per Million, c(x)	320	323	327	333	340	348	358	370	382

Source: Environmental Protection Agency

(a) Plot the points and connect them to graph the underlying function.

(b) Estimate $c(x)$ when x is 1972.

(c) What do you estimate $c(x)$ will be in the year 2002?

(d) How would you compare the rate of increase in $c(x)$ between 1960 and 1975 to the rate of increase in $c(x)$ between 1975 and 1990?

(e) During what year did the concentration of carbon dioxide reach 330 parts per million?

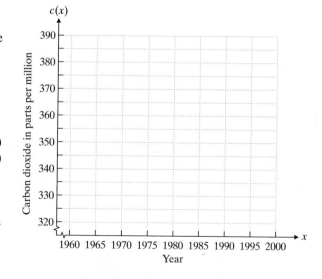

33. *Computer Manufacturing* The following table shows the relationship between the number of computers manufactured by a company each day and the company's profit for that day.

Number of Computers Made	20	30	40	50	60	70	80
Profit for the Company ($)	0	5000	8000	9000	8000	5000	0

(a) Plot the points and connect them to graph the underlying function.

(b) How many computers should be made each day to achieve the maximum profit?

(c) If the company wants to earn a profit of $8000 or more each day, how many computers should they manufacture each day?

(d) What will the profit picture be if the company manufactures eighty-two computers per day?

(e) Estimate the profit if the company manufactures forty-five computers per day.

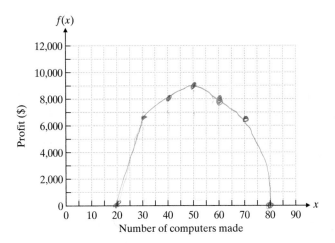

To Think About

34. *Heat Index* In hot weather, meteorologists use a heat index to indicate the relative comfort and safety of people subjected to high temperatures and humidity. The following table gives the heat index at 80°F for a relative humidity of 0% to 100%.

Relative Humidity	0%	20%	40%	60%	80%	100%
Heat Index at 80°F	73	77	79	82	86	91

Source: National Oceanic and Atmospheric Administration.

(a) Graph the heat index values at 80°F and connect the points. Is the function linear? Why or why not?

(b) Estimate the value of the function when the relative humidity is 50%.

(c) Between what levels of humidity does the heat index increase the fastest?

(d) For what level of humidity does 80°F have a heat index of 80°F?

(e) What is the significance of the heat index being only 77°F when the relative humidity is 20%?

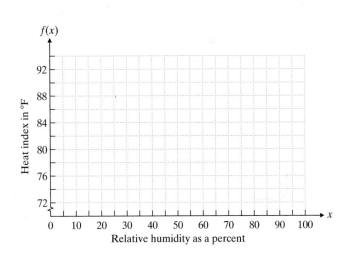

Cumulative Review

Solve for x.

35. $2(3ax - 4y) = 5(ax + 3)$

36. $0.12(x - 4) = 1.16x - 8.02$

37. $\dfrac{1}{2}(x + 2) - 5 = x - \dfrac{3}{4}(x + 4)$

38. *Ice Cover of the Earth* Approximately 3.9 million square miles of the Earth's land surface (about 10% of the entire Earth), is under a permanent ice cover. Eighty percent of the ice cover is in Antarctica, 12% is in Greenland, and 8% is distributed among various polar islands and mountain peaks. How many square feet of Antarctica is covered in ice? *Source:* National Oceanic and Atmospheric Administration.

39. *Woodchuck Hibernation* A woodchuck breathes only ten times per hour while hibernating. How many times will it breathe during its hibernation months of December, January, and February (during a leap year)?

40. *Purchasing a Set of Speakers* Charlene is at a store trying to decide between two sets of speakers for her stereo. The first set costs $475 per pair with a subwoofer thrown in for free. The subwoofer enhances the bass tones and is valued at $120. The second pair of speakers costs $525 and also includes a subwoofer. The difference is that this subwoofer is valued at $140. The salesperson tells Charlene that he will give her a portable CD player (valued at $99) if she buys the first set of speakers. If Charlene chooses the second set of speakers, he will give her a 15% discount. If the speakers sound the same, which deal is better?

Putting Your Skills to Work

Using Functions to Manage a Cherry Orchard

Imagine that you had to efficiently manage a cherry orchard. One of your first concerns would be to plant enough trees so that you have large yield. However, there is some advice against just planting more and more trees. While it may seem that more trees will always produce more fruit, one must also take into consideration the demand on the land that more trees would make.

A commercial cherry grower estimates that when 30 trees are planted per acre, each tree can produce about 50 pounds of cherries per season. For each additional tree planted per acre, the yield per tree is reduced by 1 pound.

Problems for Individual Investigation

1. Complete the following table of the yield per tree for several different planting densities:

D = Density (number of trees per acre)	30	31	32	33	34	35	40	45
Y = Yield (yield per tree in pounds)								

2. Based on the pattern above, if there were 55 trees per acre, how many pounds of cherries per tree would you expect at harvest time?

3. Someone proposed that this data can be described by the formula $Y = 50 - (D - 30)$, where Y is the yield per tree in pounds and D is the density of the planting in trees per acre. Does this formula agree with the data in your table?

4. Simplify the formula from your answer to 3.

Problems for Group Investigation and Cooperative Activity

While it is very necessary to consider the cherry yield per tree, of more interest to the farmer is the cherry yield per acre.

5. Modify the formula in problem 4 to describe the number of pounds of cherries produced per acre. Let H stand for the yield of cherries, in pounds, per acre.

6. Using the equation you obtained in exercise 5, fill in the following chart with the yield per acre.

D (number of trees per acre)	0	10	20	30	40	50	60	70	80
H (yield per acre (in pounds))									

7. (Use the data from the preceding chart to graph the equation you obtained in exercise 5.) From your graph, how many trees should be planted per acre to achieve the highest yield?

8. Using your answer for problem 7, what is the maximum income for one acre of land planted in cherry trees if the cherry farmers receive $0.47 a pound from the wholesaler?

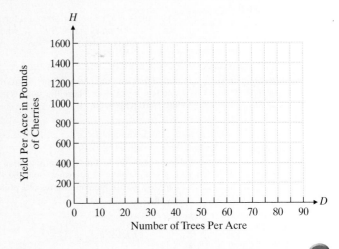

Topic	Procedure	Examples
Graphing straight lines, p. 127.	An equation of the form $$Ax + By = C$$ is a linear equation and has a graph that is a straight line. To graph such an equation, plot any three points—two give the line and the third checks it. (Where possible, use the x- and y-intercepts.)	$$2x + 3y = 12$$ Form is $Ax + By = C$, so it is a straight line. Let $x = 0$; then $y = 4$. Let $y = 0$; then $x = 6$. Let $x = 3$; then $y = 2$. | x	y
---	---
0	4
6	0
3	2	
Intercepts, p. 127.	A line crosses the x-axis at its x-intercept $(a, 0)$. A line crosses the y-axis at its y-intercept $(0, b)$.	Find the intercepts of $7x - 2y = -14$. If $x = 0$, $-2y = -14$ and $y = 7$. The y-intercept is $(0, 7)$. If $y = 0$, $7x = -14$ and $x = -2$. The x-intercept is $(-2, 0)$.
Slope, p. 135.	The *slope* m of the straight line that contains the points (x_1, y_1) and (x_2, y_2) is defined by $$m = \frac{y_2 - y_1}{x_2 - x_1}, \quad \text{where } x_2 \neq x_1.$$	Find the slope of the line passing through $(6, -1)$ and $(-4, -5)$. $$m = \frac{-5 - (-1)}{-4 - 6} = \frac{-5 + 1}{-4 - 6} = \frac{-4}{-10} = \frac{2}{5}$$
Zero slope, p. 136.	All horizontal lines have *zero slope*. They can be described by equations of the form $y = b$, where b is a real number.	What is the slope of $2y = 8$? This equation can be simplified to $y = 4$. It is a horizontal line; the slope is zero.
Undefined slope, p. 136.	The slopes of all vertical lines are undefined. The lines can be described by equations of the form $x = a$, where a is a real number.	What is the slope of $5x = -15$? This equation can be simplified to $x = -3$. It is a vertical line. This line has an undefined slope.
Parallel and perpendicular lines, pp. 137 and 138.	Two distinct lines with nonzero slopes m_1 and m_2, are **1.** parallel if $m_1 = m_2$. **2.** perpendicular if $m_1 = -\dfrac{1}{m_2}$.	Find the slope of a line parallel to $y = -\dfrac{3}{2}x + 6$. $$m = -\frac{3}{2}$$ Find the slope of a line perpendicular to $$y = -4x + 7, \quad m = \frac{1}{4}$$
Standard form, p. 126.	The equation of a line is in standard form when it is written as $Ax + By = C$, where A, B, and C are real numbers.	Place this equation in standard form: $y = -5(x + 6)$. $$y = -5x - 30$$ $$5x + y = -30$$
Slope–intercept form, p. 143.	The slope–intercept form of the equation of a line is $y = mx + b$, where the slope is m and the y-intercept is $(0, b)$.	Find the slope and y-intercept of $y = -\dfrac{7}{3}x + \dfrac{1}{4}$. The slope is $-\dfrac{7}{3}$; the y-intercept is $\left(0, \dfrac{1}{4}\right)$.
Point–slope form, p. 145.	The point–slope form of the equation of a line is $y - y_1 = m(x - x_1)$, where m is the slope and (x_1, y_1) are the coordinates of a point on the line.	Find an equation of the line passing through the points $(6, 0)$ and $(3, 4)$. $$m = \frac{4 - 0}{3 - 6} = -\frac{4}{3}$$ Then use the point–slope form. $$y - 0 = -\frac{4}{3}(x - 6)$$ $$y = -\frac{4}{3}x + 8$$

Topic	Procedure	Examples
Graphing the solution of a linear inequality in two variables, p. 154.	1. Replace the inequality symbol by an equals sign. This equation will be a boundary for the desired region. 2. Graph the line obtained in step 1, using a solid line if the original inequality contains a $\leq$ or $\geq$ symbol and a dashed line if the original inequality contains a $<$ or $>$ symbol. 3. Pick any point that does not lie on the boundary line. Substitute the coordinates into the original inequality. If you obtain a true inequality, shade the region on the side of the line containing the point. If you obtain a false inequality, shade the region on the side of the line opposite the test point.	Graph the solution of $3y - 4x \geq 6$. 1. The boundary line is $3y - 4x = 6$. 2. The line passes through $(0, 2)$ and $(-1.5, 0)$. The inequality includes the boundary; draw a solid line. 3. Pick $(0, 0)$ as a test point. $$3y - 4x \geq 6$$ $$3(0) - 4(0) \overset{?}{\geq} 6$$ $$0 - 0 \overset{?}{\geq} 6$$ $$0 \quad 6$$ Our test point fails. We shade the side that does not contain $(0, 0)$.
Finding the domain and the range of a relation, p. 160.	The set of all the first items of the ordered pairs in a relation is called the domain. The set of all the second items of the ordered pairs in a relation is called range.	Find the domain and range of the relation $$A = \{(5, 6), (1, 6), (3, 4), (2, 3)\},$$ $$\text{Domain} = \{1, 2, 3, 5\}$$ $$\text{Range} = \{3, 4, 6\}$$
Determining whether a relation is a function, p. 162.	A function is a relation in which no two different ordered pairs have the same first coordinate.	Determine whether each of the following are functions. $$B = \{(6, 7), (3, 0), (6, 4)\}$$ $$C = \{(-9, 3), (16, 4), (9, 3)\}$$ B is not a function. Two different pairs have the same first coordinate. They are $(6, 7)$ and $(6, 4)$. C is a function. There are no different pairs with the same first coordinate.
Determining whether a graph represents the graph of a function, pp. 162, 164.	The graph of a function will have no two different ordered pairs with the same first coordinate.	This is a function. This is not a function. There are at least two different ordered pairs with the same first coordinate.

Topic	Procedure	Examples
Function notation, **p. 164.**	Use function notation to evaluate the function at a given value. Replace the x by the quantity within the parentheses and then simplify.	$$f(x) = 2x^2 - 3x + 4$$ Find $f(-2)$. $$f(-2) = 2(-2)^2 - 3(-2) + 4$$ $$= 2(4) - 3(-2) + 4$$ $$= 8 + 6 + 4 = 18$$
Graphing functions, **p. 170.**	Prepare a table of ordered pairs (if one is not provided) that satisfy the function equation. Graph these ordered pairs. Connect the ordered pairs by a line or curve.	Graph $f(x) = |x| - 2$. First make a table. Then graph the ordered pairs and connect them. | x	$f(x)$
---	---
-2	0
-1	-1
0	-2
1	-1
2	0	

Chapter 3 Review Problems

Graph the straight line determined by each of the following equations.

1. $y = -\dfrac{1}{4}x - 1$

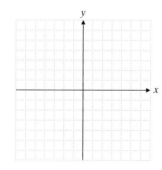

2. $y = -\dfrac{3}{2}x + 5$

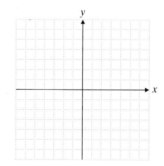

3. $y - 2x + 4 = 0$

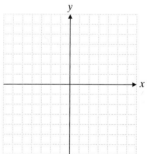

4. $7x = x + 2y$

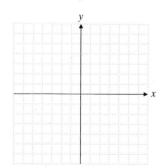

5. $-3y + 2 = -8y - 13$

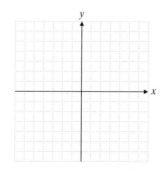

6. $5x - 6 = -2x + 8$

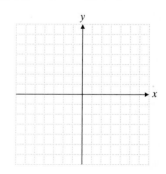

Find the slope, if possible, of the line connecting

7. $\left(\frac{1}{2}, 3\right)$ and $\left(\frac{3}{2}, 1\right)$

8. $(-8, -4)$ and $(2, -5)$

9. $(-3, 6)$ and $(-3, 1.8)$

10. $(7.5, -1)$ and $(0.3, -1)$

11. Find the slope of the line connecting $(-8, -4)$ and $(2, -3)$.

12. Find the slope of a line perpendicular to the line passing through $\left(\frac{2}{3}, \frac{1}{3}\right)$ and $(4, 2)$.

13. A line has a slope of $\frac{2}{3}$ and a y-intercept of $(0, -4)$. Write its equation in standard form.

14. Find the standard form of the equation of the line that passes through $\left(\frac{1}{2}, -2\right)$ and has slope -4.

15. Find the standard form of the equation of the line that passes through $(-3, 1)$ and has slope 0.

16. ***Computer Company Profit*** A microcomputer company's profit in dollars is given by the equation $P = 140x - 2000$, where x is the number of micro-computers sold each day. **(a)** What is the slope of the equation $P = 140x - 2000$? **(b)** How many microcomputers must be sold each day for the company to make a profit?

In Exercises 17–20, find the equation of the line satisfying the conditions given. Write your answer in standard form.

17. A line passing through $(5, 6)$ and $\left(-1, -\frac{1}{2}\right)$

18. A line that has an undefined slope and passes through $(-6, 5)$

19. A line perpendicular to $7x + 8y - 12 = 0$ and passing through $(-2, 5)$

20. A line parallel to $3x - 2y = 8$ and passing through $(5, 1)$

In exercises 21–23, find the slope and y-intercept of each of the following lines. Use these to write the equation of the line.

21.
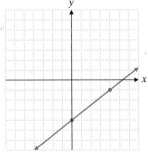

22.
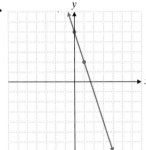

23.
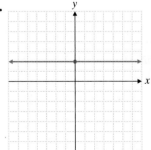

Graph the region described by the inequality.

24. $y < 2x + 4$

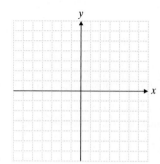

25. $y < 3x + 1$

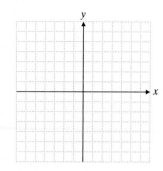

26. $y > -\dfrac{1}{2}x + 3$

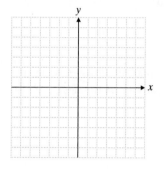

27. $y > -\dfrac{2}{3}x + 1$

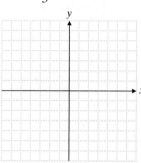

28. $3x + 4y \le -12$

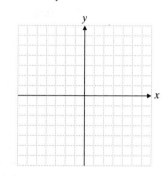

29. $x \le 3y$

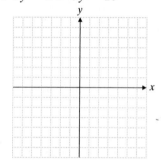

30. $5x + 3y \le -15$

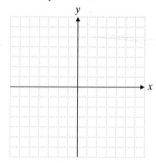

31. $3x - 5 < 7$

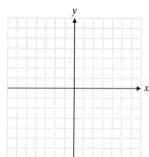

32. $5y - 2 > 3y - 10$

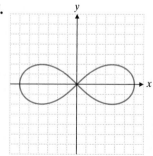

What is the domain and range of each relation? Is the relation a function?

33. $B = \{(-20, 18), (-18, 16), (-16, 14), (-12, 18)\}$

34. $A = \{(0, 0), (1, 1), (2, 4), (3, 9), (1, 16)\}$

Determine whether the graph represents a function.

35.

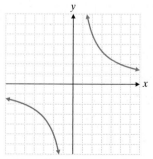

36.

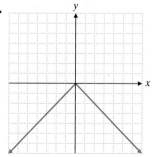

37.

38. If $f(x) = 3x - 8$, find $f(-2)$ and $f(-3)$.

39. If $g(x) = 2x^2 - 3x - 5$, find $g(-3)$ and $g(2)$.

40. If $h(x) = x^3 + 2x^2 - 5x + 8$, find $h(-1)$.

41. If $p(x) = |-6x - 3|$, find $p(3)$.

Graph each function.

42. $f(x) = 2|x - 1|$

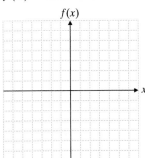

43. $g(x) = x^2 - 5$

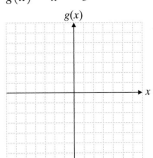

44. $h(x) = x^3 + 3$

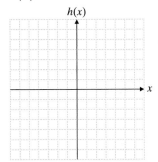

Plot the points and connect them to graph the underlying function. Estimate the value of $f(x)$ when $x = -2$.

45.

x	f(x)
−1	5
−3	−3
−4	−4
−5	−3
−6	0
−7	5

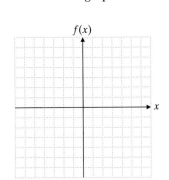

46.

x	f(x)
−3	4
−1	2
0	1
1	0
2	−1
3	0
4	1

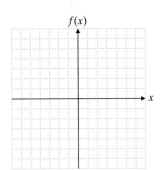

Complete the table of values for each function.

47. $f(x) = -\dfrac{4}{5}x + 3$

x	f(x)
−5	
0	
10	

48. $f(x) = 2x^2 - 3x + 4$

x	f(x)
−3	
0	
4	

49. $f(x) = 3x^3 - 4$

x	f(x)
−1	
0	
2	

50. $f(x) = \dfrac{7}{2x + 3}$

x	f(x)
−2	
0	
2	

Mixed Practice

51. Find the slope of the line connecting $(3, -5)$ and $(-5, -6)$.

52. A line n is perpendicular to the line passing through $(-4, 2)$ and $(1, 10)$. What is the slope of line n?

53. A line p is parallel to the line passing through $(4.5, 8)$ and $(2.5, -6)$. What is the slope of line p?

54. Find the standard form of the equation of the line that has slope $\frac{5}{6}$ and y-intercept $(0, -5)$.

55. Find the equation of the line parallel to $y = 5x - 2$ and passing through $(4, 10)$. Write the answer in slope–intercept form.

56. Find the slope–intercept form of the equation of the line that passes through $(5, 6)$ and $(-7, 3)$.

57. Find the equation of the line perpendicular to $3x - 6y = 9$ and passing through $(-2, -1)$. Write the answer in standard form.

58. Find the equation of a vertical line passing through $(5, 6)$.

Applications

59. *Car Rental Costs* The cost of renting a full-size sedan at Ocean City Rentals is $35 per day plus $0.15 per mile that the car is driven. Express the cost of renting a full-size sedan at Ocean City Rentals as a function $f(x)$, where x is the number of miles the car is driven.

60. *Population Growth of a Town* The population of South Hadley was 15,000 in 1990. Since then the town has grown by 500 people each year. Express the population as a function $f(x)$, where x is the number of years since 1990.

61. *Fish Population in a Stream* A biologist in Montana has determined that a certain mountain stream initially had 18,000 fish on June 1. The number of fish in the stream has decreased by 65 fish each day since June 1. Express the number of fish in this stream as a function $f(x)$, where x is the number of days after June 1.

Remember to use your Chapter Test Prep Video CD to see the worked-out solutions to the test problems you want to review.

Graph each line. Plot at least three points.

1. $y = \dfrac{1}{3}x - 2$

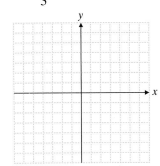

2. $2x - 3 = 1$

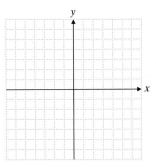

3. $5x + 3y = 9$

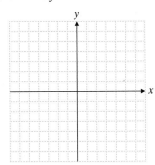

4. $2x + 3y = -10$

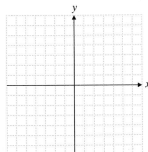

5. Find the slope of the line passing through $(2, -3)$ and $\left(\dfrac{1}{2}, -6\right)$.

6. Find the slope of a line passing through $(-7, 5)$ and $(6, 5)$.

7. Find the slope of the line $9x + 7y = 13$.

8. Write the standard form equation of the line that is perpendicular to $6x - 7y - 1 = 0$ and passes through $(0, -2)$.

9. Write the standard form of the equation of the line that passes through $(5, -2)$ and $(-3, -1)$.

10. Write the equation of the horizontal line passing through $\left(-\dfrac{1}{3}, 2\right)$.

11. Write the equation of the line with slope -5 and y-intercept $(0, -8)$. Write the answer in slope–intercept form.

Graph the regions.

12. $y \geq -4x$

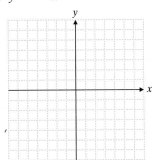

13. $4x - 2y < -6$

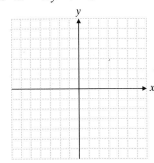

1. _____

2. _____

3. _____

4. _____

5. _____

6. _____

7. _____

8. _____

9. _____

10. _____

11. _____

12. _____

13. _____

14. What are the domain and range of the following relation?

$$A = \{(0,0), (1,1), (1,-1), (2,4), (2,-4)\}$$

14. _____

15. If $f(x) = 2x - 3$, find $f\left(\dfrac{3}{4}\right)$.

16. If $g(x) = \dfrac{1}{2}x^2 + 3$, find $g(-4)$.

15. _____

17. If $h(x) = \left| -\dfrac{2}{3}x + 4 \right|$, find $h(-9)$.

18. If $p(x) = -2x^3 + 3x^2 + x - 4$, find $p(-2)$.

16. _____

Graph each function.

19. $g(x) = 5 - x^2$

17. _____

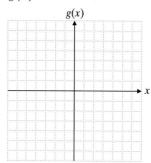

20. $h(x) = x^3 - 4$

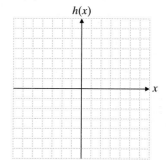

21. The following table describes the approximate distance $f(x)$ in miles that you can see across the ocean on a clear day if you are x feet above the water. Plot these points and connect them to graph the function.

18. _____

Height in Feet, x	0	3	9	15
Distance in Miles, $f(x)$	0	6	54	150

Based on your graph, how many miles can you see if you are 4 feet above the water?

19. _____

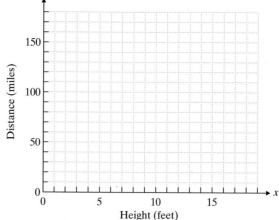

20. _____

21. _____

Approximately one-half of this test covers the content of Chapters 1 and 2. The remainder covers the content of Chapter 3.

1. Name the property that justifies $(-6) + 6 = 0$.

2. Evaluate the expression: $3(4 - 6)^2 + \sqrt{16} + 12 \div (-3)$

3. Simplify $(3x^2 y^{-3})^{-4}$.

4. Simplify $5x(2x - 3y) - 3(x^2 + 4)$.

5. Write using scientific notation: 0.000437

6. Solve and graph your solution: $3(x - 2) > 6$ or $5 - 3(x + 1) > 8$

$\longleftarrow\!\!+\!\!+\!\!+\!\!+\!\!+\!\!+\!\!+\!\!+\!\!+\!\!+\!\!+\!\!+\!\!\longrightarrow$

7. Solve for x: $2a - x = \dfrac{1}{3}(6x - y)$

▲ **8.** A plastic insulator is made in a rectangular shape with a perimeter of 92 centimeters. The length is 1 centimeter longer than double the width. Find the dimensions of the insulator.

9. Sharim invested $3000 for 1 year. Part was invested at 5% simple interest and part at 8% simple interest. At the end of 1 year, he had earned $189 in interest. How much had he invested at each rate?

▲ **10.** Find the area of the semicircle below. Round your answer to the nearest hundredth. (Use $\pi = 3.14$.)

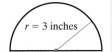
$r = 3$ inches

11. Graph the line $4x - 6y = 10$. Plot at least three points.

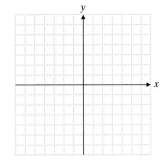

12. Find the slope of the line passing through $(6, 5)$ and $(-2, 1)$.

13. Write the standard form of the equation of the line that passes through $(5, 1)$ and $(4, 3)$.

14. Write the standard form of the equation of the line that passes through $(-2, -3)$ and is perpendicular to $y = \dfrac{2}{3}x - 4$.

15. What are the domain and range of the relation $\{(3, 7), (5, 8), (\frac{1}{2}, -1), (2, 2)\}$? Is the relation a function?

16. Find $f(-3)$ for $f(x) = -2x^2 - 4x + 1$.

1. _____

2. _____

3. _____

4. _____

5. _____

6. _____

7. _____

8. _____

9. _____

10. _____

11. _____

12. _____

13. _____

14. _____

15. _____

16. _____

193

17. _____

Graph the following functions.

17. $p(x) = -\frac{1}{3}x + 2$

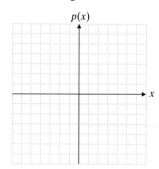

18. _____

18. $h(x) = |x - 2|$

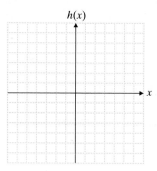

19. _____

19. $r(x) = \frac{3}{x}$

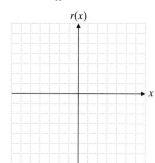

20. _____

20. $f(x) = x^2 - 3$

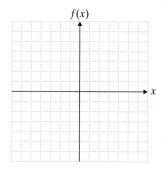

21. _____

21. Graph the region. $y \le -\frac{3}{2}x + 3$

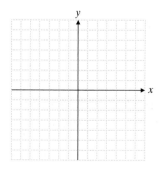

22. Complete the following table of values for $f(x)$. $f(x) = -2x^3 + 4$

x	f(x)
-2	
0	
3	

22. _____

Just before the new interstate highway opened, the traffic on Route 1 was measured by the highway department. They found that an average of 32,500 cars traveled on Route 1 each weekday. Since Interstate I-95 was completed, in 1990, the daily traffic on Route 1 has decreased by about 1400 cars each year.

23. _____

23. Express the average number of cars driven each weekday on Route 1 as a function $f(x)$ of the number of years x since 1990.

24. Using your answer from Exercise 23, predict how many cars will travel each weekday on Route 1 in the year 2005.

24. _____

CHAPTER

4

Cell phones have become increasingly popular in recent years. You almost always see someone talking on his or her cell phone in public, even at the beach. For many of us, it is important that other people can contact us wherever we are. Having a cell phone available adds a sense of security in case an emergency should arise. However, choosing a cell phone plan can be confusing. Use your math skills to determine which phone plans are most economical in a variety of situations. Turn to page 237 to see if you can make wise mathematical decisions.

Systems of Linear Equations and Inequalities

Student Learning Objectives

After studying this section, you will be able to:

1 Determine whether an ordered pair is a solution to a system of two linear equations.

2 Solve a system of two linear equations by the graphing method.

3 Solve a system of two linear equations by the substitution method.

4 Solve a system of two linear equations by the addition (elimination) method.

5 Identify systems of linear equations that do not have a unique solution.

6 Choosing an appropriate method to solve a system of linear equations algebraically.

1 ## Determining Whether an Ordered Pair Is a Solution to a System of Two Linear Equations

In Chapter 3 we found that a linear equation containing two variables, such as $4x + 3y = 12$, has an unlimited number of ordered pairs (x, y) that satisfy it. For example, $(3, 0)$, $(0, 4)$, and $(-3, 8)$ all satisfy the equation $4x + 3y = 12$. We call *two* linear equations in two unknowns a **system of two linear equations in two variables.** Many such systems have exactly one solution. A **solution to a system** of two linear equations in two variables is an *ordered pair* that is a solution to *each* equation.

EXAMPLE 1   Determine whether $(3, -2)$ is a solution to the following system.

$$x + 3y = -3$$
$$4x + 3y = \ \ 6$$

Solution   We will begin by substituting $(3, -2)$ into the first equation to see whether the ordered pair is a solution to the first equation.

$$3 + 3(-2) \overset{?}{=} -3$$
$$3 - 6 \overset{?}{=} -3$$
$$-3 = -3 \ \checkmark$$

Likewise, we will determine whether $(3, -2)$ is a solution to the second equation.

$$4(3) + 3(-2) \overset{?}{=} 6$$
$$12 - 6 \overset{?}{=} 6$$
$$6 = 6 \ \checkmark$$

Since $(3, -2)$ is a solution to each equation in the system, it is a solution to the system itself.

It is important to remember that we cannot confirm that a particular ordered pair is in fact the solution to a system of two equations unless we have checked to see whether the solution satisfies both equations. Merely checking one equation is not sufficient. Determining whether an ordered pair is a solution to a system of equations requires that we verify that the solution satisfies *both* equations.

Practice Problem 1   Determine whether $(-3, 4)$ is a solution to the following system.

$$2x + 3y = 6$$
$$3x - 4y = 7$$

NOTE TO STUDENT: Fully worked-out solutions to all of the Practice Problems can be found at the back of the text starting at page SP-1

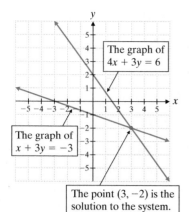

The graph of $4x + 3y = 6$

The graph of $x + 3y = -3$

The point $(3, -2)$ is the solution to the system.

2 ## Solving a System of Two Linear Equations by the Graphing Method

We can verify the solution to a system of linear equations by graphing each equation. If the lines intersect, the system has a unique solution. The point of intersection lies on both lines. Thus, it is a solution to each equation and the solution to the system. We will illustrate this by graphing the equations in Example 1. Notice that the coordinates of the point of intersection are $(3, -2)$. The solution to the system is $(3, -2)$.

This example shows that we can find the solution to a system of linear equations by graphing each line and determining the point of intersection.

EXAMPLE 2   Solve this system of equations by graphing.

$$2x + 3y = 12$$
$$x - y = 1$$

Solution   Using the methods that we developed in Chapter 3, we graph each line and determine the point at which the two lines intersect.

Finding the solution by the graphing method does not always lead to an accurate result, however, because it involves visual estimation of the point of intersection. Also, our plotting of one or more of the lines could be off slightly. Thus, we verify that our answer is correct by substituting $x = 3$ and $y = 2$ into the system of equations.

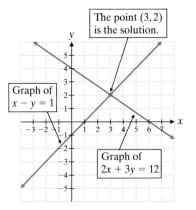

The point $(3, 2)$ is the solution.

Graph of $x - y = 1$

Graph of $2x + 3y = 12$

$x - y = 1$	$2x + 3y = 12$
$3 - 2 \overset{?}{=} 1$	$2(3) + 3(2) \overset{?}{=} 12$
$1 = 1$ ✓	$12 = 12$ ✓

Thus, we have verified that the solution to the system is $(3, 2)$.

Practice Problem 2   Solve this system of equations by graphing. Check your solution.

$$3x + 2y = 10$$
$$x - y = 5$$

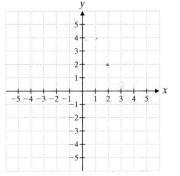

Many times when we graph a system, we find that the two straight lines intersect at one point. However, it is possible for a given system to have as its graph two parallel lines. In such a case there is no solution because there is no point that lies on both lines (i.e., no ordered pair that satisfies both equations). Such a system of equations is said to be **inconsistent.** Another possibility is that when we graph each equation in the system, we obtain one line. In such a case there are an infinite number of solutions. Any point that lies on the first line will also lie on the second line (i.e., any ordered pair). A system of equations in two variables is said to have **dependent equations** if it has infinitely many solutions. We will discuss these situations in more detail after we have developed algebraic methods for solving a system of equations.

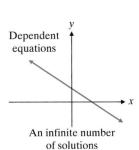

Inconsistent system of equations

No solution

Dependent equations

An infinite number of solutions

 ### Solving a System of Two Linear Equations by the Substitution Method

An algebraic method of solving a system of linear equations in two variables is the **substitution method.** To use this method, we choose one equation and solve for one variable. It is usually best to solve for a variable that has a coefficient of $+1$ or -1. This will help us avoid introducing fractions. When we solve for one variable, we obtain an expression that contains the other variable. We *substitute* this expression into the second equation. Then we have one equation with one unknown, which we can easily solve. Once we know the value of this variable, we can substitute it into one of the original equations to find the value of the other variable.

Graphing Calculator

Solving Systems of Linear Equations

We can solve systems of equations graphically by using a graphing calculator. For example, to solve the system of equations in Example 2, first rewrite each equation in slope–intercept form.

$$y = -\frac{2}{3}x + 4$$
$$y = x - 1$$

Then graph $y_1 = -\frac{2}{3}x + 4$ and $y_2 = x - 1$ on the same screen.

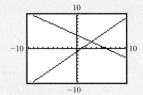

Next, use the Zoom and Trace features to find the intersection of the two lines.

Some graphing calculators have a command to find and calculate the intersection.

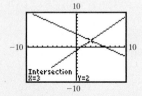

Try the following:

1. $y_1 = -3x + 9$
   $y_2 = 4x - 5$
2. $y_1 = -0.48x + 4.79$
   $y_2 = 1.52x - 2.98$

EXAMPLE 3 Find the solution to the following system of equations. Use the substitution method.

$$x + 3y = -7 \quad \textbf{(1)}$$
$$4x + 3y = -1 \quad \textbf{(2)}$$

Solution We can work with equation **(1)** or equation **(2)**. Let's choose equation **(1)** because x has a coefficient of 1. Now let us solve for x. This gives us equation **(3)**.

$$x = -7 - 3y \quad \textbf{(3)}$$

Now we substitute this expression for x into equation **(2)** and solve the equation for y.

$$4x + 3y = -1 \quad \textbf{(2)}$$
$$4(-7 - 3y) + 3y = -1$$
$$-28 - 12y + 3y = -1$$
$$-28 - 9y = -1$$
$$-9y = -1 + 28$$
$$-9y = 27$$
$$y = -3$$

Now we substitute $y = -3$ into equation **(1)** or **(2)** to find x. Let's use **(1)**:

$$x + 3(-3) = -7$$
$$x - 9 = -7$$
$$x = -7 + 9$$
$$x = 2$$

Therefore, our solution is the ordered pair $(2, -3)$.
Check. We must verify the solution in both of the *original* equations.

$$x + 3y = -7 \qquad\qquad 4x + 3y = -1$$
$$2 + 3(-3) \overset{?}{=} -7 \qquad 4(2) + 3(-3) \overset{?}{=} -1$$
$$2 - 9 \overset{?}{=} -7 \qquad\qquad 8 - 9 \overset{?}{=} -1$$
$$-7 = -7 \ \checkmark \qquad\qquad -1 = -1 \ \checkmark$$

Practice Problem 3 Use the substitution method to solve this system.

$$2x - y = 7$$
$$3x + 4y = -6$$

We summarize the substitution method here.

HOW TO SOLVE A SYSTEM OF TWO LINEAR EQUATIONS BY THE SUBSTITUTION METHOD

1. Choose one of the two equations and solve for one variable in terms of the other variable.
2. Substitute this expression from step 1 into the *other* equation.
3. You now have one equation with one variable. Solve this equation for that variable.
4. Substitute this value for the variable into one of the original equations to obtain a value for the second variable.
5. Check the solution in both original equations.

Optional Graphing Calculator Note. Before solving the system in Example 3 with a graphing calculator, you will first need to solve each equation for y. Equation **(1)** can be written as $y_1 = -\frac{1}{3}x - \frac{7}{3}$ or as $y_1 = \frac{-x - 7}{3}$. Likewise, equation **(2)** can be written as $y_2 = -\frac{4}{3}x - \frac{1}{3}$ or as $y_2 = \frac{-4x - 1}{3}$.

EXAMPLE 4   Solve the following system of equations.

$$\frac{1}{2}x - \frac{1}{4}y = -\frac{3}{4} \quad \textbf{(1)}$$
$$3x - 2y = -6 \quad \textbf{(2)}$$

Solution   First clear equation **(1)** of fractions by multiplying each term by 4.

$$4\left(\frac{1}{2}x\right) - 4\left(\frac{1}{4}y\right) = 4\left(-\frac{3}{4}\right)$$
$$2x - y = -3 \quad \textbf{(3)}$$

The new system is as follows:

$$2x - y = -3 \quad \textbf{(3)}$$
$$3x - 2y = -6 \quad \textbf{(2)}$$

Step 1   Let's solve equation **(3)** for y. We select this because the y-variable has a coefficient of -1.

$$-y = -3 - 2x$$
$$y = 3 + 2x$$

Step 2   Substitute this expression for y into equation **(2)**.

$$3x - 2(3 + 2x) = -6$$

Step 3   Solve this equation for x.

$$3x - 6 - 4x = -6$$
$$-6 - x = -6$$
$$-x = -6 + 6$$
$$-x = 0$$
$$x = 0$$

Step 4   Substitute $x = 0$ into equation **(2)**.

$$3(0) - 2y = -6$$
$$-2y = -6$$
$$y = 3$$

So our solution is $(0, 3)$.

Step 5   We must verify the solution in both original equations.

$$\frac{1}{2}x - \frac{1}{4}y = -\frac{3}{4} \qquad\qquad 3x - 2y = -6$$
$$\frac{0}{2} - \frac{3}{4} \overset{?}{=} -\frac{3}{4} \qquad\qquad 3(0) - 2(3) \overset{?}{=} -6$$
$$-\frac{3}{4} = -\frac{3}{4} \checkmark \qquad\qquad -6 = -6 \checkmark$$

NOTE TO STUDENT: Fully worked-out solutions to all of the Practice Problems can be found at the back of the text starting at page SP-1

Practice Problem 4   Use the substitution method to solve this system.

$$\frac{1}{2}x + \frac{2}{3}y = 1$$

$$\frac{1}{3}x + y = -1$$

4 Solving a System of Two Linear Equations by the Addition Method

Another way to solve a system of two linear equations in two variables is to add the two equations so that a variable is eliminated. This technique is called the **addition method** or the **elimination method.** We usually have to multiply one or both of the equations by suitable factors so that we obtain opposite coefficients on one variable (either x or y) in the equations.

EXAMPLE 5   Solve the following system by the addition method.

$$5x + 8y = -1 \quad \textbf{(1)}$$
$$3x + y = 7 \quad \textbf{(2)}$$

Solution   We can eliminate either the x- or the y-variable. Let's choose y. We multiply equation **(2)** by -8.

$$-8(3x) + (-8)(y) = -8(7)$$
$$-24x - 8y = -56 \quad \textbf{(3)}$$

We now add equations **(1)** and **(3).**

$$5x + 8y = -1 \quad \textbf{(1)}$$
$$\underline{-24x - 8y = -56} \quad \textbf{(3)}$$
$$-19x = -57$$

We solve for x.

$$x = \frac{-57}{-19} = 3$$

Now we substitute $x = 3$ into equation **(2)** (or equation **(1)**).

$$3(3) + y = 7$$
$$9 + y = 7$$
$$y = -2$$

Our solution is $(3, -2)$.

$$\textit{Check.} \quad 5(3) + 8(-2) \stackrel{?}{=} -1$$
$$15 + (-16) \stackrel{?}{=} -1$$
$$-1 = -1$$
$$3(3) + (-2) \stackrel{?}{=} 7$$
$$9 + (-2) \stackrel{?}{=} 7$$
$$7 = 7 \quad \checkmark$$

Practice Problem 5   Use the addition method to solve this system.

$$-3x + y = 5$$
$$2x + 3y = 4$$

For convenience, we summarize the addition method here.

HOW TO SOLVE A SYSTEM OF TWO LINEAR EQUATIONS BY THE ADDITION (ELIMINATION) METHOD

1. Arrange each equation in the form $ax + by = c$. (Remember that a, b, and c can be any real numbers.)

2. Multiply one or both equations by appropriate numbers so that the coefficients of one of the variables are opposites.

3. Add the two equations from step 2 so that one variable is eliminated.

4. Solve the resulting equation for the remaining variable.

5. Substitute this value into one of the *original* equations and solve to find the value of the other variable.

6. Check the solution in both of the original equations.

EXAMPLE 6   Solve the following system by the addition method.

$$3x + 2y = -8 \quad \textbf{(1)}$$
$$2x + 5y = 2 \quad \textbf{(2)}$$

Solution   To eliminate the variable x, we multiply equation **(1)** by 2 and equation **(2)** by -3. We now have the following equivalent system.

$$6x + 4y = -16$$
$$\underline{-6x - 15y = -6}$$
$$-11y = -22 \quad \text{Add the equations.}$$
$$y = 2 \quad \text{Solve for } y.$$

Substitute $y = 2$ into equation **(1)**.

$$3x + 2(2) = -8$$
$$3x + 4 = -8$$
$$3x = -12$$
$$x = -4$$

The solution to the system is $(-4, 2)$.

Check. Verify that this solution is correct.

Note. We could have easily eliminated the variable y in Example 6 by multiplying equation **(1)** by 5 and equation **(2)** by -2. Try it. Is the solution the same? Why?

Practice Problem 6   Use the addition (elimination) method to solve this system.

$$5x + 4y = 23$$
$$7x - 3y = 15$$

Identifying Systems of Linear Equations That Do Not Have a Unique Solution

So far we have examined only those systems that have one solution. But other systems must also be considered. These systems can best be illustrated with graphs. In general, the system of equations

$$ax + by = c$$
$$dx + ey = f$$

may have one solution, no solution, or an infinite number of solutions.

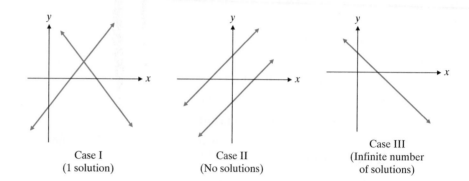

Case I	Case II	Case III
(1 solution)	(No solutions)	(Infinite number of solutions)

Case I: *One solution.* The two graphs intersect at one point, which is the solution. We say that the equations are **independent.** It is a **consistent system** of equations. There is a point (an ordered pair) *consistent* with both equations.

Case II: *No solution.* The two graphs are parallel and so do not intersect. We say that the system of equations is **inconsistent** because there is no point consistent with both equations.

Case III: *An infinite number of solutions.* The graphs of each equation yield the same line. Every ordered pair on this line is a solution to both of the equations. We say that the equations are **dependent.**

EXAMPLE 7   If possible, solve the system.

$$2x + 8y = 16 \quad \textbf{(1)}$$
$$4x + 16y = -8 \quad \textbf{(2)}$$

Solution   To eliminate the variable y, we'll multiply equation **(1)** by -2.

$$-2(2x) + (-2)(8y) = (-2)(16)$$
$$-4x - 16y = -32 \quad \textbf{(3)}$$

We now have the following equivalent system.

$$-4x - 16y = -32 \quad \textbf{(3)}$$
$$4x + 16y = -8 \quad \textbf{(2)}$$

When we add equations **(3)** and **(2)**, we get

$$0 = -40,$$

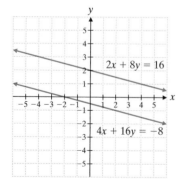

which, of course, is false. Thus, we conclude that this system of equations is inconsistent, and **there is no solution.** Therefore, equations **(1)** and **(2)** do not intersect, as we can see on the accompanying graph.

If we had used the substitution method to solve this system, we still would have obtained a false statement. When you try to solve an inconsistent system of linear equations by any method, you will always obtain a mathematical equation that is not true.

Practice Problem 7   If possible, solve the system.

$$4x - 2y = 6$$
$$-6x + 3y = 9$$

NOTE TO STUDENT: Fully worked-out solutions to all of the Practice Problems can be found at the back of the text starting at page SP-1

EXAMPLE 8   If possible, solve the system.

$$0.5x - 0.2y = 1.3 \quad \textbf{(1)}$$
$$-1.0x + 0.4y = -2.6 \quad \textbf{(2)}$$

Solution   Although we could work directly with the decimals, it is easier to multiply each equation by the appropriate power of 10 (10, 100, and so on) so that the coefficients of the new system are integers. Therefore, we will multiply equations **(1)** and **(2)** by 10 to obtain the following equivalent system.

$$5x - 2y = 13 \quad \textbf{(3)}$$
$$-10x + 4y = -26 \quad \textbf{(4)}$$

We can eliminate the variable y by multiplying each term of equation **(3)** by 2.

$$10x - 4y = 26 \quad \textbf{(3)}$$
$$\underline{-10x + 4y = -26} \quad \textbf{(4)}$$
$$0 = 0 \quad \text{Add the equations.}$$

This statement is always true; it is an **identity.** Hence, the two equations are dependent, and there are an infinite number of solutions. Any solution satisfying equation **(1)** will also satisfy equation **(2)**. For example, $(3, 1)$ is a solution to equation **(3)**. (Prove this.) Hence, it must also be a solution to equation **(4)**. (Prove it). Thus, the equations actually describe the same line, as you can see on the graph.

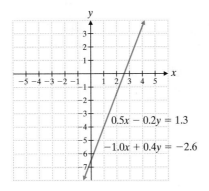

Practice Problem 8   If possible, solve the system.

$$0.3x - 0.9y = 1.8$$
$$-0.4x + 1.2y = -2.4$$

6 Choosing an Appropriate Method to Solve a System of Linear Equations Algebraically

At this point we will review the algebraic methods for solving systems of linear equations and discuss the advantages and disadvantages of each method.

Method	Advantage	Disadvantage
Substitution	Works well if one or more variables has a coefficient of 1 or -1.	Often becomes difficult to use if no variable has a coefficient of 1 or -1.
Addition	Works well if equations have fractional or decimal coefficients. Works well if no variable has a coefficient of 1 or -1.	None

EXAMPLE 9 Select a method and solve each system of equations.

(a) $\quad x + y = 3080$
$\quad\;\; 2x + 3y = 8740$

(b) $\quad 5x - 2y = 19$
$\quad\;\; -3x + 7y = 35$

Solution

(a) Since there are x- and y-values that have coefficients of 1, we will select the substitution method.

$$y = 3080 - x \quad\quad \text{Solve the first equation for } y.$$
$$2x + 3(3080 - x) = 8740 \quad\quad \text{Substitute the expression into the second equation.}$$
$$2x + 9240 - 3x = 8740 \quad\quad \text{Remove parentheses.}$$
$$-1x = -500 \quad\quad \text{Simplify.}$$
$$x = 500 \quad\quad \text{Divide each side by } -1.$$

Substitute $x = 500$ into the equation obtained in step one.

$$y = 3080 - 500$$
$$y = 2580 \quad\quad \text{Simplify.}$$

The solution is $(500, 2580)$.

(b) Because none of the x- and y-variables have a coefficient of 1 or -1, we select the addition method. We choose to eliminate the y-variable. Thus, we would like the coefficients of y to be -14 and 14.

$$7(5x) - 7(2y) = 7(19) \quad\quad \text{Multiply each term of the first equation by 7.}$$
$$2(-3x) + 2(7y) = 2(35) \quad\quad \text{Multiply each term of the second equation by 2.}$$
$$35x - 14y = 133 \quad\quad \text{We now have an equivalent system of equations.}$$
$$\underline{-6x + 14y = \;\;70}$$
$$29x \qquad\quad = 203 \quad\quad \text{Add the two equations.}$$
$$x = 7 \quad\quad \text{Divide each side by 29.}$$

Substitute $x = 7$ into one of the original equations. We will use the first equation.

$$5(7) - 2y = 19$$
$$35 - 2y = 19 \quad\quad \text{Solve for } y.$$
$$-2y = -16$$
$$y = 8$$

The solution is $(7, 8)$.

Practice Problem 9   Select a method and solve each system of equations.

(a) $3x + 5y = 1485$

$\quad x + 2y = 564$

(b) $7x + 6y = 45$

$\quad 6x - 5y = -2$

NOTE TO STUDENT: Fully worked-out solutions to all of the Practice Problems can be found at the back of the text starting at page SP-1

TO THINK ABOUT: Two Linear Equations with Two Variables   Now is a good time to look back over what we have learned. When you graph a system of two linear equations, what possible kinds of graphs will you obtain?

What will happen when you try to solve a system of two linear equations using algebraic methods? How many solutions are possible in each case? The following chart may help you to organize your answers to these questions.

Graph	Number of Solutions	Algebraic Interpretation
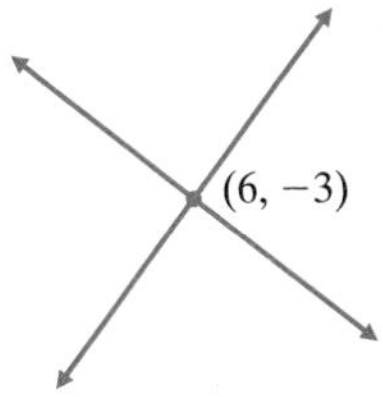 Two lines intersect at one point — $(6, -3)$	**One unique solution**	You obtain one value for x and one value for y. For example, $x = 6, \quad y = -3$.
 Parallel lines	**No solution**	You obtain an equation that is inconsistent with known facts. For example, $0 = 6$. The system of equations is inconsistent.
 Lines coincide	**Infinite number of solutions**	You obtain an equation that is always true. For example, $8 = 8$. The equations are dependent.

4.1 EXERCISES

Student Solutions Manual | CD/Video | PH Math Tutor Center | MathXL®Tutorials on CD | MathXL® | MyMathLab® | Interactmath.com

Verbal and Writing Skills

1. Explain what happens when a system of two linear equations is inconsistent. What effect does it have in obtaining a solution? What would the graph of such a system look like?

2. Explain what happens when a system of two linear equations has dependent equations. What effect does it have in obtaining a solution? What would the graph of such a system look like?

3. How many possible solutions can a system of two linear equations in two unknowns have?

4. When you have graphed a system of two linear equations in two unknowns, how do you determine the solution of the system?

Determine whether the given ordered pair is a solution to the system of equations.

5. $\left(\dfrac{3}{2}, -1\right)$ $\begin{aligned} 4x + 1 &= 6 - y \\ 2x - 5y &= 8 \end{aligned}$

6. $\left(-4, \dfrac{2}{3}\right)$ $\begin{aligned} 2x - 3(y - 5) &= 5 \\ 6y &= x + 8 \end{aligned}$

Solve the system of equations by graphing. Check your solution.

7. $\begin{aligned} 3x + y &= 2 \\ 2x - y &= 3 \end{aligned}$

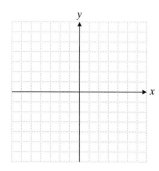

8. $\begin{aligned} 3x + y &= 5 \\ 2x - y &= 5 \end{aligned}$

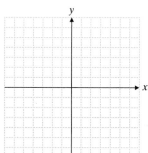

9. $\begin{aligned} 2x + 3y &= 6 \\ 2x + y &= -2 \end{aligned}$

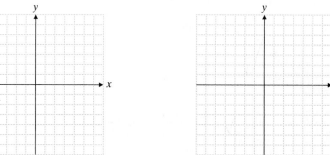

10. $\begin{aligned} 2x + 3y &= -6 \\ x - 3y &= 6 \end{aligned}$

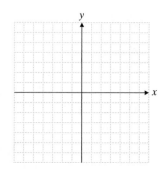

11. $\begin{aligned} y &= -x + 3 \\ x + y &= -\dfrac{2}{3} \end{aligned}$

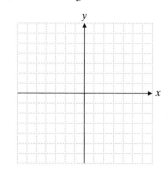

12. $\begin{aligned} y &= \dfrac{1}{3}x - 2 \\ -x + 3y &= 9 \end{aligned}$

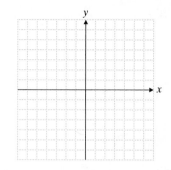

13. $y = -2x + 5$
$3y + 6x = 15$

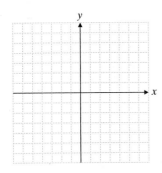

14. $x - 3 = 2y + 1$
$y - \dfrac{x}{2} = -2$

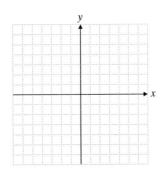

Find the solution to each system by the substitution method. Check your answers for Exercises 15–18.

15. $3x + 2y = -17$
$2x + y = 3$

16. $5x - 2y = 8$
$3x - y = 7$

17. $-x + 3y = -8$
$2x - y = 6$

18. $10x + 3y = 8$
$2x + y = 2$

19. $2x - \dfrac{1}{2}y = -3$
$\dfrac{x}{5} + 2y = \dfrac{19}{5}$

20. $x - \dfrac{3}{2}y = 1$
$2x - 7y = 10$

21. $\dfrac{3}{2}x - \dfrac{1}{10}y = 6$
$\dfrac{1}{2}x + \dfrac{3}{10}y = 2$

22. $\dfrac{1}{6}x - y = \dfrac{1}{2}$
$x + \dfrac{7}{11}y = 3$

Find the solution to each system by the addition (elimination) method. Check your answers for Exercises 23–26.

23. $9x + 2y = 2$
$3x + 5y = 5$

24. $12x - 5y = -7$
$4x + 2y = 5$

25. $6s - 3t = 1$
$5s + 6t = 15$

26. $2s + 3t = 5$
$3s - 6t = 18$

27. $\dfrac{7}{2}x + \dfrac{5}{2}y = -4$
$3x + \dfrac{2}{3}y = 1$

28. $\dfrac{4}{3}x - y = 4$
$\dfrac{3}{4}x - y = \dfrac{1}{2}$

29. $1.6x + 1.5y = 1.8$
$0.4x + 0.3y = 0.6$

30. $2.5x + 0.6y = 0.2$
$0.5x - 1.2y = 0.7$

Mixed Practice

If possible, solve each system of equations. Use any method. If there is not a unique solution to a system, state a reason.

31. $7x - y = 6$
$3x + 2y = 22$

32. $8x - y = 17$
$4x + 3y = 33$

33. $3x + 4y = 8$
$5x + 6y = 10$

34. $7x + 5y = -25$
$3x + 7y = -1$

35. $2x + y = 4$
$\dfrac{2}{3}x + \dfrac{1}{4}y = 2$

36. $2x + 3y = 16$
$5x - \dfrac{3}{4}y = 7$

37. $0.2x = 0.1y - 1.2$
$2x - y = 6$

38. $0.1x - 0.6 = 0.3y$
$0.3x + 0.1y + 2.2 = 0$

39. $5x - 7y = 12$
$-10x + 14y = -24$

40. $3x - 11y = 9$
$-9x + 33y = 18$

41. $0.8x + 0.9y = 1.3$
$0.6x - 0.5y = 4.5$

42. $0.6y = 0.9x + 1$
$3x = 2y - 4$

43. $\dfrac{4}{5}b = \dfrac{1}{5} + a$
$15a - 12b = 4$

44. $3a - 2b = \dfrac{3}{2}$
$\dfrac{3a}{2} = \dfrac{3}{4} + b$

45. $\dfrac{3}{8}x + y = 14$
$2x - \dfrac{7}{4}y = 18$

46. $\dfrac{4}{5}x - y = 7$
$x - \dfrac{4}{3}y = 8$

47. $3.2x - 1.5y = -3$
$0.7x + y = 2$

48. $3x - 0.2y = 1$
$1.1x + 0.4y = -2$

49. $3 - (2x + 1) = y + 6$
$x + y + 5 = 1 - x$

50. $2(y - 3) = x + 3y$
$x + 2 = 3 - y$

To Think About

51. *Bathroom Tile* Wayne Burton is having some tile replaced in his bathroom. He has obtained an estimate from two tile companies. Old World Tile gave an estimate of $200 to remove the old tile and $50 per hour to place new tile on the wall. Modern Bathroom Headquarters gave an estimate of $300 to remove the old tile and $30 per hour to place new tile on the wall.

(a) Create a cost equation for each company where y is the total cost of the tile work and x is the number of hours of labor. Write a system of equations.

(b) Graph the two equations using the values $x = 0, 4,$ and 8.

(c) Determine from your graph how many hours of installing new tile will be required for the two companies to cost the same.

(d) Determine from your graph which company costs less to remove old tile and to install new tile if the time needed to install new tile is 6 hours.

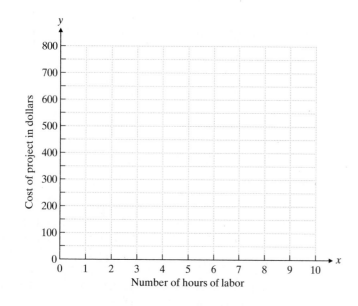

52. ***Moving Furniture*** Jeff and Shelley are planning to move some furniture to their daughter's house. Seaside Movers quoted a price of $100 for the truck and $40 per hour for the movers. Beverly Rapid Mover quoted a price of $50 for the truck and $50 per hour for the movers.
    (a) Create a cost equation for each company where y is the total cost of the move and x is the number of hours of labor. Write a system of equations.
    (b) Graph the two equations using the values $x = 0, 4,$ and 7.
    (c) Determine from your graph how many hours of moving would be required for the two companies to cost the same.
    (d) Determine from your graph which company costs less to conduct the move if the total number of hours needed for moving is 3 hours.

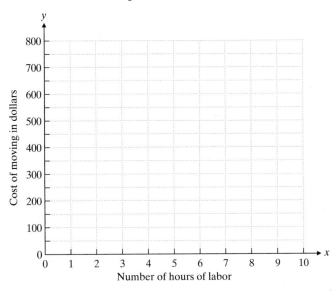

Optional Graphing Calculator Problems

On your graphing calculator, graph each system of equations on the same set of axes. Find the point of intersection to the nearest hundredth.

53. $y_1 = -1.7x + 3.8$
    $y_2 = 0.7x - 2.1$

54. $y_1 = -0.81x + 2.3$
    $y_2 = 1.6x + 0.8$

55. $0.5x + 1.1y = 5.5$
    $-3.1x + 0.9y = 13.1$

56. $5.86x + 6.22y = -8.89$
    $-2.33x + 4.72y = -10.61$

Cumulative Review

57. ***Winter Road Salt*** Nine million tons of salt are applied to American highways for road deicing each year. The cost of buying and applying the salt totals $200 million. How much is this per pound? Round your answer to the nearest cent.

58. ***City Parking Space*** Four-fifths of the automobiles that enter the city of Boston during rush hour will have to park in private or municipal parking lots. If there are 273,511 private or municipal lot spaces filled by cars entering the city during rush hour every morning, how many cars enter the city during rush hour? Round your answer to the nearest car.

Student Learning Objectives

After studying this section, you will be able to:

1 Determine whether an ordered triple is the solution to a system of three linear equations in three variables.

2 Find the solution to a system of three linear equations in three variables if none of the coefficients is zero.

3 Find the solution to a system of three linear equations in three variables if some of the coefficients are zero.

1 Determining Whether an Ordered Triple Is the Solution to a System of Three Linear Equations in Three Variables

We are now going to study **systems of three linear equations in three variables** (unknowns). A **solution** to a system of three linear equations in three unknowns is an **ordered triple** of real numbers (x, y, z) that satisfies each equation in the system.

EXAMPLE 1 Determine whether $(2, -5, 1)$ is the solution to the following system.

$$3x + y + 2z = 3$$
$$4x + 2y - z = -3$$
$$x + y + 5z = 2$$

Solution How can we prove that $(2, -5, 1)$ is a solution to this system? We will substitute $x = 2$, $y = -5$, and $z = 1$ into each equation. If a true statement occurs each time, $(2, -5, 1)$ is a solution to each equation and hence, solution to the system. For the first equation:

$$3(2) + (-5) + 2(1) \stackrel{?}{=} 3$$
$$6 - 5 + 2 \stackrel{?}{=} 3$$
$$3 = 3. \checkmark$$

For the second equation:

$$4(2) + 2(-5) - 1 \stackrel{?}{=} -3$$
$$8 - 10 - 1 \stackrel{?}{=} -3$$
$$-3 = -3. \checkmark$$

For the third equation:

$$2 + (-5) + 5(1) \stackrel{?}{=} 2$$
$$2 - 5 + 5 \stackrel{?}{=} 2$$
$$2 = 2. \checkmark$$

Since we obtained three true statements, the ordered triple $(2, -5, 1)$ is a solution to the system.

Practice Problem 1 Determine whether $(3, -2, 2)$ is a solution to this system.

$$2x + 4y + z = 0$$
$$x - 2y + 5z = 17$$
$$3x - 4y + z = 19$$

NOTE TO STUDENT: Fully worked-out solutions to all of the Practice Problems can be found at the back of the text starting at page SP-1

TO THINK ABOUT: Graphs in Three Variables
Can we graph an equation in three variables? How? What would the graph look like? What would the graph of the system in Example 1 look like? Describe the graph of the solution.

2 Finding the Solution to a System of Three Linear Equations in Three Variables If None of the Coefficients Is Zero

One way to solve a system of three equations with three variables is to obtain from it a system of two equations in two variables; in other words, we eliminate one variable from both equations. We can then use the methods of Section 4.1 to solve the resulting system. You can find the third variable (the one that was eliminated) by substituting the two variables that you have found into one of the original equations.

EXAMPLE 2 Find the solution to (that is, solve) the following system of equations.

$$-2x + 5y + z = 8 \quad (1) \checkmark$$
$$-x + 2y + 3z = 13 \quad (2)$$
$$x + 3y - z = 5 \quad (3) \checkmark$$

Solution Let's eliminate z because it can be done easily by adding equations **(1)** and **(3)**.

$$-2x + 5y + z = 8 \quad (1)$$
$$\underline{x + 3y - z = 5} \quad (3)$$
$$-x + 8y \quad\quad = 13 \quad (4)$$

Now we need to choose a *different pair* from the original system of equations and once again eliminate the same variable. In other words, we have to use equations **(1)** and **(2)** or equations **(2)** and **(3)** and eliminate z. Let's multiply each term of equation **(3)** by 3 (and call it equation **(6)**) and add the result to equation **(2)**.

$$-x + 2y + 3z = 13 \quad (2)$$
$$\underline{3x + 9y - 3z = 15} \quad (6)$$
$$2x + 11y \quad\quad = 28 \quad (5)$$

We now can solve the resulting system of two linear equations.

$$-x + 8y = 13 \quad (4)$$
$$2x + 11y = 28 \quad (5)$$

Multiply each term of equation **(4)** by 2.

$$-2x + 16y = 26$$
$$\underline{2x + 11y = 28}$$
$$27y = 54 \quad \text{Add the equations.}$$
$$y = 2 \quad \text{Solve for } y.$$

Substituting $y = 2$ into equation **(4)**, we have the following:

$$-x + 8(2) = 13$$
$$-x = -3$$
$$x = 3.$$

Now substitute $x = 3$ and $y = 2$ into one of the original equations (any one will do) to solve for z. Let's use equation **(1)**.

$$-2x + 5y + z = 8$$
$$-2(3) + 5(2) + z = 8$$
$$-6 + 10 + z = 8$$
$$z = 4$$

The solution to the system is $(3, 2, 4)$.

Check. Verify that $(3, 2, 4)$ satisfies *each* of the three *original* equations.

Practice Problem 2   Solve this system.

$$x + 2y + 3z = 4$$
$$2x + y - 2z = 3$$
$$3x + 3y + 4z = 10$$

Here's a summary of the procedure that we just used.

> **HOW TO SOLVE A SYSTEM OF THREE LINEAR EQUATIONS IN THREE VARIABLES**
>
> 1. Use the addition method to eliminate any variable from any pair of equations. (The choice of variable is arbitrary.)
> 2. Use appropriate steps to eliminate the *same variable* from a *different pair* of equations. (If you don't eliminate the same variable, you will still have three unknowns.)
> 3. Solve the resulting system of two equations in two variables.
> 4. Substitute the values obtained in step 3 into one of the three original equations. Solve for the remaining variable.
> 5. Check the solution in all of the original equations.

It is helpful to write all equations in the form $Ax + By + Cz = D$ before using this five-step method.

③ Finding the Solution to a System of Three Linear Equations in Three Variables If Some of the Coefficients Are Zero

If a system of three linear equations in three variables contains one or more equations of the form $Ax + By + Cz = 0$, where one of the values of A, B, or C is zero, then we will slightly modify our approach to solving the system. We will select one equation that contains only two variables. Then we will take the remaining system of two equations and eliminate the variable that was missing in the equation that we selected.

EXAMPLE 3   Solve the system.

$$4x + 3y + 3z = 4 \quad (1)$$
$$3x \quad\quad + 2z = 2 \quad (2)$$
$$2x - 5y \quad\quad = -4 \quad (3)$$

Solution   Note that equation **(2)** has no y-term and equation **(3)** has no z-term. Obviously, that makes our work easier. Let's work with equations **(2)** and **(1)** to obtain an equation that contains only x and y.

Step 1   Multiply equation **(1)** by 2 and equation **(2)** by -3 to obtain the following system.

$$8x + 6y + 6z = 8 \quad (4)$$
$$\underline{-9x \quad\quad - 6z = -6} \quad (5)$$
$$-x + 6y \quad\quad = 2 \quad (6)$$

Step 2   This step is already done, since equation **(3)** has no z-term.

Step 3   Now we can solve the system formed by equations **(3)** and **(6)**.

$$2x - 5y = -4 \quad (3)$$
$$-x + 6y = 2 \quad (6)$$

If we multiply each term of equation **(6)** by 2, we obtain the system

$$2x - 5y = -4$$
$$\underline{-2x + 12y = \quad 4}$$
$$7y = \quad 0 \quad \text{Add.}$$
$$y = \quad 0. \quad \text{Solve for } y.$$

Substituting $y = 0$ in equation **(6)**, we find the following:

$$-x + 6(0) = \quad 2$$
$$-x = \quad 2$$
$$x = -2.$$

Step 4   To find z, we substitute $x = -2$ and $y = 0$ into one of the original equations containing z. Since equation **(2)** has only two variables, let's use it.

$$3x + 2z = 2$$
$$3(-2) + 2z = 2$$
$$2z = 8$$
$$z = 4$$

The solution to the system is $(-2, 0, 4)$.

Check. Verify this solution by substituting these values into equations **(1)**, **(2)**, and **(3)**.

Practice Problem 3   Solve the system.

$$2x + y + z = 11$$
$$4y + 3z = -8$$
$$x - 5y \qquad = 2$$

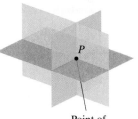

Point of
intersection

(a)

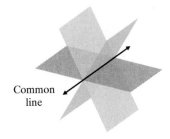

Common
line

(b)

(c)

(d)

(e)

A linear equation in three variables is a plane in three-dimensional space. A system of linear equations in three variables is three planes. The solution to the system is the set of points at which all three planes intersect. There are three possible results. The three planes may intersect at one point. (See figure **(a)** in the margin.)

This point is described by an ordered triple of the form (x, y, z) and lies in each plane. The three planes may intersect at a line. (See figure **(b)** in the margin.) In this case the system has an infinite number of solutions; that is, all the points on the line are solutions to the system.

Finally, all three planes may not intersect at any points. It may mean that all three planes never share any point of intersection, but that two planes intersect. (See figures **(c)**, **(d)**, and **(e)** in the margin.) In all such cases there is no solution to the system of equations.

4.2 EXERCISES

Student Solutions Manual | CD/Video | PH Math Tutor Center | MathXL®Tutorials on CD | MathXL® | MyMathLab® | Interactmath.com

1. Determine whether $(2, 1, -4)$ is a solution to the system.

$$2x - 3y + 2z = -7$$
$$x + 4y - z = 10$$
$$3x + 2y + z = 4$$

2. Determine whether $(-3, 0, 1)$ is a solution to the system.

$$2x + 5y - z = -7$$
$$x - 11y + 4z = 1$$
$$-5x + 8y - 12z = 3$$

3. Determine whether $(-1, 5, 1)$ is a solution to the system.

$$3x + 2y - z = 6$$
$$x - y - 2z = -8$$
$$4x + y + 2z = 5$$

4. Determine whether $(3, 2, 1)$ is a solution to the system.

$$x + y + 2z = 7$$
$$2x + y + z = 9$$
$$3x + 4y - 2z = 13$$

Solve each system.

5.
$$x + y + 2z = 0$$
$$2x - y - z = 1$$
$$x + 2y + 3z = 1$$

6.
$$2x + y + 3z = 2$$
$$x - y + 2z = -4$$
$$x + 3y - z = 1$$

7.
$$x + 2y - 3z = -11$$
$$-2x + y - z = -11$$
$$x + y + z = 6$$

8.
$$-5x + 3y + 2z = 1$$
$$x + y + z = 7$$
$$2x - y + z = 7$$

9.
$$8x - 5y + z = 15$$
$$3x + y - z = -7$$
$$x + 4y + z = -3$$

10.
$$-4x + y - 3z = 2$$
$$5x - 3y + 4z = 1$$
$$3x - 2y + 5z = 1$$

11.
$$x + 4y - z = -5$$
$$-2x - 3y + 2z = 5$$
$$x - \frac{2}{3}y + z = \frac{11}{3}$$

12.
$$x - 4y + 4z = -1$$
$$-x + \frac{y}{2} - \frac{5}{2}z = -3$$
$$-x + 3y - z = 5$$

13.
$$2x + 2z = -7 + 3y$$
$$\frac{3}{2}x + y + \frac{1}{2}z = 2$$
$$x + 4y = 10 + z$$

14.
$$8x - y = 2z$$
$$y = 3x + 4z + 2$$
$$5x + 2z = y - 1$$

15.
$$a = 8 + 3b - 2c$$
$$4a + 2b - 3c = 10$$
$$c = 10 + b - 2a$$

16.
$$a = c - b$$
$$3a - 2b + 6c = 1$$
$$c = 4 - 3b - 7a$$

17. $0.2a + 0.1b + 0.2c = 0.1$
$0.3a + 0.2b + 0.4c = -0.1$
$0.6a + 1.1b + 0.2c = 0.3$

18. $-0.1a + 0.2b + 0.3c = 0.1$
$0.2a - 0.6b + 0.3c = 0.5$
$0.3a - 1.2b - 0.4c = -0.4$

Find the solution for each system of equations. Round your answers to five decimal places.

19. $x - 4y + 4z = -3.72186$
$-x + 3y - z = 5.98115$
$2x - y + 5z = 7.93645$

20. $4x + 2y + 3z = 9$
$9x + 3y + 2z = 3$
$2.987x + 5.027y + 3.867z = 18.642$

Solve each system.

21. $x + y = 1$
$y - z = -3$
$2x + 3y + z = 1$

22. $y - 2z = 5$
$2x + z = -1$
$3x + y - z = 4$

23. $-y + 2z = 1$
$x + y + z = 2$
$-x + 3z = 2$

24. $-2x + y - 3z = 0$
$-2y - z = -1$
$x + 2y - z = 5$

25. $x - 2y + z = 0$
$-3x - y = -6$
$y - 2z = -7$

26. $x + 2z = 0$
$3x + 3y + z = 6$
$6y + 5z = -3$

27. $\dfrac{a}{2} - b + c = 8$

$\dfrac{3}{2}a + b + 2c = 0$

$a + c = 2$

28. $2a + b + \dfrac{c}{3} = -2$

$\dfrac{a}{3} + \dfrac{b}{3} = -1$

$3b + c = 0$

Try to solve the system of equations. Explain your result in each case.

29.
$$\begin{aligned} 2x + y &= -3 \\ 2y + 16z &= -18 \\ -7x - 3y + 4z &= 6 \end{aligned}$$

30.
$$\begin{aligned} 6x - 2y + 2z &= 2 \\ 4x + 8y - 2z &= 5 \\ -2x - 4y + z &= -2 \end{aligned}$$

31.
$$\begin{aligned} 3x + 3y - 3z &= -1 \\ 4x + y - 2z &= 1 \\ -2x + 4y - 2z &= -8 \end{aligned}$$

32.
$$\begin{aligned} -3x + 4y - z &= -4 \\ x + 2y + z &= 4 \\ -12x + 16y - 4z &= -16 \end{aligned}$$

Cumulative Review

33. Solve for x: $|3 - 2x| = 5$

34. Write using scientific notation: 76,300,000

35. Find the standard form of the equation of the line that passes through $(1, 4)$ and $(-2, 3)$.

36. Find the standard form of the equation of the line that is perpendicular to $y = -\frac{2}{3}x + 4$ and passes through $(-4, 2)$.

37. *Australia Ranch* A rancher in Australia has 346 horses, 545 sheep, and 601 cattle. He wants to purchase more animals so that he has 80% more cattle than horses and 74% more sheep than horses. How many animals of each type will he have to buy? How many animals of each type will he have after his purchases? (Round all answers to the nearest whole number.)

38. *Boat Traveling in a River Current* The current in a river moves at a speed of 3.5 miles per hour. A boat travels with the current 48 miles downstream in a total of 3 hours. What would the speed of the boat have been if there had been no current?

How are you doing with your homework assignments in Sections 4.1 to 4.2? Do you feel you have mastered the material so far? Do you understand the concepts you have covered? Before you go further in the textbook, take some time to do each of the following problems.

4.1

1. Solve by the substitution method.

$4x - y = -1$
$3x + 2y = 13$

2. Solve by the addition method.

$3x + 2y = 9$
$5x + 4y = 13$

Find the solution to each system of equations by any method. If there is no single solution to a system, state the reason.

3. $5x - 2y = 27$
$3x - 5y = -18$

4. $7x + 3y = 15$
$\frac{1}{3}x - \frac{1}{2}y = 2$

5. $2x = 3 + y$
$3y = 6x - 9$

6. $0.2x + 0.7y = -1$
$0.5x + 0.6y = -0.2$

7. $6x - 9y = 15$
$-4x + 6y = 8$

4.2

8. Determine whether $(-1, -2, 3)$ is a solution to the system.

$3x + 2y - 5z = -22$
$2x + 3y + 7z = 13$
$x + y + 2z = 6$

Find the solution to each system of equations.

9. $5x - 2y + z = -1$
$3x + y - 2z = 6$
$-2x + 3y - 5z = 7$

10. $2x - y + 3z = -1$
$5x + y + 6z = 0$
$2x - 2y + 3z = -2$

11. $x + y + 2z = 9$
$3x + 2y + 4z = 16$
$2y + z = 10$

12. $x - 2z = -5$
$y - 3z = -3$
$2x - z = -4$

Now turn to page SA-17 for the answers to each of these problems. Each answer also includes a reference to the objective in which the problem is first taught. If you missed any of these problems, you should stop and review the Examples and Practice Problems in the referenced objective. A little review now will help you master the material in the upcoming sections of the text.

1. _____
2. _____
3. _____
4. _____
5. _____
6. _____
7. _____
8. _____
9. _____
10. _____
11. _____
12. _____

Student Learning Objectives

After studying this section, you will be able to:

1 Solve applications requiring the use of a system of two linear equations in two unknowns.

2 Solve applications requiring the use of a system of three linear equations in three unknowns.

Allosaurus

1 Solving Applications Requiring the Use of a System of Two Linear Equations in Two Unknowns

We will now examine how a system of linear equations can assist us in solving applied exercises.

EXAMPLE 1   For the paleontology lecture on campus, advance tickets cost $5 and tickets at the door cost $6. The ticket sales this year came to $4540. The department chairman wants to raise prices next year to $7 for advance tickets and $9 for tickets at the door. He said that if exactly the same number of people attend next year, the ticket sales at these new prices will total $6560. If he is correct, how many tickets were sold in advance this year? How many tickets were sold at the door?

Solution

1. ***Understand the problem.*** Since we are looking for the number of tickets sold, we let

$$x = \text{number of tickets bought in advance and}$$
$$y = \text{number of tickets bought at the door.}$$

2. ***Write a system of two equations in two unknowns.*** If advance tickets cost $5, then the total sales will be $5x$; similarly, total sales of door tickets will be $6y$. Since the total sales of both types of tickets was $4540, we have

$$5x + 6y = 4540.$$

By the same reasoning, we have

$$7x + 9y = 6560.$$

Thus, our system is as follows:

$$5x + 6y = 4540 \quad \textbf{(1)}$$
$$7x + 9y = 6560. \quad \textbf{(2)}$$

3. ***Solve the system of equations and state the answer.*** We will multiply each term of equation **(1)** by -3 and each term of equation **(2)** by 2 to obtain the following equivalent system.

$$-15x - 18y = -13{,}620 \quad \textbf{(3)}$$
$$\underline{14x + 18y = 13{,}120} \quad \textbf{(4)}$$
$$-x = -500$$

Therefore, $x = 500$. Substituting $x = 500$ into equation **(1)**, we have the following:

$$5(500) + 6y = 4540$$
$$6y = 2040$$
$$y = 340.$$

Thus, 500 advance tickets and 340 door tickets were sold.

4. ***Check.*** We need to check our answers. Do they seem reasonable?

Would 500 advance tickets at $5 and 340 door tickets at $6 yield $4540?

$$5(500) + 6(340) \stackrel{?}{=} 4540$$
$$2500 + 2040 \stackrel{?}{=} 4540$$
$$4540 = 4540 \quad \checkmark$$

Would 500 advance tickets at $7 and 340 door tickets at $9 yield $6560?

$$7(500) + 9(340) \stackrel{?}{=} 6560$$
$$3500 + 3060 \stackrel{?}{=} 6560$$
$$6560 = 6560 \quad \checkmark$$

Practice Problem 1   Coach Perez purchased baseballs at $6 each and bats at $21 each last week for the college baseball team. The total cost of the purchase was $318. This week he noticed that the same items are on sale. Baseballs are now $5 each and bats are $17. He found that if he made the same purchase this week, it would cost only $259. How many baseballs and how many bats did he buy last week?

NOTE TO STUDENT: Fully worked-out solutions to all of the Practice Problems can be found at the back of the text starting at page SP-1

EXAMPLE 2   An electronics firm makes two types of switching devices. Type A takes 4 minutes to make and requires $3 worth of materials. Type B takes 5 minutes to make and requires $5 worth of materials. When the production manager reviewed the latest batch, he found that it took 35 hours to make these switches with a materials cost of $1900. How many switches of each type were produced for this latest batch?

Solution

1. **Understand the problem.** We are given a lot of information, but the major concern is to find out how many of the type A devices and the type B devices were produced. This becomes our starting point to define the variables we will use.

   Let A = the number of type A devices produced and

         B = the number of type B devices produced.

2. **Write a system of two equations.** How should we construct the equations? What relationships exist between our variables (or unknowns)? According to the problem, the devices are related by time and by cost. So we set up one equation in terms of time (minutes in this case) and one in terms of cost (dollars). Each type A took 4 minutes to make, each type B took 5 minutes to make, and the total time was 2100 minutes. Each type A used $3 worth of materials, each type B used $5 worth of materials, and the total material cost was $1900. We can gather this information in a table. Making a table will help us form the equations.

	Type *A* Devices	Type *B* Devices	Total
Number of Minutes	4A	5B	2100
Cost of Materials	3A	5B	1900

   $$4A + 5B = 2100$$
   $$3A + 5B = 1900$$

   Therefore, we have the following system.

   $$4A + 5B = 2100 \quad \textbf{(1)}$$
   $$3A + 5B = 1900 \quad \textbf{(2)}$$

3. **Solve the system of equations and state the answers.** Multiplying equation **(2)** by -1 and adding the equations, we find the following:

   $$4A + 5B = 2100$$
   $$\underline{-3A - 5B = -1900}$$
   $$A = 200.$$

   Substituting $A = 200$ into equation **(1)**, we have the following:

   $$800 + 5B = 2100$$
   $$5B = 1300$$
   $$B = 260.$$

   Thus, 200 type A devices and 260 type B devices were produced.

4. **Check.** If each type A requires 4 minutes and each type B requires 5 minutes, does this amount to a total time of 2100 minutes?

$$4A + 5B = 2100$$
$$4(200) + 5(260) \overset{?}{=} 2100$$
$$800 + 1300 \overset{?}{=} 2100$$
$$2100 = 2100 \quad \checkmark$$

If each type A costs \$3 and each type B costs \$5, does this amount to a total cost of \$1900?

$$3A + 5B = 1900$$
$$3(200) + 5(260) \overset{?}{=} 1900$$
$$600 + 1300 \overset{?}{=} 1900$$
$$1900 = 1900 \quad \checkmark$$

NOTE TO STUDENT: *Fully worked-out solutions to all of the Practice Problems can be found at the back of the text starting at page SP-1*

Practice Problem 2   A furniture company makes both small and large chairs. It takes 30 minutes of machine time and 1 hour and 15 minutes of labor to build the small chair. The large chair requires 40 minutes of machine time and 1 hour and 20 minutes of labor. The company has 57 hours of labor time and 26 hours of machine time available each day. If all available time is used, how many chairs of each type can the company make?

When we encounter motion problems involving rate, time, or distance, it is useful to recall the formula $D = RT$ or distance = (rate)(time).

EXAMPLE 3   An airplane travels between two cities that are 1500 miles apart. The trip against the wind takes 3 hours. The return trip with the wind takes $2\frac{1}{2}$ hours. What is the speed of the plane in still air (in other words, how fast would the plane travel if there were no wind)? What is the speed of the wind?

Solution

1. **Understand the problem.** Our unknowns are the speed of the plane in still air and the speed of the wind.

Let

$$x = \text{the speed of the plane in still air and}$$
$$y = \text{the speed of the wind.}$$

Let's make a sketch to help us see how these speeds are related to one another. When we travel against the wind, the wind is slowing us down. Since the wind speed opposes the plane's speed in still air, we must subtract: $x - y$.

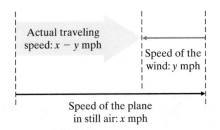

When we travel with the wind, the wind is helping us travel forward. Thus, the wind speed is added to the plane's speed in still air, and we add: $x + y$.

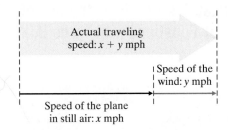

2. ***Write a system of two equations.*** To help us write our equations, we organize the information in a chart. The chart will be based on the formula $RT = D$, which is (rate)(time) = distance.

	R	·	T	=	D
Flying against the wind	$x - y$		3		1500
Flying with the wind	$x + y$		2.5		1500

Using the rows of the chart, we obtain a system of equations.

$$(x - y)(3) = 1500$$
$$(x + y)(2.5) = 1500$$

If we remove the parentheses, we will obtain the following system.

$$3x - 3y = 1500 \quad \textbf{(1)}$$
$$2.5x + 2.5y = 1500 \quad \textbf{(2)}$$

3. ***Solve the system of equations and state the answer.*** It will be helpful to clear equation **(2)** of decimal coefficients. Although we could multiply each term by 10, doing so will result in large coefficients on x and y. For this equation, multiplying by 2 is a better choice.

$$3x - 3y = 1500 \quad \textbf{(1)}$$
$$5x + 5y = 3000 \quad \textbf{(3)}$$

If we multiply equation **(1)** by 5 and equation **(3)** by 3, we will obtain the following system.

$$15x - 15y = 7500$$
$$\underline{15x + 15y = 9000}$$
$$30x \quad\quad = 16{,}500$$
$$x = 550$$

Graphing Calculator

Exploration

A visual interpretation of two equations in two unknowns is sometimes helpful. Study Example 3. Graph the two equations.

$$3x - 3y = 1500$$
$$2.5x + 2.5y = 1500$$

What is the significance of the point of intersection? If you were an air traffic controller, how would you interpret the linear equation $3x - 3y = 1500$? Why would this be useful? How would you interpret $2.5x + 2.5y = 1500$? How would this be useful?

Substituting this result in equation **(1)**, we obtain the following:

$$3(550) - 3y = 1500$$
$$1650 - 3y = 1500$$
$$-3y = -150$$
$$y = 50.$$

Thus, the speed of the plane in still air is 550 miles per hour, and the speed of the wind is 50 miles per hour.

4. Check. The check is left to the student.

Practice Problem 3   An airplane travels west from city A to city B against the wind. It takes 3 hours to travel 1950 kilometers. On the return trip the plane travels east from city B to city C, a distance of 1600 kilometers in a time of 2 hours. On the return trip the plane travels with the wind. What is the speed of the plane in still air? What is the speed of the wind?

2 Solving Applications Requiring the Use of a System of Three Linear Equations with Three Unknowns

EXAMPLE 4   A trucking firm has three sizes of trucks. The biggest truck holds 10 tons of gravel, the next size holds 6 tons, and the smallest holds 4 tons. The firm's manager has fifteen trucks available to haul 104 tons of gravel. However, to reduce fuel costs she wants to use two more of the fuel-efficient 10-ton trucks than the 6-ton trucks. Her assistant tells her that she has two more 10-ton trucks than 6-ton trucks available. How many trucks of each type should she use?

Solution

1. **Understand the problem.** Since we need to find three things (the numbers of 10-ton trucks, 6-ton trucks, and 4-ton trucks), it would be helpful to have three variables. Let

$$x = \text{the number of 10-ton trucks used,}$$
$$y = \text{the number of 6-ton trucks used, and}$$
$$z = \text{the number of 4-ton trucks used.}$$

2. **Write a system of three equations.** We know that fifteen trucks will be used; hence, we have the following:

$$x + y + z = 15. \quad \textbf{(1)}$$

How can we get our second equation? Well, we also know the *capacity* of each truck type, and we know the total tonnage to be hauled. The first type of truck hauls 10 tons, the second type 6 tons, and the third type 4 tons, and the total tonnage is 104 tons. Hence, we can write the following:

$$10x + 6y + 4z = 104. \quad \textbf{(2)}$$

We still need one more equation. What other given information can we use? The problem states that the manager wants to use two more 10-ton trucks than 6-ton trucks. Thus, we have the following:

$$x = 2 + y. \quad \textbf{(3)}$$

(We could also have written $x - y = 2$.) Hence, our system of equations is as follows:

$$x + y + z = 15 \quad \textbf{(1)}$$
$$10x + 6y + 4z = 104 \quad \textbf{(2)}$$
$$x - y = 2. \quad \textbf{(3)}$$

3. *Solve the system of equations and state the answers.* Equation **(3)** doesn't contain the variable z. Let's work with equations **(1)** and **(2)** to eliminate z. First, we multiply equation **(1)** by -4 and add it to equation **(2)**.

$$
\begin{array}{ll}
-4x - 4y - 4z = -60 & \textbf{(4)} \\
\underline{10x + 6y + 4z = 104} & \textbf{(2)} \\
6x + 2y = 44 & \textbf{(5)}
\end{array}
$$

Make sure you understand how we got equation **(5)**. Dividing each term of equation **(5)** by 2 and adding to equation **(3)** gives the following:

$$
\begin{array}{ll}
3x + y = 22 & \textbf{(6)} \\
\underline{x - y = 2} & \textbf{(3)} \\
4x = 24 & \\
x = 6. &
\end{array}
$$

For $x = 6$, equation **(3)** yields the following:

$$
6 - y = 2
$$
$$
4 = y.
$$

Now we substitute the known x- and y-values into equation **(1)**.

$$
6 + 4 + z = 15
$$
$$
z = 5
$$

Thus, the manager needs six 10-ton trucks, four 6-ton trucks, and five 4-ton trucks.

4. *Check.* The check is left to the student.

Practice Problem 4   A factory uses three machines to wrap boxes for shipment. Machines A, B, and C can wrap 260 boxes in 1 hour. If machine A runs 3 hours and machine B runs 2 hours, they can wrap 390 boxes. If machine B runs 3 hours and machine C runs 4 hours, 655 boxes can be wrapped. How many boxes per hour can each machine wrap?

NOTE TO STUDENT: Fully worked-out solutions to all of the Practice Problems can be found at the back of the text starting at page SP-1

Developing Your Study Skills

Applications or Word Problems

Applications, or word problems, are the very life of mathematics! They are the reason for doing mathematics because they teach you how to put into use the mathematical skills you have developed. Learning mathematics without ever doing word problems is similar to learning all the skills of a sport without ever playing a game or learning all the notes on an instrument without ever playing a song.

The key to success is practice. Make yourself work through as many exercises as you can. You may not be able to do them all correctly at first, but keep trying. Do not give up when you reach a difficult exercise. If you cannot solve it, just try another one. Then go back and try the "difficult" one again later.

A misconception among students when they begin studying word problems is that each one is different. At first the exercises may seem this way, but as you practice more and more, you will begin to see the similarities, the different "types." You will see patterns, which will enable you to solve exercises of a given type more easily.

Math XL

Applications

Use a system of two linear equations to solve each exercise.

1. The sum of 2 numbers is 87. If twice the smaller number is subtracted from the larger number, the result is 12. Find the two numbers.

2. The difference between 2 numbers is 15. If twice the larger number is added to 3 times the smaller, the result is 90. Find the 2 numbers.

3. *Temporary Employment Agency* An employment agency specializing in temporary construction help pays heavy equipment operators $140 per day and general laborers $90 per day. If thirty-five people were hired and the payroll was $3950, how many heavy equipment operators were employed? How many laborers?

4. *Broadway Ticket Prices* A Broadway performance of *The Phantom of the Opera* had a paid attendance of 320 people. Balcony tickets cost $42, and orchestra tickets cost $64. Ticket sales receipts totaled $16,630. How many tickets of each type were sold?

5. *Amtrak Train Tickets* Ninety-eight passengers rode in an Amtrak train from Boston to Denver. Tickets for regular coach seats cost $120. Tickets for sleeper car seats cost $290. The receipts for the trip totaled $19,750. How many passengers purchased each type of ticket?

6. *Farm Operations* The Tupper Farm has 450 acres of land allotted for raising corn and wheat. The cost to cultivate corn is $42 per acre. The cost to cultivate wheat is $35 per acre. The Tuppers have $16,520 available to cultivate these crops. How many acres of each crop should the Tuppers plant?

7. *Computer Training for Managers* A large company wants to train its managers to use new word processing and spreadsheet software. Experienced managers can learn the word processor in 2 hours and the spreadsheet in 3 hours. Newly hired managers require 5 hours to learn the word processor and 8 hours to learn the spreadsheet. The company can afford to pay for 140 hours of word processing instruction and 215 hours of spreadsheet instruction. How many of each type of manager can the company train?

8. *Radar Detector Manufacturing* Ventex makes auto radar detectors. Ventex has found that its basic model requires 3 hours of manufacturing for the inside components and 2 hours for the housing and controls. Its advanced model requires 5 hours to manufacture the inside components and 3 hours for the housing and controls. This week, the production division has available 1050 hours for producing inside components and 660 hours for housing and controls. How many detectors of each type can be made?

9. *Farm Management* A farmer has several packages of fertilizer for his new grain crop. The old packages contain 50 pounds of long-term-growth supplement and 60 pounds of weed killer. The new packages contain 65 pounds of long-term-growth supplement and 45 pounds of weed killer. Using past experience, the farmer estimates that he needs 3125 pounds of long-term-growth supplement and 2925 pounds of weed killer for the fields. How many old packages of fertilizer and how many new packages of fertilizer should he use?

10. *Hospital Dietician* A staff hospital dietician has two prepackaged mixtures of vitamin additives available for patients. Mixture 1 contains 5 grams of vitamin C and 3 grams of niacin; mixture 2 contains 6 grams of vitamin C and 5 grams of niacin. On an average day she needs 87 grams of niacin and 117 grams of vitamin C. How many packets of each mixture will she need?

11. *Coffee and Snack Expenses* On Monday, Harold picked up three doughnuts and four large coffees for the office staff. He paid $4.91. On Tuesday, Melinda picked up five doughnuts and six large coffees for the office staff. She paid $7.59. What is the cost of one doughnut? What is the cost of one large coffee?

12. *Advertising Costs* A local department store is preparing four-color sales brochures to insert into the *Salem Evening News*. The printer has a fixed charge to set up the printing of the brochure and a specific per-copy amount for each brochure printed. He quoted a price of $1350 for printing five thousand brochures and a price of $1750 for printing seven thousand brochures. What is the fixed charge to set up the printing? What is the per-copy cost for printing a brochure?

13. *Airspeed* Against the wind a small plane flew 210 miles in 1 hour and 10 minutes. The return trip took only 50 minutes. What was the speed of the wind? What was the speed of the plane in still air?

14. *Aircraft Operation* Against the wind a commercial airline in South America flew 630 miles in 3 hours and 30 minutes. With a tailwind the return trip took 3 hours. What was the speed of the wind? What was the speed of the plane in still air?

15. *Fishing and Boating* Don Williams uses his small motorboat to go 8 miles upstream to his favorite fishing spot. Against the current, the trip takes $\frac{2}{3}$ hour. With the current, the trip takes $\frac{1}{2}$ hour. How fast can the boat travel in still water? What is the speed of the current?

16. *Canoe Trip* It took Linda and Alice 4 hours to travel 24 miles downstream by canoe on Indian River. The next day they traveled for 6 hours upstream for 18 miles. What was the rate of the current? What was their average speed in still water?

17. *Basketball* Tim Duncan scored 32 points in an NBA basketball game without scoring any 3-point shots. He scored 21 times. He made several free throws worth 1 point each and several regular shots from the floor, which were worth 2 points each. How many free throws did he make? How many 2-point shots did he make?

18. *Basketball* Shaquille O'Neal scored 38 points in a recent basketball game. He scored no free throws, but he made a number of 2-point shots and 3-point shots. He scored 16 times. How many 2-point baskets did he make? How many 3-point baskets did he make?

19. *Telephone Charges* Nick's telephone company charges $0.05 per minute for weekend calls and $0.08 for calls made on weekdays. This month Nick was billed for 625 minutes. The charge for these minutes was $43.40. How many minutes did he talk on weekdays and how many minutes did he talk on the weekend?

20. *Office Supply Costs* A new catalog from an office supply company shows that some of its prices will increase next month. This month, copier paper costs $2.70 per ream and printer cartridges cost $15.50. If Chris submits his order this month, the cost will be $462. Next month, when paper will cost $3.00 per ream and the cartridges will cost $16, his order would cost $495. How many reams of paper and how many printer cartridges are in his order?

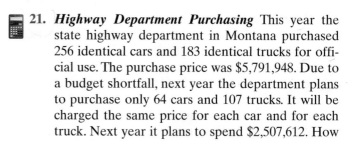

21. *Highway Department Purchasing* This year the state highway department in Montana purchased 256 identical cars and 183 identical trucks for official use. The purchase price was $5,791,948. Due to a budget shortfall, next year the department plans to purchase only 64 cars and 107 trucks. It will be charged the same price for each car and for each truck. Next year it plans to spend $2,507,612. How much does the department pay for each car and for each truck?

22. *Concert Ticket Prices* A recent concert at Gordon College had a paid audience of 987 people. Advance tickets were $9.95 and tickets at the door were $12.95. A total of $10,738.65 was collected in ticket sales. How many of each type of ticket were sold?

Use a system of three linear equations to solve Exercises 23–30.

23. *Bookstore Supplies* Zoe bought 12 items at the college bookstore. The items cost a total of $20.70. The pens cost $1.20 each, the notebooks were $3.00 each and the highlighters cost $0.90. She bought 2 more pens than highlighters. How many of each item did she buy?

24. *Student Survey* A survey was conducted among college seniors who owned cars. A total of 235 students was surveyed. Three times as many students owned compact cars as SUVs. Eleven more students owned fullsize cars than owned compact cars. How many of each type of vehicle were owned by those seniors surveyed?

25. *High School Play* A total of three hundred people attended the high school play. The admission prices were $5 for adults, $3 for high school students, and $2 for any children not yet in high school. The ticket sales totaled $1010. The school principal suggested that next year they raise prices to $7 for adults, $4 for high school students, and $3 for children not yet in high school. He said that if exactly the same number of people attend next year, the ticket sales at the higher prices will total $1390. How many adults, high school students, and children not yet in high school attended this year?

26. *CPR Training* The college conducted a CPR training class for students, faculty, and staff. Faculty were charged $10, staff were charged $8, and students were charged $2 to attend the class. A total of four hundred people came. The receipts for all who attended totaled $2130. The college president remarked that if he had charged faculty $15 and staff $10 and let students come free, the receipts this year would have been $2425. How many students, faculty, and staff came to the CPR training class?

27. *City Subway Token Costs* A total of twelve thousand passengers normally ride the green line of the MBTA during the morning rush hour. The token prices for a ride are $0.25 for children under 12, $1 for adults, and $0.50 for senior citizens, and the revenue from these riders is $10,700. If the token prices were raised to $0.35 for children under 12 and $1.50 for adults, and the senior citizen price were unchanged, the expected revenue from these riders would be $15,820. How many riders in each category normally ride the green line during the morning rush hour?

28. *Commission Sales* The owner of Danvers Ford found that he sold a total of 520 cars, Freestars, and Explorers last year. He paid the sales staff a commission of $100 for every car, $200 for every Freestar, and $300 for every Explorer sold. The total of these commissions last year was $87,000. In the coming year he is contemplating an increase so that the commission will be $150 for every car and $250 for every Freestar, with no change in the commission for Explorer sales. If the sales are the same this year as they were last year, the commissions will total $106,500. How many vehicles in each category were sold last year?

29. *Pizza Costs* The Essex House of Pizza delivered twenty pepperoni pizzas to Gordon College on the first night of final exams. The cost of these pizzas totaled $181. A small pizza costs $5 and contains 3 ounces of pepperoni. A medium pizza costs $9 and contains 4 ounces of pepperoni. A large pizza costs $12 and contains 5 ounces of pepperoni. The owner of the pizza shop used 5 pounds 2 ounces of pepperoni in making these twenty pizzas. How many pizzas of each size were delivered to Gordon College?

30. *Roast Beef Sandwich Costs* One of the favorite meeting places for local college students is Nick's Roast Beef in Beverly, Massachusetts. Last night from 8 P.M. to 9 P.M. Nick served twenty-four roast beef sandwiches. He sliced 15 pounds 8 ounces of roast beef to make these sandwiches and collected $82 for them. The medium roast beef sandwich has 6 ounces of beef and costs $2.50. The large roast beef sandwich has 10 ounces of beef and costs $3. The extra large roast beef sandwich has 14 ounces of beef and costs $4.50. How many of each size of roast beef sandwich did Nick sell from 8 P.M. to 9 P.M.?

31. *Packing Fruit* Sunshine Fruit Company packs three types of gift boxes of oranges, pink grapefruit, and white grapefruit. Box *A* contains 10 oranges, 3 pink grapefruit, and 3 white grapefruit. Box *B* contains 5 oranges, 2 pink grapefruit, and 3 white grapefruit. Box *C* contains 4 oranges, 1 pink grapefruit, and 2 white grapefruit. The shipping manager has available 51 oranges, 16 pink grapefruit, and 23 white grapefruit. How many gift boxes of each type can she prepare?

32. *Shipping Packages* A company packs three types of packages in a large shipping box. Package type *A* weighs 0.5 pound, package type *B* weighs 0.25 pound, and package type *C* weighs 1.0 pound. Seventy-five packages weighing a total of 42 pounds were placed into the box. The number of type *A* packages was three less than the total number of types *C* and *B* combined. How many packages of each type were placed into the box?

Solve each of the following exercises by using three equations and three unknowns. Recall that the sum of the measures of the angles of a triangle is 180°.

▲ **33.** *Jet Airliner Construction* An engineer constructed a triangular piece of metal for the wheel housing of a jet airliner. He measured each angle of the piece of metal. The sum of the measures of the first angle and the second angle is half the measure of the third angle. The sum of the measures of the second angle and the third angle is eight times the measure of the first angle. Find the measure of each angle.

▲ **34.** *Land Survey* In Caribou, Maine, Nancy Alberto owns a triangular field. She and her husband measured each angle of the field. The sum of the measures of the first and second angles is 100° more than the measure of the third angle. If the measure of the third angle is subtracted from the measure of the second angle, the result is 30° greater than the measure of the first angle. Find the measure of each angle.

To Think About

Use a system of four linear equations and four unknowns to solve the following problem.

35. *Clinical Test* A scientist at the University of Chicago is performing an experiment to determine how to increase the life span of mice through a controlled diet. The mice need 134 grams of carbohydrates, 150 grams of protein, 178 grams of fat, and 405 grams of moisture during the length of the experiment. The food is available in four packets, as shown in the table. How many packets of each type should the scientist use?

	Packet			
Contents	A	B	C	D
Carbohydrates	42	20	0	10
Protein	20	10	20	0
Fat	34	0	10	20
Moisture	50	35	30	40

Cumulative Review

Solve for the variable indicated.

36. $\frac{1}{3}(4 - 2x) = \frac{1}{2}x - 3$

37. $0.06x + 0.15(0.5 - x) = 0.04$

38. $2(y - 3) - (2y + 4) = -6y$

39. Solve for x: $6a(2x - 3y) = 7ax - 3$

4.4 SYSTEMS OF LINEAR INEQUALITIES

1 Graphing a System of Linear Inequalities

We learned how to graph a linear inequality in two variables in Section 3.4. We call two linear inequalities in two variables a **system of linear inequalities in two variables.** We now consider how to graph such a system. The solution to a system of inequalities is the intersection of the solution sets of individual inequalities.

Student Learning Objective

After studying this section, you will be able to:

1 Graph a system of linear inequalities.

EXAMPLE 1 Graph the solution of the system.

$$y \leq -3x + 2$$
$$-2x + y \geq -1$$

Solution

In this example, we will first graph each inequality separately. The graph of $y \leq -3x + 2$ is the region on or below the line $y = -3x + 2$.

The graph of $-2x + y \geq -1$ consists of the region on or above the line $-2x + y = -1$.

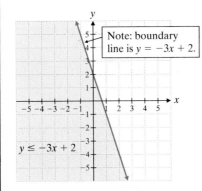

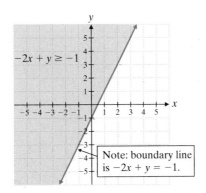

We will now place these graphs on one rectangular coordinate system. The darker shaded region is the intersection of the two graphs. Thus, the solution to the system of two inequalities is the darker shaded region and its boundary lines.

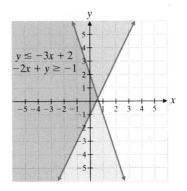

PRACTICE PROBLEM 1

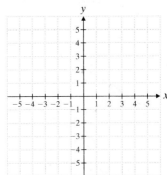

Practice Problem 1 Graph the solution of the system.

$$-2x + y \leq -3$$
$$x + 2y \geq 4$$

NOTE TO STUDENT: Fully worked-out solutions to all of the Practice Problems can be found at the back of the text starting at page SP-1

Usually we sketch the graphs of the individual inequalities on one set of axes. We will illustrate that concept with the following example.

229

EXAMPLE 2 Graph the solution of the system.

$$y < 4$$

$$y > \frac{3}{2}x - 2$$

Solution The graph of $y < 4$ is the region below the line $y = 4$. It does not include the line since we have the $<$ symbol. Thus, we use a dashed line to indicate that the boundary line is not part of the answer. The graph of $y > \frac{3}{2}x - 2$ is the region above the line $y = \frac{3}{2}x - 2$. Again, we use the dashed line to indicate that the boundary line is not part of the answer. The final solution is the darker shaded region. The solution does *not* include the dashed boundary lines.

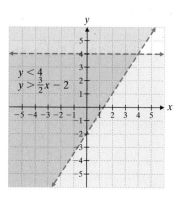

PRACTICE PROBLEM 2

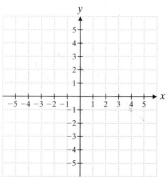

NOTE TO STUDENT: *Fully worked-out solutions to all of the Practice Problems can be found at the back of the text starting at page SP-1*

Practice Problem 2 Graph the solution of the system.

$$y > -1$$

$$y < -\frac{3}{4}x + 2$$

There are times when we require the exact location of the point where two boundary lines intersect. In these cases the boundary points are labeled on the final sketch of the solution.

EXAMPLE 3 Graph the solution to the following system of inequalities. Find the coordinates of any point where boundary lines intersect.

$$x + y \le 5$$

$$x + 2y \le 8$$

$$x \ge 0$$

$$y \ge 0$$

Solution The graph of $x + y \le 5$ is the region on and below the line $x + y = 5$. The graph of $x + 2y \le 8$ is the region on and below the line $x + 2y = 8$. We solve the system containing the equations $x + y = 5$ and $x + 2y = 8$ to find that their point of intersection is $(2, 3)$. The graph of $x \ge 0$ is the y-axis and all the region to the right of the y-axis. The graph of $y \ge 0$ is the x-axis and all the region above the x-axis. Thus, the solution to the system is the shaded region and its boundary lines. The boundary lines intersect at four points.

These points are called the **vertices** of the solution. Thus, the vertices of the solution are $(0, 0)$, $(0, 4)$, $(2, 3)$, and $(5, 0)$.

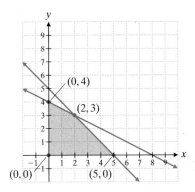

Practice Problem 3 Graph the solution to the system of inequalities. Find the vertices of the solution.

$$x + y \leq 6$$
$$3x + y \leq 12$$
$$x \geq 0$$
$$y \geq 0$$

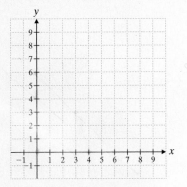

Graphing Calculator

 Graphing Systems of Linear Inequalities

On some graphing calculators, you can graph a system of linear inequalities. To graph the system in Example 1, first rewrite it as follows:

$$y \leq -3x + 2$$
$$y \geq 2x - 1.$$

Enter each expression into the Y = editor of your graphing calculator and then select the appropriate direction for shading.
 Display:

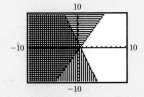

 Notice that one inequality is shaded vertically and the other is shaded horizontally. The intersection is the solution. Note also that the graphing calculator will not indicate whether the boundary of the solution region is included in the solution.

Developing Your Study Skills

How to Review for an Exam

Reviewing adequately for an exam enables you to bring together the concepts you have learned over several sections. For your review, you will need to:

1. Reread your textbook. Make a list of any terms, rules, or formulas you need to know for the exam. Be sure you understand them all.

2. Reread your notes. Go over returned homework and quizzes. Redo the exercises you missed.

3. Practice some of each type of exercise covered in the chapter(s) you are to be tested on.

4. Use the end-of-chapter materials provided in your textbook. Read carefully through the Chapter Organizer. Take the Chapter Test. When you are finished, check your answers. Redo any exercises you missed.

5. Get help if any concepts give you difficulty.

Student Solutions Manual · CD/Video · PH Math Tutor Center · MathXL®Tutorials on CD · MathXL® · MyMathLab® · Interactmath.com

Verbal and Writing Skills

1. In the graph of the system $y > 3x + 1$ and $y < -2x + 5$, would the boundary lines be solid or dashed? Why?

2. In the graph of the system $y \geq -6x + 3$ and $y \leq -4x - 2$, would the boundary lines be solid or dashed? Why?

3. Stephanie wanted to know if the point $(3, -4)$ lies in the region that is a solution for $y < -2x + 3$ and $y > 5x - 3$. How could she determine if this is true?

4. John wanted to know if the point $(-5, 2)$ lies in the region that is a solution for $x + 2y < 3$ and $-4x + y > 2$. How could he determine if this is true?

Graph the solution for each of the following systems.

5. $y \geq 2x - 1$
 $x + y \leq 6$

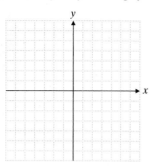

6. $y \geq x - 3$
 $x + y \geq 2$

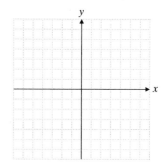

7. $y \geq -4x$
 $y \geq 3x - 2$

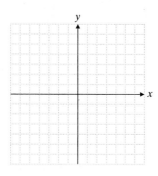

8. $y \geq x$
 $y \leq -x + 2$

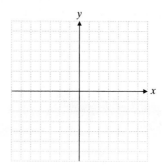

9. $y \geq 2x - 3$
 $y \leq \dfrac{2}{3}x$

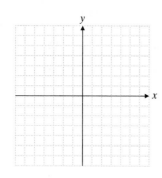

10. $y \leq \dfrac{1}{2}x - 3$
 $y \geq -\dfrac{1}{2}x$

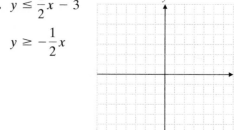

11. $x - y \geq -1$
   $-3x - y \leq \ \ 4$

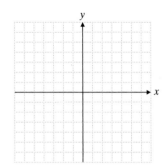

12. $3x - y \leq 3$
    $-x + y \leq 4$

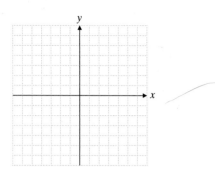

13. $x + 2y < 6$
      $y < 3$

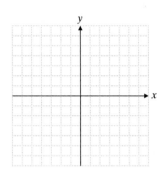

14. $x + 3y < 12$
      $y > 4$

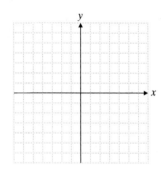

15. $y < \ \ 4$
   $x > -2$

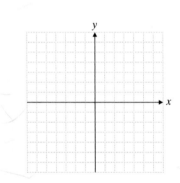

16. $y > -3$
   $x < \ \ 2$

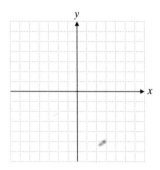

Mixed Practice

Graph the solution for each of the following systems.

17. $x - 4y \geq -4$
    $3x + y \leq \ \ 3$

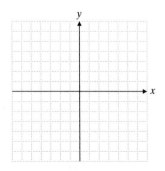

18. $5x - 2y \leq \ \ 10$
     $x - \ \ y \geq -1$

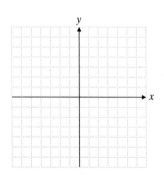

19. $3x + 2y < 6$
$3x + 2y > -6$

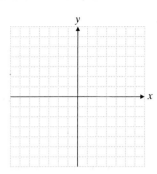

20. $2x - y < 2$
$2x - y > -2$

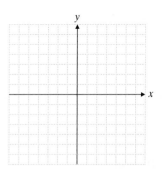

Graph the solution to the following systems of inequalities. Find the vertices of the solution.

21. $x + y \leq 5$
$2x - y \geq 1$

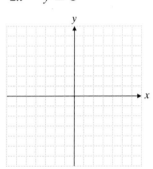

22. $x + y \geq 2$
$y + 4x \leq -1$

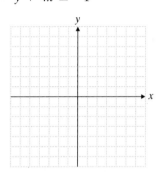

23. $x + 3y \leq 12$
$y < x$

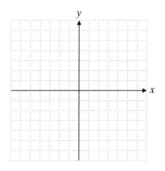

24. $x + 2y \leq 4$
$y < -x$

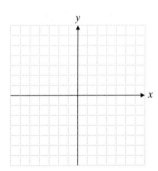

25. $x + y \geq 1$
$x - y \geq 1$
$x \geq 3$

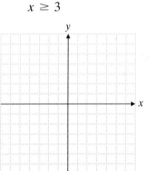

26. $x - y \leq 2$
$x + y \leq 2$
$x \geq -2$

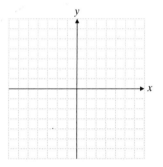

To Think About

Graph the region determined by each of the following systems.

27. $y \leq 3x + 6$
$4y + 3x \leq 3$
$x \geq -2$
$y \geq -3$

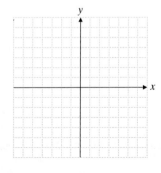

28. $-x + y \leq 100$
$x + 3y \leq 150$
$x \geq -80$
$y \geq 20$

(Hint: Use a
scale of each
square = 20 units
on both axes.)

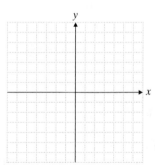

Applications

Hint: In Exercises 29 and 30, if the coordinates of a boundary point contain fractions, it is wise to obtain the point of intersection algebraically rather than graphically.

29. ***Hospital Staffing Levels*** The equation that represents the proper level of medical staffing in the cardiac care unit of a local hospital is $N \le 2D$, where N is the number of nurses on duty and D is the number of doctors on duty. In order to control costs, the equation $4N + 3D \le 20$ is appropriate. The number of doctors and nurses on duty at any time cannot be negative, so $N \ge 0$ and $D \ge 0$.

  (a) Graph the region satisfying all of the medical requirements for the cardiac care unit. Use the following special graph grid, where D is measured on the horizontal axis and N is measured on the vertical axis.

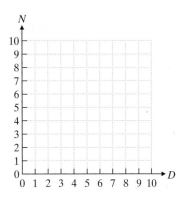

  (b) If there are three doctors and two nurses on duty in the cardiac care unit, are all of the medical requirements satisfied?

  (c) If there is one doctor and four nurses on duty in the cardiac care unit, are all of the medical requirements satisfied?

30. ***Traffic Control*** The equation that represents the proper traffic control and emergency vehicle response availability in the city of Salem is $P + 3F \le 18$, where P is the number of police cars on active duty and F is the number of fire trucks that have left the firehouse and are involved in a response to a call. In order to comply with staffing limitations, the equation $4P + F \le 28$ is appropriate. The number of police cars on active duty and the number of fire trucks that have left the firehouse cannot be negative, so $P \ge 0$ and $F \ge 0$.

  (a) Graph the regions satisfying all of the availability and staffing limitation requirements for the city of Salem. Use the following special graph grid, where P is measured on the horizontal axis and F is measured on the vertical axis.

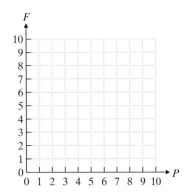

  (b) If four police cars are on active duty and four fire trucks have left the firehouse in response to a call, are all of the requirements satisfied?

  (c) If two police cars are on active duty and six fire trucks have left the firehouse in response to a call, are all of the requirements satisfied?

Cumulative Review

31. Find the slope of the line passing through $(3, -4)$ and $(-1, -2)$.

32. What is the slope and the y-intercept of the line defined by the equation $3x + 4y = -8$?

33. Evaluate for $x = 2$, $y = -1$:
$-3x^2y - x^2 + 5y^2$

34. Simplify: $2x - 2\{y + 3(x - y)\}$

35. *Movie Theatre Receipts* The Cape Cod Cinema took in $23,400 on two rainy days and five sunny days. The next week the cinema took in $25,800 on four rainy days and three sunny days. What is the average amount of money taken in when a film is shown at the Cape Cod Cinema on a rainy day? On a sunny day?

36. *Building Bicycle Trails* Illinois is establishing a number of new bicycle trails throughout the state. When a group of volunteers worked three days and a group of experienced professionals worked four days, they were able to establish 389 feet of bicycle trails. When the same group of volunteers worked five days and the same group of experienced professionals worked seven days, they were able to establish 670 feet of bicycle trails. How many feet per day are established by the group of volunteers? By the group of experienced professionals?

37. *Commission Sales* Hector sells televisions at a local department store. He earns $200 per week plus a commission of 5% on his total sales. His brother Fernando sells automobile tires at a tire store. His salary is $100 per week plus a commission of 8% on his total sales. After they had both worked several weeks, they discovered that they had sold the same dollar value in sales. However, their total earnings were quite different. Hector had earned $7400, and Fernando had earned $9200. How many weeks had each worked? What was the dollar value of the amount of sales that each had made?

38. *Office Lunch Expenses* Two weeks ago Larry went from his office to Nick's Roast Beef and bought 3 roast beef sandwiches, 2 orders of french fries, and 3 sodas for $13.85. One week ago it was Alice's turn to take the trip to Nick's. She bought 4 roast beef sandwiches, 3 orders of french fries, and 5 sodas for $20. This week Roberta made the trip. She purchased 3 roast beef sandwiches, 3 orders of french fries, and 4 sodas for $16.55. What is the cost of one roast beef sandwich? Of one order of french fries? Of one soda?

Putting Your Skills to Work

Analyzing Cell Phone Plans

Aaron and Gina are each interested in getting a cell phone, but their situations and the uses they will have for their cell phones are quite different.

Gina wants a cell phone to use in case of emergencies. Her job requires that she drive many miles each week, and she would like to have a cell phone available in case her car, which is far from new, breaks down. Aaron is self-employed and would use a cell phone to conduct business when he's occasionally away from his office during the week. He would probably use his cell phone for about 2 weekday hours each month.

Two of the cell phone plans available in their area are quite economical. Both plans offer a generous number of weekend minutes, but both Aaron and Gina plan to use their cell phones primarily during the month.

Plan A would cost $20 per month. It offers 30 free weekday minutes per month and charges $0.52/minute for any weekday minutes over 30. Plan B costs $30 per month. It also offers 30 free weekday minutes but charges $0.40 for any minutes used over 30.

Problems for Individual Investigation and Analysis

1. Write an equation that gives the monthly cost for each plan in terms of x, where x represents the number of weekday minutes more than 30 that are used.

2. Plot both equations on the axes.

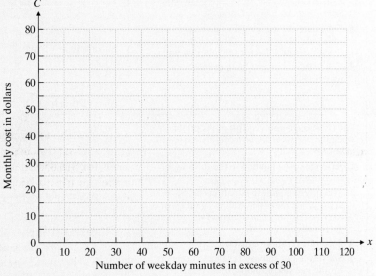

At what value of x do the equations appear to intersect? How many minutes of weekday cell phone use per month does this represent?

3. Solve the system of two equations for x algebraically. Does your answer agree with your answer to Exercise 2?

4. Which plan would be more economical for Gina? Which plan would be more economical for Aaron?

5. If Aaron uses his cell phone for 10 weekday hours during the month, what will his cell phone bill for that month be?

Problems for Group Investigation and Cooperative Study

Aaron investigates further and finds other cell phone plans that may be more economical for him. Plan C costs $40 per month. It offers 450 free weekday minutes per month and charges $0.30/minute for any weekday minutes over 450. Plan D costs $45 per month. It offers 600 free weekday minutes and charges $0.25 for additions minutes.

6. Of Plans B, C, or D, which is most economical for Aaron, given his estimated phone use?

7. If Aaron's business expands and he finds that he needs to use his phone about 8 weekday hours per month, which phone plan will be most economical?

Chapter 4 Organizer

Topic	Procedure	Examples
Finding a solution to a system of equations by the graphing method, p. 196.	1. Graph the first equation. 2. Graph the second equation. 3. Approximate from your graph where the two lines intersect, if they intersect at one point. 4. If the lines are parallel, there is no solution. If the lines coincide, there are an infinite number of solutions.	Solve by graphing: $x + y = 6$ $\qquad\qquad 2x - y = 6$ Graph each line. The solution is $(4, 2)$.
Solving a system of two linear equations by the substitution method, p. 198.	The substitution method is most appropriate when *at least one variable has a coefficient of 1 or −1.* 1. Solve for one variable in one of the equations. 2. In the other equation, replace that variable with the expression you obtained in step 1. 3. Solve the resulting equation. 4. Substitute the numerical value you obtain for a variable into one of the original equations and solve for the other variable. 5. Check the solution in both original equations.	Solve: $\quad 2x + y = 11 \quad$ **(1)** $\qquad\quad x + 3y = 18 \quad$ **(2)** $y = 11 - 2x$ from equation **(1)**. Substitute into **(2)**. $$x + 3(11 - 2x) = 18$$ $$x + 33 - 6x = 18$$ $$-5x = -15$$ $$x = 3$$ Substitute $x = 3$ into $2x + y = 11$. $$2(3) + y = 11$$ $$y = 5$$ The solution is $(3, 5)$.
Solving a system of two linear equations by the addition method, p. 200.	The addition method is most appropriate when the variables *all have coefficients other than 1 or −1.* 1. Arrange each equation in the form $ax + by = c$. 2. Multiply one or both equations by appropriate numerical values so that when the two resulting equations are added, one variable is eliminated. 3. Solve the resulting equation. 4. Substitute the numerical value you obtain for the variable into one of the original equations. 5. Solve this equation to find the other variable.	Solve: $\quad 2x + 3y = 5 \quad$ **(1)** $\qquad -3x - 4y = -2 \quad$ **(2)** Multiply equation **(1)** by 3 and **(2)** by 2. $$6x + 9y = 15$$ $$\underline{-6x - 8y = -4}$$ $$y = 11$$ Substitute $y = 11$ into equation **(1)**. $$2x + 3(11) = 5$$ $$2x + 33 = 5$$ $$2x = -28$$ $$x = -14$$ The solution is $(-14, 11)$.
Inconsistent system of equations, p. 202.	If there is *no solution* to a system of linear equations, the system of equations is inconsistent. When you try to solve an inconsistent system, you obtain an equation that is not true, such as $0 = 5$.	Solve: $\quad 4x + 3y = 10 \quad$ **(1)** $\qquad -8x - 6y = 5 \quad$ **(2)** Multiply equation **(1)** by 2 and add to **(2)**. $$8x + 6y = 20$$ $$\underline{-8x - 6y = 5}$$ $$0 = 25$$ But $0 \neq 25$. Thus, there is no solution. The system of equations is inconsistent.

Topic	Procedure	Examples
Dependent equations, p. 202.	If there are an *infinite number of solutions* to a system of linear equations, at least one pair of equations is dependent. When you try to solve a system that contains dependent equations, you will obtain an equation that is always true (such as $0 = 0$ or $3 = 3$). These equations are called *identities*.	Attempt to solve the system. $x - 2y = -5$ (1) $-3x + 6y = 15$ (2) Multiply equation (1) by 3 and add to (2). $3x - 6y = -15$ $\underline{-3x + 6y = \quad 15}$ $0 = \quad 0$ There are an infinite number of solutions. The equations are dependent.
Solving a system of three linear equations by algebraic methods, p. 211.	If there is one solution to a system of three linear equations in three unknowns, it may be obtained in the following manner. 1. Choose two equations from the system. 2. Multiply one or both of the equations by the appropriate constants so that by adding the two equations together, one variable can be eliminated. 3. Choose a *different* pair of the three original equations and eliminate the *same* variable using the procedure of step 2. 4. Solve the system formed by the two equations resulting from steps 2 and 3 for both variables. 5. Substitute the two values obtained in step 4 into one of the original three equations to find the third variable.	Solve: $2x - y - 2z = -1$ (1) $\quad\quad\quad x - 2y - z = \quad 1$ (2) $\quad\quad\quad x + y + z = \quad 4$ (3) Add equations (2) and (3) to eliminate z. $2x - y = 5$ (4) Multiply equation (3) by 2 and add to (1). $4x + y = 7$ (5) Add equations (4) and (5). $2x - y = 5$ $\underline{4x + y = 7}$ $6x = 12$ $x = 2$ Substitute $x = 2$ into equation (5). $4(2) + y = \quad 7$ $y = -1$ Substitute $x = 2, y = -1$ into equation (3). $2 + (-1) + z = 4$ $z = 3$ The solution is $(2, -1, 3)$.
Graphing the solution to a system of inequalities in two variables, p. 229.	1. Determine the region that satisfies each individual inequality. 2. Shade the common region that satisfies all the inequalities.	Graph: $3x + 2y \leq 10$ $\quad\quad\quad -1x + 2y \geq 2$ 1. $3x + 2y \leq 10$ can be graphed more easily as $y \leq -\dfrac{3}{2}x + 5$. We draw a solid line and shade the region below it. $-1x + 2y \geq 2$ can be graphed more easily as $y \geq \dfrac{1}{2}x + 1$. We draw a solid line and shade the region above it. 2. The common region is shaded.

Chapter 4 Review Problems

Solve the following systems by graphing.

1. $x + 2y = 8$
    $x - y = 2$

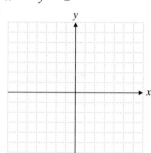

2. $x + y = 2$
    $3x - y = 6$

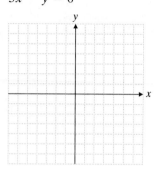

3. $2x + y = 6$
    $3x + 4y = 4$

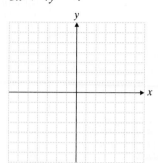

Solve the following systems by substitution.

4. $3x - 2y = -9$
   $2x + y = 1$

5. $-6x - y = 1$
   $3x - 4y = 31$

6. $4x - 3y = 15$
   $7x - y = 5$

7. $-7x + y = -4$
   $5x + 2y = 11$

Solve the following systems by addition.

8. $-2x + 5y = -12$
   $3x + y = 1$

9. $-3x + 4y = 9$
   $5x + 3y = -15$

10. $7x - 4y = 2$
   $6x - 5y = -3$

11. $5x + 2y = 3$
   $7x + 5y = -20$

Solve by any appropriate method. If there is no unique solution, state why.

12. $x = 3 - 2y$
    $3x + 6y = 8$

13. $x + 5y = 10$
    $y = 2 - \dfrac{1}{5}x$

14. $7x + 6y = -10$
    $2x + y = 0$

15. $3x + 4y = 1$
    $9x - 2y = -4$

16. $x + \dfrac{1}{3}y = 1$
    $\dfrac{1}{4}x - \dfrac{3}{4}y = -\dfrac{9}{4}$

17. $\dfrac{2}{3}x + y = 1$
    $\dfrac{1}{3}x + y = \dfrac{5}{6}$

18. $3a + 8b = 0$
$9a + 2b = 11$

19. $3a + 5b = -2$
$10b = -6a - 4$

20. $x + 3 = 3y + 1$
$1 - 2(x - 2) = 6y + 1$

21. $10(x + 1) - 13 = -8y$
$4(2 - y) = 5(x + 1)$

22. $0.3x - 0.2y = 0.7$
$-0.6x + 0.4y = 0.3$

23. $0.2x - 0.1y = 0.8$
$0.1x + 0.3y = 1.1$

Solve by an appropriate method.

24. $3x - 2y - z = 3$
$2x + y + z = 1$
$-x - y + z = -4$

25. $-2x + y - z = -7$
$x - 2y - z = 2$
$6x + 4y + 2z = 4$

26. $2x + 5y + z = 3$
$x + y + 5z = 42$
$2x + y = 7$

27. $x + 2y + z = 5$
$3x - 8y = 17$
$2y + z = -2$

28. $2x - 4y + 3z = 0$
$x - 2y - 5z = 13$
$5x + 3y - 2z = 19$

29. $5x + 2y + 3z = 10$
$6x - 3y + 4z = 24$
$-2x + y + 2z = 2$

30. $3x + 2y = 7$
$2x + 7z = -26$
$5y + z = 6$

31. $x - y = 2$
$5x + 7y - 5z = 2$
$3x - 5y + 2z = -2$

Use a system of linear equations to solve each of the following exercises.

32. *Commercial Airline* A plane flies 720 miles against the wind in 3 hours. The return trip with the wind takes only $2\frac{1}{2}$ hours. Find the speed of the wind. Find the speed of the plane in still air.

33. *Football* Two football teams scored a total of 11 times during Saturday's game. They scored a number of touchdowns for 7 points each and several field goals at 3 points each. Altogether they scored 65 points. How many touchdowns and how many field goals did they score?

34. *Temporary Help Expenses* When the circus came to town last year, they hired general laborers at $70 per day and mechanics at $90 per day. They paid $1950 for this temporary help for one day. This year they hired exactly the same number of people of each type, but they paid $80 for general laborers and $100 for mechanics for the one day. This year they paid $2200 for temporary help. How many general laborers did they hire? How many mechanics did they hire?

35. *Circus Ticket Prices* A total of 590 tickets were sold for the circus matinee performance. Children's admission tickets were $6, and adult tickets were $11. The ticket receipts for the matinee performance were $4790. How many children's tickets were sold? How many adult tickets were sold?

36. *Baseball Equipment* A baseball coach bought two hats, five shirts, and four pairs of pants for $129. His assistant purchased one hat, one shirt, and two pairs of pants for $42. The next week the coach bought two hats, three shirts, and one pair of pants for $63. What was the cost of each item?

37. *Math Exam Scores* Jess, Chris, and Nick scored a total of 249 points on their last math exam. Jess's score was 20 points higher than Chris's score. Twice Nick's score was 6 more than the sum of Chris's and Jess's scores. What was each of their score on the exam?

38. *Food Costs* Four jars of jelly, three jars of peanut butter, and five jars of honey cost $9.80. Two jars of jelly, two jars of peanut butter, and one jar of honey cost $4.20. Three jars of jelly, four jars of peanut butter, and two jars of honey cost $7.70. Find the cost for one jar of jelly, one jar of peanut butter, and one jar of honey.

39. *Transportation Logistics* The church youth group is planning a trip to Mount Washington. A total of 127 people need rides. The church has available buses that hold forty passengers, and several parents have volunteered station wagons that hold eight passengers or sedans that hold five passengers. The youth leader is planning to use nine vehicles to transport the people. One parent said that if they didn't use any buses, tripled the number of station wagons, and doubled the number of sedans, they would be able to transport 126 people. How many buses, station wagons, and sedans are they planning to use if they use nine vehicles?

Mixed Practice

Solve by any appropriate method.

40. $-x - 5z = -5$
$13x + 2z = 2$

41. $x - y = 1$
$5x + y = 7$

42. $2x + 5y = 4$
$5x - 7y = -29$

43. $\dfrac{x}{2} - 3y = -6$
$\dfrac{4}{3}x + 2y = 4$

44. $\frac{3}{5}x - y = 6$

    $x + \frac{y}{3} = 10$

45. $\frac{x + 1}{5} = y + 2$

    $\frac{2y + 7}{3} = x - y$

46. $3(2 + x) = y + 1$

    $5(x - y) = -7 - 3y$

47. $7(x + 3) = 2y + 25$

    $3(x - 6) = -2(y + 1)$

48. $0.3x - 0.4y = 0.9$

    $0.2x - 0.3y = 0.4$

49. $1.2x - y = 1.6$

    $x + 1.5y = 6$

50. $x - \frac{y}{2} + \frac{1}{2}z = -1$

    $2x \qquad + \frac{5}{2}z = -1$

    $\frac{3}{2}y + 2z = 1$

51. $2x - 3y + 2z = 0$

    $x + 2y - z = 2$

    $2x + y + 3z = -1$

52. $x - 4y + 4z = -1$

    $2x - y + 5z = -3$

    $x - 3y + z = 4$

53. $x - 2y + z = -5$

    $2x + z = -10$

    $y - z = 15$

Solve each of the following systems of linear inequalities by graphing.

54. $x - y \le 3$

    $y \le -\frac{1}{4}x + 2$

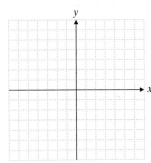

55. $-2x + 3y < 6$

    $y > -2$

56. $x + y > 1$

    $2x - y < 5$

57. $x + y \ge 4$

    $y \le x$

    $x \le 6$

Remember to use your Chapter Test Prep Video CD to see the worked-out solutions to the test problems you want to review.

Solve each system of equations. If there is no solution to the system, give a reason. In exercises 3–9, you may use any method.

1. Solve using the substitution method:
$$x - y = 3$$
$$2x - 3y = -1$$

2. Solve using the addition method:
$$3x + 2y = 1$$
$$5x + 3y = 3$$

3.
$$5x - 3y = 3$$
$$7x + y = 25$$

4.
$$\frac{1}{4}a - \frac{3}{4}b = -1$$
$$\frac{1}{3}a + b = \frac{5}{3}$$

5.
$$\frac{1}{3}x + \frac{5}{6}y = 2$$
$$\frac{3}{5}x - y = -\frac{7}{5}$$

6.
$$8x - 3y = 5$$
$$-16x + 6y = 8$$

7.
$$3x + 5y - 2z = -5$$
$$2x + 3y - z = -2$$
$$2x + 4y + 6z = 18$$

8.
$$3x + 2y = 0$$
$$2x - y + 3z = 8$$
$$5x + 3y + z = 4$$

9.
$$x + 5y + 4z = -3$$
$$x - y - 2z = -3$$
$$x + 2y + 3z = -5$$

Use a system of linear equations to solve the following exercises.

10. A plane flew 1000 miles with a tailwind in 2 hours. The return trip against the wind took $2\frac{1}{2}$ hours. Find the speed of the wind and the speed of the plane in still air.

11. The math club is selling items with the college logo to raise money. Sam bought 4 pens, a mug, and a T-shirt for $20.00. Alicia bought 2 pens and 2 mugs for $11.00. Ramon bought 6 pens, a mug, and 2 T-shirts for $33.00. What was the price of each pen, mug, and T-shirt?

12. Sue Miller had to move some supplies to Camp Cherith for the summer camp program. She rented a Portland Rent-A-Truck in April for 5 days and drove 150 miles. She paid $180 for the rental in April. Then in May she rented the same truck again for 7 days and drove 320 miles. She paid $274 for the rental in May. How much does Portland Rent-A-Truck charge for a daily rental of the truck? How much do they charge per mile?

Solve the following systems of linear inequalities by graphing.

13.
$$x + 2y \le 6$$
$$-2x + y \ge -2$$

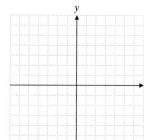

14.
$$3x + y > 8$$
$$x - 2y > 5$$

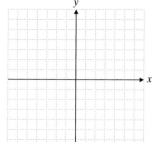

1. _____

2. _____

3. _____

4. _____

5. _____

6. _____

7. _____

8. _____

9. _____

10. _____

11. _____

12. _____

13. _____

14. _____

Approximately one half of this test covers the content of Chapters 1–3. The remainder covers the content of Chapter 4.

1. State what property is illustrated. $7 + 0 = 7$

2. Evaluate. $\sqrt{25} + (2 - 3)^3 + 20 \div (-10)$

3. Simplify $(5x^{-2})(3x^{-4}y^2)$.

4. Simplify $2x - 4[x - 3(2x + 1)]$.

5. Solve for P: $A = P(3 + 4rt)$

6. Solve for x: $\dfrac{1}{4}x + 5 = \dfrac{1}{3}(x - 2)$

7. Graph the line $4x - 8y = 10$. Plot at least three points.

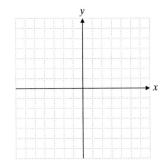

8. Find the slope of the line passing through $(6, -1)$ and $(-4, -2)$.

Solve the following linear inequalities and graph the solution on a number line.

9. $4x + 3 - 13x - 7 < 2(3 - 4x)$

10. $\dfrac{2x - 1}{3} \le 7$ and $2(x + 1) \ge 12$

11. Find the standard form of the equation of the line passing through $(2, -3)$ and perpendicular to $5x + 6y = -2$.

▲ **12.** A triangle has a perimeter of 69 meters. The second side is 7 meters longer than the first side. The third side is 6 meters shorter than double the length of the first side. Find the length of each side.

13. Victor invests $6000 in a bank. Part is invested at 7% simple interest and part at 9% simple interest. In 1 year Victor earns $510 in interest. How much did he invest at each amount?

1. _____

2. _____

3. _____

4. _____

5. _____

6. _____

7. _____

8. _____

9. _____

10. _____

11. _____

12. _____

13. _____

14. _____

14. Solve the system.

$$5x + 2y = 2$$
$$4x + 3y = -4$$

15. Solve the system.

$$\frac{1}{2}x - 3y = 5$$
$$\frac{1}{4}x + y = 0$$

15. _____

16. Patricia bought five shirts and eight pairs of slacks for $345 at Super Discount Center. Joanna bought seven of the same shirts and three pairs of the same slacks at the same store, and her total was $237. How much did a shirt cost? How much did a pair of slacks cost?

16. _____

17. Solve the system.

$$7x - 6y = 17$$
$$3x + y = 18$$

17. _____

18. Solve the system.

$$x + 3y + z = 5$$
$$2x - 3y - 2z = 0$$
$$x - 2y + 3z = -9$$

18. _____

19. What happens when you attempt to solve the system below? Why is this?

$$-5x + 6y = 2$$
$$10x - 12y = -4$$

20. Solve the following system of inequalities by graphing.

$$x - y \geq -4$$
$$x + 2y \geq 2$$

19. _____

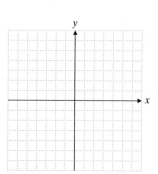

20. _____

Occupations change over time. A hundred years ago, every village and town in the United States had a blacksmith. Today it is difficult to find the few remaining blacksmiths in our country. Forty years ago, there were very few jobs in the computer area. Today, computer-related occupations are among the fastest growing. Which occupations are growing at the fastest rate? Where will there be a demand for workers in the future? Turn to page 297 to find the answers to these questions and to use functions to determine how many of some of these jobs will be available a few years from now.

Polynomials

5.1 INTRODUCTION TO POLYNOMIALS AND POLYNOMIAL FUNCTIONS: ADDING, SUBTRACTING, AND MULTIPLYING

Student Learning Objectives

After studying this section, you will be able to:

1 Identify types and degrees of polynomials.

2 Evaluate polynomial functions.

3 Add and subtract polynomials.

4 Multiply two binomials by FOIL.

5 Multiply two binomials $(a + b)(a - b)$.

6 Multiply two binomials $(a - b)^2$ or $(a + b)^2$.

7 Multiply polynomials with more than two terms.

1 Identifying Types and Degrees of Polynomials

A **polynomial** is an algebraic expression of one or more terms. A **term** is a number, a variable raised to a nonnegative integer power, or a product of numbers and variables raised to nonnegative integer powers. There must be no division by a variable. Three types of polynomials that you will see often are **monomials, binomials,** and **trinomials.**

1. A **monomial** has *one* term.
2. A **binomial** has *two* terms.
3. A **trinomial** has *three* terms.

Here are some examples of polynomials.

Number of Variables	Monomials	Binomials	Trinomials	Other Polynomials
One Variable	$8x^3$	$2y^2 + 3y$	$5x^2 + 2x - 6$	$x^4 + 2x^3 - x^2 + 9$
Two Variables	$6x^2y$	$3x^2 - 5y^3$	$8x^2 + 5xy - 3y^2$	$x^3y + 5xy^2 + 3xy - 7y^5$
Three Variables	$12uvw^3$	$11a^2b + 5c^2$	$4a^2b^4 + 7c^4 - 2a^5$	$3c^2 + 4c - 8d + 2e - e^2$

The following are *not* polynomials.

$$2x^{-3} + 5x^2 - 3 \qquad 4ab^{\frac{1}{2}} \qquad \frac{2}{x} + \frac{3}{y}$$

TO THINK ABOUT: Understanding Polynomials Give a reason each expression is not a polynomial.

Polynomials are also classified by degree. The **degree of a term** is the sum of the exponents of its variables. The **degree of a polynomial** is the degree of the highest-degree term in the polynomial. If the polynomial has no variable, then it has degree zero.

EXAMPLE 1 Name the type of polynomial and give its degree.

(a) $5x^6 + 3x^2 + 2$
(b) $7x + 6$
(c) $5x^2y + 3xy^3 + 6xy$
(d) $7x^4y^5$

Solution

(a) This is a trinomial of degree 6.
(b) This is a binomial of degree 1.
(c) This is a trinomial of degree 4.
(d) This is a monomial of degree 9.

Remember that if a variable has no exponent, the exponent is understood to be 1.

NOTE TO STUDENT: Fully worked-out solutions to all of the Practice Problems can be found at the back of the text starting at page SP-1

Practice Problem 1 State the type of polynomial and give its degree.

(a) $3x^5 - 6x^4 + x^2$
(b) $5x^2 + 2$
(c) $3ab + 5a^2b^2 - 6a^4b$
(d) $16x^4y^6$

Some polynomials contain only one variable. A **polynomial in x** is an expression of the form

$$a_n x^n + a_{n-1} x^{n-1} + a_{n-2} x^{n-2} + \cdots + a_0$$

where n is a nonnegative integer and the constants $a_n, a_{n-1}, a_{n-2}, \ldots, a_0$ are real numbers. We usually write polynomials in **descending order** of the variable. That is, the exponents on the variables decrease from left to right. For example, the polynomial $4x^5 - 2x^3 + 6x^2 + 5x - 8$ is written in descending order.

2 Evaluating Polynomial Functions

A **polynomial function** is a function that is defined by a polynomial.
For example,

$$p(x) = 5x^2 - 3x + 6 \quad \text{and} \quad p(x) = 2x^5 - 3x^3 + 8x - 15$$

are both polynomial functions.
To evaluate a polynomial function, we use the skills developed in Section 3.5.

EXAMPLE 2   Evaluate the polynomial function $p(x) = -3x^3 + 2x^2 - 5x + 6$ for **(a)** $p(-3)$ and **(b)** $p(6)$.

Solution

(a) $p(-3) = -3(-3)^3 + 2(-3)^2 - 5(-3) + 6$
$\qquad\qquad = -3(-27) + 2(9) - 5(-3) + 6$
$\qquad\qquad = 81 + 18 + 15 + 6$
$\qquad\qquad = 120$

(b) $p(6) = -3(6)^3 + 2(6)^2 - 5(6) + 6$
$\qquad\qquad = -3(216) + 2(36) - 5(6) + 6$
$\qquad\qquad = -648 + 72 - 30 + 6$
$\qquad\qquad = -600$

Practice Problem 2   Evaluate the polynomial function $p(x) = 2x^4 - 3x^3 + 6x - 8$ for **(a)** $p(-2)$ **(b)** $p(5)$.

3 Adding and Subtracting Polynomials

We can add and subtract polynomials by combining like terms as we learned in Section 1.5.

EXAMPLE 3   Add: $(5x^2 - 3x - 8) + (-3x^2 - 7x + 9)$

Solution

$5x^2 - 3x - 8 - 3x^2 - 7x + 9$ We remove the parentheses and combine like terms.
$\quad = 2x^2 - 10x + 1$

Practice Problem 3   Add: $(-7x^2 + 5x - 9) + (2x^2 - 3x + 5)$

To subtract real numbers, we add the opposite of the second number to the first. Thus, for real numbers a and b, we have $a - (+b) = a + (-b)$. Similarly for polynomials, to subtract polynomials we add the opposite of the second polynomial to the first.

EXAMPLE 4   Subtract: $(-5x^2 - 19x + 15) - (3x^2 - 4x + 13)$

Solution

$(-5x^2 - 19x + 15) + (-3x^2 + 4x - 13)$ We add the opposite of the second polynomial to the first polynomial.

$= -8x^2 - 15x + 2$

Practice Problem 4   Subtract: $(2x^2 - 14x + 9) - (-3x^2 + 10x + 7)$

NOTE TO STUDENT: Fully worked-out solutions to all of the Practice Problems can be found at the back of the text starting at page SP-1

4 Multiplying Two Binomials by FOIL

The FOIL method for multiplying two binomials has been developed to help you keep track of the order of the terms to be multiplied. The acronym FOIL means the following:

F	First
O	Outer
I	Inner
L	Last

That is, we multiply the first terms, then the outer terms, then the inner terms, and finally, the last terms.

EXAMPLE 5   Multiply: $(5x + 2)(7x - 3)$

Solution

First   Last   First + Outer + Inner + Last

$(5x + 2)(7x - 3) = 35x^2 - 15x + 14x - 6$

Inner
Outer

$= 35x^2 - x - 6$

Practice Problem 5   Multiply: $(7x + 3)(2x - 5)$

EXAMPLE 6   Multiply: $(7x^2 - 8)(2x - 3)$

Solution   First   Last

$(7x^2 - 8)(2x - 3) = 14x^3 - 21x^2 - 16x + 24$

Inner
Outer

Note that in this case we were not able to combine the inner and outer products.

Practice Problem 6   Multiply: $(3x^2 - 2)(5x - 4)$

5 **Multiplying** $(a + b)(a - b)$

Products of the form $(a + b)(a - b)$ deserve special attention.

$$(a + b)(a - b) = a^2 - ab + ab - b^2 = a^2 - b^2$$

Notice that the middle terms, $-ab$ and $+ab$, when combined equal zero. The product is the difference of two squares, $a^2 - b^2$. This is always true when you multiply binomials of the form $(a + b)(a - b)$. You should memorize the following formula.

$$(a + b)(a - b) = a^2 - b^2$$

EXAMPLE 7   Multiply: $(2a - 9b)(2a + 9b)$

Solution $(2a - 9b)(2a + 9b) = (2a)^2 - (9b)^2 = 4a^2 - 81b^2$

Of course, we could have used the FOIL method, but recognizing the special product allowed us to save time.

Practice Problem 7   Multiply: $(7x - 2y)(7x + 2y)$

6 **Multiplying** $(a - b)^2$ **or** $(a + b)^2$

Another special product is the square of a binomial.

$$(a - b)^2 = (a - b)(a - b) = a^2 - ab - ab + b^2 = a^2 - 2ab + b^2$$

Once you understand the pattern, you should memorize these two formulas.

$$(a - b)^2 = a^2 - 2ab + b^2 \qquad (a + b)^2 = a^2 + 2ab + b^2$$

This procedure is also called **expanding a binomial.** *Note:* $(a - b)^2 \neq a^2 - b^2$ and $(a + b)^2 \neq a^2 + b^2$.

EXAMPLE 8   Multiply.   **(a)** $(5a - 8b)^2$ **(b)** $(3u + 11v^2)^2$

Solution

(a) $(5a - 8b)^2 = (5a)^2 - 2(5a)(8b) + (8b)^2 = 25a^2 - 80ab + 64b^2$

(b) Here $a = 3u$ and $b = 11v^2$.

$$(3u + 11v^2)^2 = (3u)^2 + 2(3u)(11v^2) + (11v^2)^2$$
$$= 9u^2 + 66uv^2 + 121v^4$$

Practice Problem 8   Multiply.   **(a)** $(4u + 5v)^2$ **(b)** $(7x^2 - 3y^2)^2$

7 **Multiplying Polynomials with More Than Two Terms**

The distributive property is the basis for multiplying polynomials. Recall that

$$a(b + c) = ab + ac.$$

We can use this property to multiply a polynomial by a monomial.

$$3xy(5x^3 + 2x^2 - 4x + 1) = 3xy(5x^3) + 3xy(2x^2) - 3xy(4x) + 3xy(1)$$
$$= 15x^4y + 6x^3y - 12x^2y + 3xy$$

A similar procedure can be used to multiply two binomials.

$$(3x + 5)(6x + 7) = (3x + 5)6x + (3x + 5)7$$ We use the distributive property again.

$$= (3x)(6x) + (5)(6x) + (3x)(7) + (5)(7)$$
$$= 18x^2 + 30x + 21x + 35$$
$$= 18x^2 + 51x + 35$$

The multiplication of a binomial and a trinomial is more involved. One way to multiply two polynomials is to write them vertically, as we do when multiplying two- and three-digit numbers. We then multiply them in the usual way.

EXAMPLE 9   Multiply: $(4x^2 - 2x + 3)(-3x + 4)$

Solution

$$
\begin{array}{r}
4x^2 - 2x + 3 \\
-3x + 4 \\
\hline
16x^2 - 8x + 12 \\
-12x^3 + 6x^2 - 9x \\
\hline
-12x^3 + 22x^2 - 17x + 12
\end{array}
$$

Multiply $(4x^2 - 2x + 3)(+4)$.
Multiply $(4x^2 - 2x + 3)(-3x)$.
Add the two products.

NOTE TO STUDENT: Fully worked-out solutions to all of the Practice Problems can be found at the back of the text starting at page SP-1

Practice Problem 9   Multiply: $(2x^2 - 3x + 1)(x^2 - 5x)$

Another way to multiply polynomials is to multiply horizontally. We redo Example 9 in the following example.

EXAMPLE 10   Multiply horizontally: $(4x^2 - 2x + 3)(-3x + 4)$

Solution   By the distributive law, we have the following:

$$(4x^2 - 2x + 3)(-3x + 4) = (4x^2 - 2x + 3)(-3x) + (4x^2 - 2x + 3)(4)$$
$$= -12x^3 + 6x^2 - 9x + 16x^2 - 8x + 12$$
$$= -12x^3 + 22x^2 - 17x + 12.$$

In actual practice you will find that you can do some of these steps mentally.

Practice Problem 10   Multiply horizontally: $(2x^2 - 3x + 1)(x^2 - 5x)$.

Developing Your Study Skills

Taking an Exam

Allow yourself plenty of time to get to your exam. You may even find it helpful to arrive a little early in order to collect your thoughts and ready yourself. This will help you feel more relaxed.

After you receive your exam, you will find it helpful to do the following.

1. Take two or three moderately deep breaths. Inhale; then exhale slowly. You will feel your entire body begin to relax.

2. Write down on the back of the exam any formulas or ideas that you need to remember.

3. Look over the entire test quickly in order to pace yourself and use your time wisely. Notice how many

points each exercise is worth. Spend more time on items of greater worth.

4. Read directions carefully and be sure to answer all questions clearly. Keep your work neat and easy to read.

5. Ask your instructor about anything that is not clear to you.

6. Work the exercises and answer the questions that are easiest for you first. Then go back to the more difficult ones.

7. Do not get bogged down on one exercise for too long because it may jeopardize your chances of finishing other exercises. Leave the tough exercise and go back to it when you have time later.

8. Check your work. This will help you catch minor errors.

9. Stay calm if others leave before you do. You are entitled to use the full amount of allotted time.

5.1 EXERCISES

| Student Solutions Manual | CD/ Video | PH Math Tutor Center | MathXL®Tutorials on CD | MathXL® | MyMathLab® | Interactmath.com |

Name the type of polynomial and give its degree.

1. $2x^2 - 5x + 3$

2. $7x^3 + 6x^2 - 2$

3. $-3.2a^4bc^3$

4. $26.8a^3bc^2$

5. $\frac{3}{5}m^3n - \frac{2}{5}mn$

6. $\frac{2}{7}m^2n^2 + \frac{1}{2}mn^2$

For the polynomial function $p(x) = 5x^2 - 9x - 12$ evaluate the following:

7. $p(3)$

8. $p(-4)$

For the polynomial function $g(x) = -3x^3 - x^2 + 4x + 2$, evaluate the following:

9. $g(2)$

10. $g(-1)$

For the polynomial function $h(x) = 2x^4 - x^3 + 2x^2 - 4x - 3$ evaluate the following:

11. $h(-1)$

12. $h(3)$

Add or subtract the following polynomials as indicated.

13. $(x^2 + 3x - 2) + (-2x^2 - 5x + 1) + (x^2 - x - 5)$

14. $(2x^2 - 5x - 1) + (3x^2 - 7x + 3) + (-5x^2 + x + 1)$

15. $(7m^3 + 4m^2 - m + 2.5) - (-3m^3 + 5m + 3.8)$

16. $(3x^3 + 2x^2 - 8x - 9.2) - (-5x^3 + x^2 - x - 12.7)$

17. $(5a^3 - 2a^2 - 6a + 8) + (5a + 6) - (-a^2 - a + 2)$

18. $(a^5 + 3a^2) + (2a^4 - a^3 - 3a^2 + 2) - (a^4 + 3a^3 - 5)$

19. $\left(\frac{1}{2}x^2 - 7x\right) + \left(\frac{1}{3}x^2 + \frac{1}{4}x\right)$

20. $\left(\frac{1}{5}x^2 + 9x\right) + \left(\frac{4}{5}x^2 - \frac{1}{6}x\right)$

21. $(2.3x^3 - 5.6x^2 - 2) - (5.5x^3 - 7.4x^2 + 2)$

22. $(5.9x^3 + 3.4x^2 - 7) - (2.9x^3 - 9.6x^2 + 3)$

Multiply.

23. $(5x + 8)(2x + 9)$

24. $(6x + 7)(3x + 2)$

25. $(5w + 2d)(3a - 4b)$

26. $(7a + 8b)(5d - 8w)$

27. $(3x - 2y)(-4x + y)$

28. $(-9x - 5y)(3a + 2y)$

29. $(7r - s^2)(-4a - 11s^2)$

30. $(-3r - 2s^2)(5r - 6s^2)$

Multiply mentally. See Examples 7 and 8.

31. $(5x - 8y)(5x + 8y)$

32. $(2a - 7b)(2a + 7b)$

33. $(5a - 2b)^2$

34. $(6a + 5b)^2$

35. $(7m - 1)^2$

36. $(5r + 3)^2$

37. $(4 + 3x^2)(4 - 3x^2)$

38. $(7 - 5x^3)(7 + 5x^3)$

39. $(3m^3 + 1)^2$

40. $(4r^3 - 5)^2$

254 Chapter 5 Polynomials

Multiply.

41. $2x(3x^2 - 5x + 1)$

42. $-5x(x^2 - 6x - 2)$

43. $-\dfrac{1}{3}xy(2x - 6y + 15)$

44. $\dfrac{3}{5}xy^3(x - 10y + 4)$

45. $(2x - 3)(x^2 - x + 1)$

46. $(4x + 1)(2x^2 + x + 1)$

47. $(3x^2 - 2xy - 6y^2)(2x - y)$

48. $(5x^2 + 3xy - 7y^2)(3x - 2y)$

49. $\left(\dfrac{3}{2}x^2 - x + 1\right)(x^2 + 2x - 6)$

50. $\left(\dfrac{2}{3}x^2 + 5x - 2\right)(2x^2 - 3x + 9)$

51. $(5a^3 - 3a^2 + 2a - 4)(a - 3)$

52. $(2b^3 - 5b^2 - 4b + 1)(2b - 1)$

First multiply any two binomials in the exercise; then multiply the result by the third binomial.

53. $(x + 2)(x - 3)(2x - 5)$

54. $(x - 6)(x + 2)(3x + 2)$

55. $(a + 3)(2 - a)(4 - 3a)$

56. $(6 - 5a)(a + 1)(2 - 3a)$

Applications

▲ **57.** *Geometry* The area of the base of a rectangular box measures $2x^2 + 5x + 8$ cm². The height of the box measures $3x + 5$ cm. Find the volume of the box.

▲ **58.** *Geometry* A rectangular garden has $3n^2 + 4n + 7$ flowers planted in each row. The garden has $2n + 5$ rows. Find the number of flowers in the garden.

Antimalaria Medication The concentration of a certain antimalaria medication, in parts per million after time t, in hours, is given by the polynomial $p(t) = -0.03t^2 + 78$.

59. Find the concentration after 3 hours.

60. Find the concentration after 30 hours.

61. Find the concentration after 50 hours.

62. Find the concentration after 50.9 hours.

Cumulative Review

63. Solve: $\dfrac{1}{2}x + 4 \le \dfrac{2}{3}(x - 3) + 1$

64. Solve: $2(x + 1) - 3 = 4 - (x + 5)$

65. *Airline Operations* An American Airlines jet is cruising at an altitude of 31,000 feet on an approach to Logan Airport. The tower instructs the jet to descend to 8000 feet for the final approach. The plane is descending at a rate of 2500 feet per minute. How long will it take the jet to reach the 8000-foot altitude?

66. *Quality Control* A certain company produces small "reminder" notepads. Each pad has eighty square $2'' \times 2''$ sheets. One day at the factory, eighty thousand reminder notepads were produced. Unfortunately, due to mechanical error, every fifth page of every notepad was defective. How many total pages were defective?

5.2 DIVIDING POLYNOMIALS

① Dividing a Polynomial by a Monomial

The easiest type of polynomial division occurs when the divisor is a monomial. We perform this type of division just as if we were dividing numbers. First we write the indicated division as the sum of separate fractions, and then we reduce each fraction (if possible).

EXAMPLE 1

Divide: $(15x^3 - 10x^2 + 40x) \div 5x$

Solution

$$\frac{15x^3 - 10x^2 + 40x}{5x} = \frac{15x^3}{5x} - \frac{10x^2}{5x} + \frac{40x}{5x}$$
$$= 3x^2 - 2x + 8$$

Practice Problem 1 Divide: $(-16x^4 + 16x^3 + 8x^2 + 64x) \div 8x$

② Dividing a Polynomial by a Polynomial

When we divide polynomials by binomials or trinomials, we perform long division. This is much like the long division method for dividing numbers. The polynomials must be in descending order.

First we write the problem in the form of long division.

$$2x + 3\overline{)6x^2 + 17x + 12}$$

The divisor is $2x + 3$; the dividend is $6x^2 + 17x + 12$. Now we divide the first term of the dividend $(6x^2)$ by the first term of the divisor $(2x)$.

$$\boxed{3x} \quad \boxed{6x^2 \div 2x = 3x}$$
$$2x + 3\overline{)6x^2 + 17x + 12}$$

Now we multiply $3x$ (the first term of the quotient) by the divisor $2x + 3$.

$$\begin{array}{r} 3x \\ 2x + 3\overline{)6x^2 + 17x + 12} \\ 6x^2 + 9x \end{array} \longleftarrow \boxed{\text{The product of } 3x(2x + 3).}$$

Next, just as in long division with numbers, we subtract and bring down the next monomial.

$$\begin{array}{r} 3x \\ 2x + 3\overline{)6x^2 + 17x + 12} \\ 6x^2 + 9x \\ \hline 8x + 12 \end{array}$$

$\boxed{\text{Subtract } 6x^2 + 9x \text{ from } 6x^2 + 17x.}$

$\boxed{\text{Bring down the next monomial.}}$

Now we divide the first term of this binomial $(8x)$ by the first term of the divisor $(2x)$.

$$\begin{array}{r} 3x + \boxed{4} \quad \boxed{8x \div 2x = 4} \\ 2x + 3\overline{)6x^2 + 17x + 12} \\ 6x^2 + 9x \\ \hline 8x + 12 \\ 8x + 12 \longleftarrow \boxed{\text{The product of } 4(2x + 3).} \\ \hline 0 \end{array}$$

Note that we then multiplied $(2x + 3)(4)$ and subtracted, just as we did before. We continue this process until the remainder is zero. Thus, we find that
$$\frac{6x^2 + 17x + 12}{2x + 3} = 3x + 4.$$

Student Learning Objectives

After studying this section, you will be able to:

① Divide a polynomial by a monomial.

② Divide a polynomial by a polynomial.

NOTE TO STUDENT: Fully worked-out solutions to all of the Practice Problems can be found at the back of the text starting at page SP-1

 Graphing Calculator

Verifying Answers When Dividing Polynomials

One way to verify that the division was performed correctly is to graph

$$y_1 = 3x + 4$$

and $y_2 = \dfrac{6x^2 + 17x + 12}{2x + 3}$.

If the graphs appear to coincide, then we have an independent verification that

$$\frac{6x^2 + 17x + 12}{2x + 3} = 3x + 4.$$

255

DIVIDING A POLYNOMIAL BY A BINOMIAL OR TRINOMIAL

1. Write the division problem in long division form. Write both polynomials in descending order; write missing terms with a coefficient of zero.
2. Divide the *first* term of the divisor into the first term of the dividend. The result is the first term of the quotient.
3. Multiply the first term of the quotient by *every* term in the divisor.
4. Write the product under the dividend (align like terms) and subtract.
5. Treat this difference as a new dividend. Repeat steps 2 through 4. Continue until the remainder is zero or a polynomial of lower degree than the *first term* of the divisor.
6. If there is a remainder, write it as the numerator of a fraction with the divisor as the denominator. Add this fraction to the quotient.

EXAMPLE 2   Divide: $(6x^3 + 7x^2 + 3) \div (3x - 1)$

Solution   There is no x-term in the dividend, so we write $0x$.

$$
\begin{array}{r}
2x^2 + 3x + 1 \\
3x - 1 \overline{\smash{)}\, 6x^3 + 7x^2 + 0x + 3} \\
\underline{6x^3 - 2x^2} \\
9x^2 + 0x \\
\underline{9x^2 - 3x} \\
3x + 3 \\
\underline{3x - 1} \\
4
\end{array}
$$

Note that we calculate $7x^2 - (-2x^2)$ to obtain $9x^2$.

Note that we calculate $0x - (-3x)$ to obtain $3x$.

The quotient is $2x^2 + 3x + 1$ with a remainder of 4. We may write this as

$$2x^2 + 3x + 1 + \frac{4}{3x - 1}$$

Check: $(3x - 1)(2x^2 + 3x + 1) + 4 \stackrel{?}{=} 6x^3 + 7x^2 + 3$

$$6x^3 + 7x^2 - 0x - 1 + 4 \stackrel{?}{=} 6x^3 + 7x^2 + 3$$

$$6x^3 + 7x^2 + 3 = 6x^3 + 7x^2 + 3 \quad \checkmark$$

Practice Problem 2   Divide: $(8x^3 - 10x^2 - 9x + 14) \div (4x - 3)$

EXAMPLE 3   Divide $\dfrac{64x^3 - 125}{4x - 5}$.

Solution   This fraction is equivalent to the problem $(64x^3 - 125) \div (4x - 5)$.
Note that two terms are missing in the dividend. We write them with zero coefficients.

$$
\begin{array}{r}
16x^2 + 20x + 25 \\
4x - 5 \overline{\smash{)}\, 64x^3 + 0x^2 + 0x - 125} \\
\underline{64x^3 - 80x^2} \\
80x^2 + 0x \\
\underline{80x^2 - 100x} \\
100x - 125 \\
\underline{100x - 125} \\
0
\end{array}
$$

Note that $0x^2 - (-80x^2) = 80x^2$.

Note that $0x - (-100x) = 100x$.

The quotient is $16x^2 + 20x + 25$.

   Check: Verify that $(4x - 5)(16x^2 + 20x + 25) = 64x^3 - 125$.

Practice Problem 3   Divide: $(8x^3 + 27) \div (2x + 3)$

NOTE TO STUDENT: Fully worked-out solutions to all of the Practice Problems can be found at the back of the text starting at page SP-1

EXAMPLE 4   Divide: $(7x^3 - 10x - 7x^2 + 2x^4 + 8) \div (2x^2 + x - 2)$

Solution   Arrange the dividend in descending order before dividing.

$$2x^4 \div 2x^2 = x^2$$

$$
\begin{array}{r}
x^2 + 3x - 4 \\
2x^2 + x - 2 \overline{)\,2x^4 + 7x^3 - 7x^2 - 10x + 8} \\
\underline{2x^4 + x^3 - 2x^2} \\
6x^3 - 5x^2 - 10x \\
\underline{6x^3 + 3x^2 - 6x} \\
-8x^2 - 4x + 8 \\
\underline{-8x^2 - 4x + 8} \\
0
\end{array}
$$

Note that
$(7x^3 - 7x^2) - (x^3 - 2x^2)$
$= 7x^3 - 7x^2 - x^3 + 2x^2$.

Note that
$(-5x^2 - 10x) - (3x^2 - 6x)$
$= -5x^2 - 10x - 3x^2 + 6x$.

The quotient is $x^2 + 3x - 4$.

   Check: Verify that $(2x^2 + x - 2)(x^2 + 3x - 4) =$
$2x^4 + 7x^3 - 7x^2 - 10x + 8$.

Practice Problem 4   Divide: $(x^4 - 3x^3 + 3x + 4) \div (x^2 - 1)$

Developing Your Study Skills

Taking Notes in a Lecture Class

An important part of studying mathematics is taking notes. In order to take meaningful notes, you must be an active listener. Keep your mind on what the instructor is saying, and be ready with questions whenever you do not understand something.

   If you have previewed the lesson material, you will be prepared to take good notes. The important concepts will seem somewhat familiar. You will have a better idea of what needs to be written down. If you frantically try to write all that the instructor says or copy all the examples done in class, you may find your notes to be nearly worthless when you are home alone. You may find that you are unable to make sense of what you have written.

   Write down *important* ideas and examples as the instructor lectures, making sure that you are listening and following the logic. Include any helpful hints or suggestions that your instructor gives you or refers to in your text. You will be amazed at how easily these are forgotten if they are not written down.

Successful note taking requires active listening and processing. Stay alert in class. You will realize the advantages of taking your own notes over copying those of someone else.

Taking Notes in an Online Class

In an online class, you will need notes to help you focus on three things.

1. What are the important types of problems I need to be able to identify?
2. How do I solve each of these types of problems?
3. Is there anything in this section that gives me difficulty?

The best thing you can do to help you is to study very carefully each sample example in a section and make a few notes (one or two sentences) about each sample example. If you find difficulty with a particular example place a * next to the example and spend some extra time going over that example. You may need to talk to the instructor, a tutor, or a classmate to help you master that particular kind of problem.

Math XL

Student Solutions Manual   CD/Video   PH Math Tutor Center   MathXL®Tutorials on CD   MathXL®   MyMathLab®   Interactmath.com

Divide.

1. $(24x^2 - 8x - 44) \div 4$

2. $(18x^2 - 63x + 81) \div 9$

3. $(27x^4 - 9x^3 + 63x^2) \div 9x$

4. $(22x^4 + 33x^3 - 121x^2) \div 11x$

5. $\dfrac{8b^4 - 6b^3 - b^2}{2b^2}$

6. $\dfrac{12w^4 - 6w^3 + w^2}{3w^2}$

7. $\dfrac{18a^3b^2 + 12a^2b^2 - 4ab^2}{2ab^2}$

8. $\dfrac{25m^5n - 10m^4n + 15m^3n}{5m^3n}$

Divide. Check your answers for Exercises 11–16.

9. $(5x^2 - 17x + 6) \div (x - 3)$

10. $(6x^2 - 31x + 5) \div (x - 5)$

11. $(15x^2 + 23x + 4) \div (5x + 1)$

12. $(12x^2 + 11x + 2) \div (4x + 1)$

13. $(28x^2 - 29x + 6) \div (4x - 3)$

14. $(30x^2 - 17x + 2) \div (5x - 2)$

15. $(x^3 - x^2 + 11x - 1) \div (x + 1)$

16. $(x^3 + 2x^2 - 3x + 2) \div (x + 1)$

17. $(2x^3 - x^2 - 7) \div (x - 2)$

18. $(4x^3 - 6x - 11) \div (2x - 4)$

19. $\dfrac{4x^3 - 6x^2 - 3}{2x + 1}$

20. $\dfrac{9x^3 + 5x - 4}{3x - 1}$

21. $\dfrac{2x^4 - x^3 + 16x^2 - 4}{2x - 1}$

22. $\dfrac{9x^4 - 13x^2 - 19x + 15}{3x - 4}$

23. $\dfrac{6t^4 - 5t^3 - 8t^2 + 16t - 8}{3t^2 + 2t - 4}$

24. $\dfrac{2t^4 + 5t^3 - 11t^2 - 20t + 12}{t^2 + t - 6}$

Applications

▲ **25.** *Space Station* For the space station an engineer has designed a new rectangular solar panel that has an area of $18x^3 - 21x^2 + 11x - 2$ square meters. The length of the solar panel is $6x^2 - 5x + 2$ meters. What is the width of the solar panel?

▲ **26.** *Space Station* For the space station an engineer has designed a new rectangular solar panel that has an area of $8x^3 + 22x^2 - 29x + 6$ square meters. The length of the solar panel is $2x^2 + 7x - 2$ meters. What is the width of the solar panel?

Optional Graphing Calculator Problems

 If you have a graphing calculator, verify the following:

27. $\dfrac{2x^2 - x - 10}{2x - 5} = x + 2$

28. $\dfrac{4x^3 + 12x^2 + 7x - 3}{2x + 3} = 2x^2 + 3x - 1$

Cumulative Review

29. Find the slope of the line passing through $\left(-\dfrac{1}{3}, 0\right)$ and $\left(\dfrac{1}{2}, -1\right)$.

30. Find the slope of a line which is perpendicular to $3y - 2x = 7$.

Solve for x.

31. $2(x + 5) - 3y = 5x - (2 - y)$

32. $\dfrac{2x + 4}{3} - \dfrac{y}{2} = x - 2y + 1$

To Think About

Soft Drinks *Sylvia likes Coca-Cola, but she doesn't like root beer or grape soda. Curt likes 7-Up and root beer but not orange soda. Fritz likes ginger ale and Coca-Cola, but does not like grape soda or 7-Up.*

33. If one person likes grape soda, who is it?

34. If one person likes orange soda, who could it be?

Student Learning Objective

After studying this section, you will be able to:

 Use synthetic division to divide polynomials.

 ## Using Synthetic Division to Divide Polynomials

When dividing a polynomial by a binomial of the form $x + b$ you may find a procedure known as **synthetic division** quite efficient. Notice the following division exercises. The right-hand problem is the same as the left, but without the variables.

$$
\begin{array}{r}
3x^2 - 2x + 2 \\
x + 3\overline{)3x^3 + 7x^2 - 4x + 3} \\
\underline{3x^3 + 9x^2} \\
-2x^2 - 4x \\
\underline{-2x^2 - 6x} \\
2x + 3 \\
\underline{2x + 6} \\
-3
\end{array}
\qquad
\begin{array}{r}
3\;-2\;\;\;2 \\
1 + 3\overline{)3\;\;\;7\;-4\;\;\;3} \\
\underline{3\;\;\;9} \\
-2\;-4 \\
\underline{-2\;-6} \\
2\;\;\;3 \\
\underline{2\;\;\;6} \\
-3
\end{array}
$$

Eliminating the variables makes synthetic division efficient, and we can make the procedure simpler yet. Note that the colored numbers $(3, -2, \text{and } 2)$ appear twice in the previous example, once in the quotient and again in the subtraction. Synthetic division makes it possible to write each number only once. Also, in synthetic division we change the subtraction that division otherwise requires to addition. We do this by dropping the 1, which is the coefficient of x in the divisor, and taking the opposite of the second number in the divisor. In our first example, this means dropping the 1 and changing 3 to -3. The following steps detail synthetic division.

Step 1

$$
\begin{array}{c|cccc}
-3 & 3 & 7 & -4 & 3 \\
\hline
& 3
\end{array}
$$

Divisor, without the 1 and opposite sign   Dividend, without variables

Step 2

$$
\begin{array}{c|cccc}
-3 & 3 & 7 & -4 & 3 \\
& & -9 & & \\
\hline
& 3 & -2 & &
\end{array}
$$

Multiply $(-3)(3) = -9$ and add $7 + (-9) = -2$.

Step 3

$$
\begin{array}{c|cccc}
-3 & 3 & 7 & -4 & 3 \\
& & -9 & 6 & \\
\hline
& 3 & -2 & 2 &
\end{array}
$$

Multiply $(-3)(-2) = 6$ and add $-4 + 6 = 2$.

Step 4

$$
\begin{array}{c|cccc}
-3 & 3 & 7 & -4 & 3 \\
& & -9 & 6 & -6 \\
\hline
& 3 & -2 & 2 & \boxed{-3}
\end{array}
$$

Multiply $(-3)(2) = -6$ and add $3 + (-6) = -3$.

$3x^2 - 2x + 2 +$ remainder of -3 *Replace the variables in descending order. The degree of the quotient should be one less than the degree of the dividend.*

The result is read from the bottom row. Our answer is $3x^2 - 2x + 2 + \dfrac{-3}{x + 3}$.

EXAMPLE 1

Divide by synthetic division: $(3x^3 - x^2 + 4x + 8) \div (x + 2)$

Solution

$$
\begin{array}{r|rrrr}
-2 & 3 & -1 & 4 & 8 \\
   &   & -6 & 14 & -36 \\
\hline
   & 3 & -7 & 18 & \underline{-28}
\end{array}
$$

The quotient is $3x^2 - 7x + 18 + \dfrac{-28}{x + 2}$.

Practice Problem 1   Divide by synthetic division:
$(x^3 - 3x^2 + 4x - 5) \div (x + 3)$

When a term is missing in the sequence of descending powers of x, we use a zero to indicate the coefficient of that term.

EXAMPLE 2

Divide by synthetic division:

$$(3x^4 - 21x^3 + 31x^2 - 25) \div (x - 5)$$

Solution   Since $b = -5$, we use 5 as the divisor for synthetic division.

$$
\begin{array}{r|rrrrr}
5 & 3 & -21 & 31 & 0 & -25 \\
  &   & 15 & -30 & 5 & 25 \\
\hline
  & 3 & -6 & 1 & 5 & \underline{0}
\end{array}
$$
Note that the remainder is zero.

The quotient is $3x^3 - 6x^2 + x + 5$.

Practice Problem 2   Divide by synthetic division:
$(2x^4 - x^2 + 5x - 12) \div (x - 3)$.

NOTE TO STUDENT: Fully worked-out solutions to all of the Practice Problems can be found at the back of the text starting at page SP-1

EXAMPLE 3

Divide by synthetic division:

$$(3x^4 - 4x^3 + 8x^2 - 5x - 5) \div (x - 2)$$

Solution

$$
\begin{array}{r|rrrrr}
2 & 3 & -4 & 8 & -5 & -5 \\
  &   & 6 & 4 & 24 & 38 \\
\hline
  & 3 & 2 & 12 & 19 & \underline{33}
\end{array}
$$

The quotient is $3x^3 + 2x^2 + 12x + 19 + \dfrac{33}{x - 2}$.

Practice Problem 3   Divide by synthetic division:
$(2x^4 - 9x^3 + 5x^2 + 13x - 3) \div (x - 3)$

Divide by synthetic division.

1. $(2x^2 - 11x - 8) \div (x - 6)$

2. $(2x^2 - 15x - 23) \div (x - 9)$

3. $(3x^3 + x^2 - x + 4) \div (x + 1)$

4. $(3x^3 + 10x^2 + 6x - 4) \div (x + 2)$

5. $(x^3 + 7x^2 + 17x + 15) \div (x + 3)$

6. $(3x^3 - x^2 + 4x + 8) \div (x + 2)$

7. $(7x^3 + 6x^2 - 40x - 15) \div (x - 2)$

8. $(6x^3 - 4x^2 - 42x + 7) \div (x - 3)$

9. $(x^3 - 2x^2 + 8) \div (x + 2)$

10. $(2x^3 + 7x^2 - 5) \div (x + 3)$

11. $(6x^4 + 13x^3 + 35x - 24) \div (x + 3)$

12. $(x^4 - 2x^3 - 11x^2 + 34) \div (x + 2)$

13. $(2x^4 + 3x^3 + x^2 + 2x + 5) \div (x + 1)$

14. $(x^4 - 4x^3 + 5x^2 - 6x + 1) \div (x - 3)$

15. $(3x^5 + x - 1) \div (x + 1)$

16. $(2x^4 - x + 3) \div (x - 2)$

17. $(7x^5 - x^3 + 3x^2 + 2) \div (x + 1)$

18. $(5x^5 + 12x^4 + 25x^2 + 8) \div (x + 3)$

19. $(x^6 - 5x^3 + x^2 + 12) \div (x + 1)$

20. $(x^6 - 4) \div (x + 1)$

 21. $(x^3 + 2.5x^2 - 3.6x + 5.4) \div (x - 1.2)$

22. $(x^3 - 4.2x^2 - 8.8x + 3.7) \div (x + 1.8)$

To Think About

23. When the quotient
$(2x^4 + 12x^3 + ax^2 - 5x + 75) \div (x + 5)$ is simplified, there is no remainder. What is the value of a?

24. When the quotient
$(x^4 + 3x^3 - 2x^2 + bx + 5) \div (x + 3)$ is simplified, there is no remainder. What is the value of b?

How do we use synthetic division when the divisor is in the form $ax + b$? We divide the divisor by a to get $x + \dfrac{b}{a}$. Then, after performing the synthetic division, we divide each term of the result by a. The number that is the remainder does not change. To divide $(2x^3 + 7x^2 - 5x - 4) \div (2x + 1)$, we would use $-\dfrac{1}{2}\Big|$ 2 7 −5 −4 and then divide each term of the result by 2.

In Exercises 25 and 26, divide by synthetic division.

25. $(4x^3 - 6x^2 + 6) \div (2x + 3)$

26. $(2x^3 - 3x^2 + 6x + 4) \div (2x + 1)$

27. When the divisor is of the form $ax + b$, why does the method discussed above work? What are we really doing when we divide the divisor and the quotient by the value a?

28. Why do we not have to divide the remainder by a when using this method?

Cumulative Review

A total of 21 people were killed and 150 people injured in the Great Boston Molasses Flood in January 1919. A molasses storage tank burst and spilled 2 million gallons of molasses through the streets of Boston.

▲ **29.** *Boston Molasses Flood* How many cubic feet of molasses were contained in the 2-million-gallon molasses tank? (Use 1 gallon $\approx$ 0.134 cubic feet.)

▲ **30.** *Boston Molasses Flood* At one point the moving flood of molasses appeared as a huge cylindrically shaped object with a radius of 200 feet. At that point how deep was the molasses? (Round to the nearest tenth.)

31. If $p(x) = 2x^4 - 3x^2 + 6x - 1$, find $p(-3)$.

Student Learning Objectives

After studying this section, you will be able to:

1 Factor out the greatest common factor from a polynomial.

2 Factor a polynomial by the grouping method.

We learned to multiply polynomials in Section 5.1. When two or more algebraic expressions (monomials, binomials, and so on) are multiplied, each expression is called a **factor.**

In the rest of this chapter, we will learn how to find the factors of a polynomial. **Factoring** is the opposite of multiplication and is an extremely important mathematical technique.

1 Factoring Out the Greatest Common Factor

To factor out a common factor, we make use of the distributive property.

$$ab + ac = a(b + c)$$

The **greatest common factor** is simply the largest factor that is common to all terms of the expression. It must contain

1. The largest possible common factor of the numerical coefficients and
2. The largest possible common variable factor

EXAMPLE 1 Factor out the greatest common factor.

(a) $7x^2 - 14x$ **(b)** $40a^3 - 20a^2$

Solution

(a) $7x^2 - 14x = 7 \cdot x \cdot x - 7 \cdot 2 \cdot x = 7x(x - 2)$
Be careful. The greatest common factor is $7x$, not 7.

(b) $40a^3 - 20a^2 = 20a^2(2a - 1)$
The greatest common factor is $20a^2$.

Suppose we had written $10a(4a^2 - 2a)$ or $10a(2a)(2a - 1)$ as our answer. Although we have factored the expression, we have not found the *greatest* common factor.

Practice Problem 1 Factor out the greatest common factor.

(a) $19x^3 - 38x^2$ **(b)** $100a^4 - 50a^2$

NOTE TO STUDENT: Fully worked-out solutions to all of the Practice Problems can be found at the back of the text starting at page SP-1

EXAMPLE 2 Factor out the greatest common factor.

(a) $9x^2 - 18xy - 15y^2$ **(b)** $4a^3 - 12a^2b^2 - 8ab^3 + 6ab$

Solution

(a) $9x^2 - 18xy - 15y^2 = 3(3x^2 - 6xy - 5y^2)$
The greatest common factor is 3.

(b) $4a^3 - 12a^2b^2 - 8ab^3 + 6ab = 2a(2a^2 - 6ab^2 - 4b^3 + 3b)$
The greatest common factor is $2a$.

Practice Problem 2 Factor out the greatest common factor.

(a) $21x^3 - 18x^2y + 24xy^2$ **(b)** $12xy^2 - 14x^2y + 20x^2y^2 + 36x^3y$

How do you know whether you have factored correctly? You can do two things to verify your answer.

1. Examine the polynomial in the parentheses. Its terms should not have any remaining common factors.
2. Multiply the two factors. You should obtain the original expression.

In each of the remaining examples, you will be asked to **factor** a polynomial (i.e., to find the factors that, when multiplied, give the polynomial as a product). For each of these examples, this will require you to factor out the greatest common factor.

EXAMPLE 3 Factor $6x^3 - 9x^2y - 6x^2y^2$. Check your answer.

Solution $6x^3 - 9x^2y - 6x^2y^2 = 3x^2(2x - 3y - 2y^2)$

Check:

1. $(2x - 3y - 2y^2)$ has no common factors. If it did, we would know that we had not factored out the *greatest* common factor.

2. Multiply the two factors.

$$3x^2(2x - 3y - 2y^2) = 6x^3 - 9x^2y - 6x^2y^2$$

Observe that we do obtain the original polynomial.

Practice Problem 3 Factor $9a^3 - 12a^2b^2 - 15a^4$. Check your answer.

The greatest common factor need not be a monomial. It may be a binomial or even a trinomial. For example, note the following:

$$5a(x + 3) + 2(x + 3) = (x + 3)(5a + 2)$$
$$5a(x + 4y) + 2(x + 4y) = (x + 4y)(5a + 2)$$

The common factors are binomials.

EXAMPLE 4 Factor.

(a) $2x(x + 5) - 3(x + 5)$
(b) $5a(a + b) - 2b(a + b) - 1(a + b)$

Solution

(a) $2x(x + 5) - 3(x + 5) = (x + 5)(2x - 3)$ The common factor is $x + 5$.
(b) $5a(a + b) - 2b(a + b) - 1(a + b) = (a + b)(5a - 2b - 1)$

The common factor is $a + b$.

Practice Problem 4 Factor $7x(x + 2y) - 8y(x + 2y) - (x + 2y)$.

② Factoring by Grouping

Because the common factors in Example 4 were grouped inside parentheses, it was easy to pick them out. However, this rarely happens, so we have to learn how to manipulate expressions to find the greatest common factor.

Polynomials of four terms can often be factored by the method of Example 4(a). However, the parentheses are not always present in the original problem. When they are not present, we look for a way to remove a common factor from the first two terms. We then factor out a common factor from the first two terms and a common factor from the second two terms. Then we can find the greatest common factor of the original expression.

EXAMPLE 5 Factor $ax + 2ay + 2bx + 4by$.

Solution

Remove the greatest common factor (a) from the first two terms.

$$ax + 2ay + 2bx + 4by = a(x + 2y) + 2b(x + 2y)$$

Remove the greatest common factor ($2b$) from the last two terms.

Now we can see that $(x + 2y)$ is a common factor.

$$a(x + 2y) + 2b(x + 2y) = (x + 2y)(a + 2b)$$

NOTE TO STUDENT: *Fully worked-out solutions to all of the Practice Problems can be found at the back of the text starting at page SP-1*

Practice Problem 5 Factor $bx + 5by + 2wx + 10wy$.

EXAMPLE 6 Factor $2x^2 - 18y - 12x + 3xy$.

Solution First write the polynomial in this order: $2x^2 - 12x + 3xy - 18y$

Remove the greatest common factor ($2x$) from the first two terms.

$$2x^2 - 12x + 3xy - 18y = 2x(x - 6) + 3y(x - 6) = (x - 6)(2x + 3y)$$

Remove the greatest common factor ($3y$) from the last two terms.

Practice Problem 6 Factor $5x^2 - 12y + 4xy - 15x$.

If a problem can be factored by this method, we must rearrange the order of the four terms whenever necessary so that the first two terms do have a common factor.

EXAMPLE 7 Factor $xy - 6 + 3x - 2y$.

Solution

$xy + 3x - 2y - 6$ Rearrange the terms so that the first two terms have a common factor and the last two terms have a common factor.

$= x(y + 3) - 2(y + 3)$ Factor out a common factor of x from the first two terms and -2 from the second two terms.

$= (y + 3)(x - 2)$ Factor out the common binomial factor $y + 3$.

Practice Problem 7 Factor $xy - 12 - 4x + 3y$.

| EXAMPLE 8 | Factor $2x^3 + 21 - 7x^2 - 6x$. Check your answer by |

multiplication.

Solution

$2x^3 - 7x^2 - 6x + 21$ Rearrange the terms.

$= x^2(2x - 7) - 3(2x - 7)$ Factor out a common factor from each group
                               of two terms.

$= (2x - 7)(x^2 - 3)$ Factor out the common binomial factor $2x - 7$.

Check:

$(2x - 7)(x^2 - 3) = 2x^3 - 6x - 7x^2 + 21$ Multiply the two binomials.

$= 2x^3 + 21 - 7x^2 - 6x$ Rearrange the terms.

The product is identical to the original expression.

Practice Problem 8   Factor $2x^3 - 15 - 10x + 3x^2$.

Developing Your Study Skills

Reading the Textbook

Your homework time each day should begin with a careful reading of the section(s) assigned in your textbook. Usually, much time and effort have gone into the selection of a particular text, and your instructor has decided that this is the book that will help you become successful in this mathematics class. Textbooks are expensive, but they can be a wise investment if you take advantage of them by reading them.

Reading a mathematics textbook is unlike reading the types of books that you may find in your literature, history, psychology, or sociology courses. Mathematics texts are technical books that provide you with exercises to practice. Learning from a mathematics text requires slow and careful reading of each word, which takes time and effort.

Begin reading your textbook with a paper and pencil in hand. As you come across a new definition or concept, underline it in the text and/or write it down in your notebook. Whenever you encounter an unfamiliar term, look it up and make a note of it. When you come to an example, work through it step-by-step. Be sure to read each word and to follow directions carefully.

Notice the helpful hints the author provides. They guide you to correct solutions and prevent you from making errors. Take advantage of these pieces of expert advice.

Be sure that you understand what you are reading. Make a note of any of the things that you do not understand, and ask your instructor about them. Do not hurry through the material. Learning mathematics takes time.

5.4 EXERCISES

Student Solutions Manual | CD/ Video | PH Math Tutor Center | MathXL®Tutorials on CD | MathXL® | MyMathLab® | Interactmath.com

Factor. (Be sure to factor out the greatest common factor.)

1. $80 - 10y$

2. $16x - 16$

3. $5a^2 - 25a$

4. $7a^2 - 14a$

5. $3c^2x^3 - 9cx - 6c$

6. $5a^2b^4 + 15ab - 30a$

7. $30y^4 + 24y^3 + 18y^2$

8. $16y^5 - 24y^4 - 40y^3$

9. $15ab^2 + 5ab - 10a^3b$

10. $-12x^2y - 18xy + 6x$

11. $12xy^3 - 24x^3y^2 + 36x^2y^4 - 60x^4y^3$

12. $15a^3b^3 + 6a^4b^3 - 9a^2b^3 + 30a^5b^3$

13. $3x(x + y) - 2(x + y)$

14. $5a(a + 3b) + 4(a + 3b)$

15. $5b(a - 3b) + 8(-3b + a)$

16. $4y(x - 5y) - 3(-5y + x)$

17. $3x(a + 5b) + (a + 5b)$

Hint: Is the expression in the first parentheses equal to the expression in the second parentheses in exercises 15 and 16?

18. $2w(s - 3t) - (s - 3t)$

19. $2a^2(3x - y) - 5b^3(3x - y)$

20. $7a^3(5a + 4) - 2(5a + 4)$

21. $3x(5x + y) - 8y(5x + y) - (5x + y)$

22. $4w(y - 8x) + 5z(y - 8x) + (y - 8x)$

23. $2a(a - 6b) - 3b(a - 6b) - 2(a - 6b)$

24. $3a(a + 4b) - 5b(a + 4b) - 9(a + 4b)$

Factor.

25. $x^3 + 5x^2 + 3x + 15$

26. $x^3 + 8x^2 + 2x + 16$

27. $2x + 6 - 3ax - 9a$

28. $2bc + 4b - 5c - 10$

29. $ab - 4a + 12 - 3b$

30. $2m^2 - 8mn - 5m + 20n$

31. $5x - 30 - 2xy + 12y$

32. $4x - 20 - 3xy + 15y$

33. $9y + 2x - 6 - 3xy$

34. $10y + 3x - 6 - 5xy$

35. $yz^2 - 15 - 3z^2 + 5y$

36. $ad^4 - 4ab - d^4 + 4b$

37. $s^3r - t - s^2 + srt$

38. $28a^2x - 15b + 12bx - 35a^2$

Applications

39. ***Stacked Oranges*** The total number of oranges stacked in a pile of x rows is given by the polynomial $\frac{1}{3}x^3 + \frac{1}{2}x^2 + \frac{1}{6}x$. Write this polynomial in factored form.

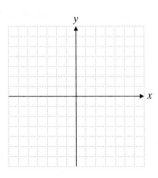

▲ **40.** ***Geometry*** The volume of the box pictured below is given by the polynomial $4x^3 + 2x^2 - 6x$. Write this polynomial in factored form.

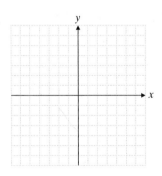

Cumulative Review

Graph the equations in 41 and 42.

41. $6x - 2y = -12$

42. $y = \frac{2}{3}x - 2$

43. Find the slope of the line passing through $(6, -1)$ and $(2, 3)$.

44. Find the slope and y-intercept of $2y + 6x = -3$.

45. ***Test Scores*** A student scored 82 on a test that had only 5-point fill-in questions and 4-point multiple-choice questions. The student had 4 fill-in questions wrong. The maximum possible score on the test was 102 points. If the entire test had 22 questions, how many multiple-choice questions did the student answer correctly?

46. ***Swimming Laps*** Reynaldo swims in a lap pool that is 50 meters long. He wants to swim at least 1 mile each day. How many laps (up and down the distance of the pool) will he have to swim to reach his goal of 1 mile per day? (Use 1 mile = 1.61 kilometers.)

Student Learning Objectives

After studying this section, you will be able to:

1 Factor trinomials of the form $x^2 + bx + c$.

2 Factor trinomials of the form $ax^2 + bx + c$.

1 Factoring Trinomials of the Form $x^2 + bx + c$

If we multiply $(x + 4)(x + 5)$, we obtain $x^2 + 9x + 20$. But suppose that we already have the polynomial $x^2 + 9x + 20$ and need to factor it. In other words, suppose we need to find the expressions that, when multiplied, give us the polynomial. Let's use this example to find a general procedure.

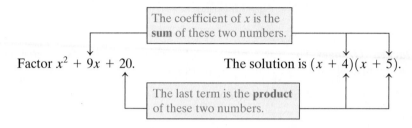

The coefficient of x is the **sum** of these two numbers.

Factor $x^2 + 9x + 20$.   The solution is $(x + 4)(x + 5)$.

The last term is the **product** of these two numbers.

FACTORING TRINOMIALS OF THE FORM $x^2 + bx + c$

1. The answer has the form $(x + m)(x + n)$, where m and n are real numbers.
2. The numbers m and n are chosen so that
   (a) $m \cdot n = c$ and
   (b) $m + n = b$.

If the last term of the trinomial is positive and the middle term is negative, the two numbers m and n will be negative numbers.

EXAMPLE 1   Factor $x^2 - 14x + 24$.

Solution   We want to find two numbers whose product is 24 and whose sum is -14. They will both be negative numbers.

Factor Pairs of 24	Sum of the Factors
$(-24)(-1)$	$-24 - 1 = -25$
$(-12)(-2)$	$-12 - 2 = -14$ ✓
$(-6)(-4)$	$-6 - 4 = -10$
$(-8)(-3)$	$-8 - 3 = -11$

The numbers whose product is 24 and whose sum is -14 are -12 and -2. Thus,

$$x^2 - 14x + 24 = (x - 12)(x - 2).$$

Practice Problem 1   Factor $x^2 - 10x + 21$.

NOTE TO STUDENT: Fully worked-out solutions to all of the Practice Problems can be found at the back of the text starting at page SP-1

If the last term of the trinomial is negative, the two numbers m and n will be opposite in sign.

EXAMPLE 2   Factor $x^2 + 11x - 26$.

Solution   We want to find two numbers whose product is -26 and whose sum is 11. One number will be positive and the other negative.

Factor Pairs of -26	Sum of the Factors
$(-26)(1)$	$-26 + 1 = -25$
$(26)(-1)$	$26 - 1 = 25$
$(-13)(2)$	$-13 + 2 = -11$
$(13)(-2)$	$13 - 2 = 11$ ✓

The numbers whose product is -26 and whose sum is 11 are -2 and 13. Thus,

$$x^2 + 11x - 26 = (x + 13)(x - 2).$$

Practice Problem 2 Factor $x^2 - 13x - 48$.

Sometimes we can make a substitution that makes a polynomial easier to factor as shown in the following example.

EXAMPLE 3 Factor $x^4 - 2x^2 - 24$.

Solution We need to recognize that we can write this as $(x^2)^2 - 2(x^2) - 24$. We can make this polynomial easier to factor if we substitute y for x^2. Then we have $y^2 + (-2)y + (-24)$. So the factors will be $(y + m)(y + n)$. The two numbers whose product is -24 and whose sum is -2 are -6 and 4. Therefore, we have $(y - 6)(y + 4)$. But $y = x^2$, so our answer is

$$x^4 - 2x^2 - 24 = (x^2 - 6)(x^2 + 4).$$

Practice Problem 3 Factor $x^4 + 9x^2 + 8$.

FACTS ABOUT SIGNS

Suppose $x^2 + bx + c = (x + m)(x + n)$. We know certain facts about m and n.

1. m and n have the same sign if c is positive. (*Note:* We did *not* say that they will have the same sign as c.)
   (a) They are positive if b is positive.
   (b) They are negative if b is negative.

2. m and n have opposite signs if c is negative. The larger number is positive if b is positive and negative if b is negative.

If you understand these sign facts, continue on to Example 4. If not, review Examples 1 through 3.

EXAMPLE 4 Factor. **(a)** $y^2 + 5y - 36$ **(b)** $x^4 - 4x^2 - 12$

Solution

(a) $y^2 + 5y - 36 = (y + 9)(y - 4)$ The larger number (9) is positive because $b = 5$ is positive.

(b) $x^4 - 4x^2 - 12 = (x^2 - 6)(x^2 + 2)$ The larger number (6) is negative because $b = -4$ is negative.

Practice Problem 4 Factor.

(a) $a^2 + 2a - 48$ **(b)** $x^4 + 2x^2 - 15$

Does the order in which we write the factors make any difference? In other words, is it true that $x^2 + bx + c = (x + n)(x + m)$? Since multiplication is commutative,

$$x^2 + bx + c = (x + n)(x + m) = (x + m)(x + n).$$

The order of the factors is not important.

We can also factor trinomials that have more than one variable.

NOTE TO STUDENT: *Fully worked-out solutions to all of the Practice Problems can be found at the back of the text starting at page SP-1*

EXAMPLE 5   Factor.   **(a)** $x^2 - 21xy + 20y^2$ **(b)** $x^2 + 4xy - 21y^2$

Solution   **(a)** $x^2 - 21xy + 20y^2 = (x - 20y)(x - y)$

The last terms in each factor contain the variable y.

(b) $x^2 + 4xy - 21y^2 = (x + 7y)(x - 3y)$

Practice Problem 5   Factor.   **(a)** $x^2 - 16xy + 15y^2$ **(b)** $x^2 + xy - 42y^2$

If the terms of a trinomial have a common factor, you should remove the greatest common factor from the terms first. Then you will be able to follow the factoring procedure we have used in the previous examples.

EXAMPLE 6   Factor $3x^2 - 30x + 48$.

Solution   The factor 3 is common to all three terms of the polynomial. Factoring out the 3 gives us the following:

$$3x^2 - 30x + 48 = 3(x^2 - 10x + 16)$$

Now we continue to factor the trinomial in the usual fashion.

$$3(x^2 - 10x + 16) = 3(x - 8)(x - 2)$$

Practice Problem 6   Factor $4x^2 - 44x + 72$.

2 Factoring Trinomials of the Form $ax^2 + bx + c$

Using the Grouping Number Method. One way to factor a trinomial $ax^2 + bx + c$ is to write it as four terms and factor it by grouping as we did in Section 5.4. For example, the trinomial $2x^2 + 11x + 12$ can be written as $2x^2 + 3x + 8x + 12$.

$$2x^2 + 3x + 8x + 12 = x(2x + 3) + 4(2x + 3)$$
$$= (2x + 3)(x + 4)$$

We can factor all factorable trinomials of the form $ax^2 + bx + c$ in this way. Use the following procedure.

> **GROUPING NUMBER METHOD FOR FACTORING TRINOMIALS OF THE FORM $ax^2 + bx + c$**
>
> 1. Obtain the grouping number ac.
> 2. Find the factor pair of the grouping number whose sum is b.
> 3. Use those two factors to write bx as the sum of two terms.
> 4. Factor by grouping.

EXAMPLE 7   Factor $2x^2 + 19x + 24$.

Solution

1. The grouping number is $(a)(c) = (2)(24) = 48$.
2. The factor pairs of 48 are as follows:

$$48 \cdot 1 \qquad 12 \cdot 4$$
$$24 \cdot 2 \qquad 8 \cdot 6$$
$$16 \cdot 3.$$

Now $b = 19$ so we want the factor pair of 48 whose sum is 19. Therefore, we select the factors 16 and 3.

3. We use the numbers 16 and 3 to write $19x$ as the sum of $16x$ and $3x$.

$$2x^2 + 19x + 24 = 2x^2 + 16x + 3x + 24$$

4. Factor by grouping.

$$2x^2 + 16x + 3x + 24 = 2x(x + 8) + 3(x + 8)$$
$$= (x + 8)(2x + 3)$$

Practice Problem 7 Factor $3x^2 + 2x - 8$. ■

EXAMPLE 8 Factor $6x^2 + 7x - 5$.

Solution

1. The grouping number is -30.
2. We want the factor pair of -30 whose sum is 7.

$$-30 = (-30)(1) \qquad -30 = (5)(-6)$$
$$= (30)(-1) \qquad\qquad = (-5)(6)$$
$$= (15)(-2) \qquad\qquad = (3)(-10)$$
$$= (-15)(2) \qquad\qquad = (-3)(10)$$

3. Since $-3 + 10 = 7$, use -3 and 10 to write $6x^2 + 7x - 5$ with four terms.

$$6x^2 + 7x - 5 = 6x^2 - 3x + 10x - 5$$

4. Factor by grouping.

$$6x^2 - 3x + 10x - 5 = 3x(2x - 1) + 5(2x - 1)$$
$$= (2x - 1)(3x + 5)$$

Practice Problem 8 Factor $10x^2 - 9x + 2$. ■

If the three terms have a common factor, then prior to using the four-step grouping number procedure, we first factor out the greatest common factor from the terms of the trinomial.

EXAMPLE 9 Factor $6x^3 - 26x^2 + 24x$.

Solution First we factor out the greatest common factor $2x$ from each term.

$$6x^3 - 26x^2 + 24x = 2x(3x^2 - 13x + 12)$$

Next we follow the four steps to factor $3x^2 - 13x + 12$.

1. The grouping number is 36.
2. We want the factor pair of 36 whose sum is -13. The two factors are -4 and -9.
3. We use -4 and -9 to write $3x^2 - 13x + 12$ with four terms.

$$3x^2 - 13x + 12 = 3x^2 - 4x - 9x + 12$$

4. Factor by grouping. Remember that we first factored out the factor $2x$. This factor must be part of the answer.

$$2x(3x^2 - 4x - 9x + 12) = 2x[x(3x - 4) - 3(3x - 4)]$$
$$= 2x(3x - 4)(x - 3)$$

Practice Problem 9 Factor $9x^3 - 15x^2 - 6x$. ■

Using the Trial-and-Error Method. Another way to factor trinomials of the form $ax^2 + bx + c$ is by trial and error. This method has an advantage if the grouping number is large and we would have to list many factors. In the trial-and-error method, we try different values and see which ones can be multiplied out to obtain the original expression.

If the last term is negative, there are many more sign possibilities.

EXAMPLE 10   Factor by trial and error $10x^2 - 49x - 5$.

Solution   The first terms in the factors could be $(10x)$ and (x) or $(5x)$ and $(2x)$. The second terms could be $(+1)$ and (-5) or (-1) and $(+5)$. We list all the possibilities and look for one that will yield a middle term of $-49x$.

Possible Factors	Middle Term of Product
$(2x - 1)(5x + 5)$	$+5x$
$(2x + 1)(5x - 5)$	$-5x$
$(2x + 5)(5x - 1)$	$+23x$
$(2x - 5)(5x + 1)$	$-23x$
$(10x - 5)(x + 1)$	$+5x$
$(10x + 5)(x - 1)$	$-5x$
$(10x - 1)(x + 5)$	$+49x$
$(10x + 1)(x - 5)$	$-49x$

Thus,

$$10x^2 - 49x - 5 = (10x + 1)(x - 5)$$

As a check, it is always a good idea to multiply the two binomials to see whether you obtain the original expression.

$$(10x + 1)(x - 5) = 10x^2 - 50x + 1x - 5.$$
$$= 10x^2 - 49x - 5$$

NOTE TO STUDENT: Fully worked-out solutions to all of the Practice Problems can be found at the back of the text starting at page SP-1

Practice Problem 10   Factor by trial and error $8x^2 - 6x - 5$.

EXAMPLE 11   Factor by trial and error $6x^4 + x^2 - 12$.

Solution   The first term of each factor must contain x^2. Suppose that we try the following:

Possible Factors	Middle Term of Product
$(2x^2 - 3)(3x^2 + 4)$	$-x^2$

The middle term we get is $-x^2$, but we need its opposite, x^2. In this case, we just need to reverse the signs of -3 and 4. Do you see why? Therefore,

$$6x^4 + x^2 - 12 = (2x^2 + 3)(3x^2 - 4).$$

Practice Problem 11   Factor by trial and error $6x^4 + 13x^2 - 5$.

5.5 EXERCISES

| Student Solutions Manual | CD/ Video | PH Math Tutor Center | MathXL®Tutorials on CD | MathXL® | MyMathLab® | Interactmath.com |

Factor each polynomial.

1. $x^2 + 8x + 7$ **2.** $x^2 + 12x + 11$ **3.** $x^2 - 9x + 14$ **4.** $x^2 - 8x + 12$

5. $x^2 - 10x + 24$ **6.** $x^2 - 9x + 18$ **7.** $a^2 + 4a - 45$ **8.** $a^2 + 2a - 35$

9. $x^2 - xy - 42y^2$ **10.** $x^2 + xy - 30y^2$ **11.** $x^2 - 15xy + 14y^2$ **12.** $x^2 + 10xy + 9y^2$

13. $x^4 - 3x^2 - 40$ **14.** $x^4 + 6x^2 + 5$ **15.** $x^4 + 16x^2y^2 + 63y^4$ **16.** $x^4 - 6x^2 - 55$

Factor out the greatest common factor from the terms of the trinomial. Then factor the remaining trinomial.

17. $2x^2 + 26x + 44$ **18.** $2x^2 + 30x + 52$ **19.** $x^3 + x^2 - 20x$ **20.** $x^3 - 4x^2 - 45x$

Factor each polynomial. You may use the grouping number method or the trial-and-error method.

21. $2x^2 - x - 1$ **22.** $3x^2 + x - 2$ **23.** $6x^2 - 7x - 5$ **24.** $5x^2 - 13x - 28$

25. $3a^2 - 8a + 5$ **26.** $6a^2 + 11a + 3$ **27.** $8a^2 + 14a - 9$ **28.** $3a^2 - 20a + 12$

29. $2x^2 + 13x + 15$ **30.** $5x^2 - 8x - 4$ **31.** $3x^4 - 8x^2 - 3$ **32.** $6x^4 + 7x^2 - 5$

33. $6x^2 + 35xy + 11y^2$ **34.** $5x^2 + 12xy + 7y^2$ **35.** $7x^2 + 11xy - 6y^2$ **36.** $4x^2 - 13xy + 3y^2$

Factor out the greatest common factor from the terms of the trinomial. Then factor the remaining trinomial.

37. $4x^3 + 4x^2 - 15x$ **38.** $8x^3 + 6x^2 - 9x$ **39.** $10x^4 + 15x^3 + 5x^2$ **40.** $16x^4 + 48x^3 + 20x^2$

Mixed Practice

Factor each polynomial.

41. $x^2 - 2x - 63$ **42.** $x^2 + 6x - 40$ **43.** $6x^2 + x - 2$ **44.** $5x^2 + 17x + 6$

45. $x^2 - 20x + 51$ **46.** $x^2 - 20x + 99$ **47.** $15x^2 + x - 2$ **48.** $12x^2 - 5x - 3$

49. $2x^2 + 4x - 96$ **50.** $3x^2 + 9x - 84$ **51.** $18x^2 + 21x + 6$ **52.** $24x^2 + 26x + 6$

53. $27ax^2 + 99ax - 36a$ **54.** $77bx^2 - 44bx - 33b$ **55.** $6x^3 + 26x^2 - 20x$ **56.** $12x^3 - 14x^2 + 4x$

57. $3x^4 - 2x^2 - 5$ **58.** $6x^4 - 13x^2 - 5$ **59.** $9a^2 - 18ab - 7b^2$ **60.** $13a^2 - 8ab - 5b^2$

61. $x^6 - 10x^3 - 39$ **62.** $x^6 - 3x^3 - 70$ **63.** $4x^3y + 2x^2y - 2xy$ **64.** $9x^3y + 24x^2y - 9xy$

Applications

▲ **65.** *Tree Reforestation* A plan has been made in northern Maine to replace trees harvested by paper mills. The proposed planting zone is in the shape of a giant rectangle with an area of $30x^2 + 19x - 5$ square feet. Use your factoring skills to determine a possible configuration of the number of rows of trees and the number of trees to be placed in each row.

▲ **66.** *Tree Reforestation* A plan has been made in northern Washington to replace trees harvested by paper mills. The proposed planting zone is in the shape of a giant rectangle with an area of $12x^2 + 20x - 25$ square feet. Use your factoring skills to determine a possible configuration of the number of rows of trees and the number of trees to be placed in each row.

Cumulative Review

▲ **67.** Find the area of a circle of radius 3 inches.

68. Solve for b: $A = \dfrac{1}{3}(3b + 4a)$

69. *Steepness of a Hill* A state college campus has a paved road called a "loop" that encircles the entire campus. There are two very steep hills at each end of the loop.
(a) Find the slope (pitch) of the hill that rises 48 yards vertically over a horizontal distance of 156 yards.
(b) Does this hill violate the city ordinance requiring all roads to have a slope that does not exceed 30%?

70. Graph $6x + 4y = -12$.

71. *Bicycle Shop Operation* John and Carolyn Ciukaj have opened a new bicycle shop in Beverly. They want to have at least 120 bicycle racks and bicycle helmets in stock. Their wholesale cost for bicycle racks averages around $60 and for bicycle helmets around $70. They have available $7950 in capital to pay for bicycle racks and bicycle helmets. How many of each should they stock?

72. *Airplane Seating* A large commercial jetliner flies from Atlanta to San Francisco. The jetliner can carry a total of 184 passengers. There are two types of seats on the aircraft: first class and coach. The number of coach seats is sixteen more than six times the number of first class seats. How many of each type of seat are there on the airplane?

How are you doing with your homework assignments in Sections 5.1 to 5.5? Do you feel you have mastered the material so far? Do you understand the concepts you have covered? Before you go further in the textbook, take some time to do each of the following problems.

5.1

Simplify.

1. $(5x^2 - 3x + 2) + (-3x^2 - 5x - 8) - (x^2 + 3x - 10)$

2. $(x^2 - 3x - 4)(2x - 3)$

3. $(5a - 8)(a - 7)$

4. $(2y - 3)(2y + 3)$

5. $(3x^2 + 4)^2$

6. Evaluate the polynomial function $p(x) = 2x^3 - 5x^2 - 6x + 1$ for $p(-3)$.

5.2 and 5.3

Divide.

7. $(25x^3y^2 - 30x^2y^3 - 50x^2y^2) \div 5x^2y^2$

8. $(3y^3 - 5y^2 + 2y - 1) \div (y - 2)$

9. $(2x^4 + 9x^3 + 8x^2 - 9x - 10) \div (2x + 5)$

10. Use synthetic division to do the following:
$(2x^4 + 10x^3 + 11x^2 - 6x - 9) \div (x + 3)$

5.4

Factor completely.

11. $24a^3b^2 + 36a^4b^2 - 60a^3b^3$

12. $3x(4x - 3y) - 2(4x - 3y)$

13. $10wx + 6xz - 15yz - 25wy$

14. $10a^2 - 8ab - 5ab + 4b^2$

5.5

Factor completely.

15. $x^2 - 7x + 10$

16. $4y^2 - 4y - 15$

17. $28x^2 - 19xy + 3y^2$

18. $2x^2 + 17x + 35$

19. $3x^2 - 6x - 72$

20. $8x^2 - 18x + 9$

Now turn to page SA-21 for the answers to each of these problems. Each answer also includes a reference to the objective in which the problem is first taught. If you missed any of these problems, you should stop and review the Examples and Practice Problems in the referenced objective. A little review now will help you master the material in the upcoming sections of the text.

1. _____
2. _____
3. _____
4. _____
5. _____
6. _____
7. _____
8. _____
9. _____
10. _____
11. _____
12. _____
13. _____
14. _____
15. _____
16. _____
17. _____
18. _____
19. _____
20. _____

1 Factoring the Difference of Two Squares

Recall the special product formula: $(a + b)(a - b) = a^2 - b^2$. We can use it now as a factoring formula.

> **FACTORING THE DIFFERENCE OF TWO SQUARES**
>
> $$a^2 - b^2 = (a + b)(a - b)$$

EXAMPLE 1 Factor $x^2 - 16$.

Solution In this case $a = x$ and $b = 4$ in the formula.

$$\begin{array}{cccccc} a^2 & - & b^2 & = & (a + b) & (a - b) \\ \downarrow & & \downarrow & & \downarrow\ \downarrow & \downarrow \\ (x)^2 & - & (4)^2 & = & (x + 4) & (x - 4) \end{array}$$

Practice Problem 1 Factor $x^2 - 9$.

EXAMPLE 2 Factor $25x^2 - 36$.

Solution In each case we will use the formula $a^2 - b^2 = (a + b)(a - b)$.

$$25x^2 - 36 = (5x)^2 - (6)^2 = (5x + 6)(5x - 6)$$

Practice Problem 2 Factor $64x^2 - 121y^2$.

EXAMPLE 3 Factor $100w^4 - 9z^4$.

Solution $100w^4 - 9z^4 = (10w^2)^2 - (3z^2)^2 = (10w^2 + 3z^2)(10w^2 - 3z^2)$

Practice Problem 3 Factor $49x^2 - 25y^4$.

Whenever possible, a common factor should be factored out in the first step. Then the formula can be applied.

EXAMPLE 4 Factor $75x^2 - 3$.

Solution We factor out a common factor of 3 from each term.

$$\begin{aligned} 75x^2 - 3 &= 3(25x^2 - 1) \\ &= 3(5x + 1)(5x - 1) \end{aligned}$$

Practice Problem 4 Factor $7x^2 - 28$.

NOTE TO STUDENT: Fully worked-out solutions to all of the Practice Problems can be found at the back of the text starting at page SP-1

2 Factoring Perfect Square Trinomials

Recall the formulas for squaring a binomial.

$$(a - b)^2 = a^2 - 2ab + b^2$$
$$(a + b)^2 = a^2 + 2ab + b^2$$

We can use these formulas to factor a perfect square trinomial.

PERFECT SQUARE FACTORING FORMULAS

$$a^2 - 2ab + b^2 = (a - b)^2$$
$$a^2 + 2ab + b^2 = (a + b)^2$$

Recognizing these special cases will save you a lot of time when factoring. How can we recognize a perfect square trinomial?

1. The first and last terms are perfect squares. (The numerical values are $1, 4, 9, 16, 25, 36, \ldots$, and the variables have an exponent that is an even whole number.)
2. The middle term is twice the product of the values that, when squared, give the first and last terms.

EXAMPLE 5   Factor $25x^2 - 20x + 4$.

Solution   Is this trinomial a perfect square? Yes.

1. The first and last terms are perfect squares.
$$25x^2 - 20x + 4 = (5x)^2 - 20x + (2)^2$$

2. The middle term is twice the product of the value $5x$ and the value 2. In other words, $2(5x)(2) = 20x$.
$$(5x)^2 - 2(5x)(2) + (2)^2 = (5x - 2)^2$$

Therefore, we can use the formula $a^2 - 2ab + b^2 = (a - b)^2$. Thus,
$$25x^2 - 20x + 4 = (5x - 2)^2.$$

Practice Problem 5   Factor $9x^2 - 30x + 25$.

EXAMPLE 6   Factor $16x^2 - 24x + 9$.

Solution

1. The first and last terms are perfect squares: $16x^2 = (4x)^2$ and $9 = (3)^2$.
2. The middle term is twice the product $(4x)(3)$. Therefore, we have the following:
$$a^2 - 2ab + b^2 = (a - b)^2$$
$$16x^2 - 24x + 9 = (4x)^2 - 2(4x)(3) + (3)^2$$
$$16x^2 - 24x + 9 = (4x - 3)^2.$$

Practice Problem 6   Factor $25x^2 - 70x + 49$.

EXAMPLE 7   Factor $200x^2 + 360x + 162$.

Solution   First we factor out the common factor of 2.

$$200x^2 + 360x + 162 = 2(100x^2 + 180x + 81)$$
$$a^2 + 2ab + b^2 = (a + b)^2$$
$$2[100x^2 + 180x + 81] = 2[(10x)^2 + (2)(10x)(9) + (9)^2]$$
$$= 2(10x + 9)^2$$

Practice Problem 7   Factor $242x^2 + 88x + 8$.

> **EXAMPLE 8**   Factor.   **(a)** $x^4 + 14x^2 + 49$ **(b)** $9x^4 + 30x^2y^2 + 25y^4$
>
> **Solution**
>
> **(a)** $x^4 + 14x^2 + 49 = (x^2)^2 + 2(x^2)(7) + (7)^2$
> $$= (x^2 + 7)^2$$
>
> **(b)** $9x^4 + 30x^2y^2 + 25y^4 = (3x^2)^2 + 2(3x^2)(5y^2) + (5y^2)^2$
> $$= (3x^2 + 5y^2)^2$$

Practice Problem 8   Factor.

(a) $49x^4 + 28x^2 + 4$ **(b)** $36x^4 + 84x^2y^2 + 49y^4$

③ Factoring the Sum or Difference of Two Cubes

There are also special formulas for factoring cubic binomials. We see that the factors of $x^3 + 27$ are $(x + 3)(x^2 - 3x + 9)$, and that the factors of $x^3 - 64$ are $(x - 4)(x^2 + 4x + 16)$. Therefore, we can generalize this pattern and derive the following factoring formulas.

> **SUM AND DIFFERENCE OF CUBES FACTORING FORMULAS**
> $$a^3 + b^3 = (a + b)(a^2 - ab + b^2)$$
> $$a^3 - b^3 = (a - b)(a^2 + ab + b^2)$$

> **EXAMPLE 9**   Factor $125x^3 + y^3$.
>
> **Solution**   Here $a = 5x$ and $b = y$.
> $$a^3 + b^3 = (a + b)(a^2 - ab + b^2)$$
> $$125x^3 + y^3 = (5x)^3 + (y)^3 = (5x + y)(25x^2 - 5xy + y^2)$$

Practice Problem 9   Factor $8x^3 + 125y^3$.

> **EXAMPLE 10**   Factor $64x^3 - 27$.
>
> **Solution**   Here $a = 4x$ and $b = 3$.
> $$a^3 - b^3 = (a - b)(a^2 + ab + b^2)$$
> $$64x^3 - 27 = (4x)^3 - (3)^3 = (4x - 3)(16x^2 + 12x + 9)$$

Practice Problem 10   Factor $64x^3 - 125y^3$.

> **EXAMPLE 11**   Factor $125w^3 - 8z^6$.
>
> **Solution**   Here $a = 5w$ and $b = 2z^2$.
> $$a^3 - b^3 = (a - b)(a^2 + ab + b^2)$$
> $$125w^3 - 8z^6 = (5w)^3 - (2z^2)^3 = (5w - 2z^2)(25w^2 + 10wz^2 + 4z^4)$$

NOTE TO STUDENT: Fully worked-out solutions to all of the Practice Problems can be found at the back of the text starting at page SP-1

Practice Problem 11   Factor $27w^3 - 125z^6$.

EXAMPLE 12   Factor $250x^3 - 2$.

Solution   First we must factor out the common factor of 2.

$$250x^3 - 2 = 2(125x^3 - 1)$$
$$= 2(5x - 1)\underbrace{(25x^2 + 5x + 1)}$$
$$\uparrow$$

Note that this trinomial cannot be factored.

Practice Problem 12   Factor $54x^3 - 16$.

What should you do if a polynomial is the difference of two cubes *and* the difference of two squares? Usually, it's easier to use the difference of two squares formula first. Then apply the difference of two cubes formula.

EXAMPLE 13   Factor $x^6 - y^6$.

Solution   We can write this binomial as $(x^2)^3 - (y^2)^3$ or as $(x^3)^2 - (y^3)^2$. Therefore, we can use either the difference of two cubes formula or the difference of two squares formula. It's usually better to use the difference of two squares formula first, so we'll do that.

$$x^6 - y^6 = (x^3)^2 - (y^3)^2$$

Here $a = x^3$ and $b = y^3$. Therefore,

$$(x^3)^2 - (y^3)^2 = (x^3 + y^3)(x^3 - y^3).$$

Now we use the sum of two cubes formula for the first factor and the difference of two cubes formula for the second factor.

$$x^3 + y^3 = (x + y)(x^2 - xy + y^2)$$
$$x^3 - y^3 = (x - y)(x^2 + xy + y^2)$$

Hence,

$$x^6 - y^6 = (x + y)(x^2 - xy + y^2)(x - y)(x^2 + xy + y^2).$$

Practice Problem 13   Factor $64a^6 - 1$.

You'll see these special cases of factoring often. You should be very familiar with the following formulas. Be sure you understand how to use each one of them.

SPECIAL CASES OF FACTORING

Difference of Two Squares

$$a^2 - b^2 = (a + b)(a - b)$$

Perfect Square Trinomial

$$a^2 - 2ab + b^2 = (a - b)^2$$
$$a^2 + 2ab + b^2 = (a + b)^2$$

Sum and Difference of Cubes

$$a^3 + b^3 = (a + b)(a^2 - ab + b^2)$$
$$a^3 - b^3 = (a - b)(a^2 + ab + b^2)$$

Verbal and Writing Skills

1. How do you determine if a factoring problem will use the difference of two squares?

2. How do you determine if a factoring problem will use the perfect square trinomial formula?

3. How do you determine if a factoring problem will use the sum of two cubes formula?

4. How do you determine if a factoring problem will use the difference of two cubes formula?

Use the difference of two squares formula to factor. Be sure to factor out any common factors.

5. $a^2 - 64$

6. $y^2 - 49$

7. $16x^2 - 81$

8. $4x^2 - 25$

9. $64x^2 - 1$

10. $81x^2 - 1$

11. $49m^2 - 9n^2$

12. $36x^2 - 25y^2$

13. $100y^2 - 81$

14. $49y^2 - 144$

15. $1 - 81x^2y^2$

16. $1 - 49x^2y^2$

17. $32x^2 - 18$

18. $50x^2 - 8$

19. $5x - 20x^3$

20. $49x^3 - 36x$

Use the perfect square trinomial formulas to factor. Be sure to factor out any common factors.

21. $9x^2 - 6x + 1$

22. $16y^2 - 8y + 1$

23. $49x^2 - 14x + 1$

24. $100y^2 - 20y + 1$

25. $81w^2 + 36wt + 4t^2$

26. $25w^2 + 20wt + 4t^2$

27. $36x^2 + 60xy + 25y^2$

28. $64x^2 + 48xy + 9y^2$

29. $8x^2 + 24x + 18$

30. $128x^2 + 32x + 2$

31. $3x^3 - 24x^2 + 48x$

32. $50x^3 - 20x^2 + 2x$

Use the sum and difference of cubes formulas to factor. Be sure to factor out any common factors.

33. $x^3 - 27$

34. $x^3 - 8$

35. $x^3 + 125$

36. $x^3 + 64$

37. $64x^3 - 1$

38. $125x^3 - 1$

39. $125x^3 - 8$

40. $27x^3 - 64$

41. $1 - 27x^3$

42. $1 - 8x^3$

43. $64x^3 + 125$

44. $27x^3 + 125$

45. $64s^6 + t^6$

46. $125s^6 + t^6$

47. $6y^3 - 6$

48. $80y^3 - 10$

49. $3x^3 - 24$ **50.** $54y^2 - 2$ **51.** $x^5 - 8x^2y^3$ **52.** $x^5 - 27x^2y^3$

Mixed Practice

Factor by the methods of this section.

53. $25w^4 - 1$ **54.** $16m^4 - 25$ **55.** $b^4 + 6b^2 + 9$ **56.** $a^4 - 10a^2 + 25$

57. $49m^6 - 81$ **58.** $4 - 9m^6$ **59.** $36y^6 - 60y^3 + 25$ **60.** $100n^6 - 140n^3 + 49$

61. $2a^8 - 50$ **62.** $12z^8 - 27$ **63.** $125m^3 + 8n^3$

64. $64z^3 - 27w^3$ **65.** $24a^3 - 3b^3$ **66.** $54w^3 + 250$

67. $4w^2 - 20wz + 25z^2$ **68.** $81x^4 - 36x^2 + 4$ **69.** $36a^2 - 81b^2$

70. $400x^4 - 36y^2$ **71.** $16x^4 - 81y^4$ **72.** $256x^4 - 1$

73. $125m^6 + 8$ **74.** $27n^6 + 125$

Try to factor the following four exercises by using the formulas for the perfect square trinomial. Why can't the formulas be used? Then factor each exercise correctly using an appropriate method.

75. $25x^2 + 25x + 4$ **76.** $16x^2 + 40x + 9$

77. $4x^2 - 15x + 9$ **78.** $36x^2 - 65x + 25$

Applications

▲ **79.** *Carpentry* Find the area of a maple cabinet surface that is constructed by a carpenter as a large square with sides of $4x$ feet and has a square cut out region whose sides are y feet. Factor the expression.

▲ **80.** *Base of a Lamp* A copper base for a lamp consists of a large circle of radius $2y$ inches with a cut out area in the center of radius x inches. Write an expression for the area of this copper base. Write your answer in factored form.

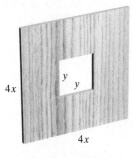

Cumulative Review

81. *Marketing Salary* The average beginning annual salary y offered to a student graduating with a bachelor's degree in marketing can be approximated by the equation $y = 1200x + 27{,}000$, where x is the number of years since 1996. The average beginning annual salary y offered to a student graduating with a bachelor's degree in mathematics can be approximated by the equation $y = 3200x + 29{,}000$, where x in the number of years since 1996. If this trend continues, in what year will a student graduating with a bachelor's degree in math be offered $20,000 more per year than a student graduating with a bachelor's degree in marketing? *Source:* U.S. Census Bureau.

▲ **83.** *Geometry* A triangular circuit board has a perimeter of 66 centimeters. The first side is two-thirds as long as the second side. The third side is 14 centimeters shorter than the second side. Find the length of each side.

82. *Investments* Belinda invested $4000 in mutual funds. In 1 year she earned $398. Part was invested at 8% simple interest and the remainder at 11% simple interest. How much did she invest at each rate?

84. *Compact Disc Player Prices* Three friends each bought a portable compact disc player. The total for the three purchases was $858. Melinda paid $110 more than Hector. Alice paid $86 less than Hector. How much did each person pay?

Developing Your Study Skills

Keep Trying

You may be one of those students who has had much difficulty with mathematics in the past and who is sure that you cannot do well in this course. Perhaps you are thinking, "I have never been any good at mathematics" or "I have always hated mathematics" or "Math scares me" or "I have not had any math for so long that I have forgotten it all." You may even have picked up on the label "math anxiety" and attached it to yourself.

It is time for you to reprogram your thinking. Replace those negative thoughts with more positive ones. You need to say things like, "I will give this math class my best shot" or "I can learn mathematics if I work at it" or "I will try to do better than I have done in previous math classes." You will be pleasantly surprised at the difference a positive attitude makes!

We live in a highly technical world, and you cannot afford to give up on the study of mathematics. Dropping mathematics may prevent you from entering certain career fields that you find interesting. You may not have to take math courses at as high a level as calculus, but such courses as finite math, college algebra, and trigonometry may be necessary. Learning mathematics can open new doors for you.

Learning mathematics is a process that takes time and effort. You will find that regular study and daily practice are necessary to strengthen your skills and to help you grow academically. This process will lead you toward success in mathematics. Then, as you become more successful, your confidence in your ability to do mathematics will grow.

5.7 FACTORING A POLYNOMIAL COMPLETELY

① Factoring Factorable Polynomials

Not all polynomials have the convenient form of one of the special formulas. Most do not. The following procedure will help you handle these common cases. You must practice this procedure until you can *recognize the various forms* and *determine which factoring method to use*.

Student Learning Objectives

After studying this section, you will be able to:

 Factor any factorable polynomial.

 Recognize polynomials that are prime.

COMPLETELY FACTORING A POLYNOMIAL

1. Check for a common factor. Factor out the greatest common factor (if there is one) before doing anything else.

2. **(a)** If the remaining polynomial has two terms, try to factor it as one of the following.
   (1) The difference of two squares: $a^2 - b^2 = (a + b)(a - b)$
   (2) The difference of two cubes: $a^3 - b^3 = (a - b)(a^2 + ab + b^2)$
   (3) The sum of two cubes: $a^3 + b^3 = (a + b)(a^2 - ab + b^2)$

   (b) If the remaining polynomial has three terms, try to factor it as one of the following.
   (1) A perfect square trinomial: $a^2 + 2ab + b^2 = (a + b)^2$ or $a^2 - 2ab + b^2 = (a - b)^2$
   (2) A general trinomial of the form $x^2 + bx + c$ or the form $ax^2 + bx + c$

   (c) If the remaining polynomial has four terms, try to factor by grouping.

3. Check to see whether the factors can be factored further.

EXAMPLE 1 Factor completely.

(a) $2x^2 - 18$ **(b)** $27x^4 - 8x$

(c) $27x^2 + 36xy + 12y^2$ **(d)** $2x^2 - 100x + 98$

(e) $6x^3 + 11x^2 - 10x$ **(f)** $5ax + 5ay - 20x - 20y$

Solution

(a) $2x^2 - 18 = 2(x^2 - 9)$ Factor out the common factor.
$\qquad\qquad = 2(x + 3)(x - 3)$ Use $a^2 - b^2 = (a + b)(a - b)$.

(b) $27x^4 - 8x = x(27x^3 - 8)$ Factor out the common factor.
$\qquad\qquad = x(3x - 2)(9x^2 + 6x + 4)$ Use $a^3 - b^3 = (a - b)(a^2 + ab + b^2)$.

(c) $27x^2 + 36xy + 12y^2 = 3(9x^2 + 12xy + 4y^2)$ Factor out the common factor.
$\qquad\qquad = 3(3x + 2y)^2$ Use $(a + b)^2 = a^2 + 2ab + b^2$.

(d) $2x^2 - 100x + 98 = 2(x^2 - 50x + 49)$ Factor out the common factor.
$\qquad\qquad = 2(x - 49)(x - 1)$ The trinomial has the form $x^2 + bx + c$.

(e) $6x^3 + 11x^2 - 10x = x(6x^2 + 11x - 10)$ Factor out the common factor.
$\qquad\qquad = x(3x - 2)(2x + 5)$ The trinomial has the form $ax^2 + bx + c$.

(f) $5ax + 5ay - 20x - 20y = 5(ax + ay - 4x - 4y)$ Factor out the common factor.

$= 5[a(x + y) - 4(x + y)]$ Factor by grouping.

$= 5(x + y)(a - 4)$

NOTE TO STUDENT: Fully worked-out solutions to all of the Practice Problems can be found at the back of the text starting at page SP-1

Practice Problem 1   Factor completely.

(a) $7x^5 + 56x^2$

(b) $125x^2 + 50xy + 5y^2$

(c) $12x^2 - 75$

(d) $3x^2 - 39x + 126$

(e) $6ax + 6ay + 18bx + 18by$

(f) $6x^3 - x^2 - 12x$

2 Recognizing Polynomials That Are Prime

Can all polynomials be factored? No. Many polynomials cannot be factored. If a polynomial cannot be factored using rational numbers, it is said to be **prime.**

EXAMPLE 2   If possible, factor $6x^2 + 10x + 3$.

Solution   The trinomial has the form $ax^2 + bx + c$. The grouping number is 18. If the trinomial can be factored, we must find two numbers whose product is 18 and whose sum is 10.

Factor Pairs of 18	Sum of the Factors
(18)(1)	19
(6)(3)	9
(9)(2)	11

There are no numbers meeting the necessary conditions. Thus, the polynomial is prime. (If you use the trial-and-error method, try all the possible factors and show that none of them has a product with a middle term of $10x$.)

Practice Problem 2   If possible, factor $3x^2 - 10x + 4$.

EXAMPLE 3   If possible, factor $25x^2 + 49$.

Solution   Unless there is a common factor that can be factored out, binomials of the form $a^2 + b^2$ cannot be factored. Therefore, $25x^2 + 49$ is prime.

Practice Problem 3   If possible, factor $16x^2 + 81$.

Verbal and Writing Skills

1. In any factoring problem the first step is
_____.

2. If $x^2 + bx + c = (x + e)(x + f)$ and c is positive and b is negative, what can you say about the signs of e and of f?

3. If you were asked to factor a problem of the form $49x^2 + 9y^2$, how would you know immediately that this polynomial is prime?

4. If you were asked to factor a problem of the form $x^2 + 9x + 12$, how would you know very quickly that this polynomial is prime?

Factor, if possible. These problems will require only one step.

5. $3xy - 6yz$

6. $33a - 44a^2b$

7. $y^2 + 7y - 18$

8. $b^2 - 7b + 12$

9. $3x^2 - 8x + 5$

10. $3x^2 + 4x - 7$

11. $ax - 2xy + 3aw - 6wy$

12. $ax - 3xy + 4aw - 12wy$

13. $8x^3 - 125y^3$

14. $27x^3 + 64y^3$

15. $x^2 + 2xy - xz$

16. $2x^2 - 3xy + xz$

17. $x^2 + 16$

18. $4x^4 + 25$

19. $64y^2 - 25z^2$

20. $81m^2 - 49n^2$

Mixed Practice

Factor if possible. Be sure to factor completely.

21. $6x^2 - 23x - 4$

22. $5x^2 + x - 4$

23. $3x^2 - x - 1$

24. $5x^2 - x - 2$

25. $x^3 - 11x^2 + 30x$

26. $x^3 + 4x^2 - 21x$

27. $25x^2 - 40x + 16$

28. $9r^2 + 48r + 64$

29. $6a^2 - 6a - 36$

30. $9a^2 + 18a - 72$

31. $3x^2 - 3x - xy + y$

32. $xb - x - yb + y$

33. $81a^4 - 1$

34. $1 - 16x^4$

35. $2x^5 - 16x^3 - 18x$

36. $2x^4 - 2x^2 - 24$ **37.** $8a^3b - 50ab^3$ **38.** $50x^2y^2 - 32y^2$

39. $4x^2 - 8x - 6$ **40.** $10x^2 + 5x + 5$

Applications

Cattle Farming   *A cattle pen is constructed with solid wood walls. The pen is divided into four rectangular compartments. Each compartment is x feet long and y feet wide. The walls are x − 10 feet high.*

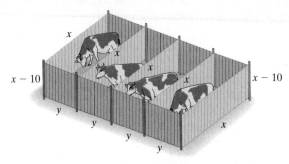

▲ **41.** Find the total surface area of the walls used in the cattle pen. Express the answer in factored form and in the form with the factors multiplied.

▲ **42.** The rancher who owns the cattle pen wants to increase the length x by 3 feet. Find the new total surface area of the walls that would be used in this enlarged cattle pen. Express the answer in factored form and in the form with the factors multiplied.

Cumulative Review

Solve the following inequalities.

43. $3x - 2 \le -5 + 2(x - 3)$ **44.** $|2 + 5x - 3| < 2$

45. $\left| \dfrac{1}{3}(5 - 4x) \right| > 4$ **46.** $x - 4 \ge 7$ or $4x + 1 \le 17$

Political Party Campaign Expenses   *Use the bar graph at the right to answer the following.*

47. What was the average value of the net receipts for a 2-year period for the Republican Party?

48. What was the average percent of increase from one 2-year period to the next for the Democratic Party? (*Hint:* First find the percents of increase from 1985–86 to 1989–90, from 1989–90 to 1993–94, from 1993–94 to 1997–98, and from 1997–98 to 2001–02. Then average the four values.)

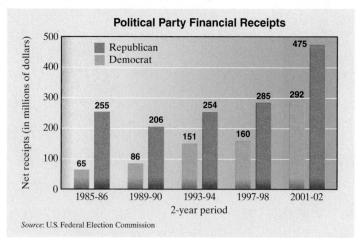

49. Assuming the average percent of increase obtained in Exercise 48, what would be the expected net receipts for the Democratic party in the 2005–2006 period?

50. If the percent of increase from 1997–98 to 2001–2002 was the same as that from 2001–2002 to 2005–2006, what would be the expected net receipts for the Republican party in the 2005–2006 period?

5.8 Solving Equations and Applications Using Polynomials

1 Factoring to Find the Roots of a Quadratic Equation

Up until now, we have solved only first-degree equations. In this section we will solve quadratic, or second-degree, equations.

DEFINITION

A second-degree equation of the form $ax^2 + bx + c = 0$, where a, b, c are real numbers and $a \neq 0$, is a **quadratic equation.** $ax^2 + bx + c = 0$ is the **standard form** of a quadratic equation.

Before solving a quadratic equation, we will first write the equation in standard form. Although it is not necessary that a, b, and c be integers, the equation is usually written this way.

The key to solving quadratic equations by factoring is called the **zero factor property.** When we multiply two real numbers, the resulting product will be zero if one or both of the factors is zero. Thus, if the product of two real numbers is zero, at least one of the factors must be zero. We state this property of real numbers formally.

ZERO FACTOR PROPERTY

For all real numbers a and b,

$$\text{if } a \cdot b = 0, \text{ then } a = 0, b = 0, \text{ or both } = 0.$$

EXAMPLE 1 Solve the equation $x^2 + 15x = 100$.

Solution When we say "solve the equation" or "find the roots," we mean "find the values of x that satisfy the equation."

$x^2 + 15x - 100 = 0$	Subtract 100 from both sides so that one side is 0.
$(x + 20)(x - 5) = 0$	Factor the trinomial.
$x + 20 = 0$ or $x - 5 = 0$	Set each factor equal to 0.
$x = -20 \qquad x = 5$	Solve each equation.

Check: Use the *original* equation $x^2 + 15x = 100$.

$$x = -20: \quad (-20)^2 + 15(-20) \overset{?}{=} 100$$
$$400 - 300 \overset{?}{=} 100$$
$$100 = 100 \quad \checkmark$$
$$x = 5: \quad (5)^2 + 15(5) \overset{?}{=} 100$$
$$25 + 75 \overset{?}{=} 100$$
$$100 = 100 \quad \checkmark$$

Thus, 5 and -20 are both roots of the quadratic equation $x^2 + 15x = 100$.

Practice Problem 1 Find the roots of $x^2 + x = 56$.

For convenience, on the next page we will list the steps we have employed to solve the quadratic equation.

NOTE TO STUDENT: Fully worked-out solutions to all of the Practice Problems can be found at the back of the text starting at page SP-1

Graphing Calculator

Finding Roots

Not all quadratic equations are as easy to factor as the one in Example 2. Suppose you are asked to find the roots of

$$10x(x + 1) = 83x + 12{,}012.$$

This can be written in the form

$$10x^2 - 73x - 12{,}012 = 0$$

and factored to obtain the following:

$$(5x + 156)(2x - 77) = 0$$

$$x = -\frac{156}{5} \text{ and } x = \frac{77}{2}.$$

In decimal form the solutions are -31.2 and 38.5.

However, you can use the graphing calculator to graph

$$y = 10x^2 - 73x - 12{,}012.$$

By setting an appropriate viewing window and then using the Zoom and Trace features of your calculator, you can find the two places where $y = 0$.

Some calculators have a command that will find the zeros (roots) of a graph.

In a similar fashion graph each equation using the form $y = ax^2 + bx + c$ and use the graph to find the roots.

(a)

$$10x^2 - 189x - 12{,}834 = 0$$

(b)

$$10x(x + 2) = 11{,}011 - 193x$$

---

1. Rewrite the quadratic equation in standard form (so that one side of the equation is 0) and, if possible, *factor* the quadratic expression.
2. Set each factor equal to zero.
3. Solve the resulting equations to find both roots. (A quadratic equation has two roots.)
4. Check your solutions.

It is extremely important to remember that when you are placing the quadratic equation in standard form, one side of the equation must be zero. Several algebraic operations may be necessary to obtain that desired result before you can factor the polynomial.

EXAMPLE 2 Find the roots of $6x^2 + 4 = 7(x + 1)$.

Solution

$$6x^2 + 4 = 7x + 7 \qquad \text{Apply the distributive property.}$$
$$6x^2 - 7x - 3 = 0 \qquad \text{Rewrite the equation in standard form.}$$
$$(2x - 3)(3x + 1) = 0 \qquad \text{Factor the trinomial.}$$
$$2x - 3 = 0 \quad \text{or} \quad 3x + 1 = 0 \qquad \text{Set each factor equal to 0.}$$
$$2x = 3 \qquad\qquad 3x = -1 \qquad \text{Solve the equations.}$$
$$x = \frac{3}{2} \qquad\qquad x = -\frac{1}{3}$$

Check: Use the *original* equation $6x^2 + 4 = 7(x + 1)$.

$$x = \frac{3}{2}: \quad 6\left(\frac{3}{2}\right)^2 + 4 \overset{?}{=} 7\left(\frac{3}{2} + 1\right)$$
$$6\left(\frac{9}{4}\right) + 4 \overset{?}{=} 7\left(\frac{5}{2}\right)$$
$$\frac{27}{2} + 4 \overset{?}{=} \frac{35}{2}$$
$$\frac{27}{2} + \frac{8}{2} \overset{?}{=} \frac{35}{2}$$
$$\frac{35}{2} = \frac{35}{2} \quad ✓$$

It checks, so $\frac{3}{2}$ is a root. Verify that $-\frac{1}{3}$ is also a root.

If you are using a calculator to check your roots, you can complete the check more rapidly using the decimal values 1.5 and -0.33333333. The latter value is approximate, so some round-off error is expected.

Practice Problem 2 Find the roots of $12x^2 - 11x + 2 = 0$.

EXAMPLE 3 Find the roots of $3x^2 - 5x = 0$.

Solution

$$3x^2 - 5x = 0 \qquad \text{The equation is already in standard form.}$$
$$x(3x - 5) = 0 \qquad \text{Factor.}$$
$$x = 0 \quad \text{or} \quad 3x - 5 = 0 \qquad \text{Set each factor equal to 0.}$$
$$3x = 5$$
$$x = \frac{5}{3}$$

Check: Verify that 0 and $\frac{5}{3}$ are roots of $3x^2 - 5x = 0$.

NOTE TO STUDENT: Fully worked-out solutions to all of the Practice Problems can be found at the back of the text starting at page SP-1

Practice Problem 3   Find the roots of $7x^2 - 14x = 0$.

EXAMPLE 4   Solve $9x(x - 1) = 3x - 4$.

Solution
$$9x^2 - 9x = 3x - 4 \quad \text{Remove parentheses.}$$
$$9x^2 - 9x - 3x + 4 = 0 \quad \text{Get 0 on one side.}$$
$$9x^2 - 12x + 4 = 0 \quad \text{Combine like terms.}$$
$$(3x - 2)^2 = 0 \quad \text{Factor.}$$

$$3x - 2 = 0 \quad \text{or} \quad 3x - 2 = 0$$
$$3x = 2 \qquad\qquad 3x = 2$$
$$x = \frac{2}{3} \qquad\qquad x = \frac{2}{3}$$

We obtain one solution twice. This value is called a **double root.**

Practice Problem 4   Solve $16x(x - 2) = 8x - 25$.

The zero factor property can be extended to a polynomial equation of degree greater than 2. In the following example, we will find the three roots of a third-degree polynomial equation.

EXAMPLE 5   Solve $2x^3 = 24x - 8x^2$.

Solution
$$2x^3 + 8x^2 - 24x = 0 \qquad\qquad \text{Get 0 on one side of the equation.}$$
$$2x(x^2 + 4x - 12) = 0 \qquad\qquad \text{Factor out the common factor } 2x.$$
$$2x(x + 6)(x - 2) = 0 \qquad\qquad \text{Factor the trinomial.}$$
$$2x = 0 \quad x + 6 = 0 \quad x - 2 = 0 \quad \text{Zero factor property.}$$
$$x = 0 \qquad x = -6 \qquad x = 2 \quad \text{Solve for } x.$$

The solutions are 0, −6, and 2.

Practice Problem 5   Solve $3x^3 + 6x^2 = 45x$.

2 **Solving Applications That Involve a Factorable Quadratic Equation**

Some applied exercises lead to a factorable quadratic equation. Using the methods developed in this section, we can solve these types of exercises.

▲ **EXAMPLE 6** A racing sailboat has a triangular sail. Find the base and altitude of the triangular sail that has an area of 35 square meters and a base that is 3 meters shorter than the altitude.

Solution

1. *Understand the problem.* We draw a sketch and recall the formula for the area of a triangle.

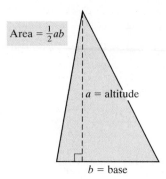

Let x = the length of the altitude in meters.

Then $x - 3$ = the length of the base in meters.

2. *Write an equation.*

$$A = \frac{1}{2}ab$$

$$35 = \frac{1}{2}x(x - 3) \qquad \text{Replace } A \text{ (area) by 35, } a \text{ (altitude) by } x, \text{ and } b \text{ (base) by } x - 3.$$

3. *Solve the equation and state the answer.*

$$70 = x(x - 3) \qquad \text{Multiply each side by 2.}$$
$$70 = x^2 - 3x \qquad \text{Remove parentheses.}$$
$$0 = x^2 - 3x - 70 \qquad \text{Subtract 70 from each side.}$$
$$0 = (x - 10)(x + 7)$$
$$x = 10 \quad or \quad x = -7$$

The altitude of a triangle must be a positive number, so we disregard -7. Thus,

$$\text{altitude} = x = 10 \text{ meters and}$$
$$\text{base} = x - 3 = 7 \text{ meters.}$$

The altitude of the triangular sail measures 10 meters, and the base of the sail measures 7 meters.

4. *Check.* Is the base 3 meters shorter than the altitude?

$$10 - 3 = 7 \quad \checkmark$$

Is the area of the triangle 35 square meters?

$$A = \frac{1}{2}ab$$

$$A = \frac{1}{2}(10)(7) = 5(7) = 35 \quad \checkmark$$

▲ **Practice Problem 6** A racing sailboat has a triangular sail. Find the base and the altitude of the triangular sail if the area is 52 square feet and the altitude is 5 feet longer than the base.

▲ **EXAMPLE 7**   A car manufacturer uses a square panel that holds fuses. The square panel that is used this year has an area that is 72 square centimeters greater than the area of the square panel used last year. The length of each side of the new panel is 3 centimeters more than double the length of last year's panel. Find the dimensions of each panel.

Solution

1. *Understand the problem.* We draw a sketch of each square panel and recall that the area of each panel is obtained by squaring its side.

Let x = the length in centimeters of last year's panel.
Then $2x + 3$ = the length in centimeters of this year's panel.

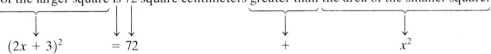

The area of this year's panel is 72 square centimeters greater than the area of last year's panel.

2. *Write an equation.*

Area of the larger square is 72 square centimeters greater than the area of the smaller square.

$$(2x + 3)^2 \qquad = 72 \qquad\qquad + \qquad\qquad x^2$$

3. *Solve the equation and state the answer.*

$$4x^2 + 12x + 9 = 72 + x^2 \quad \text{Remove parentheses.}$$
$$4x^2 + 12x + 9 - x^2 - 72 = 0 \quad \text{Get 0 on one side of the equation.}$$
$$3x^2 + 12x - 63 = 0 \quad \text{Simplify.}$$
$$3(x^2 + 4x - 21) = 0 \quad \text{Factor out the common factor of 3.}$$
$$3(x + 7)(x - 3) = 0 \quad \text{Factor the trinomial.}$$
$$x + 7 = 0 \qquad x - 3 = 0 \quad \text{Use the zero factor property.}$$
$$x = -7 \qquad\quad x = 3 \quad \text{Solve for } x.$$

A fuse box cannot measure -7 centimeters, so we reject the negative answer. We use $x = 3$; then $2x + 3 = 2(3) + 3 = 6 + 3 = 9$.
Thus, the old fuse panel measured 3 centimeters on a side. The new fuse panel measures 9 centimeters on a side.

4. *Check.* Verify that each answer is correct.

▲ **Practice Problem 7**   Last year Grandpa Jones had a small square garden. This year he has a square garden that is 112 square feet larger in area. Each side of the new garden is 2 feet longer than triple each side of the old garden. Find the dimensions of each garden. ◼

Developing Your Study Skills

Getting Help

Getting the right kind of help at the right time can be a key ingredient in being successful in mathematics. Even if you have gone to class on a regular basis, taken careful notes, methodically read your textbook, and diligently done your homework—all of which means making every effort possible to learn the mathematics—you may find that you are still having difficulty. If this is the case, then you need to seek help. Make an appointment with your instructor to find out what help is available to you. The instructor, tutoring services, a mathematics lab, videotapes, and computer software may be among the resources you can draw on.

Once you discover the resources available in your school, you need to take advantage of them. Do not put it off, or you will find yourself getting behind. You cannot afford that. When studying mathematics, you must keep up with your work.

5.8 EXERCISES

| Student Solutions Manual | CD/ Video | PH Math Tutor Center | MathXL®Tutorials on CD | MathXL® | MyMathLab® | Interactmath.com |

Find all the roots and check your answers.

1. $x^2 - x - 6 = 0$ **2.** $x^2 + 5x - 14 = 0$ **3.** $5x^2 - 6x = 0$ **4.** $3x^2 + 5x = 0$

5. $25x^2 - 36 = 0$ **6.** $4x^2 - 9 = 0$ **7.** $3x^2 - 2x - 8 = 0$ **8.** $4x^2 - 13x + 3 = 0$

9. $8x^2 - 3 = 2x$ **10.** $4x^2 - 15 = 4x$ **11.** $8x^2 = 11x - 3$ **12.** $5x^2 = 11x - 2$

13. $x^2 + \dfrac{5}{3}x = \dfrac{2}{3}x$ **14.** $x^2 - \dfrac{5}{2}x = \dfrac{x}{2}$ **15.** $25x^2 + 10x + 1 = 0$ **16.** $36x^2 - 12x + 1 = 0$

Find all the roots and check your answers.

17. $x^3 + 5x^2 + 6x = 0$ **18.** $x^3 + 11x^2 + 18x = 0$ **19.** $\dfrac{x^3}{6} - 8x = \dfrac{x^2}{3}$ **20.** $\dfrac{x^3}{24} - \dfrac{x^2}{12} = x$

21. $3x^3 - 10x = 17x$ **22.** $5x^3 - 8x = 12x$ **23.** $3x^3 + 15x^2 = 42x$ **24.** $2x^3 + 6x^2 = 20x$

25. $\dfrac{7x^2 - 3}{2} = 2x$ **26.** $\dfrac{3x^2 + 3x}{2} = \dfrac{2}{3}$ **27.** $2(x + 3) = -3x + 2(x^2 - 3)$

28. $2(x^2 - 4) - 3x = 4x - 11$ **29.** $7x^2 + 6 = 2x^2 + 2(4x + 3)$ **30.** $11x^2 - 3x + 1 = 2(x^2 - 5x) + 1$

To Think About

31. The equation $2x^2 - 3x + c = 0$ has a solution of $-\dfrac{1}{2}$. What is the value of c? What is the other solution to the equation?

32. The equation $x^2 + bx - 12 = 0$ has a solution of -4. What is the value of b? What is the other solution to the equation?

Applications

Solve the following applied exercises.

▲ **33.** ***Warning Road Sign*** An orange triangular warning sign by the side of the road has an area of 180 square inches. The base of the sign is 2 inches longer than the altitude. Find the measurements of the base and altitude.

▲ **34.** ***Baseball Banner*** A triangular Boston Red Sox banner has an area of 150 square inches. The altitude of the triangle is 3 times the base. Find the measurements of the base and altitude.

▲ **35.** ***Neon Billboard*** The area of a triangular neon billboard advertising the local mall is 104 square feet. The base of the triangle is 2 feet longer than triple the length of the altitude.
　(a) What are the dimensions of the triangular billboard in feet?
　(b) What are the dimensions of the triangular billboard in yards?

▲ **36.** ***Entertainment Platform*** During halftime at the Super Bowl, one of the performers will sing on a triangular platform that measures 119 square yards. The base of the triangular stage is 6 yards longer than four times the length of the altitude. What are the dimensions of the triangular stage?

▲ **37.** ***Desk Telephone*** The area of the base of a rectangular desk telephone is 896 square centimeters. The length of the rectangular telephone is 4 centimeters longer than its width.
　(a) What are the length and width, in centimeters, of the desk telephone?
　(b) What are the length and width, in millimeters, of the desk telephone?

▲ **38.** ***Mouse Pad*** The area of a rectangular mouse pad is 480 square centimeters. Its length is 16 centimeters shorter than twice its width.
　(a) What are the length and width of the mouse pad in centimeters?
　(b) What are the length and width of the mouse pad in millimeters?

▲ **39.** ***Turkish Rug*** A rare square Turkish rug, belonging to the family of President John F. Kennedy, was auctioned off. The area of the Turkish rug in square feet is 96 more than its perimeter in feet. Find the length in feet of the side.

▲ **40.** ***Circus*** The backstage dressing room of the most famous circus in the world is in the shape of a square. The area of the square dressing room in square feet is 165 more than its perimeter in feet. Find the length of the side.

▲ **41.** ***Cereal Box*** The volume of a rectangular solid can be written as $V = LWH$, where L is the length of the solid, W is the width, and H is the height. A box of cereal has a width of 2 inches. Its height is 2 inches longer than its length. If the volume of the box is 198 cubic inches, what are the length and height of the box?

▲ **42.** ***College Catalog*** The volume of a rectangular solid can be written as $V = LWH$, where L is the length of the solid, W is the width, and H is the height. The North Shore Community College catalog is $\frac{1}{2}$ inch wide. Its height is 3 inches longer than its length. The volume of the catalog is 27 cubic inches. What are the height and length of the catalog?

▲ **43.** *Moon Exploration* In planning for the first trip to the moon, NASA surveyed a rectangular area of 54 square miles. The length of the rectangle was 3 miles less than double the width. What were the dimensions of this potential landing area?

▲ **44.** *Ocean Fishing Area* In the northern Atlantic Ocean, a certain fishing area is in the shape of a rectangle. The length of the rectangle is 7 miles longer than double its width. The area of this rectangle is 85 square miles. What are the dimensions of this rectangle?

▲ **45.** *Television Tower* Comcast has a square target that receives signal transmissions from a television tower. They recently enlarged the square target so that its area is 176 square centimeters more than the area of the old target. The new target is a square and has a side that is 1 cm longer than double the side of the old target. What are the dimensions of the old and the new targets?

▲ **46.** *Vegetable Garden* The Parad family has a square vegetable garden. They increased the size of the garden so that the new, larger square garden is 84 square feet larger than the old garden. The new garden has a side that is 2 feet longer than double the length of the side of the old garden. What are the dimensions of the old and the new gardens?

Manufacturing A manufacturer finds that the profit in dollars for manufacturing n units is $P = 2n^2 - 19n - 10$. (Assume that n is a positive integer.) Use this formula for Exercises 47–50.

47. How many units are produced when the profit is $410?

48. How many units are produced when the profit is $0?

49. How many units are produced when there is a loss of $52 ($P = -52$)?

50. How many units are produced when there is a loss of $34 ($P = -34$)?

Mutual Funds The number N of mutual funds in the United States can be approximated by the equation $N = 28x^2 + 80x + 560$, where x is the number of years since 1980. Use this equation to solve Exercises 51–54. Source: U.S. Bureau of Economic Analysis.

51. Approximately how many mutual funds were there in the year 2000?

52. Approximately how many mutual funds were there in the year 2003?

53. In what year were there 668 mutual funds?

54. In what year were there 4160 mutual funds?

Cumulative Review

Simplify. Do not leave negative exponents in your answers.

55. $(2x^3y^2)^3(5xy^2)^2$

56. $\dfrac{(2a^3b^2)^3}{16a^5b^8}$

57. Solve the system:
$x - 2y = 8$
$x + y = -1$

58. Find the equation of the line passing through $(2, 6)$ and $(1, 4)$.

Putting Your Skills to Work

Using Mathematical Equations to Predict Employment

In 1900 there was a blacksmith for every 100 families in the United States. Since that time, the number of blacksmiths in our country has been decreasing. The number of blacksmiths can be estimated by the equation $f(x) = -7x^2 - 1250x + 200,000$, where x is the number of years since 1900 and $f(x)$ is the number of blacksmiths. This equation is only valid for the years 1900 to 2000. (*Source*: Artist Blacksmiths Association of North America)

Problems for Individual Investigation and Analysis

1. How many blacksmiths were there in 1900? In 1930?

2. How many blacksmiths were there in 1940? In 1960?

3. How many blacksmiths were there in 1990? What was the percent of decrease from 1900 to 1990? (Use your answer from problem 1 to help answer the question.)

4. How many blacksmiths were there in 2000? What was the percent of decrease from 1940 to 2000? (Use your answer from problem 2 to help answer the question.)

Problems for Group Investigation and Cooperative Study

Desktop publishers rank among the 10 fastest growing occupations. They use computer software to combine text, data, photographs, charts, and other elements to produce a wide variety of publication-ready materials, including books, calendars, magazines, newsletters and newspapers, and promotional materials. They may be responsible for converting text and graphics to an Internet-ready format. Job-seekers with certification or degrees will have the best job opportunities.

A function that gives the predicted number of jobs for computer support specialists up to the year 2010 is $f(x) = 0.11x^3 - 1.9x^2 + 9.4x + 26$, where x represents the number of years since 1998 and f is expressed in thousands. Use a calculator or computer to assist you in answering these questions. (*Source*: U.S. Bureau of Labor Statistics) Round your answers to the nearest thousand.

5. Find the number of jobs and expected number of jobs in the following years: 2000, 2002, 2004, 2006, 2008, 2010.

6. Use the values you found in problem 5 to graph the function.

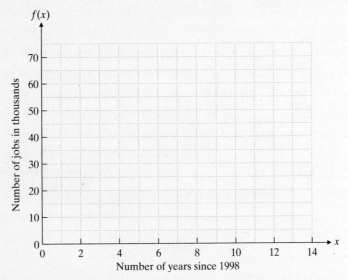

7. If the trend continues, how many jobs for desktop publishers will there be in 2012?

8. What factors could affect whether jobs in this area will continue to increase at the current rate?

Chapter 5 Organizer

Topic	Procedure	Examples
Adding and subtracting polynomials, p. 249.	Combine like terms following the rules of signs.	$(5x^2 - 6x - 8) + (-2x^2 - 5x + 3) = 3x^2 - 11x - 5$ $(3a^2 - 2ab - 5b^2) - (-7a^2 + 6ab - b^2)$ $= (3a^2 - 2ab - 5b^2) + (7a^2 - 6ab + b^2)$ $= 10a^2 - 8ab - 4b^2$
Multiplying polynomials, p. 250.	**1.** Multiply each term of the first polynomial by each term of the second polynomial. **2.** Combine like terms.	$2x^2(3x^2 - 5x - 6) = 6x^4 - 10x^3 - 12x^2$ $(3x + 4)(2x - 7) = 6x^2 - 21x + 8x - 28$ $\qquad\qquad\qquad\;\; = 6x^2 - 13x - 28$ $(x - 3)(x^2 + 5x + 8)$ $= x^3 + 5x^2 + 8x - 3x^2 - 15x - 24$ $= x^3 + 2x^2 - 7x - 24$
Division of a polynomial by a monomial, p. 255.	**1.** Write the division as the sum of separate fractions. **2.** If possible, reduce the separate fractions.	$(16x^3 - 24x^2 + 56x) \div (-8x)$ $= \dfrac{16x^3}{-8x} + \dfrac{-24x^2}{-8x} + \dfrac{56x}{-8x}$ $= -2x^2 + 3x - 7$
Dividing a polynomial by a binomial or a trinomial, p. 256.	**1.** Write the division exercise in long division form. Write both polynomials in descending order; write any missing terms with a coefficient of zero. **2.** Divide the *first* term of the divisor into the first term of the dividend. The result is the first term of the quotient. **3.** Multiply the first term of the quotient by *every* term in the divisor. **4.** Write this product under the dividend (align like terms) and subtract. **5.** Treat this difference as a new dividend. Repeat steps 2 to 4. Continue until the remainder is zero or is a polynomial of lower degree than the *first term* of the divisor. **6.** If there is a remainder, write it as the numerator of a fraction with the divisor as the denominator. Add this fraction to the quotient.	Divide $(6x^3 + 5x^2 - 2x + 1) \div (3x + 1)$. $$\begin{array}{r} 2x^2 + x - 1 \\ 3x + 1\overline{)6x^3 + 5x^2 - 2x + 1} \\ \underline{6x^3 + 2x^2} \\ 3x^2 - 2x \\ \underline{3x^2 + x} \\ -3x + 1 \\ \underline{-3x - 1} \\ 2 \end{array}$$ The quotient is $2x^2 + x - 1 + \dfrac{2}{3x + 1}$.
Synthetic division, p. 260.	Synthetic division can be used if the divisor is in the form $(x + b)$. **1.** Write the coefficients of the terms in descending order of the dividend. Write any missing terms with a coefficient of zero. **2.** The divisor will be of the form $x + b$. Write down the opposite of b to the left. **3.** Bring down the first coefficient to the bottom row. **4.** Multiply the coefficient in the bottom row by the opposite of b and add it to the coefficient above it in the top row. Write the result in the second row. **5.** Add the values in the top and second rows and place the result in the bottom row. **6.** Repeat steps 3 and 4 until the bottom row is filled.	Divide $(3x^5 - 2x^3 + x^2 - x + 7) \div (x + 2)$. $$\begin{array}{r|rrrrrr} -2 & 3 & 0 & -2 & 1 & -1 & 7 \\ & & -6 & 12 & -20 & 38 & -74 \\ \hline & 3 & -6 & 10 & -19 & 37 & \underline{|-67} \end{array}$$ The quotient is $3x^4 - 6x^3 + 10x^2 - 19x + 37 + \dfrac{-67}{x + 2}$.

Topic	Procedure	Examples
Factoring out a common factor, p. 264.	Remove the greatest common factor from each term. Many factoring problems are two steps, of which this is the first.	$5x^3 - 25x^2 - 10x = 5x(x^2 - 5x - 2)$ $20a^3b^2 - 40a^4b^3 + 30a^3b^3 = 10a^3b^2(2 - 4ab + 3b)$
Factoring the difference of two squares, p. 278.	$a^2 - b^2 = (a + b)(a - b)$	$9x^2 - 1 = (3x + 1)(3x - 1)$ $8x^2 - 50 = 2(4x^2 - 25) = 2(2x + 5)(2x - 5)$
Factoring a perfect square trinomial, p. 279.	$a^2 + 2ab + b^2 = (a + b)^2$ $a^2 - 2ab + b^2 = (a - b)^2$	$16x^2 + 40x + 25 = (4x + 5)^2$ $18x^2 + 120xy + 200y^2 = 2(9x^2 + 60xy + 100y^2)$ $\qquad\qquad\qquad\qquad = 2(3x + 10y)^2$ $4x^2 - 36x + 81 = (2x - 9)^2$ $25a^3 - 10a^2b + ab^2 = a(25a^2 - 10ab + b^2)$ $\qquad\qquad\qquad\qquad = a(5a - b)^2$
Factoring the sum and difference of two cubes, p. 280.	$a^3 + b^3 = (a + b)(a^2 - ab + b^2)$ $a^3 - b^3 = (a - b)(a^2 + ab + b^2)$	$8x^3 + 27 = (2x + 3)(4x^2 - 6x + 9)$ $250x^3 + 2y^3 = 2(125x^3 + y^3)$ $\qquad\qquad\qquad = 2(5x + y)(25x^2 - 5xy + y^2)$ $27x^3 - 64 = (3x - 4)(9x^2 + 12x + 16)$ $125y^4 - 8y = y(125y^3 - 8)$ $\qquad\qquad\qquad = y(5y - 2)(25y^2 + 10y + 4)$
Factoring trinomials of the form $x^2 + bx + c$, p. 270.	The factors will be of the form $(x + m)(x + n)$, where $m \cdot n = c$ and $m + n = b$.	$x^2 - 7x + 12 = (x - 4)(x - 3)$ $3x^2 - 36x + 60 = 3(x^2 - 12x + 20)$ $\qquad\qquad\qquad = 3(x - 2)(x - 10)$ $x^2 + 2x - 15 = (x + 5)(x - 3)$ $2x^2 - 44x - 96 = 2(x^2 - 22x - 48)$ $\qquad\qquad\qquad = 2(x - 24)(x + 2)$
Factoring trinomials of the form $ax^2 + bx + c$, p. 272.	Use the trial-and-error method or the grouping number method.	$2x^2 + 7x + 3 = (2x + 1)(x + 3)$ $8x^2 - 26x + 6 = 2(4x^2 - 13x + 3)$ $\qquad\qquad\qquad = 2(4x - 1)(x - 3)$ $7x^2 + 20x - 3 = (7x - 1)(x + 3)$ $5x^3 - 18x^2 - 8x = x(5x^2 - 18x - 8)$ $\qquad\qquad\qquad = x(5x + 2)(x - 4)$
Factoring by grouping, p. 266.	1. Make sure that the first two terms have a common factor and the last two terms have a common factor; otherwise, regroup the terms. 2. Factor out the common factor in each group. 3. Factor out the common binomial factor.	$6xy - 8y + 3xw - 4w$ $= 2y(3x - 4) + w(3x - 4)$ $= (3x - 4)(2y + w)$
Solving a quadratic equation by factoring, p. 289.	1. Rewrite the equation in standard form. 2. Factor, if possible. 3. Set each factor equal to 0. 4. Solve each of the resulting equations.	Solve $(x + 3)(x - 2) = 5(x + 3)$. $x^2 - 2x + 3x - 6 = 5x + 15$ $x^2 + x - 6 = 5x + 15$ $x^2 + x - 5x - 6 - 15 = 0$ $x^2 - 4x - 21 = 0$ $(x - 7)(x + 3) = 0$ $x - 7 = 0 \quad x + 3 = 0$ $x = 7 \qquad x = -3$

Perform the indicated operations.

1. $(x^2 - 3x + 5) + (-2x^2 - 7x + 8)$

2. $(-4x^2y - 7xy + y) + (5x^2y + 2xy - 9y)$

3. $(-6x^2 + 7xy - 3y^2) - (5x^2 - 3xy - 9y^2)$

4. $(-13x^2 + 9x - 14) - (-2x^2 - 6x + 1)$

5. $(3x - 1) - (2 - 8x) + (x + 7)$

6. $(x^2 + 4) + (3x - 5) - (2x^2 - x)$

For the polynomial function $p(x) = 3x^3 - 2x^2 - 6x + 1$ find the following.

7. $p(-4)$

8. $p(-1)$

9. $p(3)$

For the polynomial function $g(x) = -2x^4 + x^3 - 5x - 2$ find the following.

10. $g(2)$

11. $g(-3)$

12. $g(0)$

For the polynomial function $h(x) = -x^3 - 6x^2 + 12x - 4$ find the following.

13. $h(3)$

14. $h(-2)$

15. $h(0)$

Multiply.

16. $3xy(x^2 - xy + y^2)$

17. $(3x^2 + 1)(2x - 1)$

18. $(5x^2 + 3)^2$

19. $(x - 3)(2x - 5)(x + 2)$

20. $(x^2 - 3x + 1)(-2x^2 + x - 2)$

21. $(3x - 5)(3x^2 + 2x - 4)$

22. $(6xy - 7)(6xy + 7)$

23. $(5a - 2b^2)(3a - 4b^2)$

Divide.

24. $(25x^3y - 15x^2y - 100xy) \div (-5xy)$

25. $(12x^2 - 16x - 4) \div (3x + 2)$

26. $(2x^3 + x^2 - x + 1) \div (2x + 3)$

27. $(3y^3 - 2y + 5) \div (y - 3)$

28. $(15a^4 - 3a^3 + 4a^2 + 4) \div (3a^2 - 1)$

29. $(2x^4 - x^2 + 6x + 3) \div (x - 1)$

30. $(2x^4 - 13x^3 + 16x^2 - 9x + 20) \div (x - 5)$

31. $(3x^4 + 5x^3 - x^2 + x - 2) \div (x + 2)$

Remove the greatest common factor.

32. $6a^2b - 3ab^2 - 3ab$

33. $x^5 - 3x^4 + 2x^2$

34. $12mn - 8m$

Factor using the grouping method.

35. $2x + 6 - xy - 3y$

36. $8x^2y + x^2b + 8y + b$

37. $3ab - 15a - 2b + 10$

Factor the trinomials.

38. $x^2 - 9x - 22$

39. $4x^2 - 5x - 6$

40. $6x^2 + 5x - 21$

Factor using one of the formulas from Section 5.6.

41. $100x^2 - 49$

42. $4x^2 - 28x + 49$

43. $8a^3 - 27$

Mixed Practice

Factor, if possible. Be sure to factor completely.

44. $9x^2 - 121$

45. $5x^2 - 11x + 2$

46. $x^3 + 8x^2 + 12x$

47. $x^2 - 8wy + 4xw - 2xy$

48. $36x^2 + 25$

49. $2x^2 - 7x - 3$

50. $x^2 + 6xy - 27y^2$

51. $27x^4 - x$

52. $21a^2 + 20ab + 4b^2$

53. $-3a^3b^3 + 2a^2b^4 - a^2b^3$

54. $a^4b^4 + a^3b^4 - 6a^2b^4$

55. $3x^4 - 5x^2 - 2$

56. $9a^2b + 15ab - 14b$

57. $2x^2 + 7x - 6$

58. $3x^2 + 5x + 4$

59. $4y^4 - 13y^3 + 9y^2$

60. $y^4 + 2y^3 - 35y^2$

61. $4x^2y^2 - 12x^2y - 8x^2$

62. $3x^4 - 7x^2 - 6$

63. $a^2 + 5ab^3 + 4b^6$

64. $3x^2 - 12 - 8x + 2x^3$

65. $2x^4 - 12x^2 - 54$

66. $8a + 8b - 4bx - 4ax$

67. $8x^4 + 34x^2y^2 + 21y^4$

68. $4x^3 + 10x^2 - 6x$

69. $2a^2x - 15ax + 7x$

70. $16x^4y^2 - 56x^2y + 49$

71. $128x^3y - 2xy$

72. $5xb - 28y + 4by - 35x$

73. $27abc^2 - 12ab$

Solve the following equations.

74. $5x^2 - 9x - 2 = 0$

75. $2x^2 - 11x + 12 = 0$

76. $(2x - 1)(3x - 5) = 20$

77. $7x^2 = 21x$

78. $3x^2 + 14x + 3 = -1 + 4(x + 1)$

79. $x^3 + 7x^2 = -12x$

Use a quadratic equation to solve each of the following exercises.

▲ **80.** *Geometry* The area of a triangle is 77 square meters. The altitude of the triangle is 3 meters longer than the base of the triangle. Find the base and the altitude of the triangle.

▲ **81.** *Parks* A rectangular park has an area of 40 square miles. The length of the rectangle is 2 miles less than triple the width. Find the dimensions of the park.

82. *Manufacturing Profit* The hourly profit in dollars made by a scientific calculator manufacturing plant is given by the equation $P = 3x^2 - 7x - 10$, where x is the number of calculators assembled in 1 hour. Find the number of calculators that should be made in 1 hour if the hourly profit is to be $30.

▲ **83.** *Sound Insulators* A square sound insulator is constructed for a restaurant. It does not provide enough insulation, so a larger square is constructed. The larger square has 24 square yards more insulation. The side of the larger square is 3 yards longer than double the side of the smaller square. Find the dimensions of each square.

Remember to use your Chapter Test Prep Video CD to see the worked-out solutions to the test problems you want to review.

Combine.

1. $(3x^2y - 2xy^2 - 6) + (5 + 2xy^2 - 7x^2y)$

2. $(5a^2 - 3) - (2 + 5a) - (4a - 3)$

Multiply.

3. $-2x(x + 3y - 4)$

4. $(2x - 3y^2)^2$

5. $(x - 2)(2x^2 + x - 1)$

Divide.

6. $(-15x^3 - 12x^2 + 21x) \div (-3x)$

7. $(2x^4 - 7x^3 + 7x^2 - 9x + 10) \div (2x - 5)$

8. $(x^3 - x^2 - 5x + 2) \div (x + 2)$

Use Synthetic division to perform this division.

9. $(x^4 + x^3 - x - 3) \div (x + 1)$

Factor, if possible.

10. $121x^2 - 25y^2$

11. $9x^2 + 30xy + 25y^2$

12. $x^3 - 26x^2 + 48x$

13. $4x^3y + 8x^2y^2 + 4x^2y$

14. $x^2 - 6wy + 3xy - 2wx$

15. $2x^2 - 3x + 2$

1. _____

2. _____

3. _____

4. _____

5. _____

6. _____

7. _____

8. _____

9. _____

10. _____

11. _____

12. _____

13. _____

14. _____

15. _____

16. _____

16. $18x^2 + 3x - 15$

17. _____

17. $54a^4 - 16a$

18. _____

18. $9x^5 - 6x^3y + xy^2$

19. _____

19. $3x^4 + 17x^2 + 10$

20. _____

20. $3x - 10ay + 6y - 5ax$

Evaluate the following if $p(x) = -2x^3 - x^2 + 6x - 10$.
21. $p(2)$

21. _____

22. $p(-3)$

22. _____

Solve the following equations.
23. $x^2 = 5x + 14$

23. _____

24. $3x^2 - 11x - 4 = 0$

24. _____

25. $7x^2 + 6x = 8x$

25. _____

Use a quadratic equation to solve the following exercise.

▲ **26.** The area of a triangular road sign is 70 square inches. The altitude of the triangle is 4 inches less than the base of the triangle. Find the altitude and the base of the triangle.

26. _____

Approximately one-half of this test covers the content of Chapters 1–4. The remainder covers the content of Chapter 5.

1. What property is illustrated by the equation $3(5 \cdot 2) = (3 \cdot 5)2$?

2. Evaluate $\dfrac{2 + 6(-2)}{(2 - 4)^3 + 3}$.

3. Simplify: $7x - 3\{1 + 2(x - y)\}$

4. Solve for x: $5x + 7y = 2$

5. Solve for x. $\dfrac{1}{2}(x - 3) + \dfrac{x}{5} = x - 1$

6. Find the slope of the line passing through $(-2, -3)$ and $(1, 5)$.

7. Graph $y = -\dfrac{2}{3}x + 4$.

8. Graph the region $3x - 4y \geq -12$.

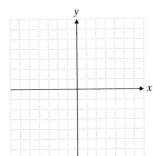

9. Solve the inequality. $-3(x + 2) < 5x - 2(4 + x)$

▲ **10.** What are the dimensions of a rectangle with a perimeter of 46 meters if the length is 5 meters longer than twice the width?

Multiply and simplify your answer.

11. Evaluate: $3x^3 - 2xy - x^2 + 5$ for $x = 1$, $y = 0$

12. Simplify: $a\{ab - 2b(a + 4)\}$

13. $(5x - 2)(2x^2 - 3x - 4)$

1. _____

2. _____

3. _____

4. _____

5. _____

6. _____

7. _____

8. _____

9. _____

10. _____

11. _____

12. _____

13. _____

305

14. _____

15. _____

16. _____

17. _____

18. _____

19. _____

20. _____

21. _____

22. _____

23. _____

24. _____

25. _____

26. _____

27. _____

28. _____

Divide.

14. $(-21x^3 + 14x^2 - 28x) \div (7x)$

15. $(2x^3 - 3x^2 + 3x - 4) \div (x - 2)$

Factor, if possible.

16. $2x^3 - 10x^2$

17. $64x^2 - 49$

18. $3x^2 - 2x - 8$

19. $25x^2 + 60x + 36$

20. $3x^2 - 15x - 42$

21. $2x^2 + 24x + 40$

22. $16x^2 + 9$

23. $6x^3 + 11x^2 + 3x$

24. $27x^4 + 64x$

25. $2x - 6 - 5xy + 15y$

Solve for x.

26. $3x^2 - 4x - 4 = 0$

27. $x^2 - 8x = 33$

▲ **28.** A hospital has paved a triangular parking lot for emergency helicopter landings. The area of the triangle is 68 square meters. The altitude of the triangle is 1 meter longer than double the base of the triangle. Find the altitude and the base of this triangular region.

CHAPTER

6

Are you considering buying a car? Automobiles are a big investment. In deciding what type of car you can afford to buy and own, there are many factors to consider—the price of the car; the amount of money you have available as a down payment; and insurance, maintenance, and gas costs. If you plan to finance the cost of the car, the interest rates available will figure into your decision. Turn to page 343 to find a formula for computing your monthly payment and to use your math skills to compare some offers from car dealers.

Rational Expressions and Equations

307

6.1 RATIONAL EXPRESSIONS AND FUNCTIONS: SIMPLIFYING, MULTIPLYING, AND DIVIDING

Student Learning Objectives

After studying this section, you will be able to:

1 Simplify a rational expression or function.

2 Simplify the product of two or more rational expressions.

3 Simplify the quotient of two or more rational expressions.

Graphing Calculator

Finding Domains

To find the domain in Example 1, graph
$$y = \frac{x - 7}{x^2 + 8x - 20}$$ on a
graphing calculator. Since graphing calculators try to connect the points in a graph, it may not be easy to see where the expression is not defined.

Display:

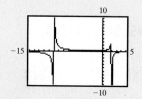

Since the domain will be values except those that make the denominator equal to zero, we can graph $y = x^2 + 8x - 20$ in order to identify the values that are not in the domain.

Display:

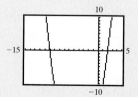

(continues on next page)

---

1 ## Simplifying a Rational Expression or Function

You may recall from Chapter 1 that a *rational number* is an exact quotient $\frac{a}{b}$ of two integers a and b with $b \neq 0$. A **rational expression** is an expression of the form $\frac{P}{Q}$, where P and Q are polynomials and Q is not zero. For example,

$$\frac{7}{x + 2} \quad \text{and} \quad \frac{x + 5}{x - 3}$$

are rational expressions. Since the denominator cannot equal zero, the first expression is undefined if $x + 2 = 0$. This occurs when $x = -2$. We say that x can be any real number except -2. Look at the second expression. When is this expression undefined? Why? For what values of x is the expression defined? Note that the numerator can be zero because any fraction $\frac{0}{a}(a \neq 0)$ is just 0.

A function defined by a rational expression is a **rational function.** The domain of a rational function is the set of values that can be used to replace the variable. Thus, the domain of $f(x) = \frac{7}{x + 2}$ is all real numbers except -2. The domain of $g(x) = \frac{x + 5}{x - 3}$ is all real numbers except 3.

EXAMPLE 1 Find the domain of the function $f(x) = \frac{x - 7}{x^2 + 8x - 20}$.

Solution The domain will be all real numbers except those that make the denominator equal to zero. What value(s) will make the denominator equal to zero? We determine this by solving the equation $x^2 + 8x - 20 = 0$.

$$(x + 10)(x - 2) = 0 \quad \text{Factor.}$$

$$x + 10 = 0 \quad \text{or} \quad x - 2 = 0 \quad \text{Use the zero factor property.}$$

$$x = -10 \qquad x = 2 \quad \text{Solve for } x.$$

The domain of $y = f(x)$ is all real numbers except -10 and 2.

Practice Problem 1 Find the domain of $f(x) = \frac{4x + 3}{x^2 - 9x - 22}$.

We have learned that we can simplify fractions (or reduce them to lowest terms) by factoring the numerator and denominator into prime factors and dividing by the common factors. For example,

$$\frac{15}{25} = \frac{3 \cdot \cancel{5}}{5 \cdot \cancel{5}} = \frac{3}{5}.$$

We can do this by using the **basic rule of fractions.**

BASIC RULE OF FRACTIONS

For any polynomials a, b, and c,

$$\frac{ac}{bc} = \frac{a}{b}, \qquad \text{where } b \text{ and } c \neq 0.$$

(continued)

Use the Zoom and Trace features or the zero command to find where $y = 0$.

Try to find the domain of the following function. Round any values where the function is not defined to the nearest tenth.

$$y = \frac{x + 9}{2.1x^2 + 5.2x - 3.1}$$

What we are doing is factoring out a common factor of 1 $\left(\frac{c}{c} = 1\right)$. We have

$$\frac{ac}{bc} = \frac{a}{b} \cdot \frac{c}{c} = \frac{a}{b} \cdot 1 = \frac{a}{b}.$$

Note that c must be a factor of the numerator *and* the denominator. Thus, the basic rule of fractions simply says that we may divide out a *common* factor from the numerator and the denominator. This nonzero factor can be a number or an algebraic expression.

EXAMPLE 2   Simplify $\dfrac{2a^2 - ab - b^2}{a^2 - b^2}$.

Solution

$$\frac{(2a + b)(a - b)}{(a + b)(a - b)} = \frac{2a + b}{a + b} \cdot 1 = \frac{2a + b}{a + b}$$

As you become more familiar with this basic rule, you won't have to write out every step. We did so here to show the application of the rule. We cannot simplify this fraction any further.

Practice Problem 2   Simplify $\dfrac{x^2 - 36y^2}{x^2 - 3xy - 18y^2}$.

NOTE TO STUDENT: Fully worked-out solutions to all of the Practice Problems can be found at the back of the text starting at page SP-1

EXAMPLE 3   Simplify $\dfrac{2x - 3y}{2x^2 - 7xy + 6y^2}$.

Solution

$$\frac{(2x - 3y)1}{(2x - 3y)(x - 2y)} = \frac{1}{x - 2y} \qquad \text{\textit{Note:} Do you see why it is necessary to have a 1 in the numerator of the answer?}$$

Practice Problem 3   Simplify $\dfrac{9x^2y}{3xy^2 + 6x^2y}$.

EXAMPLE 4   Simplify $\dfrac{2x^2 + 2x - 12}{x^3 + 7x^2 + 12x}$.

Solution

$$\frac{2x^2 + 2x - 12}{x^3 + 7x^2 + 12x} = \frac{2(x^2 + x - 6)}{x(x^2 + 7x + 12)} = \frac{2(x + 3)(x - 2)}{x(x + 3)(x + 4)} = \frac{2(x - 2)}{x(x + 4)}$$

We usually leave the answer in factored form.

NOTE TO STUDENT: *Fully worked-out solutions to all of the Practice Problems can be found at the back of the text starting at page SP-1*

Practice Problem 4   Simplify $\dfrac{2x^2 - 8x - 10}{2x^2 - 20x + 50}$.

Be alert for situations in which one factor in the numerator is the opposite of another factor in the denominator. In such cases you should factor -1 from one of the factors.

EXAMPLE 5   Simplify $\dfrac{-2x + 14y}{x^2 - 5xy - 14y^2}$.

Solution

$$\frac{-2x + 14y}{x^2 - 5xy - 14y^2} = \frac{-2(x - 7y)}{(x + 2y)(x - 7y)} \qquad \text{Factor } -2 \text{ from each term of the numerator and then factor the denominator.}$$

$$= \frac{-2}{x + 2y} \qquad \text{Use the basic rule of fractions.}$$

Practice Problem 5   Simplify $\dfrac{-3x + 6y}{x^2 - 7xy + 10y^2}$.

EXAMPLE 6   Simplify $\dfrac{25y^2 - 16x^2}{8x^2 - 14xy + 5y^2}$.

$$\frac{(5y + 4x)(5y - 4x)}{(4x - 5y)(2x - y)} = \frac{(5y + 4x)(5y - 4x)}{-1(-4x + 5y)(2x - y)} = \frac{5y + 4x}{-1(2x - y)} = \frac{5y + 4x}{y - 2x}$$

> Observe that $4x - 5y = -1(-4x + 5y)$.

Practice Problem 6   Simplify $\dfrac{7a^2 - 23ab + 6b^2}{4b^2 - 49a^2}$.

② Simplifying the Product of Two or More Rational Expressions

Multiplication of rational expressions follows the same rule as multiplication of integer fractions. However, it is particularly helpful to use the basic rule of fractions to simplify whenever possible.

MULTIPLYING RATIONAL EXPRESSIONS

For any polynomials a, b, c, and d,

$$\frac{a}{b} \cdot \frac{c}{d} = \frac{ac}{bd}, \qquad \text{where } b \text{ and } d \neq 0.$$

> **EXAMPLE 7**   Multiply $\dfrac{2x^2 - 4x}{x^2 - 5x + 6} \cdot \dfrac{x^2 - 9}{2x^4 + 14x^3 + 24x^2}$.

Solution   We first use the basic rule of fractions; that is, we factor (if possible) the numerator and denominator and divide out common factors.

$$\frac{2x(x - 2)}{(x - 2)(x - 3)} \cdot \frac{(x + 3)(x - 3)}{2x^2(x^2 + 7x + 12)} = \frac{2x(x - 2)(x + 3)(x - 3)}{(2x)x(x - 2)(x - 3)(x + 3)(x + 4)}$$

$$= \frac{2x}{2x} \cdot \frac{1}{x} \cdot \frac{x - 2}{x - 2} \cdot \frac{x + 3}{x + 3} \cdot \frac{x - 3}{x - 3} \cdot \frac{1}{x + 4}$$

$$= 1 \cdot \frac{1}{x} \cdot 1 \cdot 1 \cdot 1 \cdot \frac{1}{x + 4}$$

$$= \frac{1}{x(x + 4)} \quad \text{or} \quad \frac{1}{x^2 + 4x}$$

Although either form of the answer is correct, we usually use the factored form.

Practice Problem 7   Multiply $\dfrac{2x^2 + 5x + 2}{4x^2 - 1} \cdot \dfrac{2x^2 + x - 1}{x^2 + x - 2}$.

> **EXAMPLE 8**   Multiply $\dfrac{7x + 7y}{4ax + 4ay} \cdot \dfrac{8a^2x^2 - 8b^2x^2}{35ax^3 - 35bx^3}$.

Solution $\dfrac{7(x + y)}{4a(x + y)} \cdot \dfrac{8x^2(a^2 - b^2)}{35x^3(a - b)} = \dfrac{7(x + y)}{4a(x + y)} \cdot \dfrac{8x^2(a + b)(a - b)}{35x^3(a - b)}$

$$= \frac{\cancel{7}\,\cancel{(x + y)}}{\cancel{4}a\cancel{(x + y)}} \cdot \frac{\overset{2}{\cancel{8}}x^2(a + b)\cancel{(a - b)}}{\underset{5x}{\cancel{35}}\,\cancel{x^3}\,\cancel{(a - b)}}$$

$$= \frac{2(a + b)}{5ax} \quad \text{or} \quad \frac{2a + 2b}{5ax}$$

Note that we shortened our steps by not writing out every factor of 1 as we did in Example 7. Either way is correct.

Practice Problem 8   Multiply $\dfrac{9x + 9y}{5ax + 5ay} \cdot \dfrac{10ax^2 - 40x^2b^2}{27ax^2 - 54bx^2}$.

③ Simplifying the Quotient of Two or More Rational Expressions

When we divide fractions, we take the **reciprocal** of the second fraction and then multiply the fractions. (Remember that the reciprocal of a fraction $\dfrac{m}{n}$ is $\dfrac{n}{m}$. Thus, the reciprocal of $\dfrac{2}{3}$ is $\dfrac{3}{2}$, and the reciprocal of $\dfrac{3x}{11y^2}$ is $\dfrac{11y^2}{3x}$.) We divide rational expressions in the same way.

DIVIDING RATIONAL EXPRESSIONS

For any polynomials a, b, c, and d,

$$\frac{a}{b} \div \frac{c}{d} = \frac{a}{b} \cdot \frac{d}{c}, \qquad \text{where } b, c, \text{ and } d \neq 0.$$

EXAMPLE 9   Divide $\dfrac{4x^2 - y^2}{x^2 + 4xy + 4y^2} \div \dfrac{4x - 2y}{3x + 6y}$.

Solution   We take the reciprocal of the second fraction and multiply the fractions.

$$\frac{4x^2 - y^2}{x^2 + 4xy + 4y^2} \cdot \frac{3x + 6y}{4x - 2y} = \frac{(2x + y)\cancel{(2x - y)}}{\cancel{(x + 2y)}(x + 2y)} \cdot \frac{3\cancel{(x + 2y)}}{2\cancel{(2x - y)}}$$

$$= \frac{3(2x + y)}{2(x + 2y)} \quad \text{or} \quad \frac{6x + 3y}{2x + 4y}$$

Practice Problem 9   Divide $\dfrac{8x^3 + 27y^3}{64x^3 - y^3} \div \dfrac{4x^2 - 9y^2}{16x^2 + 4xy + y^2}$.

EXAMPLE 10   Divide $\dfrac{24 + 10x - 4x^2}{2x^2 + 13x + 15} \div (2x - 8)$.

Solution   We take the reciprocal of the second fraction and multiply the fractions. The reciprocal of $(2x - 8)$ is $\dfrac{1}{2x - 8}$.

$$\frac{-4x^2 + 10x + 24}{2x^2 + 13x + 15} \cdot \frac{1}{2x - 8} = \frac{-2(2x^2 - 5x - 12)}{(2x + 3)(x + 5)} \cdot \frac{1}{2(x - 4)}$$

$$= \frac{\overset{-1}{\cancel{-2}}\cancel{(x - 4)}\cancel{(2x + 3)}}{\cancel{(2x + 3)}(x + 5)} \cdot \frac{1}{\underset{1}{\cancel{2}}\cancel{(x - 4)}}$$

$$= \frac{-1}{x + 5} \quad \text{or} \quad -\frac{1}{x + 5}$$

NOTE TO STUDENT: Fully worked-out solutions to all of the Practice Problems can be found at the back of the text starting at page SP-1

Practice Problem 10   Divide $\dfrac{4x^2 - 9}{2x^2 + 11x + 12} \div (-6x + 9)$.

6.1 EXERCISES

| Student Solutions Manual | CD/ Video | PH Math Tutor Center | MathXL®Tutorials on CD | MathXL® | MyMathLab® | interactmath.com |

Find the domain of each of the following rational functions.

1. $f(x) = \dfrac{5x + 6}{2x - 6}$

2. $f(x) = \dfrac{3x - 8}{4x + 20}$

3. $g(x) = \dfrac{-7x + 2}{x^2 - 5x - 36}$

4. $g(x) = \dfrac{-8x + 9}{x^2 + 10x - 24}$

Simplify completely.

5. $\dfrac{-18x^4 y}{12x^2 y^6}$

6. $\dfrac{-7xy^2}{28x^5 y}$

7. $\dfrac{3x^3 - 24x^2}{6x - 48}$

8. $\dfrac{10x^2 + 15x}{35x^2 - 5x}$

9. $\dfrac{9x^2}{12x^2 - 15x}$

10. $\dfrac{20x^2}{28x^2 - 12x}$

11. $\dfrac{5x^2 y^2 - 15xy^2}{10x^3 y - 20x^3 y^2}$

12. $\dfrac{6m^2 n^4 + 2mn^3}{4mn^2 - 8n^2}$

13. $\dfrac{2x + 10}{2x^2 - 50}$

14. $\dfrac{6x^2 - 15x}{3x}$

15. $\dfrac{2y^2 - 8}{2y + 4}$

16. $\dfrac{x + 2}{7x^2 - 28}$

17. $\dfrac{2x^2 - x^3 - x^4}{x^4 - x^3}$

18. $\dfrac{30x - x^2 - x^3}{x^3 - x^2 - 20x}$

19. $\dfrac{2y^2 + y - 10}{4 - y^2}$

20. $\dfrac{36 - b^2}{3b^2 - 16b - 12}$

Multiply.

21. $\dfrac{-8mn^5}{3m^4 n^3} \cdot \dfrac{9m^3 n^3}{6mn}$

22. $\dfrac{35x^2 y^6}{10x^7 y^2} \cdot \dfrac{-15y^3}{21xy}$

23. $\dfrac{3a^2}{a^2 + 4a + 4} \cdot \dfrac{a^2 - 4}{3a}$

24. $\dfrac{8x^2}{x^2 - 9} \cdot \dfrac{x^2 + 6x + 9}{16x^3}$

25. $\dfrac{x^2 + 5x + 7}{x^2 - 5x + 6} \cdot \dfrac{3x - 6}{x^2 + 5x + 7}$

26. $\dfrac{x - 5}{10x - 2} \cdot \dfrac{25x^2 - 1}{x^2 - 10x + 25}$

27. $\dfrac{x^2 - 5xy - 24y^2}{x - y} \cdot \dfrac{x^2 + 6xy - 7y^2}{x + 3y}$

28. $\dfrac{x - 3y}{x^2 + 3xy - 18y^2} \cdot \dfrac{x^2 + xy - 30y^2}{x - 5y}$

29. $\dfrac{y^2 - 3y - 10}{2y^2 - y - 1} \cdot \dfrac{2y^2 + 11y + 5}{2y^2 - 50}$

30. $\dfrac{6a^2 - a - 2}{a^2 - 4a - 5} \cdot \dfrac{3a^2 - 3}{3a^2 - 5a + 2}$

31. $\dfrac{x^3 - 125}{x^5 y} \cdot \dfrac{x^3 y^2}{x^2 + 5x + 25}$

32. $\dfrac{3a^3 b^2}{8a^3 - b^3} \cdot \dfrac{4a^2 + 2ab + b^2}{12ab^4}$

Divide.

33. $\dfrac{2mn - m}{15m^3} \div \dfrac{2n - 1}{3m^2}$

34. $\dfrac{3y + 12}{8y^3} \div \dfrac{9y + 36}{16y^3}$

35. $\dfrac{b^2 - 6b + 9}{5b^2 - 16b + 3} \div \dfrac{6b - 3}{15b - 3}$

36. $\dfrac{a^2 - 3a}{a^2 - 9} \div \dfrac{a^3 - a^2}{4a^2 - a - 3}$

37. $\dfrac{x^2 - xy - 6y^2}{x^2 + 2} \div (x^2 + 2xy)$

38. $\dfrac{x^2 - 5x + 4}{2x - 8} \div (3x^2 - 3x)$

Mixed Practice

Perform the indicated operation. When no operation is indicated, simplify the rational expression completely.

39. $\dfrac{7x}{y^2} \div 21x^3$

40. $\dfrac{10m^4}{9a^2} \cdot 3a^2 b$

41. $\dfrac{3x^2 - 2x}{6x - 4}$

42. $\dfrac{4x^2 - 3x}{20x - 15}$

43. $\dfrac{x^2 y - 49y}{x^2 y^3} \cdot \dfrac{3x^2 y - 21xy}{x^2 - 14x + 49}$

44. $\dfrac{x^2 + 6x + 9}{2x^2 y - 18y} \div \dfrac{6xy + 18y}{3x^2 y - 27y}$

45. $\dfrac{x^2 - 9x + 18}{x^2 y^3} \div \dfrac{x^2 - 8x + 12}{x^4 y^2}$

46. $\dfrac{x^2 - 7x + 10}{xy^6} \div \dfrac{x^2 - 11x + 30}{x^2 y^5}$

47. $\dfrac{a^2 - a - 12}{2a^2 + 5a - 12}$

48. $\dfrac{3y^2 - 3y - 36}{2y^2 - y - 3}$

Optional Graphing Calculator Problems

Find the domain of the following functions. Round any values that you exclude to the nearest tenth.

49. $f(x) = \dfrac{2x + 5}{3.6x^2 + 1.8x - 4.3}$

50. $f(x) = \dfrac{5x - 4}{1.6x^2 - 1.3x - 5.9}$

Applications

Tropical Fish *The total number of many tropical fish will grow and then hold constant in a pond or aquarium. The total number of fish will depend on the size of the pond or aquarium. At the Reading Mandarin Restaurant a total of 90 tropical fish were placed in an aquarium. For the next 12 months the total number of fish could be predicted by the equation*

$$P(x) = \frac{90(1 + 1.5x)}{1 + 0.5x},$$

where P is the number of fish and x is the number of months since the fish were placed in the aquarium.

51. One month after the fish are first placed in the aquarium, what is the total number of fish?

52. Three months after the fish are first placed in the aquarium, what is the total number of fish?

53. Six months after the fish are first placed in the aquarium, what is the total number of fish?

54. One year after the fish are first placed in the aquarium, what is the total number of fish? Round your answer to the nearest whole number.

55. Using your answers for problems 51 to 54 graph the function $P(x)$.

56. Based on your graph, what would you estimate to be the total number of fish eleven months after the fish are first placed in the aquarium?

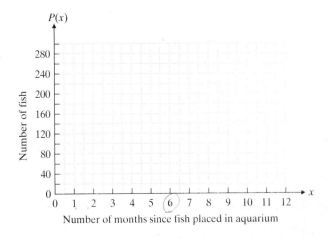

Cumulative Review

Graph the straight line. Plot at least three points.

57. $y = -\frac{3}{2}x + 4$

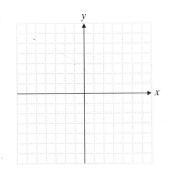

58. $6x - 3y = -12$

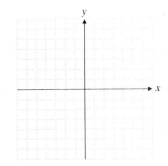

59. Solve the system: $2x - y = 8$
$x + 3y = 4$

60. Solve: $3x - (x + 5) < 7x + 10$

61. ***Directory Assistance*** A directory assistance operator averages two inquiries per minute. If she works a 38-hour week and takes 1 hour each day for lunch (away from the phone), how many inquiries will she have answered in 1 week?

62. ***Hair Stylist*** A hair stylist is able to service 21 customers in a 7-hour shift if he only *cuts* hair. The same hair stylist is able to service 14 customers in a 7-hour shift if he *cuts and colors* hair. He is able to service 18 customers in a 7-hour shift if he only *colors* hair. Can he feasibly service 12 cuts, 13 cuts and color, and 5 colors during one shift?

63. ***Availability of Donor Organs*** There is a great shortage of available donor organs in the United States. As of the spring of 2000, there were 69,399 patients waiting for organ transplants. Each organ donor helps an average of six people live longer, more productive lives. Every year 37% of the people waiting for a donor organ die. The number of patients waiting for an organ transplant has risen 7% per year since the spring of 1997. How many people have died waiting for a donor organ from the spring of 1997 to the spring of 2000? *Source:* National Center for Health Statistics.

6.2 ADDING AND SUBTRACTING RATIONAL EXPRESSIONS

Student Learning Objectives

After studying this section, you will be able to:

1 Find the LCD of two or more rational expressions.

2 Add or subtract two or more rational expressions.

1 Finding the LCD of Two or More Rational Expressions

Recall that if we wish to add or subtract fractions, the denominators must be the same. If the denominators are not the same, we use the basic rule of fractions to rewrite one or both fractions so that the denominations are the same. For example,

$$\frac{3}{7} + \frac{2}{7} = \frac{5}{7} \quad \text{and} \quad \frac{2}{3} + \frac{3}{4} = \frac{2}{3} \cdot \frac{4}{4} + \frac{3}{4} \cdot \frac{3}{3} = \frac{8}{12} + \frac{9}{12} = \frac{17}{12}.$$

How did we know that 12 was the least common denominator (LCD)? The least common denominator of two or more fractions is the product of the different prime factors in each denominator. If a factor is repeated, we use the highest power that appears on that factor in the denominators.

$$3 = 3$$
$$4 = 2 \cdot 2$$
$$\text{LCD} = 3 \cdot 2 \cdot 2 = 12$$

This same technique is used to add or subtract rational expressions.

HOW TO FIND THE LCD

1. Factor each denominator completely into prime factors.
2. List all the different prime factors.
3. The LCD is the product of these factors, each of which is raised to the highest power that appears on that factor in the denominators.

Now we'll do some sample problems.

EXAMPLE 1 Find the LCD of the rational expressions $\dfrac{7}{x^2 - 4}$ and $\dfrac{2}{x - 2}$.

Solution

Step 1 We factor each denominator completely (into prime factors).
$$x^2 - 4 = (x + 2)(x - 2)$$
$$x - 2 \text{ cannot be factored.}$$

Step 2 We list all the *different* prime factors. The different factors are $x + 2$ and $x - 2$.

Step 3 Since no factor occurs more than once in the denominators, the LCD is the product $(x + 2)(x - 2)$.

Practice Problem 1 Find the LCD of the rational expressions
$$\frac{8}{x^2 - x - 12} \text{ and } \frac{3}{x - 4}.$$

EXAMPLE 2 Find the LCD of the rational expressions $\dfrac{7}{12xy^2}$ and $\dfrac{4}{15x^3y}$.

Solution

Step 1 We factor each denominator.
$$12xy^2 = 2 \cdot 2 \cdot 3 \cdot x \cdot y \cdot y$$
$$15x^3y = 3 \cdot 5 \cdot x \cdot x \cdot x \cdot y$$

Step 2 Our LCD will require each of the different factors that appear in either denominator. They are 2, 3, 5, x, and y.

Step 3 The factor 2 and the factor y each occur twice in $12xy^2$.
The factor x occurs three times in $15x^3y$.
Thus, the LCD $= 2 \cdot 2 \cdot 3 \cdot 5 \cdot x \cdot x \cdot x \cdot y \cdot y$
$= 60x^3y^2$

Practice Problem 2 Find the LCD. $\dfrac{2}{15x^3y^2}$ and $\dfrac{13}{25xy^3}$.

NOTE TO STUDENT: Fully worked-out solutions to all of the Practice Problems can be found at the back of the text starting at page SP-1

② Adding or Subtracting Two or More Rational Expressions

We can add and subtract rational expressions with the same denominator just as we do in arithmetic: We simply add or subtract the numerators.

> **ADDITION AND SUBTRACTION OF RATIONAL EXPRESSIONS**
>
> For any polynomials a, b, and c we have the following:
>
> $$\frac{a}{b} + \frac{c}{b} = \frac{a+c}{\cdot b}, \qquad b \neq 0$$
>
> $$\frac{a}{b} - \frac{c}{b} = \frac{a-c}{b}, \qquad b \neq 0.$$

EXAMPLE 3   Subtract: $\dfrac{5x+2}{(x+3)(x-4)} - \dfrac{6x}{(x+3)(x-4)}$

Solution $\dfrac{5x+2}{(x+3)(x-4)} - \dfrac{6x}{(x+3)(x-4)} = \dfrac{-x+2}{(x+3)(x-4)}$

Practice Problem 3   Subtract: $\dfrac{4x}{(x+6)(2x-1)} - \dfrac{3x+1}{(x+6)(2x-1)}$

If the two rational expressions have different denominators, we first need to find the LCD. Then we must rewrite each fraction as an equivalent fraction that has the LCD as the denominator.

EXAMPLE 4   Add the rational expressions: $\dfrac{7}{(x+2)(x-2)} + \dfrac{2}{x-2}$

Solution   The LCD $= (x+2)(x-2)$.
  Before we can add the fractions, we must rewrite our fractions as fractions with the LCD. The first fraction needs no change, but the second fraction does.

$$\frac{7}{(x+2)(x-2)} + \frac{2}{x-2} \cdot \frac{x+2}{x+2}$$

Since $\dfrac{x+2}{x+2} = 1$, we have not changed the *value* of the fraction. We are simply writing it in an equivalent form. Thus, we now have the following:

$$\frac{7}{(x+2)(x-2)} + \frac{2(x+2)}{(x+2)(x-2)} = \frac{7+2(x+2)}{(x+2)(x-2)}$$

$$= \frac{7+2x+4}{(x+2)(x-2)} = \frac{2x+11}{(x+2)(x-2)}$$

Practice Problem 4   Add the rational expressions: $\dfrac{8}{(x-4)(x+3)} + \dfrac{3}{x-4}$

EXAMPLE 5   Add: $\dfrac{4}{x+6} + \dfrac{5}{6x}$

Solution   The LCD $= 6x(x+6)$. We need to multiply the fractions by 1 to obtain two equivalent fractions that have the LCD for the denominator. We multiply the first fraction by $1 = \dfrac{6x}{6x}$. We multiply the second fraction by $1 = \dfrac{x+6}{x+6}$.

$$\frac{4(6x)}{(x+6)(6x)} + \frac{5(x+6)}{6x(x+6)}$$

$$\frac{24x}{6x(x+6)} + \frac{5x+30}{6x(x+6)} = \frac{29x+30}{6x(x+6)}$$

Practice Problem 5   Add: $\dfrac{5}{x+4} + \dfrac{3}{4x}$

EXAMPLE 6   Add: $\dfrac{7}{2x^2y} + \dfrac{3}{xy^2}$

Solution   You should be able to see that the LCD of these fractions is $2x^2y^2$.

$$\frac{7}{2x^2y} \cdot \frac{y}{y} + \frac{3}{xy^2} \cdot \frac{2x}{2x} = \frac{7y}{2x^2y^2} + \frac{6x}{2x^2y^2} = \frac{6x+7y}{2x^2y^2}$$

Practice Problem 6   Add: $\dfrac{7}{4ab^3} + \dfrac{1}{3a^3b^2}$

   When two rational expressions are subtracted, we must be very careful with the signs in the numerator of the second fraction.

EXAMPLE 7   Subtract: $\dfrac{2}{x^2+3x+2} - \dfrac{4}{x^2+4x+3}$

Solution
$$x^2+3x+2 = (x+1)(x+2)$$
$$x^2+4x+3 = (x+1)(x+3)$$

Therefore, the LCD is $(x+1)(x+2)(x+3)$. We now have the following:

$$\frac{2}{(x+1)(x+2)} \cdot \frac{x+3}{x+3} - \frac{4}{(x+1)(x+3)} \cdot \frac{x+2}{x+2}$$

$$= \frac{2x+6}{(x+1)(x+2)(x+3)} - \frac{4x+8}{(x+1)(x+2)(x+3)} = \frac{2x+6-4x-8}{(x+1)(x+2)(x+3)}$$

$$= \frac{-2x-2}{(x+1)(x+2)(x+3)} = \frac{-2\cancel{(x+1)}}{\cancel{(x+1)}(x+2)(x+3)}$$

$$= \frac{-2}{(x+2)(x+3)}$$

   Study this problem carefully. Be sure you understand the reason for each step. You'll see this type of problem often.

Practice Problem 7   Subtract: $\dfrac{4x+2}{x^2+x-12} - \dfrac{3x+8}{x^2+6x+8}$

The following example involves repeated factors in the denominator. Read it through carefully to be sure you understand each step.

EXAMPLE 8   Subtract: $\dfrac{2x+1}{25x^2+10x+1}-\dfrac{6x}{25x+5}$

Solution

Step 1 Factor each denominator into prime factors.

$$25x^2+10x+1=(5x+1)^2$$
$$25x+5=5(5x+1)$$

Step 2 The different factors are 5 and $5x+1$. However, $5x+1$ appears to the first power *and* to the second power. So we need step 3.

Step 3 We must use the *highest* power of each factor. In this example the highest power of the factor 5 is 1, and the highest power of the factor $5x+1$ is 2. So we use $(5x+1)^2$. Thus, the LCD is $5(5x+1)^2$.

So first we write our problem in factored form.

$$\frac{2x+1}{(5x+1)^2}-\frac{6x}{5(5x+1)}$$

Next, we must multiply our fractions by the appropriate factor to change them to equivalent fractions with the LCD.

$$\frac{2x+1}{(5x+1)^2}\cdot\frac{5}{5}-\frac{6x}{5(5x+1)}\cdot\frac{5x+1}{5x+1}=\frac{5(2x+1)-6x(5x+1)}{5(5x+1)^2}$$

$$=\frac{10x+5-30x^2-6x}{5(5x+1)^2}$$

$$=\frac{-30x^2+4x+5}{5(5x+1)^2}$$

Practice Problem 8   Subtract: $\dfrac{7x-3}{4x^2+20x+25}-\dfrac{3x}{4x+10}$

CAUTION: Adding and subtracting rational expressions is somewhat difficult. You should take great care in selecting the LCD. Students sometimes make careless errors when picking the LCD.

Likewise, great care should be taken to copy correctly all + and − signs. It is very easy to make a sign error when combining the equivalent fractions. Try to work very neatly and very carefully. A little extra diligence will result in greater accuracy.

Verbal and Writing Skills

1. Explain how to find the LCD of the fractions $\dfrac{3}{5xy}$ and $\dfrac{11}{y^3}$.

2. Explain how to find the LCD of the fractions $\dfrac{8}{7xy}$ and $\dfrac{3}{x^2}$.

Find the LCD.

3. $\dfrac{3}{x-1}, \dfrac{4}{x^2-2x+1}$

4. $\dfrac{5}{x-6}, \dfrac{7}{x^2-12x+36}$

5. $\dfrac{7}{2m^3n}, \dfrac{3}{2mn^2}$

6. $\dfrac{5}{7ab^5}, \dfrac{2}{7a^3b^3}$

7. $\dfrac{5x}{(3x+4)^3}, \dfrac{2x-1}{(x+2)(3x+4)^2}$

8. $\dfrac{9x}{(x-5)(x+2)^2}, \dfrac{11xy}{(x+2)^3}$

9. $\dfrac{15xy}{3x^2+2x}, \dfrac{17y}{18x^2+9x-2}$

10. $\dfrac{8x}{3x^2-4x}, \dfrac{10xy}{3x^2+5x-12}$

Add or subtract and simplify your answers.

11. $\dfrac{3}{x+4} + \dfrac{2}{x^2-16}$

12. $\dfrac{7}{x^2-x-6} + \dfrac{5}{x-3}$

13. $\dfrac{12}{5x^2} + \dfrac{2}{5xy}$

14. $\dfrac{7}{4ab} + \dfrac{3}{4b^2}$

15. $\dfrac{3}{x^2-7x+12} + \dfrac{5}{x^2-4x}$

16. $\dfrac{7}{x^2-1} + \dfrac{5}{3x^2+3x}$

17. $\dfrac{6x}{2x-5} + 4$

18. $\dfrac{15}{7a+3} + 5$

19. $\dfrac{-5y}{y^2-1} + \dfrac{6}{y^2-2y+1}$

20. $\dfrac{7y}{y^2+6y+9} + \dfrac{5}{y^2-9}$

Mixed Practice

Add or subtract and simplify your answers.

21. $\dfrac{a+5}{a^2-4} + \dfrac{a-3}{2a-4}$

22. $\dfrac{4b}{b^2-b-6} + \dfrac{b-1}{3b-9}$

23. $\dfrac{5}{x-4} - \dfrac{3}{x+1}$

24. $\dfrac{7}{x+2} - \dfrac{4}{2x-3}$

25. $\dfrac{1}{x^2-x-2} - \dfrac{3}{x^2+2x+1}$

26. $\dfrac{3x}{x^2+3x-10} - \dfrac{2x}{x^2+x-6}$

27. $\dfrac{4y}{y^2+3y+2} - \dfrac{y-3}{y+2}$

28. $\dfrac{3y^2}{y^2-1} - \dfrac{y+2}{y+1}$

29. $a+3+\dfrac{2}{3a-5}$

30. $a-2+\dfrac{3}{2a+1}$

Applications

31. *Artificial Lung* If an artificial lung company man-ufactures more than five machines per day, the revenue function in thousands of dollars to manu-facture and sell x machines is given by

$$R(x) = \frac{80-24x}{2-x}.$$

The cost function in thousands of dollars to manu-facture x machines is given by

$$C(x) = \frac{60-12x}{3-x}.$$

Determine the profit function in thousands of dol-lars for this company when more than five ma-chines per day are manufactured by obtaining $P(x) = R(x) - C(x)$.

32. *Artificial Heart* If an artificial heart company manufactures more than five machines per day, the revenue function in thousands of dollars to manu-facture and sell x machines is given by

$$R(x) = \frac{150-38x}{3-x}.$$

The cost function in thousands of dollars to manu-facture x machines is given by

$$C(x) = \frac{120-25x}{2-x}.$$

Determine the profit function in thousands of dol-lars for this company when more than five ma-chines per day are manufactured by obtaining $P(x) = R(x) - C(x)$.

33. *Artificial Lung* Determine the daily profit of the artificial lung company in Exercise 31 if ten ma-chines per day are manufactured. Round your an-swer to the nearest dollar.

34. *Artificial Heart* Determine the daily profit of the artificial heart company in Exercise 32 if twenty machines per day are manufactured. Round your answer to the nearest dollar.

To Think About

Perform the operations indicated.

35. $3x - 2 + \dfrac{5x}{3x - 2} + \dfrac{2x^2}{(3x - 2)^2}$

36. $\left[x + 1 + \dfrac{1}{x - 1} \right] \div \left[\dfrac{1}{x} + \dfrac{1}{x - 1} \right]$

Cumulative Review

37. Evaluate: $\dfrac{|2 - 7| - 3 \cdot 4}{2\sqrt{16} - 1^2 \cdot 4 + 5 \cdot 2}$

38. Write using scientific notation: 0.000351

39. *Volume of Letters* Charlie Chaplin was one of the most celebrated men of the 1920s and 1930s. During a visit to London, England, he received 73,000 letters in just 2 days. If he had had five people reading his letters, and each person was able to read ten letters per hour, how long would it have taken his assistants to read these letters?

40. *New Prescription Drugs* There were 985 new prescription drugs introduced into the United States between 1940 and 1976. Of these, the United States originated 630.
   (a) What percentage of these prescription drugs were formulated in the United States?

   (b) If 7% of the new prescription drugs came from Switzerland, how many actual medications would have been formulated there? *Source: National Center for Health Statistics.*

41. *Purchase Price of Automobiles* Alreda, Tony, and Melissa each purchased a car. The total cost of the cars was $26,500. Alreda purchased a used car that cost $1500 more than the one Tony purchased. Melissa purchased a car that cost $1000 more than double the cost of Tony's car. How much did each car cost?

42. *Chemical Mixtures* A chemist at Argonne Laboratories must combine a mixture that is 15% acid with a mixture that is 30% acid to obtain 60 liters of a mixture that is 20% acid. How much of each kind should he use?

43. *Boat Ride in the River* A Four Winns speedboat traveled up the Hudson River against the current a distance of 75 kilometers in 5 hours. The return trip with the current took only 3 hours. Find the speed of the current and the speed of the speedboat in still water.

44. *Restaurant Inspections* The Health Department of Springfield recently inspected fifteen thousand restaurants. A total of 22% of the restaurants passed inspection. The remainder had one or more health violations. Of the restaurants that did not pass inspection, 3200 had only a cleanliness violation (e.g., vents, stores, floors, or walls). Of the restaurants that did not pass, 3500 had only a vermin violation (e.g., flies, rodents, or other insects). How many restaurants had both a cleanliness violation and a vermin violation?

1 Simplifying Complex Rational Expressions

A **complex rational expression** is a large fraction that has at least one rational expression in the numerator, in the denominator, or in both the numerator and the denominator. The following are three examples of complex rational expressions.

$$\frac{7 + \dfrac{1}{x}}{x + 2}, \qquad \frac{2}{\dfrac{x}{y} + 3}, \qquad \frac{\dfrac{a + b}{7}}{\dfrac{1}{x} + \dfrac{1}{x + a}}$$

There are two ways to simplify complex rational expressions. You can use whichever method you like.

EXAMPLE 1 Simplify $\dfrac{x + \dfrac{1}{x}}{\dfrac{1}{x} + \dfrac{3}{x^2}}$.

Solution

Method 1

1. Simplify numerator and denominator.

$$x + \frac{1}{x} = \frac{x^2 + 1}{x}$$
$$\frac{1}{x} + \frac{3}{x^2} = \frac{x + 3}{x^2}$$

2. Divide the numerator by the denominator.

$$\frac{\dfrac{x^2 + 1}{x}}{\dfrac{x + 3}{x^2}} = \frac{x^2 + 1}{x} \div \frac{x + 3}{x^2}$$

$$= \frac{x^2 + 1}{x} \cdot \frac{x^2}{x + 3}$$

$$= \frac{x^2 + 1}{\cancel{x}} \cdot \frac{\cancel{x^2}^{\,x}}{x + 3}$$

$$= \frac{x(x^2 + 1)}{x + 3}$$

3. The result is already simplified.

Method 2

1. Find the LCD of all the fractions in the numerator and denominator. The LCD is x^2.

2. Multiply the numerator and denominator by the LCD. Use the distributive property.

$$\frac{x + \dfrac{1}{x}}{\dfrac{1}{x} + \dfrac{3}{x^2}} \cdot \frac{x^2}{x^2} = \frac{x^3 + x}{x + 3}$$

3. The result is already simplified, but we will write it in factored form.

$$\frac{x^3 + x}{x + 3} = \frac{x(x^2 + 1)}{x + 3}$$

Practice Problem 1 Simplify $\dfrac{y + \dfrac{3}{y}}{\dfrac{2}{y^2} + \dfrac{5}{y}}$.

Student Learning Objective

After studying this section, you will be able to:

1 Simplify complex rational expressions.

Graphing Calculator

 Exploration

You can verify the answer for Example 1 on a graphing calculator. Graph y_1 and y_2 on the same set of axes.

$$y_1 = \frac{x + \dfrac{1}{x}}{\dfrac{1}{x} + \dfrac{3}{x^2}}$$

$$y_2 = \frac{x(x^2 + 1)}{x + 3}$$

The domain of y_1 is more restrictive than that of y_2. If we use the domain of y_1, then the graphs of y_1 and y_2 should be identical.

Show on your graphing calculator whether y_1 is or is not equivalent to y_2 in each of the following:

1. $y_1 = \dfrac{1 + \dfrac{3}{x + 2}}{1 + \dfrac{6}{x - 1}}$

   $y_2 = \dfrac{x - 1}{x + 2}$

2. $y_1 = \dfrac{\dfrac{1}{x + 1} - \dfrac{1}{x}}{\dfrac{1}{x}}$

   $y_2 = \dfrac{1}{x + 1}$

NOTE TO STUDENT: Fully worked-out solutions to all of the Practice Problems can be found at the back of the text starting at page SP-1

> ## METHOD 1: COMBINING FRACTIONS IN BOTH NUMERATOR AND DENOMINATOR
>
> 1. Simplify the numerator and denominator, if possible, by combining quantities to obtain one fraction in the numerator and one fraction in the denominator.
> 2. Divide the numerator by the denominator (that is, multiply the numerator by the reciprocal of the denominator).
> 3. Simplify the expression.

EXAMPLE 2 Simplify $\dfrac{\dfrac{1}{2x+6}+\dfrac{3}{2}}{\dfrac{3}{x^2-9}+\dfrac{x}{x-3}}$.

Solution

Method 1

1. Simplify the numerator.

$$\frac{1}{2x+6}+\frac{3}{2}=\frac{1}{2(x+3)}+\frac{3}{2}$$

$$=\frac{1}{2(x+3)}+\frac{3(x+3)}{2(x+3)}$$

$$=\frac{1+3x+9}{2(x+3)}$$

$$=\frac{3x+10}{2(x+3)}$$

Simplify the denominator.

$$\frac{3}{x^2-9}+\frac{x}{x-3}=\frac{3}{(x+3)(x-3)}+\frac{x}{x-3}$$

$$=\frac{3}{(x+3)(x-3)}+\frac{x(x+3)}{(x+3)(x-3)}$$

$$=\frac{x^2+3x+3}{(x+3)(x-3)}$$

2. Divide the numerator by the denominator.

$$\frac{3x+10}{2(x+3)}\div\frac{x^2+3x+3}{(x+3)(x-3)}=\frac{3x+10}{2\cancel{(x+3)}}\cdot\frac{\cancel{(x+3)}(x-3)}{x^2+3x+3}$$

$$=\frac{(3x+10)(x-3)}{2(x^2+3x+3)}$$

3. Simplify. The answer is already simplified.

Before we continue the example, we state Method 2 in the following box.

METHOD 2: MULTIPLYING EACH TERM OF THE NUMERATOR AND DENOMINATOR BY THE LCD OF ALL INDIVIDUAL FRACTIONS

1. Find the LCD of all the rational expressions in the numerator and denominator.
2. Multiply the numerator and denominator of the complex fraction by the LCD.
3. Simplify the result.

We will now proceed to do Example 2 by Method 2.

Method 2

1. To find the LCD, we factor.

$$\frac{\dfrac{1}{2x+6}+\dfrac{3}{2}}{\dfrac{3}{x^2-9}+\dfrac{x}{x-3}} = \frac{\dfrac{1}{2(x+3)}+\dfrac{3}{2}}{\dfrac{3}{(x+3)(x-3)}+\dfrac{x}{x-3}}$$

The LCD of the two fractions in the numerator and the two fractions in the denominator is

$$2(x+3)(x-3).$$

2. Multiply the numerator and the denominator by the LCD.

$$\frac{\dfrac{1}{2(x+3)}+\dfrac{3}{2}}{\dfrac{3}{(x+3)(x-3)}+\dfrac{x}{x-3}} \cdot \frac{2(x+3)(x-3)}{2(x+3)(x-3)}$$

$$=\frac{\dfrac{1}{\cancel{2(x+3)}}\cdot\cancel{2(x+3)}(x-3)+\dfrac{3}{\cancel{2}}\cdot\cancel{2}(x+3)(x-3)}{\dfrac{3}{\cancel{(x+3)(x-3)}}\cdot 2\cancel{(x+3)(x-3)}+\dfrac{x}{\cancel{x-3}}\cdot 2(x+3)\cancel{(x-3)}}$$

$$=\frac{x-3+3(x+3)(x-3)}{6+2x(x+3)}$$

$$=\frac{3x^2+x-30}{2x^2+6x+6}=\frac{(3x+10)(x-3)}{2(x^2+3x+3)}$$

3. Simplify. The answer is already simplified.

Whether we use Method 1 or Method 2, we can leave the answer in factored form, or we can multiply it out to obtain

$$\frac{3x^2+x-30}{2x^2+6x+6}.$$

Practice Problem 2 Simplify $\dfrac{\dfrac{4}{16x^2-1}+\dfrac{3}{4x+1}}{\dfrac{x}{4x-1}+\dfrac{5}{4x+1}}$.

NOTE TO STUDENT: Fully worked-out solutions to all of the Practice Problems can be found at the back of the text starting at page SP-1

EXAMPLE 3

Simplify by Method 1: $\dfrac{x+3}{\dfrac{9}{x}-x}$

Solution

$$\dfrac{x+3}{\dfrac{9}{x}-\dfrac{x}{1}\cdot\dfrac{x}{x}}=\dfrac{x+3}{\dfrac{9}{x}-\dfrac{x^2}{x}}=\dfrac{\dfrac{x+3}{1}}{\dfrac{9-x^2}{x}}=\dfrac{x+3}{1}\div\dfrac{9-x^2}{x}=\dfrac{x+3}{1}\cdot\dfrac{x}{9-x^2}$$

$$=\dfrac{\cancel{(x+3)}}{1}\cdot\dfrac{x}{\cancel{(3+x)}(3-x)}=\dfrac{x}{3-x}$$

NOTE TO STUDENT: Fully worked-out solutions to all of the Practice Problems can be found at the back of the text starting at page SP-1

Practice Problem 3   Simplify by Method 1: $\dfrac{4+x}{x-\dfrac{16}{x}}$

EXAMPLE 4

Simplify by Method 2: $\dfrac{\dfrac{3}{x+2}+\dfrac{1}{x}}{\dfrac{3}{y}-\dfrac{2}{x}}$

Solution   The LCD of the numerator is $x(x+2)$. The LCD of the denominator is xy. Thus, the LCD of the complex fraction is $xy(x+2)$.

$$\dfrac{\dfrac{3}{x+2}+\dfrac{1}{x}}{\dfrac{3}{y}-\dfrac{2}{x}}\cdot\dfrac{xy(x+2)}{xy(x+2)}=\dfrac{3xy+xy+2y}{3x(x+2)-2y(x+2)}$$

$$=\dfrac{4xy+2y}{(x+2)(3x-2y)}$$

$$=\dfrac{2y(2x+1)}{(x+2)(3x-2y)}$$

Practice Problem 4   Simplify by Method 2: $\dfrac{\dfrac{7}{y+3}-\dfrac{3}{y}}{\dfrac{2}{y}+\dfrac{5}{y+3}}$

Student Solutions Manual   CD/Video   PH Math Tutor Center   MathXL®Tutorials on CD   MathXL®   MyMathLab®   Interactmath.com

Simplify the complex fractions by any method.

1. $\dfrac{\dfrac{7}{x}}{\dfrac{3}{xy}}$

2. $\dfrac{-\dfrac{2}{ab^2}}{\dfrac{5}{a}}$

3. $\dfrac{\dfrac{2x}{x+5}}{\dfrac{x^2}{x-1}}$

4. $\dfrac{\dfrac{3y+2}{7y^3}}{\dfrac{y-1}{y^2}}$

5. $\dfrac{1-\dfrac{6}{5y}}{\dfrac{3}{y}+1}$

6. $\dfrac{2-\dfrac{3}{4y}}{\dfrac{1}{2y}+1}$

7. $\dfrac{\dfrac{y}{6}-\dfrac{1}{2y}}{\dfrac{3}{2y}-\dfrac{1}{y}}$

8. $\dfrac{\dfrac{1}{3y}+\dfrac{1}{6y}}{\dfrac{1}{2y}+\dfrac{3}{4y}}$

9. $\dfrac{\dfrac{2}{y^2-9}}{\dfrac{3}{y+3}+1}$

10. $\dfrac{\dfrac{2}{y+4}}{\dfrac{3}{y-4}-\dfrac{1}{y^2-16}}$

11. $\dfrac{\dfrac{3}{2x+4}+\dfrac{1}{2}}{\dfrac{2}{x^2-4}+\dfrac{x}{x+2}}$

12. $\dfrac{\dfrac{5}{2x+8}+\dfrac{3}{2}}{\dfrac{3}{x^2-16}+\dfrac{1}{x+4}}$

13. $\dfrac{-8}{\dfrac{6x}{x-1}-4}$

14. $\dfrac{6}{2x-\dfrac{10}{x-4}}$

15. $\dfrac{\dfrac{1}{2x+3}+\dfrac{2}{4x^2+12x+9}}{\dfrac{5}{2x^2+3x}}$

16. $\dfrac{\dfrac{3}{5x-2}-\dfrac{2}{25x^2-4}}{\dfrac{7x}{5x^2-2x}}$

17. $\dfrac{\dfrac{3}{x-y}+\dfrac{1}{2}}{1-\dfrac{x}{x-y}}$

18. $\dfrac{\dfrac{5}{y+2}-1}{\dfrac{1}{2}+\dfrac{3}{y+2}}$

19. $\dfrac{\dfrac{1}{x-a}-\dfrac{1}{x}}{a}$

20. $\dfrac{\dfrac{1}{x+a}-\dfrac{1}{x}}{a}$

To Think About

Simplify the complex fractions.

21. $1 - \dfrac{1}{1 - \dfrac{1}{y - 2}}$

22. $\dfrac{x}{1 + \dfrac{1}{x}} + \dfrac{2x}{2 + \dfrac{2}{x}}$

Applications

23. *Amount of Electric Current* The amount of electrical current in amps in a certain alternating current circuit can be described by the expression

$$\dfrac{\dfrac{x^2 - 1}{6x^2 + 3x}}{\dfrac{x - 1}{2x^2}},$$

where x is the number of milliseconds since the power has started to flow through the circuit. Simplify this expression.

24. *Sound Traveling from a Moving Object* If an object is traveling toward you at x miles per hour and is emitting a sound (like a train whistle or an ambulance siren) of pitch y, then the actual pitch that you hear is given by the expression

$$\dfrac{y}{1 - \dfrac{x}{770}}.$$

Simplify this expression.

Cumulative Review

Solve for x.

25. $|2 - 3x| = 4$

26. $\left|\dfrac{1}{2}(5 - x)\right| = 5$

27. $|7x - 3 - 2x| < 6$

28. $|0.6x + 0.3| \geq 1.2$

29. *Interstate Highway Cost* The price of building an interstate highway project in the late 1970s near New York City was $4000 per inch.
   (a) How much money was spent per mile?

   (b) Now a new interstate highway connector will be built in the same area beginning in 2002 for $660 million per mile. If the land acquisition cost was $570 million, how many miles can be built if the total budget limit is $4,860,000,000?

30. *Moving Expenses* Jan Robbins and her family are being relocated by her corporation from Dallas to Stockholm, Sweden. The moving company charges $2.50 per kilogram for airfreight and $1.30 per kilogram for belongings shipped by ocean freighter. Jan needs to ship 5600 kilograms of her belongings to Stockholm. Her corporation will pay only $9380 for the shipments. How many kilograms should she ship by airfreight and how many kilograms should she ship by ocean freighter?

How are you doing with your homework assignments in Sections 6.1 to 6.3? Do you feel you have mastered the material so far? Do you understand the concepts you have covered? Before you go further in the textbook, take some time to do each of the following problems.

6.1

Simplify.

1. $\dfrac{49x^2 - 9}{7x^2 + 4x - 3}$

2. $\dfrac{x^2 - 4x - 21}{x^2 + x - 56}$

3. $\dfrac{2x^3 - 5x^2 - 3x}{x^3 - 8x^2 + 15x}$

4. $\dfrac{6a - 30}{3a + 3} \cdot \dfrac{9a^2 + a - 8}{2a^2 - 15a + 25}$

5. $\dfrac{5x^3y^2}{x^2y + 10xy^2 + 25y^3} \div \dfrac{2x^4y^5}{3x^3 - 75xy^2}$

6. $\dfrac{8x^3 + 1}{4x^2 + 4x + 1} \cdot \dfrac{6x + 3}{4x^2 - 2x + 1}$

6.2

Add or subtract. Simplify your answers.

7. $\dfrac{x}{3x - 6} - \dfrac{4}{3x}$

8. $\dfrac{2}{x + 5} + \dfrac{3}{x - 5} + \dfrac{7x}{x^2 - 25}$

9. $\dfrac{y + 1}{y^2 + y - 12} - \dfrac{y - 3}{y^2 + 7y + 12}$

10. $\dfrac{x + 1}{x + 4} + \dfrac{4 - x^2}{x^2 - 16}$

6.3

Simplify.

11. $\dfrac{\dfrac{1}{12x} + \dfrac{5}{3x}}{\dfrac{2}{3x^2}}$

12. $\dfrac{\dfrac{x}{4x^2 - 1}}{3 - \dfrac{2}{2x + 1}}$

13. $\dfrac{\dfrac{5}{x} + 3}{\dfrac{6}{x} - 2}$

14. $\dfrac{\dfrac{x}{x + 2} + \dfrac{5}{x}}{\dfrac{x + 2}{x} + \dfrac{3}{x + 2}}$

Now turn to page SA-24 for the answers to each of these problems. Each answer also includes a reference to the objective in which the problem is first taught. If you missed any of these problems, you should stop and review the Examples and Practice Problems in the referenced objective. A little review now will help you master the material in the upcoming sections of the text.

1. _____

2. _____

3. _____

4. _____

5. _____

6. _____

7. _____

8. _____

9. _____

10. _____

11. _____

12. _____

13. _____

14. _____

6.4 RATIONAL EQUATIONS

Student Learning Objectives

After studying this section, you will be able to:

1 Solve a rational equation that has a solution and be able to check the solution.

2 Identify those rational equations that have no solution.

Graphing Calculator

Solving Rational Equations

To solve Example 2 on your graphing calculator, find the point of intersection of

$$y_1 = \frac{2}{3x + 6}$$

and $y_2 = \frac{1}{6} - \frac{1}{2x + 4}$.

Use the Zoom and Trace features or the intersection command to find that the solution is 5.00 (to the nearest hundredth). What two difficulties do you observe in the graph that make this exploration more challenging? How can these be overcome?

Use this method to find the solutions to the following equations on your graphing calculator. (Round your answers to the nearest hundredth.)

1. $\dfrac{3}{x + 3} + \dfrac{5}{x + 4}$
$= \dfrac{12x + 19}{x^2 + 7x + 12}$

2. $\dfrac{x - 2.84}{x + 1.12} = \dfrac{x - 5.93}{x + 5.06}$

NOTE TO STUDENT: Fully worked-out solutions to all of the Practice Problems can be found at the back of the text starting at page SP-1

330

① Solving a Rational Equation

A **rational equation** is an equation that has one or more rational expressions as terms. To solve a rational equation, we find the LCD of all fractions in the equation and multiply each side of the equation by the LCD. We then solve the resulting linear equation.

EXAMPLE 1 Solve $\dfrac{9}{4} - \dfrac{1}{2x} = \dfrac{4}{x}$. Check your solution.

Solution First we multiply each side of the equation by the LCD, which is $4x$.

$$4x\left(\frac{9}{4} - \frac{1}{2x}\right) = 4x\left(\frac{4}{x}\right)$$

$$4x\left(\frac{9}{4}\right) - \overset{2}{4x}\left(\frac{1}{2x}\right) = 4x\left(\frac{4}{x}\right) \quad \text{Use the distributive property.}$$

$$9x - 2 = 16 \qquad \text{Simplify.}$$

$$9x = 18 \qquad \text{Collect like terms.}$$

$$x = 2 \qquad \text{Divide each side by the coefficient of } x.$$

Check:

$$\frac{9}{4} - \frac{1}{2(2)} \overset{?}{=} \frac{4}{2}$$

$$\frac{9}{4} - \frac{1}{4} \overset{?}{=} 2$$

$$\frac{8}{4} \overset{?}{=} 2$$

$$2 = 2 \quad ✓$$

Practice Problem 1 Solve and check $\dfrac{4}{3x} + \dfrac{x + 1}{x} = \dfrac{1}{2}$.

Usually, we combine the first two steps of the exercise and show only the step of multiplying each term of the equation by the LCD. We will follow this approach in the remaining examples in this section.

This is another illustration of the need to understand a mathematical principle rather than merely copying down a step without understanding it. Because we understand the distributive property, we can move directly to simplifying a rational equation by multiplying each term of the equation by the LCD.

EXAMPLE 2 Solve $\dfrac{2}{3x + 6} = \dfrac{1}{6} - \dfrac{1}{2x + 4}$.

Solution

$$\frac{2}{3(x + 2)} = \frac{1}{6} - \frac{1}{2(x + 2)} \quad \text{Factor each denominator.}$$

$$\overset{2}{6(x + 2)}\left[\frac{2}{3(x + 2)}\right] = 6(x + 2)\left[\frac{1}{6}\right] - \overset{3}{6(x + 2)}\left[\frac{1}{2(x + 2)}\right] \quad \begin{array}{l}\text{Multiply} \\ \text{each term} \\ \text{by the LCD} \\ 6(x + 2).\end{array}$$

$$4 = x + 2 - 3 \qquad \text{Simplify.}$$

$$4 = x - 1 \qquad \text{Collect like terms.}$$

$$5 = x \qquad \text{Solve for } x.$$

Check: Verify that 5 is the solution.

Practice Problem 2 Solve and check $\dfrac{1}{3x - 9} = \dfrac{1}{2x - 6} - \dfrac{5}{6}$.

EXAMPLE 3   Solve $\dfrac{y^2 - 10}{y^2 - y - 20} = 1 + \dfrac{7}{y - 5}$.

Solution

$\dfrac{y^2 - 10}{(y - 5)(y + 4)} = 1 + \dfrac{7}{y - 5}$ Factor each denominator. Multiply each term by the LCD $(y - 5)(y + 4)$.

$\cancel{(y - 5)}\cancel{(y + 4)}\left[\dfrac{y^2 - 10}{\cancel{(y - 5)}\cancel{(y + 4)}}\right] = (y - 5)(y + 4)\,(1) + \cancel{(y - 5)}(y + 4)\left[\dfrac{7}{\cancel{(y - 5)}}\right]$

$y^2 - 10 = (y - 5)(y + 4)(1) + 7(y + 4)$ Divide out common factors.

$y^2 - 10 = y^2 - y - 20 + 7y + 28$ Simplify.

$y^2 - 10 = y^2 + 6y + 8$ Collect like terms.

$-10 = 6y + 8$ Subtract y^2 from each side.

$-18 = 6y$ Add -8 to each side.

$-3 = y$ Divide each side by the coefficient of y.

Check: $\dfrac{(-3)^2 - 10}{(-3)^2 - (-3) - 20} \overset{?}{=} 1 + \dfrac{7}{-3 - 5}$

$\dfrac{9 - 10}{9 + 3 - 20} \overset{?}{=} 1 + \dfrac{7}{-8}$

$\dfrac{-1}{-8} \overset{?}{=} 1 - \dfrac{7}{8}$

$\dfrac{1}{8} = \dfrac{1}{8}$ ✓

Practice Problem 3   Solve $\dfrac{y^2 + 4y - 2}{y^2 - 2y - 8} = 1 + \dfrac{4}{y - 4}$.

② Identifying Equations with No Solution

Some rational equations have no solution. This can happen in two distinct ways. In the first case, when you attempt to solve the equation, you obtain a contradiction, such as $0 = 1$. This occurs because the variable "drops out" of the equation. No solution can be obtained. In the second case, we may solve an equation to get an *apparent* solution, but it may not satisfy the original equation. We call the apparent solution an **extraneous solution.** An equation that yields an extraneous solution has no solution.

Case 1: The Variable Drops Out. In this case, when you attempt to solve the equation, the coefficient of the variable term becomes zero. Thus, you are left with a statement such as $0 = 1$, which is false. In such a case you know that there is no value for the variable that could make $0 = 1$; hence, there cannot be any solution.

EXAMPLE 4   Solve $\dfrac{z + 1}{z^2 - 3z + 2} + \dfrac{3}{z - 1} = \dfrac{4}{z - 2}$.

Solution

$\dfrac{z + 1}{(z - 2)(z - 1)} + \dfrac{3}{z - 1} = \dfrac{4}{z - 2}$ Factor to find the LCD $(z - 2)(z - 1)$. Then multiply each term by the LCD.

$\cancel{(z - 2)}\cancel{(z - 1)}\left[\dfrac{z + 1}{\cancel{(z - 2)}\cancel{(z - 1)}}\right] + (z - 2)\cancel{(z - 1)}\left[\dfrac{3}{\cancel{z - 1}}\right] = \cancel{(z - 2)}(z - 1)\left[\dfrac{4}{\cancel{(z - 2)}}\right]$

$$z + 1 + 3(z - 2) = 4(z - 1) \quad \text{Divide out common factors.}$$
$$z + 1 + 3z - 6 = 4z - 4 \quad \text{Simplify.}$$
$$4z - 5 = 4z - 4 \quad \text{Collect like terms.}$$
$$4z - 4z = -4 + 5 \quad \text{Obtain variable terms on one side and constant values on the other.}$$
$$0 = 1$$

Of course, $0 \neq 1$. Therefore, no value of z makes the original equation true. Hence, the equation has **no solution.**

Practice Problem 4 Solve $\dfrac{2x - 1}{x^2 - 7x + 10} + \dfrac{3}{x - 5} = \dfrac{5}{x - 2}$.

Case 2: The Obtained Value of the Variable Leads to a Denominator of Zero

EXAMPLE 5 Solve $\dfrac{4y}{y + 3} - \dfrac{12}{y - 3} = \dfrac{4y^2 + 36}{y^2 - 9}$.

Solution

$$\frac{4y}{y + 3} - \frac{12}{y - 3} = \frac{4y^2 + 36}{(y + 3)(y - 3)} \quad \begin{array}{l}\text{Factor each denominator to find the LCD } (y + 3)(y - 3). \text{ Multiply} \\ \text{each term by the LCD.}\end{array}$$

$$\cancel{(y + 3)}(y - 3)\left[\frac{4y}{\cancel{y + 3}}\right] - (y + 3)\cancel{(y - 3)}\left[\frac{12}{\cancel{y - 3}}\right] = \cancel{(y + 3)}\,\cancel{(y - 3)}\left[\frac{4y^2 + 36}{\cancel{(y + 3)}\,\cancel{(y - 3)}}\right]$$

$$4y(y - 3) - 12(y + 3) = 4y^2 + 36 \quad \text{Divide out common factors.}$$
$$4y^2 - 12y - 12y - 36 = 4y^2 + 36 \quad \text{Remove parentheses.}$$
$$4y^2 - 24y - 36 = 4y^2 + 36 \quad \text{Collect like terms.}$$
$$-24y - 36 = 36 \quad \text{Subtract } 4y^2 \text{ from each side.}$$
$$-24y = 72 \quad \text{Add 36 to each side.}$$
$$y = \frac{72}{-24} \quad \text{Divide each side by } -24.$$
$$y = -3$$

Check: $\quad \dfrac{4(-3)}{-3 + 3} - \dfrac{12}{-3 - 3} \overset{?}{=} \dfrac{4(-3)^2 + 36}{(-3)^2 - 9}$

$$\frac{-12}{0} - \frac{12}{-6} \overset{?}{=} \frac{36 + 36}{0}$$

You cannot divide by zero. Division by zero is not defined. A value of a variable that makes a denominator in the original equation zero is not a solution to the equation. Thus, this equation has **no solution.**

TO THINK ABOUT: Quick Solution Checks Sometimes you may find that you do not have sufficient time for a complete check, but you still wish to make sure that you do not have a "no solution" situation. In those instances you can do a quick analysis to be sure that your obtained value for the variable does not make a denominator zero. If you were solving the equation $\dfrac{4x - 1}{x^2 + 5x - 14} = \dfrac{1}{x - 2} - \dfrac{2}{x + 7}$, you would know immediately that you could not have 2 or -7 as a solution. Do you see why?

Practice Problem 5 Solve and check $\dfrac{y}{y - 2} - 3 = 1 + \dfrac{2}{y - 2}$.

Student Solutions Manual   CD/Video   PH Math Tutor Center   MathXL®Tutorials on CD   MathXL®   MyMathLab®   Interactmath.com

Solve the equations and check your solutions. If there is no solution, say so.

1. $\dfrac{2}{x} + \dfrac{3}{2x} = \dfrac{7}{6}$

2. $\dfrac{1}{x} + \dfrac{2}{3x} = \dfrac{1}{3}$

3. $3 - \dfrac{2}{x} = \dfrac{1}{4x}$

4. $\dfrac{5}{3x} + 2 = \dfrac{1}{x}$

5. $\dfrac{5}{2x + 3} + \dfrac{1}{x} = \dfrac{3}{x}$

6. $\dfrac{4}{3x - 1} + \dfrac{1}{2x} = \dfrac{3}{2x}$

7. $\dfrac{2}{y} = \dfrac{5}{y - 3}$

8. $\dfrac{10}{3x - 8} = \dfrac{2}{x}$

9. $\dfrac{y + 6}{y + 3} - 2 = \dfrac{3}{y + 3}$

10. $4 - \dfrac{8x}{x + 1} = \dfrac{8}{x + 1}$

11. $\dfrac{1}{3x} - \dfrac{2}{x} = \dfrac{-5}{x + 4}$

12. $\dfrac{2}{x} - \dfrac{1}{5x} = \dfrac{3}{2x - 3}$

13. $\dfrac{2x + 3}{x + 3} = \dfrac{2x}{x + 1}$

14. $\dfrac{3x}{x - 2} = \dfrac{3x + 5}{x - 1}$

15. $\dfrac{3}{y^2 - 1} = \dfrac{6}{y^2 - y}$

16. $\dfrac{5}{2x^2 - 4x} = \dfrac{2}{x^2 - 4}$

Mixed Practice

Solve the equations and check your solutions. If there is no solution, say so.

17. $\dfrac{3}{2x - 1} + \dfrac{3}{2x + 1} = \dfrac{8x}{4x^2 - 1}$

18. $\dfrac{6}{x} - \dfrac{3}{x^2 - x} = \dfrac{7}{x - 1}$

19. $\dfrac{5}{y - 3} + 2 = \dfrac{3}{3y - 9}$

20. $\dfrac{2}{3} + \dfrac{5}{y - 4} = \dfrac{y + 6}{3y - 12}$

21. $1 - \dfrac{10}{z - 3} = \dfrac{-5}{3z - 9}$

22. $\dfrac{3}{2} + \dfrac{2}{2z - 8} = \dfrac{1}{z - 4}$

23. $\dfrac{8}{3x + 2} - \dfrac{7x + 4}{3x^2 + 5x + 2} = \dfrac{2}{x + 1}$

24. $\dfrac{1}{x - 1} - \dfrac{2x + 1}{2x^2 + 5x - 7} = \dfrac{6}{2x + 7}$

25. $\dfrac{4}{z^2 - 9} = \dfrac{2}{z^2 - 3z}$

26. $\dfrac{z^2 + 16}{z^2 - 16} = \dfrac{z}{z + 4} - \dfrac{4}{z - 4}$

27. $\dfrac{2x + 3}{2} + \dfrac{1}{x + 1} = x$

28. $\dfrac{3x - 4}{3} - x = -\dfrac{4}{x + 3}$

Verbal and Writing Skills

29. In what situations will a rational equation have no solution?

30. What does "extraneous solution" mean? What must we do to determine whether a solution is an extraneous solution?

Optional Graphing Calculator Problems

Solve each equation. Round your answers to the nearest tenth.

31. $\dfrac{153.8}{x^2 + 4.9x - 39.56} = \dfrac{75.3}{x + 9.2} + \dfrac{84.2}{x - 4.3}$

32. $\dfrac{5}{x + 3.6} - \dfrac{4.2}{x - 7.6} = \dfrac{3.3}{x^2 - 4x - 27.36}$

Cumulative Review

Factor completely.

33. $7x^2 - 63$

34. $2x^2 + 20x + 50$

35. $64x^3 - 27y^3$

36. $3x^2 - 13x + 14$

Effect of Marital Counseling *A recent study from Rutgers University showed that the number of couples who intend to divorce decreases rapidly if those couples undergo marital counseling. Those who stay in counseling less than 1 year have a 10% chance of not getting divorced. Couples who remain in counseling at least 1 year but less than 2 years have a 25% chance of not getting divorced. Couples who remain in counseling 2 years or more have a 78% chance of not getting divorced. Suppose 160,000 couples in the Midwest have marriage problems and intend to divorce. Of these couples, 15% refuse to attend counseling, 10% decide to attend counseling for less than 1 year, 50% decide to attend counseling for at least a year but less than 2 years, and 25% decide to attend counseling for 2 years or more.*

37. How many couples will attend counseling for at least 1 year but less than 2 years? How many of these couples will likely remain married?

38. How many of the original 160,000 couples will likely remain married? (Assume that all couples refusing to attend counseling do in fact get divorced.)

① Solving a Formula for a Particular Variable

In science, economics, business, and mathematics, we use formulas that contain rational expressions. We often have to solve these formulas for a specific variable in terms of the other variables.

EXAMPLE 1 Solve for a: $\dfrac{1}{f} = \dfrac{1}{a} + \dfrac{1}{b}$

Solution This formula is used in optics in the study of light passing through a lens. It relates the focal length f of the lens to the distance a of an object from the lens and the distance b of the image from the lens.

$$ab\!\!\not{f}\left[\frac{1}{\not{f}}\right] = \not{a}bf\left[\frac{1}{\not{a}}\right] + a\not{b}f\left[\frac{1}{\not{b}}\right] \qquad \text{Multiply each term by the LCD } abf.$$

$$ab = bf + af \qquad \text{Simplify.}$$

$$ab - af = bf \qquad \begin{array}{l}\text{Obtain all the terms containing the}\\ \text{variable } a \text{ on one side of the equation.}\end{array}$$

$$a(b - f) = bf \qquad \text{Factor.}$$

$$a = \frac{bf}{b - f} \qquad \text{Divide each side by } b - f.$$

Practice Problem 1 Solve for t: $\dfrac{1}{t} = \dfrac{1}{c} + \dfrac{1}{d}$

This formula relates the total amount of time t in hours that is required for two workers to complete a job working together if one worker can complete it alone in c hours and the other worker in d hours.

EXAMPLE 2 The gravitational force F between two masses m_1 and m_2 a distance d apart is represented by the formula

$$F = \frac{Gm_1m_2}{d^2}.$$

Solution Solve for m_2.

The subscripts on the variable m mean that m_1 and m_2 are *different*. (The m stands for "mass.")

$$F = \frac{Gm_1m_2}{d^2}$$

$$d^2\,[F] = \not{d^2}\left[\frac{Gm_1m_2}{\not{d^2}}\right] \qquad \text{Multiply each side by the LCD } d^2.$$

$$d^2 F = Gm_1m_2 \qquad \text{Simplify.}$$

$$\frac{d^2 F}{Gm_1} = \frac{Gm_1m_2}{Gm_1} \qquad \text{Divide each side by the coefficient of } m_2, \text{ which is } Gm_1.$$

$$\frac{d^2 F}{Gm_1} = m_2$$

Practice Problem 2 The number of telephone calls C between two cities of populations p_1 and p_2 that are a distance d apart may be represented by the formula

$$C = \frac{Bp_1p_2}{d^2}.$$

Solve this equation for p_1.

NOTE TO STUDENT: Fully worked-out solutions to all of the Practice Problems can be found at the back of the text starting at page SP-1

2 Solving Advanced Exercises Involving Ratio and Rate

You have already encountered the simple idea of proportions in a previous course. In that course you learned that a **proportion** is an equation that says that two ratios are equal.

For example, if Wendy's car traveled 180 miles on 7 gallons of gas, how many miles can the car travel on 11 gallons of gas? You probably remember that you can solve this type of exercise quickly if you let x be the number of miles the car can travel. Then to find x you solve the proportion $\dfrac{7}{180} = \dfrac{11}{x}$. Rounded to the nearest mile, the answer is 283 miles. Can you obtain that answer? So far in Chapters 1–5 in this book in the Cumulative Review Exercises, there have been several ratio and proportion exercises for you to solve at this elementary level.

Now we proceed with more advanced exercises in which the ratio of two quantities is more difficult to establish. Study carefully the next two examples. See whether you can follow the reasoning in each case.

EXAMPLE 3 A company plans to employ 910 people with a ratio of two managers for every eleven workers. How many managers should be hired? How many workers?

Solution If we let x = the number of managers, then $910 - x$ = the number of workers. We are given the ratio of managers to workers, so let's set up our proportion in that way.

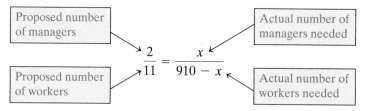

The LCD is $11(910 - x)$. Multiplying by the LCD, we get the following:

$$\cancel{11}(910 - x)\left[\dfrac{2}{\cancel{11}}\right] = 11\cancel{(910 - x)}\left[\dfrac{x}{\cancel{910 - x}}\right]$$

$$2(910 - x) = 11x$$
$$1820 - 2x = 11x$$
$$1820 = 13x$$
$$140 = x$$
$$910 - x = 910 - 140 = 770$$

The number of managers needed is 140. The number of workers needed is 770.

NOTE TO STUDENT: Fully worked-out solutions to all of the Practice Problems can be found at the back of the text starting at page SP-1

Practice Problem 3 Western University has 168 faculty. The university always maintains a student-to-faculty ratio of 21:2. How many students should they enroll to maintain that ratio?

The next example concerns **similar triangles**. These triangles have corresponding angles that are equal and corresponding sides that are *proportional* (not equal). Similar triangles are frequently used to determine distances that cannot be conveniently measured. For example, in the following sketch, x and X are corresponding angles, y and Y are corresponding angles, and s and S are corresponding sides.

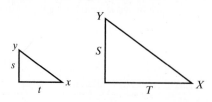

Hence, angle x = angle X, angle y = angle Y, and side s is proportional to side S. (Again, note that we did not say that side s is equal to side S.) Also, side t is proportional to side T. So, really, one triangle is just a magnification of the other triangle. If the sides are proportional, they are in the same ratio.

▲ **EXAMPLE 4** A helicopter is hovering an unknown distance above an 850-foot building. A man watching the helicopter is 500 feet from the base of the building and 11 feet from a flagpole that is 29 feet tall. The man's line of sight to the helicopter is directly above the flagpole, as you can see in the sketch. How far above the building is the helicopter? Round your answer to the nearest foot.

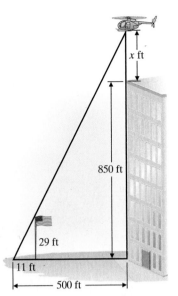

Solution

1. *Understand the problem.*
   Can you see the two triangles in the diagram in the margin? For convenience, we separate them out in the sketch on the right. We want to find the distance x. Are the triangles similar? The angles at the bases of the triangles are equal. (Why?) It follows, then, that the top angles must also be equal. (Remember that the angles of any triangle add up to 180°.) Since the angles are equal, the triangles are similar and the sides are proportional.

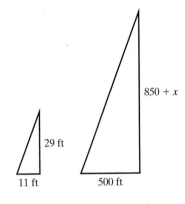

2. *Write an equation.*
   We can set up our proportion like this:

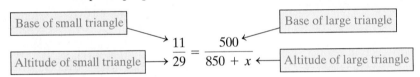

3. *Solve the equation and state the answer.*
   The LCD is $29(850 + x)$.

$$29(850 + x)\left[\frac{11}{29}\right] = 29(850 + x)\left[\frac{500}{850 + x}\right]$$

$$11(850 + x) = 29(500)$$
$$9350 + 11x = 14{,}500$$
$$11x = 5150$$
$$x = \frac{5150}{11} = 468.\overline{18}$$

So the helicopter is about 468 feet above the building.

▲ **Practice Problem 4** Solve the exercise in Example 4 for a man watching 450 feet from the base of a 900-foot building as shown in the figure in the margin. The flagpole is 35 feet tall, and the man is 10 feet from the flagpole.

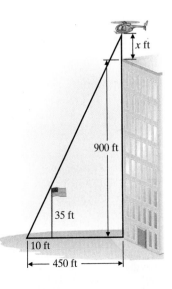

We will see some challenging exercises involving similar triangles in Exercises 55 and 56.

We will sometimes encounter exercises in which two or more people or machines are working together to complete a certain task. These types of exercises are sometimes called *work problems*. In general, these types of exercises can be analyzed by using the following concept.

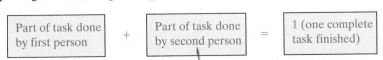

We will also use a general idea about the rate at which something is done. If Robert can do a task in 3 hours, then he can do $\frac{1}{3}$ of the task in 1 hour. If Susan can do the same task in 2 hours, then she can do $\frac{1}{2}$ of the task in 1 hour. In general, if a person can do a task in t hours, then that person can do $\frac{1}{t}$ of the task in 1 hour.

EXAMPLE 5 Robert can paint the kitchen in 3 hours. Susan can paint the kitchen in 2 hours. How long will it take Robert and Susan to paint the kitchen if they work together?

Solution

1. Understand the problem.

Robert can paint $\frac{1}{3}$ of the kitchen in 1 hour.

Susan can paint $\frac{1}{2}$ of the kitchen in 1 hour.

We do not know how long it will take them working together, so we let x = the number of hours it takes them to paint the kitchen working together. To assist us, we will construct a table that relates the data. We will use the concept that (rate)(time) = fraction of task done.

	Rate of Work per Hour	Time Worked in Hours	Fraction of Task Done
Robert	$\frac{1}{3}$	x	$\frac{x}{3}$
Susan	$\frac{1}{2}$	x	$\frac{x}{2}$

2. Write an equation.

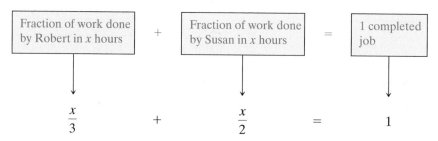

Fraction of work done by Robert in x hours	+	Fraction of work done by Susan in x hours	=	1 completed job
$\frac{x}{3}$	+	$\frac{x}{2}$	=	1

3. Solve the equation and state the answer.

Multiply each side of the equation by the LCD and use the distributive property.

$$6\left(\frac{x}{3}\right) + 6\left(\frac{x}{2}\right) = 6(1)$$
$$2x + 3x = 6$$
$$5x = 6$$
$$x = \frac{6}{5} = 1\frac{1}{5} = 1.2 \text{ hours}$$

Working together, Robert and Susan can paint the kitchen in 1.2 hours.

NOTE TO STUDENT: *Fully worked-out solutions to all of the Practice Problems can be found at the back of the text starting at page SP-1*

Practice Problem 5 Alfred can mow the huge lawn at his Vermont farm with his new lawn mower in 4 hours. His young son can mow the lawn with the old lawn mower in 5 hours. If they work together, how long will it take them to mow the lawn?

Solve for the indicated variable.

1. $x = \dfrac{y - b}{m}$; for m

2. $m = \dfrac{y - b}{x}$; for x

3. $\dfrac{1}{f} = \dfrac{1}{a} + \dfrac{1}{b}$; for b

4. $\dfrac{1}{f} = \dfrac{1}{a} + \dfrac{1}{b}$; for f

5. $\dfrac{V}{lh} = w$, for h

6. $r = \dfrac{I}{pt}$, for p

7. $F = \dfrac{xy + xz}{2}$; for x

8. $A = \dfrac{ha + hb}{2}$; for h

9. $\dfrac{r^3}{V} = \dfrac{3}{4\pi}$; for V

10. $\dfrac{V}{\pi r^2} = h$; for r^2

11. $\dfrac{E}{e} = \dfrac{R + r}{r}$; for e

12. $\dfrac{E}{e} = \dfrac{R + r}{r}$, for r

13. $\dfrac{P_1 V_1}{T_1} = \dfrac{P_2 V_2}{T_2}$; for T_1

14. $\dfrac{P_1 V_1}{T_1} = \dfrac{P_2 V_2}{T_2}$; for T_2

15. $\dfrac{S - 2lw}{2w + 2l} = h$, for w

16. $\dfrac{S - 2lw}{2w + 2l} = h$, for l

Mixed Practice

Solve for the indicated variable.

17. $E = T_1 - \dfrac{T_1}{T_2}$; for T_1

18. $E = T_1 - \dfrac{T_1}{T_2}$; for T_2

19. $m = \dfrac{y_2 - y_1}{x_2 - x_1}$; for x_1

20. $m = \dfrac{y_2 - y_1}{x_2 - x_1}$; for x_2

21. $\dfrac{2D - at^2}{2t} = V$; for D

22. $\dfrac{2D - at^2}{2t} = V$; for a

23. $Q = \dfrac{kA(t_1 - t_2)}{L}$; for t_2

24. $Q = \dfrac{kA(t_1 - t_2)}{L}$; for A

25. $\dfrac{T_2 W}{T_2 - T_1} = q$; for T_2

26. $d = \dfrac{LR_2}{R_2 + R_1}$; for R_1

27. $\dfrac{s - s_0}{v_0 + gt} = t$; for v_0

28. $\dfrac{A - P}{Pr} = t$; for P

Round your answers to four decimal places.

 29. Solve for T: $\dfrac{1.98V}{1.96V_0} = 0.983 + 5.936(T - T_0)$

30. Solve for r_1: $\dfrac{1}{R} = \dfrac{1}{r_1} + \dfrac{1}{0.368} + \dfrac{1}{0.736}$

Applications

31. *Map Scale* On a map of Nigeria, the cities of Benin City and Onitsha are 6.5 centimeters apart. The map scale shows that 3 centimeters = 55 kilometers on the map. How far apart are the two cities?

32. *Scale Drawing* The lobby of the new athletic center at Mansfield Community College is shown on a blueprint made by the architect. The scale drawing is 6 inches wide and 12 inches long. The blueprint has a scale of 3:100. Find the length and width of the lobby measured in inches. Find the length and width of the lobby measured in feet.

33. *Speed Conversion* A speed of 60 miles per hour (mph) is equivalent to 88 kilometers per hour (km/h). On a trip through Nova Scotia, Lora passes a sign stating that the speed limit is 80 km/h. What is the speed limit in miles per hour?

34. *Currency Conversion* Miguel traveled to Italy on business for a week and will now travel to Australia for a week. He wants to convert the 60 euros he now has to Australian dollars. 0.84 euros are currently equivalent to 1.4 Australian dollars. How many Australian dollars will he receive for his 60 euros?

In Exercises 35–54, round your answers to the nearest hundredth, unless otherwise directed.

35. *Grizzly Bear Population* Thirty-five grizzly bears were captured and tagged by wildlife personnel in the Yukon and then released back into the wild. Later that same year, fifty were captured, and twenty-two had tags. Estimate the number of grizzly bears in that part of the Yukon. (Round this answer to the nearest whole number.)

36. *Alligator Population* Thirty alligators were captured in the Louisiana bayou (swamp) and tagged by National Park officials and then put back into the bayou. One month later, fifty alligators were captured from the same part of the bayou, eighteen of which had been tagged. Estimate to the nearest whole number how many alligators are in that part of Louisiana.

37. *Composition of Russian Navy* A ship in the Russian navy has a ratio of two officers for every seven seamen. The crew of the ship totals 117 people. How many officers and how many seamen are on this ship?

38. *Logging Company Employees* A Maine logging company has a total of 210 people at the north camp. The ratio of workers is three new employees for every eleven experienced employees. How many new employees are there at the north camp? How many experienced employees are there at the north camp?

39. *Cell Phone Employees* Southwestern Cell Phones has a total of 187 employees in the marketing and sales offices in Dallas. For every four people employed in marketing, there are thirteen people employed in sales. How many people in the Dallas office work in marketing? How many people in the Dallas office work in sales?

40. *Education* At Elmwood University there are nine men for every five women on the faculty. If Elmwood University employs 182 faculty members, how many are men and how many are women?

▲ **41.** *Picture Framing* Rose is carefully framing an 8 inch × 10 inch family photo so that the length and width of the frame have the same ratio as the photo. She is using a 54 inch length piece of frame. What are the length and width of the frame?

▲ **42.** *Enlarging Photographs* Becky DeWitt is a photographer at Photographics. She wants to enlarge a photograph that measures 3 inches wide and 5 inches long to an oversize photograph with the same width-to-length ratio. The perimeter of the new photograph is 115.2 inches. What are the length and the width of the new oversize photograph?

43. *E-Mail Software* A college recently installed new e-mail software on all of its computers. A poll was taken to find out how many of the faculty and staff preferred the new software. The ratio of those preferring the new software to the old software was 3:11. If there are 280 faculty and staff at the college, how many prefer the new software?

44. *Police Force Staff* The ratio of detectives to patrol officers at Center City is 2:9. The police force has 187 detectives and patrol officers. How many are detectives? How many are patrol officers?

45. *Boat Distribution* The Salem harbormaster said that last year the ratio of powerboats to sailboats moored in the harbor was 4:9. If a total of seventy-eight boats was moored in the harbor, how many boats were powerboats? How many boats were sailboats?

46. *Exercise Plan* When Jino D'Alessandro retired, his doctor told him to run 2 miles for every 7 miles he walks. If he plans to cover 63 miles in total each week, how many miles should he walk? How many miles should he run?

▲ **47.** *Shadow Length* A 12-foot marble statue in Italy casts a shadow that is 15 feet long. At the same time of day, a wall casts a shadow that is 8 feet long. How high is the wall?

▲ **48.** *Shadow Length* A 3-foot-tall child casts a shadow that is 4.8 feet long. At the same time of day, a building casts a shadow that extends 177 feet. How tall is the building?

49. *Nutrition Guidelines* A nutritionist told Maggie that for health reasons, she should eat fish seven times for every four times she eats red meat. If over the next 5 months she eats red meat 112 times, how many times should she eat fish?

50. *Mattress Stock* The manager of a mattress store knows that he should stock five queen-size mattresses for every eleven double-size mattresses. Next week he will have 448 mattresses in stock that are either queen-size or doubles. How many of each type should he have?

51. *Snow Plowing* Matt can plow out all the driveways on his street in Duluth, Minnesota, with his new four-wheel-drive truck in 6 hours. Using a snow blade on a lawn tractor, his neighbor can plow out the same number of driveways in 9 hours. How long would it take them to do the work together?

52. *Oil Change Time* The new mechanic at Speedy Lube can perform thirty oil changes in 4 hours. His assistant can perform the same number of oil changes in 6 hours. How long would it take them to do the work together?

53. *Swimming Pool* Houghton College has just built a new Olympic-size swimming pool. One large pipe with cold water and one small pipe with hot water are used to fill the pool. When both pipes are used, the pool can be filled in 2 hours. If only the cold water pipe is used, the pool can be filled in 3 hours. How long would it take to fill the pool with just the hot water pipe?

54. *Splitting Firewood* A lumberjack and his cousin Fred can split a cord of seasoned oak firewood in 3 hours. The lumberjack can split the wood alone without any help from Fred in 4 hours. How long would it take Fred if he worked without the lumberjack?

To Think About

River Width   *To find the width of a river, a hiking club laid out a triangular pattern of measurements. See the figure. Use your knowledge of similar triangles to solve Exercises 55 and 56.*

▲ **55.** If any observer stands at the point shown in the figure, then $a = 2$ feet, $b = 5$ feet, and $c = 116$ feet. What is the width of the river?

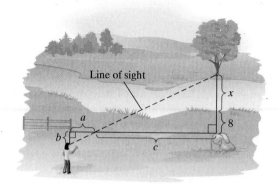

Line of sight

▲ **56.** What is the width of the river if $a = 3$ feet, $b = 8$ feet, and $c = 297$ feet?

Cumulative Review

57. Find the slope and the y-intercept of the line: $7x - 3y = 8$

58. Find the slope of the line passing through $(-1, 2.5)$ and $(1.5, 2)$

59. Find the equation of the line that is perpendicular to $y = 3x - 7$ passing through $(6, 1)$. Express the solution in slope–intercept form.

60. Solve the system: $x - 3y = 8$
$$2x + y = 9$$

61. ***Scholarships*** A philanthropic organization cofounded by Theodore J. Forstmann, the billionaire financier, and John T. Walton, the son of Wal-Mart founder Sam Walton, offered $50 million in partial and full scholarships for a certain fall term to 7500 underpriviledged students. Full scholarships were twice as large as partial scholarships, and there were an equal number of full scholarships and partial scholarships. What was the average partial scholarship award per student and the average full scholarship award per student? Round your answers to the nearest cent.

Developing Your Study Skills

Why Is Review Necessary?

You master a course in mathematics by learning the concepts one at a time. There are basic concepts like addition, subtraction, multiplication, and division of whole numbers, which are considered the foundation upon which all of mathematics is built. These must be mastered first. The study of mathematics is built upon this foundation, each concept supporting the next. The process is a carefully designed procedure, and no steps can be skipped. A student of mathematics needs to realize the importance of this building process to succeed.

Because new concepts depend on those previously learned, students often need to take time to review. Reviewing at the right time on the right concepts can strengthen previously learned skills and make progress possible.

Timely, periodic review of previously learned mathematical concepts is absolutely necessary in order to master new concepts. You may have forgotten a concept or grown a bit rusty in applying it. Reviewing is the answer. Make use of any review sections in your textbook, whether they are assigned or not. Look back to previous chapters whenever you have forgotten how to do something. Study the examples and practice some exercises to refresh your understanding.

Be sure that you understand and can perform the computations of each new concept. This will enable you to move successfully on to the next ones.

Remember that mathematics is a step-by-step building process. Learn one concept at a time, skipping none, and reinforce and strengthen skills with review whenever necessary.

Putting Your Skills to Work

Calculating Monthly Payments When Purchasing a Vehicle

How much will a new car cost?
The average cost of a new car in 1990 was $14,371. In 2003 the average cost of a new car had risen to $22,360 (*Source:* U.S. Bureau of Economic Analysis.)

Therefore, a major consideration in purchasing a new vehicle is the amount of the monthly payment you will be making. How much money do you have in your monthly budget for a car payment? The amount of the monthly payment is a function of the cost of the car, the size of your down payment, the finance rate and the period of the loan. The formula

$$M = \frac{P\left(\frac{r}{12}\right)}{1 - \left(1 + \frac{r}{12}\right)^{-m}}$$ computes the monthly payment

where P is the amount of money being financed (principal), r is the finance rate expressed as a decimal, and m is the number of monthly payments.

Problems for Individual Investigation and Analysis

Hadley Motors is advertising a used Nissan Sentra for $10,388 with financing of 3.9% over 36 months. They also offer a financing rate of 6.9% over 72 months.

1. Use the formula above to find your monthly payment on the Sentra if you make a $2000 down payment and want to pay off the car in 36 months.

2. What is your monthly payment if you choose to finance over 72 months, given the same down payment?

3. Not including the down payment, what is the total cost of the car if you finance over 36 months? Over 72 months?

4. What is the total finance cost if you pay over 36 months? Over 72 months?

Problems for Group Investigation and Cooperative Study

The Monarch Automobile Dealership is offering a new Ford Ranger truck for $17,500. The dealership offers several plans for acquiring the truck. To buy the truck, a 10% down payment is required. One option is to receive a $2000 rebate, taken before the down payment is figured, and finance the balance over 72 months at 5.9%. Another option is to finance the car over 48 months at 0% in lieu of the rebate. Or, the truck can be leased for 39 months for $209 per month and no down payment is required.

5. Find the monthly payment for each of the financing plans.

6. Find the total cost of the truck for each of the finance plans.

7. What are the advantages of each of the financing plans?

8. What are the advantages and disadvantages of leasing?

Topic	Procedure	Examples
Simplifying rational expressions, p. 308.	1. Factor the numerator and denominator, if possible. 2. Any factor that is common to both numerator and denominator can be divided out. This is an application of the basic rule of fractions. *Basic rule of fractions:* For any polynomials $a, b,$ and c (where $b \neq 0$ and $c \neq 0$), $$\frac{ac}{bc} = \frac{a}{b}.$$	Simplify $\dfrac{6x^2 - 14x + 4}{6x^2 - 20x + 6}$. $$\frac{2(3x^2 - 7x + 2)}{2(3x^2 - 10x + 3)} = \frac{\cancel{2}\,\cancel{(3x-1)}(x-2)}{\cancel{2}\,\cancel{(3x-1)}(x-3)}$$ $$= \frac{x-2}{x-3}$$
Multiplying rational expressions, p. 310.	1. Factor all numerators and denominators, if possible. 2. Any factor that is common to a numerator of one fraction and the denominator of the same fraction or any fraction that is multiplied by it can be divided out. 3. Write the product of the remaining factors in the numerator. Write the product of the remaining factors in the denominator. $$\frac{a}{b} \cdot \frac{c}{d} = \frac{ac}{bd}$$	Multiply: $\dfrac{x^2 - 4x}{6x - 12} \cdot \dfrac{3x^2 - 6x}{x^3 + 3x^2}$ $$\frac{\cancel{x}(x-4)}{\underset{2}{\cancel{6}\cancel{(x-2)}}} \cdot \frac{\overset{1}{\cancel{3}}\cancel{x}\cancel{(x-2)}}{\cancel{x^2}(x+3)} = \frac{x-4}{2(x+3)} \text{ or } \frac{x-4}{2x+6}$$
Dividing rational expressions, p. 311.	1. Invert the second fraction and multiply it by the first fraction. $$\frac{a}{b} \div \frac{c}{d} = \frac{a}{b} \cdot \frac{d}{c}$$ 2. Apply the steps for multiplying rational expressions.	Divide: $\dfrac{6x^2 - 5x - 6}{24x^2 + 13x - 2} \div \dfrac{4x^2 + x - 3}{8x^2 + 7x - 1}$ $$\frac{6x^2 - 5x - 6}{24x^2 + 13x - 2} \cdot \frac{8x^2 + 7x - 1}{4x^2 + x - 3}$$ $$= \frac{\cancel{(3x+2)}(2x-3)}{\cancel{(3x+2)}\cancel{(8x-1)}} \cdot \frac{\cancel{(8x-1)}\cancel{(x+1)}}{\cancel{(x+1)}(4x-3)} = \frac{2x-3}{4x-3}$$
Adding rational expressions, p. 317.	1. If all fractions have a common denominator, add the numerators and place the result over the common denominator. $$\frac{a}{c} + \frac{b}{c} = \frac{a+b}{c}$$ 2. If the fractions do not have a common denominator, factor the denominators (if necessary) and determine the least common denominator (LCD). 3. Rewrite each fraction as an equivalent fraction with the LCD as the denominator. 4. Add the numerators and place the result over the common denominator. 5. Simplify, if possible.	Add: $\dfrac{7x}{x^2 - 9} + \dfrac{x+2}{x+3}$ $$\frac{7x}{(x+3)(x-3)} + \frac{x+2}{x+3}$$ The LCD $= (x+3)(x-3)$. $$\frac{7x}{(x+3)(x-3)} + \frac{x+2}{x+3} \cdot \frac{x-3}{x-3}$$ $$= \frac{7x}{(x+3)(x-3)} + \frac{x^2 - x - 6}{(x+3)(x-3)}$$ $$= \frac{x^2 + 6x - 6}{(x+3)(x-3)}$$
Subtracting rational expressions, p. 317.	Follow the procedure for adding rational expressions through step 3. Then subtract the second numerator from the first and place the result over the common denominator. $$\frac{a}{c} - \frac{b}{c} = \frac{a-b}{c}$$ Simplify, if possible.	Subtract: $\dfrac{4x}{3x - 2} - \dfrac{5x}{x + 4}$ The LCD $= (3x-2)(x+4)$. $$\frac{4x}{3x-2} \cdot \frac{x+4}{x+4} - \frac{5x}{x+4} \cdot \frac{3x-2}{3x-2}$$ $$= \frac{4x^2 + 16x}{(3x-2)(x+4)} - \frac{15x^2 - 10x}{(x+4)(3x-2)}$$ $$= \frac{-11x^2 + 26x}{(3x-2)(x+4)}$$

Topic	Procedure	Examples
Simplifying a complex rational expression by Method 1, p. 324.	**1.** Simplify the numerator and denominator, if possible, by combining quantities to obtain one fraction in the numerator and one fraction in the denominator. **2.** Divide the numerator by the denominator. (That is, multiply the numerator by the reciprocal of the denominator.) **3.** Simplify the result.	Simplify by Method 1: $\dfrac{4 - \dfrac{1}{x^2}}{\dfrac{2}{x} + \dfrac{1}{x^2}}$ **Step 1:** $\dfrac{\dfrac{4x^2}{x^2} - \dfrac{1}{x^2}}{\dfrac{2x}{x^2} + \dfrac{1}{x^2}} = \dfrac{\dfrac{4x^2 - 1}{x^2}}{\dfrac{2x + 1}{x^2}}$ **Step 2:** $\dfrac{4x^2 - 1}{x^2} \cdot \dfrac{x^2}{2x + 1}$ **Step 3:** $\dfrac{\cancel{(2x + 1)}(2x - 1)}{\cancel{x^2}} \cdot \dfrac{\cancel{x^2}}{\cancel{2x + 1}} = 2x - 1$
Simplifying a complex rational expression by Method 2, p. 325.	**1.** Find the LCD of the rational expressions in the numerator and the denominator. **2.** Multiply the numerator and the denominator of the complex fraction by the LCD. **3.** Simplify the results.	Simplify by Method 2: $\dfrac{4 - \dfrac{1}{x^2}}{\dfrac{2}{x} - \dfrac{1}{x^2}}$ **Step 1:** The LCD of the rational expressions is x^2. **Step 2:** $\dfrac{\left[4 - \dfrac{1}{x^2}\right]x^2}{\left[\dfrac{2}{x} - \dfrac{1}{x^2}\right]x^2} = \dfrac{4(x^2) - \left(\dfrac{1}{x^2}\right)(x^2)}{\left(\dfrac{2}{x}\right)(x^2) - \left(\dfrac{1}{x^2}\right)(x^2)}$ $= \dfrac{4x^2 - 1}{2x - 1}$ **Step 3:** $\dfrac{(2x + 1)\cancel{(2x - 1)}}{\cancel{(2x - 1)}} = 2x + 1$
Solving rational equations, p. 330.	**1.** Determine the LCD of all denominators in the equation. **2.** Multiply each term in the equation by the LCD. **3.** Simplify and remove parentheses. **4.** Collect any like terms. **5.** Solve for the variable. If the variable term drops out, there is no solution. **6.** Check your answer. Be sure that the value you obtained does not make any denominator in the original equation 0. If so, there is no solution.	Solve $\dfrac{4}{y - 1} + \dfrac{-y + 5}{3y^2 - 4y + 1} = \dfrac{9}{3y - 1}$. The LCD $= (y - 1)(3y - 1)$. $\cancel{(y - 1)}(3y - 1)\left[\dfrac{4}{\cancel{y - 1}}\right]$ $+ (y - 1)(3y - 1)\left[\dfrac{-y + 5}{(y - 1)(3y - 1)}\right]$ $= (y - 1)\cancel{(3y - 1)}\left[\dfrac{9}{\cancel{3y - 1}}\right]$ $4(3y - 1) + (-y) + 5 = 9(y - 1)$ $12y - 4 - y + 5 = 9y - 9$ $11y + 1 = 9y - 9$ $11y - 9y = -9 - 1$ $2y = -10$ $y = -5$ *Check:* $\dfrac{4}{-5 - 1} + \dfrac{-(-5) + 5}{3(-5)^2 - 4(-5) + 1} \overset{?}{=} \dfrac{9}{3(-5) - 1}$ $\dfrac{4}{-6} + \dfrac{10}{96} \overset{?}{=} \dfrac{9}{-16}$ $-\dfrac{9}{16} = -\dfrac{9}{16}$ ✓

Topic	Procedure	Examples
Solving formulas containing rational expressions for a specified variable, p. 335.	1. Remove any parentheses. 2. Multiply each term of the equation by the LCD. 3. Add a quantity to or subtract a quantity from each side of the equation so that only terms containing the desired variable are on one side of the equation while all other terms are on the other side. 4. If there are two or more unlike terms containing the desired variable, remove that variable as a common factor. 5. Divide each side of the equation by the coefficient of the desired variable. 6. Simplify, if possible.	Solve for n: $v = c\left(1 - \dfrac{t}{n}\right)$ $v = c - \dfrac{ct}{n}$ $n(v) = n(c) - n\left(\dfrac{ct}{n}\right)$ $nv = nc - ct$ $nv - nc = -ct$ $n(v - c) = -ct$ $n = \dfrac{-ct}{v - c}$ or $\dfrac{ct}{c - v}$
Using advanced ratios to solve applied problems, p. 336.	1. Determine a given ratio in the problem for which both values are known. 2. Use variables to describe each of the quantities in the other ratio. 3. Set the two ratios equal to each other to form an equation. 4. Solve the resulting equation. Determine both quantities in the second ratio.	A new navy cruiser has a crew of 304 people. For every thirteen seamen there are three officers. How many officers and how many seamen are in the crew? The ratio of seamen to officers is: $\dfrac{13}{3}$ Let x be the number of officers. Because the crew totals 304, it follows that $304 - x$ represents the number of seamen. $\begin{array}{l}\text{seamen} \rightarrow \\ \text{officers} \rightarrow\end{array} \dfrac{13}{3} = \dfrac{304 - x}{x} \begin{array}{l}\leftarrow \text{seamen} \\ \leftarrow \text{officers}\end{array}$ $13x = 3(304 - x)$ $13x = 912 - 3x$ $16x = 912$ $x = 57$ $304 - x = 247$ There are 57 officers and 247 seamen.

Chapter 6 Review Problems

Simplify.

1. $\dfrac{6x^3 - 9x^2}{12x^2 - 18x}$

2. $\dfrac{15x^4}{5x^2 - 20x}$

3. $\dfrac{26x^3y^2}{39xy^4}$

4. $\dfrac{42a^4bc^3}{24a^7b}$

5. $\dfrac{2x^2 - 5x + 3}{3x^2 + 2x - 5}$

6. $\dfrac{ax + 2a - bx - 2b}{3x^2 - 12}$

7. $\dfrac{4x^2 - 1}{x^2 - 4} \cdot \dfrac{2x^2 + 4x}{4x + 2}$

8. $\dfrac{3y}{4xy - 6y^2} \cdot \dfrac{2x - 3y}{12xy}$

9. $\dfrac{y^2 + 8y - 20}{y^2 + 6y - 16} \cdot \dfrac{y^2 + 3y - 40}{y^2 + 6y - 40}$

10. $\dfrac{3x^3y}{x^2 + 7x + 12} \cdot \dfrac{x^2 + 8x + 15}{6xy^2}$

11. $\dfrac{2x + 12}{3x - 15} \div \dfrac{2x^2 - 6x - 20}{x^2 - 10x + 25}$

12. $\dfrac{2a^4b^5}{6x^2 + x - 1} \div \dfrac{10a^5b^2}{3x^2 + 5x - 2}$

13. $\dfrac{9y^2 - 3y - 2}{6y^2 - 13y - 5} \div \dfrac{3y^2 + 10y - 8}{2y^2 + 13y + 20}$

14. $\dfrac{4a^2 + 12a + 5}{2a^2 - 7a - 13} \div (4a^2 + 2a)$

Add or subtract the rational expressions and simplify your answers.

15. $\dfrac{x - 5}{2x + 1} - \dfrac{x + 1}{x - 2}$

16. $\dfrac{5}{4x} + \dfrac{-3}{x + 4}$

17. $\dfrac{2y - 1}{12y} - \dfrac{3y + 2}{9y}$

18. $\dfrac{4}{y + 5} + \dfrac{3y + 2}{y^2 - 25}$

19. $\dfrac{4y}{y^2 + 2y + 1} + \dfrac{3}{y^2 - 1}$

20. $\dfrac{y^2 - 4y - 19}{y^2 + 8y + 15} - \dfrac{2y - 3}{y + 5}$

21. $\dfrac{a}{5 - a} - \dfrac{2}{a + 3} + \dfrac{2a^2 - 2a}{a^2 - 2a - 15}$

22. $\dfrac{2}{x^2 + 8x + 16} - \dfrac{x}{2x^2 + 9x + 4}$

23. $5b - 1 - \dfrac{b+2}{b+3}$

24. $\dfrac{1}{x} + \dfrac{3}{2x} + 3 + 2x$

Simplify the complex rational expressions.

25. $\dfrac{\dfrac{5}{x} + 1}{1 - \dfrac{25}{x^2}}$

26. $\dfrac{\dfrac{4}{x+3}}{\dfrac{2}{x-2} - \dfrac{1}{x^2+x-6}}$

27. $\dfrac{\dfrac{y}{y+1} + \dfrac{1}{y}}{\dfrac{y}{y+1} - \dfrac{1}{y}}$

28. $\dfrac{\dfrac{10}{a+2} - 5}{\dfrac{4}{a+2} - 2}$

29. $\dfrac{\dfrac{2}{x+4} - \dfrac{1}{x^2+4x}}{\dfrac{3}{2x+8}}$

30. $\dfrac{\dfrac{y^2}{y^2-x^2} - 1}{x + \dfrac{xy}{x-y}}$

31. $\dfrac{\dfrac{2x+1}{x-1}}{1 + \dfrac{x}{x+1}}$

32. $\dfrac{\dfrac{3}{x} - \dfrac{2}{x+1}}{\dfrac{5}{x^2+5x+4} - \dfrac{1}{x+4}}$

Solve for the variable and check your solutions. If there is no solution, say so.

33. $\dfrac{3}{2} = 1 - \dfrac{1}{x-1}$

34. $\dfrac{3}{7} + \dfrac{4}{x+1} = 1$

35. $\dfrac{7}{2x-3} + \dfrac{3}{x+2} = \dfrac{6}{2x-3}$

36. $\dfrac{9}{5x-2} - \dfrac{2}{x} = \dfrac{1}{x}$

37. $\dfrac{5}{2a} = \dfrac{2}{a} - \dfrac{1}{12}$

38. $\dfrac{1}{2a} = \dfrac{2}{a} - \dfrac{3}{10}$

39. $\dfrac{1}{y} + \dfrac{1}{2y} = 2$

40. $\dfrac{5}{y^2} + \dfrac{7}{y} = \dfrac{6}{y^2}$

41. $\dfrac{a+2}{2a+6} = \dfrac{3}{2} - \dfrac{3}{a+3}$

42. $\dfrac{5}{a+5} + \dfrac{a+4}{2a+10} = \dfrac{3}{2}$

43. $\dfrac{3x-23}{2x^2-5x-3} + \dfrac{2}{x-3} = \dfrac{5}{2x+1}$

44. $\dfrac{2x-10}{2x^2-5x-3} + \dfrac{1}{x-3} = \dfrac{3}{2x+1}$

Solve for the variable indicated.

45. $\dfrac{N}{V} = \dfrac{m}{M+N}$; for M

46. $m = \dfrac{y-y_0}{x-x_0}$; for x

47. $\dfrac{1}{f} = \dfrac{1}{a} + \dfrac{1}{b}$; for a

48. $S = \dfrac{V_1 t + V_2 t}{2}$; for t

49. $d = \dfrac{LR_2}{R_2 + R_1}$; for R_2

50. $\dfrac{S-P}{Pr} = t$; for r

Mixed Practice

Simplify.

51. $\dfrac{x^2 - x - 42}{x^2 - 2x - 35}$

52. $\dfrac{2x^2 - 5x - 3}{x^2 - 9} \cdot \dfrac{2x^2 + 5x - 3}{2x^2 + 5x + 2}$

53. $\dfrac{-2x-1}{x+4} + 4x + 3$

54. $\dfrac{\dfrac{1}{x^2 - 3x + 2}}{\dfrac{3}{x-2} - \dfrac{2}{x-1}}$

55. Solve for x.

$$\frac{5}{2} - \frac{3}{x+1} = \frac{2-x}{x+1}$$

Solve the following exercises. If necessary, round your answers to the nearest hundredth.

56. *Calculator Supply* The campus bookstore at Boston University ordered 253 scientific and graphing calculators for the spring semester. For every seven scientific calculators they ordered they also ordered four graphing calculators. How many scientific calculators did they order? How many graphing calculators did they order?

57. *New Homes* Walter Johnson built a new development of 112 homes in Naperville, Illinois. For every three one-story homes, he built thirteen two-story homes. How many one-story homes did he build? How many two-story homes did he build?

▲ 58. *Photograph Enlargement* Jill VanderWoude decided to enlarge a photograph that measures 5 inches wide and 7 inches long into a poster-size photograph with a perimeter of 168 inches. The new photograph will maintain the same width-to-length ratio. How wide will the enlarged photograph be? How long will the enlarged photograph be?

59. *Swimming Pool* How long will it take a pump to empty a 4900-gallon swimming pool if the same pump can empty a 3500-gallon swimming pool in 4 hours?

60. *Rabbit Population* In a sanctuary a sample of one hundred wild rabbits is tagged and released by the wildlife management team. In a few weeks, after they have mixed with the general rabbit population, a sample of forty rabbits is caught, and eight have a tag. Estimate the population of rabbits in the sanctuary.

61. *Police Statistics* The ratio of officers to state troopers is 2:9. If there are 154 men and women on the force, how many are officers?

62. *Maritime Chart* The scale on a maritime sailing chart shows that 2 centimeters is equivalent to 7 nautical miles. A boat captain lays out a course on the chart that is 3.5 cm long. How many nautical miles will this be?

▲ 63. *Shadow Length* A 7-foot-tall tree casts a shadow that is 6 feet long. At the same time of day, a building casts a shadow that is 156 feet long. How tall is the building?

64. *Painting Time* In the summer it takes Dominic 12 hours to paint the barn. It takes his young son 18 hours to paint the barn. How many hours would it take if they both worked together?

65. *Jacuzzi* If the hot water faucet at Mike's house is left on, it takes 15 minutes to fill the jacuzzi. If the cold water faucet is left on, it takes 10 minutes to fill the jacuzzi. How many minutes would it take if both faucets are left on?

Instant Messenger e-Mail   *The growth in instant messenger e-mail has been more rapid in the last few years than the growth of regular e-mail.*

The bar graph below shows the growth in usage of instant messenger e-mail. Use this bar graph to answer Exercises 66–69. Round all answers to the nearest million.

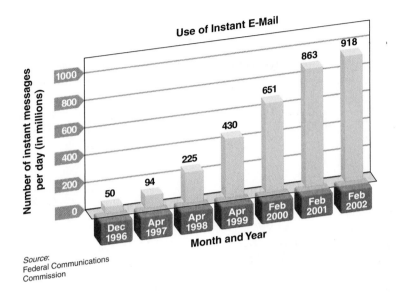

Source:
Federal Communications
Commission

66. If the increase in usage per month that occurred over the period from February 2001 to February 2002 continues at the same increase per month, what would be the expected number of daily instant messages in February 2003?

67. If the increase in usage per month that occurred from February 2000 to February 2002 is proportional to the increase in usage from February 2002 to February 2004, what would be the expected number of daily instant messages in February 2004?

68. If the increase in usage per month that occurred from April 1999 to February 2000 is one-half as great as the increase in usage from February 2000 to December 2000, what would be the expected number of daily instant messages in December 2000?

69. If the increase in usage per month that occurred from April 1997 to April 1999 is double the increase in usage from April 1999 to April 2001, what would be the expected number of daily instant messages in April 2001?

Remember to use your Chapter Test Prep Video CD to see the worked-out solutions to the test problems you want to review.

Simplify.

1. $\dfrac{x^3 + 3x^2 + 2x}{x^3 - 2x^2 - 3x}$

2. $\dfrac{-25p^4qr^3}{45pqr^6}$

3. $\dfrac{2y^2 + 7y - 4}{y^2 + 2y - 8} \cdot \dfrac{2y^2 - 8}{3y^2 + 11y + 10}$

4. $\dfrac{4 - 2x}{3x^2 - 2x - 8} \div \dfrac{2x^2 + x - 1}{9x + 12}$

5. $\dfrac{3}{x} - \dfrac{2}{x + 1}$

6. $\dfrac{2}{x^2 + 5x + 6} + \dfrac{3x}{x^2 + 6x + 9}$

7. $\dfrac{\dfrac{4}{y + 2} - 2}{5 - \dfrac{10}{y + 2}}$

8. $\dfrac{\dfrac{1}{x} - \dfrac{3}{x + 2}}{\dfrac{2}{x^2 + 2x}}$

Solve for the variable and check your answers. If no solution exists, say so.

9. $\dfrac{7}{4} = \dfrac{x + 4}{x}$

10. $2 + \dfrac{x}{x + 4} = \dfrac{3x}{x - 4}$

11. $\dfrac{1}{2y + 4} - \dfrac{1}{6} = \dfrac{-2}{3y + 6}$

12. $2 + \dfrac{3}{x} = \dfrac{2x}{x - 1}$

13. Solve for W: $h = \dfrac{S - 2WL}{2W + 2L}$

14. Solve for h: $\dfrac{3V}{\pi h} = r^2$

15. A total of 286 employees at Kaiser Telecommunication Systems were eligible this year for a high-performance bonus. The company president announced that for every three employees who got the bonus, nineteen employees did not. If the president was correct, how many employees got the high-performance bonus? How many did not get the high-performance bonus?

▲ **16.** The Newbury Elementary School had a rectangular playground that measured 500 feet wide and 850 feet long. When the school had a major addition to its property, it was decided to increase the playground to a large rectangular shape with the same proportional ratio that had a perimeter of 8100 feet. What is the width and the length of the new playground?

1. _____

2. _____

3. _____

4. _____

5. _____

6. _____

7. _____

8. _____

9. _____

10. _____

11. _____

12. _____

13. _____

14. _____

15. _____

16. _____

Approximately one-half of this test covers the content of Chapters 1–5. The remainder covers the content of Chapter 6.

1. Simplify: $(3x^{-2}y)^2(x^4y^{-3})$

2. Solve for x: $\dfrac{2}{3}(3x - 1) = \dfrac{2}{5}x + 3$

3. Graph the straight line $-6x + 2y = -12$.

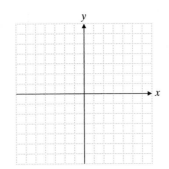

4. Find the standard form of the equation of the line parallel to $5x - 6y = 8$ that passes through $(-1, -3)$.

5. Brenda invested $7000 in two accounts at the bank. One account earns 5% simple interest. The other earns 8% simple interest. She earned $539 interest in 1 year. How much was invested at each rate?

6. Solve for x and graph. $3(2 - 6x) > 4(x + 1) + 24$

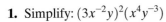

7. Evaluate $2x^2 - 3x - 4y^2$ when $x = -2$ and $y = 3$.

8. Solve for x: $|3x - 4| \le 10$

Factor.

9. $8x^3 - 125y^3$ **10.** $81x^3 - 90x^2y + 25xy^2$

Solve for x.

11. $x^2 + 20x + 36 = 0$ **12.** $3x^2 - 11x - 4 = 0$

1. _____

2. _____

3. _____

4. _____

5. _____

6. _____

7. _____

8. _____

9. _____

10. _____

11. _____

12. _____

Simplify.

13. _____

13. $\dfrac{7x^2 - 28}{x^2 + 6x + 8}$

14. _____

14. $\dfrac{2x^2 + x - 1}{2x^2 - 9x + 4} \cdot \dfrac{3x^2 - 12x}{6x + 15}$

15. _____

15. $\dfrac{x^3 + 2x^2}{3x - 21} \div \dfrac{2x^3 + 5x^2 + 2x}{x - 7}$

16. _____

16. $\dfrac{7}{3x - 3} - \dfrac{x + 2}{x^2 - 1}$

17. _____

17. $\dfrac{\dfrac{1}{2x + 1} + 1}{4 - \dfrac{3}{4x^2 - 1}}$

18. _____

18. $\dfrac{3}{x - 6} + \dfrac{4}{x + 4}$

Solve for the variable and check your answers.

19. _____

19. $\dfrac{1}{2x + 3} - \dfrac{4}{4x^2 - 9} = \dfrac{3}{2x - 3}$

20. _____

20. $\dfrac{1}{4x} - \dfrac{3}{2x} = \dfrac{5}{8}$

21. _____

21. Solve for b: $\quad H = \dfrac{3b + 2x}{5 - 4b}$

22. _____

22. The mayor of Chicago decided that in the coming year the city police force should be deployed in a particular ratio. He decided that for every three police officers who patrol on foot, there should be eleven police officers patrolling in squad cars. If 3234 police officers are normally on duty during the day, how many are patrolling on foot, and how many are patrolling in squad cars?

CHAPTER

7

Most of us enjoy listening to a variety of types of music and musical instruments. Did you know that tuning musical instruments involves math? This is especially significant when tuning stringed instruments such as a piano. Turn to page 409 to learn about math and music and use your math skills to solve some musical problems.

Rational Exponents and Radicals

7.1 RATIONAL EXPONENTS

Student Learning Objectives

After studying this section, you will be able to apply the laws of exponents to:

1 Simplify expressions with rational exponents.

2 Add expressions with rational exponents.

3 Factor expressions with rational exponents.

NOTE TO STUDENT: *Fully worked-out solutions to all of the Practice Problems can be found at the back of the text starting at page SP-1*

1 Simplifying Expressions with Rational Exponents

Before studying this section, you may need to review Section 1.4. For convenience, we list the rules of exponents that we learned there.

$$x^m x^n = x^{m+n} \qquad x^0 = 1$$

$$\frac{x^m}{x^n} = x^{m-n} \qquad (x^m)^n = x^{mn}$$

$$x^{-n} = \frac{1}{x^n} \qquad (xy)^n = x^n y^n$$

$$\frac{x^{-n}}{y^{-m}} = \frac{y^m}{x^n} \qquad \left(\frac{x}{y}\right)^n = \frac{x^n}{y^n}$$

To ensure that you understand these rules, study Example 1 carefully and work Practice Problem 1.

EXAMPLE 1 Simplify $\left(\dfrac{5xy^{-3}}{2x^{-4}y}\right)^{-2}$.

Solution

$$\left(\frac{5xy^{-3}}{2x^{-4}y}\right)^{-2} = \frac{(5xy^{-3})^{-2}}{(2x^{-4}y)^{-2}} \qquad \left(\frac{x}{y}\right)^n = \frac{x^n}{y^n}.$$

$$= \frac{5^{-2}x^{-2}(y^{-3})^{-2}}{2^{-2}(x^{-4})^{-2}y^{-2}} \qquad (xy)^n = x^n y^n.$$

$$= \frac{5^{-2}x^{-2}y^6}{2^{-2}x^8 y^{-2}} \qquad (x^m)^n = x^{mn}.$$

$$= \frac{5^{-2}}{2^{-2}} \cdot \frac{x^{-2}}{x^8} \cdot \frac{y^6}{y^{-2}}$$

$$= \frac{2^2}{5^2} \cdot x^{-2-8} \cdot y^{6+2} \qquad \frac{x^{-n}}{y^{-m}} = \frac{y^m}{x^n}, \frac{x^m}{x^n} = x^{m-n}.$$

$$= \frac{4}{25} x^{-10} y^8$$

The answer can also be written as $\dfrac{4y^8}{25x^{10}}$. Explain why.

Practice Problem 1 Simplify $\left(\dfrac{3x^{-2}y^4}{2x^{-5}y^2}\right)^{-3}$.

SIDELIGHT: Example 1 Follow-Up Deciding when to use the rule $\dfrac{x^{-n}}{y^{-m}} = \dfrac{y^m}{x^n}$ is entirely up to you. In Example 1, we could have begun by writing

$$\left(\frac{5xy^{-3}}{2x^{-4}y}\right)^{-2} = \left(\frac{5x \cdot x^4}{2y \cdot y^3}\right)^{-2} = \left(\frac{5x^5}{2y^4}\right)^{-2}.$$

Complete the steps to simplify this expression.

Likewise, in the fourth step in Example 1, we could have written

$$\frac{5^{-2}x^{-2}y^6}{2^{-2}x^8 y^{-2}} = \frac{2^2 y^6 y^2}{5^2 x^8 x^2}.$$

Complete the steps to simplify this expression. Are the two answers the same as the answer in Example 1? Why or why not?

We generally begin to simplify a rational expression with exponents by raising a power to a power because sometimes negative powers become positive. The order in which you use the rules of exponents is up to you. Work carefully. Keep track of your exponents and where you are as you simplify the rational expression.

These rules for exponents can also be extended to include rational exponents—that is, exponents that are fractions. As you recall, rational numbers are of the form $\frac{a}{b}$, where a and b are integers and b does not equal zero. We will write fractional exponents using diagonal lines. Thus, we will write $\frac{5}{6}$ as 5/6 and $\frac{a}{b}$ as a/b throughout this chapter when writing fractional exponents. For now we restrict the base to *positive* real numbers. Later we will talk about negative bases.

EXAMPLE 2 Simplify.

(a) $(x^{2/3})^4$ **(b)** $\dfrac{x^{5/6}}{x^{1/6}}$ **(c)** $x^{2/3} \cdot x^{-1/3}$ **(d)** $5^{3/7} \cdot 5^{2/7}$

Solution We will not write out every step or every rule of exponents that we use. You should be able to follow the solutions.

(a) $(x^{2/3})^4 = x^{(2/3)(4/1)} = x^{8/3}$ **(b)** $\dfrac{x^{5/6}}{x^{1/6}} = x^{5/6-1/6} = x^{4/6} = x^{2/3}$

(c) $x^{2/3} \cdot x^{-1/3} = x^{2/3-1/3} = x^{1/3}$ **(d)** $5^{3/7} \cdot 5^{2/7} = 5^{3/7+2/7} = 5^{5/7}$

Practice Problem 2 Simplify.

(a) $(x^4)^{3/8}$ **(b)** $\dfrac{x^{3/7}}{x^{2/7}}$ **(c)** $x^{-7/5} \cdot x^{4/5}$

Sometimes fractional exponents will not have the same denominator. Remember that you need to change the fractions to equivalent fractions with the same denominator when the rule of exponents requires you to add or to subtract them.

EXAMPLE 3 Simplify. Express your answers with positive exponents only.

(a) $(2x^{1/2})(3x^{1/3})$ **(b)** $\dfrac{18x^{1/4}y^{-1/3}}{-6x^{-1/2}y^{1/6}}$

Solution

(a) $(2x^{1/2})(3x^{1/3}) = 6x^{1/2+1/3} = 6x^{3/6+2/6} = 6x^{5/6}$

(b) $\dfrac{18x^{1/4}y^{-1/3}}{-6x^{-1/2}y^{1/6}} = -3x^{1/4-(-1/2)}y^{-1/3-1/6}$

$= -3x^{1/4+2/4}y^{-2/6-1/6}$

$= -3x^{3/4}y^{-3/6}$

$= -3x^{3/4}y^{-1/2}$

$= \dfrac{-3x^{3/4}}{y^{1/2}}$

Practice Problem 3 Simplify. Express your answers with positive exponents only.

(a) $(-3x^{1/4})(2x^{1/2})$ **(b)** $\dfrac{13x^{1/12}y^{-1/4}}{26x^{-1/3}y^{1/2}}$

EXAMPLE 4 Multiply and simplify $-2x^{5/6}(3x^{1/2} - 4x^{-1/3})$.

Solution We will need to be very careful when we add the exponents for x as we use the distributive property. Study each step of the following example. Be sure you understand each operation.

$$-2x^{5/6}(3x^{1/2} - 4x^{-1/3}) = -6x^{5/6+1/2} + 8x^{5/6-1/3}$$
$$= -6x^{5/6+3/6} + 8x^{5/6-2/6}$$
$$= -6x^{8/6} + 8x^{3/6}$$
$$= -6x^{4/3} + 8x^{1/2}$$

NOTE TO STUDENT: Fully worked-out solutions to all of the Practice Problems can be found at the back of the text starting at page SP-1

Practice Problem 4 Multiply and simplify $-3x^{1/2}(2x^{1/4} + 3x^{-1/2})$.

Sometimes we can use the rules of exponents to simplify numerical values raised to rational powers.

EXAMPLE 5 Evaluate: **(a)** $(25)^{3/2}$ **(b)** $(27)^{2/3}$

Solution

(a) $(25)^{3/2} = (5^2)^{3/2} = 5^{2/1 \cdot 3/2} = 5^3 = 125$

(b) $(27)^{2/3} = (3^3)^{2/3} = 3^{3/1 \cdot 2/3} = 3^2 = 9$

Practice Problem 5 Evaluate: **(a)** $(4)^{5/2}$ **(b)** $(27)^{4/3}$

---

② Adding Expressions with Rational Exponents

Adding expressions with rational exponents may require several steps. Sometimes this involves removing negative exponents. For example, to add $2x^{-1/2} + x^{1/2}$, we begin by writing $2x^{-1/2}$ as $\dfrac{2}{x^{1/2}}$. This is a rational expression. Recall that to add rational expressions we need to have a common denominator. Take time to look at the steps needed to write $2x^{-1/2} + x^{1/2}$ as one term.

EXAMPLE 6 Write as one fraction with positive exponents. $2x^{-1/2} + x^{1/2}$

Solution $2x^{-1/2} + x^{1/2} = \dfrac{2}{x^{1/2}} + \dfrac{x^{1/2} \cdot x^{1/2}}{x^{1/2}} = \dfrac{2}{x^{1/2}} + \dfrac{x^1}{x^{1/2}} = \dfrac{2 + x}{x^{1/2}}$

Practice Problem 6 Write as one fraction with only positive exponents.
$3x^{1/3} + x^{-1/3}$

3 Factoring Expressions with Rational Exponents

To factor expressions, we need to be able to recognize common factors. If the terms of the expression contain exponents, we look for the same exponential factor in each term. For example, in the expression $6x^5 + 4x^3 - 8x^2$, the common factor of each term is $2x^2$. Thus, we can factor out the common factor $2x^2$ from each term. The expression then becomes $2x^2(3x^3 + 2x - 4)$.

We do exactly the same thing when we factor expressions with rational exponents. The key is to identify the exponent of the common factor. In the expression $6x^{3/4} + 4x^{1/2} - 8x^{1/4}$, the common factor is $2x^{1/4}$. Thus, we factor the expression $6x^{3/4} + 4x^{1/2} - 8x^{1/4}$ as $2x^{1/4}(3x^{1/2} + 2x^{1/4} - 4)$. We do not always need to factor out the greatest common factor. In the following examples we simply factor out a common factor.

EXAMPLE 7   Factor out the common factor of $2x$ from $2x^{3/2} + 4x^{5/2}$.

Solution   We rewrite the exponent of each term so that we can see that each term contains the factor $2x$ or $2x^{2/2}$.

$$2x^{3/2} + 4x^{5/2} = 2x^{2/2+1/2} + 4x^{2/2+3/2}$$

$$= 2(x^{2/2})(x^{1/2}) + 4(x^{2/2})(x^{3/2})$$

$$= 2x(x^{1/2} + 2x^{3/2})$$

Practice Problem 7   Factor out the common factor of $4y$ from $4y^{3/2} - 8y^{5/2}$.

For convenience we list here the properties of exponents that we have discussed in this section, as well as the property $x^0 = 1$.

> When x and y are **positive real numbers** and a and b are **rational numbers:**
>
> $$x^a x^b = x^{a+b} \qquad \frac{x^a}{x^b} = x^{a-b} \qquad x^0 = 1$$
>
> $$x^{-a} = \frac{1}{x^a} \qquad \frac{x^{-a}}{y^{-b}} = \frac{y^b}{x^a}$$
>
> $$(x^a)^b = x^{ab} \qquad (xy)^a = x^a y^a \qquad \left(\frac{x}{y}\right)^a = \frac{x^a}{y^a}$$

7.1 EXERCISES

    Math XP

Student Solutions Manual | CD/ Video | PH Math Tutor Center | MathXL®Tutorials on CD | MathXL® | MyMathLab® | Interactmath.com

Simplify.

Express your answer with positive exponents.

1. $\left(\dfrac{3xy^{-1}}{z^2}\right)^4$

2. $\left(\dfrac{3xy^{-2}}{x^3}\right)^2$

3. $\left(\dfrac{2a^{-1}b^3}{-3b^2}\right)^3$

4. $\left(\dfrac{-a^{-2}b}{5b^2}\right)^2$

5. $\left(\dfrac{2x^2}{y}\right)^{-3}$

6. $\left(\dfrac{4x}{y^3}\right)^{-3}$

7. $\left(\dfrac{3xy^{-2}}{y^3}\right)^{-2}$

8. $\left(\dfrac{5x^{-2}y}{x^4}\right)^{-2}$

9. $(x^{3/4})^2$

10. $(x^{5/6})^3$

11. $(y^{12})^{2/3}$

12. $(y^2)^{5/2}$

13. $\dfrac{x^{3/5}}{x^{1/5}}$

14. $\dfrac{y^{6/7}}{y^{3/7}}$

15. $\dfrac{x^{7/12}}{x^{1/12}}$

16. $\dfrac{x^{7/8}}{x^{3/8}}$

17. $\dfrac{x^3}{x^{1/2}}$

18. $\dfrac{x^2}{x^{1/3}}$

19. $x^{1/7} \cdot x^{3/7}$

20. $x^{3/5} \cdot x^{1/5}$

21. $a^{3/8} \cdot a^{1/2}$

22. $b^{2/5} \cdot b^{2/15}$

23. $y^{3/5} \cdot y^{-1/10}$

24. $y^{7/10} \cdot y^{-1/5}$

Write each expression with positive exponents.

25. $x^{-3/4}$

26. $x^{-5/6}$

27. $a^{-5/6}b^{1/3}$

28. $2a^{-1/6}b^{3/4}$

29. $6^{-1/2}$

30. $4^{-1/3}$

31. $2a^{-1/4}$

32. $-5y^{-2/3}$

Evaluate or simplify the numerical expressions.

33. $(27)^{2/3}$

34. $(16)^{3/4}$

35. $(4)^{3/2}$

36. $(9)^{3/2}$

37. $(-8)^{5/3}$

38. $(-27)^{5/3}$

39. $(-27)^{2/3}$

40. $(-64)^{2/3}$

Mixed Practice

Simplify and express your answers with positive exponents. Evaluate or simplify the numerical expressions.

41. $(x^{1/2}y^{1/3})(x^{1/3}y^{2/3})$

42. $(x^{-1/3}y^{2/3})(x^{1/3}y^{1/4})$

43. $(7x^{1/3}y^{1/4})(-2x^{1/4}y^{-1/6})$

44. $(8x^{-1/5}y^{1/3})(-3x^{-1/4}y^{1/6})$

45. $6^2 \cdot 6^{-2/3}$

46. $11^{1/2} \cdot 11^3$

47. $\dfrac{2x^{1/5}}{x^{-1/2}}$

48. $\dfrac{3y^{2/3}}{y^{-1/4}}$

49. $\dfrac{-20x^2y^{-1/5}}{5x^{-1/2}y}$

50. $\dfrac{12x^{-2/3}y}{-6xy^{-3/4}}$

51. $\left(\dfrac{8a^2b^6}{a^{-1}b^3}\right)^{1/3}$

52. $\left(\dfrac{16a^5b^{-2}}{a^{-1}b^{-6}}\right)^{1/2}$

53. $(-3x^{2/5}y^{3/2}z^{1/3})^2$

54. $(5x^{-1/2}y^{1/3}z^{4/5})^3$

55. $x^{2/3}(x^{4/3} - x^{1/5})$

56. $y^{-2/3}(y^{2/3} + y^{3/2})$

57. $m^{7/8}(m^{-1/2} + 2m)$

58. $m^{2/3}(m^{-1/2} + 3m)$

59. $(8)^{-1/3}$

60. $(100)^{-1/2}$

61. $(49)^{-3/2}$

62. $(16)^{-3/4}$

63. $(81)^{3/4} + (25)^{1/2}$

64. $9^{3/2} + 4^{1/2}$

Write each expression as one fraction with positive exponents.

65. $3y^{1/2} + y^{-1/2}$

66. $2y^{1/3} + y^{-2/3}$

67. $x^{-1/3} + 6^{4/3}$

68. $5^{-1/4} + x^{-1/2}$

Factor out the common factor of 2a.

69. $10a^{5/4} - 4a^{8/5}$

70. $6a^{4/3} - 8a^{3/2}$

Factor out the common factor of 3x.

71. $6x^{7/4} - 15x^{3/2}$

72. $21x^{13/8} - 12x^{4/3}$

To Think About

73. What is the value of a if $x^a \cdot x^{1/4} = x^{-1/8}$?

74. What is the value of b if $x^b \div x^{1/3} = x^{-1/12}$?

Applications

Radius and Volume of a Sphere *The radius needed to create a sphere with a given volume V can be approximated by the equation* $r = 0.62(V)^{1/3}$. *Find the radius of the spheres with the following volumes.*

▲ **75.** 27 cubic meters

▲ **76.** 64 cubic meters

Radius and Volume of a Cone *The radius required for a cone to have a volume V and a height h is given by the equation*

$$r = \left(\frac{3V}{\pi h}\right)^{1/2}$$

Find the necessary radius to have a cone with the properties below. Use $\pi \approx 3.14$.

▲ **77.** $V = 314$ cubic feet and $h = 12$ feet.　　　　▲ **78.** $V = 3140$ cubic feet and $h = 30$ feet.

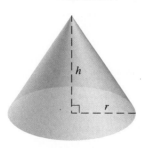

Cumulative Review

Solve for x.

79. $-4(x + 1) = \frac{1}{3}(3 - 2x)$

Solve for b.

80. $A = \frac{h}{2}(a + b)$

Dosage of Medication for a Child *Giving a young patient the wrong amount of medication can have serious and even fatal consequences. A formula used by doctors, nurses, and pharmacists to verify the correct dosage of a prescription drug for a child is*

$$y = \frac{ax}{a + 12},$$

where y = the child dosage, x = the adult dosage, and a = the age of the child in years.

81. If the adult dosage of a medication is 400 milligrams, how much should a 7-year-old child receive? Round your answer to the nearest milligram.

82. The adult dosage of a medication is 250 milligrams, and a certain child was assigned the correct dosage level of 75 milligrams. How old was the child? Round your answer to the nearest year.

1 Evaluating Radical Expressions and Functions

In Section 1.3 we studied simple radical expressions called square roots. The **square root** of a number is a value that when multiplied by itself is equal to the original number. That is, since $3 \cdot 3 = 9$, 3 is a square root of 9. But $(-3) \cdot (-3) = 9$, so -3 is also a square root. We call the positive square root the **principal square root.**

The symbol $\sqrt{}$ is called a **radical sign.** We use it to denote positive square roots (and positive higher-order roots also). A negative square root is written $-\sqrt{}$. Thus, we have the following:

$$\sqrt{9} = 3 \qquad -\sqrt{9} = -3$$
$$\sqrt{64} = 8 \qquad (\text{because } 8 \cdot 8 = 64)$$
$$\sqrt{121} = 11 \qquad (\text{because } 11 \cdot 11 = 121)$$

Because $\sqrt{9} = \sqrt{3 \cdot 3} = \sqrt{3^2} = 3$, we can say the following:

> **DEFINITION OF SQUARE ROOT**
>
> If x is a nonnegative real number, then $\sqrt{x}$ is the *nonnegative* (or principal) *square root* of x; in other words, $\left(\sqrt{x}\right)^2 = x$.

Note that x must be *nonnegative*. Why? Suppose we want to find $\sqrt{-36}$. We must find a number that when multiplied by itself gives -36. Is there one? No, because

$$6 \cdot 6 = 36 \quad \text{and}$$
$$(-6)(-6) = 36.$$

So there is no real number that we can square to get -36.

We call $\sqrt[n]{x}$ a **radical expression.** The $\sqrt{}$ symbol is the radical sign, the x is the **radicand,** and the n is the **index** of the radical. When no number for n appears in the radical expression, it is understood that 2 is the index, which means that we are looking for the square root. For example, in the radical expression $\sqrt{25}$, with no number given for the index n we take the index to be 2. Thus, $\sqrt{25}$ is the principal square root of 25.

We can extend the notion of square root to **higher-order roots,** such as cube roots, fourth roots, and so on. A **cube root** of a number is a value that when cubed is equal to the original number. The index n of the radical is 3, and the radical used is $\sqrt[3]{}$. Similarly, a **fourth root** of a number is a value that when raised to the fourth power is equal to the original number. The index n of the radical is 4, and the radical used is $\sqrt[4]{}$. Thus, we have the following:

$$\sqrt[3]{27} = 3 \qquad \text{because } 3 \cdot 3 \cdot 3 = 3^3 = 27.$$
$$\sqrt[3]{8} = 2 \qquad \text{because } 2 \cdot 2 \cdot 2 = 2^3 = 8.$$
$$\sqrt[4]{81} = 3 \qquad \text{because } 3 \cdot 3 \cdot 3 \cdot 3 = 3^4 = 81.$$
$$\sqrt[5]{32} = 2 \qquad \text{because } 2 \cdot 2 \cdot 2 \cdot 2 \cdot 2 = 2^5 = 32.$$
$$\sqrt[3]{-64} = -4 \quad \text{because } (-4)(-4)(-4) = (-4)^3 = -64.$$

You should be able to see a pattern here.

$$\sqrt[3]{27} = \sqrt[3]{3^3} = 3$$
$$\sqrt[4]{81} = \sqrt[4]{3^4} = 3$$
$$\sqrt[5]{32} = \sqrt[5]{2^5} = 2$$
$$\sqrt[6]{729} = \sqrt[6]{3^6} = 3$$
$$\sqrt[3]{-64} = \sqrt[3]{(-4)^3} = -4$$

In these cases, we see that $\sqrt[n]{x^n} = x$. We now give the following definition.

DEFINITION OF HIGHER-ORDER ROOTS

1. If x is a *nonnegative* real number, then $\sqrt[n]{x}$ is a nonnegative nth root and has the property that

$$\left(\sqrt[n]{x}\right)^n = x.$$

2. If x is a *negative* real number, then

   (a) $\left(\sqrt[n]{x}\right)^n = x$ when n is an *odd integer*.

   (b) $\left(\sqrt[n]{x}\right)^n$ is *not* a real number when n is an *even integer*.

EXAMPLE 1 If possible, find the root of each negative number. If there is no real number root, say so.

(a) $\sqrt[3]{-216}$ (b) $\sqrt[5]{-32}$ (c) $\sqrt[4]{-16}$ (d) $\sqrt[6]{-64}$

Solution

(a) $\sqrt[3]{-216} = \sqrt[3]{(-6)^3} = -6$

(b) $\sqrt[5]{-32} = \sqrt[5]{(-2)^5} = -2$

(c) $\sqrt[4]{-16}$ is not a real number because n is even and x is negative.

(d) $\sqrt[6]{-64}$ is not a real number because n is even and x is negative.

Practice Problem 1 If possible, find the roots. If there is no real number root, say so.

(a) $\sqrt[3]{216}$ (b) $\sqrt[5]{32}$ (c) $\sqrt[3]{-8}$ (d) $\sqrt[4]{-81}$

Because the symbol $\sqrt{x}$ represents exactly one real number for all real numbers x that are nonnegative, we can use it to define the **square root function** $f(x) = \sqrt{x}$.

This function has a domain of all real numbers x that are greater than or equal to zero.

EXAMPLE 2 Find the indicated function values of the function $f(x) = \sqrt{2x + 4}$.

Round your answers to the nearest tenth when necessary.

(a) $f(-2)$ (b) $f(6)$ (c) $f(3)$

Solution

(a) $f(-2) = \sqrt{2(-2) + 4} = \sqrt{-4 + 4} = \sqrt{0} = 0$ The square root of zero is zero.

(b) $f(6) = \sqrt{2(6) + 4} = \sqrt{12 + 4} = \sqrt{16} = 4$

(c) $f(3) = \sqrt{2(3) + 4} = \sqrt{6 + 4} = \sqrt{10} \approx 3.2$ We use a calculator or a square root table to approximate $\sqrt{10}$.

NOTE TO STUDENT: Fully worked-out solutions to all of the Practice Problems can be found at the back of the text starting at page SP-1

Practice Problem 2 Find the indicated values of the function $f(x) = \sqrt{4x - 3}$. Round your answers to the nearest tenth when necessary.

(a) $f(3)$ (b) $f(4)$ (c) $f(7)$

EXAMPLE 3 Find the domain of the function $f(x) = \sqrt{3x - 6}$.

Solution We know that the expression $3x - 6$ must be nonnegative. That is, $3x - 6 \geq 0$.

$$3x - 6 \geq 0$$
$$3x \geq 6$$
$$x \geq 2$$

Thus, the domain is all real numbers x where $x \geq 2$.

Practice Problem 3 Find the domain of the function $f(x) = \sqrt{0.5x + 2}$.

EXAMPLE 4 Graph the function $f(x) = \sqrt{x + 2}$. Use the values $f(-2), f(-1), f(0), f(1), f(2)$, and $f(7)$. Round your answers to the nearest tenth when necessary.

Solution We show the table of values here.

x	$f(x)$
-2	0
-1	1
0	1.4
1	1.7
2	2
7	3

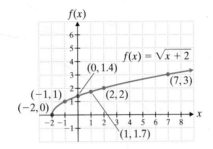

Practice Problem 4 Graph the function $f(x) = \sqrt{3x - 9}$. Use the values $f(3), f(4), f(5), f(6)$, and $f(15)$.

② Changing Radical Expressions to Expressions with Rational Exponents

Now we want to extend our definition of roots to rational exponents. By the laws of exponents we know that

$$x^{1/2} \cdot x^{1/2} = x^{1/2+1/2} = x^1 = x.$$

Since $x^{1/2}x^{1/2} = x$, $x^{1/2}$ must be a square root of x. That is, $x^{1/2} = \sqrt{x}$. Is this true? By the definition of square root, $(\sqrt{x})^2 = x$. Does $(x^{1/2})^2 = x$? Using the law of exponents we have

$$\left(x^{1/2}\right)^2 = x^{(1/2)(2)} = x^1 = x.$$

We conclude that

$$x^{1/2} = \sqrt{x}.$$

In the same way we can write the following:

$$x^{1/3} \cdot x^{1/3} \cdot x^{1/3} = x \qquad x^{1/3} = \sqrt[3]{x}$$
$$x^{1/4} \cdot x^{1/4} \cdot x^{1/4} \cdot x^{1/4} = x \qquad x^{1/4} = \sqrt[4]{x}$$
$$\vdots \qquad\qquad \vdots$$
$$\underbrace{x^{1/n} \cdot x^{1/n} \cdot \ \cdots \ \cdot x^{1/n}}_{n \text{ factors}} = x \qquad x^{1/n} = \sqrt[n]{x}$$

Therefore, we are ready to define fractional exponents in general.

DEFINITION

If n is a positive integer and x is a nonnegative real number, then

$$x^{1/n} = \sqrt[n]{x}.$$

EXAMPLE 5 Change to rational exponents and simplify. Assume that all variables are nonnegative real numbers.

(a) $\sqrt[4]{x^4}$ (b) $\sqrt[5]{(32)^5}$

Solution

(a) $\sqrt[4]{x^4} = (x^4)^{1/4} = x^{4/4} = x^1 = x$

(b) $\sqrt[5]{(32)^5} = (32^5)^{1/5} = 32^{5/5} = 32^1 = 32$

NOTE TO STUDENT: *Fully worked-out solutions to all of the Practice Problems can be found at the back of the text starting at page SP-1*

Practice Problem 5 Change to rational exponents and simplify. Assume that all variables are nonnegative real numbers.

(a) $\sqrt[3]{x^3}$ (b) $\sqrt[4]{y^4}$

EXAMPLE 6 Replace all radicals with rational exponents.

(a) $\sqrt[3]{x^2}$ (b) $\left(\sqrt[5]{w}\right)^7$

Solution

(a) $\sqrt[3]{x^2} = (x^2)^{1/3} = x^{2/3}$ (b) $\left(\sqrt[5]{w}\right)^7 = (w^{1/5})^7 = w^{7/5}$

Practice Problem 6 Replace all radicals with rational exponents.

(a) $\sqrt[4]{x^3}$ (b) $\sqrt[5]{(xy)^7}$

EXAMPLE 7 Evaluate or simplify. Assume that all variables are positive.

(a) $\sqrt[5]{32x^{10}}$ (b) $\sqrt[3]{125x^9}$ (c) $(16x^4)^{3/4}$

Solution

(a) $\sqrt[5]{32x^{10}} = (2^5 x^{10})^{1/5} = 2x^2$

(b) $\sqrt[3]{125x^9} = [(5)^3 x^9]^{1/3} = 5x^3$

(c) $(16x^4)^{3/4} = (2^4 x^4)^{3/4} = 2^3 x^3 = 8x^3$

Practice Problem 7 Evaluate or simplify. Assume that all variables are nonnegative.

(a) $\sqrt[4]{81x^{12}}$ (b) $\sqrt[3]{27x^6}$ (c) $(32x^5)^{3/5}$

Changing Expressions with Rational Exponents to Radical Expressions

Sometimes we need to change an expression with rational exponents to a radical expression. This is especially helpful because the value of the radical form of an expression is sometimes more recognizable. For example, because of our experience with radicals, we know that $\sqrt{25} = 5$. It is not as easy to see that $25^{1/2} = 5$. Therefore, we simplify expressions with rational exponents by first rewriting them as radical expressions. Recall that

$$x^{1/n} = \sqrt[n]{x}.$$

Again, using the laws of exponents, we know that

$$x^{m/n} = (x^m)^{1/n} = (x^{1/n})^m,$$

when x is nonnegative. We can make the following general definition.

> **DEFINITION**
>
> For positive integers m and n and any real number x for which $x^{1/n}$ is defined,
>
> $$x^{m/n} = \left(\sqrt[n]{x}\right)^m = \sqrt[n]{x^m}.$$
>
> If it is also true that $x \neq 0$, then
>
> $$x^{-m/n} = \frac{1}{x^{m/n}} = \frac{1}{\left(\sqrt[n]{x}\right)^m} = \frac{1}{\sqrt[n]{x^m}}.$$

EXAMPLE 8 Change to radical form.

(a) $(xy)^{5/7}$ **(b)** $w^{-2/3}$ **(c)** $3x^{3/4}$ **(d)** $(3x)^{3/4}$

Solution

(a) $(xy)^{5/7} = \sqrt[7]{(xy)^5} = \sqrt[7]{x^5 y^5}$ or $(xy)^{5/7} = \left(\sqrt[7]{xy}\right)^5$

(b) $w^{-2/3} = \dfrac{1}{w^{2/3}} = \dfrac{1}{\sqrt[3]{w^2}}$ or $w^{-2/3} = \dfrac{1}{w^{2/3}} = \dfrac{1}{\left(\sqrt[3]{w}\right)^2}$

(c) $3x^{3/4} = 3\sqrt[4]{x^3}$ or $3x^{3/4} = 3\left(\sqrt[4]{x}\right)^3$

(d) $(3x)^{3/4} = \sqrt[4]{(3x)^3} = \sqrt[4]{27x^3}$ or $(3x)^{3/4} = \left(\sqrt[4]{3x}\right)^3$

Practice Problem 8 Change to radical form.

(a) $(xy)^{3/4}$ **(b)** $y^{-1/3}$ **(c)** $(2x)^{4/5}$ **(d)** $2x^{4/5}$

EXAMPLE 9 Change to radical form and evaluate.

(a) $125^{2/3}$ **(b)** $(-16)^{5/2}$ **(c)** $144^{-1/2}$

Solution

(a) $125^{2/3} = \left(\sqrt[3]{125}\right)^2 = (5)^2 = 25$

(b) $(-16)^{5/2} = \left(\sqrt{-16}\right)^5$; however, $\sqrt{-16}$ is not a real number. Thus, $(-16)^{5/2}$ is not a real number.

(c) $144^{-1/2} = \dfrac{1}{144^{1/2}} = \dfrac{1}{\sqrt{144}} = \dfrac{1}{12}$

 Graphing Calculator

Rational Exponents

To evaluate Example 9(a) on a graphing calculator we use

125 $\boxed{\wedge}$ $\boxed{(}$ $\boxed{2}$ $\boxed{\div}$ $\boxed{3}$ $\boxed{)}$

$\boxed{\text{enter}}$.

Note the need to include the parentheses around the quantity 2/3. Try each part of Example 9 on your graphing calculator.

NOTE TO STUDENT: Fully worked-out solutions to all of the Practice Problems can be found at the back of the text starting at page SP-1

Practice Problem 9 Change to radical form and evaluate.

(a) $8^{2/3}$ (b) $(-8)^{4/3}$ (c) $100^{-3/2}$

 ## Evaluating Higher-Order Radicals Containing a Variable Radicand That Represents Any Real Number (Including a Negative Real Number)

We now give a definition of higher-order radicals that works for all radicals, no matter what their signs are.

DEFINITION

For all real numbers x (including negative real numbers),

$$\sqrt[n]{x^n} = |x| \quad \text{when } n \text{ is an } even \text{ positive integer, and}$$

$$\sqrt[n]{x^n} = x \quad \text{when } n \text{ is an } odd \text{ positive integer.}$$

EXAMPLE 10   Evaluate; x may be any real number.

(a) $\sqrt[3]{(-2)^3}$ (b) $\sqrt[4]{(-2)^4}$ (c) $\sqrt[5]{x^5}$ (d) $\sqrt[6]{x^6}$

Solution

(a) $\sqrt[3]{(-2)^3} = -2$ because the index is odd .

(b) $\sqrt[4]{(-2)^4} = |-2| = 2$ because the index is even .

(c) $\sqrt[5]{x^5} = x$ because the index is odd .

(d) $\sqrt[6]{x^6} = |x|$ because the index is even .

Practice Problem 10   Evaluate; y and w may be any real numbers.

(a) $\sqrt[5]{(-3)^5}$ (b) $\sqrt[4]{(-5)^4}$ (c) $\sqrt[4]{w^4}$ (d) $\sqrt[7]{y^7}$

EXAMPLE 11   Simplify. Assume that x and y may be any real numbers.

(a) $\sqrt{49x^2}$ (b) $\sqrt[4]{81y^{16}}$ (c) $\sqrt[3]{27x^6y^9}$

Solution

(a) We observe that the index is an even positive number. We will need the absolute value. $\sqrt{49x^2} = 7|x|$

(b) Again, we need the absolute value. $\sqrt[4]{81y^{16}} = 3|y^4|$

Since we know that $3y^4$ is positive (anything to the fourth power will be positive), we can write $3|y^4|$ without the absolute value symbol. Thus, $\sqrt[4]{81y^{16}} = 3y^4$.

(c) The index is an odd integer. The absolute value is never needed in such a case. $\sqrt[3]{27x^6y^9} = \sqrt[3]{(3)^3(x^2)^3(y^3)^3} = 3x^2y^3$

Practice Problem 11   Simplify. Assume that x and y may be any real numbers.

(a) $\sqrt{36x^2}$ (b) $\sqrt[4]{16y^8}$ (c) $\sqrt[3]{125x^3y^6}$

7.3 SIMPLIFYING, ADDING, AND SUBTRACTING RADICALS

1. Simplifying a Radical by Using the Product Rule

When we simplify a radical, we want to get an equivalent expression with the smallest possible quantity in the radicand. We can use the product rule for radicals to simplify radicals.

> **PRODUCT RULE FOR RADICALS**
>
> For all nonnegative real numbers a and b and positive integers n,
> $$\sqrt[n]{a}\,\sqrt[n]{b} = \sqrt[n]{ab}.$$

You should be able to derive the product rule from your knowledge of the laws of exponents. We have

$$\sqrt[n]{a}\,\sqrt[n]{b} = a^{1/n}b^{1/n} = (ab)^{1/n} = \sqrt[n]{ab}.$$

Throughout the remainder of this chapter, assume that all variables in any radicand represent nonnegative numbers, unless a specific statement is made to the contrary.

EXAMPLE 1 Simplify $\sqrt{32}$.

Solution 1: $\sqrt{32} = \sqrt{16 \cdot 2} = \sqrt{16}\sqrt{2} = 4\sqrt{2}$

Solution 2: $\sqrt{32} = \sqrt{4 \cdot 8} = \sqrt{4}\sqrt{8} = 2\sqrt{8} = 2\sqrt{4 \cdot 2} = 2\sqrt{4}\sqrt{2} = 4\sqrt{2}$

Although we obtained the same answer both times, the first solution is much shorter. You should try to use the largest factor that is a perfect square when you use the product rule.

Practice Problem 1 Simplify $\sqrt{20}$.

EXAMPLE 2 Simplify $\sqrt{48}$.

Solution $$\sqrt{48} = \sqrt{16}\sqrt{3} = 4\sqrt{3}$$

Practice Problem 2 Simplify $\sqrt{27}$.

EXAMPLE 3 Simplify. **(a)** $\sqrt[3]{16}$ **(b)** $\sqrt[3]{-81}$

Solution

(a) $\sqrt[3]{16} = \sqrt[3]{8}\sqrt[3]{2} = 2\sqrt[3]{2}$ **(b)** $\sqrt[3]{-81} = \sqrt[3]{-27}\sqrt[3]{3} = -3\sqrt[3]{3}$

Practice Problem 3 Simplify. **(a)** $\sqrt[3]{24}$ **(b)** $\sqrt[3]{-108}$

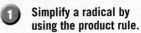
NOTE TO STUDENT: Fully worked-out solutions to all of the Practice Problems can be found at the back of the text starting at page SP-1

EXAMPLE 4 Simplify $\sqrt[4]{48}$.

Solution $\sqrt[4]{48} = \sqrt[4]{16}\sqrt[4]{3} = 2\sqrt[4]{3}$

Practice Problem 4 Simplify $\sqrt[4]{64}$.

EXAMPLE 5 Simplify. **(a)** $\sqrt{27x^3y^4}$ **(b)** $\sqrt[3]{16x^4y^3z^6}$

Solution

(a) $\sqrt{27x^3y^4} = \sqrt{9 \cdot 3 \cdot x^2 \cdot x \cdot y^4} = \sqrt{9x^2y^4}\sqrt{3x}$ Factor out the perfect squares.

$$= 3xy^2\sqrt{3x}$$

(b) $\sqrt[3]{16x^4y^3z^6} =$

$\sqrt[3]{8 \cdot 2 \cdot x^3 \cdot x \cdot y^3 \cdot z^6} = \sqrt[3]{8x^3y^3z^6}\sqrt[3]{2x}$ Factor out the perfect cubes.

$$= 2xyz^2\sqrt[3]{2x}$$ Why is z^6 a perfect cube?

Practice Problem 5 Simplify. **(a)** $\sqrt{45x^6y^7}$ **(b)** $\sqrt{27a^7b^8c^9}$

---

2 Adding and Subtracting Like Radical Terms

Only like radicals can be added or subtracted. Two radicals are **like radicals** if they have the same radicand and the same index. $2\sqrt{5}$ and $3\sqrt{5}$ are like radicals. $2\sqrt{5}$ and $2\sqrt{3}$ are not like radicals; $2\sqrt{5}$ and $2\sqrt[3]{5}$ are not like radicals. When we combine radicals, we combine like terms by using the distributive property.

EXAMPLE 6 Combine $2\sqrt{5} + 3\sqrt{5} - 4\sqrt{5}$.

Solution $2\sqrt{5} + 3\sqrt{5} - 4\sqrt{5} = (2 + 3 - 4)\sqrt{5} = 1\sqrt{5} = \sqrt{5}$

NOTE TO STUDENT: Fully worked-out solutions to all of the Practice Problems can be found at the back of the text starting at page SP-1

Practice Problem 6 Combine $19\sqrt{xy} + 5\sqrt{xy} - 10\sqrt{xy}$.

Sometimes when you simplify radicands, you may find you have like radicals.

EXAMPLE 7 Combine $5\sqrt{3} - \sqrt{27} + 2\sqrt{48}$.

Solution $5\sqrt{3} - \sqrt{27} + 2\sqrt{48} = 5\sqrt{3} - \sqrt{9}\sqrt{3} + 2\sqrt{16}\sqrt{3}$

$$= 5\sqrt{3} - 3\sqrt{3} + 2(4)\sqrt{3}$$
$$= 5\sqrt{3} - 3\sqrt{3} + 8\sqrt{3}$$
$$= 10\sqrt{3}$$

Practice Problem 7 Combine $4\sqrt{2} - 5\sqrt{50} - 3\sqrt{98}$.

EXAMPLE 8   Combine $6\sqrt{x} + 4\sqrt{12x} - \sqrt{75x} + 3\sqrt{x}$.

Solution

$$6\sqrt{x} + 4\sqrt{12x} - \sqrt{75x} + 3\sqrt{x} = 6\sqrt{x} + 4\sqrt{4}\sqrt{3x} - \sqrt{25}\sqrt{3x} + 3\sqrt{x}$$

$$= 6\sqrt{x} + 8\sqrt{3x} - 5\sqrt{3x} + 3\sqrt{x}$$

$$= 6\sqrt{x} + 3\sqrt{x} + 8\sqrt{3x} - 5\sqrt{3x}$$

$$= 9\sqrt{x} + 3\sqrt{3x}$$

Practice Problem 8   Combine $4\sqrt{2x} + \sqrt{18x} - 2\sqrt{125x} - 6\sqrt{20x}$.

EXAMPLE 9   Combine $2\sqrt[3]{81x^3y^4} + 3xy\sqrt[3]{24y}$.

Solution

$$2\sqrt[3]{81x^3y^4} + 3xy\sqrt[3]{24y} = 2\sqrt[3]{27x^3y^3}\sqrt[3]{3y} + 3xy\sqrt[3]{8}\sqrt[3]{3y}$$

$$= 2(3xy)\sqrt[3]{3y} + 3xy(2)\sqrt[3]{3y}$$

$$= 6xy\sqrt[3]{3y} + 6xy\sqrt[3]{3y}$$

$$= 12xy\sqrt[3]{3y}$$

Practice Problem 9   Combine $3x\sqrt[3]{54x^4} - 3\sqrt[3]{16x^7}$.

Student Solutions Manual | CD/ Video | PH Math Tutor Center | MathXL®Tutorials on CD | MathXL® | MyMathLab® | Interactmath.com

Simplify. Assume that all variables are nonnegative real numbers.

1. $\sqrt{8}$ **2.** $\sqrt{12}$ **3.** $\sqrt{18}$ **4.** $\sqrt{75}$

5. $\sqrt{28}$ **6.** $\sqrt{54}$ **7.** $\sqrt{44}$ **8.** $\sqrt{90}$

9. $\sqrt{9x^3}$ **10.** $\sqrt{16x^5}$ **11.** $\sqrt{40a^6b^7}$

12. $\sqrt{45a^3b^8}$ **13.** $\sqrt{90x^3yz^4}$ **14.** $\sqrt{24xy^8z^3}$

15. $\sqrt[3]{8}$ **16.** $\sqrt[3]{125}$ **17.** $\sqrt[3]{40}$

18. $\sqrt[3]{128}$ **19.** $\sqrt[3]{54a^2}$ **20.** $\sqrt[3]{24m}$

21. $\sqrt[3]{8a^3b^8}$ **22.** $\sqrt[3]{125a^6b^2}$ **23.** $\sqrt[3]{24x^6y^{11}}$

24. $\sqrt[3]{40x^7y^{26}}$ **25.** $\sqrt[4]{81kp^{23}}$ **26.** $\sqrt[4]{16k^{12}p^{18}}$

27. $\sqrt[5]{-32x^5y^6}$ **28.** $\sqrt[5]{-243x^4y^{10}}$

To Think About

29. $\sqrt[4]{1792} = a\sqrt[4]{7}$. What is the value of a? **30.** $\sqrt[3]{3072} = b\sqrt[3]{6}$. What is the value of b?

Combine.

31. $4\sqrt{5} + 8\sqrt{5}$ **32.** $3\sqrt{13} + 7\sqrt{13}$ **33.** $4\sqrt{3} + \sqrt{7} - 5\sqrt{7}$

34. $2\sqrt{6} + \sqrt{2} - 5\sqrt{6}$ **35.** $3\sqrt{18} - \sqrt{2}$ **36.** $\sqrt{40} - \sqrt{10}$

37. $4\sqrt{12} + \sqrt{27}$ **38.** $5\sqrt{75} + \sqrt{48}$ **39.** $\sqrt{8} + \sqrt{50} - 2\sqrt{72}$

40. $\sqrt{45} + \sqrt{80} - 3\sqrt{20}$

41. $-2\sqrt{50} + \sqrt{32} - 3\sqrt{8}$

42. $-\sqrt{12} + 2\sqrt{48} - \sqrt{75}$

43. $-5\sqrt{45} + 6\sqrt{20} + 3\sqrt{5}$

44. $-7\sqrt{10} + 4\sqrt{40} - 8\sqrt{90}$

Combine. Assume that all variables represent nonnegative real numbers.

45. $3\sqrt{48x} - 2\sqrt{12x}$

46. $5\sqrt{27x} - 4\sqrt{75x}$

47. $\sqrt{98x} + \sqrt{8x} + 5\sqrt{32x}$

48. $\sqrt{18x} + \sqrt{50x} + 6\sqrt{72x}$

49. $\sqrt{44} - 3\sqrt{63x} + 4\sqrt{28x}$

50. $\sqrt{63x} - \sqrt{54x} + \sqrt{24x}$

51. $\sqrt{200x^3} - x\sqrt{32x}$

52. $\sqrt{75a^3} + a\sqrt{12a}$

53. $\sqrt[3]{16} + 3\sqrt[3]{54}$

54. $\sqrt[3]{128} - 4\sqrt[3]{16}$

55. $-2\sqrt[3]{125x^3y^4} + 3y^2\sqrt[3]{8x^3}$

56. $3x\sqrt[3]{x^4y} - 2\sqrt[3]{x^3y^5}$

To Think About

 57. Use a calculator to show that
$$\sqrt{48} + \sqrt{27} + \sqrt{75} = 12\sqrt{3}.$$

58. Use a calculator to show that
$$\sqrt{98} + \sqrt{50} + \sqrt{128} = 20\sqrt{2}.$$

Applications

Electric Current   *We can approximate the amount of current in amps I (amperes) drawn by an appliance in the home using the formula*

$$I = \sqrt{\frac{P}{R}},$$

where P is the power measured in watts and R is the resistance measured in ohms. In Exercises 59 and 60, round your answers to three decimal places.

59. What is the current I if $P = 500$ watts and $R = 10$ ohms?

60. What is the current I if $P = 480$ watts and $R = 8$ ohms?

Period of a Pendulum   *The **period** of a pendulum is the amount of time it takes the pendulum to make one complete swing back and forth. If the length of the pendulum L is measured in feet, then its period T measured in seconds is given by the formula*

$$T = 2\pi\sqrt{\frac{L}{32}}.$$

Use $\pi \approx 3.14$ for exercises 61–62.

61. Find the period of a pendulum if its length is 8 feet.

62. A person suspended on a rope swinging back and forth acts like a human pendulum. What is the period of a person swinging on a rope that is 128 feet long?

Cumulative Review

Factor completely.

63. $16x^3 - 56x^2y + 49xy^2$

64. $81x^2y - 25y$

FDA Recommendations for Phosphorus   *The FDA recommends that an adult's minimum daily intake of the mineral phosphorus be 1 gram. A small serving of scallops (six average-size scallops) has 0.2 gram of phosphorus, while one small serving of skim milk (1 cup) has 0.25 gram of phosphorus.*

65. If the number of servings of scallops and the number of servings of skim milk totals 4.5 servings, how many of each would you need to meet the minimum daily requirement of phosphorus?

66. If you eat only scallops, how many servings would you need to obtain the minimum daily requirement of phosphorus? If you drink only skim milk, how many servings would you need to obtain the daily requirement for phosphorus?

Population Change   *According to the United Nations, since the collapse of communism in 1989, there has been a marked drop in the fertility rate throughout Eastern and Central Europe. The population of the former Soviet Union, which was 307 million in the year 2000, could fall to about 250 million by the year 2050. Source: United Nations Statistics Division.*

67. What is the projected percent of decrease in the population during the 50 years from 2000 to 2050?

68. What is the projected percent of decrease in the population per year during these 50 years?

69. If this trend is slowed so that there are 280 million people in the former Soviet Union in 2050, what will be the projected percent of decrease in the population from 2000 to 2050?

7.4 MULTIPLYING AND DIVIDING RADICALS

1 Multiplying Radical Expressions

We use the product rule for radicals to multiply radical expressions. Recall that $\sqrt[n]{a}\sqrt[n]{b} = \sqrt[n]{ab}$.

EXAMPLE 1   Multiply $\left(3\sqrt{2}\right)\left(5\sqrt{11x}\right)$.

Solution $\left(3\sqrt{2}\right)\left(5\sqrt{11x}\right) = (3)(5)\sqrt{2 \cdot 11x} = 15\sqrt{22x}$

Practice Problem 1   Multiply $\left(-4\sqrt{2}\right)\left(-3\sqrt{13x}\right)$.

EXAMPLE 2   Multiply $\sqrt{6x}\left(\sqrt{3} + \sqrt{2x} + \sqrt{5}\right)$.

Solution

$$\sqrt{6x}\left(\sqrt{3} + \sqrt{2x} + \sqrt{5}\right) = \left(\sqrt{6x}\right)\left(\sqrt{3}\right) + \left(\sqrt{6x}\right)\left(\sqrt{2x}\right) + \left(\sqrt{6x}\right)\left(\sqrt{5}\right)$$
$$= \sqrt{18x} + \sqrt{12x^2} + \sqrt{30x}$$
$$= \sqrt{9}\sqrt{2x} + \sqrt{4x^2}\sqrt{3} + \sqrt{30x}$$
$$= 3\sqrt{2x} + 2x\sqrt{3} + \sqrt{30x}$$

Practice Problem 2   Multiply $\sqrt{2x}\left(\sqrt{5} + 2\sqrt{3x} + \sqrt{8}\right)$.

To multiply two binomials containing radicals, we can use the distributive property. Most students find that the FOIL method is helpful in remembering how to find the four products.

EXAMPLE 3   Multiply $\left(\sqrt{2} + 3\sqrt{5}\right)\left(2\sqrt{2} - \sqrt{5}\right)$.

Solution   By FOIL:
$$\left(\sqrt{2} + 3\sqrt{5}\right)\left(2\sqrt{2} - \sqrt{5}\right) = 2\sqrt{4} - \sqrt{10} + 6\sqrt{10} - 3\sqrt{25}$$
$$= 4 + 5\sqrt{10} - 15$$
$$= -11 + 5\sqrt{10}$$

By the distributive property:
$$\left(\sqrt{2} + 3\sqrt{5}\right)\left(2\sqrt{2} - \sqrt{5}\right) = \left(\sqrt{2} + 3\sqrt{5}\right)\left(2\sqrt{2}\right) - \left(\sqrt{2} + 3\sqrt{5}\right)\sqrt{5}$$
$$= \left(\sqrt{2}\right)\left(2\sqrt{2}\right) + \left(3\sqrt{5}\right)\left(2\sqrt{2}\right) - \left(\sqrt{2}\right)\left(\sqrt{5}\right) - \left(3\sqrt{5}\right)\left(\sqrt{5}\right)$$
$$= 2\sqrt{4} + 6\sqrt{10} - \sqrt{10} - 3\sqrt{25}$$
$$= 4 + 5\sqrt{10} - 15$$
$$= -11 + 5\sqrt{10}$$

Practice Problem 3   Multiply $\left(\sqrt{7} + 4\sqrt{2}\right)\left(2\sqrt{7} - 3\sqrt{2}\right)$.

EXAMPLE 4   Multiply $\left(7 - 3\sqrt{2}\right)\left(4 - \sqrt{3}\right)$.

Solution $\left(7 - 3\sqrt{2}\right)\left(4 - \sqrt{3}\right) = 28 - 7\sqrt{3} - 12\sqrt{2} + 3\sqrt{6}$

Practice Problem 4   Multiply $\left(2 - 5\sqrt{5}\right)\left(3 - 2\sqrt{2}\right)$.

Student Learning Objectives

After studying this section, you will be able to:

1. Multiply radical expressions.

2. Divide radical expressions.

3. Simplify radical expressions by rationalizing the denominator.

NOTE TO STUDENT: Fully worked-out solutions to all of the Practice Problems can be found at the back of the text starting at page SP-1

EXAMPLE 5   Multiply $\left(\sqrt{7} + \sqrt{3x}\right)^2$.

Solution

Method 1: We can use the FOIL method or the distributive property.

$$\left(\sqrt{7} + \sqrt{3x}\right)\left(\sqrt{7} + \sqrt{3x}\right) = \sqrt{49} + \sqrt{21x} + \sqrt{21x} + \sqrt{9x^2}$$
$$= 7 + \sqrt{21x} + \sqrt{21x} + 3x$$
$$= 7 + 2\sqrt{21x} + 3x$$

Method 2: We could also use the Chapter 5 formula.

$$(a + b)^2 = a^2 + 2ab + b^2,$$

where $a = \sqrt{7}$ and $b = \sqrt{3x}$. Then

$$\left(\sqrt{7} + \sqrt{3x}\right)^2 = \left(\sqrt{7}\right)^2 + 2\sqrt{7}\sqrt{3x} + \left(\sqrt{3x}\right)^2$$
$$= 7 + 2\sqrt{21x} + 3x$$

Practice Problem 5   Multiply $\left(\sqrt{5x} + \sqrt{10}\right)^2$. Use the approach that seems easiest to you.

EXAMPLE 6   Multiply.

(a) $\sqrt[3]{3x}\left(\sqrt[3]{x^2} + 3\sqrt[3]{4y}\right)$ **(b)** $\left(\sqrt[3]{2y} + \sqrt[3]{4}\right)\left(2\sqrt[3]{4y^2} - 3\sqrt[3]{2}\right)$

Solution

(a) $\sqrt[3]{3x}\left(\sqrt[3]{x^2} + 3\sqrt[3]{4y}\right) = \left(\sqrt[3]{3x}\right)\left(\sqrt[3]{x^2}\right) + 3\left(\sqrt[3]{3x}\right)\left(\sqrt[3]{4y}\right)$
$$= \sqrt[3]{3x^3} + 3\sqrt[3]{12xy}$$
$$= x\sqrt[3]{3} + 3\sqrt[3]{12xy}$$

(b) $\left(\sqrt[3]{2y} + \sqrt[3]{4}\right)\left(2\sqrt[3]{4y^2} - 3\sqrt[3]{2}\right) = 2\sqrt[3]{8y^3} - 3\sqrt[3]{4y} + 2\sqrt[3]{16y^2} - 3\sqrt[3]{8}$
$$= 2(2y) - 3\sqrt[3]{4y} + 2\sqrt[3]{8}\sqrt[3]{2y^2} - 3(2)$$
$$= 4y - 3\sqrt[3]{4y} + 4\sqrt[3]{2y^2} - 6$$

NOTE TO STUDENT: Fully worked-out solutions to all of the Practice Problems can be found at the back of the text starting at page SP-1

Practice Problem 6   Multiply.

(a) $\sqrt[3]{2x}\left(\sqrt[3]{4x^2} + 3\sqrt[3]{y}\right)$ **(b)** $\left(\sqrt[3]{7} + \sqrt[3]{x^2}\right)\left(2\sqrt[3]{49} - \sqrt[3]{x}\right)$

---

2   **Dividing Radical Expressions**

We can use the laws of exponents to develop a rule for dividing two radicals.

$$\sqrt[n]{\frac{a}{b}} = \left(\frac{a}{b}\right)^{1/n} = \frac{a^{1/n}}{b^{1/n}} = \frac{\sqrt[n]{a}}{\sqrt[n]{b}}$$

This quotient rule is very useful. We now state it more formally.

QUOTIENT RULE FOR RADICALS

For all nonnegative real numbers a, all positive real numbers b, and positive integers n,

$$\frac{\sqrt[n]{a}}{\sqrt[n]{b}} = \sqrt[n]{\frac{a}{b}}.$$

Sometimes it will be best to change $\sqrt[n]{\dfrac{a}{b}}$ to $\dfrac{\sqrt[n]{a}}{\sqrt[n]{b}}$, whereas at other times it will be best to change $\dfrac{\sqrt[n]{a}}{\sqrt[n]{b}}$ to $\sqrt[n]{\dfrac{a}{b}}$. To use the quotient rule for radicals, you need to have good number sense. You should know your squares up to 15^2 and your cubes up to 5^3.

EXAMPLE 7 Divide. **(a)** $\dfrac{\sqrt{48}}{\sqrt{3}}$ **(b)** $\sqrt[3]{\dfrac{125}{8}}$ **(c)** $\dfrac{\sqrt{28x^5y^3}}{\sqrt{7x}}$

Solution

(a) $\dfrac{\sqrt{48}}{\sqrt{3}} = \sqrt{\dfrac{48}{3}} = \sqrt{16} = 4$

(b) $\sqrt[3]{\dfrac{125}{8}} = \dfrac{\sqrt[3]{125}}{\sqrt[3]{8}} = \dfrac{5}{2}$

(c) $\dfrac{\sqrt{28x^5y^3}}{\sqrt{7x}} = \sqrt{\dfrac{28x^5y^3}{7x}} = \sqrt{4x^4y^3} = 2x^2y\sqrt{y}$

Practice Problem 7 Divide. **(a)** $\dfrac{\sqrt{75}}{\sqrt{3}}$ **(b)** $\sqrt[3]{\dfrac{27}{64}}$ **(c)** $\dfrac{\sqrt{54a^3b^7}}{\sqrt{6b^5}}$

③ Simplifying Radical Expressions by Rationalizing the Denominator

Recall that to simplify a radical we want to get the smallest possible quantity in the radicand. Whenever possible, we find the square root of a perfect square. Thus, to simplify $\sqrt{\dfrac{7}{16}}$ we have

$$\sqrt{\dfrac{7}{16}} = \dfrac{\sqrt{7}}{\sqrt{16}} = \dfrac{\sqrt{7}}{4}.$$

Notice that the denominator does not contain a square root. The expression $\dfrac{\sqrt{7}}{4}$ is in simplest form.

Let's look at $\sqrt{\dfrac{16}{7}}$. We have

$$\sqrt{\dfrac{16}{7}} = \dfrac{\sqrt{16}}{\sqrt{7}} = \dfrac{4}{\sqrt{7}}.$$

Notice that the denominator contains a square root. If an expression contains a square root in the denominator, it is not considered to be simplified. How can we rewrite $\dfrac{4}{\sqrt{7}}$ as an equivalent expression that does not contain the $\sqrt{7}$ in the denominator? Since $\sqrt{7}\sqrt{7} = 7$, we can multiply the numerator and the denominator by the radical in the denominator.

$$\dfrac{4}{\sqrt{7}} \cdot \dfrac{\sqrt{7}}{\sqrt{7}} = \dfrac{4\sqrt{7}}{\sqrt{49}} = \dfrac{4\sqrt{7}}{7}$$

This expression is considered to be in simplest form. We call this process rationalizing the denominator.

Rationalizing the denominator is the process of transforming a fraction with one or more radicals in the denominator into an equivalent fraction without a radical in the denominator.

EXAMPLE 8 Simplify by rationalizing the denominator $\dfrac{3}{\sqrt{2}}$.

Solution
$$\frac{3}{\sqrt{2}} = \frac{3}{\sqrt{2}} \cdot \frac{\sqrt{2}}{\sqrt{2}} \quad \text{Since } \frac{\sqrt{2}}{\sqrt{2}} = 1.$$

$$= \frac{3\sqrt{2}}{\sqrt{4}} \qquad \text{Product rule for radicals.}$$

$$= \frac{3\sqrt{2}}{2}$$

NOTE TO STUDENT: Fully worked-out solutions to all of the Practice Problems can be found at the back of the text starting at page SP-1

Practice Problem 8 Simplify by rationalizing the denominator $\dfrac{7}{\sqrt{3}}$.

We can rationalize the denominator either before or after we simplify the denominator.

EXAMPLE 9 Simplify $\dfrac{3}{\sqrt{12x}}$.

Solution

Method 1: First we simplify the radical in the denominator, and then we multiply in order to rationalize the denominator.

$$\frac{3}{\sqrt{12x}} = \frac{3}{\sqrt{4}\sqrt{3x}} = \frac{3}{2\sqrt{3x}} \cdot \frac{\sqrt{3x}}{\sqrt{3x}} = \frac{3\sqrt{3x}}{2(3x)} = \frac{\sqrt{3x}}{2x}$$

Method 2: We can multiply numerator and denominator by a value that will make the radicand in the denominator a perfect square (i.e., rationalize the denominator).

$$\frac{3}{\sqrt{12x}} = \frac{3}{\sqrt{12x}} \cdot \frac{\sqrt{3x}}{\sqrt{3x}}$$

$$= \frac{3\sqrt{3x}}{\sqrt{36x^2}} \qquad \text{Since } \sqrt{12x}\sqrt{3x} = \sqrt{36x^2}.$$

$$= \frac{3\sqrt{3x}}{6x} = \frac{\sqrt{3x}}{2x}$$

Practice Problem 9 Simplify $\dfrac{8}{\sqrt{20x}}$.

If the radicand has a fraction, it is not considered to be simplified. We can use the quotient rule for radicals and then rationalize the denominator to simplify the radical. We have already rationalized denominators when they contain square roots. Now we will rationalize denominators when they contain radical expressions that are cube roots or higher-order roots.

EXAMPLE 10   Simplify $\sqrt[3]{\dfrac{2}{3x^2}}$.

Solution

Method 1: $\sqrt[3]{\dfrac{2}{3x^2}} = \dfrac{\sqrt[3]{2}}{\sqrt[3]{3x^2}}$ Quotient rule for radicals.

$= \dfrac{\sqrt[3]{2}}{\sqrt[3]{3x^2}} \cdot \dfrac{\sqrt[3]{9x}}{\sqrt[3]{9x}}$ Multiply the numerator and denominator by an appropriate value so that the new denominator will be a perfect cube.

$= \dfrac{\sqrt[3]{18x}}{\sqrt[3]{27x^3}}$ Observe that we can evaluate the cube root in the denominator.

$= \dfrac{\sqrt[3]{18x}}{3x}$

Method 2: $\sqrt[3]{\dfrac{2}{3x^2}} = \sqrt[3]{\dfrac{2}{3x^2} \cdot \dfrac{9x}{9x}}$

$= \sqrt[3]{\dfrac{18x}{27x^3}}$

$= \dfrac{\sqrt[3]{18x}}{\sqrt[3]{27x^3}}$

$= \dfrac{\sqrt[3]{18x}}{3x}$

Practice Problem 10   Simplify $\sqrt[3]{\dfrac{6}{5x}}$.

If the denominator of a radical expression contains a sum or difference with radicals, we multiply the numerator and denominator by the *conjugate* of the denominator. For example, the conjugate of $x + \sqrt{y}$ is $x - \sqrt{y}$; similarly, the conjugate of $x - \sqrt{y}$ is $x + \sqrt{y}$. What is the conjugate of $3 + \sqrt{2}$? It is $3 - \sqrt{2}$. How about $\sqrt{11} + \sqrt{xyz}$? It is $\sqrt{11} - \sqrt{xyz}$.

CONJUGATES

The expressions $a + b$ and $a - b$, where a and b represent any algebraic term, are called **conjugates.** Each expression is the conjugate of the other expression.

Multiplying by conjugates is simply an application of the formula

$$(a + b)(a - b) = a^2 - b^2.$$

For example,

$$\left(\sqrt{x} + \sqrt{y}\right)\left(\sqrt{x} - \sqrt{y}\right) = \left(\sqrt{x}\right)^2 - \left(\sqrt{y}\right)^2 = x - y.$$

EXAMPLE 11   Simplify $\dfrac{5}{3 + \sqrt{2}}$.

Solution

$$\frac{5}{3 + \sqrt{2}} = \frac{5}{3 + \sqrt{2}} \cdot \frac{3 - \sqrt{2}}{3 - \sqrt{2}}$$

Multiply the numerator and denominator by the conjugate of $3 + \sqrt{2}$.

$$= \frac{15 - 5\sqrt{2}}{(3)^2 - (\sqrt{2})^2}$$

$$= \frac{15 - 5\sqrt{2}}{9 - 2} = \frac{15 - 5\sqrt{2}}{7}$$

NOTE TO STUDENT: Fully worked-out solutions to all of the Practice Problems can be found at the back of the text starting at page SP-1

Practice Problem 11   Simplify $\dfrac{4}{2 + \sqrt{5}}$.

EXAMPLE 12   Simplify $\dfrac{\sqrt{7} + \sqrt{3}}{\sqrt{7} - \sqrt{3}}$.

Solution   The conjugate of $\sqrt{7} - \sqrt{3}$ is $\sqrt{7} + \sqrt{3}$.

$$\frac{\sqrt{7} + \sqrt{3}}{\sqrt{7} - \sqrt{3}} \cdot \frac{\sqrt{7} + \sqrt{3}}{\sqrt{7} + \sqrt{3}} = \frac{\sqrt{49} + 2\sqrt{21} + \sqrt{9}}{(\sqrt{7})^2 - (\sqrt{3})^2}$$

$$= \frac{7 + 2\sqrt{21} + 3}{7 - 3}$$

$$= \frac{10 + 2\sqrt{21}}{4}$$

$$= \frac{2(5 + \sqrt{21})}{2 \cdot 2}$$

$$= \frac{\cancel{2}(5 + \sqrt{21})}{\cancel{2} \cdot 2}$$

$$= \frac{5 + \sqrt{21}}{2}$$

Practice Problem 12   Simplify $\dfrac{\sqrt{11} + \sqrt{2}}{\sqrt{11} - \sqrt{2}}$.

Multiply and simplify. Assume that all variables represent nonnegative numbers.

1. $\sqrt{5}\sqrt{7}$

2. $\sqrt{11}\sqrt{6}$

3. $(5\sqrt{2})(-6\sqrt{5})$

4. $(-7\sqrt{3})(-4\sqrt{10})$

5. $(2\sqrt{6})(-3\sqrt{2})$

6. $(-4\sqrt{5})(2\sqrt{10})$

7. $(-3\sqrt{y})(\sqrt{5x})$

8. $(\sqrt{2x})(-7\sqrt{3y})$

9. $(3x\sqrt{2x})(-2\sqrt{10xy})$

10. $(4\sqrt{3a})(a\sqrt{6ab})$

11. $5\sqrt{a}(3\sqrt{b} - 5)$

12. $-\sqrt{x}(5\sqrt{y} + 3)$

13. $-2\sqrt{y}(\sqrt{2x} - 3\sqrt{5})$

14. $3\sqrt{7a}(\sqrt{b} + 8\sqrt{3})$

15. $-\sqrt{a}(\sqrt{a} - 2\sqrt{b})$

16. $4\sqrt{xy}(3\sqrt{x} - 10\sqrt{xy})$

17. $7\sqrt{x}(2\sqrt{3} - 5\sqrt{x})$

18. $3\sqrt{y}(4\sqrt{6} + 11\sqrt{y})$

19. $(3 - \sqrt{2})(8 + \sqrt{2})$

20. $(\sqrt{5} + 4)(\sqrt{5} - 1)$

21. $(2\sqrt{3} + \sqrt{2})(2\sqrt{3} - 4\sqrt{2})$

22. $(3\sqrt{3} + \sqrt{5})(\sqrt{3} - 2\sqrt{5})$

23. $(\sqrt{7} + 4\sqrt{5x})(2\sqrt{7} + 3\sqrt{5x})$

24. $(\sqrt{6} + 3\sqrt{3y})(5\sqrt{6} + 2\sqrt{3y})$

25. $(\sqrt{3} + 2\sqrt{2})(\sqrt{5} + \sqrt{3})$

26. $(3\sqrt{5} + \sqrt{3})(\sqrt{2} + 2\sqrt{5})$

27. $(\sqrt{5} - 2\sqrt{6})^2$

28. $(\sqrt{3} + 4\sqrt{7})^2$

29. $(6 - 5\sqrt{a})^2$

30. $(3\sqrt{b} + 4)^2$

31. $(\sqrt{3x + 4} + 3)^2$

32. $(\sqrt{2x + 1} - 2)^2$

33. $(\sqrt[3]{x^2})(3\sqrt[3]{4x} - 4\sqrt[3]{x^5})$

34. $(2\sqrt[3]{x})(\sqrt[3]{4x^2} - \sqrt[3]{14x})$

35. $(\sqrt[3]{3} + \sqrt[3]{2})(\sqrt[3]{9} - \sqrt[3]{4})$

36. $(\sqrt[3]{4} - \sqrt[3]{6})(\sqrt[3]{2} + \sqrt[3]{36})$

Divide and simplify. Assume that all variables represent positive numbers.

37. $\sqrt{\dfrac{49}{25}}$

38. $\sqrt{\dfrac{16}{36}}$

39. $\sqrt{\dfrac{12x}{49y^6}}$

40. $\sqrt{\dfrac{27a^4}{64x^2}}$

41. $\sqrt[3]{\dfrac{8x^5y^6}{27}}$

42. $\sqrt[3]{\dfrac{125a^3b^4}{64}}$

43. $\dfrac{\sqrt[3]{5y^8}}{\sqrt[3]{27x^3}}$

44. $\dfrac{\sqrt[3]{9y^{10}}}{\sqrt[3]{64x^6}}$

Simplify by rationalizing the denominator.

45. $\dfrac{3}{\sqrt{2}}$

46. $\dfrac{5}{\sqrt{7}}$

47. $\sqrt{\dfrac{4}{3}}$

48. $\sqrt{\dfrac{25}{2}}$

49. $\dfrac{1}{\sqrt{5y}}$

50. $\dfrac{1}{\sqrt{3x}}$

51. $\dfrac{\sqrt{14a}}{\sqrt{2y}}$

52. $\dfrac{\sqrt{15a}}{\sqrt{3b}}$

53. $\dfrac{\sqrt{2}}{\sqrt{6x}}$

54. $\dfrac{\sqrt{5y}}{\sqrt{10x}}$

55. $\dfrac{x}{\sqrt{5}-\sqrt{2}}$

56. $\dfrac{y}{\sqrt{7}+\sqrt{3}}$

57. $\dfrac{2y}{\sqrt{6}+\sqrt{5}}$

58. $\dfrac{3x}{\sqrt{10}-\sqrt{2}}$

59. $\dfrac{\sqrt{x}}{\sqrt{3x}+\sqrt{2}}$

60. $\dfrac{\sqrt{x}}{\sqrt{5}+\sqrt{2x}}$

61. $\dfrac{\sqrt{5}+\sqrt{3}}{\sqrt{5}-\sqrt{3}}$

62. $\dfrac{\sqrt{11}-\sqrt{5}}{\sqrt{11}+\sqrt{5}}$

63. $\dfrac{\sqrt{3x}-2\sqrt{y}}{\sqrt{3x}+\sqrt{y}}$

64. $\dfrac{\sqrt{x}+\sqrt{y}}{\sqrt{x}-2\sqrt{y}}$

Mixed Practice

Simplify each of the following.

65. $3\sqrt{8}-5\sqrt{50}+\sqrt{98}$

66. $2\sqrt{48}-4\sqrt{27}+\sqrt{75}$

67. $\left(3\sqrt{2}-5\sqrt{3}\right)\left(\sqrt{2}+2\sqrt{3}\right)$

68. $\left(5\sqrt{6}-3\sqrt{2}\right)\left(\sqrt{6}+2\sqrt{2}\right)$

69. $\dfrac{9}{\sqrt{8x}}$

70. $\dfrac{5}{\sqrt{12x}}$

71. $\dfrac{\sqrt{5}+1}{\sqrt{5}+2}$

72. $\dfrac{\sqrt{2}-1}{2\sqrt{2}+1}$

To Think About

73. A student rationalized the denominator of $\dfrac{\sqrt{6}}{2\sqrt{3}-\sqrt{2}}$ and obtained $\dfrac{\sqrt{3}+3\sqrt{2}}{5}$. Find a decimal approximation of each expression. Are the decimals equal? Did the student do the work correctly?

74. A student rationalized the denominator of $\dfrac{\sqrt{5}}{\sqrt{5}+\sqrt{3}}$ and obtained $\dfrac{5-\sqrt{15}}{2}$. Find a decimal approximation of each expression. Are the decimals equal? Did the student do the work correctly?

In calculus, students are sometimes required to rationalize the numerator of an expression. In this case the numerator will not have a radical in the answer. Rationalize the numerator in each of the following:

75. $\dfrac{\sqrt{3}+2\sqrt{7}}{8}$

76. $\dfrac{\sqrt{5}-4\sqrt{3}}{6}$

Applications

Fertilizer Costs *The cost of fertilizing a lawn is $0.18 per square foot. Find the cost to fertilize each of the following triangular lawns in problems 77 and 78. Round your answers to the nearest cent.*

▲ **77.** The base of the triangle is $\sqrt{21}$ feet, and the altitude is $\sqrt{50}$ feet.

▲ **78.** The base of the triangle is $\sqrt{17}$ feet, and the altitude is $\sqrt{40}$ feet.

▲ **79.** ***Pacemaker Control Panel*** A medical doctor has designed a pacemaker that has a rectangular control panel. This rectangle has a width of $\sqrt{x}+3$ millimeters and a length of $\sqrt{x}+5$ millimeters. Find the area in square millimeters of this rectangle.

▲ **80.** ***FBI Listening Device*** An FBI agent has designed a secret listening device that has a rectangular base. The rectangle has a width of $\sqrt{x}+7$ centimeters and a length of $\sqrt{x}+11$ centimeters. Find the area in square centimeters of this rectangle.

Cumulative Review

Solve the system for x and y.

81. $2x+3y=13$
$5x-2y=4$

Solve the system for x, y, and z.

82. $3x-y-z=5$
$2x+3y-z=-16$
$x+2y+2z=-3$

83. *Caffeine in Coffee and Tea* A cup of strong coffee contains about 200 milligrams of caffeine. A cup of strong tea contains 80 milligrams of caffeine. Juanita used to drink 1 cup of each every day. However, she resolved on January 1 to reduce her intake of caffeine to less than 18 milligrams per day. On January 2 she cut her consumption of both coffee and tea in half. Three days later she again cut her consumption in half. If she continues this pattern of reduction, on what day will she reach her goal?

84. *Caffeine in Coffee and Tea* Juanita's husband, Carlos, has several cups of coffee and tea each day. On January 1, he had 11 cups in total and consumed a total of 1480 milligrams of caffeine. Using the information in Exercise 83, find out how many cups of coffee and how many cups of tea he consumed. If he cut his consumption of coffee and tea in half on January 2 and continues to cut his consumption of coffee and tea in half every 4 days after that, when will he reach his goal of fewer than 24 milligrams per day?

Price of Engagement Rings Tiffany & Co., one of the most famous diamond merchants, offers engagement rings in a variety of price ranges. These data are displayed in the bar graph below. The data indicate the price of all diamond rings sold during a recent year.

85. What percent of the rings sold for $23,000 or less?

86. If Tiffany & Co. sold 85,000 diamond engagement rings last year, what is the number of rings that cost more than $5,000?

Price of Diamond Engagement Rings

Percentage of rings in each price range

40% 30% 20% 10% 0%

15%   31%   26%   19%   9%

Cost of diamond engagement rings sold at Tiffany & Co. (in dollars)

500 – 5000   5001 – 15,000   15,001 – 23,000   23,001 – 50,000   More than 50,000

Source:
U.S. Bureau of
Economic Analysis

How are you doing with your homework assignments in Sections 7.1 to 7.4? Do you feel you have mastered the material so far? Do you understand the concepts you have covered? Before you go further in the textbook, take some time to do each of the following problems.

Leave all answers with positive exponents. Assume all variables are nonnegative.

7.1

1. Multiply and simplify your answer. $(-3x^{1/4}y^{1/2})(-2x^{-1/2}y^{1/3})$

Simplify.

2. $(-4x^{-1/4}y^{1/3})^3$

3. $\dfrac{-18x^{-2}y^2}{-3x^{-5}y^{1/3}}$

4. $\left(\dfrac{27x^2y^{-5}}{x^{-4}y^4}\right)^{2/3}$

7.2

Evaluate.

5. $27^{-4/3}$

6. $\sqrt[5]{-243}$

7. $\sqrt{169} + \sqrt[3]{-64}$

8. $\sqrt{49x^6y^{20}}$

9. $\sqrt[3]{27a^{12}b^6c^{15}}$

10. Replace the radical with a rational Exponent. $\left(\sqrt[6]{4x}\right)^5$

7.3

11. Simplify. $\sqrt[4]{16x^{20}y^{28}}$

12. Simplify $\sqrt[3]{32x^8y^{15}}$

13. Combine like terms. $\sqrt{44} - 2\sqrt{99} + 7\sqrt{11}$

14. Combine like terms where possible. $3\sqrt{48y^3} - 2\sqrt[3]{16} + 3\sqrt[3]{54} - 5y\sqrt{12y}$

7.4

15. Multiply and simplify $\left(3\sqrt{3} - 5\sqrt{6}\right)\left(\sqrt{12} - 3\sqrt{6}\right)$.

16. Rationalize the denominator and simplify your answer. $\dfrac{6}{\sqrt{20x}}$

17. Rationalize the denominator and simplify your answer. $\dfrac{\sqrt{2} + \sqrt{3}}{\sqrt{2} - \sqrt{3}}$

Now turn to page SA-27 for the answers to each of these problems. Each answer also includes a reference to the objective in which the problem is first taught. If you missed any of these problems, you should stop and review the Examples and Practice Problems in the referenced objective. A little review now will help you master the material in the upcoming sections of the text.

1. _____

2. _____

3. _____

4. _____

5. _____

6. _____

7. _____

8. _____

9. _____

10. _____

11. _____

12. _____

13. _____

14. _____

15. _____

16. _____

17. _____

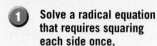
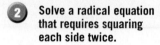
 Solving a Radical Equation by Squaring Each Side Once

A **radical equation** is an equation with a variable in one or more of the radicals. $3\sqrt{x} = 8$ and $\sqrt{3x - 1} = 5$ are radical equations. We solve radical equations by raising each side of the equation to the appropriate power. In other words, we square both sides if the radicals are square roots, cube both sides if the radicals are cube roots, and so on. Once we have done this, solving for the unknown becomes routine.

Sometimes after we square each side, we obtain a quadratic equation. In this case we collect all terms on one side and use the zero factor method that we developed in Section 5.8. After solving the equation, *always* check your answers to see whether extraneous solutions have been introduced.

We will now generalize this rule because it is very useful in higher-level mathematics courses.

> **RAISING EACH SIDE OF AN EQUATION TO A POWER**
>
> If $y = x$, then $y^n = x^n$, for all natural numbers n.

EXAMPLE 1   Solve $\sqrt{2x + 9} = x + 3$.

Solution

$$\left(\sqrt{2x + 9}\right)^2 = (x + 3)^2 \qquad \text{Square each side.}$$
$$2x + 9 = x^2 + 6x + 9 \qquad \text{Simplify.}$$
$$0 = x^2 + 4x \qquad \text{Collect all terms on one side.}$$
$$0 = x(x + 4) \qquad \text{Factor.}$$
$$x = 0 \quad \text{or} \quad x + 4 = 0 \qquad \text{Set each factor equal to zero.}$$
$$x = 0 \qquad\qquad x = -4 \qquad \text{Solve for } x.$$

Check:

For $x = 0$: $\sqrt{2(0) + 9} \stackrel{?}{=} 0 + 3$ For $x = -4$: $\sqrt{2(-4) + 9} \stackrel{?}{=} -4 + 3$
$$\sqrt{9} \stackrel{?}{=} 3 \qquad\qquad\qquad\qquad \sqrt{1} \stackrel{?}{=} -1$$
$$3 = 3 \; \checkmark \qquad\qquad\qquad\qquad 1 \neq -1$$

Therefore, 0 is the only solution to this equation.

Practice Problem 1   Solve and check your solution(s). $\sqrt{3x - 8} = x - 2$

As you begin to solve more complicated radical equations, it is important to make sure that one radical expression is alone on one side of the equation. This is often referred to as **isolating the radical term.**

EXAMPLE 2   Solve $\sqrt{10x + 5} - 1 = 2x$.

Solution

$$\sqrt{10x + 5} = 2x + 1 \qquad \text{Isolate the radical term.}$$
$$\left(\sqrt{10x + 5}\right)^2 = (2x + 1)^2 \qquad \text{Square each side.}$$
$$10x + 5 = 4x^2 + 4x + 1 \qquad \text{Simplify.}$$
$$0 = 4x^2 - 6x - 4 \qquad \text{Collect all terms on one side.}$$
$$0 = 2(2x^2 - 3x - 2) \qquad \text{Factor out the common factor.}$$
$$0 = 2(2x + 1)(x - 2) \qquad \text{Factor completely.}$$
$$2x + 1 = 0 \quad \text{or} \quad x - 2 = 0 \qquad \text{Set each factor equal to zero.}$$
$$2x = -1 \qquad\qquad x = 2 \qquad \text{Solve for } x.$$
$$x = -\frac{1}{2}$$

Check:

$$x = -\frac{1}{2}: \sqrt{10\left(-\frac{1}{2}\right) + 5} - 1 \overset{?}{=} 2\left(-\frac{1}{2}\right) \qquad x = 2: \sqrt{10(2) + 5} - 1 \overset{?}{=} 2(2)$$

$$\sqrt{-5 + 5} - 1 \overset{?}{=} -1 \qquad\qquad\qquad \sqrt{25} - 1 \overset{?}{=} 4$$

$$\sqrt{0} - 1 \overset{?}{=} -1 \qquad\qquad\qquad\qquad 5 - 1 \overset{?}{=} 4$$

$$-1 = -1 \ \checkmark \qquad\qquad\qquad\qquad\qquad 4 = 4 \ \checkmark$$

Both answers check, so $-\dfrac{1}{2}$ and 2 are roots of the equation.

Practice Problem 2   Solve and check your solution(s). $\sqrt{x + 4} = x + 4$

NOTE TO STUDENT: *Fully worked-out solutions to all of the Practice Problems can be found at the back of the text starting at page SP-1*

❷ Solving a Radical Equation by Squaring Each Side Twice

In some exercises, we must square each side twice in order to remove all the radicals. It is important to isolate at least one radical before squaring each side.

EXAMPLE 3   Solve $\sqrt{5x + 1} - \sqrt{3x} = 1$.

Solution

$$\sqrt{5x + 1} = 1 + \sqrt{3x} \qquad\quad \text{Isolate one of the radicals.}$$

$$\left(\sqrt{5x + 1}\right)^2 = \left(1 + \sqrt{3x}\right)^2 \qquad \text{Square each side.}$$

$$5x + 1 = \left(1 + \sqrt{3x}\right)\left(1 + \sqrt{3x}\right)$$

$$5x + 1 = 1 + 2\sqrt{3x} + 3x$$

$$2x = 2\sqrt{3x} \qquad\qquad \text{Isolate the remaining radical.}$$

$$x = \sqrt{3x} \qquad\qquad\quad \text{Divide each side by 2.}$$

$$(x)^2 = \left(\sqrt{3x}\right)^2 \qquad\qquad \text{Square each side.}$$

$$x^2 = 3x$$

$$x^2 - 3x = 0 \qquad\qquad\quad \text{Collect all terms on one side.}$$

$$x(x - 3) = 0 \qquad\qquad\quad \text{Factor.}$$

$$x = 0 \quad \text{or} \quad x - 3 = 0 \qquad \text{Solve for } x.$$

$$x = 3$$

Check:

$$x = 0: \ \sqrt{5(0) + 1} - \sqrt{3(0)} \overset{?}{=} 1 \qquad x = 3: \ \sqrt{5(3) + 1} - \sqrt{3(3)} \overset{?}{=} 1$$

$$\sqrt{1} - \sqrt{0} \overset{?}{=} 1 \qquad\qquad\qquad \sqrt{16} - \sqrt{9} \overset{?}{=} 1$$

$$1 = 1 \ \checkmark \qquad\qquad\qquad\qquad 1 = 1 \ \checkmark$$

Both answers check. The solutions are 0 and 3.

Practice Problem 3   Solve and check your solution(s).
$$\sqrt{2x + 5} - 2\sqrt{2x} = 1$$

We will now formalize the procedure for solving radical equations.

PROCEDURE FOR SOLVING RADICAL EQUATIONS

1. Perform algebraic operations to obtain one radical by itself on one side of the equation.
2. If the equation contains square roots, square each side of the equation. Otherwise, raise each side to the appropriate power for third- and higher-order roots.
3. Simplify, if possible.
4. If the equation still contains a radical, repeat steps 1 to 3.
5. Collect all terms on one side of the equation.
6. Solve the resulting equation.
7. Check all apparent solutions. Solutions to radical equations must be verified.

EXAMPLE 4

Solve $\sqrt{2y + 5} - \sqrt{y - 1} = \sqrt{y + 2}$.

Solution

$$\left(\sqrt{2y + 5} - \sqrt{y - 1}\right)^2 = \left(\sqrt{y + 2}\right)^2$$

$$\left(\sqrt{2y + 5} - \sqrt{y - 1}\right)\left(\sqrt{2y + 5} - \sqrt{y - 1}\right) = y + 2$$

$$2y + 5 - 2\sqrt{(y - 1)(2y + 5)} + y - 1 = y + 2$$

$$-2\sqrt{(y - 1)(2y + 5)} = -2y - 2$$

$$\sqrt{(y - 1)(2y + 5)} = y + 1 \qquad \text{Divide each side by } -2.$$

$$\left(\sqrt{2y^2 + 3y - 5}\right)^2 = (y + 1)^2 \qquad \text{Square each side.}$$

$$2y^2 + 3y - 5 = y^2 + 2y + 1$$

$$y^2 + y - 6 = 0 \qquad \text{Collect all terms on one side.}$$

$$(y + 3)(y - 2) = 0$$

$$y = -3 \quad \text{or} \quad y = 2$$

Check: Verify that 2 is a valid solution but -3 is not a valid solution.

NOTE TO STUDENT: *Fully worked-out solutions to all of the Practice Problems can be found at the back of the text starting at page SP-1*

Practice Problem 4 Solve and check your solution(s).

$$\sqrt{y - 1} + \sqrt{y - 4} = \sqrt{4y - 11}$$

Verbal and Writing Skills

1. Before squaring each side of a radical equation, what step should be taken first?

2. Why do we have to check the solutions when we solve radical equations?

Solve each radical equation. Check your solution(s).

3. $\sqrt{8x + 1} = 5$

4. $\sqrt{5x - 4} = 6$

5. $\sqrt{7x - 3} - 2 = 0$

6. $1 = 5 - \sqrt{9x - 2}$

7. $y + 1 = \sqrt{5y - 1}$

8. $\sqrt{y + 10} = y - 2$

9. $2x = \sqrt{x + 3}$

10. $3x = \sqrt{9x - 2}$

11. $2 = 5 + \sqrt{2x + 1}$

12. $12 + \sqrt{4x + 5} = 7$

13. $y - \sqrt{y - 3} = 5$

14. $\sqrt{2y - 4} + 2 = y$

15. $\sqrt{y + 1} - 1 = y$

16. $5 + \sqrt{2y + 5} = y$

17. $x - 2\sqrt{x - 3} = 3$

18. $2\sqrt{4x + 1} + 5 = x + 9$

19. $\sqrt{3x^2 - x} = x$

20. $\sqrt{5x^2 - 3x} = 2x$

21. $\sqrt[3]{2x + 3} = 2$

22. $\sqrt[3]{3x - 6} = 3$

23. $\sqrt[3]{4x - 1} = 3$

24. $\sqrt[3]{3 - 5x} = 2$

Solve each radical equation. This will usually involve squaring each side twice. Check your solutions.

25. $\sqrt{x + 4} = 1 + \sqrt{x - 3}$

26. $\sqrt{x - 5} + 1 = \sqrt{x + 2}$

27. $\sqrt{5x + 1} = 1 + \sqrt{3x}$

28. $2\sqrt{x + 4} = 1 + \sqrt{2x + 9}$

29. $\sqrt{x + 6} = 1 + \sqrt{x + 2}$

30. $\sqrt{3x + 1} - \sqrt{x - 4} = 3$

31. $\sqrt{6x + 6} = 1 + \sqrt{4x + 5}$ **32.** $\sqrt{8x + 17} = \sqrt{2x + 8} + 3$ **33.** $\sqrt{2x + 9} - \sqrt{x + 1} = 2$

34. $\sqrt{2x + 6} = \sqrt{7 - 2x} + 1$ **35.** $\sqrt{4x + 6} = \sqrt{x + 1} - \sqrt{x + 5}$ **36.** $\sqrt{3x + 4} + \sqrt{x + 5} = \sqrt{7 - 2x}$

37. $2\sqrt{x} - \sqrt{x - 5} = \sqrt{2x - 2}$ **38.** $\sqrt{3 - 2\sqrt{x}} = \sqrt{x}$

Optional Graphing Calculator Problems

Solve for x. Round your answer to four decimal places.

39. $x = \sqrt{5.326x - 1.983}$ **40.** $\sqrt[3]{5.62x + 9.93} = 1.47$

Applications

41. **Length of Skid Marks** When a car traveling on wet pavement at a speed V in miles per hour stops suddenly, it will produce skid marks of length S feet according to the formula $V = 2\sqrt{3S}$.

   (a) Solve the equation for S.

   (b) Use your result from **(a)** to find the length of the skid mark S if the car is traveling at 18 miles per hour.

42. **Flight Data Recorder** The volume V of a steel container inside a flight data recorder is defined by the equation

$$x = \sqrt{\frac{V}{5}},$$

where x is the sum of the length and the width of the container in inches and the height of the container is 5 inches.

   (a) Solve the equation for V.

   (b) Use the result from **(a)** to find the volume of the container whose length and width total 3.5 inches.

Stopping Distance   *Recently an experiment was conducted relating the speed a car is traveling and the stopping distance. In this experiment, a car is traveling on dry pavement at a constant rate of speed. From the instant that a driver recognizes the need to stop, the number of feet it takes for him to stop the car is recorded. For example, for a driver traveling at 50 miles per hour, it requires a stopping distance of 190 feet. In general, the stopping distance x in feet is related to the speed of the car y in miles per hour by the equation*

$$0.11y + 1.25 = \sqrt{3.7625 + 0.22x}.$$

Source: National Highway Traffic Safety Administration.

43. Solve this equation for x.

44. Use your answer from Exercise 43 to find what the stopping distance x would have been for a car traveling at $y = 60$ miles per hour.

To Think About

45. The solution to the equation

$$\sqrt{x^2 - 4x + c} = x - 1$$

is $x = 4$. What is the value of c?

46. The solution to the equation

$$\sqrt{x + b} - \sqrt{x} = -2$$

is $x = 16$. What is the value of b?

Cumulative Review

Simplify.

47. $(4^3 x^6)^{2/3}$

48. $(2^{-3} x^{-6})^{1/3}$

49. $\sqrt[3]{-216 x^6 y^9}$

50. $\sqrt[5]{-32 x^{15} y^5}$

▲ **51.** *Coffee Table* The area of the top of a solid rectangular coffee table measures $(4x^2 + 2x + 9)$ square centimeters. The height of the coffee table measures $(2x + 3)$ centimeters. Find the volume of the solid coffee table.

▲ **52.** *Cereal Display Case* A rectangular display case has $(2r^2 + 5r + 3)$ boxes of cereal on each shelf. The display case has $(2r + 4)$ shelves. Find the number of cereal boxes in the display case.

53. *Mississippi Paddleboat* The Mississippi Magic paddleboat can travel 12 miles per hour in still water. After traveling for 3 hours downstream with the current, it takes 5 hours to get upstream with the current and return to its original starting point. What is the speed of the current?

54. *Skiing* Louise Elton rides the ski lift for 1.75 miles to the top of Mount Gray. Once she is there, she immediately skis directly down the mountain. The ski trail winding down the mountain is 2.5 miles long. If she skis five times as fast as the lift runs and the round trip takes 45 minutes, find the rate at which she skis.

Student Learning Objectives

After studying this section, you will be able to:

1 Simplify expressions involving complex numbers.

2 Add and subtract complex numbers.

3 Multiply complex numbers.

4 Evaluate complex numbers of the form i^n.

5 Divide two complex numbers.

1 Simplifying Expressions Involving Complex Numbers

Until now we have not been able to solve an equation such as $x^2 = -4$ because there is no *real* number that satisfies this equation. However, this equation *does* have a nonreal solution. This solution is an *imaginary number*.

We define a new number:

$$i = \sqrt{-1} \text{ or } i^2 = -1.$$

Now let us use this procedure

$$\sqrt{-a} = \sqrt{-1}\sqrt{a}$$

and see if it is valid.

Then $\sqrt{-4} = \sqrt{4(-1)} = \sqrt{4}\sqrt{-1} = \sqrt{4} \cdot i = 2i.$

Thus, one solution to the equation $x^2 = -4$ is $2i$. Let's check it.

$$x^2 = -4$$
$$(2i)^2 \stackrel{?}{=} -4$$
$$4i^2 \stackrel{?}{=} -4$$
$$4(-1) \stackrel{?}{=} -4$$
$$-4 = -4 \checkmark$$

The value $-2i$ is also a solution. You should verify this.

Now we formalize our definitions and give some examples of imaginary numbers.

DEFINITION OF IMAGINARY NUMBER

The **imaginary number** i is defined as follows:

$$i = \sqrt{-1} \quad \text{and} \quad i^2 = -1.$$

The set of imaginary numbers consists of numbers of the form bi, where b is a real number and $b \neq 0$.

DEFINITION

For all positive real numbers a,

$$\sqrt{-a} = \sqrt{-1}\sqrt{a} = i\sqrt{a}.$$

EXAMPLE 1 Simplify.

(a) $\sqrt{-36}$ **(b)** $\sqrt{-17}$

Solution

(a) $\sqrt{-36} = \sqrt{-1}\sqrt{36} = (i)(6) = 6i$

(b) $\sqrt{-17} = \sqrt{-1}\sqrt{17} = i\sqrt{17}$

Practice Problem 1 Simplify. **(a)** $\sqrt{-49}$ **(b)** $\sqrt{-31}$

To avoid confusing $\sqrt{17}i$ with $\sqrt{17i}$, we write the i before the radical. That is, we write $i\sqrt{17}$.

EXAMPLE 2 Simplify $\sqrt{-45}$.

Solution

$$\sqrt{-45} = \sqrt{-1}\sqrt{45} = i\sqrt{45} = i\sqrt{9}\sqrt{5} = 3i\sqrt{5}$$

Practice Problem 2 Simplify $\sqrt{-98}$.

The rule $\sqrt{a}\sqrt{b} = \sqrt{ab}$ requires that $a \geq 0$ and $b \geq 0$. Therefore, we cannot use our product rule when the radicands are negative unless we first use the definition of $\sqrt{-1}$. Recall that

$$\sqrt{-1} \cdot \sqrt{-1} = i \cdot i = i^2 = -1.$$

EXAMPLE 3 Multiply $\sqrt{-16} \cdot \sqrt{-25}$.

Solution First we must use the definition $\sqrt{-1} = i$. Thus, we have the following:

$$\left(\sqrt{-16}\right)\left(\sqrt{-25}\right) = \left(i\sqrt{16}\right)\left(i\sqrt{25}\right)$$
$$= i^2(4)(5)$$
$$= -1(20) \qquad i^2 = -1.$$
$$= -20$$

Practice Problem 3 Multiply $\sqrt{-8} \cdot \sqrt{-2}$.

Now we formally define a complex number.

DEFINITION

A number that can be written in the form $a + bi$, where a and b are real numbers, is a **complex number.** We say that a is the **real part** and bi is the **imaginary part.**

Under this definition, every real number is also a complex number. For example, the real number 5 can be written as $5 + 0i$. Therefore, 5 is a complex number. In a similar fashion, the imaginary number $2i$ can be written as $0 + 2i$. So $2i$ is a complex number. Thus, the set of complex numbers includes the set of real numbers and the set of imaginary numbers.

DEFINITION

Two complex numbers $a + bi$ and $c + di$ are equal if and only if $a = c$ and $b = d$.

This definition means that two complex numbers are equal if and only if their real parts are equal *and* their imaginary parts are equal.

Graphing Calculator

Complex Numbers

Some graphing calculators, such as the TI-83, have a complex number mode. If your graphing calculator has this capability, you will be able to use it to do complex number operations. First you must use the Mode command to transfer selection from "Real" to "Complex" or "a + bi." To verify your status, try to find $\sqrt{-7}$ on your graphing calculator. If you obtain an approximate answer of "2.645751311 i," then your calculator is operating in the complex number mode. If you obtain "ERROR: NONREAL ANSWER," then your calculator is not operating in the complex number mode.

NOTE TO STUDENT: *Fully worked-out solutions to all of the Practice Problems can be found at the back of the text starting at page SP-1*

EXAMPLE 4 Find the real numbers x and y if $x + 3i\sqrt{7} = -2 + yi$.

Solution By our definition, the real parts must be equal, so x must be -2; the imaginary parts must also be equal, so y must be $3\sqrt{7}$.

Practice Problem 4 Find the real numbers x and y if
$-7 + 2yi\sqrt{3} = x + 6i\sqrt{3}$.

② Adding and Subtracting Complex Numbers

ADDING AND SUBTRACTING COMPLEX NUMBERS

For all real numbers a, b, c, and d,

$$(a + bi) + (c + di) = (a + c) + (b + d)i \quad \text{and}$$
$$(a + bi) - (c + di) = (a - c) + (b - d)i.$$

In other words, to combine complex numbers we add (or subtract) the real parts, and we add (or subtract) the imaginary parts.

EXAMPLE 5 Subtract $(6 - 2i) - (3 - 5i)$.

Solution

$(6 - 2i) - (3 - 5i) = (6 - 2i) + (-3 + 5i) = (6 - 3) + (-2 + 5)i = 3 + 3i$

Practice Problem 5 Subtract $(3 - 4i) - (-2 - 18i)$.

③ Multiplying Complex Numbers

As we might expect, the procedure for multiplying complex numbers is similar to the procedure for multiplying polynomials. We will see that the complex numbers obey the associative, commutative, and distributive properties.

EXAMPLE 6 Multiply $(7 - 6i)(2 + 3i)$.

Solution Use FOIL.

$$(7 - 6i)(2 + 3i) = (7)(2) + (7)(3i) + (-6i)(2) + (-6i)(3i)$$
$$= 14 + 21i - 12i - 18i^2$$
$$= 14 + 21i - 12i - 18(-1)$$
$$= 14 + 21i - 12i + 18$$
$$= 32 + 9i$$

Practice Problem 6 Multiply $(4 - 2i)(3 - 7i)$.

EXAMPLE 7   Multiply $3i(4 - 5i)$.

Solution   Use the distributive property.

$$3i(4 - 5i) = (3)(4)i + (3)(-5)i^2$$
$$= 12i - 15i^2$$
$$= 12i - 15(-1)$$
$$= 15 + 12i$$

Practice Problem 7   Multiply $-2i(5 + 6i)$.

4 Evaluating Complex Numbers of the Form i^n

How would you evaluate i^n, where n is any positive integer? What if n is a negative integer? We look for a pattern. We have defined

$$i^2 = -1.$$

We could write

$$i^3 = i^2 \cdot i = (-1)i = -i.$$

We also have the following:

$$i^4 = i^2 \cdot i^2 = (-1)(-1) = +1$$
$$i^5 = i^4 \cdot i = (+1)i = +i$$

We notice that $i^5 = i$. Let's look at i^6.

$$i^6 = i^4 \cdot i^2 = (+1)(-1) = -1$$

We begin to see a pattern that starts with i and repeats itself for i^5. Will $i^7 = -i$? Why or why not?

VALUES OF i^n

$i = i$	$i^5 = i$	$i^9 = i$
$i^2 = -1$	$i^6 = -1$	$i^{10} = -1$
$i^3 = -i$	$i^7 = -i$	$i^{11} = -i$
$i^4 = +1$	$i^8 = +1$	$i^{12} = +1$

We can use this pattern to evaluate powers of i.

EXAMPLE 8   Evaluate.   **(a)** i^{36} **(b)** i^{27}

Solution

(a) $i^{36} = (i^4)^9 = (1)^9 = 1$

(b) $i^{27} = (i^{24+3}) = (i^{24})(i^3) = (i^4)^6(i^3) = (1)^6(-i) = -i$

This suggests a quick method for evaluating powers of i. Divide the exponent by 4. i^4 raised to any power will be 1. Then use the first column of the values of i^n chart above to evaluate the remainder.

Practice Problem 8   Evaluate.   **(a)** i^{42} **(b)** i^{53}

NOTE TO STUDENT: Fully worked-out
solutions to all of the Practice Problems
can be found at the back of the text
starting at page SP-1

Graphing Calculator

 Complex Operations

When we perform complex number operations on a graphing calculator, the answer will usually be displayed as an approximate value in decimal form. Try Example 9 on your graphing calculator by entering $(7 + i) \div (3 - 2i)$. You should obtain an approximate answer of $1.461538462 + 1.307692308i$.

5 Dividing Two Complex Numbers

The complex numbers $a + bi$ and $a - bi$ are called **conjugates.** The product of two complex conjugates is always a real number.

$$(a + bi)(a - bi) = a^2 - abi + abi - b^2i^2$$
$$= a^2 - b^2(-1)$$
$$= a^2 + b^2$$

When dividing two complex numbers, we want to remove any expression involving i from the denominator. So we multiply the numerator and denominator by the conjugate of the denominator. This is just what we did when we rationalized the denominator in a radical expression.

EXAMPLE 9 Divide $\dfrac{7 + i}{3 - 2i}$.

Solution

$$\frac{(7 + i)}{(3 - 2i)} \cdot \frac{(3 + 2i)}{(3 + 2i)} = \frac{21 + 14i + 3i + 2i^2}{9 - 4i^2} = \frac{21 + 17i + 2(-1)}{9 - 4(-1)}$$

$$= \frac{21 + 17i - 2}{9 + 4}$$

$$= \frac{19 + 17i}{13} \quad \text{or} \quad \frac{19}{13} + \frac{17}{13}i$$

Practice Problem 9 Divide $\dfrac{4 + 2i}{3 + 4i}$.

EXAMPLE 10 Divide $\dfrac{3 - 2i}{4i}$.

Solution The conjugate of $0 + 4i$ is $0 - 4i$ or simply $-4i$.

$$= \frac{(3 - 2i)}{(4i)} \cdot \frac{(-4i)}{(-4i)} = \frac{-12i + 8i^2}{-16i^2} = \frac{-12i + 8(-1)}{-16(-1)}$$

$$= \frac{-8 - 12i}{16} = \frac{\cancel{4}(-2 - 3i)}{\cancel{4} \cdot 4}$$

$$= \frac{-2 - 3i}{4} \quad \text{or} \quad -\frac{1}{2} - \frac{3}{4}i$$

Practice Problem 10 Divide $\dfrac{5 - 6i}{-2i}$.

Verbal and Writing Skills

1. Does $x^2 = -9$ have a real number solution? Why or why not?

2. Describe a complex number and give an example(s).

3. Are the complex numbers $2 + 3i$ and $3 + 2i$ equal? Why or why not?

4. Describe in your own words how to add or subtract complex numbers.

Simplify. Express in terms of i.

5. $\sqrt{-25}$

6. $\sqrt{-100}$

7. $\sqrt{-50}$

8. $\sqrt{-48}$

9. $\sqrt{-\dfrac{4}{49}}$

10. $\sqrt{-\dfrac{9}{16}}$

11. $-\sqrt{-81}$

12. $-\sqrt{-36}$

13. $2 + \sqrt{-3}$

14. $5 + \sqrt{-7}$

15. $-1.5 + \sqrt{-81}$

16. $\dfrac{7}{6} + \sqrt{-64}$

17. $-3 + \sqrt{-24}$

18. $-6 - \sqrt{-32}$

19. $\left(\sqrt{-3}\right)\left(\sqrt{-2}\right)$

20. $\left(\sqrt{-5}\right)\left(\sqrt{-3}\right)$

21. $\left(\sqrt{-36}\right)\left(\sqrt{-4}\right)$

22. $\left(\sqrt{-25}\right)\left(\sqrt{-9}\right)$

Find the real numbers x and y.

23. $x - 3i = 5 + yi$

24. $x - 6i = 7 + yi$

25. $1.3 - 2.5yi = x - 5i$

26. $3.4 - 0.8i = 2x - yi$

27. $23 + yi = 17 - x + 3i$

28. $2 + x - 11i = 19 + yi$

Perform the addition or subtraction.

29. $(1 + 8i) + (-6 + 3i)$

30. $(-12 - 4i) + (6 - 9i)$

31. $\left(-\dfrac{3}{2} + \dfrac{1}{2}i\right) + \left(\dfrac{5}{2} - \dfrac{3}{2}i\right)$

32. $\left(\dfrac{3}{4} - \dfrac{3}{4}i\right) + \left(\dfrac{9}{4} + \dfrac{5}{4}i\right)$

33. $(2.8 - 0.7i) - (1.6 - 2.8i)$

34. $(5.4 + 4.1i) - (4.8 + 2.6i)$

Multiply and simplify your answers. Place in i notation before doing any other operations.

35. $(4i)(3i)$

36. $(6i)(5i)$

37. $(-7i)(6i)$

38. $(i)(-3i)$

39. $(2 + 3i)(2 - i)$

40. $(4 - 6i)(2 + i)$

41. $5i - 2(-4 + i)$

42. $12i - 6(3 + i)$

43. $2i(5i - 6)$

44. $4i(7 - 2i)$

45. $\left(\dfrac{1}{2} + i\right)^2$

46. $\left(\dfrac{1}{3} - i\right)^2$

47. $\left(i\sqrt{3}\right)\left(i\sqrt{7}\right)$

48. $\left(i\sqrt{2}\right)\left(i\sqrt{6}\right)$

49. $\left(3 + \sqrt{-2}\right)\left(4 + \sqrt{-5}\right)$

50. $\left(2 + \sqrt{-3}\right)\left(6 + \sqrt{-2}\right)$

Evaluate.

51. i^{17}

52. i^{21}

53. i^{24}

54. i^{16}

55. i^{46}

56. i^{83}

57. i^{37}

58. i^{10}

59. $i^{30} + i^{28}$

60. $i^{26} + i^{24}$

61. $i^{100} - i^7$

62. $3i^{64} - 2i^{11}$

Divide.

63. $\dfrac{2 + i}{3 - i}$

64. $\dfrac{4 + 2i}{2 - i}$

65. $\dfrac{3i}{4 + 2i}$

66. $\dfrac{-2i}{3 + 5i}$

67. $\dfrac{5 - 2i}{6i}$

68. $\dfrac{7 + 10i}{3i}$

69. $\dfrac{2}{i}$

70. $\dfrac{-5}{i}$

71. $\dfrac{7}{5 - 6i}$

72. $\dfrac{3}{4 + 2i}$

73. $\dfrac{5 - 2i}{3 + 2i}$

74. $\dfrac{6 + 3i}{6 - 3i}$

Mixed Practice

Simplify.

75. $\sqrt{-98}$

76. $\sqrt{-72}$

77. $(4 - 7i) - (-2 + 5i)$

78. $(-6 + 3i) - (4 - 8i)$

79. $(5i - 4)(6i - 2)$

80. $(2i + 3)(7i - 5)$

81. $\dfrac{2 - 3i}{2 + i}$

82. $\dfrac{4 - 3i}{5 + 2i}$

Optional Graphing Calculator Problems

Perform each operation to obtain approximate answers.

83. $(29.3 + 56.2i)^2$

84. $\dfrac{196 - 34.8i}{24.9 + 56.4i}$

Applications

The impedance Z in an alternating current circuit (like the one used in your home and in your classroom) is given by the formula $Z = V/I$, where V is the voltage and I is the current.

85. Find the value of Z if $V = 3 + 2i$ and $I = 3i$.

86. Find the value of Z if $V = 4 + 2i$ and $I = -3i$.

Cumulative Review

87. *Factory Production* A grape juice factory produces juice in three different types of containers. $x + 3$ hours per week are spent on producing juice in glass bottles. $2x - 5$ hours per week are spent on producing juice in cans. $4x + 2$ hours per week are spent on producing juice in plastic bottles. If the factory operates 105 hours per week, how much time is spent producing juice in each type of container?

88. *Donation of Computers* Citizens Bank has decided to donate its older personal computers to the Boston Public Schools. Each computer donated is worth $120 in tax-deductible dollars to the bank. In addition, the computer company supplying the bank with its new computers gives a 7% rebate to any customer donating used computers to schools. If sixty new computers are purchased at a list price of $1850 each and sixty older computers are donated to the Boston Public Schools, what is the net cost to the bank for this purchase?

We are now able to write a more specific equation,

$$L = \frac{400}{d^2}.$$

We will use this to find L when $d = 4$ meters.

$$L = \frac{400}{4^2}$$

$$L = \frac{400}{16}$$

$$L = 25 \text{ lumens}$$

Check: Does this answer seem reasonable? Would we expect to have more light if we move closer to the light source? ✓

NOTE TO STUDENT: Fully worked-out solutions to all of the Practice Problems can be found at the back of the text starting at page SP-1

Practice Problem 3 If the amount of power in an electrical circuit is held constant, the resistance in the circuit varies inversely with the square of the amount of current. If the amount of current is 0.01 ampere, the resistance is 800 ohms. What is the resistance if the amount of current is 0.04 ampere?

3 Solving Problems Using Joint or Combined Variation

Sometimes a quantity depends on the variation of two or more variables. This is called joint or **combined variation.**

EXAMPLE 4 y varies directly with x and z and inversely with d^2. When $x = 7$, $z = 3$, and $d = 4$, the value of y is 20. Find the value of y when $x = 5$, $z = 6$, and $d = 2$.

Solution We can write the equation

$$y = \frac{kxz}{d^2}.$$

To find the value of k, we substitute into the equation $y = 20$, $x = 7$, $z = 3$, and $d = 4$.

$$20 = \frac{k(7)(3)}{4^2}$$

$$20 = \frac{21k}{16}$$

$$320 = 21k$$

$$\frac{320}{21} = k$$

Now we substitute $\frac{320}{21}$ for k into our original equation.

$$y = \frac{\frac{320}{21}xz}{d^2} \quad \text{or} \quad y = \frac{320xz}{21d^2}$$

We use this equation to find y for the known values of x, z, and d. We want to find y when $x = 5$, $z = 6$, and $d = 2$.

$$y = \frac{320(5)(6)}{21(2)^2} = \frac{9600}{84}$$

$$y = \frac{800}{7}$$

Practice Problem 4 y varies directly with z and w^2 and inversely with x. $y = 20$ when $z = 3$, $w = 5$, and $x = 4$. Find y when $z = 4$, $w = 6$, and $x = 2$.

Many applied problems involve joint variation. For example, a cylindrical cement column has a safe load capacity that varies directly with the diameter raised to the fourth power and inversely with the square of its length.

Therefore, if d = diameter and l = length, the equation would be of the form

$$y = \frac{kd^4}{l^2}.$$

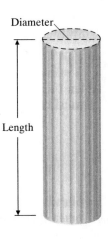

Diameter

Length

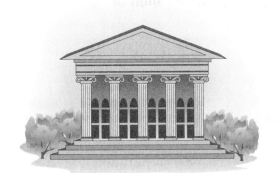

Student Solutions Manual   CD/ Video   PH Math Tutor Center   MathXL®Tutorials on CD   MathXL®   MyMathLab®   Interactmath.com

Verbal and Writing Skills

1. Give an example in everyday life of direct variation and write an equation as a mathematical model.

2. The general equation $y = kx$ means that y varies _____ with x. k is called the _____ of variation.

3. If y varies inversely with x, we write the equation _____.

4. Write a mathematical model for the following situation: The strength of a rectangular beam varies directly with its width and the square of its depth.

Round all answers to the nearest tenth unless otherwise directed.

5. If y varies directly with x and $y = 15$ when $x = 40$, find y when $x = 64$.

6. If y varies directly with x and $y = 63$ when $x = 36$, find y when $x = 84$.

7. *Pressure on a Submarine* A marine biology submarine was searching the waters for blue whales at 50 feet below the surface, where it experienced a pressure of 21 pounds per square inch (psi). If the pressure of water on a submerged object varies directly with its distance beneath the surface, how much pressure would the submarine have experienced if it had to dive to 170 feet?

8. *Spring Stretching* The distance a spring stretches varies directly with the weight of the object hung on the spring. If a 10-pound weight stretches a spring 6 inches, how far will a 35-pound weight stretch this spring?

9. *Stopping Distance* A car's stopping distance varies directly with the square of its speed. A car that is traveling 30 miles per hour can stop in 40 feet. What distance will it take to stop if it is traveling 60 miles per hour?

10. *Time of Fall in Gravitation* When an object is dropped, the distance it falls in feet varies directly with the square of the duration of the fall in seconds. An apple that falls from a tree falls 1 foot in $\frac{1}{4}$ second. How far will it fall in 1 second? How far will it fall in 2 seconds?

11. If *y* varies inversely with the square of *x*, and *y* = 10 when *x* = 2, find *y* when *x* = 0.5.

12. If *y* varies inversely with the square root of *x*, and *y* = 1.8 when *x* = 0.04, find *y* when *x* = 0.3.

13. *Gasoline Prices* During one month last summer, the price of gasoline changed frequently. One station owner noticed that the number of gallons he sold each day seemed to vary inversely with the price per gallon. If he sold 3000 gallons when the price was $2.10, how many gallons could he expect to sell if the price fell to $1.90? Round your answer to the nearest gallon.

14. *Weight of an Object* The weight of an object on the Earth's surface varies inversely with the square of its distance from the center of the Earth. An object weighs 1000 pounds on the Earth's surface. This is approximately 4000 miles from the center of the Earth. How much would an object weigh 4500 miles from the center of the Earth?

15. *Speed Detection* Police officers can detect speeding by using variation. The speed of a car varies inversely with the time it takes to cover a certain fixed distance. Between two points on a highway, a car travels 45 miles per hour in 6 seconds. What is the speed of a car that travels the same distance in 9 seconds?

16. *Electric Current* If the voltage in an electric circuit is kept at the same level, the current varies inversely with the resistance. The current measures 40 amperes when the resistance is 270 ohms. Find the current when the resistance is 100 ohms.

17. *Support Beam* The weight that can be safely supported by a 2- by 6-inch support beam varies inversely with its length. A builder finds that a support beam that is 8 feet long will support 900 pounds. Find the weight that can be safely supported by a beam that is 18 feet long.

18. *Satellite Orbit Speed* The speed that is required to maintain a satellite in a circular orbit around the Earth varies directly with the square root of the distance of the satellite from the center of the Earth. We will assume that the radius of the Earth is approximately 4000 miles. A satellite that is 100 miles above the surface of the Earth is orbiting at approximately 18,000 miles per hour. What speed would be necessary for the satellite to orbit 500 miles above the surface of the Earth? Round to nearest mile per hour.

19. *Strength of a Beam* The strength of a rectangular beam varies jointly with its width and the square of its thickness. If a beam 5 inches wide and 2 inches thick supports 400 pounds, how much can a beam of the same material that is 4 inches wide and 3.5 inches thick support?

20. *Aquarium Tank* The amount of time it takes to drain the water from a large tank in an aquarium is inversely proportional to the square of the radius of its drainage pipe. If a pipe of radius 2.4 inches could drain the tank in 5.5 minutes, how long will it take the tank to drain if a pipe of radius 3 inches is installed?

21. *Atmospheric Drag* Atmospheric drag tends to slow down moving objects. Atmospheric drag varies jointly with an object's surface area A and velocity v. If a Dodge Intrepid, traveling at a speed of 45 mph with a surface area of 37.8 square feet, experiences a drag of 222 newtons, how fast must a Dodge Caravan, with a surface area of 55 square feet, travel in order to experience a drag force of 450 newtons?

22. *Wind Generator* The force on a blade of a wind generator varies jointly with the product of the blade's area and the square of the wind velocity. The force of the wind is 20 pounds when the area is 3 square feet and the velocity is 30 feet per second. Find the force when the area is increased to 5 square feet and the velocity is reduced to 25 feet per second.

Cumulative Review

Solve each of the following equations or word problems.

23. $3x^2 - 8x + 4 = 0$

24. $4x^2 = -28x + 32$

25. *Sales Tax* In Champaign, Illinois, the sales tax is 6.25%. Donny bought an amplifier for his stereo that cost $488.75 after tax. What was the original price of the amplifier?

26. *Tennis Courts* It takes 7.5 gallons of white paint to properly paint lines on three tennis courts. How much paint is needed to paint twenty-two tennis courts?

27. *Photograph Collection* Craig Emanuel has a photography studio in Los Angeles. He wants to frame his collection of 110 antique photographs in special gold leaf and silver frames. The price of each gold leaf frame is $140, and the price of each silver frame is $95. If Craig has $13,375 for the frames, how many of each frame can he buy to decorate his studio?

▲ **28.** *Geometry* A triangular New Year's noisemaking toy has a perimeter of 50 centimeters. The first side is $\frac{4}{5}$ as long as the second side. The third side is 2 centimeters shorter than the first side. Find the length of each side.

Putting Your Skills to Work

Mathematics and Music Frequency

On a piano, the different notes are created by the piano strings' vibrating at different rates. This rate of vibration is called frequency. The frequency of the note middle A is 440 times a second (440 hertz). In general, the frequency of the next note above any given note is calculated by multiplying the frequency of the given note by $\sqrt[12]{2}$.

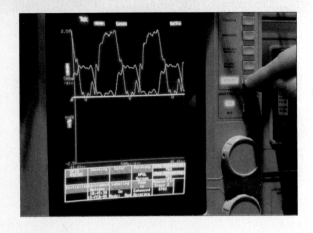

Problems for Individual Investigation and Analysis

1. Write $\sqrt[12]{2}$ using exponents instead of a radical. Use your calculator to find this value.

2. Find the frequency of A sharp (A#), the note just above A, by multiplying 440 by the value you found in exercise 1.

3. Find the frequency of G sharp (G#), the note just below A. Divide 440 by the value you found in exercise 1.

Problems for Group Investigation and Cooperative Study

If you continue the process in Exercise 2, you will find that the frequency of middle C is approximately 261.63 hertz. An octave contains 12 notes. To obtain the frequency for the note above middle C, we multiply 261.63 by $2^{\frac{1}{12}}$. To obtain the frequency of each successive note, we multiply by $2^{\frac{1}{12}}$ again. Then the note above middle C has a frequency of $261.63(2^{\frac{1}{12}})$ and the note above that has a frequency of $261.63(2^{\frac{1}{12}})(2^{\frac{1}{12}})$, etc.

4. Without computing the value of all 12 notes in the octave, use the pattern just given to find the frequency of high C, which is 12 notes above middle C.

5. What is the relationship between these two frequencies that are an octave apart? Would this relationship hold for all notes which are an octave apart?

6. Find the frequency of the C that is an octave below middle C.

7. What is the frequency of A that is an octave below middle A?

Chapter 7 Organizer

Topic	Procedure	Examples
Multiplication of variables with rational exponents, p. 357.	$$x^m x^n = x^{m+n}$$	$$(3x^{1/5})(-2x^{3/5}) = -6x^{4/5}$$
Division of variables with rational exponents, p. 357.	$$\frac{x^m}{x^n} = x^{m-n}, \quad n \neq 0, x \neq 0$$	$$\frac{-16x^{3/20}}{24x^{5/20}} = -\frac{2x^{-1/10}}{3}$$
Removing negative exponents, p. 356.	$$x^{-n} = \frac{1}{x^n}, \quad m \text{ and } n \neq 0, x \text{ and } y \neq 0$$ $$\frac{x^{-n}}{y^{-m}} = \frac{y^m}{x^n}$$	Write with positive exponents. $$3x^{-4} = \frac{3}{x^4}$$ $$\frac{2x^{-6}}{5y^{-8}} = \frac{2y^8}{5x^6}$$ $$4^{-2} = \frac{1}{4^2} = \frac{1}{16}$$
Zero exponent, p. 359.	$$x^0 = 1 \quad (\text{if } x \neq 0)$$	$$(3x^{1/2})^0 = 1$$
Raising a variable with an exponent to a power, p. 356.	$$(x^m)^n = x^{mn}$$ $$(xy)^n = x^n y^n$$ $$\left(\frac{x}{y}\right)^n = \frac{x^n}{y^n}, \quad y \neq 0$$	$$(x^{-1/2})^{-2/3} = x^{1/3}$$ $$(3x^{-2}y^{-1/2})^{2/3} = 3^{2/3}x^{-4/3}y^{-1/3}$$ $$\left(\frac{4x^{-2}}{3^{-1}y^{-1/2}}\right)^{1/4} = \frac{4^{1/4}x^{-1/2}}{3^{-1/4}y^{-1/8}}$$
Multiplication of expressions with rational exponents, p. 357.	Add exponents whenever expressions with the same base are multiplied.	$$x^{2/3}(x^{1/3} - x^{1/4}) = x^{3/3} - x^{2/3+1/4} = x - x^{11/12}$$
Higher-order roots, p. 363.	If x is a nonnegative real number, $\sqrt[n]{x}$ is a nonnegative nth root and has the property that $$\left(\sqrt[n]{x}\right)^n = x.$$ If x is a negative real number, $\left(\sqrt[n]{x}\right)^n = x$ when n is an odd integer. If x is a negative real number, $\left(\sqrt[n]{x}\right)^n$ is not a real number when n is an even integer. $\sqrt[3]{27} = 3$ because $3^3 = 27$.	$\sqrt[5]{-32} = -2$ because $(-2)^5 = -32$. $\sqrt[4]{-16}$ is *not* a real number.
Rational exponents and radicals, p. 364.	For positive integers m and n and any real number x for which $x^{1/n}$ is defined, $$x^{m/n} = \left(\sqrt[n]{x}\right)^m = \sqrt[n]{x^m}.$$ If it is also true that $x \neq 0$, then $$x^{1/n} = \sqrt[n]{x}.$$	Write as a radical: $x^{3/7} = \sqrt[7]{x^3}$, $3^{1/5} = \sqrt[5]{3}$ Write as an expression with a fractional exponent: $\sqrt[3]{w^4} = w^{3/4}$ Evaluate. $$25^{3/2} = \left(\sqrt{25}\right)^3 = (5)^3 = 125$$
Higher-order roots and absolute value, p. 368.	$\sqrt[n]{x^n} = |x|$ when n is an even positive integer. $\sqrt[n]{x^n} = x$ when n is an odd positive integer.	$\sqrt[6]{x^6} = |x|$ $\sqrt[5]{x^5} = x$
Evaluation of higher-order roots, p. 368.	Use exponent notation.	$\sqrt[5]{-32x^{15}} = \sqrt[5]{(-2)^5 x^{15}}$ $= [(-2)^5 x^{15}]^{1/5} = (-2)^1 x^3 = -2x^3$

Topic	Procedure	Examples
Simplification of radicals with the product rule, p. 371.	For nonnegative real numbers a and b and positive integers n, $$\sqrt[n]{a}\,\sqrt[n]{b} = \sqrt[n]{ab}.$$	Simplify when $x \geq 0$, $y \geq 0$. $$\sqrt{75x^3} = \sqrt{25x^2}\sqrt{3x}$$ $$= 5x\sqrt{3x}$$ $$\sqrt[3]{16x^5y^6} = \sqrt[3]{8x^3y^6}\sqrt[3]{2x^2}$$ $$= 2xy^2\sqrt[3]{2x^2}$$
Combining radicals, p. 372.	Simplify radicals and combine them if they have the same index and the same radicand.	Combine. $$2\sqrt{50} - 3\sqrt{98} = 2\sqrt{25}\sqrt{2} - 3\sqrt{49}\sqrt{2}$$ $$= 2(5)\sqrt{2} - 3(7)\sqrt{2}$$ $$= 10\sqrt{2} - 21\sqrt{2} = -11\sqrt{2}$$
Multiplying radicals, p. 377.	1. Multiply coefficients outside the radical and then multiply the radicands. 2. Simplify your answer.	$$(2\sqrt{3})(4\sqrt{5}) = 8\sqrt{15}$$ $$2\sqrt{6}(\sqrt{2} - 3\sqrt{12}) = 2\sqrt{12} - 6\sqrt{72}$$ $$= 2\sqrt{4}\sqrt{3} - 6\sqrt{36}\sqrt{2}$$ $$= 4\sqrt{3} - 36\sqrt{2}$$ $(\sqrt{2} + \sqrt{3})(2\sqrt{2} - \sqrt{3})$ By the FOIL method. $$= 2\sqrt{4} - \sqrt{6} + 2\sqrt{6} - \sqrt{9}$$ $$= 4 + \sqrt{6} - 3$$ $$= 1 + \sqrt{6}$$
Simplifying quotients of radicals with the quotient rule, p. 378.	For nonnegative real numbers a, positive real numbers b, and positive integers n, $$\sqrt[n]{\frac{a}{b}} = \frac{\sqrt[n]{a}}{\sqrt[n]{b}}.$$	$$\sqrt[3]{\frac{5}{27}} = \frac{\sqrt[3]{5}}{\sqrt[3]{27}} = \frac{\sqrt[3]{5}}{3}$$
Rationalizing denominators, p. 379.	Multiply numerator and denominator by a value that eliminates the radical in the denominator.	$$\frac{2}{\sqrt{7}} = \frac{2}{\sqrt{7}} \cdot \frac{\sqrt{7}}{\sqrt{7}} = \frac{2\sqrt{7}}{7}$$ $$\frac{3}{\sqrt{5} + \sqrt{2}} = \frac{3}{\sqrt{5} + \sqrt{2}} \cdot \frac{\sqrt{5} - \sqrt{2}}{\sqrt{5} - \sqrt{2}}$$ $$= \frac{3\sqrt{5} - 3\sqrt{2}}{(\sqrt{5})^2 - (\sqrt{2})^2} = \frac{3\sqrt{5} - 3\sqrt{2}}{5 - 2}$$ $$= \frac{3\sqrt{5} - 3\sqrt{2}}{3} = \sqrt{5} - \sqrt{2}$$
Solving radical equations, p. 388.	1. Perform algebraic operations to obtain one radical by itself on one side of the equation. 2. If the equation contains square roots, square each side of the equation. Otherwise, raise each side to the appropriate power for third- and higher-order roots. 3. Simplify, if possible. 4. If the equation still contains a radical, repeat steps 1 to 3. 5. Collect all terms on one side of the equation. 6. Solve the resulting equation. 7. Check all apparent solutions. Solutions to radical equations must be verified.	Solve. $$x = \sqrt{2x + 9} - 3$$ $$x + 3 = \sqrt{2x + 9}$$ $$(x + 3)^2 = (\sqrt{2x + 9})^2$$ $$x^2 + 6x + 9 = 2x + 9$$ $$x^2 + 6x - 2x + 9 - 9 = 0$$ $$x^2 + 4x = 0$$ $$x(x + 4) = 0$$ $$x = 0 \quad \text{or} \quad x = -4$$ *Check:* $x = 0$: $0 \overset{?}{=} \sqrt{2(0) + 9} - 3$ $$0 \overset{?}{=} \sqrt{9} - 3$$ $$0 = 3 - 3 \ \checkmark$$ $x = -4$: $-4 \overset{?}{=} \sqrt{2(-4) + 9} - 3$ $$-4 \overset{?}{=} \sqrt{1} - 3$$ $$-4 \neq -2$$ The only solution is 0.

Topic	Procedure	Examples
Simplifying imaginary numbers, p. 394.	Use $i = \sqrt{-1}$ and $i^2 = -1$ and $\sqrt{-a} = \sqrt{a}\sqrt{-1}$.	$\sqrt{-16} = \sqrt{-1}\sqrt{16} = 4i$ $\sqrt{-18} = \sqrt{-1}\sqrt{18} = i\sqrt{9}\sqrt{2} = 3i\sqrt{2}$
Adding and subtracting complex numbers, p. 396.	Combine real parts and imaginary parts separately.	$(5 + 6i) + (2 - 4i) = 7 + 2i$ $(-8 + 3i) - (4 - 2i) = -8 + 3i - 4 + 2i$ $\qquad\qquad\qquad\qquad = -12 + 5i$
Multiplying complex numbers, p. 396.	Use the FOIL method and $i^2 = -1$.	$(5 - 6i)(2 - 4i) = 10 - 20i - 12i + 24i^2$ $\qquad = 10 - 32i + 24(-1)$ $\qquad = 10 - 32i - 24$ $\qquad = -14 - 32i$
Dividing complex numbers, p. 398.	Multiply the numerator and denominator by the conjugate of the denominator.	$\dfrac{5 + 2i}{4 - i} = \dfrac{5 + 2i}{4 - i} \cdot \dfrac{4 + i}{4 + i} = \dfrac{20 + 5i + 8i + 2i^2}{16 - i^2}$ $\qquad = \dfrac{20 + 13i + 2(-1)}{16 - (-1)}$ $\qquad = \dfrac{20 + 13i - 2}{16 + 1}$ $\qquad = \dfrac{18 + 13i}{17} \quad \text{or} \quad \dfrac{18}{17} + \dfrac{13}{17}i$
Raising i to a power, p. 397.	$i^1 = i$ $i^2 = -1$ $i^3 = -i$ $i^4 = 1$	Evaluate. $i^{27} = i^{24} \cdot i^3$ $\qquad = (i^4)^6 \cdot i^3$ $\qquad = (1)^6(-i)$ $\qquad = -i$
Direct variation, p. 402.	If y varies directly with x, there is a constant of variation k such that $y = kx$. After k is determined, other values of y or x can easily be computed.	y varies directly with x. When $x = 2$, $y = 7$. $\qquad y = kx$ $\qquad 7 = k(2)$ Substitute. $\qquad k = \dfrac{7}{2}$ Solve. $\qquad y = \dfrac{7}{2}x$ What is y when $x = 18$? $\qquad y = \dfrac{7}{2}x = \dfrac{7}{2} \cdot 18 = 63$
Inverse variation, p. 403.	If y varies inversely with x, the constant k is such that $$y = \dfrac{k}{x}.$$	y varies inversely with x. When x is 5, y is 12. What is y when x is 30? $\qquad y = \dfrac{k}{x}$ $\qquad 12 = \dfrac{k}{5}$ Substitute. $\qquad k = 60$ Solve. $\qquad y = \dfrac{60}{x}$ Substitute. When $x = 30$, $y = \dfrac{60}{30} = 2$.

In all exercises assume that the variables represent positive real numbers unless otherwise stated. Simplify using only positive exponents in your answers.

1. $(3xy^{1/2})(5x^2y^{-3})$

2. $\dfrac{3x^{2/3}}{6x^{1/6}}$

3. $(25a^3b^4)^{1/2}$

4. $5^{1/4} \cdot 5^{1/2}$

5. $(2a^{1/3}b^{1/4})(-3a^{1/2}b^{1/2})$

6. $\dfrac{6x^{2/3}y^{1/10}}{12x^{1/6}y^{-1/5}}$

7. $(2x^{-1/5}y^{1/10}z^{4/5})^{-5}$

8. $\left(\dfrac{49a^3b^6}{a^{-7}b^4}\right)^{1/2}$

9. $\dfrac{(x^{3/4}y^{2/5})^{1/2}}{x^{-1/8}}$

10. $\left(\dfrac{27x^{5n}}{x^{2n-3}}\right)^{1/3}$

11. $(5^{6/5})^{10/7}$

12. Combine as one fraction containing only positive exponents. $2x^{1/3} + x^{-2/3}$

13. Factor out a common factor of $3x$ from $6x^{3/2} - 9x^{1/2}$.

In Exercises 14–46, assume that all variables represent nonnegative real numbers.

Evaluate, if possible:

14. $-\sqrt{16}$

15. $\sqrt[5]{-32}$

16. $\sqrt[6]{-20}$

17. $-\sqrt{\dfrac{1}{25}}$

18. $\sqrt{0.04}$

19. $\sqrt[4]{-256}$

20. $\sqrt[3]{-\dfrac{1}{8}}$

21. $\sqrt[3]{\dfrac{27}{64}}$

22. $64^{2/3}$

23. $125^{4/3}$

Simplify:

24. $\sqrt{81x^2y^6z^{10}}$

25. $\sqrt[3]{125a^9b^{60}}$

26. $\sqrt[3]{-8a^{12}b^{15}c^{21}}$

27. $\sqrt{49x^{22}y^2}$

Replace radicals with rational exponents:

28. $\sqrt[5]{a^2}$

29. $\sqrt[4]{y^3}$

30. $\sqrt{2b}$

31. $\sqrt[3]{6c}$

32. $\left(\sqrt[6]{ab}\right)^5$

Change to radical form:

33. $m^{1/2}$

34. $n^{1/4}$

35. $y^{3/5}$

36. $(3z)^{2/3}$

37. $(2x)^{3/7}$

Evaluate or simplify:

38. $16^{3/4}$

39. $64^{5/6}$

40. $(-27)^{2/3}$

41. $(-8)^{1/3}$

42. $\left(\dfrac{1}{9}\right)^{1/2}$

43. $(0.49)^{1/2}$

44. $\left(\dfrac{1}{16}\right)^{-1/4}$

45. $\left(\dfrac{1}{36}\right)^{-1/2}$

46. $(25a^2b^4)^{3/2}$

47. $(4a^6b^2)^{5/2}$

Combine where possible.

48. $\sqrt{50} + 2\sqrt{32} - \sqrt{8}$

49. $\sqrt{28} - 4\sqrt{7} + 5\sqrt{63}$

50. $3\sqrt{50} + 2\sqrt{75} - \sqrt{300}$

51. $\sqrt{40x^3} + x\sqrt{90x}$

52. $2\sqrt{32x} - 5x\sqrt{2} + \sqrt{18x}$

53. $3\sqrt[3]{16} - 4\sqrt[3]{54}$

Multiply and simplify.

54. $\left(5\sqrt{12}\right)\left(3\sqrt{6}\right)$

55. $\left(-2\sqrt{15}\right)\left(4x\sqrt{3}\right)$

56. $3\sqrt{x}\left(2\sqrt{8x} - 3\sqrt{48}\right)$

57. $\sqrt{5a}\left(2 - \sqrt{15a}\right)$

58. $-\sqrt{3xy}\left(\sqrt{2x} - \sqrt{6y}\right)$

59. $2\sqrt{7b}\left(\sqrt{ab} - b\sqrt{3bc}\right)$

60. $\left(5\sqrt{2} + \sqrt{3}\right)\left(\sqrt{2} - 2\sqrt{3}\right)$

61. $\left(5\sqrt{6} - 2\sqrt{2}\right)\left(\sqrt{6} - \sqrt{2}\right)$

62. $\left(2\sqrt{5} - 3\sqrt{6}\right)^2$

63. $\left(\sqrt[3]{2x} + \sqrt[3]{6}\right)\left(\sqrt[3]{4x^2} - \sqrt[3]{y}\right)$

64. Let $f(x) = \sqrt{5x + 20}$.

   (a) Find $f(16)$.

   (b) What is the domain of $f(x)$?

65. Let $f(x) = \sqrt{36 - 4x}$.

   (a) Find $f(5)$.

   (b) What is the domain of $f(x)$?

66. Let $f(x) = \sqrt{\dfrac{3}{4}x - \dfrac{1}{2}}$.

   (a) Find $f(1)$.

   (b) What is the domain of $f(x)$?

Rationalize the denominator and simplify the expression.

67. $\sqrt{\dfrac{3x^2}{y}}$

68. $\dfrac{2}{\sqrt{3y}}$

69. $\dfrac{3\sqrt{7x}}{\sqrt{21x}}$

70. $\dfrac{2}{\sqrt{6} - \sqrt{5}}$

71. $\dfrac{\sqrt{x}}{3\sqrt{x} + \sqrt{y}}$

72. $\dfrac{\sqrt{5}}{\sqrt{7} - 3}$

73. $\dfrac{2\sqrt{3} + \sqrt{6}}{\sqrt{3} + 2\sqrt{6}}$

74. $\dfrac{5\sqrt{2} - \sqrt{3}}{\sqrt{6} - \sqrt{3}}$

75. $\dfrac{3\sqrt{x} + \sqrt{y}}{\sqrt{x} - \sqrt{y}}$

76. $\dfrac{2xy}{\sqrt[3]{16xy^5}}$

77. Simplify $\sqrt{-16} + \sqrt{-45}$.

78. Find x and y. $2x - 3i + 5 = yi - 2 + \sqrt{6}$

Simplify by performing the operation indicated.

79. $(-12 - 6i) + (3 - 5i)$

80. $(2 - i) - (12 - 3i)$

81. $(7 + 3i)(2 - 5i)$

82. $(8 - 4i)^2$

83. $2i(3 + 4i)$

84. $3 - 4(2 + i)$

85. Evaluate i^{34}.

86. i^{65}

Divide.

87. $\dfrac{7 - 2i}{3 + 4i}$

88. $\dfrac{5 - 2i}{1 - 3i}$

89. $\dfrac{4 - 3i}{5i}$

90. $\dfrac{12}{3 - 5i}$

91. $\dfrac{10 - 4i}{2 + 5i}$

Solve and check your solution(s).

92. $\sqrt{3x - 2} = 5$

93. $\sqrt[3]{3x - 1} = 2$

94. $\sqrt{2x + 1} = 2x - 5$

95. $1 + \sqrt{3x + 1} = x$

96. $\sqrt{3x + 1} - \sqrt{2x - 1} = 1$

97. $\sqrt{7x + 2} = \sqrt{x + 3} + \sqrt{2x - 1}$

Round all answers to the nearest tenth.

98. If y varies directly with x, and $y = 16$ when $x = 5$, find the value of y when $x = 3$.

99. If y varies directly with x, and $y = 5$ when $x = 20$, find the value of y when $x = 50$.

100. *Stopping Distance* A car's stopping distance varies directly with the square of its speed. A car traveling on wet pavement can stop in 50 feet when traveling at 30 miles per hour. What distance will it take the car to stop if it is traveling at 55 miles per hour?

101. *Time of Falling Object* The time it takes a falling object to drop a given distance varies directly with the square root of the distance traveled. A steel ball takes 2 seconds to drop a distance of 64 feet. How many seconds will it take to drop a distance of 196 feet?

102. If y varies inversely with x, and $y = 8$ when $x = 3$, find the value of y when $x = 48$.

103. *Volume of a Gas* The volume of a gas varies inversely with the pressure of the gas on its container. If a pressure of 24 pounds per square inch corresponds to a volume of 70 cubic inches, what pressure corresponds to a volume of 100 cubic inches?

104. Suppose that y varies directly with x and inversely with the square of z. When $x = 8$ and $z = 4$, then $y = 1$. Find y when $x = 6$ and $z = 3$.

▲ **105.** *Capacity of a Cylinder* The capacity of a cylinder varies directly with the height and the square of the radius. A cylinder with a radius of 3 centimeters and a height of 5 centimeters has a capacity of 50 cubic centimeters. What is the capacity of a cylinder with a height of 9 centimeters and a radius of 4 centimeters?

Remember to use your Chapter Test Prep Video CD to see the worked-out solutions to the test problems you want to review.

Simplify.

1. $(2x^{1/2}y^{1/3})(-3x^{1/3}y^{1/6})$

2. $\dfrac{7x^3}{4x^{3/4}}$

3. $(8x^{1/3})^{3/2}$

4. Evaluate: $\left(\dfrac{4}{9}\right)^{\frac{3}{2}}$

5. Evaluate: $\sqrt[5]{-32}$

Evaluate.

6. $8^{-2/3}$

7. $16^{5/4}$

Simplify. Assume that all variables are nonnegative.

8. $\sqrt{75a^4b^9}$

9. $\sqrt{49a^4b^{10}}$

10. $\sqrt[3]{54m^3n^5}$

Combine like terms where possible.

11. $3\sqrt{24} - \sqrt{18} + \sqrt{50}$

12. $\sqrt{40x} - \sqrt{27x} + 2\sqrt{12x}$

Multiply and simplify.

13. $\left(-3\sqrt{2y}\right)\left(5\sqrt{10xy}\right)$

14. $2\sqrt{3}\left(3\sqrt{6} - 5\sqrt{2}\right)$

15. $\left(5\sqrt{3} - \sqrt{6}\right)\left(2\sqrt{3} + 3\sqrt{6}\right)$

Rationalize the denominator.

16. $\dfrac{30}{\sqrt{5x}}$

17. $\sqrt{\dfrac{xy}{3}}$

18. $\dfrac{1 + 2\sqrt{3}}{3 - \sqrt{3}}$

1. _____

2. _____

3. _____

4. _____

5. _____

6. _____

7. _____

8. _____

9. _____

10. _____

11. _____

12. _____

13. _____

14. _____

15. _____

16. _____

17. _____

18. _____

19. _____

Solve and check your solution(s).

19. $\sqrt{3x - 2} = x$

20. _____

20. $5 + \sqrt{x + 15} = x$

21. _____

21. $5 - \sqrt{x - 2} = \sqrt{x + 3}$

Simplify by using the properties of complex numbers.

22. _____

22. $(8 + 2i) - 3(2 - 4i)$

23. _____

23. $i^{18} + \sqrt{-16}$

24. $(3 - 2i)(4 + 3i)$

24. _____

25. $\dfrac{2 + 5i}{1 - 3i}$

25. _____

26. $(6 + 3i)^2$

26. _____

27. i^{43}

27. _____

28. If y varies inversely with x, and $y = 9$ when $x = 2$, find the value of y when $x = 6$.

28. _____

29. Suppose y varies directly with x and inversely with the square of z. When $x = 8$ and $z = 4$, then $y = 3$. Find y when $x = 5$ and $z = 6$.

29. _____

30. A car's stopping distance varies directly with the square of its speed. A car traveling on pavement can stop in 30 feet when traveling at 30 miles per hour. What distance will it take the car to stop if it is traveling at 50 miles per hour?

30. _____

Approximately one half of this test covers the content of Chapters 1–6. The remainder covers the content of Chapter 7.

1. Identify what property of real numbers is illustrated by the equation $7 + (2 + 3) = (7 + 2) + 3$.

2. Remove the parentheses and collect like terms: $-3a(2ab - a^3) + b(ab^2 + 4a^2)$

3. Simplify. $7(12 - 14)^3 - 7 + 3 \div (-3)$

4. Solve for x. $y = -\dfrac{3}{4}x + 2$

5. Graph $3x - 5y = 15$.

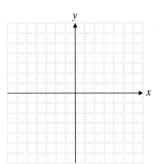

6. Factor completely $16x^2 + 24x - 16$.

7. Solve the system for x, y, and z.

$$x + 4y - z = 10$$
$$3x + 2y + z = 4$$
$$2x - 3y + 2z = -7$$

8. Combine $\dfrac{7x}{x^2 - 2x - 15} - \dfrac{2}{x - 5}$.

▲ **9.** The length of a rectangle is 3 meters longer than twice its width. The perimeter of the rectangle is 48 meters. Find the dimensions of the rectangle.

10. Solve for a: $2ax - 3b = y - 5ax$

Simplify.

11. $\dfrac{2x^{-3}y^{-4}}{4x^{-5/2}y^{7/2}}$

12. $(3x^{-1/2}y^2)^{-1/3}$

1. _____

2. _____

3. _____

4. _____

5. _____

6. _____

7. _____

8. _____

9. _____

10. _____

11. _____

12. _____

419

13. _____

13. Evaluate $64^{-1/3}$.

14. _____

14. Simplify $\sqrt[3]{40x^5y^9}$.

15. _____

15. Combine like terms. $\sqrt{80x} + 2\sqrt{45x} - 3\sqrt{20x}$

16. _____

16. Multiply and simplify. $\left(2\sqrt{3} - 5\sqrt{2}\right)\left(\sqrt{3} + 4\sqrt{2}\right)$

17. _____

17. Rationalize the denominator. $\dfrac{\sqrt{3} + 2}{2\sqrt{3} - 5}$

18. _____

18. Simplify $i^{21} + \sqrt{-16} + \sqrt{-49}$.

19. _____

19. Simplify $(3 - 4i)^2$.

20. _____

20. Simplify $\dfrac{1 + 4i}{1 + 3i}$.

Solve for x and check your solutions.

21. _____

21. $x - 3 = \sqrt{3x + 1}$

22. _____

22. $1 + \sqrt{x + 1} = \sqrt{x + 2}$

23. _____

23. If y varies directly with the square of x, and $y = 12$ when $x = 2$, find the value of y if $x = 5$.

24. _____

24. The amount of light provided by a lightbulb varies inversely with the square of the distance from the lightbulb. A lightbulb provides 120 lumens at a distance of 10 feet from the light. How many lumens are provided if the distance from the light is 15 feet?

CHAPTER

8

Eat more fresh vegetables. This seems to be the advice of doctors, dieticians, and leaders in the healthcare industry. In response it seems Americans are doing it. Recently, there has been a significant increase in the number of pounds of fresh vegetables that are consumed by the average American each year. Just how large is this increase? What are the trends for the future? Turn to page 473 and see if you can use mathematics to answer several of these questions.

Quadratic Equations and Inequalities

Student Learning Objectives

After studying this section, you will be able to:

1 Solve quadratic equations by the square root property.

2 Solve quadratic equations by completing the square.

1 Solving Quadratic Equations by the Square Root Property

Recall that an equation written in the form $ax^2 + bx + c = 0$, where a, b, and c are real numbers and $a \neq 0$, is called a **quadratic equation.** Recall also that we call this the **standard form** of a quadratic equation. We have previously solved quadratic equations using the zero-factor property. This has allowed us to factor the left side of an equation such as $x^2 - 7x + 12 = 0$ and obtain $(x - 3)(x - 4) = 0$ and then solve to find that $x = 3$ and $x = 4$. In this chapter we develop new methods of solving quadratic equations.

The first method is often called the **square root property.**

> **THE SQUARE ROOT PROPERTY**
>
> If $x^2 = a$, then $x = \pm\sqrt{a}$ for all real numbers a.

The notation $\pm\sqrt{a}$ is a shorthand way of writing "$+\sqrt{a}$ or $-\sqrt{a}$." The symbol $\pm$ is read "plus or minus." We can justify this property by using the zero-factor property. If we write $x^2 = a$ in the form $x^2 - a = 0$, we can factor it to obtain $\left(x + \sqrt{a}\right)\left(x - \sqrt{a}\right) = 0$ and thus, $x = -\sqrt{a}$ or $x = +\sqrt{a}$. This can be written more compactly as $x = \pm\sqrt{a}$.

EXAMPLE 1 Solve and check: $x^2 - 36 = 0$

Solution If we add 36 to each side, we have $x^2 = 36$.

$$x = \pm\sqrt{36}$$
$$x = \pm 6$$

Thus, the two roots are 6 and -6.

Check: $6^2 = 36$ and $(-6)^2 = 36$

Practice Problem 1 Solve and check: $x^2 - 121 = 0$

NOTE TO STUDENT: Fully worked-out solutions to all of the Practice Problems can be found at the back of the text starting at page SP-1

EXAMPLE 2 Solve $x^2 = 48$.

Solution
$$x = \pm\sqrt{48} = \pm\sqrt{16 \cdot 3}$$
$$x = \pm 4\sqrt{3}$$

The roots are $4\sqrt{3}$ and $-4\sqrt{3}$.

Practice Problem 2 Solve $x^2 = 18$.

EXAMPLE 3 Solve and check: $3x^2 + 2 = 77$

Solution
$$3x^2 + 2 = 77$$
$$3x^2 = 75$$
$$x^2 = 25$$
$$x = \pm\sqrt{25}$$
$$x = \pm 5$$

The roots are 5 and -5.

Check:

$$3(5)^2 + 2 \overset{?}{=} 77 \qquad\qquad 3(-5)^2 + 2 \overset{?}{=} 77$$
$$3(25) + 2 \overset{?}{=} 77 \qquad\qquad 3(25) + 2 \overset{?}{=} 77$$
$$75 + 2 \overset{?}{=} 77 \qquad\qquad 75 + 2 \overset{?}{=} 77$$
$$77 = 77 \;\checkmark \qquad\qquad\qquad 77 = 77 \;\checkmark$$

Practice Problem 3 Solve $5x^2 + 1 = 46$. ▪

Sometimes we obtain roots that are complex numbers.

EXAMPLE 4 Solve and check: $4x^2 = -16$

Solution

$$x^2 = -4$$
$$x = \pm\sqrt{-4}$$
$$x = \pm 2i \qquad \text{Simplify using } \sqrt{-1} = i.$$

The roots are $2i$ and $-2i$.

Check:

$$4(2i)^2 \overset{?}{=} -16 \qquad\qquad 4(-2i)^2 \overset{?}{=} -16$$
$$4(4i^2) \overset{?}{=} -16 \qquad\qquad 4(4i^2) \overset{?}{=} -16$$
$$4(-4) \overset{?}{=} -16 \qquad\qquad 4(-4) \overset{?}{=} -16$$
$$-16 = -16 \;\checkmark \qquad\qquad -16 = -16 \;\checkmark$$

Practice Problem 4 Solve and check: $3x^2 = -27$ ▪

EXAMPLE 5 Solve $(4x - 1)^2 = 5$.

Solution

$$4x - 1 = \pm\sqrt{5}$$
$$4x = 1 \pm \sqrt{5}$$
$$x = \frac{1 \pm \sqrt{5}}{4}$$

The roots are $\dfrac{1 + \sqrt{5}}{4}$ and $\dfrac{1 - \sqrt{5}}{4}$.

Practice Problem 5 Solve $(2x + 3)^2 = 7$. ▪

② Solving Quadratic Equations by Completing the Square

Often, a quadratic equation cannot be factored (or it may be difficult to factor). So we use another method of solving the equation, called **completing the square.** When we complete the square, we are changing the polynomial to a perfect square trinomial. The form of the equation then becomes $(x + d)^2 = e$.

We already know that

$$(x + d)^2 = x^2 + 2dx + d^2 .$$

Notice three things about the quadratic equation on the right-hand side.

1. The coefficient of the quadratic term (x^2) is 1.
2. The coefficient of the linear (x) term is $2d$.
3. The constant term (d^2) is the square of *half* the coefficient of the linear term.

For example, in the trinomial $x^2 + 6x + 9$, the coefficient of the linear term is 6 and the constant term is $\left(\dfrac{6}{2}\right)^2 = (3)^2 = 9$.

For the trinomial $x^2 - 10x + 25$, the coefficient of the linear term is -10 and the constant term is $\left(\dfrac{-10}{2}\right)^2 = (-5)^2 = 25$.

What number n makes the trinomial $x^2 + 12x + n$ a perfect square?

$$n = \left(\dfrac{12}{2}\right)^2 = 6^2 = 36$$

Hence, the trinomial $x^2 + 12x + 36$ is a perfect square trinomial and can be written as $(x + 6)^2$.

Now let's solve some equations.

EXAMPLE 6   Solve by completing the square and check: $x^2 + 6x + 1 = 0$

Solution

Step 1   First we rewrite the equation in the form $ax^2 + bx = c$ by adding -1 to each side of the equation. Thus, we obtain

$$x^2 + 6x = -1.$$

Step 2   We want to complete the square of $x^2 + 6x$. That is, we want to add a constant term to $x^2 + 6x$ so that we get a perfect square trinomial. We do this by taking half the coefficient of x and squaring it.

$$\left(\dfrac{6}{2}\right)^2 = 3^2 = 9$$

Adding 9 to $x^2 + 6x$ gives the perfect square trinomial $x^2 + 6x + 9$, which we factor to $(x + 3)^2$. But we cannot just add 9 to the left side of our equation unless we also add 9 to the right side. (Why?) We now have

$$x^2 + 6x + 9 = -1 + 9.$$

Step 3   Now we factor.

$$(x + 3)^2 = 8$$

Step 4   We now use the square root property.

$$(x + 3) = \pm\sqrt{8}$$
$$x + 3 = \pm 2\sqrt{2}$$

Step 5   Next we solve for x by adding -3 to each side of the equation.

$$x = -3 \pm 2\sqrt{2}$$

The roots are $-3 + 2\sqrt{2}$ and $-3 - 2\sqrt{2}$.

Step 6   We *must* check our solution in the *original* equation (not the perfect square trinomial we constructed).

$x^2 + 6x + 1 = 0$
$(-3 + 2\sqrt{2})^2 + 6(-3 + 2\sqrt{2}) + 1 \stackrel{?}{=} 0$
$9 - 12\sqrt{2} + 8 - 18 + 12\sqrt{2} + 1 \stackrel{?}{=} 0$
$18 - 18 - 12\sqrt{2} + 12\sqrt{2} \stackrel{?}{=} 0$
$0 = 0$ ✓

$x^2 + 6x + 1 = 0$
$(-3 - 2\sqrt{2})^2 + 6(-3 - 2\sqrt{2}) + 1 \stackrel{?}{=} 0$
$9 + 12\sqrt{2} + 8 - 18 - 12\sqrt{2} + 1 \stackrel{?}{=} 0$
$18 - 18 + 12\sqrt{2} - 12\sqrt{2} \stackrel{?}{=} 0$
$0 = 0$ ✓

Practice Problem 6   Solve by completing the square: $x^2 + 8x + 3 = 0$

Let us summarize for future reference the six steps we have performed to solve a quadratic equation by completing the square.

COMPLETING THE SQUARE

1. Put the equation in the form $ax^2 + bx = -c$.
2. If $a \neq 1$, divide each term of the equation by a.
3. Square half of the numerical coefficient of the linear term. Add the result to both sides of the equation.
4. Factor the left side; then take the square root of both sides of the equation.
5. Solve each resulting equation for x.
6. Check the solutions in the original equation.

EXAMPLE 7   Solve by completing the square: $3x^2 - 8x + 1 = 0$

Solution $3x^2 - 8x = -1$ Add -1 to each side.

$\dfrac{3x^2}{3} - \dfrac{8x}{3} = -\dfrac{1}{3}$ Divide each term by 3. (Remember that the coefficient of the quadratic term must be 1.)

$x^2 - \dfrac{8}{3}x + \dfrac{16}{9} = -\dfrac{1}{3} + \dfrac{16}{9}$

$\left(x - \dfrac{4}{3}\right)^2 = \dfrac{13}{9}$

$x - \dfrac{4}{3} = \pm\sqrt{\dfrac{13}{9}}$

$x - \dfrac{4}{3} = \pm\dfrac{\sqrt{13}}{3}$

$x = \dfrac{4}{3} \pm \dfrac{\sqrt{13}}{3}$

$x = \dfrac{4 \pm \sqrt{13}}{3}$

Check: For $x = \dfrac{4 + \sqrt{13}}{3}$,

$3\left(\dfrac{4 + \sqrt{13}}{3}\right)^2 - 8\left(\dfrac{4 + \sqrt{13}}{3}\right) + 1 \stackrel{?}{=} 0$

$\dfrac{16 + 8\sqrt{13} + 13}{3} - \dfrac{32 + 8\sqrt{13}}{3} + 1 \stackrel{?}{=} 0$

$\dfrac{16 + 8\sqrt{13} + 13 - 32 - 8\sqrt{13}}{3} + 1 \stackrel{?}{=} 0$

$\dfrac{29 - 32}{3} + 1 \stackrel{?}{=} 0$

$-\dfrac{3}{3} + 1 \stackrel{?}{=} 0$

$-1 + 1 = 0$ ✓

See whether you can check the solution $\dfrac{4 - \sqrt{13}}{3}$.

Practice Problem 7   Solve by completing the square: $2x^2 + 4x + 1 = 0$

NOTE TO STUDENT: Fully worked-out solutions to all of the Practice Problems can be found at the back of the text starting at page SP-1

Solve the equations by using the square root property. Express any complex numbers using i notation.

1. $x^2 = 100$

2. $x^2 = 49$

3. $3x^2 - 45 = 0$

4. $4x^2 - 68 = 0$

5. $2x^2 - 80 = 0$

6. $5x^2 - 40 = 0$

7. $x^2 = -81$

8. $x^2 = -64$

9. $x^2 + 81 = 0$

10. $x^2 + 144 = 0$

11. $(x - 3)^2 = 12$

12. $(x + 2)^2 = 18$

13. $(x + 9)^2 = 21$

14. $(x - 8)^2 = 23$

15. $(2x + 1)^2 = 7$

16. $(3x + 2)^2 = 5$

17. $(4x - 3)^2 = 36$

18. $(5x - 2)^2 = 25$

19. $(2x + 5)^2 = 49$

20. $(3x + 7)^2 = 81$

21. $3x^2 - 5 = 0$

22. $7x^2 - 2 = 0$

Solve the equations by completing the square. Simplify your answers. Express any complex numbers using i notation.

23. $x^2 + 10x + 5 = 0$

24. $x^2 + 6x + 2 = 0$

25. $x^2 - 8x = 17$

26. $x^2 - 12x = 4$

27. $x^2 - 14x = -48$

28. $x^2 - 18x = -56$

29. $\dfrac{x^2}{2} + \dfrac{5}{2}x = 2$

30. $\dfrac{x^2}{3} - \dfrac{x}{3} = 3$

31. $2y^2 + 10y = -11$

32. $7x^2 + 4x - 5 = 0$

33. $3x^2 + 10x - 2 = 0$

34. $5x^2 + 4x - 3 = 0$

35. $2y^2 - y = 6$

36. $2y^2 - y = 15$

Mixed Practice

Solve the equations by any method. Simplify your answers. Express any complex numbers using i notation.

37. $x^2 + 2x - 5 = 0$

38. $x^2 + 6x - 5 = 0$

39. $\dfrac{x^2}{2} - x = 4$

40. $\dfrac{x^2}{3} - 2x = -3$

41. $3x^2 + 1 = x$

42. $2x^2 + 5 = -3x$

43. $x^2 + 1 = x$

44. $x^2 - 2x = -7$

45. $2x^2 + 2 = 3x$

46. $3x^2 + 8x + 3 = 2$

47. Check the solution $x = -1 + \sqrt{6}$ in the equation $x^2 + 2x - 5 = 0$.

48. Check the solution $x = 2 + \sqrt{3}$ in the equation $x^2 - 4x + 1 = 0$.

Applications

Volume of a Box   *The sides of the box shown are labeled with the dimensions in feet.*

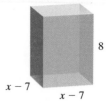

▲ **49.** What is the value of x if the volume of the box is 648 cubic feet?

▲ **50.** What is the value of x if the volume of the box is 1800 cubic feet?

8

$x - 7$

$x - 7$

Basketball   *The time a basketball player spends in the air when shooting a basket is called "the hang time." The vertical leap L measured in feet is related to the hang time t measured in seconds by the equation $L = 4t^2$.*

51. During his career as a Boston Celtics player, Larry Bird often displayed a leap of 3.1 feet. Find the hang time for that leap.

52. Shaquille O'Neal of the Los Angeles Lakers has often shown a vertical leap of 3.3 feet. Find the hang time for that leap.

Time for an Object to Fall   *The formula $D = 16t^2$ is used to approximate the distance in feet that an object falls in t seconds.*

53. A parachutist jumps from an airplane, falls 3600 feet, and then opens her parachute. For how many seconds was the parachutist falling before she opened the parachute?

54. How long would it take an object to fall to the ground from a helicopter hovering at 1936 feet above the ground?

Cumulative Review

Evaluate the expressions for the given values.

55. $\sqrt{b^2 - 4ac}$; $b = 4, a = 3, c = -4$

56. $\sqrt{b^2 - 4ac}$; $b = -5, a = 2, c = -3$

57. $5x^2 - 6x + 8$; $x = -2$

58. $2x^2 + 3x - 5$; $x = -3$

1 Solving a Quadratic Equation by Using the Quadratic Formula

The last method we'll study for solving quadratic equations is the **quadratic formula.** This method works for *any* quadratic equation.

The quadratic formula is developed from completing the square. We begin with the **standard form** of the quadratic equation.

$$ax^2 + bx + c = 0$$

To complete the square, we want the equation to be in the form $x^2 + dx = e$. Thus, we divide by a.

$$\frac{ax^2}{a} + \frac{b}{a}x + \frac{c}{a} = 0$$

$$x^2 + \frac{b}{a}x = -\frac{c}{a}$$

Now we complete the square by adding $\left(\frac{b}{2a}\right)^2$ to each side.

$$x^2 + \frac{b}{a}x + \left(\frac{b}{2a}\right)^2 = -\frac{c}{a} + \left(\frac{b}{2a}\right)^2$$

We factor the left side and write the right side as one fraction.

$$\left(x + \frac{b}{2a}\right)^2 = \frac{b^2 - 4ac}{4a^2}$$

Now we use the square root property.

$$x + \frac{b}{2a} = \pm\sqrt{\frac{b^2 - 4ac}{4a^2}}$$

We solve for x and simplify.

$$x = -\frac{b}{2a} \pm \sqrt{\frac{b^2 - 4ac}{4a^2}}$$

$$x = \frac{-b \pm \sqrt{b^2 - 4ac}}{2a}$$

This is the quadratic formula.

QUADRATIC FORMULA

For all equations $ax^2 + bx + c = 0$,

$$x = \frac{-b \pm \sqrt{b^2 - 4ac}}{2a}, \qquad \text{where } a \neq 0.$$

Student Learning Objectives

After studying this section, you will be able to:

1. Solve a quadratic equation by using the quadratic formula.

2. Use the discriminant to determine the nature of the roots of a quadratic equation.

3. Write a quadratic equation given the solutions of the equation.

EXAMPLE 1 Solve by using the quadratic formula: $x^2 + 8x = -3$

Solution The standard form is $x^2 + 8x + 3 = 0$. We substitute $a = 1$, $b = 8$, and $c = 3$.

$$x = \frac{-b \pm \sqrt{b^2 - 4ac}}{2a}$$

$$x = \frac{-8 \pm \sqrt{8^2 - 4(1)(3)}}{2(1)}$$

$$x = \frac{-8 \pm \sqrt{64 - 12}}{2} = \frac{-8 \pm \sqrt{52}}{2} = \frac{-8 \pm \sqrt{4}\sqrt{13}}{2}$$

$$x = \frac{-8 \pm 2\sqrt{13}}{2} = \frac{\cancel{2}(-4 \pm \sqrt{13})}{\cancel{2}}$$

$$x = -4 \pm \sqrt{13}$$

NOTE TO STUDENT: Fully worked-out solutions to all of the Practice Problems can be found at the back of the text starting at page SP-1

Practice Problem 1 Solve by using the quadratic formula:
$$x^2 + 5x = -1 + 2x$$

EXAMPLE 2 Solve by using the quadratic formula: $3x^2 - x - 2 = 0$

Solution Here $a = 3$, $b = -1$, and $c = -2$.

$$x = \frac{-b \pm \sqrt{b^2 - 4ac}}{2a}$$

$$x = \frac{-(-1) \pm \sqrt{(-1)^2 - 4(3)(-2)}}{2(3)}$$

$$x = \frac{1 \pm \sqrt{1 + 24}}{6} = \frac{1 \pm \sqrt{25}}{6}$$

$$x = \frac{1 + 5}{6} = \frac{6}{6} \quad \text{or} \quad x = \frac{1 - 5}{6} = -\frac{4}{6}$$

$$x = 1 \qquad\qquad\qquad x = -\frac{2}{3}$$

Practice Problem 2 Solve by using the quadratic formula:
$$2x^2 + 7x + 6 = 0$$

EXAMPLE 3 Solve by using the quadratic formula: $2x^2 - 48 = 0$

Solution This equation is equivalent to $2x^2 - 0x - 48 = 0$. Therefore, we know that $a = 2$, $b = 0$, and $c = -48$.

$$x = \frac{-b \pm \sqrt{b^2 - 4ac}}{2a}$$

$$x = \frac{-0 \pm \sqrt{(0)^2 - 4(2)(-48)}}{2(2)}$$

$$x = \frac{\pm\sqrt{384}}{4} \qquad \text{but this is not simplified}$$

$$x = \frac{\pm\sqrt{64}\sqrt{6}}{4} = \frac{\pm 8\sqrt{6}}{4}$$

$$x = \pm 2\sqrt{6}$$

Practice Problem 3   Solve by using the quadratic formula: $2x^2 - 26 = 0$

EXAMPLE 4   A small company that manufactures canoes makes a daily profit p according to the equation $p = -100x^2 + 3400x - 26{,}196$, where p is measured in dollars and x is the number of canoes made per day. Find the number of canoes that must be made each day to produce a zero profit for the company. Round your answer to the nearest whole number.

Solution   Since $p = 0$, we are solving the equation
$0 = -100x^2 + 3400x - 26{,}196$.
    In this case we have $a = -100$, $b = 3400$, and $c = -26{,}196$.
    Now we substitute these into the quadratic formula.

$$x = \frac{-b \pm \sqrt{b^2 - 4ac}}{2a}$$

$$x = \frac{-3400 \pm \sqrt{(3400)^2 - 4(-100)(-26{,}196)}}{2(-100)}$$

We will use a calculator to assist us with computation in this problem.

$$x = \frac{-3400 \pm \sqrt{11{,}560{,}000 - 10{,}478{,}400}}{-200}$$

$$x = \frac{-3400 \pm \sqrt{1{,}081{,}600}}{-200}$$

$$x = \frac{-3400 \pm 1040}{-200}$$

We now obtain two answers.

$$x = \frac{-3400 + 1040}{-200} = \frac{-2360}{-200} = 11.8 \approx 12$$

$$x = \frac{-3400 - 1040}{-200} = \frac{-4440}{-200} = 22.2 \approx 22$$

A zero profit is obtained when approximately 12 canoes are produced or when approximately 22 canoes are produced. Actually a slight profit of $204 is made when these numbers of canoes are produced. The discrepancy is due to the round-off error that occurs when we approximate. By methods that we will learn later in this chapter, the maximum profit is produced when seventeen canoes are made at the factory. We will investigate exercises of this kind later.

Practice Problem 4   A company that manufactures modems makes a daily profit p according to the equation $p = -100x^2 + 4800x - 52{,}559$, where p is measured in dollars and x is the number of modems made per day. Find the number of modems that must be made each day to produce a zero profit for the company. Round your answer to the nearest whole number.

    When a quadratic equation contains fractions, eliminate them by multiplying each term by the LCD. Then rewrite the equation in standard form before using the quadratic formula.

Chapter 8 Quadratic Equations and Inequalities

EXAMPLE 5 Solve by using the quadratic formula: $\dfrac{2x}{x+2} = 1 - \dfrac{3}{x+4}$

Solution The LCD is $(x+2)(x+4)$.

$$\frac{2}{x+2} = 1 - \frac{3}{x+4}$$

$$\frac{2x}{x+2}(x+2)(x+4) = 1(x+2)(x+4) - \frac{3}{x+4}(x+2)(x+4)$$

$$2x(x+4) = (x+2)(x+4) - 3(x+2)$$

$$2x^2 + 8x = x^2 + 6x + 8 - 3x - 6 \quad \text{Now we have an equation that is quadratic.}$$

$$2x^2 + 8x = x^2 + 3x + 2$$

$$x^2 + 5x - 2 = 0$$

Now the equation is in standard form, and we can use the quadratic formula with $a=1, b=5$, and $c=-2$.

$$x = \frac{-5 \pm \sqrt{5^2 - 4(1)(-2)}}{2(1)} = \frac{-5 \pm \sqrt{25+8}}{2}$$

$$x = \frac{-5 \pm \sqrt{33}}{2}$$

NOTE TO STUDENT: Fully worked-out solutions to all of the Practice Problems can be found at the back of the text starting at page SP-1

Practice Problem 5 Solve by using the quadratic formula: $\dfrac{1}{x} + \dfrac{1}{x-1} = \dfrac{5}{6}$

Some quadratic equations will have solutions that are not real numbers. You should use i notation to simplify the solutions of nonreal complex numbers.

EXAMPLE 6 Solve and simplify your answer: $8x^2 - 4x + 1 = 0$

Solution $a=8, b=-4$, and $c=1$.

$$x = \frac{-(-4) \pm \sqrt{(-4)^2 - 4(8)(1)}}{2(8)}$$

$$x = \frac{4 \pm \sqrt{16-32}}{16} = \frac{4 \pm \sqrt{-16}}{16}$$

$$x = \frac{4 \pm 4i}{16} = \frac{4(1 \pm i)}{16} = \frac{1 \pm i}{4}$$

Practice Problem 6 Solve by using the quadratic formula:

$$2x^2 - 4x + 5 = 0$$

You may have noticed that complex roots come in pairs. In other words, if $a + bi$ is a solution of a quadratic equation, its conjugate $a - bi$ is also a solution.

② Using the Discriminant to Determine the Nature of the Roots of a Quadratic Equation

So far we have used the quadratic formula to solve quadratic equations that had two real roots. Sometimes the roots were rational, and sometimes they were irrational. We have also solved equations like Example 6 with nonreal complex numbers. Such solutions occur when the expression $b^2 - 4ac$, the radicand in the quadratic formula, is negative.

$$x = \frac{-b \pm \sqrt{b^2 - 4ac}}{2a},$$

The expression $b^2 - 4ac$ is called the **discriminant.** Depending on the value of the discriminant and whether the discriminant is positive, zero, or negative, the roots of the quadratic equation will be rational, irrational, or complex. We summarize the types of solutions in the following table.

If the Discriminant $b^2 - 4ac$ is:	Then the Quadratic Equation $ax^2 + bx + c = 0$, where a, b, and c are integers, will have:
A positive number that is also a perfect square	Two different rational solutions (Such an equation can always be factored.)
A positive number that is not a perfect square	Two different irrational solutions
Zero	One rational solution
Negative	Two complex solutions containing i (They will be complex conjugates.)

EXAMPLE 7   What type of solutions does the equation $2x^2 - 9x - 35 = 0$ have? Do not solve the equation.

Solution $a = 2, b = -9$, and $c = -35$. Thus,

$$b^2 - 4ac = (-9)^2 - 4(2)(-35) = 361.$$

Since the discriminant is positive, the equation has two real roots.
     Since $(19)^2 = 361$, 361 is a perfect square. Thus, the equation has two different rational solutions. This type of quadratic equation can always be factored.

Practice Problem 7   Use the discriminant to find what type of solutions the equation $9x^2 + 12x + 4 = 0$ has. Do not solve the equation.

EXAMPLE 8   Use the discriminant to determine the type of solutions each of the following equations has.

(a) $3x^2 - 4x + 2 = 0$ **(b)** $5x^2 - 3x - 5 = 0$

Solution
(a) Here $a = 3, b = -4$, and $c = 2$. Thus,

$$b^2 - 4ac = (-4)^2 - 4(3)(2)$$
$$= 16 - 24 = -8$$

Since the discriminant is negative, the equation will have two complex solutions containing i.

(b) Here $a = 5, b = -3$, and $c = -5$. Thus,

$$b^2 - 4ac = (-3)^2 - 4(5)(-5)$$
$$= 9 + 100 = 109$$

Since this positive number is not a perfect square, the equation will have two different irrational solutions.

NOTE TO STUDENT: Fully worked-out solutions to all of the Practice Problems can be found at the back of the text starting at page SP-1

Practice Problem 8 Use the discriminant to determine the type of solutions each of the following equations has.

(a) $x^2 - 4x + 13 = 0$

(b) $9x^2 + 6x + 7 = 0$

3 Writing a Quadratic Equation Given the Solutions of the Equation

By using the zero-product rule in reverse, we can find a quadratic equation that contains two given solutions. To illustrate, if 3 and 7 are the two solutions, then we could write the equation $(x - 3)(x - 7) = 0$, and therefore, a quadratic equation that has these two solutions is $x^2 - 10x + 21 = 0$. This answer is not unique. Any constant multiple of $x^2 - 10x + 21 = 0$ would also have roots of 3 and 7. Thus, $2x^2 - 20x + 42 = 0$ also has roots of 3 and 7.

EXAMPLE 9 Find a quadratic equation whose roots are 5 and -2.

Solution

$$x = 5 \qquad\qquad x = -2$$
$$x - 5 = 0 \qquad\qquad x + 2 = 0$$
$$(x - 5)(x + 2) = 0$$
$$x^2 - 3x - 10 = 0$$

Practice Problem 9 Find a quadratic equation whose roots are -10 and -6.

EXAMPLE 10 Find a quadratic equation whose solutions are $3i$ and $-3i$.

Solution First we write the two equations.

$$x - 3i = 0 \quad \text{and} \quad x + 3i = 0$$
$$(x - 3i)(x + 3i) = 0$$
$$x^2 + 3ix - 3ix - 9i^2 = 0$$
$$x^2 - 9(-1) = 0 \quad \text{Use } i^2 = -1.$$
$$x^2 + 9 = 0$$

Practice Problem 10 Find a quadratic equation whose solutions are $2i\sqrt{3}$ and $-2i\sqrt{3}$.

Verbal and Writing Skills

1. How is the quadratic formula used to solve a quadratic equation?

2. The discriminant in the quadratic formula is the expression _____.

3. If the discriminant in the quadratic formula is zero, then the quadratic equation will have _____ solution(s).

4. If the discriminant in the quadratic formula is a perfect square, then the quadratic equation will have _____ solution(s).

Solve by the quadratic formula. Simplify your answers.

5. $x^2 + x - 5 = 0$

6. $x^2 - 3x - 1 = 0$

7. $2x^2 + x - 4 = 0$

8. $5x^2 - x - 1 = 0$

9. $x^2 = \frac{2}{3}x$

10. $\frac{4}{5}x^2 = x$

11. $3x^2 - x - 2 = 0$

12. $7x^2 + 4x - 3 = 0$

13. $4x^2 + 3x - 2 = 0$

14. $6x^2 - 2x - 1 = 0$

15. $3x^2 + 1 = 8$

16. $5x^2 - 1 = 5$

Simplify each equation. Then solve by the quadratic formula. Simplify your answers using i notation for nonreal complex numbers.

17. $2x(x + 3) - 3 = 4x - 2$

18. $5 + 3x(x - 2) = 4$

19. $x(x + 3) - 2 = 3x + 7$

20. $3(x^2 - 12) - 2x = 2x(x - 1)$

21. $(x - 2)(x + 1) = \dfrac{2x + 3}{2}$

22. $3x(x + 1) = \dfrac{7x + 1}{3}$

23. $\dfrac{1}{x + 2} + \dfrac{1}{x} = \dfrac{1}{3}$

24. $\dfrac{1}{y} - y = \dfrac{5}{3}$

25. $\dfrac{1}{15} + \dfrac{3}{y} = \dfrac{4}{y + 1}$

26. $\dfrac{1}{4} + \dfrac{6}{y + 2} = \dfrac{6}{y}$

27. $x(x + 4) = -12$

28. $x^2 = 2(x - 4)$

29. $2x^2 + 15 = 0$

30. $5x^2 = -3$

31. $3x^2 - 8x + 7 = 0$

32. $3x^2 - 4x + 6 = 0$

*Use the discriminant to find what type of solutions (two rational, two irrational, one rational, or two nonreal complex)
each of the following equations has. Do not solve the equation.*

33. $3x^2 + 4x = 2$

34. $4x^2 - 20x + 25 = 0$

35. $2x^2 + 10x + 8 = 0$

36. $2x^2 - 7x - 4 = 0$

37. $9x^2 + 4 = 12x$

38. $5x^2 - 8x - 2 = 0$

39. $7x(x - 1) + 15 = 10$

40. $x^2 - 3(x - 8) = 2x$

Write a quadratic equation having the given solutions.

41. $13, -2$

42. $5, -11$

43. $-5, -12$

44. $-6, -10$

45. $4i, -4i$

46. $6i, -6i$

47. $3, -\dfrac{5}{2}$

48. $-2, \dfrac{5}{6}$

Solve for x by using the quadratic formula. Approximate your answers to four decimal places.

49. $3x^2 + 5x - 9 = 0$

50. $1.2x^2 - 12.3x - 4.2 = 0$

51. $0.162x^2 + 0.094x - 0.485 = 0$

52. $20.6x^2 - 73.4x + 41.8 = 0$

Applications

53. ***Business Profit*** A company that manufactures mountain bikes makes a daily profit p according to the equation $p = -100x^2 + 4800x - 54{,}351$, where p is measured in dollars and x is the number of mountain bikes made per day. Find the number of mountain bikes that must be made each day to produce a zero profit for the company. Round your answer to the nearest whole number.

54. ***Business Profit*** A company that manufactures sport parachutes makes a daily profit p according to the equation $p = -100x^2 + 4200x - 39{,}476$, where p is measured in dollars and x is the number of parachutes made per day. Find the number of parachutes that must be made each day to produce a zero profit for the company. Round your answer to the nearest whole number.

To Think About

55. ***Business Profit*** The company described in Exercise 53 earns a maximum profit when $x = 24$. What profit does the company make per day if it produces twenty-four mountain bikes? Speculate how you could have predicted that the maximum profit occurs when $x = 24$ based on the answers you obtained in Exercise 53.

56. ***Business Profit*** The company described in Exercise 54 earns a maximum profit when $x = 21$. What profit does the company make per day if it produces twenty-one parachutes? Speculate how you could have predicted that the maximum profit occurs when $x = 21$ based on the answers you obtained in the Exercise 54.

Cumulative Review

Simplify.

57. $9x^2 - 6x + 3 - 4x - 12x^2 + 8$

58. $3y(2 - y) + \dfrac{1}{5}(10y^2 - 15y)$

▲ **59.** ***Retail Management*** Music Galaxy sells compact discs, cassettes, and everything else you could possibly want from a music supply superstore. The management plans to expand its compact disc section. Presently, it takes 50 feet of an inner security fence to enclose the rectangular section. The expansion plans call for tripling the width and doubling the length. The new CD section will need 118 feet of inner security fencing. What is the length and width of the current compact disc section?

60. ***Mountain Bike Gear*** Last year, Cecile, a professional mountain bike racer, purchased three new padded riding suits to protect her from injury and compress her muscles while riding. In addition, she purchased two pairs of racing goggles. The cost for these items was $343. This year, suits cost $10 more and goggles cost $5 more than last year. This year she purchased two new suits and three pairs of goggles for $312. How much did each suit cost last year? How much did each pair of goggles cost last year?

Student Learning Objectives

After studying this section, you will be able to:

1 Solve equations of degree greater than 2 that can be transformed into quadratic form.

2 Solve equations with fractional exponents that can be transformed into quadratic form.

1 Solving Equations of Degree Greater than 2

Some higher-order equations can be solved by writing them in the form of a quadratic equation. An equation is **quadratic in form** if we can substitute a linear term for the variable raised to the lowest power and get an equation of the form $ay^2 + by + c = 0$.

EXAMPLE 1   Solve $x^4 - 13x^2 + 36 = 0$.

Solution   Let $y = x^2$. Then $y^2 = x^4$. Thus, we obtain a new equation and solve it as follows:

$$y^2 - 13y + 36 = 0 \qquad \text{Replace } x^2 \text{ by } y \text{ and } x^4 \text{ by } y^2.$$
$$(y - 4)(y - 9) = 0 \qquad \text{Factor.}$$
$$y - 4 = 0 \quad \text{or} \quad y - 9 = 0 \qquad \text{Solve for } y.$$
$$y = 4 \qquad\qquad y = 9 \qquad \text{These are } not \text{ the roots to the original equation. We must replace } y \text{ by } x^2.$$
$$x^2 = 4 \qquad\qquad x^2 = 9$$
$$x = \pm\sqrt{4} \qquad\qquad x = \pm\sqrt{9}$$
$$x = \pm 2 \qquad\qquad x = \pm 3$$

Thus, there are *four* solutions to the original equation: $x = +2$, $x = -2$, $x = +3$, and $x = -3$. Check these values to verify that they are solutions.

Practice Problem 1   Solve $x^4 - 5x^2 - 36 = 0$.

NOTE TO STUDENT: Fully worked-out solutions to all of the Practice Problems can be found at the back of the text starting at page SP-1

EXAMPLE 2   Solve for all real roots: $2x^6 - x^3 - 6 = 0$

Solution   Let $y = x^3$. Then $y^2 = x^6$. Thus, we have the following:

$$2y^2 - y - 6 = 0 \qquad \text{Replace } x^3 \text{ by } y \text{ and } x^6 \text{ by } y^2.$$
$$(2y + 3)(y - 2) = 0 \qquad \text{Factor.}$$
$$2y + 3 = 0 \quad \text{or} \quad y - 2 = 0 \qquad \text{Solve for } y.$$
$$y = -\frac{3}{2} \qquad\qquad y = 2$$
$$x^3 = -\frac{3}{2} \quad \text{or} \quad x^3 = 2 \qquad \text{Replace } y \text{ by } x^3.$$
$$x = \sqrt[3]{-\frac{3}{2}} \qquad x = \sqrt[3]{2} \qquad \text{Take the cube root of each side of the equation.}$$
$$x = \frac{\sqrt[3]{-12}}{2} \qquad\qquad \text{Simplify } \sqrt[3]{-\frac{3}{2}} \text{ by rationalizing the denominator.}$$

Check these solutions.

Practice Problem 2   Solve for all real roots: $x^6 - 5x^3 + 4 = 0$

2 Solving Equations with Fractional Exponents

> **EXAMPLE 3**   Solve and check your solutions: $x^{2/3} - 3x^{1/3} + 2 = 0$

Solution   Let $y = x^{1/3}$. Then $y^2 = x^{2/3}$.

$$y^2 - 3y + 2 = 0 \qquad \text{Replace } x^{1/3} \text{ by } y \text{ and } x^{2/3} \text{ by } y^2.$$
$$(y - 2)(y - 1) = 0 \qquad \text{Factor.}$$

$$y - 2 = 0 \quad \text{ or } \quad y - 1 = 0$$
$$y = 2 \qquad\qquad y = 1 \qquad \text{Solve for } y.$$
$$x^{1/3} = 2 \quad \text{ or } \quad x^{1/3} = 1 \qquad \text{Replace } y \text{ by } x^{1/3}.$$
$$(x^{1/3})^3 = (2)^3 \qquad (x^{1/3})^3 = (1)^3 \qquad \text{Cube each side of the equation.}$$
$$x = 8 \qquad\qquad\quad x = 1$$

Check:

$x = 8$: $(8)^{2/3} - 3(8)^{1/3} + 2 \overset{?}{=} 0$ $x = 1$: $(1)^{2/3} - 3(1)^{1/3} + 2 \overset{?}{=} 0$

$\ \ (\sqrt[3]{8})^2 - 3(\sqrt[3]{8}) + 2 \overset{?}{=} 0$ $\ \ (\sqrt[3]{1})^2 - 3(\sqrt[3]{1}) + 2 \overset{?}{=} 0$

$\ (2)^2 - 3(2) + 2 \overset{?}{=} 0$ $\ 1 - 3 + 2 \overset{?}{=} 0$

$\ 4 - 6 + 2 \overset{?}{=} 0$ $\ 0 = 0 \ \checkmark$

$\ 0 = 0 \ \checkmark$

The exercises that appear in this section are somewhat difficult to solve. Part of the difficulty lies in the fact that the equations have different numbers of solutions. A fourth-degree equation like the one in Example 1 has four different solutions. Whereas a sixth-degree equation such as the one in Example 2 has only two solutions, some sixth-degree equations will have as many as six solutions. Although the equation that we examined in Example 3 has only two solutions, other equations with fractional exponents may have one solution or even no solution at all. It is good to take some time to carefully examine your work to determine that you have obtained the correct number of solutions.

A graphing program on a computer such as TI Interactive, Derive, or Maple can be very helpful in determining or verifying the solutions to these types of problems. Of course a graphing calculator can be most helpful, particularly in verifying the value of a solution and the number of solutions.

Optional Graphing Calculation Exploration: If you have a graphing calculator, verify the solutions for Example 3 by graphing the equation

$$y = x^{2/3} - 3x^{1/3} + 2.$$

Determine from your graph whether the curve does in fact cross the *x*-axis (that is, $y = 0$ when $x = 1$ and $x = 8$). You will have to carefully select the window so that you can see the behavior of the curve clearly. For this equation a useful window is $[-1, 12, -1, 2]$. Remember that with most graphing calculators, you will need to surround the exponents with parentheses.

Practice Problem 3   Solve and check your solutions:

$$3x^{4/3} - 5x^{2/3} + 2 = 0$$

EXAMPLE 4 Solve and check your solutions: $2x^{1/2} = 5x^{1/4} + 12$

Solution

$$2x^{1/2} - 5x^{1/4} - 12 = 0 \qquad \text{Place in standard form.}$$
$$2y^2 - 5y - 12 = 0 \qquad \text{Replace } x^{1/4} \text{ by } y \text{ and } x^{1/2} \text{ by } y^2.$$
$$(2y + 3)(y - 4) = 0 \qquad \text{Factor.}$$
$$2y = -3 \quad \text{or} \quad y = 4$$
$$y = -\frac{3}{2} \qquad \text{Solve for } y.$$
$$x^{1/4} = -\frac{3}{2} \quad \text{or} \quad x^{1/4} = 4 \qquad \text{Replace } y \text{ by } x^{1/4}.$$
$$(x^{1/4})^4 = \left(-\frac{3}{2}\right)^4 \qquad (x^{1/4})^4 = (4)^4 \qquad \text{Solve for } x.$$
$$x = \frac{81}{16} \qquad\qquad x = 256$$

Check:

$$x = \frac{81}{16}: \quad 2\left(\frac{81}{16}\right)^{1/2} - 5\left(\frac{81}{16}\right)^{1/4} - 12 \overset{?}{=} 0 \qquad\qquad x = 256: \quad 2(256)^{1/2} - 5(256)^{1/4} - 12 \overset{?}{=} 0$$
$$2\left(\frac{9}{4}\right) - 5\left(\frac{3}{2}\right) - 12 \overset{?}{=} 0 \qquad\qquad\qquad 2(16) - 5(4) - 12 \overset{?}{=} 0$$
$$\frac{9}{2} - \frac{15}{2} - 12 \overset{?}{=} 0 \qquad\qquad\qquad\qquad 32 - 20 - 12 \overset{?}{=} 0$$
$$-15 \neq 0 \qquad\qquad\qquad\qquad\qquad\qquad\qquad 0 = 0 \quad \checkmark$$

$\frac{81}{16}$ is extraneous and not a valid solution. The only valid solution is 256.

Practice Problem 4 Solve and check your solutions: $3x^{1/2} = 8x^{1/4} - 4$

Although we have covered just four basic examples here, this substitution technique can be extended to other types of equations. In each case we substitute y for an appropriate expression in order to obtain a quadratic equation. The following table lists some substitutions that would be appropriate.

If You Want to Solve:	Then You Would Use the Substitution:	Resulting Equation
$x^4 - 13x^2 + 36 = 0$	$y = x^2$	$y^2 + 13y + 36 = 0$
$2x^6 - x^3 - 6 = 0$	$y = x^3$	$2y^2 - y - 6 = 0$
$x^{2/3} - 3x^{1/3} + 2 = 0$	$y = x^{1/3}$	$y^2 - 3y + 2 = 0$
$6(x - 1)^{-2} + (x - 1)^{-1} - 2 = 0$	$y = (x - 1)^{-1}$	$6y^2 + y - 2 = 0$
$(2x^2 + x)^2 + 4(2x^2 + x) + 3 = 0$	$y = 2x^2 + x$	$y^2 + 4y + 3 = 0$
$\left(\dfrac{1}{x - 1}\right)^2 + \dfrac{1}{x - 1} - 6 = 0$	$y = \dfrac{1}{x - 1}$	$y^2 + y - 6 = 0$
$2x - 5x^{1/2} + 2 = 0$	$y = x^{1/2}$	$2y^2 - 5y + 2 = 0$

Exercises like these appear in the exercise set.

Student Solutions Manual   CD/ Video   PH Math Tutor Center   MathXL®Tutorials on CD   MathXL®   MyMathLab®   Interactmath.com

Solve. Express any nonreal complex numbers with i notation.

1. $x^4 - 9x^2 + 20 = 0$

2. $x^4 - 11x^2 + 18 = 0$

3. $x^4 + x^2 - 12 = 0$

4. $x^4 - 2x^2 - 8 = 0$

5. $3x^4 = 10x^2 + 8$

6. $5x^4 = 4x^2 + 1$

In Exercises 7–10, find all valid real roots for each equation.

7. $x^6 - 7x^3 - 8 = 0$

8. $x^6 - 3x^3 - 4 = 0$

9. $x^6 - 3x^3 = 0$

10. $x^6 + 8x^3 = 0$

Solve for real roots.

11. $x^8 = 17x^4 - 16$

12. $x^8 - 6x^4 = 0$

13. $3x^8 + 13x^4 = 10$

14. $3x^8 - 10x^4 = 8$

Solve for real roots.

15. $x^{2/3} + x^{1/3} - 12 = 0$ **16.** $x^{2/3} + 2x^{1/3} - 8 = 0$ **17.** $12x^{2/3} + 5x^{1/3} - 2 = 0$ **18.** $2x^{2/3} - 7x^{1/3} - 4 = 0$

19. $2x^{1/2} - 5x^{1/4} - 3 = 0$ **20.** $3x^{1/2} - 14x^{1/4} - 5 = 0$ **21.** $2x^{1/2} - x^{1/4} - 6 = 0$

22. $2x^{1/2} - x^{1/4} - 1 = 0$ **23.** $x^{2/5} + x^{1/5} - 2 = 0$ **24.** $2x^{2/5} + 7x^{1/5} + 3 = 0$

Mixed Practice

In each exercise make an appropriate substitution in order to obtain a quadratic equation. Find all complex values for x.

25. $x^6 - 5x^3 = 14$ **26.** $x^6 + 2x^3 = 15$

27. $(x^2 + x)^2 - 5(x^2 + x) = -6$

28. $(x^2 - 2x)^2 + 2(x^2 - 2x) = 3$

29. $x - 5x^{1/2} + 6 = 0$

30. $x - 5x^{1/2} - 36 = 0$

31. $x^{-2} + 3x^{-1} = 0$

32. $2x^{-2} - x^{-1} = 0$

To Think About

Solve. Find all valid real roots for each equation.

33. $15 - \dfrac{2x}{x - 1} = \dfrac{x^2}{x^2 - 2x + 1}$

34. $4 - \dfrac{x^3 + 1}{x^3 + 6} = \dfrac{x^3 - 3}{x^3 + 2}$

Cumulative Review

35. Solve the system: $2x + 3y = 5$
$-5x - 2y = 4$

36. Simplify: $\dfrac{5 + \frac{2}{x}}{\frac{7}{3x} - 1}$

Multiply and simplify.

37. $3\sqrt{2}\left(\sqrt{5} - 2\sqrt{6}\right)$

38. $\left(\sqrt{2} + \sqrt{6}\right)\left(3\sqrt{2} - 2\sqrt{5}\right)$

39. ***Salary and Educational Attainment*** How much greater is the average annual salary of a man than a woman at each of the four levels of educational attainment shown in the graph? Express your answers to the nearest tenth of a percent.

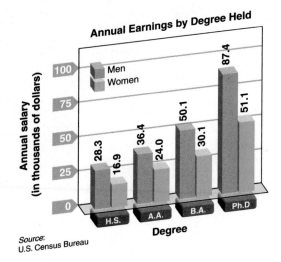

Annual Earnings by Degree Held

Source: U.S. Census Bureau

40. ***Salary and Educational Attainment*** Approximately how many extra years would an average woman with an associate's degree have to work to earn the same lifetime earnings as a male counterpart who worked for 30 years? Use the data displayed in the graph to the right. Round your answer to the nearest tenth of a year.

Profit of Broadway Performances *A Broadway musical called Miss Saigon closed in 2000. The musical was performed 4063 times, making it the sixth longest running show in Broadway history. The show cost $109 million, grossed $281.5 million, and was seen by about six million people. Cats, the longest running show in Broadway history, was performed 7451 times.*

41. What was the average profit per performance for *Miss Saigon*?

42. What was the average cost for a person to see the performance?

43. If *Cats* had the same profit per performance as *Miss Saigon*, what was the total profit of *Cats*?

How are you doing with your homework assignments in Sections 8.1 to 8.3? Do you feel you have mastered the material so far? Do you understand the concepts you have covered? Before you go further in the textbook take some time to do each of the following problems.

8.1

Simplify your answer to all problems.

1. Solve by the square root property: $2x^2 + 3 = 39$

2. Solve by the square root property: $(3x + 4)^2 = 20$

3. Solve by completing the square: $x^2 - 8x = -12$

4. Solve by completing the square: $2x^2 - 4x - 3 = 0$

8.2

Solve by the quadratic formula. Express any nonreal roots using i notation.

5. $8x^2 - 2x - 7 = 0$

6. $(x - 1)(x + 5) = 2$

7. $4x^2 = -12x - 17$

8. $5x^2 + 4x - 12 = 0$

9. $7x^2 + 9x = 14x^2 - 3x$

10. $4x^2 - 3x = -6$

11. $\dfrac{18}{x} + \dfrac{12}{x + 1} = 9$

8.3

Solve for any real roots and check your answers.

12. $x^6 - 7x^3 - 8 = 0$

13. $w^{4/3} - 6w^{2/3} + 8 = 0$

14. $x^8 = 7x^4 - 12$

15. $2x^{2/5} = 7x^{1/5} - 3$

Now turn to page SA-29 for the answers to each of these problems. Each answer also includes a reference to the objective in which the problem is first taught. If you missed any of these problems, you should stop and review the Examples and Practice Problems in the referenced objective. A little review now will help you master the material in the upcoming sections of the text.

1. _____

2. _____

3. _____

4. _____

5. _____

6. _____

7. _____

8. _____

9. _____

10. _____

11. _____

12. _____

13. _____

14. _____

15. _____

445

8.4 FORMULAS AND APPLICATIONS

Student Learning Objectives

After studying this section, you will be able to:

1. Solve a quadratic equation containing several variables.

2. Solve problems requiring the use of the Pythagorean theorem.

3. Solve applied problems requiring the use of a quadratic equation.

1 Solving a Quadratic Equation Containing Several Variables

In mathematics, physics, and engineering, we must often solve an equation for a variable in terms of other variables. You recall we solved linear equations in several variables in Section 2.2. We now examine several cases where the variable that we are solving for is squared. If the variable we are solving for is squared, and there is no other term containing that variable, then the equation can be solved using the square root property.

EXAMPLE 1 The surface area of a sphere is given by $A = 4\pi r^2$. Solve this equation for r. (You do not need to rationalize the denominator.)

Solution

$$A = 4\pi r^2$$

$$\frac{A}{4\pi} = r^2$$

$$\pm\sqrt{\frac{A}{4\pi}} = r \qquad \text{Use the square root property.}$$

$$\pm\frac{1}{2}\sqrt{\frac{A}{\pi}} = r \qquad \text{Simplify.}$$

Since the radius of a sphere must be a positive value, we use only the principal root.

$$r = \frac{1}{2}\sqrt{\frac{A}{\pi}}$$

Practice Problem 1 The volume of a cylindrical cone is $V = \frac{1}{3}\pi r^2 h$. Solve this equation for r. (You do not need to rationalize the denominator.)

NOTE TO STUDENT: Fully worked-out solutions to all of the Practice Problems can be found at the back of the text starting at page SP-1

Some quadratic equations containing many variables can be solved for one variable by factoring.

EXAMPLE 2 Solve for y: $y^2 - 2yz - 15z^2 = 0$

Solution

$$(y + 3z)(y - 5z) = 0 \qquad \text{Factor.}$$

$$y + 3z = 0 \qquad y - 5z = 0 \qquad \text{Set each factor equal to 0.}$$

$$y = -3z \qquad y = 5z \qquad \text{Solve for } y.$$

Practice Problem 2 Solve for y: $2y^2 + 9wy + 7w^2 = 0$

Sometimes the quadratic formula is required in order to solve the equation.

EXAMPLE 3   Solve for x: $2x^2 + 3wx - 4z = 0$

Solution   We use the quadratic formula where the variable is considered to be x and the letters w and z are considered constants. Thus, $a = 2$, $b = 3w$, and $c = -4z$.

$$x = \frac{-b \pm \sqrt{b^2 - 4ac}}{2a}$$

$$x = \frac{-3w \pm \sqrt{(3w)^2 - 4(2)(-4z)}}{2(2)} = \frac{-3w \pm \sqrt{9w^2 + 32z}}{4}$$

Note that this answer cannot be simplified any further.

Practice Problem 3   Solve for y: $3y^2 + 2fy - 7g = 0$

EXAMPLE 4   The formula for the curved surface area S of a right circular cone of altitude h and with base of radius r is $S = \pi r \sqrt{r^2 + h^2}$.

Solution   Solve for r^2.

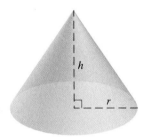

$$S = \pi r \sqrt{r^2 + h^2}$$

$$\frac{S}{\pi r} = \sqrt{r^2 + h^2} \qquad \text{Isolate the radical.}$$

$$\frac{S^2}{\pi^2 r^2} = r^2 + h^2 \qquad \text{Square both sides.}$$

$$\frac{S^2}{\pi^2} = r^4 + h^2 r^2 \qquad \text{Multiply each term by } r^2.$$

$$0 = r^4 + h^2 r^2 - \frac{S^2}{\pi^2} \qquad \text{Subtract } S^2/\pi^2.$$

This equation is quadratic in form. If we let $y = r^2$, then we have

$$0 = y^2 + h^2 y - \frac{S^2}{\pi^2}.$$

By the quadratic formula we have the following:

$$y = \frac{-h^2 \pm \sqrt{(h^2)^2 - 4(1)\left(-\dfrac{S^2}{\pi^2}\right)}}{2}$$

$$y = \frac{-h^2 \pm \sqrt{\dfrac{\pi^2 h^4}{\pi^2} + \dfrac{4S^2}{\pi^2}}}{2}$$

$$y = \frac{-h^2 \pm \dfrac{1}{\pi} \sqrt{\pi^2 h^4 + 4S^2}}{2}$$

$$y = \frac{-\pi h^2 \pm \sqrt{\pi^2 h^4 + 4S^2}}{2\pi}$$

Since $y = r^2$, we have

$$r^2 = \frac{-\pi h^2 \pm \sqrt{\pi^2 h^4 + 4S^2}}{2\pi}.$$

NOTE TO STUDENT: Fully worked-out solutions to all of the Practice Problems can be found at the back of the text starting at page SP-1

Practice Problem 4   The formula for the number of diagonals d in a polygon of n sides is $d = \dfrac{n^2 - 3n}{2}$. Solve for n.

② Solving Problems Requiring the Use of the Pythagorean Theorem

A very useful formula is the Pythagorean theorem for right triangles.

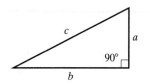

> **PYTHAGOREAN THEOREM**
>
> If c is the length of the longest side of a right triangle and a and b are the lengths of the other two sides, then $a^2 + b^2 = c^2$.

The longest side of a right triangle is called the **hypotenuse.** The other two sides are called the **legs** of the triangle.

▲ EXAMPLE 5

(a) Solve the Pythagorean theorem $a^2 + b^2 = c^2$ for a.

(b) Find the value of a if $c = 13$ and $b = 5$.

Solution

(a) $a^2 = c^2 - b^2$ Subtract b^2 from each side.

$a = \pm\sqrt{c^2 - b^2}$ Use the square root property.

Since a, b, and c must be positive numbers because they represent lengths, we use only the positive root, $a = \sqrt{c^2 - b^2}$.

(b) $a = \sqrt{c^2 - b^2}$

$a = \sqrt{(13)^2 - (5)^2} = \sqrt{169 - 25} = \sqrt{144} = 12$

Thus, $a = 12$.

▲ Practice Problem 5

(a) Solve the Pythagorean theorem for b.

(b) Find the value of b if $c = 26$ and $a = 24$.

There are many practical uses for the Pythagorean Theorem. For hundreds of years in America, it was used in land surveying and in ship navigation. So, for hundreds of years students in America learned it in public schools and were shown applications in surveying and navigation. In almost any situation where you have a right triangle and you know two sides and need to find the third side, you can use the Pythagorean Theorem.

See if you can see how it is used in Example 6. Notice that it is helpful right at the beginning of the problem to draw a picture of the right triangle and label information that you know.

▲ **EXAMPLE 8**   A triangular sign marks the edge of the rocks in Rockport Harbor. The sign has an area of 35 square meters. Find the base and altitude of this triangular sign if the base is 3 meters shorter than the altitude.

Solution   The area of a triangle is given by

$$A = \frac{1}{2}ab.$$

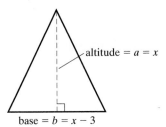

Let x = the length in meters of the altitude. Then $x - 3$ = the length in meters of the base.

$35 = \dfrac{1}{2}x(x - 3)$	Replace A(area) by 35, a (altitude) by x, and b (base) by $x - 3$.
$70 = x(x - 3)$	Multiply each side by 2.
$70 = x^2 - 3x$	Use the distributive property.
$0 = x^2 - 3x - 70$	Subtract 70 from each side.
$0 = (x - 10)(x + 7)$	
$x = 10 \quad \text{or} \quad x = -7$	

The length of a side of a triangle must be a positive number, so we disregard -7. Thus,

$$\text{altitude} = x = 10 \text{ meters and}$$
$$\text{base} = x - 3 = 7 \text{ meters.}$$

The check is left to the student.

▲ **Practice Problem 8**   The length of a rectangle is 3 feet shorter than twice the width. The area of the rectangle is 54 square feet. Find the dimensions of the rectangle.

We will now examine a few word problems that require the use of the formula distance = (rate)(time) or $d = rt$.

EXAMPLE 9   When Barbara was training for a bicycle race, she rode a total of 135 miles on Monday and Tuesday. On Monday she rode for 75 miles in the rain. On Tuesday she rode 5 miles per hour faster because the weather was better. Her total cycling time for the 2 days was 8 hours. Find her speed for each day.

Solution   We can find each distance. If Barbara rode 75 miles on Monday and a total of 135 miles during the 2 days, then she rode $135 - 75 = 60$ miles on Tuesday.

Let $x =$ the cycling rate in miles per hour on Monday. Since Barbara rode 5 miles per hour faster on Tuesday, $x + 5 =$ the cycling rate in miles per hour on Tuesday.

Since distance divided by rate is equal to time $\left(\dfrac{d}{r} = t \right)$, we can determine that the time Barbara cycled on Monday was $\dfrac{75}{x}$ and the time she cycled on Tuesday was $\dfrac{60}{x + 5}$.

Day	Distance	Rate	Time
Monday	75	x	$\dfrac{75}{x}$
Tuesday	60	$x + 5$	$\dfrac{60}{x + 5}$
Totals	135	(not used)	8

Since the total cycling time was 8 hours, we have the following:

time cycling Monday + time cycling Tuesday = 8 hours

$$\frac{75}{x} \quad + \quad \frac{60}{x + 5} \quad = 8$$

The LCD of this equation is $x(x + 5)$. Multiply each term by the LCD.

$$x(x + 5)\left(\frac{75}{x} \right) + x(x + 5)\left(\frac{60}{x + 5} \right) = x(x + 5)(8)$$

$$75(x + 5) + 60x = 8x(x + 5)$$

$$75x + 375 + 60x = 8x^2 + 40x$$

$$0 = 8x^2 - 95x - 375$$

$$0 = (x - 15)(8x + 25)$$

$$x - 15 = 0 \quad \text{or} \quad 8x + 25 = 0$$

$$x = 15 \qquad\qquad x = \frac{-25}{8}$$

We disregard the negative answer. The cyclist did not have a negative rate of speed—unless she was pedaling backward! Thus, $x = 15$. So Barbara's rate of speed on Monday was 15 mph, and her rate of speed on Tuesday was $x + 5 = 15 + 5 = 20$ mph.

Practice Problem 9   Carlos traveled in his car at a constant speed on a secondary road for 150 miles. Then he traveled 10 mph faster on a better road for 240 miles. If Carlos drove for 7 hours, find the car's speed for each part of the trip.

8.4 EXERCISES

Student Solutions Manual | CD/ Video | PH Math Tutor Center | MathXL®Tutorials on CD | MathXL® | MyMathLab® | Interactmath.com

Solve for the variable specified. Assume that all other variables are nonzero.

1. $S = 16t^2$; for t

2. $E = mc^2$; for c

3. $S = 4\pi r^2$, for r

4. $A = \dfrac{1}{2}r^2\theta$, for r

5. $3H = \dfrac{1}{2}ax^2$; for x

6. $5B = \dfrac{2}{3}hx^2$; for x

7. $4(y^2 + w) - 5 = 7R$; for y

8. $9x^2 - 2 = 3B$; for x

9. $Q = \dfrac{3mwM^2}{2c}$; for M

10. $H = \dfrac{5abT^2}{7k}$; for T

11. $V = \pi(r^2 + R^2)h$; for r

12. $H = b(a^2 + w^2)$; for w

13. $7bx^2 - 3ax = 0$, for x

14. $2x^2 - 5ax = 0$, for x

15. $P = EI - RI^2$; for I

16. $A = P(1 + r)^2$; for r

17. $10w^2 - 3qw - 4 = 0$; for w

18. $7w^2 + 5qw - 1 = 0$; for w

19. $S = 2\pi rh + \pi r^2$; for r

20. $B = 3abx^2 - 5x$; for x

21. $(a + 1)x^2 + 5x + 2w = 0$; for x

22. $(b - 2)x^2 - 3x + 5y = 0$; for x

In Exercises 23–28, use the Pythagorean theorem to find the missing side(s).

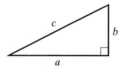

▲ **23.** $c = 6, a = 4$; find b

▲ **24.** $b = 5, c = 10$; find a

▲ **25.** $c = \sqrt{34}, b = \sqrt{19}$; find a

▲ **26.** $c = \sqrt{21}, a = \sqrt{5}$; find b

▲ **27.** $c = 12, b = 2a$; find b and a

▲ **28.** $c = 15, a = 2b$; find b and a

Applications

▲ **29.** *Brace for a Shelf* A brace for a shelf has the shape of a right triangle. Its hypotenuse is 10 inches long and the two legs are equal in length. How long are the legs of the triangle?

▲ **30.** *Tent* The two sides of a tent are equal in length and meet to form a right triangle. If the width of the floor of the tent is 8 feet, what is the length of the two sides?

▲ **31.** *College Parking Lot* Knox College is creating a new rectangular parking lot. The length is 0.07 miles longer than the width and the area of the parking lot is 0.026 square miles. Find the length and width of the parking lot.

▲ **32.** *Computer Chip* A high-tech company is producing a new rectangular computer chip. The length is 0.3 centimeters less than twice the width. The chip has an area of 1.04 square centimeters. Find the length and width of the chip.

▲ 33. **Barn** The area of a rectangular wall of a barn is 126 square feet. Its length is 4 feet longer than twice its width. Find the length and width of the wall of the barn.

▲ 34. **Tennis Court** The area of a rectangular tennis court is 140 square meters. Its length is 6 meters shorter than twice its width. Find the length and width of the tennis court.

▲ 35. **Triangular Flag** The area of a triangular flag is 72 square centimeters. Its altitude is 2 centimeters longer than twice its base. Find the lengths of the altitude and the base.

▲ 36. **Children's Playground** A children's playground is triangular in shape. Its altitude is 2 yards shorter than its base. The area of the playground is 60 square yards. Find the base and altitude of the playground.

37. **Driving Speed** Roberto drove at a constant speed in a rainstorm for 225 miles. He took a break, and the rain stopped. He then drove 150 miles at a speed that was 5 miles per hour faster than his previous speed. If he drove for 8 hours, find the car's speed for each part of the trip.

38. **Driving Speed** Benita traveled at a constant speed on an old road for 160 miles. She then traveled 5 miles per hour faster on a newer road for 90 miles. If she drove for 6 hours, find the car's speed for each part of the trip.

39. **Commuter Traffic** Bob drove from home to work at 50 mph. After work the traffic was heavier, and he drove home at 45 mph. His driving time to and from work was 1 hour and 16 minutes. How far does he live from his job?

40. **Commercial Trucking** A driver drove his heavily loaded truck from the company warehouse to a delivery point at 35 mph. He unloaded the truck and drove back to the warehouse at 45 mph. The total trip took 5 hours and 20 minutes. How far is the delivery point from the warehouse?

Federal and State Prisons *The number of inmates N (measured in thousands) in federal and state prisons in the United States can be approximated by the equation $N = 1.11x^2 + 33.39x + 304.09$, where x is the number of years since 1980. For example, when $x = 1$, $N = 338.59$. This tells us that in 1981 there were approximately 338,590 inmates in federal and state prisons. Use this equation to answer the following questions. (Source: U.S. Bureau of Justice Statistics.)*

41. How many inmates does the equation predict there were in the year 2003?

42. How many inmates does the equation predict there will be in the year 2006?

43. In what year is the number of inmates expected to be 1,744,800?

44. In what year is the number of inmates expected to be 1,832,600?

To Think About

45. Solve for w: $w = \dfrac{12b^2}{\frac{5}{2}w + \frac{7}{2}b + \frac{21}{2}}$

46. ***Investment Income*** The formula $A = P(1 + r)^2$ gives the amount A in dollars that will be obtained in 2 years if P dollars are invested at an annual compound interest rate of r. If you invest $P = \$1400$ and it grows to $\$1514.24$ in 2 years, what is the annual interest rate r?

Cumulative Review

Rationalize the denominators.

47. $\dfrac{4}{\sqrt{3x}}$

48. $\dfrac{5\sqrt{6}}{2\sqrt{5}}$

49. $\dfrac{3}{\sqrt{x} + \sqrt{y}}$

50. $\dfrac{2\sqrt{3}}{\sqrt{3} - \sqrt{6}}$

51. $\dfrac{3ab}{\sqrt[3]{8ab^2}}$

8.5 QUADRATIC FUNCTIONS

① Finding the Vertex and Intercepts of a Quadratic Function

In Section 3.6 we graphed functions such as $p(x) = x^2$ and $g(x) = (x + 2)^2$. We will now study quadratic functions in more detail.

Student Learning Objectives

After studying this section, you will be able to:

① Find the vertex and intercepts of a quadratic function.

② Graph a quadratic function.

DEFINITION OF A QUADRATIC FUNCTION

A **quadratic function** is a function of the form

$f(x) = ax^2 + bx + c$, where $a, b,$ and c are real numbers and $a \neq 0$.

Graphs of quadratic functions written in this form will be parabolas opening upward if $a > 0$ or downward if $a < 0$. The **vertex** of a parabola is the lowest point on a parabola opening upward or the highest point on a parabola opening downward. The vertex will occur at $x = \dfrac{-b}{2a}$. To find the y-value, or $f(x)$, when $x = \dfrac{-b}{2a}$, we find $f\left(\dfrac{-b}{2a}\right)$. Therefore, we can say that a quadratic function has its vertex at $\left(\dfrac{-b}{2a}, f\left(\dfrac{-b}{2a}\right)\right)$.

It is helpful to know the x-intercepts and the y-intercept when graphing a quadratic function.

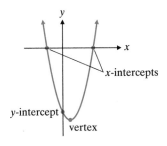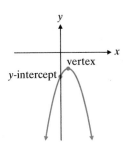

A quadratic function will always have exactly one y-intercept. However, it may have zero, one, or two x-intercepts. Why?

WHEN GRAPHING QUADRATIC FUNCTIONS OF THE FORM
$f(x) = ax^2 + bx + c, a \neq 0$

1. The coordinates of the vertex are $\left(\dfrac{-b}{2a}, f\left(\dfrac{-b}{2a}\right)\right)$.
2. The y-intercept is at $f(0)$.
3. The x-intercepts (if they exist) occur where $f(x) = 0$. They can always be found with the quadratic formula and can sometimes be found by factoring.

Since we may replace y by $f(x)$, the graph is equivalent to the graph of $y = ax^2 + bx + c$.

Finding the *x*-intercepts and the vertex

You can use a graphing calculator to find the *x*-intercepts and vertex of a quadratic function. To find the intercepts of the quadratic function

$$f(x) = x^2 - 4x + 3,$$

graph $y = x^2 - 4x + 3$ on a graphing calculator using an appropriate window.
Display:

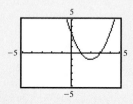

Next you can use the Trace and Zoom features or zero command of your calculator to find the *x*-intercepts.

You can also use the Trace and Zoom features to determine the vertex.

Some calculators have a feature that will calculate the maximum or minimum point on the graph. Use the feature that calculates the minimum point to find the vertex of $f(x) = x^2 - 4x + 3$.
Display:

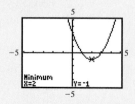

Thus, the vertex is at $(2, -1)$.

EXAMPLE 1   Find the coordinates of the vertex and the intercepts of the quadratic function $f(x) = x^2 - 8x + 15$.

Solution   For this function $a = 1, b = -8$, and $c = 15$.

Step 1   The vertex occurs at $x = \dfrac{-b}{2a}$. Thus,

$$x = \frac{-(-8)}{2(1)} = \frac{8}{2} = 4.$$

The vertex has an *x*-coordinate of 4. To find the *y*-coordinate, we evaluate $f(4)$.

$$f(4) = 4^2 - 8(4) + 15 = 16 - 32 + 15 = -1$$

Thus, the vertex is $(4, -1)$.

Step 2   The *y*-intercept is at $f(0)$. We evaluate $f(0)$ to find the *y*-coordinate when *x* is 0.

$$f(0) = 0^2 - 8(0) + 15 = 15$$

The *y*-intercept is $(0, 15)$.

Step 3   If there are *x*-intercepts, they will occur when $f(x) = 0$—that is, when $x^2 - 8x + 15 = 0$. We solve for *x*.

$$(x - 5)(x - 3) = 0$$
$$x - 5 = 0 \qquad x - 3 = 0$$
$$x = 5 \qquad\quad x = 3$$

Thus, we conclude that the *x*-intercepts are $(5, 0)$ and $(3, 0)$. We list these four important points of the function in table form.

Name	*x*	*f(x)*
Vertex	4	−1
y-intercept	0	15
x-intercept	5	0
x-intercept	3	0

NOTE TO STUDENT: Fully worked-out solutions to all of the Practice Problems can be found at the back of the text starting at page SP-1

Practice Problem 1   Find the coordinates of the vertex and the intercepts of the quadratic function $f(x) = x^2 - 6x + 5$.

② Graphing a Quadratic Function

It is helpful to find the vertex and the intercepts of a quadratic function before graphing it.

EXAMPLE 2   Find the vertex and the intercepts, and then graph the function $f(x) = x^2 + 2x - 4$.

Solution   Here $a = 1, b = 2$, and $c = -4$. Since $a > 0$, the parabola opens *upward*.

Step 1   We find the vertex.

$$x = \frac{-b}{2a} = \frac{-2}{2(1)} = \frac{-2}{2} = -1$$

$$f(-1) = (-1)^2 + 2(-1) - 4 = 1 + (-2) - 4 = -5$$

The vertex is $(-1, -5)$.

Step 2   We find the y-intercept. The y-intercept is at $f(0)$.

$$f(0) = (0)^2 + 2(0) - 4 = -4$$

The y-intercept is $(0, -4)$.

Step 3   We find the x-intercepts. The x-intercepts occur when $f(x) = 0$.
Thus, we solve $x^2 + 2x - 4 = 0$ for x. We cannot factor this equation, so we use the quadratic formula.

$$x = \frac{-b \pm \sqrt{b^2 - 4ac}}{2a} = \frac{-2 \pm \sqrt{2^2 - 4(1)(-4)}}{2(1)} = \frac{-2 \pm \sqrt{20}}{2} = -1 \pm \sqrt{5}$$

To aid our graphing, we will approximate the value of x to the nearest tenth by using a square root table or a scientific calculator.

$$1 \boxed{+/-} \boxed{+} 5 \boxed{\sqrt{}} \boxed{=} \quad 1.236068$$
$$x \approx 1.2$$

$$1 \boxed{+/-} \boxed{-} 5 \boxed{\sqrt{}} \boxed{=} \quad -3.236068$$
$$x \approx -3.2$$

The x-intercepts are approximately $(-3.2, 0)$ and $(1.2, 0)$.
We have found that the vertex is $(-1, -5)$; the y-intercept is $(0, -4)$; and the x-intercepts are approximately $(-3.2, 0)$ and $(1.2, 0)$. We connect these points by a smooth curve to graph the parabola.

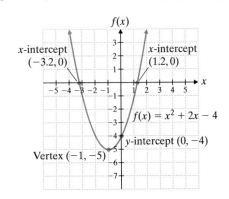

PRACTICE PROBLEM 2

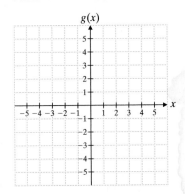

Practice Problem 2   Find the vertex and the intercepts, and then graph the function $g(x) = x^2 - 2x - 2$.

EXAMPLE 3 Find the vertex and the intercepts, and then graph the function $f(x) = -2x^2 + 4x - 3$.

Solution Here $a = -2$, $b = 4$, and $c = -3$. Since $a < 0$, the parabola opens *downward*.

The vertex occurs at $x = \dfrac{-b}{2a}$.

$$x = \frac{-4}{2(-2)} = \frac{-4}{-4} = 1$$

$$f(1) = -2(1)^2 + 4(1) - 3 = -2 + 4 - 3 = -1$$

The vertex is $(1, -1)$.

The y-intercept is at $f(0)$.

$$f(0) = -2(0)^2 + 4(0) - 3 = -3$$

The y-intercept is $(0, -3)$.

If there are any x-intercepts, they will occur when $f(x) = 0$. We use the quadratic formula to solve $-2x^2 + 4x - 3 = 0$ for x.

$$x = \frac{-4 \pm \sqrt{4^2 - 4(-2)(-3)}}{2(-2)} = \frac{-4 \pm \sqrt{-8}}{-4}$$

Because $\sqrt{-8}$ yields an imaginary number, there are no real roots. Thus, there are no x-intercepts for the graph of the function. That is, the graph does not intersect the x-axis.

We know that the parabola opens *downward*. Thus, the vertex is a maximum value at $(1, -1)$. Since this graph has no x-intercepts, we will look for three additional points to help us in drawing the graph. We try $f(2), f(3)$, and $f(-1)$.

$$f(2) = -2(2)^2 + 4(2) - 3 = -8 + 8 - 3 = -3$$
$$f(3) = -2(3)^2 + 4(3) - 3 = -18 + 12 - 3 = -9$$
$$f(-1) = -2(-1)^2 + 4(-1) - 3 = -2 - 4 - 3 = -9$$

We plot the vertex, the y-intercept, and the points $(2, -3), (3, -9)$, and $(-1, -9)$.

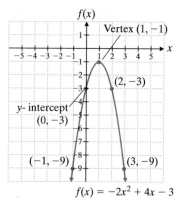

NOTE TO STUDENT: Fully worked-out solutions to all of the Practice Problems can be found at the back of the text starting at page SP-1

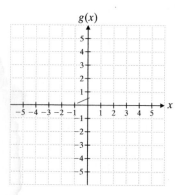

Practice Problem 3 Find the vertex and the intercepts, and then graph the function $g(x) = -2x^2 - 8x - 6$.

Find the coordinates of the vertex and the intercepts of each of the following quadratic functions. When necessary, approximate the x-intercepts to the nearest tenth.

1. $f(x) = x^2 - 2x - 8$

2. $f(x) = x^2 - 4x - 5$

3. $g(x) = -x^2 - 4x + 12$

4. $f(x) = -x^2 + 6x + 16$

5. $p(x) = 3x^2 + 12x + 3$

6. $p(x) = 2x^2 + 4x + 1$

7. $r(x) = -3x^2 - 2x - 6$

8. $f(x) = -2x^2 + 3x - 2$

9. $f(x) = 2x^2 + 2x - 4$

10. $f(x) = 5x^2 + 2x - 3$

In each of the following exercises, find the vertex, the y-intercept, and the x-intercepts (if any exist), and then graph the function.

11. $f(x) = x^2 - 6x + 8$

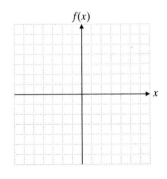

12. $f(x) = x^2 + 6x + 8$

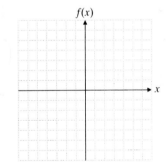

13. $g(x) = x^2 + 2x - 8$

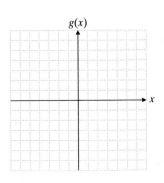

14. $g(x) = x^2 - 2x - 8$

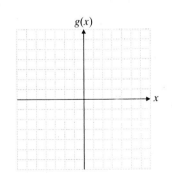

461

15. $p(x) = -x^2 + 4x - 3$

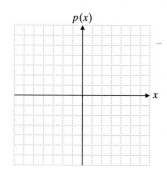

16. $p(x) = -x^2 - 4x - 3$

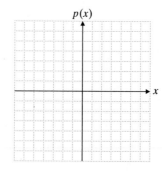

17. $r(x) = 3x^2 + 6x + 4$

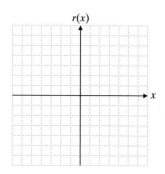

18. $r(x) = -3x^2 + 6x - 4$

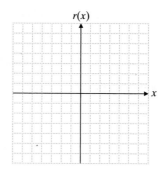

19. $f(x) = x^2 - 6x + 5$

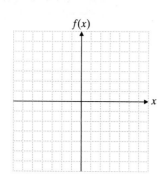

20. $g(x) = 2x^2 - 2x + 1$

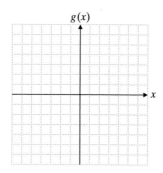

21. $f(x) = x^2 - 4x + 4$

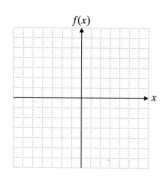

22. $g(x) = -x^2 + 6x - 9$

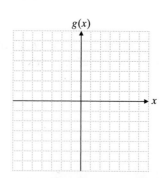

23. $f(x) = x^2 - 4$

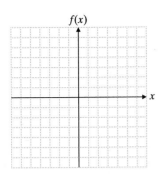

24. $r(x) = -x^2 + 1$

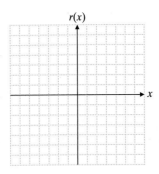

Applications

Scuba Diving   *Some sports such as scuba diving are expensive. People with larger incomes are more likely to partici-pate in such a sport. The number of people N (measured in thousands) who engage in scuba diving can be described by the function N(x) = 0.18x² − 3.18x + 102.25, where x is the mean income (measured in thousands) and x ≥ 20. Use this information to answer problems 25–30. (Source: U.S. Census Bureau.)*

25. Find $N(20)$, $N(40)$, $N(60)$, $N(80)$, and $N(100)$.

26. Use the results of Exercise 25 to graph the function from $x = 20$ to $x = 100$. You may use the graph grid provided at the bottom of the page.

27. Find $N(70)$ from your graph. Explain what $N(70)$ means.

28. Find $N(70)$ from the equation for $N(x)$. Compare your answers for Exercises 27 and 28.

29. Use your graph to determine for what value of x $N(x)$ is equal to 390. Explain what this means.

30. Use your graph to determine for what value of x $N(x)$ is equal to 560. Explain what this means.

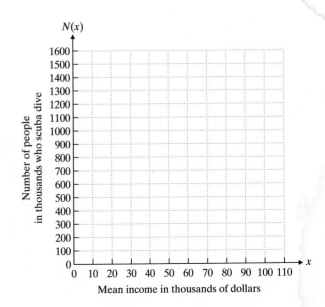

Manufacturing Profit   *The daily profit P in dollars of the Pine Tree Table Company is described by the function* $P(x) = -6x^2 + 312x - 3672$, *where x is the number of tables that are manufactured in 1 day. Use this information to answer Exercises 31–36.*

31. Find $P(16)$, $P(20)$, $P(24)$, $P(30)$, and $P(35)$.

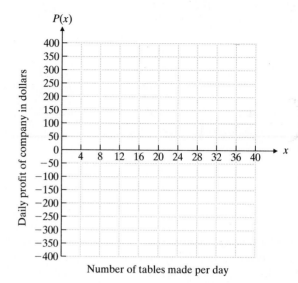

Number of tables made per day

32. Use the results of Exercise 31 to graph the function from $x = 16$ to $x = 35$.

33. The maximum profit of the company occurs at the vertex of the parabola. How many tables should be made per day in order to obtain the maximum profit for the company? What is the maximum profit?

34. How many tables per day should be made in order to obtain a daily profit of $360? Why are there two answers to this question?

35. How many tables are made per day if the company has a daily profit of zero dollars?

36. How many tables are made per day if the company has a daily profit of $288?

37. ***Softball*** Susan throws a softball upward into the air at a speed of 32 feet per second from a 40-foot platform. The distance upward that the ball travels is given by the function $d(t) = -16t^2 + 32t + 40$. What is the maximum height of the softball? How many seconds does it take to reach the ground after first being thrown upward? (Round your answer to the nearest tenth.)

38. ***Baseball*** Henry is standing on a platform overlooking a baseball stadium. It is 160 feet above the playing field. When he throws a baseball upward at 64 feet per second, the distance d from the baseball to the ground is given by the function $d(t) = -16t^2 + 64t + 160$. What is the maximum height of the baseball if he throws it upward? How many seconds does it take until the ball finally hits the ground? (Round your answer to the nearest tenth.)

Find the vertex and intercepts for each of the following. If your answers are not exact, round them to the nearest tenth.

39. $y = x^2 - 4.4x + 7.59$

40. $y = x^2 + 7.8x + 13.8$

41. Graph $y = 2.3x^2 - 5.4x - 1.6$. Find the x-intercepts to the nearest tenth.

42. Graph $y = -4.6x^2 + 7.2x - 2.3$. Find the x-intercepts to the nearest tenth.

To Think About

43. A graph of a quadratic equation of the form $y = ax^2 + bx + c$ passes through the points $(0, 2)$, $(2, 10)$, and $(-2, 34)$. What is the value of a, b, and c?

44. A graph of a quadratic equation of the form $y = ax^2 + bx + c$ has a vertex of $(3, 4)$ and passes through the point $(0, 13)$. What is the value of a, b, and c?

Cumulative Review

Solve each system.

45. $9x + 5y = 6$
$2x - 5y = -17$

46. $x + y = 16$
$95x + 143y = 1760$

47. $3x - y + 2z = 12$
$2x - 3y + z = 5$
$x + 3y + 8z = 22$

48. $7x + 3y - z = -2$
$x + 5y + 3z = 2$
$x + 2y + z = 1$

8.6 QUADRATIC INEQUALITIES IN ONE VARIABLE

1 Solving a Factorable Quadratic Inequality in One Variable

We will now solve quadratic inequalities such as $x^2 - 2x - 3 > 0$ and $2x^2 + x - 15 < 0$. A **quadratic inequality** has the form $ax^2 + bx + c < 0$ (or replace $<$ by $>$, $\leq$, or $\geq$), where a, b, and c are real numbers $a \neq 0$. We use our knowledge of solving quadratic equations to solve quadratic inequalities.

Let's solve the inequality $x^2 - 2x - 3 > 0$. We want to find the two points where the expression on the left side is equal to zero. We call these the **critical points.** To do this, we replace the inequality symbol by an equal sign and solve the resulting equation.

$$x^2 - 2x - 3 = 0$$
$$(x + 1)(x - 3) = 0 \quad \text{Factor.}$$
$$x + 1 = 0 \quad \text{or} \quad x - 3 = 0 \quad \text{Zero-product rule}$$
$$x = -1 \qquad\qquad x = 3$$

These two solutions form critical points that divide the number line into three segments.

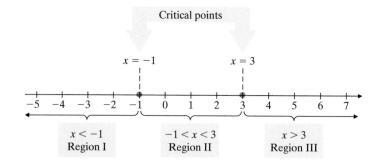

We will show as an exercise that all values of x in a given segment produce results that are greater than zero, or all values of x in a given segment produce results that are less than zero.

To solve the quadratic inequality, we pick an arbitrary test point in each region and then substitute it into the inequality to determine whether it satisfies the inequality. If one point in a region satisfies the inequality, then *all* points in the region satisfy the inequality. We will test three values of x in the expression $x^2 - 2x - 3$.

$\boxed{x < -1, \, region \, I:}$ A sample point is $x = -2$.

$$(-2)^2 - 2(-2) - 3 = 4 + 4 - 3 = 5 > 0$$

$\boxed{-1 < x < 3, \, region \, II:}$ A sample point is $x = 0$.

$$(0)^2 - 2(0) - 3 = 0 + 0 - 3 = -3 < 0$$

$\boxed{x > 3, \, region \, III:}$ A sample point is $x = 4$.

$$(4)^2 - 2(4) - 3 = 16 - 8 - 3 = 5 > 0$$

Thus, we see that $x^2 - 2x - 3 > 0$ when $x < -1$ or $x > 3$. No points in region II satisfy the inequality. The graph of the solution is shown next.

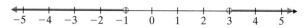

We summarize our method.

SOLVING A QUADRATIC INEQUALITY

1. Replace the inequality symbol by an equal sign. Solve the resulting equation to find the critical points.
2. Use the critical points to separate the number line into three distinct regions.
3. Evaluate the quadratic expression at a test point in each region.
4. Determine which regions satisfy the original conditions of the quadratic inequality.

EXAMPLE 1   Solve and graph $x^2 - 10x + 24 > 0$.

Solution

1. We replace the inequality symbol by an equal sign and solve the resulting equation.

$$x^2 - 10x + 24 = 0$$
$$(x - 4)(x - 6) = 0$$
$$x - 4 = 0 \quad \text{or} \quad x - 6 = 0$$
$$x = 4 \qquad\qquad x = 6$$

2. We use the critical points to separate the number line into distinct regions.

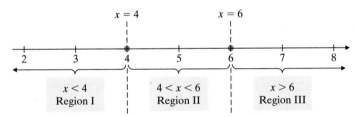

3. We evaluate the quadratic expression at a test point in each of the regions.

$$x^2 - 10x + 24$$

   | $x < 4$, *region I:* | We pick the sample point $x = 1$.

$$(1)^2 - 10(1) + 24 = 1 - 10 + 24 = 15 \ > 0$$

   | $4 < x < 6$, *region II:* | We pick the sample point $x = 5$.

$$(5)^2 - 10(5) + 24 = 25 - 50 + 24 = -1 \ < 0$$

   | $x > 6$, *region III:* | We pick the sample point $x = 7$.

$$(7)^2 - 10(7) + 24 = 49 - 70 + 24 = 3 \ > 0$$

4. We determine which regions satisfy the original conditions of the quadratic inequality.

$$x^2 - 10x + 24 > 0 \text{ when } x < 4 \text{ or when } x > 6.$$

The graph of the solution is shown next.

Practice Problem 1   Solve and graph $x^2 - 2x - 8 < 0$.

PRACTICE PROBLEM 1

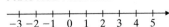

NOTE TO STUDENT: Fully worked-out solutions to all of the Practice Problems can be found at the back of the text starting at page SP-1

EXAMPLE 2   Solve and graph $2x^2 + x - 6 \leq 0$.

Solution   We replace the inequality symbol by an equal sign and solve the resulting equation.

$$2x^2 + x - 6 = 0$$
$$(2x - 3)(x + 2) = 0$$
$$2x - 3 = 0 \qquad \text{or} \quad x + 2 = 0$$
$$2x = 3 \qquad\qquad\qquad x = -2$$
$$x = \frac{3}{2} = 1.5$$

We use the critical points to separate the number line into distinct regions. The critical points are $x = -2$ and $x = 1.5$. Now we arbitrarily pick a test point in each region.

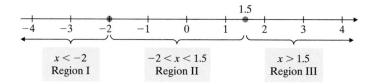

We will pick 3 values for x for the polynomial $2x^2 + x - 6$.

Region I: First we need an x value less than -2. We pick $x = -3$.

$$2(-3)^2 + (-3) - 6 = 18 - 3 - 6 = 9 \; > 0$$

Region II: Next we need an x value between -2 and 1.5. We pick $x = 0$.

$$2(0)^2 + (0) - 6 = 0 + 0 - 6 = -6 \; < 0$$

Region III: Finally we need an x value greater than 1.5. We pick $x = 2$.

$$2(2)^2 + (2) - 6 = 8 + 2 - 6 = 4 \; > 0$$

Since our inequality is $\leq$ and not just $<$, we need to include the critical points. Thus, $2x^2 + x - 6 \leq 0$ when $-2 \leq x \leq 1.5$. The graph of our solution is shown next.

Practice Problem 2   Solve and graph $3x^2 - x - 2 \geq 0$.

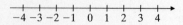

② Solving a Nonfactorable Quadratic Inequality in One Variable

If the quadratic expression in a quadratic inequality cannot be factored, then we will use the quadratic formula to obtain the critical points.

EXAMPLE 3   Solve and graph $x^2 + 4x > 6$. Round your answer to the nearest tenth.

Solution   First we write $x^2 + 4x - 6 > 0$. Because we cannot factor $x^2 + 4x - 6$, we use the quadratic formula to find the critical points.

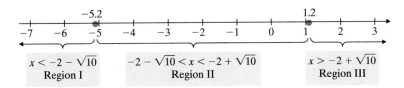

$$x = \frac{-4 \pm \sqrt{4^2 - 4(1)(-6)}}{2(1)} = \frac{-4 \pm \sqrt{16 + 24}}{2}$$

$$= \frac{-4 \pm \sqrt{40}}{2} = \frac{-4 \pm 2\sqrt{10}}{2} = -2 \pm \sqrt{10}$$

Using a calculator or our table of square roots, we find the following:

$$-2 + \sqrt{10} \approx -2 + 3.162 \approx 1.162 \text{ or about } 1.2$$
$$-2 - \sqrt{10} \approx -2 - 3.162 \approx -5.162 \text{ or about } -5.2$$

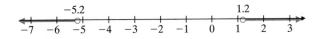

$x < -2 - \sqrt{10}$	$-2 - \sqrt{10} < x < -2 + \sqrt{10}$	$x > -2 + \sqrt{10}$
Region I	Region II	Region III

We will see where $x^2 + 4x - 6 > 0$.

Region I: $x = -6$

$$(-6)^2 + 4(-6) - 6 = 36 - 24 - 6 = 6 > 0$$

Region II: $x = 0$

$$(0)^2 + 4(0) - 6 = 0 + 0 - 6 = -6 < 0$$

Region III: $x = 2$

$$(2)^2 + 4(2) - 6 = 4 + 8 - 6 = 6 > 0$$

Thus, $x^2 + 4x > 6$ when $x^2 + 4x - 6 > 0$, and this occurs when $x < -2 - \sqrt{10}$ or $x > -2 + \sqrt{10}$. Rounding to the nearest tenth, our answer is

$$x < -5.2 \quad \text{or} \quad x > 1.2.$$

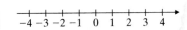

PRACTICE PROBLEM 3

Practice Problem 3   Solve and graph $x^2 + 2x < 7$. Round your answer to the nearest tenth.

Verbal and Writing Skills

1. When solving a quadratic inequality, why is it necessary to find the critical points?

2. What is the difference between solving an exercise like $ax^2 + bx + c > 0$ and an exercise like $ax^2 + bx + c \geq 0$?

Solve and graph.

3. $x^2 + x - 12 < 0$

4. $x^2 - x - 6 > 0$

5. $x^2 \geq 4$

6. $x^2 - 9 \leq 0$

7. $2x^2 + x - 3 < 0$

8. $6x^2 - 5x + 1 < 0$

Solve.

9. $x^2 + x - 20 > 0$

10. $x^2 + 3x - 28 > 0$

11. $8x^2 \leq 2x + 3$

12. $4x^2 - 10 \leq 3x$

13. $6x^2 - 5x > 6$

14. $3x^2 + 17x > -10$

15. $-2x + 30 \geq x(x + 5)$
Hint: Put variables on the right and zero on the left in your first step.

16. $55 - x^2 \geq 6x$
Hint: Put variables on the right and zero on the left in your first step.

17. $x^2 - 2x \geq -1$

18. $x^2 + 25 \geq 10x$

19. $x^2 - 4x \leq -4$ **20.** $x^2 - 6x \leq -9$

Solve each of the following quadratic inequalities if possible. Round your answers to the nearest tenth.

21. $x^2 - 2x > 4$ **22.** $x^2 + 6x > 8$ **23.** $x^2 - 6x < -7$

24. $x^2 < 2x + 1$ **25.** $2x^2 \geq x^2 - 4$ **26.** $4x^2 \geq 3x^2 - 9$

27. $5x^2 \leq 4x^2 - 1$ **28.** $x^2 - 1 \leq -17$

Applications

Projectile Flight   *In Exercises 29 and 30, a projectile is fired vertically with an initial velocity of 640 feet per second. The distance s in feet above the ground after t seconds is given by the equation* $s = -16t^2 + 640t$.

29. For what range of time (measured in seconds) will the height s be greater than 6000 feet?

30. For what range of time (measured in seconds) will the height s be less than 4800 feet?

Manufacturing Profit   *In Exercises 31 and 32, the profit of a manufacturing company is determined by the number of units x manufactured each day according to the given equation.* ***(a)*** *Find when the profit is greater than zero.* ***(b)*** *Find the daily profit when 50 units are manufactured.* ***(c)*** *Find the daily profit when 60 units are manufactured.*

31. Profit $= -20(x^2 - 220x + 2400)$ **32.** Profit $= -25(x^2 - 280x + 4000)$

Cumulative Review

33. *Test Scores* The university's synchronized swimming team will not let Mona participate unless she passes biology with a C (70 or better) average. There are six tests in the semester, and she failed the first one (with a score of 0). She decided to find a tutor. Since then, she received an 81, 92, and 80 on the next three tests. What must her minimum scores be on the last two tests to pass the course with a minimum grade of 70 and participate in synchronized swimming?

34. *Party Mixture* In a huge bowl at a college party, there are 360 ounces of mixed potato chips, peanuts, pretzels, and popcorn. There are 70 more ounces of peanuts than potato chips. There are twice as many ounces of pretzels as ounces of popcorn. There are 10 more ounces of popcorn than potato chips. How many ounces of each ingredient are in the snack mix?

Cost of a Cruise *The Circle Line Cruise is a 2-hour, 24-mile cruise around southern Manhattan (New York City). The charge for adults is $18; children 12 and under, $10; and the elderly, $16. For the 3-hour, 35-mile cruise around all of Manhattan, the charge for adults is $22; children 12 and under, $12; and the elderly, $19. The Yoffa family has come to New York for their family reunion and is planning family activities. The family has ten adults, fourteen children under 12, and five elderly members.*

35. What would it cost for all of the family to take the 2-hour trip? The 3-hour trip?

36. Six people do not take a cruise. If the rest of the family takes a 2-hour cruise, it will cost $314. If the rest of the family takes a 3-hour cruise, it will cost $380. How many adults, how many children, and how many elderly members plan to take a cruise?

Putting Your Skills to Work

Using Mathematics to Measure the Consumption of Fresh Vegetables

As medical experts find more benefits from eating fresh vegetables, the consumption of fresh vegetables for the average American has increased each year. The average number of pounds of fresh vegetables that is eaten by each person can be predicted by the quadratic function $P(x) = 0.046x^2 + 1.58x + 150.4$, where $P(x)$ is the number of pounds of fresh vegetables consumed each year by the average American and x is the number of years since 1980. (*Source:* U.S. Department of Agriculture)

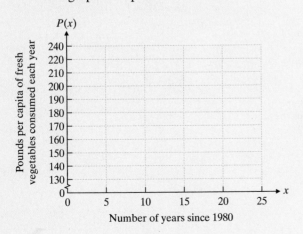

Problems for Individual Investigation and Analysis

1. Find $P(0)$ and $P(15)$.

2. Find $P(10)$ and $P(20)$.

3. Use the values you obtained in problems 1 and 2 above to graph this quadratic function.

4. Use your graph to estimate the value of $P(15)$.

Problems for Group Investigation and Analysis

5. According to the equation, by how many pounds did the annual consumption per person of fresh vegetables increase from 1990 to 2002? Round to the nearest hundredth.

6. According to the equation, by how many pounds did the annual consumption per person of fresh vegetables increase from 1980 to 1998? Round to the nearest hundredth.

7. According to the equation, in what year will the annual consumption per person of fresh vegetables reach the level of 230.7 pounds per year?

8. According to the equation, in what year will the annual consumption per person of fresh vegetables reach the level of 239.2 pounds per year?

Topic	Procedure	Examples
Solving a quadratic equation by using the square root property, p. 422.	If $x^2 = a$, then $x = \pm\sqrt{a}$.	Solve.$$2x^2 - 50 = 0$$$$2x^2 = 50$$$$x^2 = 25$$$$x = \pm\sqrt{25}$$$$x = \pm 5$$
Solving a quadratic equation by completing the square, p. 423.	1. Rewrite the equation in the form $ax^2 + bx = c$. If $a \neq 1$, divide each term of the equation by a. 2. Square half of the numerical coefficient of the linear term. Add the result to both sides of the equation. 3. Factor the left side. 4. Take the square root of both sides of the equation. 5. Solve the resulting equation for x. 6. Check the solutions in the original equation.	Solve.$$2x^2 - 4x - 1 = 0$$$$2x^2 - 4x = 1$$$$\frac{2x^2}{2} - \frac{4x}{2} = \frac{1}{2}$$$$x^2 - 2x + \underline{\quad} = \frac{1}{2} + \underline{\quad}$$$$x^2 - 2x + 1 = \frac{1}{2} + 1$$$$(x - 1)^2 = \frac{3}{2}$$$$x - 1 = \pm\sqrt{\frac{3}{2}}$$$$x - 1 = \frac{\pm\sqrt{6}}{2}$$$$x = 1 \pm \frac{1}{2}\sqrt{6}$$
Placing a quadratic equation in standard form, p. 429.	A quadratic equation in standard form is an equation of the form $ax^2 + bx + c = 0$, where a, b, and c are real numbers and $a \neq 0$. It is often necessary to remove parentheses and clear away fractions by multiplying each term of the equation by the LCD to obtain the standard form.	Rewrite in quadratic form:$$\frac{2}{x - 3} + \frac{x}{x + 3} = \frac{5}{x^2 - 9}$$$$(x + 3)(x - 3)\left[\frac{2}{x - 3}\right] + (x + 3)(x - 3)\left[\frac{x}{x + 3}\right]$$$$= (x + 3)(x - 3)\left[\frac{5}{(x + 3)(x - 3)}\right]$$$$2(x + 3) + x(x - 3) = 5$$$$2x + 6 + x^2 - 3x = 5$$$$x^2 - x + 1 = 0$$
Solve a quadratic equation by using the quadratic formula, p. 429.	If $ax^2 + bx + c = 0$, where $a \neq 0$,$$x = \frac{-b \pm \sqrt{b^2 - 4ac}}{2a}.$$ 1. Rewrite the equation in standard form. 2. Determine the values of a, b, and c. 3. Substitute the values of a, b, and c into the formula. 4. Simplify the result to obtain the values of x. 5. Any imaginary solutions to the quadratic equation should be simplified by using the definition $\sqrt{-a} = i\sqrt{a}$, where $a > 0$.	Solve.$$2x^2 = 3x - 2$$$$2x^2 - 3x + 2 = 0$$$$a = 2, b = -3, c = 2$$$$x = \frac{-(-3) \pm \sqrt{(-3)^2 - 4(2)(2)}}{2(2)}$$$$x = \frac{3 \pm \sqrt{9 - 16}}{4}$$$$x = \frac{3 \pm \sqrt{-7}}{4}$$$$x = \frac{3 \pm i\sqrt{7}}{4}$$

Topic	Procedure	Examples
Equations that can be transformed into quadratic form, p. 438.	1. Find the variable with the smallest exponent. Let this quantity be replaced by y. 2. Continue to make substitutions for the remaining variable terms based on the first substitution. (You should be able to replace the variable with the largest exponent by y^2.) 3. Solve the resulting equation for y. 4. Reverse the substitution used in step 1. 5. Solve the resulting equation for x. 6. Check your solution in the *original* equation.	Solve: $x^{2/3} - x^{1/3} - 2 = 0$ Let $y = x^{1/3}$. Then $y^2 = x^{2/3}$. $$y^2 - y - 2 = 0$$ $$(y-2)(y+1) = 0$$ $y = 2$ or $y = -1$ $x^{1/3} = 2$ or $x^{1/3} = -1$ $(x^{1/3})^3 = 2^3$ $\quad (x^{1/3})^3 = (-1)^3$ $x = 8$ $\qquad\quad x = -1$
Checking solutions for equations in quadratic form (continued), p. 439.		*Check.* $x = 8$: $\quad (8)^{2/3} - (8)^{1/3} - 2 \stackrel{?}{=} 0$ $\qquad\qquad 2^2 - 2 - 2 \stackrel{?}{=} 0$ $\qquad\qquad 4 - 4 = 0$ ✓ $x = -1$: $(-1)^{2/3} - (-1)^{1/3} - 2 \stackrel{?}{=} 0$ $\qquad\qquad (-1)^2 - (-1) - 2 \stackrel{?}{=} 0$ $\qquad\qquad 1 + 1 - 2 = 0$ ✓ Both 8 and -1 are solutions.
Solving quadratic equations containing two or more variables, p. 446.	Treat the letter to be solved for as a variable, but treat all other letters as constants. Solve the equation by factoring, by using the square root property, or by using the quadratic formula.	Solve for x. (a) $6x^2 - 11xw + 4w^2 = 0$ (b) $4x^2 + 5b = 2w^2$ (c) $2x^2 + 3xz - 10z = 0$ (a) By factoring: $$(3x - 4w)(2x - w) = 0$$ $3x - 4w = 0$ or $2x - w = 0$ $x = \dfrac{4w}{3}$ $\qquad x = \dfrac{w}{2}$ (b) Using the square root property: $4x^2 = 2w^2 - 5b$ $x^2 = \dfrac{2w^2 - 5b}{4}$ $x = \pm\sqrt{\dfrac{2w^2 - 5b}{4}} = \pm\dfrac{1}{2}\sqrt{2w^2 - 5b}$ (c) By the quadratic formula, with $a = 2, b = 3z, c = -10z$: $x = \dfrac{-3z \pm \sqrt{9z^2 + 80z}}{4}$
The Pythagorean theorem, p. 448.	In any right triangle, if c is the length of the hypotenuse and a and b are the lengths of the two legs, then $$c^2 = a^2 + b^2.$$	Find a if $c = 7$ and $b = 5$. $49 = a^2 + 25$ $49 - 25 = a^2$ $24 = a^2$ $\sqrt{24} = a$ $2\sqrt{6} = a$

Topic	Procedure	Examples
Graphing quadratic functions, p. 459.	Graph quadratic functions of the form $f(x) = ax^2 + bx + c$ with $a \neq 0$ as follows: **1.** Find the vertex at $\left(\dfrac{-b}{2a}, f\left(\dfrac{-b}{2a}\right)\right)$. **2.** Find the y-intercept, which occurs at $f(0)$. **3.** Find the x-intercepts if they exist. Solve $f(x) = 0$ for x.	Graph $f(x) = x^2 + 6x + 8$. Vertex: $$x = \frac{-6}{2} = -3$$ $$f(-3) = (-3)^2 + 6(-3) + 8 = -1$$ The vertex is $(-3, -1)$. Intercepts: $f(0) = (0)^2 + 6(0) + 8 = 8$ The y-intercept is $(0, 8)$. $$x^2 + 6x + 8 = 0$$ $$(x + 2)(x + 4) = 0$$ $$x = -2, x = -4$$ The x-intercepts are $(-2, 0)$ and $(-4, 0)$. $f(x) = x^2 + 6x + 8$
Solving quadratic inequalities in one variable, p. 466.	**1.** Replace the inequality symbol by an equal sign. Solve the resulting equation to find the critical points. **2.** Use the critical points to separate the number line into distinct regions. **3.** Evaluate the quadratic expression at a test point in each region. **4.** Determine which regions satisfy the original conditions of the quadratic inequality.	Solve and graph: $3x^2 + 5x - 2 > 0$ **1.** $3x^2 + 5x - 2 = 0$ $$(3x - 1)(x + 2) = 0$$ $$3x - 1 = 0 \qquad x + 2 = 0$$ $$x = \frac{1}{3} \qquad x = -2$$ Critical points are -2 and $\dfrac{1}{3}$. **2.** **3.** $3x^2 + 5x - 2$ *Region I:* Pick $x = -3$. $3(-3)^2 + 5(-3) - 2 = 27 - 15 - 2 = 10 > 0$ *Region II:* Pick $x = 0$. $3(0)^2 + 5(0) - 2 = 0 + 0 - 2 = -2 < 0$ *Region III:* Pick $x = 3$. $3(3)^2 + 5(3) - 2 = 27 + 15 - 2 = 40 > 0$ **4.** We know that the expression is greater than zero (that is, $3x^2 + 5x - 2 > 0$) when $$x < -2 \text{ or } x > \frac{1}{3}.$$

Solve each of the following exercises by the specified method. Simplify all answers.

Solve by the square root property.

1. $6x^2 = 24$

2. $(x + 8)^2 = 81$

Solve by completing the square.

3. $x^2 + 8x + 13 = 0$

4. $4x^2 - 8x + 1 = 0$

Solve by the quadratic formula.

5. $3x^2 - 10x + 6 = 0$

6. $x^2 - 6x - 4 = 0$

Solve by any appropriate method and simplify your answers. Express any nonreal complex solutions using i notation.

7. $4x^2 - 12x + 9 = 0$

8. $x^2 - 14 = 5x$

9. $6x^2 - 23x = 4x$

10. $2x^2 = 5x - 1$

11. $x^2 - 3x - 23 = 5$

12. $5x^2 - 10 = 0$

13. $3x^2 - 2x = 15x - 10$

14. $6x^2 + 12x - 24 = 0$

15. $7x^2 + 24 = 5x^2$

16. $3x^2 + 5x + 1 = 0$

17. $3x(3x + 2) - 2 = 3x$

18. $10x(x - 2) + 10 = 2x$

19. $\dfrac{x - 5}{x} + 9x = 1$

20. $\dfrac{4}{5}x^2 + x + \dfrac{1}{5} = 0$

21. $y + \dfrac{5}{3y} + \dfrac{17}{6} = 0$

22. $\dfrac{19}{y} - \dfrac{15}{y^2} + 10 = 0$

23. $\dfrac{15}{y^2} - \dfrac{2}{y} = 1$

24. $y - 18 + \dfrac{81}{y} = 0$

25. $(3y + 2)(y - 1) = 7(-y + 1)$

26. $y(y + 1) + (y + 2)^2 = 4$

27. $\dfrac{2x}{x + 3} + \dfrac{3x - 1}{x + 1} = 3$

28. $\dfrac{4x + 1}{2x + 5} + \dfrac{3x}{x + 4} = 2$

Determine the nature of each of the following quadratic equations. Do not solve the equation. Find the discriminant in each case and determine whether the equation has (a) one rational solution, (b) two rational solutions, (c) two irrational solutions, or (d) two nonreal complex solutions.

29. $2x^2 + 5x - 3 = 0$

30. $3x^2 - 7x - 12 = 0$

31. $3x^2 - 5x + 6 = 0$

32. $25x^2 - 20x + 4 = 0$

Write a quadratic equation having the given numbers as solutions.

33. $5, -5$

34. $3i, -3i$

35. $4\sqrt{2}, -4\sqrt{2}$

36. $-\dfrac{3}{4}, -\dfrac{1}{2}$

Solve for any valid real roots.

37. $x^4 - 6x^2 + 8 = 0$

38. $2x^6 - 5x^3 - 3 = 0$

39. $x^{2/3} - 3 = 2x^{1/3}$

40. $3x - x^{1/2} = 2$

41. $(2x - 5)^2 + 4(2x - 5) + 3 = 0$

42. $1 + 4x^{-8} = 5x^{-4}$

Solve for the variable specified. Assume that all radical expressions obtained have a positive radicand.

43. $3M = \dfrac{2A^2}{N}$; for A

44. $3t^2 + 4b = t^2 + 6ay$; for t

45. $yx^2 - 3x - 7 = 0$; for x

46. $20d^2 - xd - x^2 = 0$; for d

47. $3y^2 - 4ay + 2a = 0$; for y

48. $PV = 5x^2 + 3y^2 + 2x$; for x

Use the Pythagorean theorem to find the missing side. Assume that c is the length of the hypotenuse of a right triangle and that a and b are the lengths of the legs. Leave your answers as a radical in simplified form.

▲ **49.** $a = 3\sqrt{2}, b = 2$; find c

▲ **50.** $c = 16, b = 4$; find a

▲ **51.** *Airplane Flight* A plane is 6 miles away from an observer and exactly 5 miles above the ground. The plane is directly above a car. How far is the car from the observer? Round your answer to the nearest tenth of a mile.

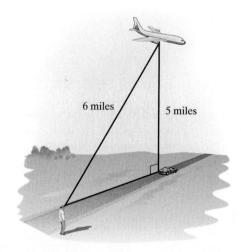

6 miles   5 miles

▲ **52.** *Geometry* The area of a triangle is 70 square centimeters. Its altitude is 6 centimeters longer than twice the length of the base. Find the dimensions of the altitude and base.

▲ **53.** *Geometry* The area of a rectangle is 203 square meters. Its length is 1 meter longer than four times its width. Find the length and width of the rectangle.

54. *Boat Fishing Trip* John rode in a motorboat for 60 miles at constant cruising speed to get to his fishing grounds. Then for 5 miles he trolled to catch fish. His trolling speed was 15 miles per hour slower than his cruising speed. The trip took 4 hours. Find his speed for each part of the trip.

55. *Car Travel in the Rain* Jessica drove at a constant speed for 200 miles. Then it started to rain. So for the next 90 miles she traveled 5 miles per hour slower. The entire trip took 6 hours of driving time. Find her speed for each part of the trip.

▲ **56.** *Garden Walkway* Mr. and Mrs. Gomez are building a rectangular garden that is 10 feet by 6 feet. Around the outside of the garden, they will build a brick walkway. They have 100 square feet of brick. How wide should they make the brick walkway? Round your answer to the nearest tenth of a foot.

▲ **57.** *Swimming Pool* The local YMCA is building a rectangular swimming pool that is 40 feet by 30 feet. The builders want to make a walkway around the pool with a nonslip cement surface. They have enough material to make 296 square feet of nonslip cement surface. How wide should the walkway be?

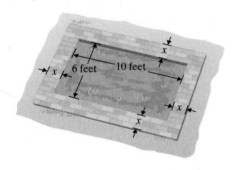

Find the vertex and the intercepts of the following quadratic functions.

58. $g(x) = -x^2 + 6x - 11$

59. $f(x) = x^2 + 10x + 25$

In each of the following exercises, find the vertex, the y-intercept, and the x-intercepts (if any exist) and then graph the function.

60. $f(x) = x^2 + 4x + 3$

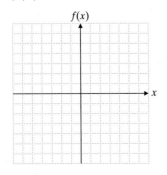

61. $f(x) = x^2 + 6x + 5$

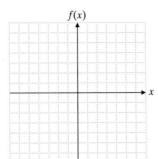

62. $f(x) = -x^2 + 6x - 5$

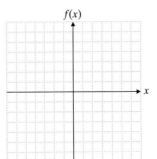

63. *Rocket Flight* A model rocket is launched upward from a platform 40 feet above the ground. The height of the rocket h is given at any time t in seconds by the function $h(t) = -16t^2 + 400t + 40$. Find the maximum height of the rocket. How long will it take the rocket to go through its complete flight and then hit the ground? (Assume that the rocket does *not* have a parachute.) Round your answer to the nearest tenth.

64. *Revenue for a Store* A salesman for an electronics store finds that in 1 month he can sell $(1200 - x)$ compact disc players that each sell for x dollars. Write a function for the revenue. What is the price x that will result in the maximum revenue for the store?

Solve and graph your solutions.

65. $x^2 + 7x - 18 < 0$

66. $x^2 + 4x - 21 < 0$

67. $x^2 - 9x + 20 > 0$

68. $x^2 - 11x + 28 > 0$

Solve each of the following if possible. Approximate, if necessary, any irrational solutions to the nearest tenth.

69. $3x^2 - 5x - 2 \le 0$

70. $2x^2 - 5x - 3 \le 0$

71. $9x^2 - 4 > 0$

72. $16x^2 - 25 > 0$

73. $4x^2 - 8x \le 12 + 5x^2$

74. $x^2 - 9x > 4 - 7x$

75. $x^2 + 13x > 16 + 7x$

76. $3x^2 - 12x > -11$

77. $4x^2 + 12x + 9 < 0$

78. $-2x^2 + 7x + 12 \le -3x^2 + x$

To Think About

79. $(x + 4)(x - 2)(3 - x) > 0$

80. $(x + 1)(x + 4)(2 - x) < 0$

Remember to use your Chapter Test Prep Video CD to see the worked-out solutions to the test problems you want to review.

Solve the quadratic equations and simplify your answers. Use i notation for any imaginary numbers.

1. $8x^2 + 9x = 0$

2. $6x^2 - 3x = 1$

3. $\dfrac{3x}{2} - \dfrac{8}{3} = \dfrac{2}{3x}$

4. $x(x - 3) - 30 = 5(x - 2)$

5. $7x^2 - 4 = 52$

6. $\dfrac{2x}{2x + 1} - \dfrac{6}{4x^2 - 1} = \dfrac{x + 1}{2x - 1}$

7. $2x^2 - 6x + 5 = 0$

8. $2x(x - 3) = -3$

Solve for any valid real roots.

9. $x^4 - 11x^2 + 18 = 0$

10. $3x^{-2} - 11x^{-1} - 20 = 0$

11. $x^{2/3} - 3x^{1/3} - 4 = 0$

1. _____

2. _____

3. _____

4. _____

5. _____

6. _____

7. _____

8. _____

9. _____

10. _____

11. _____

481

Solve for the variable specified.

12.

12. $B = \dfrac{xyw}{z^2}$; for z

13.

13. $5y^2 + 2by + 6w = 0$; for y

▲ **14.** The area of a rectangle is 80 square miles. Its length is 1 mile longer than three times its width. Find its length and width.

14.

▲ **15.** Find the hypotenuse of a right triangle if the lengths of its legs are 6 and $2\sqrt{3}$.

15.

16. Shirley and Bill paddled a canoe at a constant speed for 6 miles. They rested, had lunch, and then paddled 1 mile per hour faster for an additional 3 miles. The travel time for the entire trip was 4 hours. How fast did they paddle during each part of the trip?

16.

17. Find the vertex and the intercepts for $f(x) = -x^2 - 6x - 5$. Then graph the function.

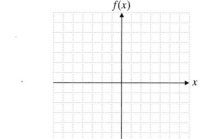

17.

18.

Solve.

18. $2x^2 + 3x \geq 27$

19.

19. $x^2 - 5x - 14 < 0$

20. Use a calculator or square root table to approximate to the nearest tenth a solution to $x^2 + 3x - 7 > 0$.

20.

Approximately one half of this test is based on the content of Chapters 1–7. The remainder is based on the content of Chapter 8.

1. Simplify: $(-3x^{-2}y^3)^4$

2. Collect like terms. $\dfrac{1}{2}a^3 - 2a^2 + 3a - \dfrac{1}{4}a^3 - 6a + a^2$

3. Solve for x: $\dfrac{1}{3}(x - 3) + 1 = \dfrac{1}{2}x - 2$

4. Graph: $6x - 3y = -12$

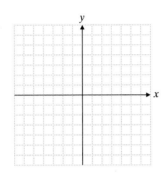

5. Write the equation of the line parallel to $2y + x = 8$ and passing through $(6, -1)$.

6. Solve the system: $\quad 7x - 3y = 1$
$\qquad\qquad\qquad -5x + 4y = 3$

7. Factor: $125x^3 - 27y^3$

8. Simplify: $\sqrt{72x^3y^6}$

9. Multiply: $\left(3 + \sqrt{2}\right)\left(\sqrt{6} + \sqrt{3}\right)$

10. Rationalize the denominator: $\dfrac{3x}{\sqrt{6}}$

Solve and simplify your answers. Use i notation for imaginary numbers.

11. $3x^2 + 12x = 26x$

12. $12x^2 = 11x - 2$

13. $44 = 3(2x - 3)^2 + 8$

14. $3 - \dfrac{4}{x} + \dfrac{5}{x^2} = 0$

1. _____

2. _____

3. _____

4. _____

5. _____

6. _____

7. _____

8. _____

9. _____

10. _____

11. _____

12. _____

13. _____

14. _____

15. _____

Solve and check.

15. $\sqrt{3x + 7} - 1 = x$

16. _____

16. $x^{2/3} + 9x^{1/3} + 18 = 0$

Solve for y.

17. _____

17. $2y^2 + 5wy - 7z = 0$

18. $3y^2 + 16z^2 = 5w$

18. _____

▲ **19.** The hypotenuse of a right triangle is $\sqrt{31}$. One leg of the triangle is 4. Find the length of the other leg.

19. _____

▲ **20.** A triangle has an area of 45 square meters. The altitude is 3 meters longer than three times the length of the base. Find each dimension.

20. _____

Exercises 21 and 22 refer to the quadratic function $f(x) = -x^2 + 8x - 12$.

21. Find the vertex and the intercepts of the function.

22. Graph the function.

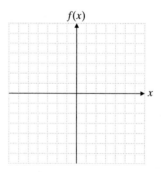

21. _____

22. _____

23. _____

Solve each of the following quadratic inequalities.

23. $6x^2 - x \le 2$

24. _____

24. $x^2 > -2x + 15$

CHAPTER

9

The United States has landed several rovers on the surface of Mars, and these rovers have sent back amazing pictures of the Red Planet. What kind of math did scientists use in planning and carrying out those missions? You can use your math skills to answer some intriguing questions about these missions to Mars. Turn to page 528.

The Conic Sections

Student Learning Objectives

After studying this section, you will be able to:

1 Find the distance between two points.

2 Find the center and radius of a circle and graph the circle if the equation is in standard form.

3 Write the equation of a circle in standard form given its center and radius.

4 Rewrite the equation of a circle in standard form.

In this chapter we'll talk about the equations and graphs of four special geometric figures—the circle, the parabola, the ellipse, and the hyperbola. These shapes are called **conic sections** because they can be formed by slicing a cone with a plane. The equation of any conic section is of degree 2.

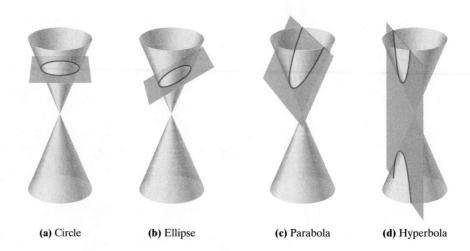

(a) Circle   **(b)** Ellipse   **(c)** Parabola   **(d)** Hyperbola

Conic sections are an important and interesting subject. They are studied along with many other things in a branch of mathematics called *analytic geometry*. Conic sections can be found in applications of physics and engineering. Satellite transmission dishes have parabolic shapes; the orbits of planets are ellipses, and the orbits of comets are hyperbolas; the path of a ball, rocket, or bullet is a parabola (if we neglect air resistance).

1 ## Finding the Distance Between Two Points

Before we investigate the conic sections, we need to know how to find the distance between two points in the *xy*-plane. We will derive a *distance formula* and use it to find the equations for the conic sections.

Recall from Chapter 1 that to find the distance between two points on the real number line, we simply find the absolute value of the difference of the values of the points. For example, the distance from -3 to 5 on the *x*-axis is

$$|5 - (-3)| = |5 + 3| = 8.$$

Remember that absolute value is another name for distance. We could have written

$$|-3 - (5)| = |-8| = 8.$$

Similarly, the distance from -3 to 5 on the *y*-axis is

$$|5 - (-3)| = 8.$$

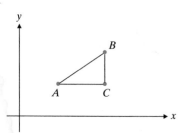

We use this simple fact to find the distance between two points in the *xy*-plane. Let $A(x_1, y_1)$ and $B(x_2, y_2)$ be points on a graph. First we draw a horizontal line through A, and then we draw a vertical line through B. (We could have drawn a horizontal line through B and a vertical line through A.) The lines intersect at point $C(x_2, y_1)$. Why are the coordinates x_2, y_1? The distance from A to C is $|x_2 - x_1|$ and from B to C $|y_2 - y_1|$.

Now, if we draw a line from A to B, we have a right triangle ABC. We can use the Pythagorean theorem to find the length (distance) of the line from A to B. By the Pythagorean theorem,

$$(AB)^2 = (AC)^2 + (BC)^2.$$

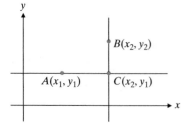

Let's rename the distance AB as d. Then

$$d^2 = (|x_2 - x_1|)^2 + (|y_2 - y_1|)^2$$

$$d = \sqrt{(x_2 - x_1)^2 + (y_2 - y_1)^2}.$$

This is the **distance formula.**

DISTANCE FORMULA

The distance between two points (x_1, y_1) and (x_2, y_2) is

$$d = \sqrt{(x_2 - x_1)^2 + (y_2 - y_1)^2}.$$

EXAMPLE 1   Find the distance between $(3, -4)$ and $(-2, -5)$.

Solution   To use the formula, we arbitrarily let $(x_1, y_1) = (3, -4)$ and $(x_2, y_2) = (-2, -5)$.

$$d = \sqrt{(x_2 - x_1)^2 + (y_2 - y_1)^2}$$
$$= \sqrt{[-2 - 3]^2 + [-5 - (-4)]^2}$$
$$= \sqrt{(-5)^2 + (-5 + 4)^2}$$
$$= \sqrt{(-5)^2 + (-1)^2}$$
$$= \sqrt{25 + 1} = \sqrt{26}$$

Practice Problem 1   Find the distance between $(-6, -2)$ and $(3, 1)$.

The choice of which point is (x_1, y_1) and which point is (x_2, y_2) is up to you. We would obtain exactly the same answer in Example 1 if $(x_1, y_1) = (-2, -5)$ and if $(x_2, y_2) = (3, -4)$. Try it for yourself and see whether you obtain the same result.

② Finding the Center and Radius of a Circle and Graphing the Circle

A **circle** is defined as the set of all points in a plane that are at a fixed distance from a point in that plane. The fixed distance is called the **radius,** and the point is called the **center** of the circle.

We can use the distance formula to find the equation of a circle. Let a circle of radius r have its center at (h, k). For any point (x, y) on the circle, the distance formula tells us that

$$\sqrt{(x - h)^2 + (y - k)^2} = r.$$

Squaring each side gives

$$(x - h)^2 + (y - k)^2 = r^2.$$

This is the equation of a circle with center at (h, k) and radius r.

Graphing Calculator

Graphing Circles

A graphing calculator is designed to graph *functions*. In order to graph a circle, you need to separate it into two halves, each of which is a function. Thus, in order to graph the circle in Example 2 on the next page, first solve for y.

$$(y - 3)^2 = 25 - (x - 2)^2$$
$$y - 3 = \pm\sqrt{25 - (x - 2)^2}$$
$$y = 3 \pm \sqrt{25 - (x - 2)^2}$$

Now graph the two functions

$$y_1 = 3 + \sqrt{25 - (x - 2)^2}$$
(the upper half of the circle)

and

$$y_2 = 3 - \sqrt{25 - (x - 2)^2}$$
(the lower half of the circle).

To get a proper-looking circle, use a "square" window setting. Window settings will vary depending on the calculator. Display:

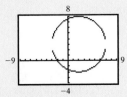

Notice that due to limitations in the calculator, it is not a perfect circle and two small gaps appear.

STANDARD FORM OF THE EQUATION OF A CIRCLE

The standard form of the equation of a circle with center at (h, k) and radius r is

$$(x - h)^2 + (y - k)^2 = r^2.$$

EXAMPLE 2 Find the center and radius of the circle $(x - 2)^2 + (y - 3)^2 = 25$. Then sketch its graph.

Solution From the equation of a circle,

$$(x - h)^2 + (y - k)^2 = r^2,$$

we see that $(h, k) = (2, 3)$. Thus, the center of the circle is at $(2, 3)$. Since $r^2 = 25$, the radius of the circle is $r = 5$.

The graph of this circle is shown on the right.

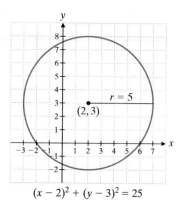

$(x - 2)^2 + (y - 3)^2 = 25$

PRACTICE PROBLEM 2

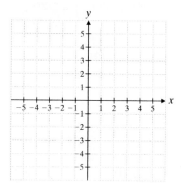

Practice Problem 2 Find the center and radius of the circle

$$(x + 1)^2 + (y + 2)^2 = 9.$$

Then sketch its graph.

③ Writing the Equation of a Circle in Standard Form Given the Center and Radius

We can write the standard form of the equation of a specific circle if we are given the center and the radius. We use the definition of the standard form of the equation of a circle to write the equation we want.

EXAMPLE 3 Write the equation of the circle with center $(-1, 3)$ and radius $\sqrt{5}$. Put your answer in standard form.

Solution We are given that $(h, k) = (-1, 3)$ and $r = \sqrt{5}$. Thus,

$$(x - h)^2 + (y - k)^2 = r^2$$

becomes the following:

$$[x - (-1)]^2 + [y - 3]^2 = \left(\sqrt{5}\right)^2$$
$$(x + 1)^2 + (y - 3)^2 = 5$$

Be careful of the signs. It is easy to make a sign error in these steps.

Practice Problem 3 Write the equation of the circle with center $(-5, 0)$ and radius $\sqrt{3}$. Put your answer in standard form.

NOTE TO STUDENT: Fully worked-out solutions to all of the Practice Problems can be found at the back of the text starting at page SP-1

4 Rewriting the Equation of a Circle in Standard Form

The standard form of the equation of a circle helps us sketch the graph of the circle. Sometimes the equation of a circle is not given in standard form, and we need to rewrite the equation.

EXAMPLE 4   Write the equation of the circle $x^2 + 2x + y^2 + 6y + 6 = 0$ in standard form. Find the radius and center of the circle and sketch its graph.

Solution   The standard form of the equation of a circle is
$$(x - h)^2 + (y - k)^2 = r^2.$$
If we multiply out the terms in the equation, we get
$$(x^2 - 2hx + h^2) + (y^2 - 2ky + k^2) = r^2.$$
Comparing this with the equation we were given,
$$(x^2 + 2x) + (y^2 + 6y) = -6,$$
suggests that we can complete the square to put the equation in standard form.

$$x^2 + 2x + \underline{\quad} + y^2 + 6y + \underline{\quad} = -6$$
$$x^2 + 2x + 1 + y^2 + 6y + 9 = -6 + 1 + 9$$
$$x^2 + 2x + 1 + y^2 + 6y + 9 = 4$$
$$(x + 1)^2 + (y + 3)^2 = 4$$

Thus, the center is at $(-1, -3)$, and the radius is 2. The sketch of the circle is shown below.

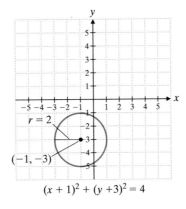

$$(x + 1)^2 + (y + 3)^2 = 4$$

Practice Problem 4   Write the equation of the circle $x^2 + 4x + y^2 + 2y - 20 = 0$ in standard form. Find the radius and center of the circle and sketch its graph.

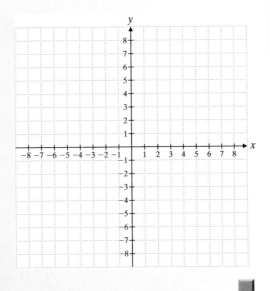

Student Solutions Manual CD/Video PH Math Tutor Center MathXL®Tutorials on CD MathXL® MyMathLab® Interactmath.com

Verbal and Writing Skills

1. Explain how you would find the distance from -2 to 4 on the y-axis.

2. Explain how you would find the distance between $(3, -1)$ and $(-4, 0)$ in the xy-plane.

3. $(x - 1)^2 + (y + 2)^2 = 9$ is the equation of a circle. Explain how to determine the center and the radius of the circle.

4. $x^2 - 6x + y^2 - 2y = 6$ is the equation of a circle. Explain how you would rewrite the equation in standard form.

Find the distance between each pair of points. Simplify your answers.

5. $(1, 6)$ and $(2, 4)$

6. $(4, 6)$ and $(7, 5)$

7. $(0, -3)$ and $(-4, 1)$

8. $(-5, 6)$ and $(0, -4)$

9. $(4, -5)$ and $(-2, -13)$

10. $(-7, 13)$ and $(-12, 1)$

11. $\left(\frac{5}{4}, -\frac{1}{3}\right)$ and $\left(\frac{1}{4}, -\frac{2}{3}\right)$

12. $\left(-1, \frac{1}{5}\right)$ and $\left(-\frac{1}{2}, \frac{11}{5}\right)$

13. $\left(\frac{1}{3}, \frac{3}{5}\right)$ and $\left(\frac{7}{3}, \frac{1}{5}\right)$

14. $\left(-\frac{1}{4}, \frac{1}{7}\right)$ and $\left(\frac{3}{4}, \frac{6}{7}\right)$

15. $(1.3, 2.6)$ and $(-5.7, 1.6)$

16. $(8.2, 3.5)$ and $(6.2, -0.5)$

Find the two values of the unknown coordinate so that the distance between the points is as given.

17. $(7, 2)$ and $(1, y)$; distance is 10

18. $(3, y)$ and $(3, -5)$; distance is 9

19. $(1.5, 2)$ and $(0, y)$; distance is 2.5

20. $\left(1, \frac{15}{2}\right)$ and $\left(x, -\frac{1}{2}\right)$; distance is 10

21. $(7, 3)$ and $(x, 6)$; distance is $\sqrt{10}$

22. $(4, 5)$ and $(2, y)$; distance is $\sqrt{5}$

Applications

Radar Detection *Use the following information to solve Exercises 23 and 24. An airport is located at point O. A short-range radar tower is located at point R. The maximum range at which the radar can detect a plane is 4 miles from point R.*

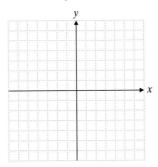

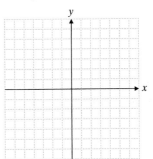

23. Assume that R is 5 miles east of O and 7 miles north of O. In other words, R is located at the point $(5, 7)$. An airplane is flying parallel to and 2 miles east of the north axis. (In other words, the plane is flying along the path $x = 2$.) What is the *shortest distance* north of the airport at which the plane can be detected by the radar tower at R? Round your answer to the nearest tenth of a mile.

24. Assume that R is 6 miles east of O and 6 miles north of O. In other words, R is located at the point $(6, 6)$. An airplane is flying parallel to and 4 miles east of the north axis. (In other words, the plane is flying along the path $x = 4$.) What is the *greatest distance* north of the airport at which the plane can still be detected by the radar tower at R? Round your answer to the nearest tenth of a mile.

Write in standard form the equation of the circle with the given center and radius.

25. center $(-3, 7)$; $r = 6$

26. center $(8, -2)$; $r = 7$

27. center $(-1.8, 0)$; $r = \dfrac{2}{5}$

28. center $\left(0, \dfrac{3}{2}\right)$; $r = \dfrac{1}{2}$

29. center $\left(\dfrac{3}{8}, 0\right)$; $r = \sqrt{3}$

30. center $(0, -1.7)$; $r = \sqrt{7}$

Give the center and radius of each circle. Then sketch its graph.

31. $x^2 + y^2 = 25$

32. $x^2 + y^2 = 9$

33. $(x - 5)^2 + (y - 3)^2 = 16$

34. $(x - 3)^2 + (y - 2)^2 = 4$ **35.** $(x + 2)^2 + (y - 3)^2 = 25$ **36.** $\left(x - \dfrac{3}{2}\right)^2 + (y + 2)^2 = 9$

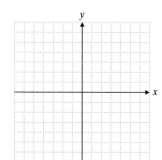

Rewrite each equation in standard form, using the approach of Example 4. Find the center and radius of each circle.

37. $x^2 + y^2 + 8x - 6y - 24 = 0$ **38.** $x^2 + y^2 + 6x - 4y - 3 = 0$

39. $x^2 + y^2 - 12x + 2y - 12 = 0$ **40.** $x^2 + y^2 + 4x - 4y + 7 = 0$

41. $x^2 + y^2 + 3x - 2 = 0$ **42.** $x^2 + y^2 - 5x - 1 = 0$

43. *Ferris Wheels* A Ferris wheel has a radius r of 25.3 feet. The height of the tower t is 31.8 feet. The distance d from the origin to the base of the tower is 44.8 feet. Find the standard form of the equation of the circle represented by the Ferris wheel.

44. *Ferris Wheels* A Ferris wheel has a radius r of 25.1 feet. The height of the tower t is 29.7 feet. The distance d from the origin to the base of the tower is 42.7 feet. Find the standard form of the equation of the circle represented by the Ferris wheel.

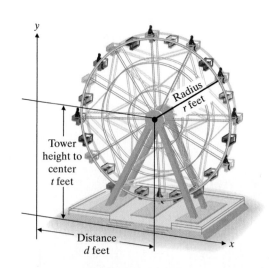

Optional Graphing Calculator Problems

Graph each circle with your graphing calculator.

45. $(x - 5.32)^2 + (y + 6.54)^2 = 47.28$

46. $x^2 + 9.56x + y^2 - 7.12y + 8.9995 = 0$

Cumulative Review

Solve the following quadratic equations by factoring.

47. $9 + \dfrac{3}{x} = \dfrac{2}{x^2}$

48. $3x^2 - 5x + 2 = 0$

Solve the following quadratic equations by using the quadratic formula.

49. $4x^2 + 2x = 1$

50. $5x^2 - 6x - 7 = 0$

▲ **51.** *Volcano Eruptions* The 1980 eruptions of Mount Saint Helens blew down or scorched 230 square miles of forest. A deposit of rock and sediments soon filled up a 20-square-mile area to an average depth of 150 feet. How many cubic feet of rock and sediments settled in this region?

52. *Volcano Eruptions* Within a 15-mile radius north of Mt. Saint Helens, the blast of its 1980 eruption traveled at up to 670 miles per hour. If an observer 15 miles north of the volcano saw the blast and attempted to run for cover, how many seconds did he have to run before the blast reached his original location?

Student Learning Objectives

After studying this section, you will be able to:

1 Graph vertical parabolas.

2 Graph horizontal parabolas.

3 Rewrite the equation of a parabola in standard form.

If we pass a plane through a cone so that the plane is parallel to but not touching a side of the cone, we form a **parabola.** A **parabola** is defined as the set of points that are the same distance from some fixed line (called the **directrix**) and some fixed point (called the **focus**) that is *not* on the line.

The shape of a parabola is a common one. For example, the cables that are used to support the weight of a bridge are in the shape of parabolas.

The simplest form for the equation is one variable = (another variable)^2. That is, $y = x^2$ or $x = y^2$. We will make a table of values for each equation, plot the points, and draw a graph. For the first equation we choose values for x and find y. For the second equation we choose values for y and find x.

$y = x^2$

x	y
−2	4
−1	1
0	0
1	1
2	4

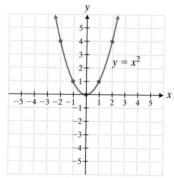

Vertical Parabola

$x = y^2$

x	y
4	−2
1	−1
0	0
1	1
4	2

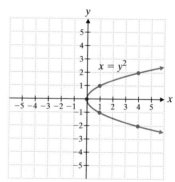

Horizontal Parabola

Notice that the graph of $y = x^2$ is symmetric about the y-axis. That is, if you folded the graph along the y-axis, the two parts of the curve would coincide. For this parabola, the y-axis is the **axis of symmetry.**

What is the axis of symmetry for the parabola $x = y^2$? Every parabola has an axis of symmetry. This axis can be *any* line; it depends on the location and orientation of the parabola in the rectangular coordinate system. The point at which the parabola crosses the axis of symmetry is the **vertex.** What are the coordinates of the vertex for $y = x^2$? For $x = y^2$?

1 Graphing Vertical Parabolas

EXAMPLE 1 Graph $y = (x - 2)^2$. Identify the vertex and the axis of symmetry.

Solution We make a table of values. We begin with $x = 2$ in the middle of the table of values because $(2 - 2)^2 = 0$. That is, when $x = 2$, $y = 0$. We then fill in the x- and y-values above and below $x = 2$. We plot the points and draw the graph.

$y = (x - 2)^2$

x	y
4	4
3	1
2	0
1	1
0	4

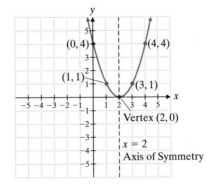

The vertex is $(2, 0)$, and the axis of symmetry is the line $x = 2$.

Practice Problem 1 Graph $y = -(x + 3)^2$. Identify the vertex and the axis of symmetry.

PRACTICE PROBLEM 1

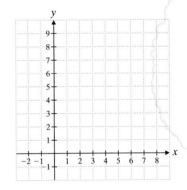

EXAMPLE 2 Graph $y = (x - 2)^2 + 3$. Find the vertex, the axis of symmetry, and the y-intercept.

Solution This graph looks just like the graph of $y = x^2$, except that it is shifted 2 units to the right and 3 units up. The vertex is $(2, 3)$. The axis of symmetry is $x = 2$. We can find the y-intercept by letting $x = 0$ in the equation. We get

$$y = (0 - 2)^2 + 3 = 4 + 3 = 7.$$

Thus, the y-intercept is $(0, 7)$.

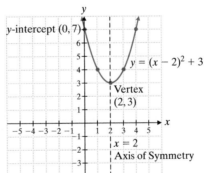

Practice Problem 2 Graph the parabola $y = (x - 6)^2 + 4$.

PRACTICE PROBLEM 2

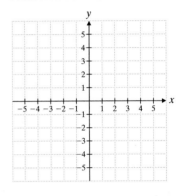

NOTE TO STUDENT: Fully worked-out solutions to all of the Practice Problems can be found at the back of the text starting at page SP-1

The examples we have studied illustrate the following properties of the standard form of the equation of a vertical parabola.

STANDARD FORM OF THE EQUATION OF A VERTICAL PARABOLA

1. The graph of $y = a(x - h)^2 + k$, where $a \neq 0$, is a vertical parabola.
2. The parabola opens upward ⌣ if $a > 0$ and downward ⌢ if $a < 0$.
3. The vertex of the parabola is (h, k).
4. The axis of symmetry is the line $x = h$.
5. The y-intercept is the point where the parabola crosses the y-axis (i.e., where $x = 0$).

We can use these properties as steps to graph a parabola. If we want greater accuracy, we should also plot a few other points.

EXAMPLE 3 Graph $y = -\frac{1}{2}(x + 3)^2 - 1$.

Solution

Step 1 The equation has the form $y = a(x - h)^2 + k$, where $a = -\frac{1}{2}, h = -3$, and $k = -1$, so it is a vertical parabola.

$$y = a(x - h)^2 + k$$

$$y = -\frac{1}{2}[x - (-3)]^2 + (-1)$$

Step 2 $a < 0$; so the parabola opens downward.

Step 3 We have $h = -3$ and $k = -1$.
Therefore, the vertex of the parabola is $(-3, -1)$.

Step 4 The axis of symmetry is the line $x = -3$.
We plot a few points on either side of the axis of symmetry. We try $x = -1$ because $(-1 + 3)^2$ is 4 and $-\frac{1}{2}(4)$ is an integer. We avoid fractions. When $x = -1, y = -\frac{1}{2}(-1 + 3)^2 - 1 = -3$. Thus, the point is $(-1, -3)$. The image of this point on the other side of the axis of symmetry is $(-5, -3)$. We now try $x = 1$. When $x = 1$, $y = -\frac{1}{2}(1 + 3)^2 - 1 = -9$. Thus, the point is $(1, -9)$. The image of this point on the other side of the axis of symmetry is $(-7, -9)$.

Step 5 When $x = 0$, we have the following:

$$y = -\frac{1}{2}(0 + 3)^2 - 1$$

$$= -\frac{1}{2}(9) - 1$$

$$= -4.5 - 1 = -5.5$$

Thus, the y-intercept is $(0, -5.5)$.
The graph is shown on the right.

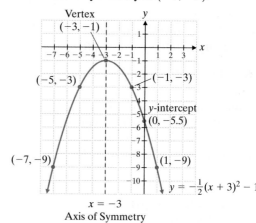

PRACTICE PROBLEM 3

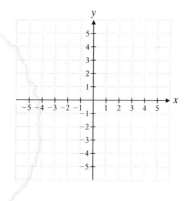

NOTE TO STUDENT: *Fully worked-out solutions to all of the Practice Problems can be found at the back of the text starting at page SP-1*

Practice Problem 3 Graph $y = \frac{1}{4}(x - 2)^2 + 3$.

② Graphing Horizontal Parabolas

Recall that the equation $x = y^2$, in which the squared term is the y-variable, describes a horizontal parabola. Horizontal parabolas open to the left or right. They are symmetric about the x-axis or about a line parallel to the x-axis. We now look at examples of horizontal parabolas.

EXAMPLE 4 Graph $x = -2y^2$.

Solution Notice that the y-term is squared. This means that the parabola is horizontal. We make a table of values, plot points, and draw the graph. To make the table of values, we choose values for y and find x. We begin with $y = 0$.

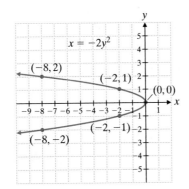

$x = -2y^2$

x	y
-8	-2
-2	-1
0	0
-2	1
-8	2

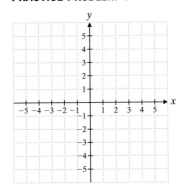

The parabola $x = -2y^2$ has its vertex at $(0, 0)$. The axis of symmetry is the x-axis.

Practice Problem 4   Graph the parabola $x = -2y^2 + 4$.

TO THINK ABOUT: Example 4 Follow-up   Compare the graphs in Example 4 and Practice Problem 4 to the graph of $x = y^2$. How are they different? How are they the same? What does the coefficient -2 in the equation $x = -2y^2$ do to the graph of the equation $x = y^2$? What does the constant 4 in $x = -2y^2 + 4$ do to the graph of $x = -2y^2$?

Now we can make the same type of observations for horizontal parabolas as we did for vertical ones.

STANDARD FORM OF THE EQUATION OF A HORIZONTAL PARABOLA

1. The graph of $x = a(y - k)^2 + h$, where $a \neq 0$, is a horizontal parabola.

2. The parabola opens to the right ⌐ if $a > 0$ and opens to the left ⌐ if $a < 0$.

3. The vertex of the parabola is (h, k).

4. The axis of symmetry is the line $y = k$.

5. The x-intercept is the point where the parabola crosses the x-axis (i.e., where $y = 0$).

EXAMPLE 5   Graph $x = (y - 3)^2 - 5$. Find the vertex, the axis of symmetry, and the x-intercept.

Solution

Step 1   The equation has the form $x = a(y - k)^2 + h$, where $a = 1$, $k = 3$, and $h = -5$, so it is a horizontal parabola.

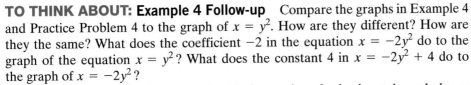

$$x = a(y - k)^2 + h$$
$$x = 1(y - 3)^2 + (-5)$$

Step 2 $a > 0$; so the parabola opens to the right.

Step 3   We have $k = 3$ and $h = -5$. Therefore, the vertex is $(-5, 3)$.

Step 4   The line $y = 3$ is the axis of symmetry.

We look for a few points on either side of the axis of symmetry. We will try y-values close to the vertex $(-5, 3)$. We try $y = 4$ and $y = 2$. When $y = 4$, $x = (4 - 3)^2 - 5 = -4$. When $y = 2$, $x = (2 - 3)^2 - 5 = -4$. Thus, the points are $(-4, 4)$ and $(-4, 2)$. (Remember to list the *x*-value first in a coordinate pair.) We try $y = 5$ and $y = 1$. When $y = 5$, $x = (5 - 3)^2 - 5 = -1$. When $y = 1$, $x = (1 - 3)^2 - 5 = -1$. Thus, the points are $(-1, 5)$ and $(-1, 1)$. You may prefer to find one point, graph it, and find its image on the other side of the axis of symmetry, as was done in Example 3. We decided to look for both pairs of points using the equation.

Step 5   When $y = 0$,

$$x = (0 - 3)^2 - 5 = 9 - 5 = 4.$$

Thus, the *x*-intercept is $(4, 0)$. We plot the points and draw the graph.

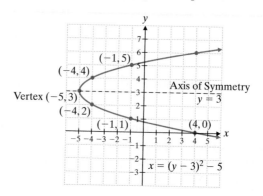

PRACTICE PROBLEM 5

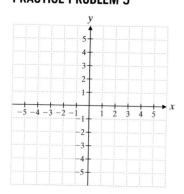

NOTE TO STUDENT: *Fully worked-out solutions to all of the Practice Problems can be found at the back of the text starting at page SP-1*

Notice that the graph also crosses the *y*-axis. You can find the *y*-intercepts by setting *x* equal to 0 and solving the resulting quadratic equation. Try it.

Practice Problem 5   Graph the parabola $x = -(y + 1)^2 - 3$. Find the vertex, the axis of symmetry, and the *x*-intercept.

③ Rewriting the Equation of a Parabola in Standard Form

So far, all the equations we have graphed have been in standard form. This rarely happens in the real world. How do you suppose we put the quadratic equation $y = ax^2 + bx + c$ in the standard form $y = a(x - h)^2 + k$? We do so by completing the square.

EXAMPLE 6   Place the equation $x = y^2 + 4y + 1$ in standard form. Then graph it.

Solution   Since the *y*-term is squared, we have a horizontal parabola. So the standard form is

$$x = a(y - k)^2 + h.$$

Now we have the following:

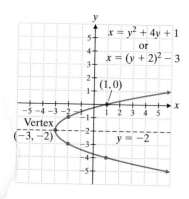

$x = y^2 + 4y + \underline{\qquad} - \underline{\qquad} + 1$ Whatever we add to the right side we must also subtract from the right side.

$= y^2 + 4y + \left(\dfrac{4}{2}\right)^2 - \left(\dfrac{4}{2}\right)^2 + 1$ Complete the square.

$= (y^2 + 4y + 4) - 3$ Simplify.

$= (y + 2)^2 - 3$ Standard form.

We see that $a = 1$, $k = -2$, and $h = -3$. Since *a* is positive, the parabola opens to the right. The vertex is $(-3, -2)$. The axis of symmetry is $y = -2$. If we let $y = 0$, we find that the *x*-intercept is $(1, 0)$. The graph is in the margin on the left.

PRACTICE PROBLEM 6

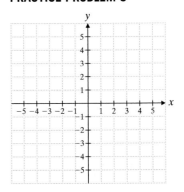

Practice Problem 6   Place the equation $x = y^2 - 6y + 13$ in standard form and graph it.

EXAMPLE 7   Place the equation $y = 2x^2 - 4x - 1$ in standard form. Then graph it.

Solution   This time the x-term is squared, so we have a vertical parabola. The standard form is

$$y = a(x - h)^2 + k.$$

We need to complete the square.

$$y = 2(x^2 - 2x + \underline{\quad}) - \underline{\quad} - 1$$
$$= 2[x^2 - 2x + (1)^2] - 2(1)^2 - 1$$
$$= 2(x - 1)^2 - 3$$

The parabola opens upward ($a > 0$), the vertex is $(1, -3)$, the axis of symmetry is $x = 1$, and the y-intercept is $(0, -1)$.

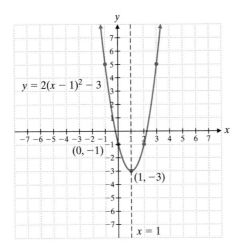

Practice Problem 7   Place $y = 2x^2 + 8x + 9$ in standard form and graph it.

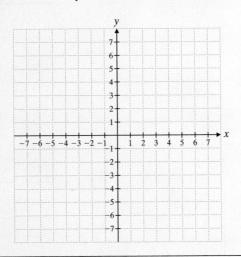

Graphing Calculator

Graphing Parabolas

Graphing horizontal parabolas such as the one in Example 6 on a graphing calculator requires dividing the curve into two halves. In this case the halves would be

$$y_1 = -2 + \sqrt{x + 3}$$

and

$$y_2 = -2 - \sqrt{x + 3}$$

Vertical parabolas can be graphed immediately on a graphing calculator. Why is this? How can you tell whether it is necessary to divide a curve into two halves? Graph the equations below on a graphing calculator. Use the quadratic formula when needed.

1. $y^2 + 8x - 4y = 28$
2. $4x^2 - 4x + 32y = 47$

Verbal and Writing Skills

1. The graph of $y = x^2$ is symmetric about the _____. The graph of $x = y^2$ is symmetric about the _____.

2. Explain how to determine the axis of symmetry of the parabola $x = \frac{1}{2}(y + 5)^2 - 1$.

3. Explain how to determine the vertex of the parabola $y = 2(x - 3)^2 + 4$.

4. How does the coefficient -6 affect the graph of the parabola $y = -6x^2$?

Graph each parabola and label the vertex. Find the y-intercept.

5. $y = -4x^2$

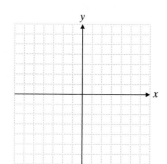

6. $y = -3x^2$

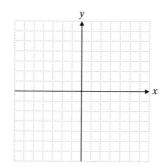

7. $y = x^2 - 6$

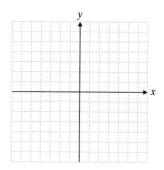

8. $y = x^2 + 2$

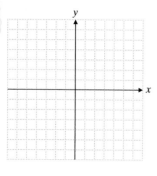

9. $y = \frac{1}{2}x^2 - 2$

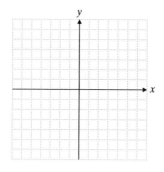

10. $y = \frac{1}{4}x^2 + 1$

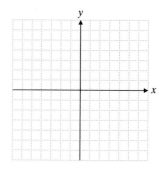

11. $y = (x - 3)^2 - 2$

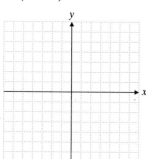

12. $y = (x - 2)^2 - 4$

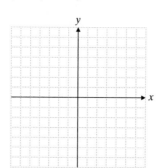

13. $y = 2(x - 1)^2 + \frac{3}{2}$

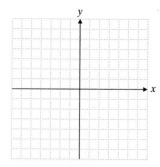

14. $y = 2(x - 2)^2 + \dfrac{5}{2}$

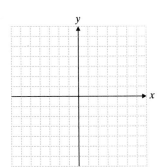

15. $y = -4\left(x + \dfrac{3}{2}\right)^2 + 5$

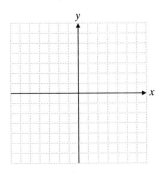

16. $y = -2\left(x + \dfrac{1}{2}\right)^2 - 1$

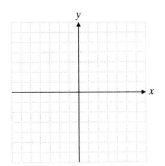

Graph each parabola and label the vertex. Find the x-intercept.

17. $x = \dfrac{1}{2}y^2$

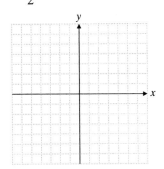

18. $x = \dfrac{2}{3}y^2$

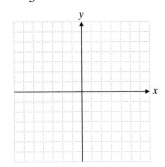

19. $x = \dfrac{1}{4}y^2 - 2$

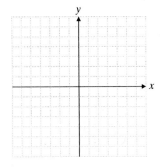

20. $x = \dfrac{1}{3}y^2 + 1$

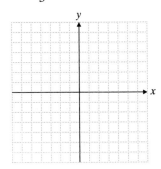

21. $x = -y^2 + 2$

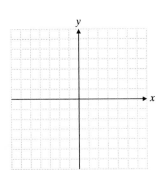

22. $x = -y^2 - 1$

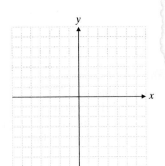

23. $x = (y - 2)^2 + 3$

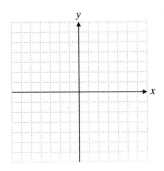

24. $x = (y - 4)^2 + 1$

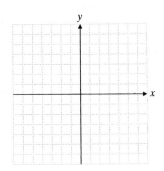

25. $x = -3(y + 1)^2 - 2$

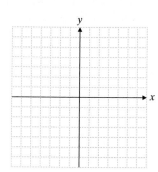

26. $x = -2(y + 3)^2 - 1$

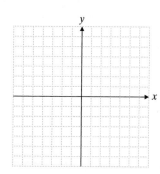

*Rewrite in standard form each equation. Determine **(a)** whether the parabola is horizontal or vertical, **(b)** the direction it opens, and **(c)** the vertex.*

27. $y = x^2 - 4x - 1$

28. $y = x^2 + 12x + 25$

29. $y = -2x^2 + 4x + 5$

30. $y = -2x^2 + 4x - 3$

31. $x = y^2 + 8y + 9$

32. $x = y^2 + 10y + 23$

Applications

33. *Satellite Dishes* Find an equation of the form $y = ax^2$ that describes the outline of a satellite dish such that the bottom of the dish passes through (0, 0), the diameter of the dish is 32 inches, and the depth of the dish is 8 inches.

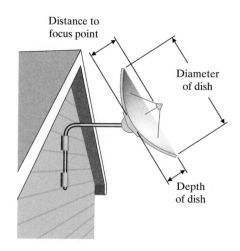

34. *Satellite Dishes* Find an equation of the form $y = ax^2$ that describes the outline of a satellite dish such that the bottom of the dish passes through (0, 0), the diameter of the dish is 24 inches, and the depth of the dish is 4 inches.

35. **Satellite Dishes** If the outline of a satellite dish is described by the equation $y = ax^2$, then the distance p from the center of the dish to the focus point of the dish is given by the equation $a = \dfrac{1}{4p}$. Find the distance p for the dish in Exercise 33.

36. **Satellite Dishes** If the outline of a satellite dish is described by the equation $y = ax^2$, then the distance p from the center of the dish to the focus point of the dish is given by the equation $a = \dfrac{1}{4p}$. Find the distance p for the dish in Exercise 34.

 Optional Graphing Calculator Problems

Find the vertex and y-intercept of each parabola. Find the two x-intercepts.

37. $y = 2x^2 + 6.48x - 0.1312$

38. $y = -3x^2 + 33.66x - 73.5063$

Applications

By writing a quadratic equation in the form $y = a(x - h)^2 + k$, we can find the maximum or minimum value of the equation and the value of x at which it occurs. Remember that the equation $y = a(x - h)^2 + k$ is a vertical parabola. For $a > 0$, the parabola opens upward. Thus, the y-coordinate of the vertex is the smallest (or minimum) value of x. Similarly, when $a < 0$, the parabola opens downward, so the y-coordinate of the vertex is the maximum value of the equation. Since the vertex occurs at (h, k), the maximum value of the equation occurs when $x = h$. Then

$$y = -a(x - h)^2 + k = a(0) + k = k.$$

For example, suppose the weekly profit of a manufacturing company in dollars is $P = -2(x - 45)^2 + 2300$ for x units manufactured. By looking at the equation, we see that the maximum profit per week is \$2300 and is attained when 45 units are manufactured. Use this approach for Exercises 39–42.

39. **Business** A company's monthly profit equation is

$$P = -3x^2 + 240x + 31{,}200,$$

where x is the number of items manufactured. Find the maximum monthly profit and the number of items that must be produced each month to attain maximum profit.

40. **Business** A company's monthly profit equation is

$$P = -2x^2 + 200x + 47{,}000,$$

where x is the number of items manufactured. Find the maximum monthly profit and the number of items that must be produced each month to attain maximum profit?

41. **Orange Grove Yield** The effective yield from a grove of orange trees is described by the equation $E = x(900 - x)$, where x is the number of orange trees per acre. What is the maximum effective yield? How many orange trees per acre should be planted to achieve the maximum yield?

42. **Drug Sensitivity** A research pharmacologist has determined that sensitivity S to a drug depends on the dosage d in milligrams, according to the equation $S = 650d - 2d^2$. What is the maximum sensitivity that will occur? What dosage will produce that maximum sensitivity?

Cumulative Review

Simplify.

43. $\sqrt{50x^3}$

44. $\sqrt[3]{40x^3y^4}$

Add.

45. $\sqrt{98x} + x\sqrt{8} - 3\sqrt{50x}$

46. $\sqrt[3]{16x^4} + 4x\sqrt[3]{2} - 8x\sqrt[3]{54}$

47. *Produce Delivery* A driver delivering eggs drove from the farm to a supermarket warehouse at 30 mph. He unloaded the eggs and drove back to the farm at 50 mph. The total trip took 2 hours and 15 minutes. How far is the farm from the supermarket warehouse?

48. *Commuting* Matthew drives from work to his home at 40 mph. One morning, an accident on the road delayed him for 15 minutes. The driving time including the delay was 56 minutes. How far does Matthew live from his job?

49. *Rose Bushes* Sir George Tipkin of Sussex has a collection of eight large English rose bushes, each having approximately 1050 buds. In normal years this type of bush produces blooms from 73% of its buds. During years of drought this figure drops to 44%. During years of heavy rainfall the figure rises to 88%. How many blooms can Sir George expect on these bushes if there is heavy rainfall this year?

50. *Rose Bushes* Last year Sir George had only six of the type of bushes described in Exercise 49. It was a drought year, and he counted 2900 blooms. Using the bloom rates given in Exercise 49, determine approximately how many buds appeared on each of these six bushes. (Round your answer to the nearest whole number.)

Suppose a plane cuts a cone at an angle so that the plane intersects all sides of the cone. If the plane is not perpendicular to the axis of the cone, the conic section that is formed is called an ellipse.

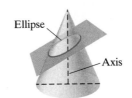

Student Learning Objectives

After studying this section, you will be able to:

1 Graph an ellipse whose center is at the origin.

2 Graph an ellipse whose center is at (h, k).

We define an **ellipse** as the set of points in a plane such that for each point in the set, the *sum* of its distances to two fixed points is constant. The fixed points are called **foci** (plural of *focus*).

We can use this definition to draw an ellipse using a piece of string tied at each end to a thumbtack. Place a pencil as shown in the drawing and draw the curve, keeping the pencil pushed tightly against the string. The two thumbtacks are the foci of the ellipse that results.

Examples of the ellipse can be found in the real world. The orbit of the Earth (and each of the other planets) is approximately an ellipse with the Sun at one focus. In the sketch at the right the Sun is located at F and the other focus of the orbit of the Earth is at F'.

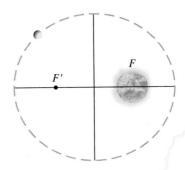

An elliptical surface has a special reflecting property. When sound, light, or some other object originating at one focus reaches the ellipse, it is reflected in such a way that it passes through the other focus. This property can be found in the United States Capitol in a famous room known as the Statuary Hall. If a person whispers at the focus of one end of this elliptically shaped room, a person at the other focus can easily hear him or her.

1 Graphing an Ellipse Whose Center Is at the Origin

The equation of an ellipse is similar to the equation of a circle. The standard form of the equation of an ellipse centered at the origin is given next.

STANDARD FORM OF THE EQUATION OF AN ELLIPSE

An ellipse with center at the origin has the equation

$$\frac{x^2}{a^2} + \frac{y^2}{b^2} = 1, \quad \text{where } a \text{ and } b > 0.$$

The **vertices** of this ellipse are at $(a, 0)$, $(-a, 0)$, $(0, b)$, and $(0, -b)$.

To plot the ellipse, we need the x- and y-intercepts.

$$\frac{x^2}{a^2} + \frac{y^2}{b^2} = 1$$

If $x = 0$, then $\dfrac{y^2}{b^2} = 1$.

$$y^2 = b^2$$
$$\pm\sqrt{y^2} = \pm\sqrt{b^2}$$
$$\pm y = \pm b \text{ or } y = \pm b$$

If $y = 0$, then $\dfrac{x^2}{a^2} = 1$.

$$x^2 = a^2$$
$$\pm\sqrt{x^2} = \pm\sqrt{a^2}$$
$$\pm x = \pm a \text{ or } x = \pm a$$

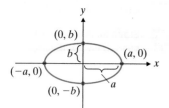

So the x-intercepts are $(a, 0)$ and $(-a, 0)$, and the y-intercepts are $(0, b)$ and $(0, -b)$ for an ellipse of the form $\dfrac{x^2}{a^2} + \dfrac{y^2}{b^2} = 1$.

A circle is a special case of an ellipse. If $a = b$, we get the following:

$$\frac{x^2}{a^2} + \frac{y^2}{a^2} = 1$$
$$x^2 + y^2 = a^2$$

This is the equation of a circle of radius a.

EXAMPLE 1 Graph $x^2 + 3y^2 = 12$. Label the intercepts.

Solution Before we can graph this ellipse, we need to rewrite the equation in standard form.

$$\frac{x^2}{12} + \frac{3y^2}{12} = \frac{12}{12} \qquad \text{Divide each side by 12.}$$

$$\frac{x^2}{12} + \frac{y^2}{4} = 1 \qquad \text{Simplify.}$$

Thus, we have the following:

$$a^2 = 12 \qquad \text{so} \qquad a = 2\sqrt{3}$$
$$b^2 = 4 \qquad \text{so} \qquad b = 2$$

The x-intercepts are $(-2\sqrt{3}, 0)$ and $(2\sqrt{3}, 0)$, and the y-intercepts are $(0, 2)$ and $(0, -2)$. We plot these points and draw the ellipse.

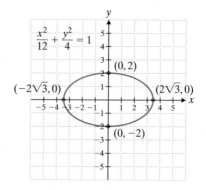

PRACTICE PROBLEM 1

NOTE TO STUDENT: *Fully worked-out solutions to all of the Practice Problems can be found at the back of the text starting at page SP-1*

Practice Problem 1 Graph $4x^2 + y^2 = 16$. Label the intercepts.

2 Graphing an Ellipse Whose Center Is at (*h, k*)

If the center of the ellipse is not at the origin but at some point whose coordinates are (h, k), then the standard form of the equation is changed.

An ellipse with center at (h, k) has the equation

$$\frac{(x - h)^2}{a^2} + \frac{(y - k)^2}{b^2} = 1,$$

where a and $b > 0$.

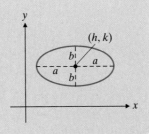

Note that a and b are *not* the x-intercepts now. Why is this? Look at the sketch. You'll see that a is the horizontal distance from the center of the ellipse to a point on the ellipse. Similarly, b is the vertical distance. Hence, when the center of the ellipse is not at the origin, the ellipse may not even cross either axis.

EXAMPLE 2 Graph $\dfrac{(x - 5)^2}{9} + \dfrac{(y - 6)^2}{4} = 1.$

Solution The center of the ellipse is $(5, 6)$, $a = 3$, and $b = 2$. Therefore, we begin at $(5, 6)$. We plot points 3 units to the left, 3 units to the right, 2 units up, and 2 units down from $(5, 6)$.

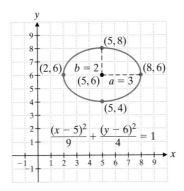

Practice Problem 2 Graph $\dfrac{(x - 2)^2}{16} + \dfrac{(y + 3)^2}{9} = 1.$

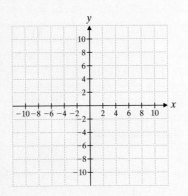

Graphing Calculator

📈 Graphing Ellipses

In order to graph the ellipse in Example 2 on a graphing calculator, we first need to solve for y.

$$\frac{(y - 6)^2}{4} = 1 - \frac{(x - 5)^2}{9}$$

$$(y - 6)^2 = 4\left[1 - \frac{(x - 5)^2}{9}\right]$$

$$y = 6 \pm 2\sqrt{1 - \frac{(x - 5)^2}{9}}$$

Is it necessary to break up the curve into two halves in order to graph the ellipse? Why or why not?

Use the above concepts to graph

$$\frac{(x - 2)^2}{9} + \frac{(y - 1)^2}{4} = 1.$$

Using the Trace feature, determine from your graph the coordinates of the two x-intercepts and the two y-intercepts. Express your answers to the nearest hundredth.

Verbal and Writing Skills

1. Explain how to determine the center of the ellipse $\dfrac{(x + 2)^2}{4} + \dfrac{(y - 3)^2}{9} = 1$.

2. Explain how to determine the x- and y-intercepts of the ellipse $\dfrac{x^2}{9} + \dfrac{y^2}{16} = 1$.

Graph each ellipse. Label the intercepts. You may need to use a scale other than 1 square = 1 unit.

3. $\dfrac{x^2}{36} + \dfrac{y^2}{4} = 1$

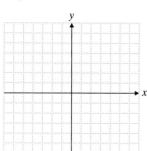

4. $\dfrac{x^2}{49} + \dfrac{y^2}{25} = 1$

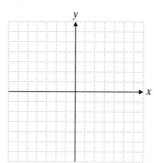

5. $\dfrac{x^2}{81} + \dfrac{y^2}{100} = 1$

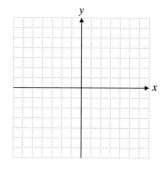

6. $\dfrac{x^2}{121} + \dfrac{y^2}{144} = 1$

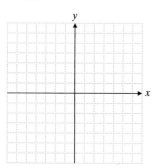

7. $4x^2 + y^2 - 36 = 0$

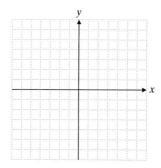

8. $9x^2 + y^2 - 9 = 0$

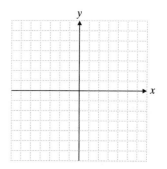

9. $x^2 + 9y^2 = 81$

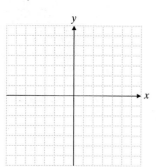

10. $4x^2 + 25y^2 = 100$

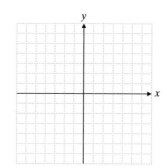

11. $x^2 + 12y^2 = 36$

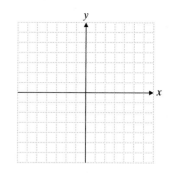

12. $2x^2 + 3y^2 = 18$

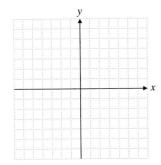

13. $\dfrac{x^2}{\dfrac{25}{4}} + \dfrac{y^2}{\dfrac{16}{9}} = 1$

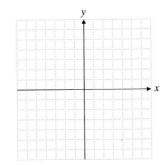

14. $\dfrac{x^2}{\dfrac{81}{4}} + \dfrac{y^2}{\dfrac{25}{16}} = 1$

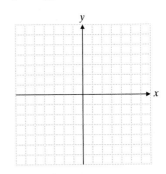

15. $121x^2 + 64y^2 = 7744$

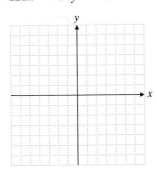

16. Write in standard form the equation of an ellipse with center at the origin, an x-intercept at $(-7, 0)$, and a y-intercept at $(0, 11)$.

17. Write in standard form the equation of an ellipse with center at the origin, an x-intercept at $(13, 0)$, and a y-intercept at $(0, -12)$.

18. Write in standard form the equation of an ellipse with center at the origin, an x-intercept at $\left(5\sqrt{2}, 0\right)$, and a y-intercept at $(0, -1)$.

19. Write in standard form the equation of an ellipse with center at the origin, an x-intercept at $(9, 0)$, and a y-intercept at $\left(0, 3\sqrt{2}\right)$.

Applications

20. ***Window Design*** The window shown in the sketch is in the shape of half of an ellipse. Find the equation for the ellipse if the center of the ellipse is at point $A = (0, 0)$.

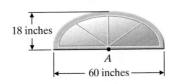

18 inches

A

60 inches

21. ***Orbit of Venus*** The orbit of Venus is an ellipse with the Sun as a focus. If we say that the center of the ellipse is at the origin, an approximate equation for the orbit is

$$\frac{x^2}{5013} + \frac{y^2}{4970} = 1,$$

where x and y are measured in millions of miles. Find the largest possible distance across the ellipse. Round your answer to the nearest million miles.

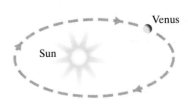

Venus

Sun

Graph each ellipse, Label the center. You may need to use a scale other than 1 square = 1 unit.

22. $\dfrac{(x-7)^2}{4} + \dfrac{(y-6)^2}{9} = 1$

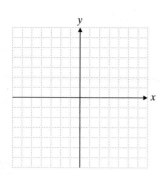

23. $\dfrac{(x-5)^2}{9} + \dfrac{(y-2)^2}{1} = 1$

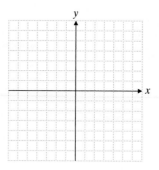

24. $\dfrac{(x+2)^2}{49} + \dfrac{y^2}{25} = 1$

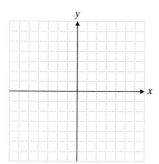

25. $\dfrac{x^2}{25} + \dfrac{(y-4)^2}{16} = 1$

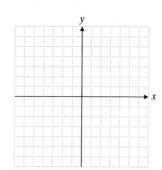

26. $\dfrac{(x+1)^2}{36} + \dfrac{(y+4)^2}{16} = 1$

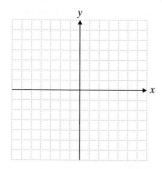

27. $\dfrac{(x+5)^2}{16} + \dfrac{(y+2)^2}{36} = 1$

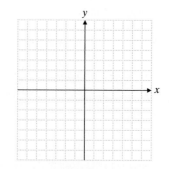

28. Write in standard form the equation of an ellipse whose vertices are $(-3, -2), (5, -2), (1, 1),$ and $(1, -5).$

29. Write in standard form the equation of an ellipse whose vertices are $(2, 3), (6, 3), (4, 7),$ and $(4, -1).$

30. For what value of a does the ellipse

$$\frac{(x + 5)^2}{4} + \frac{(y + a)^2}{9} = 1$$

pass through the point $(-4, 4)?$

31. *Pet Exercise Area* Bob's backyard is a rectangle 40 meters by 60 meters. He uses this backyard for an exercise area for his dog. He drove two posts into the ground and fastened a rope to each post, passing the rope through the metal ring on his dog's collar. When the dog pulls on the rope while running, its path is an ellipse. (See the figure.) If the dog can just reach all four sides of the rectangle, find the equation of the elliptical path.

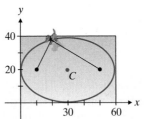

Find the four intercepts, accurate to four decimal places, for each ellipse.

32. $\dfrac{x^2}{12} + \dfrac{y^2}{19} = 1$

33. $\dfrac{(x - 3.6)^2}{14.98} + \dfrac{(y - 5.3)^2}{28.98} = 1$

To Think About

The area enclosed by the ellipse $\dfrac{x^2}{a^2} + \dfrac{y^2}{b^2} = 1$ is given by the equation $A = \pi ab$. Use the value $\pi \approx 3.1416$ to find an approximate value for each of the following answers.

34. *Mirror Design* An oval mirror has an outer boundary in the shape of an ellipse. The width of the mirror is 20 inches, and the length of the mirror is 45 inches. Find the area of the mirror. Round your answer to the nearest tenth.

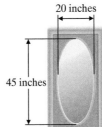

35. ***Australian Football*** In Australia a type of football is played on Aussie Rules fields. These fields are in the shape of an ellipse. Suppose the distance from A to B for the field shown is 185 meters and the distance from C to D is 154 meters. Find the area of the playing field. Round your answer to the nearest tenth.

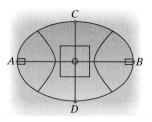

Cumulative Review

Rationalize the denominator.

36. $\dfrac{5}{\sqrt{2x} - \sqrt{y}}$

Multiply and simplify.

37. $\left(2\sqrt{3} + 4\sqrt{2}\right)\left(5\sqrt{6} - \sqrt{2}\right)$

38. ***Empire State Building*** The Empire State Building was the tallest building in the world for many years. Construction began on March 17, 1930, and the framework rose at a rate of 4.5 stories per week. How many weeks did it take to complete the framework for all 102 stories?

39. ***Empire State Building*** The top floor (the observatory on the 102nd floor) of the Empire State Building is 1224 feet above street level. To walk up the stairway of the building from street level to the 102nd floor requires climbing 1850 steps. (Once a year an official race is held to see who is the fastest to climb the stairs to the observatory.) What is the average height of one step at the Empire State Building? Round your answer to the nearest tenth of an inch.

How are you doing with your homework assignments in Sections 9.1 to 9.3? Before you go further in the textbook, take some time to do each of the following problems. Simplify all answers.

9.1

1. Write the standard form of the equation of a circle with center at $(8, -2)$ and a radius of $\sqrt{7}$.

2. Find the distance between $(-6, -2)$ and $(-3, 4)$.

3. Rewrite the equation
$x^2 + y^2 - 2x - 4y + 1 = 0$
in standard form. Find the
center and radius of the circle and
sketch its graph.

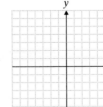

9.2

4. What is the axis of symmetry of the parabola $y = 4(x - 3)^2 + 5$?

5. Find the vertex of the parabola $y = \dfrac{1}{3}(x + 4)^2 + 6$.

Graph each parabola. Write the equation in standard form.

6. $x = (y + 1)^2 + 2$

7. $x^2 = y - 4x - 1$

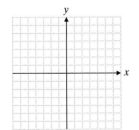

9.3

8. Write in standard form the equation of an ellipse with center at the origin, an x-intercept at $(-10, 0)$ and a y-intercept at $(0, 7)$.

Graph each ellipse. Write the equations in standard form.

9. $4x^2 + y^2 - 36 = 0$

10. $\dfrac{(x + 3)^2}{25} + \dfrac{(y - 1)^2}{16} = 1$

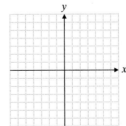

Now turn to page SA-35 for the answers to each of these problems. If you missed any of these problems, you should stop and review the Examples and Practice Problems in the referenced objective.

1. _____

2. _____

3. _____

4. _____

5. _____

6. _____

7. _____

8. _____

9. _____

10. _____

513

By cutting two branches of a cone by a plane as shown in the sketch, we obtain the two branches of a hyperbola. A comet moving with more than enough kinetic energy to escape the Sun's gravitational pull will travel in a hyperbolic path. Similarly, a rocket traveling with more than enough velocity to escape the Earth's gravitational field will follow a hyperbolic path.

We define a **hyperbola** as the set of points in a plane such that for each point in the set, the absolute value of the *difference* of its distances to two fixed points (called **foci**) is constant.

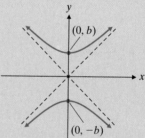

1 Graphing a Hyperbola Whose Center Is at the Origin

Notice the similarity of the definition of a hyperbola to the definition of an ellipse. If we replace the word *difference* by *sum,* we have the definition of an ellipse. Hence, we should expect that the equation of a hyperbola will be that of an ellipse with the plus sign replaced by a minus sign. And it is. If the hyperbola has its center at the origin, its equation is

$$\frac{x^2}{a^2} - \frac{y^2}{b^2} = 1 \quad \text{or} \quad \frac{y^2}{b^2} - \frac{x^2}{a^2} = 1.$$

The hyperbola has two branches. If the center of the hyperbola is at the origin and the two branches have two *x*-intercepts but no *y*-intercepts, the hyperbola is a *horizontal hyperbola,* and its **axis** is the *x*-axis. If the center of the hyperbola is at the origin and the two branches have two *y*-intercepts but no *x*-intercepts, the hyperbola is a *vertical hyperbola,* and its axis is the *y*-axis.

The points where the hyperbola intersects its axis are called the **vertices** of the hyperbola.

For hyperbolas centered at the origin, the vertices are also the intercepts.

STANDARD FORM OF THE EQUATION OF A HYPERBOLA WITH CENTER AT THE ORIGIN

Let *a* and *b* be any positive real numbers. A hyperbola with center at the origin and vertices $(-a, 0)$ and $(a, 0)$ has the equation

$$\frac{x^2}{a^2} - \frac{y^2}{b^2} = 1.$$

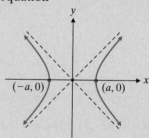

This is called a *horizontal hyperbola.*

A hyperbola with center at the origin and vertices $(0, b)$ and $(0, -b)$ has the equation

$$\frac{y^2}{b^2} - \frac{x^2}{a^2} = 1.$$

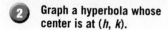

This is called a *vertical hyperbola.*

Notice that the two equations are slightly different. Be aware of this difference so that when you look at an equation you will be able to tell whether the hyperbola is horizontal or vertical.

Notice also the diagonal lines that we've drawn on the graphs of the hyperbolas. These lines are called **asymptotes.** The two branches of the hyperbola come increasingly closer to the asymptotes as the value of $|x|$ gets very large. By drawing the asymptotes and plotting the vertices, we can easily graph a hyperbola.

ASYMPTOTES OF HYPERBOLAS

The asymptotes of the hyperbolas $\dfrac{x^2}{a^2} - \dfrac{y^2}{b^2} = 1$ and $\dfrac{y^2}{b^2} - \dfrac{x^2}{a^2} = 1$ are

$$y = \frac{b}{a}x \ \text{ and } \ y = -\frac{b}{a}x.$$

Note that $\dfrac{b}{a}$ and $-\dfrac{b}{a}$ are the slopes of the asymptotes.

An easy way to find the asymptotes is to draw extended diagonal lines through the rectangle whose center is at the origin and whose corners are at (a, b), $(a, -b)$, $(-a, b)$, and $(-a, -b)$. (This rectangle is sometimes called the **fundamental rectangle.**) We draw the fundamental rectangle and the asymptotes with a dashed line because they are not part of the curve.

EXAMPLE 1   Graph $\dfrac{x^2}{25} - \dfrac{y^2}{16} = 1$.

Solution   The equation has the form $\dfrac{x^2}{a^2} - \dfrac{y^2}{b^2} = 1$, so it is a horizontal hyperbola. $a^2 = 25$, so $a = 5$; $b^2 = 16$, so $b = 4$. Since the hyperbola is horizontal, it has vertices at $(a, 0)$ and $(-a, 0)$ or $(5, 0)$ and $(-5, 0)$.

To draw the asymptotes, we construct a fundamental rectangle with corners at $(5, 4), (5, -4), (-5, 4)$, and $(-5, -4)$. We draw extended diagonal lines through the rectangle as the asymptotes. We construct each branch of the curve so that it passes through a vertex and gets closer to the asymptotes as it moves away from the origin.

PRACTICE PROBLEM 1

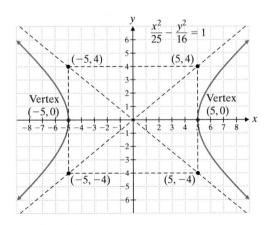

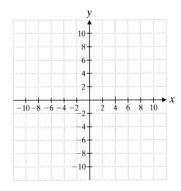

Practice Problem 1   Graph $\dfrac{x^2}{16} - \dfrac{y^2}{25} = 1$.

NOTE TO STUDENT: *Fully worked-out solutions to all of the Practice Problems can be found at the back of the text starting at page SP-1*

EXAMPLE 2   Graph $4y^2 - 7x^2 = 28$.

Solution   To find the vertices and asymptotes, we must rewrite the equation in standard form. Divide each term by 28.

$$\frac{4y^2}{28} - \frac{7x^2}{28} = \frac{28}{28}$$

$$\frac{y^2}{7} - \frac{x^2}{4} = 1$$

Thus, we have the standard form of a vertical hyperbola with center at the origin. Here $b^2 = 7$, so $b = \sqrt{7}$; $a^2 = 4$, so $a = 2$. The hyperbola has vertices at $(0, \sqrt{7})$ and $(0, -\sqrt{7})$. The fundamental rectangle has corners at $(2, \sqrt{7})$, $(2, -\sqrt{7})$, $(-2, \sqrt{7})$, and $(-2, -\sqrt{7})$. To aid us in graphing, we measure the distance $\sqrt{7}$ as approximately 2.6.

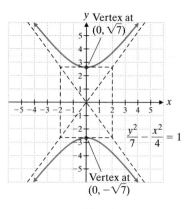

Practice Problem 2   Graph $y^2 - 4x^2 = 4$.

PRACTICE PROBLEM 2

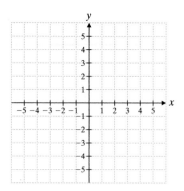

NOTE TO STUDENT: Fully worked-out solutions to all of the Practice Problems can be found at the back of the text starting at page SP-1

Graphing Calculator

Graphing Hyperbolas

Graph on your graphing calculator the hyperbola in Example 2 using

$$y_1 = \frac{\sqrt{28 + 7x^2}}{2}$$

and

$$y_2 = -\frac{\sqrt{28 + 7x^2}}{2}.$$

Do you see how we obtained y_1 and y_2?

② Graphing a Hyperbola Whose Center Is at (h, k)

If a hyperbola does not have its center at the origin but is shifted h units to the right or left and k units up or down, its equation is one of the following:

STANDARD FORM OF THE EQUATION OF A HYPERBOLA WITH CENTER AT (h, k)

Let a and b be any positive real numbers. A horizontal hyperbola with center at (h, k) and vertices $(h - a, k)$ and $(h + a, k)$ has the equation

$$\frac{(x - h)^2}{a^2} - \frac{(y - k)^2}{b^2} = 1.$$

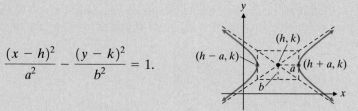

Horizontal Hyperbola

A vertical hyperbola with center at (h, k) and vertices $(h, k + b)$ and $(h, k - b)$ has the equation

$$\frac{(y - k)^2}{b^2} - \frac{(x - h)^2}{a^2} = 1.$$

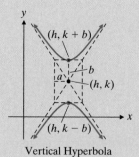

Vertical Hyperbola

EXAMPLE 3   Graph $\dfrac{(x-4)^2}{9} - \dfrac{(y-5)^2}{4} = 1$.

Solution   The center is at $(4, 5)$, and the hyperbola is horizontal. We have $a = 3$ and $b = 2$, so the vertices are $(4 \pm 3, 5)$, or $(7, 5)$ and $(1, 5)$. We can sketch the hyperbola more readily if we can draw a fundamental rectangle. Using $(4, 5)$ as the center, we construct a rectangle $2a$ units wide and $2b$ units high. We then draw and extend the diagonals of the rectangle. The extended diagonals are the asymptotes for the branches of the hyperbola.

In this example, since $a = 3$ and $b = 2$, we draw a rectangle $2a = 6$ units wide and $2b = 4$ units high with a center at $(4, 5)$. We draw extended diagonals through the rectangle. From the vertex at $(7, 5)$, we draw a branch of the hyperbola opening to the right. From the vertex at $(1, 5)$, we draw a branch of the hyperbola opening to the left. The graph of the hyperbola is shown.

Graphing Calculator

Exploration

Graph the hyperbola in Example 3 using

$$y_1 = 5 + \sqrt{\dfrac{4(x-4)^2}{9} - 4}$$

and

$$y_2 = 5 - \sqrt{\dfrac{4(x-4)^2}{9} - 4}.$$

Do you see how we obtained y_1 and y_2?

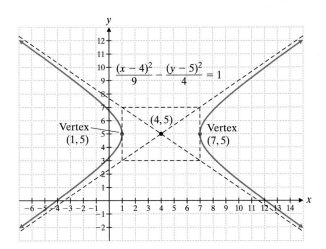

Practice Problem 3   Graph $\dfrac{(y+2)^2}{9} - \dfrac{(x-3)^2}{16} = 1$.

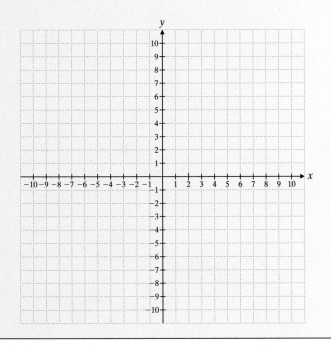

Student Solutions Manual   CD/Video   PH Math Tutor Center   MathXL®Tutorials on CD   MathXL®   MyMathLab®   Interactmath.com

Verbal and Writing Skills

1. What is the standard form of the equation of a horizontal hyperbola centered at the origin?

2. What are the vertices of the hyperbola $\dfrac{y^2}{9} - \dfrac{x^2}{4} = 1$? Is this a horizontal hyperbola or a vertical hyperbola? Why?

3. Explain in your own words how you would draw the graph of the hyperbola $\dfrac{x^2}{16} - \dfrac{y^2}{4} = 1$.

4. Explain how you determine the center of the hyperbola $\dfrac{(x-2)^2}{4} - \dfrac{(y+3)^2}{25} = 1$?

Find the vertices and graph each hyperbola. If the equation is not in standard form, write it as such.

5. $\dfrac{x^2}{4} - \dfrac{y^2}{25} = 1$

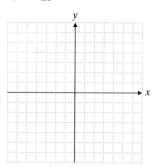

6. $\dfrac{x^2}{9} - \dfrac{y^2}{36} = 1$

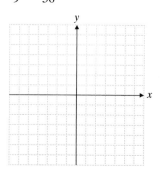

7. $\dfrac{y^2}{25} - \dfrac{x^2}{16} = 1$

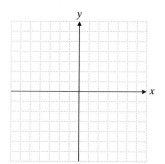

8. $\dfrac{y^2}{9} - \dfrac{x^2}{4} = 1$

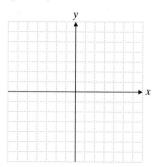

9. $4x^2 - y^2 = 64$

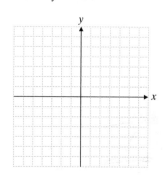

10. $x^2 - 4y^2 = 4$

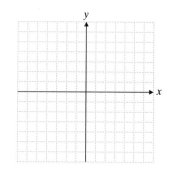

11. $8x^2 - y^2 = 16$

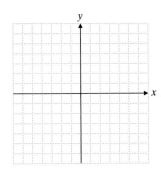

12. $12x^2 - y^2 = 36$

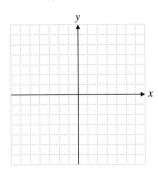

13. $4y^2 - 3x^2 = 48$

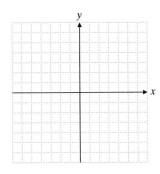

14. $8x^2 - 3y^2 = 24$

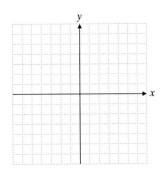

Find the equation of the hyperbola with center at the origin and with the following vertices and asymptotes.

15. Vertices at $(3, 0)$ and $(-3, 0)$;

asymptotes $y = \dfrac{4}{3}x, \ y = -\dfrac{4}{3}x$

16. Vertices at $(2, 0)$ and $(-2, 0)$;

asymptotes $y = \dfrac{3}{2}x, \ y = -\dfrac{3}{2}x$

17. Vertices $(0, 11)$ and $(0, -11)$;

asymptotes $y = \dfrac{11}{13}x, \ y = -\dfrac{11}{13}x$

18. Vertices $\left(0, \sqrt{15}\right)$ and $\left(0, -\sqrt{15}\right)$;

asymptotes $y = \dfrac{\sqrt{15}}{4}x, \ y = -\dfrac{\sqrt{15}}{4}x$

Applications

19. *Comet Orbits* Some comets have an orbit that is hyperbolic in shape with the Sun at the focus of the hyperbola. A comet is heading toward the Earth but then veers off as shown in the graph. It comes within 120 million miles of the Earth. As it travels into the distance, it moves closer and closer to the line $y = 3x$ with the Earth at the origin. Find the equation that describes the path of the comet.

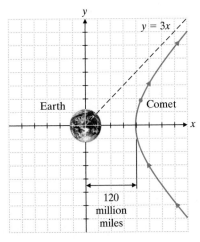

Scale on x-axis: each square is 30 million miles.
Scale on y-axis: each square is 90 million miles.

20. ***Rocket Path*** A rocket following the hyperbolic path shown in the graph turns rapidly at $(4, 0)$ and then moves closer and closer to the line $y = \dfrac{2}{3}x$ as the rocket gets farther from the tracking station at the origin. Find the equation that describes the path of the rocket if the center of the hyperbola is at $(0, 0)$.

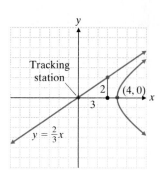

Find the center and then graph each hyperbola. Draw the axes on the grid at a convenient location. You may want to use a scale other than 1 square = 1 unit.

21. $\dfrac{(x-1)^2}{4} - \dfrac{(y+2)^2}{9} = 1$

22. $\dfrac{(x+3)^2}{16} - \dfrac{(y-1)^2}{4} = 1$

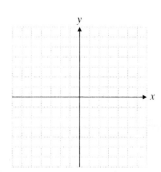

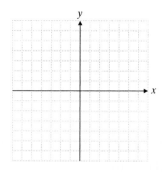

23. $\dfrac{(y+2)^2}{36} - \dfrac{(x+1)^2}{81} = 1$

24. $\dfrac{(y+1)^2}{49} - \dfrac{(x+3)^2}{81} = 1$

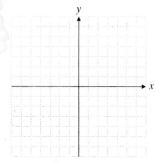

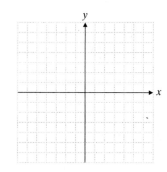

Find the center and the two vertices for each of the following hyperbolas.

25. $\dfrac{(x+6)^2}{7} - \dfrac{y^2}{3} = 1$

26. $\dfrac{x^2}{5} - \dfrac{(y+3)^2}{10} = 1$

27. A hyperbola's center is not at the origin. Its vertices are $(4, -14)$ and $(4, 0)$. One asymptote is $y = -\dfrac{7}{4}x$. Find the equation of the hyperbola.

28. A hyperbola's center is not at the origin. Its vertices are $(5, 0)$ and $(5, 14)$; one asymptote is $y = \dfrac{7}{5}x$. Find the equation of the hyperbola.

Optional Graphing Calculator Problems

29. For the hyperbola $8x^2 - y^2 = 16$, if $x = 3.5$, what are the two values of y?

30. For the hyperbola $x^2 - 12y^2 = 36$, if $x = 8.2$, what are the two values of y?

Cumulative Review

Factor completely.

31. $12x^2 + x - 6$

32. $2x^3 - 54$

Combine.

33. $\dfrac{3}{x^2 - 5x + 6} + \dfrac{2}{x^2 - 4}$

34. $\dfrac{2x}{5x^2 + 9x - 2} - \dfrac{3}{5x - 1}$

35. *Gift Contributions* A school's hockey team is giving the coach a retirement gift that costs $240. On the day that the captain of the team was collecting money, four hockey players were absent. He collected an equal amount from each player present and purchased the gift. Later the team captain said that if everyone had been present to chip in, each person would have contributed $2 less. How many people actually contributed? How many people in total are on the hockey team?

36. *Radio Airtime* A Connecticut FM radio station claims that a minimum of 104,755 songs are played every year.
   (a) Determine the number of songs played daily. (Assume that it is not a leap year.)
   (b) How much time would be left over each day for advertisements, news, sports, interviews, syndicated shows, and DJ chatter if the average song lasts 4 minutes?
   (c) What percentage of the airtime is music?

37. *Pencil Production* In 2000, approximately 2.1 billion pencils were produced in the United States by domestic manufacturers. If you add to this the number of pencils that were imported, then each American used ten pencils during the year. If there were 274 million people in the United States in 2000, how many pencils were imported to the United States in that year? *Source:* U.S. Department of Commerce

38. *Pencil Production* It is estimated that the number of pencils produced in the United States will increase by 5% from 2000 to 2005. During that time the number of imported pencils will increase by 750 million. If these figures hold true, what percent of all the pencils used in the United States in 2005 will have been imported? *Source:* U.S. Department of Commerce

Student Learning Objectives

After studying this section, you will be able to:

1 Solve a nonlinear system by the substitution method.

2 Solve a nonlinear system by the addition method.

1 Solving a Nonlinear System by the Substitution Method

Any equation that is of second degree or higher is a **nonlinear equation.** In other words, the equation is not a straight line (which is what the word *nonlinear* means) and can't be written in the form $y = mx + b$. A **nonlinear system of equations** includes at least one nonlinear equation.

The most frequently used method for solving a nonlinear system is the method of substitution. This method works especially well when one equation of the system is linear. A sketch can often be used to verify the solution(s).

EXAMPLE 1 Solve the following nonlinear system and verify your answer with a sketch.

$$x + y - 1 = 0 \qquad (1)$$

$$y - 1 = x^2 + 2x \qquad (2)$$

Solution We'll use the substitution method.

$$y = -x + 1 \qquad (3) \qquad \text{Solve for } y \text{ in equation (1).}$$

$$(-x + 1) - 1 = x^2 + 2x \qquad \text{Substitute (3) into equation (2).}$$

$$-x + 1 - 1 = x^2 + 2x$$

$$0 = x^2 + 3x \qquad \text{Solve the resulting quadratic equation.}$$

$$0 = x(x + 3)$$

$$x = 0 \quad \text{or} \quad x = -3$$

Now substitute the values for x in the equation $y = -x + 1$.

For $x = -3$: $\qquad y = -(-3) + 1 = +3 + 1 = 4$

For $x = 0$: $\qquad y = -(0) + 1 = +1 = 1$

Thus, the solutions of the system are $(-3, 4)$ and $(0, 1)$.

To sketch the system, we see that equation (2) describes a parabola. We can rewrite it in the form

$$y = x^2 + 2x + 1 = (x + 1)^2.$$

This is a parabola opening upward with its vertex at $(-1, 0)$. Equation (1) can be written as $y = -x + 1$, which is a straight line with slope $= -1$ and y-intercept $(0, 1)$.

A sketch shows the two graphs intersecting at $(0, 1)$ and $(-3, 4)$. Thus, the solutions are verified.

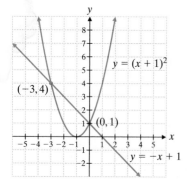

NOTE TO STUDENT: Fully worked-out solutions to all of the Practice Problems can be found at the back of the text starting at page SP-1

Practice Problem 1 Solve the system.

$$\frac{x^2}{4} - \frac{y^2}{4} = 1$$

$$x + y + 1 = 0$$

EXAMPLE 2   Solve the following nonlinear system and verify your answer with a sketch.

$$y - 2x = 0 \qquad (1)$$

$$\frac{x^2}{4} + \frac{y^2}{9} = 1 \qquad (2)$$

$$y = 2x \qquad (3) \qquad \text{Solve equation (1) for } y.$$

Solution

$$\frac{x^2}{4} + \frac{(2x)^2}{9} = 1 \qquad \text{Substitute (3) into equation (2).}$$

$$\frac{x^2}{4} + \frac{4x^2}{9} = 1 \qquad \text{Simplify.}$$

$$36\left(\frac{x^2}{4}\right) + 36\left(\frac{4x^2}{9}\right) = 36(1) \qquad \text{Clear the fractions.}$$

$$9x^2 + 16x^2 = 36$$

$$25x^2 = 36$$

$$x^2 = \frac{36}{25}$$

$$x = \pm\sqrt{\frac{36}{25}}$$

$$x = \pm\frac{6}{5} = \pm 1.2$$

For $x = +1.2$: $y = 2(1.2) = 2.4$.

For $x = -1.2$: $y = 2(-1.2) = -2.4$.

Thus, the solutions are $(1.2, 2.4)$ and $(-1.2, -2.4)$.

We recognize $\frac{x^2}{4} + \frac{y^2}{9} = 1$ as an ellipse with center at the origin and vertices $(0, 3), (0, -3), (2, 0)$, and $(-2, 0)$. When we rewrite $y - 2x = 0$ as $y = 2x$, we recognize it as a straight line with slope 2 passing through the origin. The sketch shows that the points of intersection at $(1.2, 2.4)$ and $(-1.2, -2.4)$ seem reasonable.

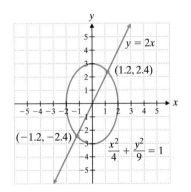

Practice Problem 2   Solve the system. Verify your answer with a sketch.

$$2x - 9 = y$$

$$xy = -4$$

Graphing Calculator

Solving Nonlinear Systems

Use a graphing calculator to solve the following system. Round your answers to the nearest tenth.

$$30x^2 + 256y^2 = 7680$$

$$3x + y - 40 = 0$$

First we will need to obtain the equations

$$y_1 = \frac{\sqrt{7680 - 30x^2}}{16},$$

$$y_2 = -\frac{\sqrt{7680 - 30x^2}}{16},$$

and

$$y_3 = 40 - 3x$$

and graph them to approximate the solutions. Be sure that your window includes enough of the graphs to find the points of intersection.

PRACTICE PROBLEM 2

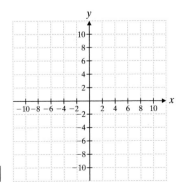

Solving a Nonlinear System by the Addition Method

Sometimes a system may be solved more readily by adding the equations together. It should be noted that some systems have no solution.

EXAMPLE 3 Solve the system.

$$4x^2 + y^2 = 1 \quad (1)$$
$$x^2 + 4y^2 = 1 \quad (2)$$

Solution Although we could use the substitution method, it is easier to use the addition method because neither equation is linear.

$$
\begin{array}{rl}
-16x^2 - 4y^2 = -4 \\
\underline{x^2 + 4y^2 = 1} \\
-15x^2 = -3
\end{array}
$$

Multiply equation (1) by -4 and add to equation (2).

$$x^2 = \frac{-3}{-15}$$

$$x^2 = \frac{1}{5}$$

$$x = \pm\sqrt{\frac{1}{5}}$$

If $x = +\sqrt{\dfrac{1}{5}}$, then $x^2 = \dfrac{1}{5}$. Substituting this value into equation (2) gives

$$\frac{1}{5} + 4y^2 = 1$$

$$4y^2 = \frac{4}{5}$$

$$y^2 = \frac{1}{5}$$

$$y = \pm\sqrt{\frac{1}{5}}$$

Similarly, if $x = -\sqrt{\dfrac{1}{5}}$, then $y = \pm\sqrt{\dfrac{1}{5}}$. It is important to determine exactly how many solutions a nonlinear system of equations actually has. In this case, we have four solutions. When x is negative, there are two values for y. When x is positive, there are two values for y. If we rationalize each expression, the four solutions are

$$\left(\frac{\sqrt{5}}{5}, \frac{\sqrt{5}}{5}\right), \left(\frac{\sqrt{5}}{5}, -\frac{\sqrt{5}}{5}\right), \left(-\frac{\sqrt{5}}{5}, \frac{\sqrt{5}}{5}\right), \text{ and } \left(-\frac{\sqrt{5}}{5}, -\frac{\sqrt{5}}{5}\right).$$

NOTE TO STUDENT: Fully worked-out solutions to all of the Practice Problems can be found at the back of the text starting at page SP-1

Practice Problem 3 Solve the system.

$$x^2 + y^2 = 12$$
$$3x^2 - 4y^2 = 8$$

Solve each of the following systems by the substitution method. Graph each equation to verify that the answer seems reasonable.

1. $y^2 = 2x$
  $y = -2x + 2$

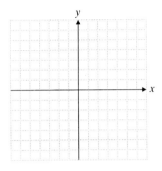

2. $y^2 = 4x$
  $y = x + 1$

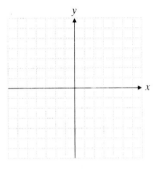

3. $x + 2y = 0$
  $x^2 + 4y^2 = 32$

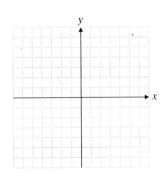

4. $y - 4x = 0$
  $4x^2 + y^2 = 20$

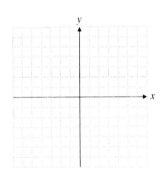

Solve each of the following systems by the substitution method.

5. $\dfrac{x^2}{1} - \dfrac{y^2}{3} = 1$
  $x + y = 1$

6. $y = (x + 3)^2 - 3$
  $2x - y + 2 = 0$

7. $x^2 + y^2 - 25 = 0$
  $3y = x + 5$

8. $x^2 + y^2 - 9 = 0$
  $2y = 3 - x$

9. $x^2 + 2y^2 = 4$
  $y = -x + 2$

10. $2x^2 + 3y^2 = 27$
  $y = x + 3$

11. $\dfrac{x^2}{4} - \dfrac{y^2}{4} = 1$
  $x + y - 4 = 0$

12. $y^2 - x^2 = 8$
  $y = 3x$

Solve each of the following systems by the addition method. Graph each equation to verify that the answer seems reasonable.

13. $2x^2 - 5y^2 = -2$
$3x^2 + 2y^2 = 35$

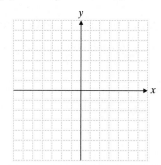

14. $2x^2 - 3y^2 = 5$
$3x^2 + 4y^2 = 16$

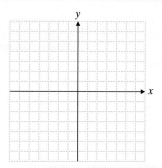

Solve each of the following systems by the addition method.

15. $x^2 + y^2 = 9$
$2x^2 - y^2 = 3$

16. $5x^2 - y^2 = 4$
$x^2 + 3y^2 = 4$

17. $x^2 + 2y^2 = 8$
$x^2 - y^2 = 1$

18. $x^2 + 4y^2 = 13$
$x^2 - 3y^2 = -8$

Mixed Practice

Solve each of the following systems by any appropriate method. If there is no real number solution, so state.

19. $x^2 + y^2 = 7$
$\dfrac{x^2}{3} - \dfrac{y^2}{9} = 1$

20. $x^2 + 2y^2 = 4$
$x^2 + y^2 = 4$

21. $2xy = 5$
$x - 4y = 3$

22. $3xy = -2$
$x + 9y = 1$

23. $xy = -6$
$2x + y = -4$

24. $xy = 1$
$3x - y + 2 = 0$

25. $x + y = 5$
$x^2 + y^2 = 4$

26. $x^2 + y^2 = 0$
$x - y = 6$

Applications

27. **Laser Beam Path** In an experiment with a laser beam, the path of a particle orbiting a central object is described by the equation $\dfrac{x^2}{49} + \dfrac{y^2}{36} = 1$, where x and y are measured in centimeters from the center of the object. The laser beam follows the path $y = 2x - 6$. Find the coordinates at which the laser will illuminate the particle (that is, when the particle will pass through the beam).

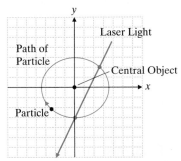

▲ **28.** **Geometry** The area of a rectangle is 540 square meters. The diagonal of the rectangle is 39 meters long. Find the dimensions of the rectangle.

Hint: Let x and y represent the length and width and write a system of two nonlinear equations.

29. **Path of a Meteor** The outline of the Earth can be considered a circle with a radius of approximately 4000 miles. Thus, if we say that the center of the Earth is located at $(0, 0)$, then an equation for this circle is $x^2 + y^2 = 16,000,000$. Suppose that an incoming meteor is approaching the Earth in a hyperbolic path. The equation describing the meteor's path is $25,000,000x^2 - 9,000,000y^2 = 2.25 \times 10^{14}$. Will the meteor strike the Earth? Why or why not? If so, locate the point (x, y) where the meteor will strike the Earth. Assume that x and y are both positive. Round your answer to three significant digits.

30. **Path of a Meteor** Suppose that a second incoming meteor is approaching the Earth in a hyperbolic path. The equation for this second meteor's path is $16,000,000x^2 - 25,000,000y^2 = 4.0 \times 10^{14}$. Will the second meteor strike the Earth? Why or why not? If so, locate the point (x, y) where the meteor will strike the Earth. Assume that x and y are both positive. Round your answer to three significant digits.

Cumulative Review

Divide.

31. $(3x^3 - 8x^2 - 33x - 10) \div (3x + 1)$

Simplify.

32. $\dfrac{6x^4 - 24x^3 - 30x^2}{3x^3 - 21x^2 + 30x}$

33. **CD-ROM Production** Ricardo is the staff accountant for a CD-ROM factory. He has determined that his monthly profit factor is given by the expression $11.5n - 290,000$. Here n represents the number of CD-ROMs manufactured each month. The profit this month was $1,187,750. How many CD-ROMs were produced this month?

34. **Speed Traps** Highway patrol officers can trap speeders by various methods. Between two certain points on a back country road in a small town in Georgia, a speeding Audi travels 55 miles per hour for 5 seconds. What is the legal speed limit there if at this speed, a car driving between those two points requires 11 seconds?

Putting Your Skills to Work

Mathematics of Navigation on Mars

Since NASA's rovers hoped to find signs of past life on Mars, they were sent to areas of Mars that were believed to hold evidence of past water. One of the rovers, Spirit, was sent to the giant Gusev Crater, which is divided by a channel. Rocks and wind complicated the landing.

Problems for Individual Investigation and Analysis

1. The target for Spirit was a landing ellipse about 12 miles wide and 60 miles long. If this ellipse is centered at the origin, find an equation for this ellipse.

2. Spirit's top speed was about 9.8 feet/minute. What is the shortest time it would take the rover to travel 1 mile? Round to the nearest hour.

3. If Spirit landed at the center of its landing ellipse, traveling at top speed, about how many days would it have taken Spirit to reach the nearest outer edge of the landing ellipse?

Problems for Group Investigation and Cooperative Study

The Martian day is about 24 hours and 40 minutes. NASA scientists in Pasadena, California, wanted to start their work day at sunrise, Mars time, each day that the rover is operating. Living on Mars time turned out to be surprisingly stressful and tiring.

4. Martian sunrise occurred at 11:36 A.M. the day after Spirit landed. After 3 days, what time did the NASA scientists start their work day?

5. After how many Earth days did the scientists' workday start at the 11:36 A.M. again?

6. If Spirit remained active for the full 92 days that the scientist had hoped for, how many hours behind Earth's time would the Martian time have been at the end of the mission? Round to the nearest tenth.

7. How many days (to the nearest tenth) behind would Martian time have been?

Chapter 9 Organizer

Topic	Procedure	Examples
Distance between two points, p. 487.	The distance d between points (x_1, y_1) and (x_2, y_2) is $$d = \sqrt{(x_2 - x_1)^2 + (y_2 - y_1)^2}.$$	Find the distance between $(-6, -3)$ and $(5, -2)$. $$d = \sqrt{[5 - (-6)]^2 + [-2 - (-3)]^2}$$ $$= \sqrt{(5 + 6)^2 + (-2 + 3)^2}$$ $$= \sqrt{121 + 1}$$ $$= \sqrt{122}$$
Standard form of the equation of a circle, p. 488.	The standard form of the equation of a circle with center at (h, k) and radius r is $$(x - h)^2 + (y - k)^2 = r^2.$$	Graph $(x - 3)^2 + (y + 4)^2 = 16$. Center at $(h, k) = (3, -4)$. Radius $= 4$.
Standard form of the equation of a vertical parabola, p. 495.	The equation of a vertical parabola with its vertex at (h, k) can be written in the form $y = a(x - h)^2 + k$. It opens upward if $a > 0$ and downward if $a < 0$. 	Graph $y = \dfrac{1}{2}(x - 3)^2 + 5$. $a = \dfrac{1}{2}$, so parabola opens upward. Vertex at $(h, k) = (3, 5)$. If $x = 0$, $y = 9.5$.
Standard form of the equation of a horizontal parabola, p. 497.	The equation of a horizontal parabola with its vertex at (h, k) can be written in the form $x = a(y - k)^2 + h$. It opens to the right if $a > 0$ and to the left if $a < 0$. 	Graph $x = \dfrac{1}{3}(y + 2)^2 - 4$. $a = \dfrac{1}{3}$, so parabola opens to the right. Vertex at $(h, k) = (-4, -2)$. If $x = 0$, $y = -2 - 2\sqrt{3} \approx -5.5$ and $y = -2 + 2\sqrt{3} \approx 1.5$.

Topic	Procedure	Examples
Standard form of the equation of an ellipse with center at (0, 0), p. 505.	An ellipse with center at the origin has the equation $$\frac{x^2}{a^2} + \frac{y^2}{b^2} = 1,$$ where $a > 0$ and $b > 0$.	Graph $\frac{x^2}{16} + \frac{y^2}{4} = 1$. $a^2 = 16$, $a = 4$; $b^2 = 4$, $b = 2$
Standard form of an ellipse with center at (h, k), p. 507.	An ellipse with center at (h, k) has the equation $$\frac{(x - h)^2}{a^2} + \frac{(y - k)^2}{b^2} = 1,$$ where $a > 0$ and $b > 0$.	Graph $\frac{(x + 2)^2}{9} + \frac{(y + 4)^2}{25} = 1$. $(h, k) = (-2, -4)$; $a = 3$, $b = 5$
Standard form of a horizontal hyperbola with center at (0, 0), p. 514.	Let a and b be positive real numbers. A horizontal hyperbola with center at the origin and vertices $(a, 0)$ and $(-a, 0)$ has the equation $$\frac{x^2}{a^2} - \frac{y^2}{b^2} = 1$$ and asymptotes $$y = \pm\frac{b}{a}x.$$	Graph $\frac{x^2}{25} - \frac{y^2}{9} = 1$. $a = 5$, $b = 3$
Standard form of a vertical hyperbola with center at (0, 0), p. 514.	Let a and b be positive real numbers. A vertical hyperbola with center at the origin and vertices $(0, b)$ and $(0, -b)$ has the equation $$\frac{y^2}{b^2} - \frac{x^2}{a^2} = 1$$ and asymptotes $$y = \pm\frac{b}{a}x.$$	Graph $\frac{y^2}{9} - \frac{x^2}{4} = 1$. $b = 3$, $a = 2$ 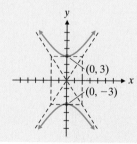

Topic	Procedure	Examples
Standard form of a horizontal hyperbola with center at (h, k), p. 516.	Let a and b be positive real numbers. A horizontal hyperbola with center at (h, k) and vertices $(h - a, k)$ and $(h + a, k)$ has the equation $$\frac{(x - h)^2}{a^2} - \frac{(y - k)^2}{b^2} = 1.$$ 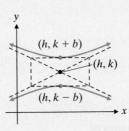	Graph $\dfrac{(x - 2)^2}{4} - \dfrac{(y - 3)^2}{25} = 1.$ Center at $(2, 3)$; $a = 2, b = 5$ 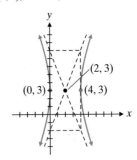
Standard form of a vertical hyperbola with center at (h, k), p. 516.	Let a and b be positive real numbers. A vertical hyperbola with center at (h, k) and vertices $(h, k + b)$ and $(h, k - b)$ has the equation $$\frac{(y - k)^2}{b^2} - \frac{(x - h)^2}{a^2} = 1.$$ 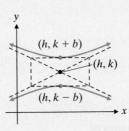	Graph $\dfrac{(y - 5)^2}{9} - \dfrac{(x - 4)^2}{4} = 1.$ Center at $(4, 5)$; $b = 3$, $a = 2$ 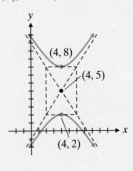
Nonlinear systems of equations, p. 522.	We can solve a nonlinear system by the substitution method or the addition method. In the addition method, we multiply one or more equations by a numerical value and then add them together so that one variable is eliminated. In the substitution method we solve one equation for one variable and substitute that expression into the other equation.	Solve by substitution. $$2x^2 + y^2 = 18$$ $$xy = 4$$ Solving the second equation for y, we have $y = \dfrac{4}{x}$. $$2x^2 + \left(\frac{4}{x}\right)^2 = 18$$ $$2x^2 + \frac{16}{x^2} = 18$$ $$2x^4 + 16 = 18x^2$$ $$2x^4 - 18x^2 + 16 = 0$$ $$x^4 - 9x^2 + 8 = 0$$ $$(x^2 - 1)(x^2 - 8) = 0$$ $x^2 - 1 = 0 \qquad x^2 - 8 = 0$ $x^2 = 1 \qquad\quad x^2 = 8$ $x = \pm 1 \qquad\quad x = \pm 2\sqrt{2}$ Since $xy = 4$, if $x = 1$, then $y = 4$. if $x = -1$, then $y = -4$. if $x = 2\sqrt{2}$, then $y = \sqrt{2}$. if $x = -2\sqrt{2}$, then $y = -\sqrt{2}$. The solutions are $(1, 4)(-1, -4), (2\sqrt{2}, \sqrt{2})$, and $(-2\sqrt{2}, -\sqrt{2})$.

In Exercises 1 and 2, find the distance between the points.

1. $(0, -6)$ and $(-3, 2)$

2. $(-7, 3)$ and $(-2, -1)$

3. Write in standard form the equation of a circle with center at $(-6, 3)$ and radius $\sqrt{15}$.

4. Write in standard form the equation of a circle with center at $(0, -7)$ and radius 5.

Rewrite each equation in standard form. Find the center and the radius of each circle.

5. $x^2 + y^2 + 2x - 6y + 5 = 0$

6. $x^2 + y^2 - 10x + 12y + 52 = 0$

Graph each parabola. Label its vertex and plot at least one intercept.

7. $x = \dfrac{1}{3}y^2$

8. $x = \dfrac{1}{2}(y - 2)^2 + 4$

9. $y = -2(x + 1)^2 - 3$

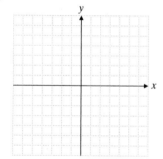

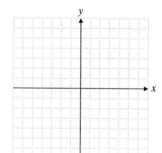

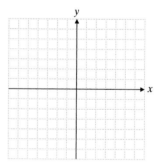

Rewrite each equation in standard form. Find the vertex and determine in which direction the parabola opens.

10. $x^2 + 6x = y - 4$

11. $x + 8y = y^2 + 10$

Graph each ellipse. Label its center and four other points.

12. $\dfrac{x^2}{4} + y^2 = 1$

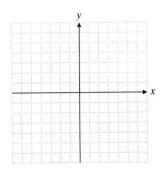

13. $16x^2 + y^2 - 32 = 0$

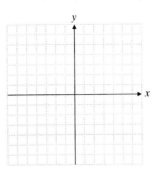

Determine the vertices and the center of each ellipse.

14. $\dfrac{(x + 5)^2}{4} + \dfrac{(y + 3)^2}{25} = 1$

15. $\dfrac{(x + 1)^2}{9} + \dfrac{(y - 2)^2}{16} = 1$

Find the center and vertices of each hyperbola and graph it.

16. $x^2 - 4y^2 - 16 = 0$

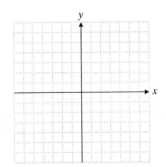

17. $3y^2 - x^2 = 27$

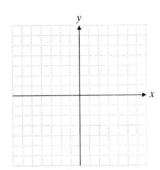

Determine the vertices and the center of each hyperbola.

18. $\dfrac{(x - 2)^2}{4} - \dfrac{(y + 3)^2}{25} = 1$

19. $9(y - 2)^2 - (x + 5)^2 - 9 = 0$

Solve each nonlinear system. If there is no real number solution, so state.

20. $x^2 + y = 9$
$\quad y - x = 3$

21. $x^2 + y^2 = 4$
$\quad x + y = 2$

22. $2x^2 + y^2 = 17$
$\quad x^2 + 2y^2 = 22$

23. $\quad xy = -2$
$\quad x^2 + y^2 = 5$

24. $3x^2 - 4y^2 = 12$
$\quad 7x^2 - y^2 = 8$

25. $y = x^2 + 1$
$\quad x^2 + y^2 - 8y + 7 = 0$

26. $2x^2 + y^2 = 18$
$\quad\quad xy = 4$

27. $y^2 - 2x^2 = 2$
$\quad 2y^2 - 3x^2 = 5$

28. $y^2 = 2x$
$\quad y = \dfrac{1}{2}x + 1$

29. $y^2 = \dfrac{1}{2}x$
$\quad y = x - 1$

Applications

30. *Searchlights* The side view of an airport searchlight is shaped like a parabola. The center of the light source of the searchlight is located 2 feet from the base along the axis of symmetry, and the opening is 5 feet across. How deep should the searchlight be? Round your answer to the nearest hundredth.

31. *Satellite Dishes* The side view of a satellite dish on Jason and Wendy's house is shaped like a parabola. The signals that come from the satellite hit the surface of the dish and are then reflected to the point where the signal receiver is located. This point is the focus of the parabolic dish. The dish is 10 feet across at its opening and 4 feet deep at its center. How far from the center of the dish should the signal receiver be placed? Round your answer to the nearest hundredth.

Remember to use your Chapter Test Prep Video CD to see the worked-out solutions to the test problems you want to review.

1. Find the distance between $(-6, -8)$ and $(-2, 5)$.

Rewrite the equation in standard form. Find the center or vertex, plot at least one other point, identify the conic, and sketch the curve.

2. $y^2 - 6y - x + 13 = 0$.

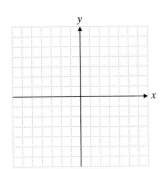

3. $x^2 + y^2 + 6x - 4y + 9 = 0$

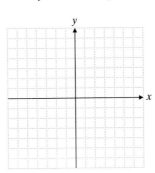

Identify and graph each conic section. Label the center and/or vertex as appropriate.

4. $\dfrac{x^2}{25} + \dfrac{y^2}{1} = 1$

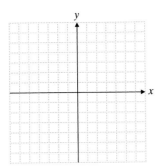

5. $\dfrac{x^2}{10} - \dfrac{y^2}{9} = 1$

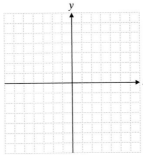

6. $y = -2(x + 3)^2 + 4$

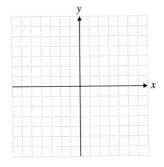

7. $\dfrac{(x + 2)^2}{16} + \dfrac{(y - 5)^2}{4} = 1$

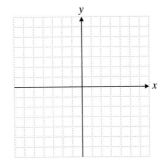

1. _____

2. _____

3. _____

4. _____

5. _____

6. _____

7. _____

535

8. _____

8. $7y^2 - 7x^2 = 28$

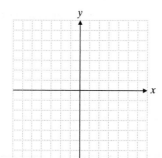

9. _____

10. _____

Find the standard form of the equation of each of the following:

9. Circle of radius $\sqrt{8}$ with its center at $(3, -5)$

11. _____

10. Ellipse with center at the origin, an x-intercept at $(3, 0)$, and a y-intercept at $(0, 5)$.

11. Parabola with its vertex at $(-7, 3)$ and that opens to the right. This parabola crosses the x-axis at $(2, 0)$. It is of the form $x = (y - k)^2 + h$.

12. _____

12. Hyperbola with center at the origin, vertices at $(3, 0)$ and $(-3, 0)$. One asymptote is $y = \dfrac{5}{3}x$.

13. _____

Solve each nonlinear system.

13. $-2x + y = 5$
$x^2 + y^2 - 25 = 0$

14. _____

14. $x^2 + y^2 = 9$
$y = x - 3$

15. _____

15. $4x^2 + y^2 - 4 = 0$
$9x^2 - 4y^2 - 9 = 0$

16. $x^2 + 2y^2 = 15$
$x^2 - y^2 = 6$

16. _____

Approximately one-half of this test covers the content of Chapters 1–8. The remainder covers the content of Chapter 9.

1. Identify the property illustrated by the equation $5(-3) = -3(5)$.

1. _____

2. Simplify. $2\{x - 3[x - 2(x + 1)]\}$

2. _____

3. Evaluate. $3(4 - 6)^3 + \sqrt{25}$

3. _____

4. Solve for p. $A = 3bt + prt$

4. _____

5. Factor. $4x^3 - 16x$

5. _____

6. Add. $\dfrac{3x}{x - 2} + \dfrac{5}{x - 1}$

6. _____

7. Solve.
$$\dfrac{3}{2x + 3} = \dfrac{1}{2x - 3} + \dfrac{2}{4x^2 - 9}$$

7. _____

8. Solve.
$$\begin{aligned} 3x - 2y - 9z &= 9 \\ x - y + z &= 8 \\ 2x + 3y - z &= -2 \end{aligned}$$

8. _____

9. Multiply and simplify. $\left(\sqrt{2} + \sqrt{3}\right)\left(2\sqrt{6} - \sqrt{3}\right)$

9. _____

10. Simplify. $\sqrt{12x^2} + 2x\sqrt{27} - \sqrt{18x}$

10. _____

Solve the following inequalities.

11. $x + 4(x + 2) > 7x + 8$

11. _____

12. $\dfrac{6(x - 4)}{5} \geq \dfrac{3(x + 2)}{4}$

12. _____

13. Find the distance between $(6, -1)$ and $(-3, -4)$.

13. _____

537

14. _____

15. _____

16. _____

17. _____

18. _____

19. _____

20. _____

21. _____

Identify and graph each equation.

14. $y = -\dfrac{1}{2}(x + 2)^2 - 3$

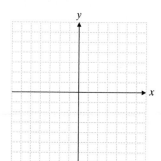

15. $25x^2 + 25y^2 = 125$

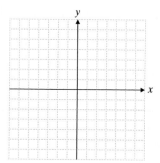

16. $16x^2 - 4y^2 = 64$

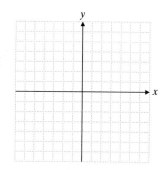

17. $\dfrac{(x - 2)^2}{25} + \dfrac{(y - 3)^2}{16} = 1$

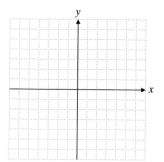

Solve the nonlinear system.

18. $y = 2x^2$
$y = 2x + 4$

19. $y = x^2 + 1$
$4y^2 = 4 - x^2$

20. $x^2 + y^2 = 25$
$x - 2y = -5$

21. $xy = -15$
$4x + 3y = 3$

CHAPTER

10

Farming throughout Africa is made more difficult by the fact that the land has been farmed for years and the yield often decreases as the soil becomes depleted. Recently, however, a new "Wonder Tree" has been developed. Now being used extensively by farmers in Zambia, this tree has enabled many farmers to increase their harvests by up to 500 percent. Can you analyze the type of increase caused by this new plant? Please turn to page 571 to find out.

Additional Properties of Functions

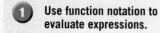

Student Learning Objectives

After studying this section, you will be able to:

1 Use function notation to evaluate expressions.

2 Use function notation to solve application exercises.

1 **Using Function Notation to Evaluate Expressions**

Function notation is useful in solving a number of interesting exercises. Suppose you wanted to skydive from an airplane. Your instructor tells you that you must wait 20 seconds before you pull the cord to open the parachute. How far will you fall in that time?

The approximate distance an object in free fall travels when there is no initial downward velocity is given by the distance function $d(t) = 16t^2$, where time t is measured in seconds and distance $d(t)$ is measured in feet. (Neglect air resistance.)

How far will a person in free fall travel in 20 seconds if he or she leaves the airplane with no downward velocity? We want to find the distance $d(20)$ where $t = 20$ seconds. To find the distance, we substitute 20 for t in the function $d(t) = 16t^2$.

$$d(20) = 16(20)^2 = 16(400) = 6400$$

Thus, a person in free fall will travel approximately 6400 feet in 20 seconds.

Now suppose that person waits longer than he or she should to pull the parachute cord. How will this affect the distance he or she falls? Suppose the person falls for a seconds beyond the 20-second mark before pulling the parachute cord.

$$\begin{aligned} d(20 + a) &= 16(20 + a)^2 \\ &= 16(20 + a)(20 + a) \\ &= 16(400 + 40a + a^2) \\ &= 6400 + 640a + 16a^2 \end{aligned}$$

Thus, if the person waited 5 seconds too long before pulling the cord, he or she would fall the following distance.

$$\begin{aligned} d(20 + 5) &= 6400 + 640(5) + 16(5)^2 \\ &= 6400 + 3200 + 16(25) \\ &= 6400 + 3200 + 400 \\ &= 10,000 \text{ feet} \end{aligned}$$

This is 3600 feet farther than the person would have fallen in 20 seconds. Obviously, a delay of 5 seconds could have life or death consequences.

We now revisit a topic that we first discussed in chapter 3, evaluating a function for particular values of the variable.

EXAMPLE 1 If $g(x) = 5 - 3x$, find the following:

(a) $g(a)$ **(b)** $g(a + 3)$ **(c)** $g(a) + g(3)$

Solution

(a) $g(a) = 5 - 3a$ **(b)** $g(a + 3) = 5 - 3(a + 3) = 5 - 3a - 9 = -4 - 3a$

(c) This exercise requires us to find each addend separately. Then we add them together.

$$g(a) = 5 - 3a$$
$$g(3) = 5 - 3(3) = 5 - 9 = -4$$

Thus,
$$\begin{aligned} g(a) + g(3) &= (5 - 3a) + (-4) \\ &= 5 - 3a - 4 \\ &= 1 - 3a \end{aligned}$$

Notice that $g(a + 3) \neq g(a) + g(3)$.

NOTE TO STUDENT: Fully worked-out solutions to all of the Practice Problems can be found at the back of the text starting at page SP-1

Practice Problem 1 If $g(x) = \dfrac{1}{2}x - 3$, find the following:

(a) $g(a)$ **(b)** $g(a + 4)$ **(c)** $g(a) + g(4)$

TO THINK ABOUT: Understanding Function Notation
Is $g(a + 4) = g(a) + g(4)$? Why or why not?

EXAMPLE 2 If $p(x) = 2x^2 - 3x + 5$, find the following:

(a) $p(-2)$ **(b)** $p(a)$ **(c)** $p(3a)$ **(d)** $p(a - 2)$

Solution

(a) $p(-2) = 2(-2)^2 - 3(-2) + 5$ **(b)** $p(a) = 2(a)^2 - 3(a) + 5 = 2a^2 - 3a + 5$
$= 2(4) - 3(-2) + 5$
$= 8 + 6 + 5$
$= 19$

(c) $p(3a) = 2(3a)^2 - 3(3a) + 5$ **(d)** $p(a - 2) = 2(a - 2)^2 - 3(a - 2) + 5$
$= 2(9a^2) - 3(3a) + 5$ $= 2(a - 2)(a - 2) - 3(a - 2) + 5$
$= 18a^2 - 9a + 5$ $= 2(a^2 - 4a + 4) - 3(a - 2) + 5$
$= 2a^2 - 8a + 8 - 3a + 6 + 5$
$= 2a^2 - 11a + 19$

Practice Problem 2 If $p(x) = -3x^2 + 2x + 4$, find the following:

(a) $p(-3)$ **(b)** $p(a)$ **(c)** $p(2a)$ **(d)** $p(a - 3)$

EXAMPLE 3 If $r(x) = \dfrac{4}{x + 2}$, find   **(a)** $r(a + 3)$ **(b)** $r(a)$

(c) $r(a + 3) - r(a)$. Express the last result as one fraction.

Solution

(a) $r(a + 3) = \dfrac{4}{a + 3 + 2} = \dfrac{4}{a + 5}$ **(b)** $r(a) = \dfrac{4}{a + 2}$

(c) $r(a + 3) - r(a) = \dfrac{4}{a + 5} - \dfrac{4}{a + 2}$

To express this as one fraction, we note that the LCD $= (a + 5)(a + 2)$.

$$r(a + 3) - r(a) = \dfrac{4(a + 2)}{(a + 5)(a + 2)} - \dfrac{4(a + 5)}{(a + 2)(a + 5)} = \dfrac{4a + 8}{(a + 5)(a + 2)} - \dfrac{4a + 20}{(a + 5)(a + 2)}$$

$$= \dfrac{4a - 4a + 8 - 20}{(a + 5)(a + 2)} = \dfrac{-12}{(a + 5)(a + 2)}$$

Practice Problem 3 If $r(x) = \dfrac{-3}{x + 1}$, find   **(a)** $r(a + 2)$ **(b)** $r(a)$
(c) $r(a + 2) - r(a)$. Express the last result as one fraction.

EXAMPLE 4 Let $f(x) = 3x - 7$. Find $\dfrac{f(x + h) - f(x)}{h}$.

Solution First

$$f(x + h) = 3(x + h) - 7 = 3x + 3h - 7$$

and

$$f(x) = 3x - 7.$$

So

$$f(x + h) - f(x) = (3x + 3h - 7) - (3x - 7)$$
$$= 3x + 3h - 7 - 3x + 7$$
$$= 3h.$$

Therefore, $\dfrac{f(x + h) - f(x)}{h} = \dfrac{3h}{h} = 3.$

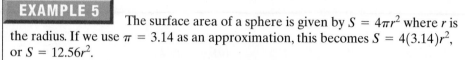

Practice Problem 4 Suppose that $g(x) = 2 - 5x$. Find $\dfrac{g(x + h) - g(x)}{h}$.

② Using Function Notation to Solve Application Exercises

▲ **EXAMPLE 5** The surface area of a sphere is given by $S = 4\pi r^2$ where r is the radius. If we use $\pi = 3.14$ as an approximation, this becomes $S = 4(3.14)r^2$, or $S = 12.56r^2$.

(a) Write the surface area of a sphere as a function of radius r.

(b) Find the surface area of a sphere with a radius of 3 centimeters.

(c) Suppose that an error is made and the radius is calculated to be $(3 + e)$ centimeters. Find an expression for the surface area as a function of the error e.

(d) Evaluate the surface area for $r = (3 + e)$ centimeters when $e = 0.2$. Round your answer to the nearest hundredth of a centimeter. What is the difference in the surface area due to the error in measurement?

Solution

(a) $S(r) = 12.56r^2$

(b) $S(3) = 12.56(3)^2 = (12.56)(9) = 113.04$ square centimeters

(c) $S(e) = 12.56(3 + e)^2$
$= 12.56(3 + e)(3 + e)$
$= 12.56(9 + 6e + e^2)$
$= 113.04 + 75.36e + 12.56e^2$

(d) If an error in measure is made so that the radius is calculated to be $r = (3 + e)$ centimeters, where $e = 0.2$, we can use the function generated in part **(c)**.

$$S = 113.04 + 75.36e + 12.56e^2$$
$$S = 113.04 + 75.36(0.2) + 12.56(0.2)^2$$
$$S = 113.04 + 15.072 + 0.5024$$
$$S = 128.6144$$

Rounding, we have $S = 128.61$ square centimeters.

Thus, if the radius of 3 centimeters was incorrectly calculated as 3.2 centimeters, the surface area would be approximately $128.61 - 113.04 = 15.57$ square centimeters too large.

▲ **Practice Problem 5** The surface area of a cylinder of height 8 meters and radius r is given by $S = 16\pi r + 2\pi r^2$.

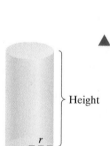
Height

(a) Write the surface area of a cylinder of height 8 meters (using $\pi = 3.14$) and radius r as a function of r.

(b) Find the surface area if the radius is 2 meters.

(c) Suppose that an error is made and the radius is calculated to be $(2 + e)$ meters. Find an expression for the surface area as a function of the error e.

(d) Evaluate the surface area for $r = (2 + e)$ meters when $e = 0.3$. Round your answer to the nearest hundredth of a meter. What is the difference in the surface area due to the error in measurement?

NOTE TO STUDENT: Fully worked-out solutions to all of the Practice Problems can be found at the back of the text starting at page SP-1

10.1 EXERCISES

| Student Solutions Manual | CD/ Video | PH Math Tutor Center | MathXL®Tutorials on CD | MathXL® | MyMathLab® | Interactmath.com |

For the function $f(x) = 3x - 5$, find the following.

1. $f\left(-\dfrac{2}{3}\right)$ **2.** $f(1.5)$ **3.** $f(a - 4)$ **4.** $f(b + 3)$

For the function $g(x) = \dfrac{1}{2}x - 3$, find the following.

5. $g(4) + g(a)$ **6.** $g(6) + g(b)$ **7.** $g(4a) - g(a)$ **8.** $g(6b) + g(-2b)$

9. $g(2a - 4)$ **10.** $g(3a + 1)$ **11.** $g(a^2) - g\left(\dfrac{2}{5}\right)$ **12.** $g(b^2) - g\left(\dfrac{4}{3}\right)$

If $p(x) = 3x^2 + 4x - 2$, find the following.

13. $p(-2)$ **14.** $p(-3)$ **15.** $p\left(\dfrac{1}{2}\right)$ **16.** $p(2.5)$

17. $p(a + 1)$ **18.** $p(b - 1)$ **19.** $p\left(-\dfrac{2a}{3}\right)$ **20.** $p\left(-\dfrac{3a}{2}\right)$

If $h(x) = \sqrt{x + 5}$, find the following.

21. $h(4)$ **22.** $h(-5)$ **23.** $h(7)$

24. $h(19)$ **25.** $h(a^2 - 1)$ **26.** $h(a^2 + 4)$

27. $h(-2b)$ **28.** $h(-7b)$ **29.** $h(4a - 1)$

30. $h(4a + 3)$ **31.** $h(b^2 + b)$ **32.** $h(b^2 + b - 5)$

If $r(x) = \dfrac{7}{x - 3}$, find the following and write your answers as one fraction.

33. $r(7)$ **34.** $r(-4)$ **35.** $r(3.5)$

36. $r(-0.5)$ **37.** $r(a^2)$ **38.** $r(3b^2)$

39. $r(a + 2)$ **40.** $r(a - 3)$

41. $r\left(\dfrac{1}{2}\right) + r(8)$ **42.** $r\left(\dfrac{5}{3}\right) + r(7)$

Find $\dfrac{f(x + h) - f(x)}{h}$ *for the following functions.*

43. $f(x) = 2x - 3$

44. $f(x) = 5 - 2x$

45. $f(x) = x^2 - x$

46. $f(x) = 2x^2$

Applications

47. *Wind Generators* A turbine wind generator produces P kilowatts of power for wind speed w (measured in miles per hour) according to the equation $P = 2.5w^2$.
 (a) Write the number of kilowatts P as a function of w.
 (b) Find the power in kilowatts when the wind speed is $w = 20$ miles per hour.
 (c) Suppose that an error is made and the speed of the wind is calculated to be $(20 + e)$ miles per hour. Find an expression for the power as a function of error e.
 (d) Evaluate the power for $w = (20 + e)$ miles per hour when $e = 2$.

▲ **48. *Geometry*** The area of a circle is $A = \pi r^2$.
 (a) Write the area of a circle as the function of the radius r. Use $\pi = 3.14$.
 (b) Find the area of a circle with a radius of 4.0 feet.
 (c) Suppose that an error is made and the radius is calculated to be $(4 + e)$ feet. Find an expression for the area as a function of error e.
 (d) Evaluate the area for $r = (4 + e)$ feet when $e = 0.4$. Round your answer to the nearest hundredth.

Lead Levels in Air Because of the elimination of lead in automobile gasoline and increased use of emission controls in automobiles and industrial operations, the amount of lead in the air in the United States has shown a marked decrease. The percent of lead in the air $p(x)$ expressed in terms of 1984 levels is given in the line graph. The variable x indicates the number of years since 1984. The function value $p(x)$ indicates the amount of lead that remains in the air in selected regions of the United States, expressed as a percent of the amount of lead in the air in 1984.

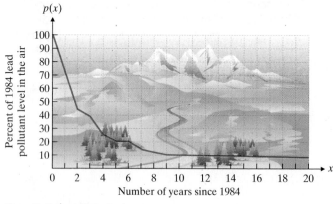

Source: Environmental Protection Agency.

49. If a new function were defined as $p(x) - 13$, what would happen to the function values associated with x? Find $p(3) - 13$.

50. If a new function were defined as $p(x + 2)$, what would happen to the function values associated with x? Find $p(x + 2)$ when $x = 4$.

If $f(x) = 3x^2 - 4.6x + 1.23$, find each of the following functions to the nearest thousandth.

51. $f(0.026a)$

52. $f(3.56a)$

53. $f(a + 2.23)$

54. $f(a - 0.152)$

▲ **55.** *Geometry* A rope 20 feet long is cut into two un-equal pieces. Each piece is used to form a square. Write a function $A(x)$ that expresses the total area enclosed by the two squares. Assume that the shorter piece of rope is x feet long. Evaluate $A(2)$, $A(5)$, and $A(8)$.

▲ **56.** *Geometry* Assume that the smaller piece of rope in Exercise 55 is used to form a circle and the longer piece is used to form a square. Write a function $A(x)$ that expresses the total area enclosed by the circle and the square. Evaluate $A(3)$ and $A(9)$. Round your answers to the nearest hundredth.

Cumulative Review

Solve for x.

57. $\dfrac{7}{6} + \dfrac{5}{x} = \dfrac{3}{2x}$

58. $\dfrac{1}{6} - \dfrac{2}{3x + 6} = \dfrac{1}{2x + 4}$

▲ **59.** *Planet Volume* The diameter of Mercury is 3031 miles, while that of Earth is 7927 miles. How many times greater is the volume of Earth compared to the volume of Mercury?

▲ **60.** *Planet Volume* The radius of Uranus is 14,584 miles. The radius of Jupiter is 43,348 miles. How many times greater is the volume of Jupiter compared to the volume of Uranus?

Student Learning Objectives

After studying this section, you will be able to:

1 Use the vertical line test to determine whether a graph represents a function.

2 Graph a function of the form $f(x + h) + k$ by means of horizontal and vertical shifts of the graph of $f(x)$.

1 Using the Vertical Line Test to Determine Whether a Graph Represents a Function

Not every graph we observe is that of a function. By definition, a function must have no ordered pairs that have the same first coordinates and different second coordinates. A graph that includes the points $(4, 2)$ and $(4, -2)$, for example, would not be the graph of a function. Thus, the graph of $x = y^2$ would not be the graph of a function.

If any vertical line crosses a graph of a relation in more than one place, the relation is not a function. If no such line exists, the relation is a function.

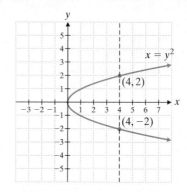

VERTICAL LINE TEST

If any vertical line intersects the graph of a relation more than once, the relation is not a function. If no such line exists, the relation is a function.

In the following sketches, we observe that the dashed vertical line crosses the curve of a function no more than once. The dashed vertical line crosses the curve of a relation that is *not* a function more than once.

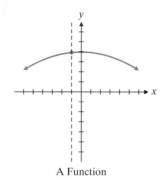

A Function

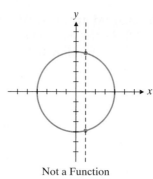

A Function

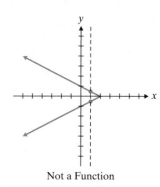

Not a Function

Not a Function

EXAMPLE 1 Determine whether each of the following is the graph of a function

(a)

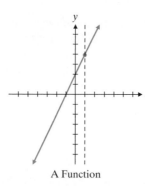

(b)

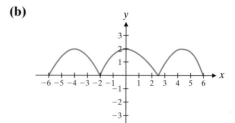

Solution

(a) By the vertical line test, this relation is not a function.

(b) By the vertical line test, this relation is a function.

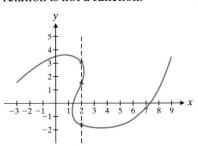

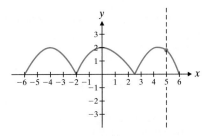

Practice Problem 1 Does this graph represent a function? Why or why not?

NOTE TO STUDENT: *Fully worked-out solutions to all of the Practice Problems can be found at the back of the text starting at page SP-1*

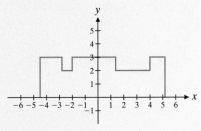

---

② Graphing a Function of the Form $f(x + h) + k$ by Means of Horizontal and Vertical Shifts of the Graph of $f(x)$

The graphs of some functions are simple vertical shifts of the graphs of similar functions.

EXAMPLE 2 Graph the functions on one coordinate plane.

$$f(x) = x^2 \quad \text{and} \quad h(x) = x^2 + 2$$

Solution First we make a table of values for $f(x)$ and for $h(x)$.

x	$f(x) = x^2$
-2	4
-1	1
0	0
1	1
2	4

x	$h(x) = x^2 + 2$
-2	6
-1	3
0	2
1	3
2	6

Now we graph each function on the same coordinate plane.

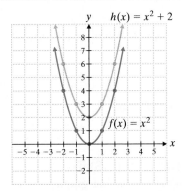

Notice that the graph of $h(x)$ is the graph of $f(x)$ moved 2 units upward.

PRACTICE PROBLEM 2

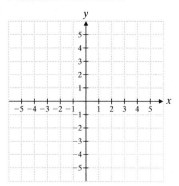

TO THINK ABOUT: Vertical Shifts What would the graph of $j(x) = x^2 - 3$ look like? Verify by making a table of values and drawing a graph of the function.

Practice Problem 2 Graph the functions on one coordinate plane.
$$f(x) = x^2 \quad \text{and} \quad h(x) = x^2 - 5$$

We have the following general summary.

> **VERTICAL SHIFTS**
>
> Suppose that k is a positive number.
>
> 1. To obtain the graph of $f(x) + k$, shift the graph of $f(x)$ up k units.
> 2. To obtain the graph of $f(x) - k$, shift the graph of $f(x)$ down k units.

Now we turn to the topic of horizontal shifts.

Graphing Calculator

 Exploration

Most graphing calculators have an absolute value function (abs). Use this function to graph $f(x)$ and $h(x)$ on one coordinate plane.

$$f(x) = |0.5x|$$
$$h(x) = |0.5x + 3.5| - 1.75$$

Describe how we shift the graph of $f(x)$ to obtain that of $h(x)$. Use your calculator to find the approximate coordinates of the point where $f(x)$ and $h(x)$ intersect. Find the coordinates to the nearest hundredth.

EXAMPLE 3 Graph the functions on one coordinate plane.
$$f(x) = |x| \quad \text{and} \quad p(x) = |x - 3|$$

Solution First we make a table of values for $f(x)$ and $p(x)$.

x	$f(x) = |x|$
-2	2
-1	1
0	0
1	1
2	2
3	3
4	4

x	$p(x) = |x - 3|$
-2	5
-1	4
0	3
1	2
2	1
3	0
4	1

Now we graph each function on the same coordinate plane.

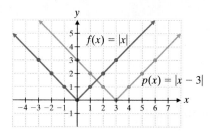

Notice that the graph of $p(x)$ is the graph of $f(x)$ shifted 3 units to the right.

Practice Problem 3 Graph the functions on one coordinate plane.
$$f(x) = |x| \quad \text{and} \quad p(x) = |x + 2|$$

PRACTICE PROBLEM 3

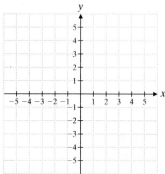

TO THINK ABOUT: Horizontal Shifts What would the graph of $h(x) = (x - 3)^2$ look like? What would the graph of $j(x) = (x + 2)^2$ look like? Verify by making tables of values and drawing the graphs.

Now we can write the following general summary.

HORIZONTAL SHIFTS

Suppose that h is a positive number.

1. To obtain the graph of $f(x - h)$, shift the graph of $f(x)$ to the right h units.
2. To obtain the graph of $f(x + h)$, shift the graph of $f(x)$ to the left h units.

Some graphs will involve both horizontal and vertical shifts.

EXAMPLE 4 Graph the functions on one coordinate plane.

$$f(x) = x^3 \quad \text{and} \quad h(x) = (x - 3)^3 - 2$$

Solution First we make a table of values for $f(x)$ and graph the function.

x	$f(x)$
-2	-8
-1	-1
0	0
1	1
2	8

Next we recognize that $h(x)$ will have a similar shape, but the curve will be shifted 3 units to the *right* and 2 units *downward*. We draw the graph of $h(x)$ using these shifts.

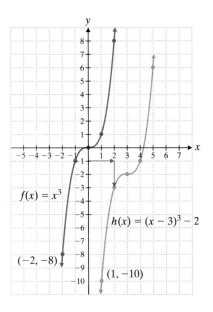

TO THINK ABOUT: Shifting Points The point $(-2, -8)$ has been shifted 3 units to the right and 2 units down to the point $(-2 + 3, -8 + (-2))$ or $(1, -10)$. The point $(-1, -1)$ is a point on $f(x)$. Use the same reasoning to find the image of $(-1, -1)$ on the graph of $h(x)$. Verify by checking the graphs.

Practice Problem 4 Graph the functions on one coordinate plane.

$$f(x) = x^3 \quad \text{and} \quad h(x) = (x + 4)^3 + 3$$

Graphing Calculator

Exploration

Graph $f(x)$ and $h(x)$ on one coordinate plane.

$$f(x) = x^3$$
$$h(x) = x^3 - 6x^2 + 12x - 4$$

How must the graph of $f(x)$ be shifted horizontally and vertically to produce the graph of $h(x)$? Use your Zoom feature to find the positive value of x where $h(x)$ crosses the x-axis. Find this value to the nearest hundredth.

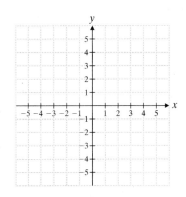

NOTE TO STUDENT: Fully worked-out solutions to all of the Practice Problems can be found at the back of the text starting at page SP-1

All the functions that we have sketched in this section so far have had a domain of all real numbers. Some functions have a restricted domain.

EXAMPLE 5 Graph the functions on one coordinate plane. State the domain of each function.

$$f(x) = \frac{4}{x} \quad \text{and} \quad g(x) = \frac{4}{x+3} + 1$$

Solution First we make a table of values for $f(x)$. The domain of $f(x)$ is all real numbers, where $x \neq 0$. Note that $f(x)$ is not defined when $x = 0$ since we cannot divide by 0.

x	f(x)
-4	-1
-2	-2
-1	-4
$-\frac{1}{2}$	-8
$\frac{1}{2}$	8
1	4
2	2
4	1

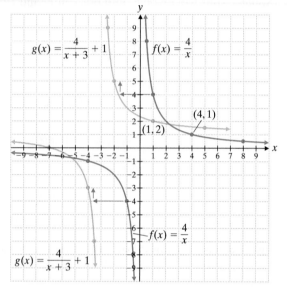

We draw $f(x)$ and note key points. From the equation, we see that the graph of $g(x)$ is 3 units to the left of and 1 unit above $f(x)$. We can find the image of each of the key points on $f(x)$ as a guide in graphing $g(x)$. For example, the image of $(4, 1)$ is $(4 - 3, 1 + 1)$ or $(1, 2)$.

Each point on $f(x)$ is shifted

$$\Leftarrow \text{ 3 units left and}$$
$$\Uparrow \text{ 1 unit up}$$

to form the graph of $g(x)$.

What is the domain of $g(x)$? Why? $g(x)$ contains the denominator $x + 3$. But $x + 3 \neq 0$. Therefore, $x \neq -3$. The domain of $g(x)$ is all real numbers, where $x \neq -3$.

NOTE TO STUDENT: Fully worked-out solutions to all of the Practice Problems can be found at the back of the text starting at page SP-1

Practice Problem 5 Graph the functions on one coordinate plane.

$$f(x) = \frac{2}{x} \quad \text{and} \quad g(x) = \frac{2}{x+1} - 2$$

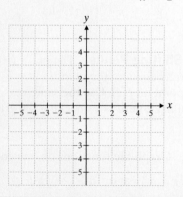

Verbal and Writing Skills

1. Does $f(x + 2) = f(x) + f(2)$? Why or why not? Give an example.

2. Explain what the vertical line test is and why it works.

3. To obtain the graph of $f(x) + k$, shift the graph of $f(x)$ _____ k units.

4. To obtain the graph of $f(x - h)$, shift the graph of $f(x)$ _____ h units.

Determine whether or not each graph represents a function.

5.

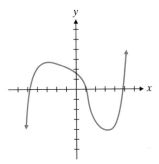

6.

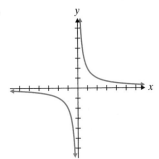

7.

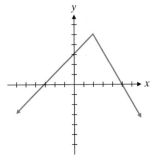

8.

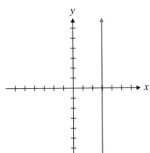

9.

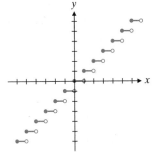

10.

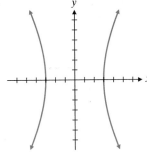

Hint: The open circle means that the function value does not exist at that point.

11.

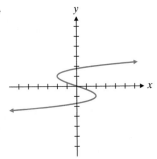

12.

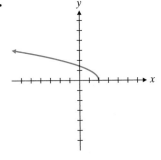

13.

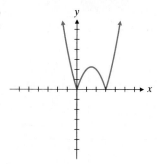

14.

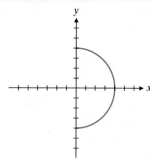

For each of Exercises 15–30, graph the two functions on one coordinate plane.

15. $f(x) = x^2$
   $h(x) = x^2 - 3$

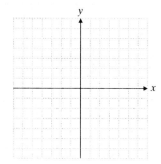

16. $f(x) = x^2$
   $h(x) = x^2 + 4$

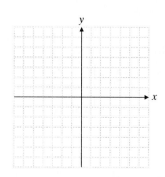

17. $f(x) = x^2$
   $p(x) = (x + 1)^2$

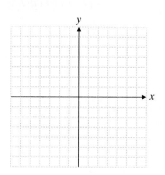

18. $f(x) = x^2$
   $p(x) = (x - 2)^2$

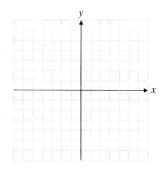

19. $f(x) = x^2$
   $g(x) = (x - 2)^2 + 1$

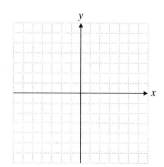

20. $f(x) = x^2$
   $g(x) = (x + 1)^2 - 2$

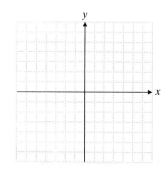

21. $f(x) = x^3$
   $r(x) = x^3 - 1$

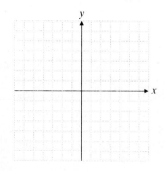

22. $f(x) = x^3$
   $r(x) = x^3 + 2$

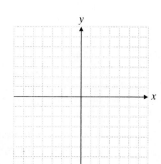

23. $f(x) = |x|$
   $s(x) = |x + 4|$

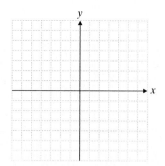

24. $f(x) = |x|$
   $s(x) = |x - 2|$

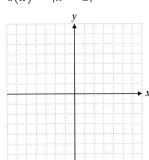

25. $f(x) = |x|$
   $t(x) = |x - 3| - 4$

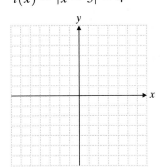

26. $f(x) = |x|$
   $t(x) = |x + 1| + 2$

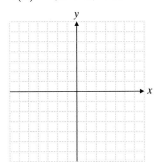

27. $f(x) = x^3$
   $j(x) = (x - 3)^3 + 3$

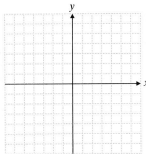

28. $f(x) = x^3$
   $j(x) = (x + 3)^3 + 1$

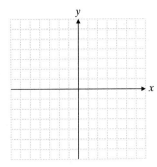

29. $f(x) = \dfrac{3}{x}$
   $g(x) = \dfrac{3}{x} - 2$

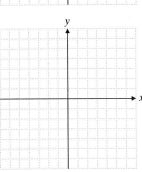

30. $f(x) = \dfrac{2}{x}$
   $g(x) = \dfrac{2}{x} + 3$

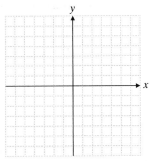

Optional Graphing Calculator Problems

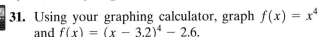

31. Using your graphing calculator, graph $f(x) = x^4$ and $f(x) = (x - 3.2)^4 - 2.6$.

32. Using your graphing calculator, graph $f(x) = x^7$ and $f(x) = (x + 1.3)^7 + 3.3$.

Cumulative Review

Simplify each expression. Assume that all variables are positive.

33. $\sqrt{12} + 3\sqrt{50} - 4\sqrt{27}$

34. $\left(\sqrt{5x} + \sqrt{2}\right)^2$

35. Rationalize the denominator: $\dfrac{\sqrt{5} - 2}{\sqrt{5} + 1}$

36. *Calorie Counting* Roy is on a diet and can have a maximum of 615 calories for lunch. A tuna fish sandwich on whole wheat bread has 315 calories, and 12 fluid ounces of a fruit juice soft drink has 120 calories. How many french fries can he eat if there are 10 calories in one french fry?

37. *Finder's Fees* For a fee, the American Scholarship Search Company helped 28,560 students locate various study grants and fellowships. They collected $13,623,120 in finder's fees from these students. The company advertised in their literature that they collect a finder's fee of only $250 per student. How much did they overcharge each student?

1. _____

2. _____

3. _____

4. _____

5. _____

6. _____

7. _____

8. _____

9. _____

10. _____

11. _____

12. _____

13. _____

14. _____

15. _____

16. _____

How are you doing with your homework assignments in Sections 10.1 to 10.2? Do you feel you have mastered the material so far? Do you understand the concepts you have covered? Before you go further in the textbook, take some time to do each of the following problems.

10.1   *For the function $f(x) = 2x - 6$, find the following:*

1. $f(-3)$ **2.** $f(a)$ **3.** $f(2a)$ **4.** $f(a + 2)$

For $f(x) = 5x^2 + 2x - 3$, find the following:

5. $f(-2)$ **6.** $f(a)$ **7.** $f(a + 1)$ **8.** $f(3a)$

For $f(x) = \dfrac{3x}{x + 2}$, find the following:

9. $f(a) + f(a - 2)$. Express your answer as one fraction.

10. $f(3a) - f(3)$. Express your answer as one fraction.

10.2   *Determine whether or not each graph represents a function.*

11.

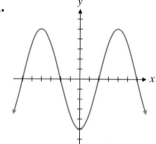

12.

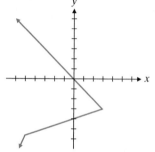

Graph the given functions on one coordinate plane.

13. $f(x) = |x|$ and $s(x) = |x - 3|$.

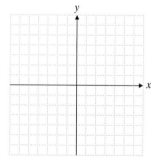

14. $f(x) = x^2$ and $h(x) = (x + 2)^2 + 3$.

15. Graph $f(x) = \dfrac{4}{x + 2}$.

16. If you were to graph $g(x) = \dfrac{4}{x + 2} - 2$, how would it compare to the graph of $f(x)$ in Problem 15?

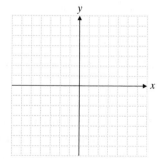

Now turn to page SA-40 for the answers to each of these problems. Each answer also includes a reference to the objective in which the problem is first taught. If you missed any of these problems, you should stop and review the Examples and Practice Problems in the referenced objective. A little review now will help you master the material in the upcoming sections of the text.

10.3 ALGEBRAIC OPERATIONS ON FUNCTIONS

1 Finding the Sum, Difference, Product, and Quotient of Two Functions

When two functions are given, new functions can be formed by combining them, as defined in the box:

> **IF f REPRESENTS ONE FUNCTION AND g REPRESENTS A SECOND FUNCTION, WE CAN DEFINE NEW FUNCTIONS AS FOLLOWS:**
>
> Sum of Functions $\qquad (f + g)(x) = f(x) + g(x)$
>
> Difference of Functions $\qquad (f - g)(x) = f(x) - g(x)$
>
> Product of Functions $\qquad (fg)(x) = f(x) \cdot g(x)$
>
> Quotient of Functions $\qquad \left(\dfrac{f}{g}\right)(x) = \dfrac{f(x)}{g(x)}, g(x) \neq 0$

EXAMPLE 1 Suppose that $f(x) = 3x^2 - 3x + 5$ and $g(x) = 5x - 2$.

(a) Find $(f + g)(x)$. **(b)** Evaluate $(f + g)(x)$ when $x = 3$.

Solution

(a) $(f + g)(x) = f(x) + g(x)$
$$= (3x^2 - 3x + 5) + (5x - 2)$$
$$= 3x^2 - 3x + 5 + 5x - 2$$
$$= 3x^2 + 2x + 3$$

(b) To evaluate $(f + g)(x)$ when $x = 3$, we write $(f + g)(3)$ and use the formula obtained in **(a)**.

$$(f + g)(x) = 3x^2 + 2x + 3$$
$$(f + g)(3) = 3(3)^2 + 2(3) + 3$$
$$= 3(9) + 2(3) + 3$$
$$= 27 + 6 + 3 = 36$$

Practice Problem 1 Given $f(x) = 4x + 5$ and $g(x) = 2x^2 + 7x - 8$, find the following: **(a)** $(f + g)(x)$ **(b)** $(f + g)(4)$

NOTE TO STUDENT: Fully worked-out solutions to all of the Practice Problems can be found at the back of the text starting at page SP-1

EXAMPLE 2 Given $f(x) = x^2 - 5x + 6$ and $g(x) = 2x - 1$, find the following: **(a)** $(fg)(x)$ **(b)** $(fg)(-4)$

Solution

(a) $(fg)(x) = f(x) \cdot g(x)$
$$= (x^2 - 5x + 6)(2x - 1)$$
$$= 2x^3 - 10x^2 + 12x - x^2 + 5x - 6$$
$$= 2x^3 - 11x^2 + 17x - 6$$

(b) To evaluate $(fg)(x)$ when $x = -4$, we write $(fg)(-4)$ and use the formula obtained in **(a)**.

$$(fg)(x) = 2x^3 - 11x^2 + 17x - 6$$
$$(fg)(-4) = 2(-4)^3 - 11(-4)^2 + 17(-4) - 6$$
$$= 2(-64) - 11(16) + 17(-4) - 6$$
$$= -128 - 176 - 68 - 6$$
$$= -378$$

Practice Problem 2   Given $f(x) = 3x + 2$ and $g(x) = x^2 - 3x - 4$, find the following: **(a)** $(fg)(x)$ **(b)** $(fg)(2)$

When finding the quotient of a function, we must be careful to avoid division by zero. Thus, we always specify any values of x that must be eliminated from the domain.

EXAMPLE 3   Given $f(x) = 3x + 1$, $g(x) = 2x - 1$, and $h(x) = 9x^2 + 6x + 1$, find the following:

(a) $\left(\dfrac{f}{g}\right)(x)$ **(b)** $\left(\dfrac{f}{h}\right)(x)$ **(c)** $\left(\dfrac{f}{h}\right)(-2)$

Solution

(a) $\left(\dfrac{f}{g}\right)(x) = \dfrac{3x + 1}{2x - 1}$

The denominator of the quotient can never be zero. Since $2x - 1 \neq 0$, we know that $x \neq \frac{1}{2}$.

(b) $\left(\dfrac{f}{h}\right)(x) = \dfrac{3x + 1}{9x^2 + 6x + 1} = \dfrac{3x + 1}{(3x + 1)(3x + 1)} = \dfrac{1}{3x + 1}$

Since $3x + 1 \neq 0$, we know that $x \neq -\frac{1}{3}$.

(c) To find $\left(\dfrac{f}{h}\right)(-2)$, we must evaluate $\left(\dfrac{f}{h}\right)(x)$ when $x = -2$.

$$\left(\dfrac{f}{h}\right)(x) = \dfrac{1}{3x + 1}$$

$$\left(\dfrac{f}{h}\right)(-2) = \dfrac{1}{(3)(-2) + 1} = \dfrac{1}{-6 + 1} = -\dfrac{1}{5}$$

NOTE TO STUDENT: Fully worked-out solutions to all of the Practice Problems can be found at the back of the text starting at page SP-1

Practice Problem 3   Given $p(x) = 5x^2 + 6x + 1$, $h(x) = 3x - 2$, and $g(x) = 5x + 1$, find the following:

(a) $\left(\dfrac{g}{h}\right)(x)$ **(b)** $\left(\dfrac{g}{p}\right)(x)$ **(c)** $\left(\dfrac{g}{h}\right)(3)$

② Finding the Composition of Two Functions

Suppose that the music section of a department store finds that the number of sales of compact discs (CDs) on a given day is generally equal to 25% of the number of people who visit the store on that day. Thus, if $x =$ the number of people who visit the store, then the sales S can be modeled by the equation $S(x) = 0.25x$.

Suppose that the average CD in the store sells for $15. Then if $S =$ the number of CD sales on a given day, the income for that day can be modeled by the equation $P(S) = 15S$. Suppose that eighty people came into the store.

$$S(x) = 0.25x$$
$$S(80) = 0.25(80) = 20$$

Thus, twenty CDs would be sold.

If twenty CDs were sold and the average price of a CD is $15, then we would have the following:

$$P(S) = 15S$$
$$P(20) = 15(20) = 300$$

That is, the income from the sales of CDs would be $300.

Let us analyze the functions we have described and record a few values of x, $S(x)$, and $P(S)$.

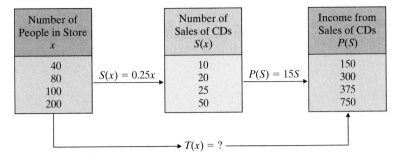

Number of People in Store x		Number of Sales of CDs $S(x)$		Income from Sales of CDs $P(S)$
40		10		150
80	$S(x) = 0.25x$	20	$P(S) = 15S$	300
100		25		375
200		50		750

$T(x) = ?$

Is there a function $T(x)$ that describes the income from CD sales as a function of x, the number of people who visit the store?

The number of sales is

$$S(x) = 0.25x.$$

Thus, $0.25x$ is the number of sales.

If we replace S in $P(S) = 15S$ by $S(x)$, we have

$$P[S(x)] = P(0.25x) = 15(0.25x) = 3.75x.$$

Thus, the formula $T(x)$ that describes the income in terms of the number of visitors is

$$T(x) = 3.75x.$$

Is this correct? Let us check by finding $T(200)$. From our table the result should be 750.

$$\text{If} \qquad T(x) = 3.75x,$$
$$\text{then} \qquad T(200) = 3.75(200) = 750.$$

Thus, we have found a function T that is the composition of the functions P and S: $T(x) = P[S(x)]$.

We now state a definition of the composition of one function with another.

The **composition** of the functions f and g, denoted $f \circ g$, is defined as follows: $(f \circ g)(x) = f[g(x)]$. The domain of $f \circ g$ is the set of all x values in the domain of g such that $g(x)$ is in the domain of f.

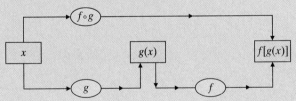

EXAMPLE 4   Given $f(x) = 3x - 2$ and $g(x) = 2x + 5$, find $f[g(x)]$.

Solution

$$
\begin{aligned}
f[g(x)] &= f(2x + 5) && \text{Substitute } g(x) = 2x + 5. \\
&= 3(2x + 5) - 2 && \text{Apply the formula for } f(x). \\
&= 6x + 15 - 2 && \text{Remove parentheses.} \\
&= 6x + 13 && \text{Simplify.}
\end{aligned}
$$

Practice Problem 4   Given $f(x) = 2x - 1$ and $g(x) = 3x - 4$, find $f[g(x)]$.

In most situations $f[g(x)]$ and $g[f(x)]$ are not the same.

EXAMPLE 5   Given $f(x) = \sqrt{x-4}$ and $g(x) = 3x+1$, find the following:

(a) $f[g(x)]$ **(b)** $g[f(x)]$

Solution

(a) $f[g(x)] = f[3x+1]$ Substitute $g(x) = 3x+1$.

$= \sqrt{(3x+1)-4}$ Apply the formula for $f(x)$.

$= \sqrt{3x+1-4}$ Remove parentheses.

$= \sqrt{3x-3}$ Simplify.

(b) $g[f(x)] = g[\sqrt{x-4}]$ Substitute $f(x) = \sqrt{x-4}$.

$= 3(\sqrt{x-4})+1$ Apply the formula for $g(x)$.

$= 3\sqrt{x-4}+1$ Remove parentheses.

We note that $g[f(x)] \neq f[g(x)]$.

Practice Problem 5 Given $f(x) = 2x^2 - 3x + 1$ and $g(x) = x + 2$, find the following:

(a) $f[g(x)]$ **(b)** $g[f(x)]$

EXAMPLE 6   Given $f(x) = 2x$ and $g(x) = \dfrac{1}{3x-4}$, $x \neq \dfrac{4}{3}$, find the following:

(a) $(f \circ g)(x)$ **(b)** $(f \circ g)(2)$

Solution

(a) $(f \circ g)(x) = f[g(x)] = f\left[\dfrac{1}{3x-4}\right]$ Substitute $g(x) = \dfrac{1}{3x-4}$.

$= 2\left(\dfrac{1}{3x-4}\right)$ Apply the formula for $f(x)$.

$= \dfrac{2}{3x-4}$ Simplify.

(b) $(f \circ g)(2) = \dfrac{2}{3(2)-4} = \dfrac{2}{6-4} = \dfrac{2}{2} = 1$

Practice Problem 6 Given $f(x) = 3x + 1$ and $g(x) = \dfrac{2}{x-3}$, find the following:

(a) $(g \circ f)(x)$ **(b)** $(g \circ f)(-3)$

Graphing Calculator

Composition of Functions

You can formulate the composition of functions on most graphing calculators by using the y-variable function (Y-VARS). To do Example 6 on most graphing calculators, you would use the following equations.

$y_1 = \dfrac{1}{3x-4}$

$y_2 = 2(y_1)$

To find the function value, you can use the TableSet command to let $x = 2$. Then enter Table and you will see displayed $y_1 = 0.5$, which represents $g(2) = 0.5$, and $y_2 = 1$, which represents $f[g(2)] = 1$.

NOTE TO STUDENT: Fully worked-out solutions to all of the Practice Problems can be found at the back of the text starting at page SP-1

10.3 EXERCISES

| Student Solutions Manual | CD/ Video | PH Math Tutor Center | MathXL®Tutorials on CD | MathXL® | MyMathLab® | Interactmath.com |

For the following functions, find **(a)** $(f + g)(x)$, **(b)** $(f - g)(x)$, **(c)** $(f + g)(2)$, *and* **(d)** $(f - g)(-1)$.

1. $f(x) = -2x + 3, g(x) = 2 + 4x$

2. $f(x) = 3x + 4, g(x) = 1 - 2x$

3. $f(x) = 3x^2 - x, g(x) = 5x + 2$

4. $f(x) = 6 - 5x, g(x) = x^2 - 7x - 3$

5. $f(x) = x^3 - \frac{1}{2}x^2 + x, g(x) = x^2 - \frac{x}{4} - 5$

6. $f(x) = 2.4x^2 + x - 3.5, g(x) = 1.1x^3 - 2.2x$

7. $f(x) = -5\sqrt{x + 6}, g(x) = 8\sqrt{x + 6}$

8. $f(x) = 3\sqrt{3 - x}, g(x) = -5\sqrt{3 - x}$

For the following functions, find **(a)** $(fg)(x)$ *and* **(b)** $(fg)(-3)$.

9. $f(x) = 2x - 3, g(x) = -2x^2 - 3x + 1$

10. $f(x) = x^2 - 3x + 2, g(x) = 1 - x$

11. $f(x) = \frac{2}{x^2}, g(x) = x^2 - x$

12. $f(x) = \frac{6x}{x - 1}, g(x) = \frac{x}{2}$

13. $f(x) = \sqrt{-2x + 1}, g(x) = -3x$

14. $f(x) = 4x, g(x) = \sqrt{3x + 10}$

For the following functions, find **(a)** $\left(\frac{f}{g}\right)(x)$ *and* **(b)** $\left(\frac{f}{g}\right)(2)$.

15. $f(x) = x - 6, g(x) = 3x$

16. $f(x) = 3x, g(x) = 4x - 1$

17. $f(x) = x^2 - 1, g(x) = x - 1$

18. $f(x) = x, g(x) = x^2 - 5x$

19. $f(x) = x^2 + 10x + 25, g(x) = x + 5$

20. $f(x) = 4x^2 + 4x + 1, g(x) = 2x + 1$

21. $f(x) = 4x - 1, g(x) = 4x^2 + 7x - 2$

22. $f(x) = 3x + 2, g(x) = 3x^2 - x - 2$

Let $f(x) = 3x + 2, g(x) = x^2 - 2x$, *and* $h(x) = \frac{x - 2}{3}$. *Find the following:*

23. $(f - g)(x)$

24. $(g - f)(x)$

25. $\left(\frac{g}{h}\right)(x)$

26. $\left(\frac{g}{h}\right)(5)$

27. $(fg)(-1)$

28. $(gh)(3)$

29. $\left(\frac{g}{f}\right)(-1)$

30. $\left(\frac{g}{f}\right)(x)$

Find $f[g(x)]$ for each of the following:

31. $f(x) = 2 - 3x, g(x) = 2x + 5$

32. $f(x) = 3x + 2, g(x) = 4x - 1$

33. $f(x) = 2x^2 + 5, g(x) = x - 1$

34. $f(x) = 7 - x^2, g(x) = x - 2$

35. $f(x) = 8 - 5x, g(x) = x^2 + 3$

36. $f(x) = 6 - x, g(x) = 5 - 2x - x^2$

37. $f(x) = \dfrac{7}{2x - 3}, g(x) = x + 2$

38. $f(x) = \dfrac{11}{6 - x}, g(x) = x + 3$

39. $f(x) = |x + 3|, g(x) = 2x - 1$

40. $f(x) = \left|\dfrac{1}{2}x - 5\right|, g(x) = 4x + 6$

Let $f(x) = x^2 + 2, g(x) = 3x + 5, h(x) = \dfrac{1}{x}$, and $p(x) = \sqrt{x - 1}$. Find each of the following:

41. $f[g(x)]$

42. $g[h(x)]$

43. $g[f(x)]$

44. $h[g(x)]$

45. $g[f(0)]$

46. $h[g(0)]$

47. $(p \circ f)(x)$

48. $(f \circ h)(x)$

49. $(g \circ h)(\sqrt{2})$

50. $(f \circ p)(x)$

51. $(p \circ f)(-5)$

52. $(f \circ p)(7)$

Applications

53. **Temperature Scales** Consider the Celsius function $C(F) = \dfrac{5F - 160}{9}$, which converts degrees Fahrenheit to degrees Celsius. A different temperature scale, called the Kelvin scale, is used by many scientists in their research. The Kelvin scale is similar to the Celsius scale, but it begins at absolute zero (the coldest possible temperature, which is around $-273°$C). To convert a Celsius temperature to a temperature on the Kelvin scale, we use the function $K(C) = C + 273$. Find $K[C(F)]$, which is the composite function that defines the temperature in Kelvins in terms of the temperature in degrees Fahrenheit.

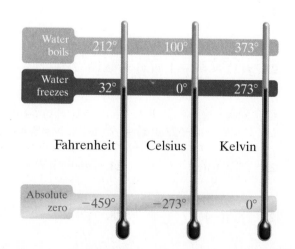

54. ***Business*** Suppose the dollar cost to produce n items in a factory is $c(n) = 5n + 4$. Furthermore, the number of items n produced in x hours is $n(x) = 3x$. Find $c[n(x)]$, which is the composite function that defines the dollar cost in terms of the number of hours of production x.

▲ **55.** ***Water Pollution*** The volume of polluted water emitted from a discharge pipe from a factory located on the ocean is shaped in a cone. The radius r of the cone of polluted water at the end of each day is given by the equation $r(h) = 3.5h$, where h is the number of hours the factory operated that day. The volume function that defines this cone is $v(r) = 31.4r^2$, where r is the radius of the cone measured in feet. Find $v[r(h)]$, which is the composite function that defines the volume of polluted water in terms of the number of hours h the factory has run. How large is the volume at the end of the day if the factory has been running for 8 hours?

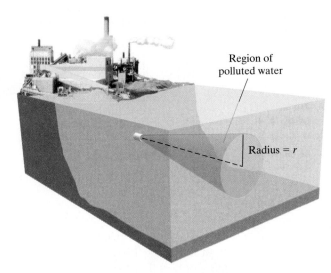

Region of polluted water

Radius = r

▲ **56.** ***Oil Slicks*** An oil tanker with a ruptured hull is leaking oil off the coast of Africa. There is no wind or significant current, so the oil slick is spreading in a circle whose radius is defined by the function $r(t) = 3t$, where t is the time in minutes since the tanker began to leak. The area of the slick for any given radius is approximately determined by the function $a(r) = 3.14r^2$ where r is the radius of the circle measured in feet. Find $a[r(t)]$, which is the composite function that defines the area of the oil slick in terms of the minutes t since the beginning of the leak. How large is the area after 20 minutes?

Cumulative Review

Factor each of the following:

57. $36x^2 - 12x + 1$

58. $25x^4 - 1$

59. $x^4 - 10x^2 + 9$

60. $3x^2 - 7x + 2$

61. ***Radio Airtime*** Juanita has an hour-long radio show called "Talk of the Town." During the broadcast she has to break for twenty commercials. She has commercial tapes that are 30 seconds long and other commercial tapes that are 60 seconds long. How many of each type should she play in order to achieve the goal of 14 minutes of commercials?

62. ***Alumni Dinner*** The West Shore Community College Alumni Dinner was held for 470 people. A children's plate cost $4.50, and an adult's plate cost $7.50. The receipts for the dinner were $3180. How many children attended the dinner? How many adults attended the dinner?

10.4 INVERSE OF A FUNCTION

Student Learning Objectives

After studying this section, you will be able to:

1 Determine whether a function is a one-to-one function.

2 Find the inverse function for a given function.

3 Graph a function and its inverse function.

Americans driving in Canada or Mexico need to be able to convert miles per hour to kilometers per hour and vice versa.

If someone is driving at 55 miles per hour, how fast is he or she going in kilometers per hour?

Approximate Value in Miles per Hour	Approximate Value in Kilometers per Hour
35	56
40	64
45	72
50	80
55	88
60	96
65	104

A function f that converts from miles per hour to an approximate value in kilometers per hour is $f(x) = 1.6x$.

For example, $f(40) = 1.6(40) = 64$.

This tells us that 40 miles per hour is approximately equivalent to 64 kilometers per hour.

We can come up with a function that does just the opposite—that is, that converts kilometers per hour to an approximate value in miles per hour. This function is $f^{-1}(x) = 0.625x$.

For example, $f^{-1}(64) = 0.625(64) = 40$.

This tells us that 64 kilometers per hour is approximately equivalent to 40 miles per hour.

Miles per hour   **Kilometers per hour**

$40 \longrightarrow \quad f(x) = 1.6x \quad \longrightarrow 64$

$40 \longleftarrow \quad f^{-1}(x) = 0.625x \quad \longleftarrow 64$

We call a function f^{-1} that reverses the domain and range of a function f the **inverse function** f.

Most American cars have numbers showing kilometers per hour in smaller print on the car speedometer. Unfortunately, these numbers are usually hard to read. If we made a list of several function values of f and several inverse function values of f^{-1}, we could create a conversion scale like the one below that we could use if we should travel to Mexico or Canada with an American car.

The original function that we studied converts miles per hour to kilometers per hour. The corresponding inverse function converts kilometers per hour to miles per hour. How do we find inverse functions? Do all functions have inverse functions? These are questions we want to explore in this section.

Determining Whether a Function Is a One-to-One Function

First we state that not all functions have inverse functions. To have an inverse that is a function, a function must be one-to-one. This means that for every value of y, there is only one value of x. Or, in the language of ordered pairs, no ordered pairs have the same second coordinate.

DEFINITION OF A ONE-TO-ONE FUNCTION

A **one-to-one function** is a function in which no ordered pairs have the same second coordinate.

TO THINK ABOUT: Relationship of One-to-One and Inverses Why must a function be one-to-one in order to have an inverse that is a function?

EXAMPLE 1 Indicate whether the following functions are one-to-one.

(a) $M = \{(1, 3), (2, 7), (5, 8), (6, 12)\}$ **(b)** $P = \{(1, 4), (2, 9), (3, 4), (4, 18)\}$

Solution

(a) M is a function because no ordered pairs have the same first coordinate. M is also a one-to-one function because no ordered pairs have the same second coordinate.

(b) P is a function, but it is not one-to-one because the ordered pairs $(1, 4)$ and $(3, 4)$ have the same second coordinate.

Practice Problem 1

(a) Is the function $A = \{(-2, -6), (-3, -5), (-1, 2), (3, 5)\}$ one-to-one?
(b) Is the function $B = \{(0, 0), (1, 1), (2, 4), (3, 9), (-1, 1)\}$ one-to-one?

NOTE TO STUDENT: Fully worked-out solutions to all of the Practice Problems can be found at the back of the text starting at page SP-1

By examining the graph of a function, we can quickly tell whether it is one-to-one. If any horizontal line crosses the graph of a function in more than one place, the function is not one-to-one. If no such line exists, then the function is one-to-one.

HORIZONTAL LINE TEST

If any horizontal line intersects the graph of a function more than once, the function is not one-to-one. If no such line exists, the function is one-to-one.

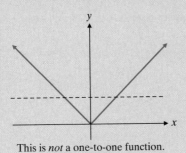

This is *not* a one-to-one function.

EXAMPLE 2 Determine whether the functions graphed are one-to-one functions.

(a)

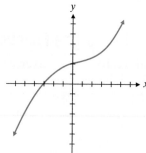

(b)

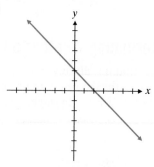

(c)

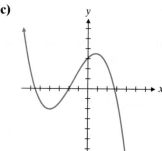

(d)

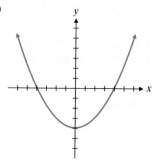

Solution The graphs of **(a)** and **(b)** represent one-to-one functions. Horizontal lines cross the graphs at most once.

(a)

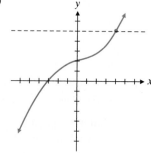

(b)

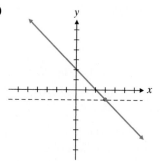

The graphs of **(c)** and **(d)** do not represent one-to-one functions. A horizontal line exists that crosses the graphs more than once.

(c)

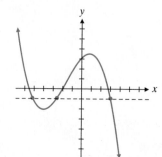

(d)

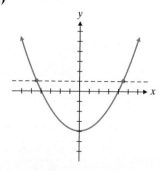

Practice Problem 2   Do the following graphs of a function represent a one-to-one function? Why or why not?

(a)

(b)

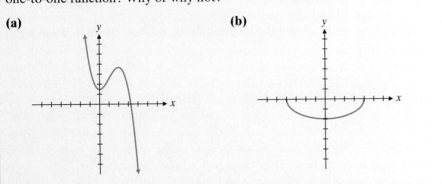

NOTE TO STUDENT: Fully worked-out solutions to all of the Practice Problems can be found at the back of the text starting at page SP-1

2 Finding the Inverse Function for a Given Function

How do we find the inverse of a function? If we have a list of ordered pairs, we simply interchange the coordinates of each ordered pair. In Example 1, we said that M has an inverse. What is it?

$$M = \{(1, 3), (2, 7), (5, 8), (6, 12)\}$$

The inverse of M, written M^{-1}, is

$$M^{-1} = \{(3, 1), (7, 2), (8, 5), (12, 6)\}.$$

Now do you see why a function must be one-to-one in order to have an inverse that is a function? Let's look at the function P from Example 1.

$$P = \{(1, 4), (2, 9), (3, 4), (4, 18)\}$$

If P had an inverse, it would be

$$P^{-1} = \{(4, 1), (9, 2), (4, 3), (18, 4)\}.$$

But we have two ordered pairs with the same first coordinate. Therefore, P^{-1} is not a function (in other words, the inverse function does not exist).

A number of real-world situations are described by functions that have inverses. Consider the function defined by the ordered pairs (year, U.S. budget in trillions of dollars). Some function values are

$$F = \{(2005, 2.40), (2000, 1.83), (1995, 1.52),$$
$$(1990, 1.25), (1985, 0.95), (1980, 0.59)\}.$$

In this case the inverse of the function is

$$F^{-1} = \{(2.40, 2005), (1.83, 2000), (1.52, 1995),$$
$$(1.25, 1990), (0.95, 1985), (0.59, 1980)\}.$$

By the way, F^{-1} does *not* mean $\frac{1}{F}$. Here the -1 simply means "inverse."

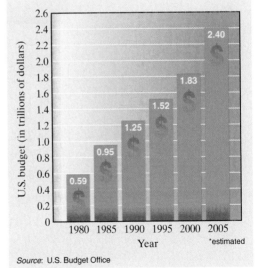

Source: U.S. Budget Office

EXAMPLE 3   Determine the inverse function of the function

$$F = \{(6, 1), (12, 2), (13, 5), (14, 6)\}.$$

Solution   The inverse function of F is
$$F^{-1} = \{(1, 6), (2, 12), (5, 13), (6, 14)\}.$$

Practice Problem 3   Find the inverse of the one-to-one function
$$B = \{(1, 2), (7, 8), (8, 7), (10, 12)\}.$$

Suppose that a function is given in the form of an equation. How do we find the inverse? Since, by definition, we interchange the ordered pairs to find the inverse of a function, this means that the x-values of the function become the y-values of the inverse function and vice versa.

Four steps will help us find the inverse of a one-to-one function when we are given its equation.

FINDING THE INVERSE OF A ONE-TO-ONE FUNCTION

1. Replace $f(x)$ with y.
2. Interchange x and y.
3. Solve for y in terms of x.
4. Replace y with $f^{-1}(x)$.

EXAMPLE 4 Find the inverse of $f(x) = 7x - 4$.

Solution

Step 1 $y = 7x - 4$ Replace $f(x)$ with y.

Step 2 $x = 7y - 4$ Interchange the variables x and y.

Step 3 $x + 4 = 7y$ Solve for y in terms of x.

$\dfrac{x + 4}{7} = y$

Step 4 $f^{-1}(x) = \dfrac{x + 4}{7}$ Replace y with $f^{-1}(x)$.

NOTE TO STUDENT: *Fully worked-out solutions to all of the Practice Problems can be found at the back of the text starting at page SP-1*

Practice Problem 4 Find the inverse of the function $g(x) = 4 - 6x$.

Let's see whether this technique works on the opening example in which we converted miles per hour to approximate values in kilometers per hour.

EXAMPLE 5 Find the inverse function of $f(x) = \dfrac{9}{5}x + 32$, which converts Celsius temperature (x) into equivalent Fahrenheit temperature.

Solution

Step 1 $y = \dfrac{9}{5}x + 32$ Replace $f(x)$ with y.

Step 2 $x = \dfrac{9}{5}y + 32$ Interchange x and y.

Step 3 $5(x) = 5\left(\dfrac{9}{5}\right)y + 5(32)$ Solve for y in terms of x.

$5x = 9y + 160$

$5x - 160 = 9y$

$\dfrac{5x - 160}{9} = \dfrac{9y}{9}$

$\dfrac{5x - 160}{9} = y$

Step 4 $f^{-1}(x) = \dfrac{5x - 160}{9}$ Replace y with $f^{-1}(x)$.

Note: Our inverse function $f^{-1}(x)$ will now convert Fahrenheit temperature to Celsius temperature. For example, suppose we wanted to know what is the Celsius temperature that corresponds to 86°F.

$$f^{-1}(86) = \frac{5(86) - 160}{9} = \frac{270}{9} = 30$$

This tells us that a temperature of 86°F corresponds to a temperature of 30°C.

Practice Problem 5   Find the inverse function of $f(x) = 0.75 + 0.55(x - 1)$, which gives the cost of a telephone call for any call over 1 minute if the telephone company charges 75 cents for the first minute and 55 cents for each minute thereafter. Here $x = $ the number of minutes.

③ Graphing a Function and its Inverse Function

The graph of a function and its inverse are symmetric about the line $y = x$. Why do you think that this is so?

EXAMPLE 6   If $f(x) = 3x - 2$, find $f^{-1}(x)$. Graph f and f^{-1} on the same set of axes. Draw the line $y = x$ as a dashed line for reference.

Solution

$$f(x) = 3x - 2$$
$$y = 3x - 2$$
$$x = 3y - 2$$
$$x + 2 = 3y$$
$$\frac{x + 2}{3} = y$$
$$f^{-1}(x) = \frac{x + 2}{3}$$

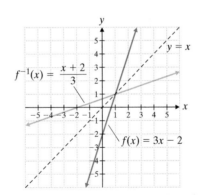

Now we graph each line.

Again we see that the graphs of f and f^{-1} are symmetric about the line $y = x$. If we folded the graph paper along the line $y = x$, the graph of f would touch the graph of f^{-1}. Try it. Redraw the functions on a separate piece of graph paper. Fold the graph paper on the line $y = x$.

Practice Problem 6   If $f(x) = -\dfrac{1}{4}x + 1$, find $f^{-1}(x)$. Graph f and f^{-1} on the same coordinate plane. Draw the line $y = x$ as a dashed line for reference.

Graphing Calculator

Inverse Functions

If f and g are inverse functions, then their graphs will be symmetric about the line $y = x$. Use a square window setting when graphing. For instance, to check the answer in Example 6, graph

$$y_1 = 3x - 2, \quad y_2 = \frac{x + 2}{3},$$

and $y_3 = x$.

Display:

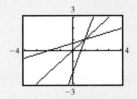

In each of the following cases, graph f, g, and $y = x$ in an appropriate square window to determine whether f and g are inverses of each other.

1. $f(x) = \left(\frac{2}{3}\right)x - 2$
   $g(x) = \left(\frac{3}{2}\right)x + 3$

2. $f(x) = (x - 1)^3$
   $g(x) = \sqrt[3]{x} + 1$

PRACTICE PROBLEM 6

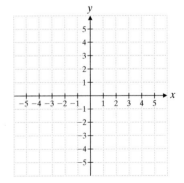

Verbal and Writing Skills

Complete the following:

1. A one-to-one function is a function in which no ordered pairs _____ .

2. If any horizontal line intersects the graph of a function more than once, the function _____ .

3. The graphs of a function f and its inverse f^{-1} are symmetric about the line _____ .

4. Do all functions have inverse functions? Why or why not?

5. Does the graph of a horizontal line represent a function? Why or why not? Does it represent a one-to-one function? Explain.

6. Does the graph of a vertical line represent a function? Why or why not? Does it represent a one-to-one function? Explain.

Indicate whether each function is one-to-one.

7. $B = \{(0, 1), (1, 0), (10, 0)\}$

8. $A = \{(-6, -2), (6, 2), (3, 4)\}$

9. $F = \left\{\left(\frac{2}{3}, 2\right), \left(3, -\frac{4}{5}\right), \left(-\frac{2}{3}, -2\right), \left(-3, \frac{4}{5}\right)\right\}$

10. $C = \{(12, 3), (-6, 1), (6, 3)\}$

11. $E = \{(1, 2.8), (3, 6), (-1, -2.8), (2.8, 1)\}$

12. $F = \{(6, 5), (-6, -5), (5, 6), (-5, -6)\}$

Indicate whether each graph represents a one-to-one function.

13.

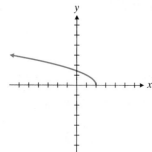

14.

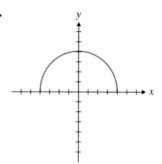

15.

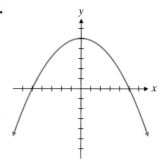

16.

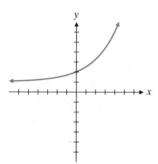

17.

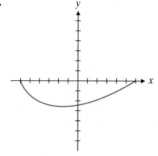

18.

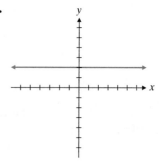

Find the inverse of each one-to-one function. Graph the function and its inverse on one coordinate plane.

19. $J = \{(8, 2), (1, 1), (0, 0), (-8, -2)\}$

20. $K = \{(-7, 1), (6, 2), (3, -1), (2, 5)\}$

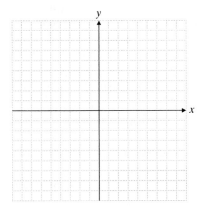

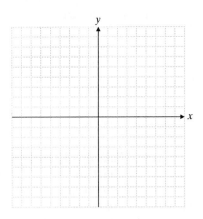

Find the inverse of each function.

21. $f(x) = 4x - 5$

22. $f(x) = 6 - 2x$

23. $f(x) = x^3 + 7$

24. $f(x) = 3 - x^3$

25. $f(x) = -\dfrac{4}{x}$

26. $f(x) = \dfrac{3}{x}$

27. $f(x) = \dfrac{4}{x - 5}$

28. $f(x) = \dfrac{7}{3 - x}$

Find the inverse of each function. Graph the function and its inverse on one coordinate plane. Graph the line $y = x$ as a dashed line.

29. $g(x) = 2x + 5$

30. $f(x) = 3x + 4$

31. $h(x) = \dfrac{1}{2}x - 2$

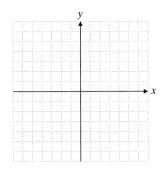

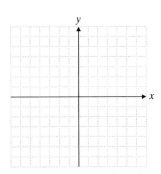

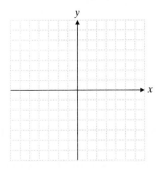

32. $p(x) = \dfrac{2}{3}x - 4$

33. $r(x) = -3x - 1$

34. $k(x) = 3 - 2x$

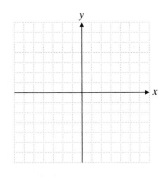

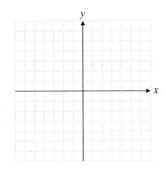

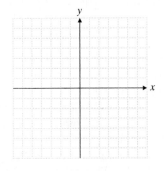

Applications

35. *Currency Conversion* Sean went to Ireland before the new Euro currency was introduced. He brought Irish pounds back with him from Dublin and went to the bank to take advantage of the favorable exchange rate and convert his leftover Irish pounds to U.S. dollars. The bank teller informed Sean that the conversion rate was US $1.437 for one Irish pound. The bank charges a fee of US $4.00 for each transaction. The function used to convert Irish pounds to U.S. dollars is given by $f(x) = 1.437x - 4$, where x is the number of Irish pounds. Find the inverse function of f. What is the significance of the inverse function? If Sean wanted to change U.S. dollars to Irish pounds, could he use this inverse function? Why or why not?

36. *Currency Conversion* Manuela went to Spain before the new Euro currency was introduced. She brought Spanish pesetas back with her from Madrid and went to the bank to take advantage of the favorable exchange rate and convert her left-over Spanish pesetas to U.S. dollars. The bank teller informed Manuela that the conversion rate was US $0.0063 for one Spanish peseta. The bank charges a fee of US $5.00 for each transaction. The function used to convert Spanish pesetas to U.S. dollars is given by $f(x) = 0.0063x - 5$, where x is the number of Spanish pesetas. Find the inverse function of f. What is the significance of the inverse function? If Manuela wanted to change U.S. dollars to Spanish pesetas, could she use this inverse function? Why or why not?

To Think About

37. Can you find an inverse function for the function $f(x) = 2x^2 + 3$? Why or why not?

38. Can you find an inverse function for the function $f(x) = |3x + 4|$? Why or why not?

For every function f and its inverse, f^{-1}, it is true that $f[f^{-1}(x)] = x$ and $f^{-1}[f(x)] = x$. Show that this is true for each pair of inverse functions.

39. $f(x) = 2x + \dfrac{3}{2}, f^{-1}(x) = \dfrac{1}{2}x - \dfrac{3}{4}$

40. $f(x) = -3x - 10, f^{-1}(x) = \dfrac{-x - 10}{3}$

Cumulative Review

Solve for x.

41. $x = \sqrt{15 - 2x}$

42. $x^{2/3} + 7x^{1/3} + 12 = 0$

43. *Red Blood Cells* The average male human has more blood than the average female. In addition, each cubic centimeter of blood in males is usually richer in red blood cells. Each cubic centimeter of blood in men contains from 4.6 million to 6.2 million red blood cells, compared with 4.2 million to 5.4 million for women. Using the lower numbers, determine the ratio of red blood cells in men to red blood cells in women.

44. *Overtime Pay* Catherine earns $17 per hour for a 40-hour week as an on-call nurse for City Hospital. She earns time and a half for every hour over 40 hours she works in one week. If Catherine made $1011.50 last week, how many overtime hours did she work?

45. *Canadian Forests* Forests are a dominant feature of Canada. Ten percent of all the world's forests lie in Canada. One out of every sixteen people in the labor force in Canada works in a job that relates to forests. If the labor force in Canada in 2000 was 12,800,000 people, how many people worked in a job related to forests in that year? *Source:* Canada Board of Tourism.

46. *Finland* The population of Finland in 2003 was approximately 5,184,000. By the year 2005 it is projected to drop to 4,170,000. What is the percent of decrease in population? Round to the nearest tenth. *Source:* United Nations Statistical Division

Putting Your Skills to Work

Using Mathematical Functions to Study Farming Productivity

In Zambia, many farmers depend on crop production for their livelihood. In the past, farmers often faced a smaller crop each successive year due to the depletion of the land. Recently a new "Wonder Tree" has been developed. This tree is planted near traditional crops. This fast growing, flowering tree can loosen compacted soil because it has a powerful and large root system. The nitrogen from the fallen leaves of this tree enrich the soil. The result is an amazing increase in the size and quality of the crops. This tree is very effective when planted near a crop of cabbage plants.

World Vision, a world relief organization, has distributed the seeds for this plant throughout Zambia and has provided the support for farmer training sessions. Already over 22,000 farmers have received formal training in how to use this new plant to increase crop productivity. Many of the farmers have seen an increase of 500% in the volume of the crop produced using the same land as before.

Problems for Individual Investigation and Analysis

Here is a hypothetical example of how the effectiveness of the "Wonder Tree" could help an individual cabbage farmer.

Hypothetical Sample Crop Production Increase

Time	Production Initially	One Year	18 Months	Two Years	30 Months
Number of Pounds of Cabbage	700	1800	2700	3100	3500

1. If a farmer was initially producing 120 pounds of cabbages per year, how many could he expect to produce one year after planting the wonder tree? Round to the nearest pound.

2. If a farmer was initially producing 200 pounds of cabbages per year, how many pounds could he expect to produce two years after planting the wonder tree? Round to the nearest pound.

If a farmer starts with an initial production of 700 pounds, the production of cabbages in pounds in this hypothetical example can be approximated by the production function

$$P(x) = 0.027x^4 - 1.88x^3 + 41.25x^2 - 178.89x + 700,$$

where x is the number of months since the wonder trees have been planted. This function is only valid for values of $x < 32$ months. Use this function to answer questions 3–4. Round all answers to the nearest pound.

3. If a farmer is initially producing 700 pounds of cabbages, how many pounds will he produce 9 months after planting the wonder trees?

4. If a farmer is initially producing 700 pounds of cabbages, how many pounds will he produce 15 months after planting the wonder trees?

Problems for Group Investigation and Cooperative Learning

The function can be made more general for the number of pounds that are initially produced.

The production of cabbages in pounds in this hypothetical example can be approximated by the production function

$$P(x) = 0.027x^4 - 1.88x^3 + 41.25x^2 - 178.89x + b,$$

where b is the number of pounds of cabbages initially planted and x is the number of months since the wonder trees have been planted.

5. If a farmer starts with 400 pounds of cabbages, how many pounds will he produce 21 months after the planting of the wonder trees?

6. If a farmer starts with 1100 pounds of cabbages, how many pounds will he produce 28 months after the planting of the wonder trees?

7. If a farmer starts with 2000 pounds of cabbages, how many pounds will he produce 10 months after the planting of the wonder trees?

8. If a farmer starts with 2500 pounds of cabbages, how many pounds will he produce 11 months after the planting of the wonder trees?

Chapter 10 Organizer

Topic	Procedure	Examples
Relations, functions, and one-to-one functions, pp. 546, and 563.	A relation is any set of ordered pairs. A function is a relation in which no ordered pairs have the same first coordinate. A one-to-one function is a function in which no ordered pairs have the same second coordinate.	Is $\{(3,6), (2,8), (9,1), (4,6)\}$ a one-to-one function? No, since $(3,6)$ and $(4,6)$ have the same second coordinate.
Vertical line test, p. 546.	If any vertical line intersects the graph of a relation more than once, the relation is not a function. If no such line exists, the relation is a function.	 Does this graph represent a function? No, because a vertical line intersects the curve more than once.
Horizontal line test, p. 563.	If any horizontal line intersects the graph of a function more than once, the function is not one-to-one. If no such line exists, the function is one-to-one.	 Does this graph represent a one-to-one function? Yes, any horizontal line will cross this function at most once.
Finding function values, p. 540.	Replace the variable by the quantity inside the parentheses. Simplify the result.	If $f(x) = 2x^2 + 3x - 4$, then we have the following: $$f(-2) = 2(-2)^2 + 3(-2) - 4$$ $$= 8 - 6 - 4 = -2$$ $$f(a) = 2a^2 + 3a - 4$$ $$f(a + 2) = 2(a + 2)^2 + 3(a + 2) - 4$$ $$= 2(a^2 + 4a + 4) + 3a + 6 - 4$$ $$= 2a^2 + 8a + 8 + 3a + 6 - 4$$ $$= 2a^2 + 11a + 10$$ $$f(3a) = 2(3a)^2 + 3(3a) - 4$$ $$= 2(9a^2) + 9a - 4$$ $$= 18a^2 + 9a - 4$$
Vertical shifts of the graph of function, p. 548.	If $k > 0$: **1.** The graph of $y = f(x) + k$ is shifted k units *upward* from the graph of $y = f(x)$.	Graph $f(x) = x^2$ and $g(x) = x^2 + 3$.

Topic	Procedure	Examples
Vertical shifts of the graph of function (continued)	**2.** The graph of $y = f(x) - k$ is shifted k units *downward* from the graph of $y = f(x)$.	Graph $f(x) = \lvert x \rvert$ and $g(x) = \lvert x \rvert - 2$.
Horizontal shifts of the graph of function, p. 548.	If $h > 0$: **1.** The graph of $y = f(x - h)$ is shifted h units to the *right* of the graph of $y = f(x)$.	Graph $f(x) = x^2$ and $g(x) = (x - 3)^2$.
	2. The graph of $y = f(x + h)$ is shifted h units to the *left* of the graph of $y = f(x)$.	Graph $f(x) = x^3$ and $g(x) = (x + 4)^3$.
Sum, difference, product, and quotient of functions, p. 555.	**1.** $(f + g)(x) = f(x) + g(x)$ **2.** $(f - g)(x) = f(x) - g(x)$ **3** $(f \cdot g)(x) = f(x) \cdot g(x)$ **4.** $\left(\dfrac{f}{g}\right)(x) = \dfrac{f(x)}{g(x)}, g(x) \neq 0$	If $f(x) = 2x + 3$ and $g(x) = 3x - 4$, then we have the following: **1.** $(f + g)(x) = (2x + 3) + (3x - 4)$ $= 5x - 1$ **2.** $(f - g)(x) = (2x + 3) - (3x - 4)$ $= 2x + 3 - 3x + 4$ $= -x + 7$ **3.** $(f \cdot g)(x) = (2x + 3)(3x - 4)$ $= 6x^2 + x - 12$ **4.** $\left(\dfrac{f}{g}\right)(x) = \dfrac{2x + 3}{3x - 4}, x \neq \dfrac{4}{3}$
Composition of functions, p. 556.	The composition of functions f and g is written as $(f \circ g)(x) = f[g(x)]$. To find $f[g(x)]$ do the following: **1.** Replace $g(x)$ by its equation. **2.** Apply the formula for $f(x)$ to this expression. **3.** Simplify the results. Usually, $f[g(x)] \neq g[f(x)]$.	If $f(x) = x^2 - 5$ and $g(x) = -3x + 4$, find $f[g(x)]$ and $g[f(x)]$. $f[g(x)] = f[-3x + 4]$ $= (-3x + 4)^2 - 5$ $= 9x^2 - 24x + 16 - 5$ $= 9x^2 - 24x + 11$ $g[f(x)] = g[x^2 - 5]$ $= -3(x^2 - 5) + 4$ $= -3x^2 + 15 + 4$ $= -3x^2 + 19$

Topic	Procedure	Examples
Finding the inverse of a function defined by a set of ordered pairs, p. 565.	Reverse the order of the coordinates of each ordered pair from (a, b) to (b, a).	Find the inverse of $A = \{(5, 6), (7, 8), (9, 10)\}$. $A^{-1} = \{(6, 5), (8, 7), (10, 9)\}$
Finding the inverse of a function defined by an equation, p. 566.	Any one-to-one function has an inverse function. To find the inverse f^{-1} of a one-to-one function f, do the following: **1.** Replace $f(x)$ with y. **2.** Interchange x and y. **3.** Solve for y in terms of x. **4.** Replace y with $f^{-1}(x)$.	Find the inverse of $f(x) = -\frac{2}{3}x + 4$. $$y = -\frac{2}{3}x + 4$$ $$x = -\frac{2}{3}y + 4$$ $$3x = -2y + 12$$ $$3x - 12 = -2y$$ $$\frac{3x - 12}{-2} = y$$ $$-\frac{3}{2}x + 6 = y$$ $$f^{-1}(x) = -\frac{3}{2}x + 6$$
Graphing the inverse of a function, p. 567.	Graph the line $y = x$ as a dashed line for reference. **1.** Graph $f(x)$. **2.** Graph $f^{-1}(x)$. The graphs of f and f^{-1} are symmetric about the line $y = x$.	$f(x) = 2x + 3$ $f^{-1}(x) = \dfrac{x - 3}{2}$ Graph f and f^{-1} on the same set of axes.

Chapter 10 Review Problems

For the function $f(x) = \dfrac{1}{2}x + 3$, find the following:

1. $f(a - 1)$ **2.** $f(a + 2)$ **3.** $f(a - 1) - f(a)$

4. $f(a + 2) - f(a)$ **5.** $f(b^2 - 3)$ **6.** $f(4b^2 + 1)$

For the function $p(x) = -2x^2 + 3x - 1$, find the following:

7. $p(-3)$ **8.** $p(-1)$ **9.** $p(2a) + p(-2)$

10. $p(-3a) + p(1)$ **11.** $p(a + 2)$ **12.** $p(a - 3)$

For the function $h(x) = |2x - 1|$, *find the following:*

13. $h(0)$ **14.** $h(-5)$ **15.** $h\left(\frac{1}{4}a\right)$

16. $h\left(\frac{3}{2}a\right)$ **17.** $h(a^2 + a)$ **18.** $h(2a^2 - 3a)$

For the function $r(x) = \dfrac{3x}{x + 4}$, $x \neq -4$, *find the following. In each case, write your answers as one fraction, if possible.*

19. $r(5)$ **20.** $r(-6)$ **21.** $r(2a - 5)$

22. $r(1 - a)$ **23.** $r(3) + r(a)$ **24.** $r(a) + r(-2)$

Find $\dfrac{f(x + h) - f(x)}{h}$ *for the following:*

25. $f(x) = 7x - 4$ **26.** $f(x) = 6x - 5$ **27.** $f(x) = 2x^2 - 5x$ **28.** $f(x) = 2x - 3x^2$

Examine each of the following graphs. **(a)** *Does the graph represent a function?* **(b)** *Does the graph represent a one-to-one function?*

29.

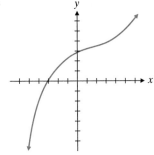

30.

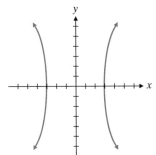

31.

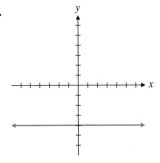

32.

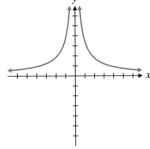

33.

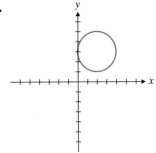

34.

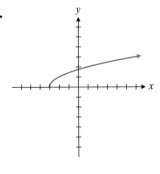

Graph each pair of functions on one set of axes.

35. $f(x) = x^2$

$g(x) = (x + 2)^2 + 4$

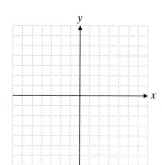

36. $f(x) = |x|$

$g(x) = |x + 3|$

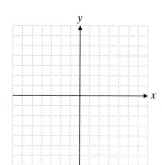

37. $f(x) = |x|$

$g(x) = |x - 4|$

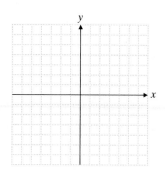

38. $f(x) = |x|$

$h(x) = |x| + 3$

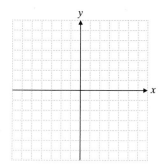

39. $f(x) = |x|$

$h(x) = |x| - 2$

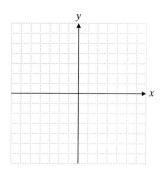

40. $f(x) = x^3$

$r(x) = (x + 3)^3 + 1$

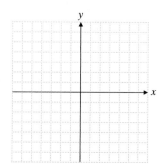

41. $f(x) = x^3$

$r(x) = (x - 1)^3 + 5$

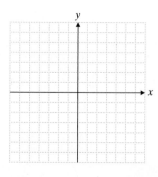

42. $f(x) = \dfrac{2}{x}, x \neq 0$

$r(x) = \dfrac{2}{x + 3} - 2, x \neq -3$

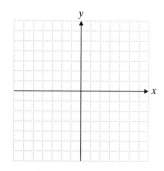

43. $f(x) = \dfrac{4}{x}, x \neq 0$

$r(x) = \dfrac{4}{x + 2}, x \neq -2$

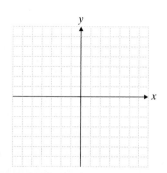

Given the functions,

$$f(x) = 3x + 5; \qquad g(x) = \frac{2}{x}, \qquad x \neq 0; \qquad s(x) = \sqrt{x - 2}, \qquad x \geq 2;$$

$$h(x) = \frac{x + 1}{x - 4}, \qquad x \neq 4; \qquad p(x) = 2x^2 - 3x + 4; \qquad \text{and} \qquad t(x) = -\frac{1}{2}x - 3,$$

find each of the following:

44. $(f + t)(x)$

45. $(f + p)(x)$

46. $(p - f)(x)$

47. $(t - f)(x)$

48. $(p - f)(2)$

49. $(t - f)(-3)$

50. $(fg)(x)$

51. $(tp)(x)$

52. $\left(\dfrac{g}{h}\right)(x)$

53. $\left(\dfrac{g}{f}\right)(x)$

54. $\left(\dfrac{g}{h}\right)(-2)$

55. $\left(\dfrac{g}{f}\right)(-3)$

56. $p[f(x)]$

57. $t[s(x)]$

58. $s[p(x)]$

59. $s[t(x)]$

60. $s[p(2)]$

61. $s[t(-18)]$

62. Show that $f[g(x)] \neq g[f(x)]$.

63. Show that $p[g(x)] \neq g[p(x)]$.

For each set, determine (a) the domain, (b) the range, (c) whether the set defines a function, and (d) whether the set defines a one-to-one function. (Hint: You can review the concept of domain and range in Section 3.5 if necessary.)

64. $B = \{(3, 7), (7, 3), (0, 8), (0, -8)\}$

65. $A = \{(100, 10), (200, 20), (300, 30), (400, 10)\}$

66. $D = \left\{\left(\frac{1}{2}, 2\right), \left(\frac{1}{4}, 4\right), \left(-\frac{1}{3}, -3\right), \left(4, \frac{1}{4}\right)\right\}$

67. $C = \{(12, 6), (0, 6), (0, -1), (-6, -12)\}$

68. $F = \{(3, 7), (2, 1), (0, -3), (1, 1)\}$

69. $E = \{(0, 1), (1, 2), (2, 9), (-1, -2)\}$

Find the inverse of each of the following functions.

70. $A = \left\{\left(3, \frac{1}{3}\right), \left(-2, -\frac{1}{2}\right), \left(-4, -\frac{1}{4}\right), \left(5, \frac{1}{5}\right)\right\}$

71. $B = \{(1, 10), (3, 7), (12, 15), (10, 1)\}$

72. $f(x) = -\frac{3}{4}x + 2$

73. $g(x) = -8 - 4x$

74. $h(x) = \frac{6}{x + 5}$

75. $j(x) = \frac{-7}{2 - x}$

76. $p(x) = \sqrt[3]{x + 1}$

77. $r(x) = x^3 + 2$

Find the inverse of each function. Graph the function and its inverse on one coordinate plane. Then on that same set of axes, graph the line $y = x$ as a dashed line.

78. $f(x) = \frac{-x - 2}{3}$

79. $f(x) = -\frac{3}{4}x + 1$

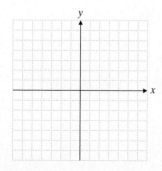

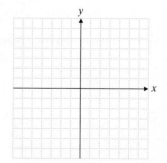

Remember to use your Chapter Test Prep Video CD to see the worked-out solutions to the test problems you want to review.

For the function $f(x) = \dfrac{3}{4}x - 2$, *find the following:*

1. $f(-8)$ **2.** $f(3a)$ **3.** $f(a) - f(2)$

For the function $f(x) = 3x^2 - 2x + 4$, *find the following:*

4. $f(-6)$ **5.** $f(a + 1)$

6. $f(a) + f(1)$ **7.** $f(-2a) - 2$

Look at each graph below. **(a)** *Does the graph represent a function?* **(b)** *Does the graph represent a one-to-one function?*

8.
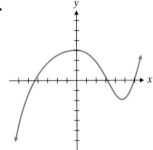

9.
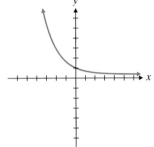

Graph each pair of functions on one coordinate plane.

10. $f(x) = x^2$
$g(x) = (x - 1)^2 + 3$

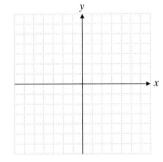

11. $f(x) = |x|$
$g(x) = |x + 1| + 2$

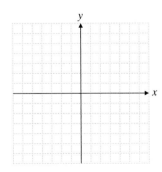

12. If $f(x) = 3x^2 - x - 6$ and $g(x) = -2x^2 + 5x + 7$, find the following:
    (a) $(f + g)(x)$ **(b)** $(f - g)(x)$ **(c)** $(f - g)(-2)$

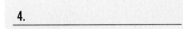

1. _____

2. _____

3. _____

4. _____

5. _____

6. _____

7. _____

8. (a) _____

   (b) _____

9. (a) _____

   (b) _____

10. _____

11. _____

12. (a) _____

    (b) _____

    (c) _____

579

13. If $f(x) = \dfrac{3}{x}$, $x \neq 0$ and $g(x) = 2x - 1$, find the following:

   (a) $(fg)(x)$ **(b)** $\left(\dfrac{f}{g}\right)(x)$ **(c)** $f[g(x)]$

14. If $f(x) = \dfrac{1}{2}x - 3$ and $g(x) = 4x + 5$, find the following:

   (a) $(f \circ g)(x)$ **(b)** $(g \circ f)(x)$ **(c)** $(f \circ g)\left(\dfrac{1}{4}\right)$

Look at the following functions. **(a)** *Is the function one-to-one?* **(b)** *If so, find the inverse of the function.*

15. $B = \{(1, 8), (8, 1), (9, 10), (-10, 9)\}$

16. $A = \{(1, 5), (2, 1), (4, -7), (0, 7), (1, -5)\}$

17. Determine the inverse of $f(x) = \sqrt[3]{2x - 1}$.

18. Find f^{-1}. Graph f and its inverse f^{-1} on one coordinate plane. Graph $y = x$ as a dashed line for a reference.

$$f(x) = -3x + 2$$

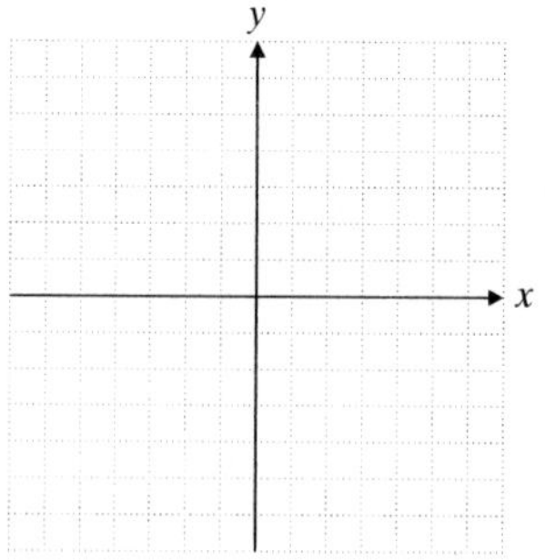

19. Given that $f(x) = \dfrac{3}{7}x + \dfrac{1}{2}$ and that $f^{-1}(x) = \dfrac{14x - 7}{6}$, find $f^{-1}[f(x)]$.

20. Find $\dfrac{f(x + h) - f(x)}{h}$ for $f(x) = 7 - 8x$.

Left margin answer blanks:

13. (a) _____

(b) _____

(c) _____

14. (a) _____

(b) _____

(c) _____

15. (a) _____

(b) _____

16. (a) _____

(b) _____

17. _____

18. _____

19. _____

20. _____

Cumulative Test for Chapters 1–10

Approximately one-half of this test covers the content of Chapters 1–9. The remainder covers the content of Chapter 10.

1. Simplify. $3x\{2y - 3[x + 2(x + 2y)]\}$

2. Evaluate $3y - 2xy - x^2$ for $x = -1$, $y = 3$.

3. Solve for x. $\dfrac{1}{2}(x - 2) = \dfrac{1}{3}(x + 10) - 2x$

4. Factor completely. $16x^4 - 1$

5. Multiply: $(x - 1)(2x^2 + x - 5)$

6. Solve for x. $\dfrac{3x}{x^2 - 4} = \dfrac{2}{x + 2} + \dfrac{4}{2 - x}$

7. Find the equation of a line with slope $= -3$ that passes through the point $(2, -1)$.

8. Solve for (x, y).
$$3x + 2y = 5$$
$$7x + 5y = 11$$

9. Simplify. $\sqrt{18x^5y^6z^3}$

10. Multiply. $\left(\sqrt{2} + \sqrt{3}\right)\left(2\sqrt{2} - 4\sqrt{3}\right)$

11. Find the distance between $(6, -1)$ and $(-3, -4)$.

12. Factor $12x^2 - 11x + 2$.

13. Factor: $x^3 - 5x^2 - 14x$

14. Write the standard form of the equation of a circle with radius $2\sqrt{2}$ and center $(0, -5)$.

15. If $f(x) = 3x^2 - 2x + 1$, find the following:
    (a) $f(-2)$
    (b) $f(a - 2)$
    (c) $f(a) + f(-2)$

1. _____

2. _____

3. _____

4. _____

5. _____

6. _____

7. _____

8. _____

9. _____

10. _____

11. _____

12. _____

13. _____

14. _____

15. (a) _____

(b) _____

(c) _____

581

16. _____

17. (a) _____

(b) _____

(c) _____

18. (a) _____

(b) _____

(c) _____

19. _____

20. (a) _____

(b) _____

(c) _____

21. (a) _____

(b) _____

22. _____

16. Graph $f(x) = x^3$ and $g(x) = (x + 2)^3 + 4$ on one coordinate plane.

17. If $f(x) = 2x^2 - 5x - 6$ and $g(x) = 5x + 3$, find the following:
  (a) $(fg)(x)$
  (b) $\left(\dfrac{f}{g}\right)(x)$
  (c) $f[g(x)]$

18. $A = \{(3, 6), (1, 8), (2, 7), (4, 4)\}$
  (a) Is A a function?
  (b) Is A a one-to-one function?
  (c) Find A^{-1}.

19. Find the inverse function for $f(x) = \sqrt[3]{7x - 3}$.

20. $f(x) = 5x^3 - 3x^2 - 6$
  (a) Find $f(5)$.
  (b) Find $f(-3)$.
  (c) Find $f(2a)$.

21. (a) Find the inverse function for $f(x) = -\dfrac{2}{3}x + 2$.
  (b) Graph f and f^{-1} on one coordinate plane. Graph $y = x$ as a reference.

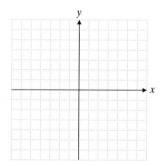

22. Find $f[f^{-1}(x)]$ using your results from Exercise 21.

CHAPTER

11

A dream trip to Hawaii. A lovely hotel a short walk from the beach. A rental car to drive across the island. Many Americans hope to take a such a trip in their lifetime. Would they go to a travel agent? Would they shop online for these things? It probably does not surprise you that more and more Americans are making such purchases on the Internet. See if you can determine how rapidly online purchases are growing. Turn to page 627 to examine some of these facts.

Logarithmic and Exponential Functions

Student Learning Objectives

After studying this section, you will be able to:

1 Graph an exponential function.

2 Solve elementary exponential equations.

3 Solve applications requiring the use of an exponential equation.

Graphing Calculator

 Exponential Functions

Using a graphing calculator, graph the function $f(x) = 4^x$. Then on the same screen, graph $g(x) = 4^{-x}$. How are these two graphs related?

Using a graphing calculator, graph the function $f(x) = 5^x$. Using the graph of $f(x) = 5^x$, draw the graph of $g(x) = 5^{-x}$ with pencil and paper. Graph $g(x) = 5^{-x}$ on your calculator to verify. What conclusions can you make about the graphs of $f(x) = a^x$ and $g(x) = a^{-x}$?

PRACTICE PROBLEM 1

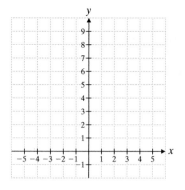

NOTE TO STUDENT: *Fully worked-out solutions to all of the Practice Problems can be found at the back of the text starting at page SP-1*

---

1 Graphing an Exponential Function

We have defined exponential equations $a^x = b$ for any rational number x. For example,

$$2^{-2} = \frac{1}{4},$$

$$2^{1/2} = \sqrt{2}, \text{ and}$$

$$2^{1.7} = 2^{17/10} = \sqrt[10]{2^{17}}.$$

We can also define such equations when x is an irrational number, such as π or $\sqrt{2}$. However, we will leave this definition for a more-advanced course.

We define an **exponential function** for all real values of x as follows:

DEFINITION OF EXPONENTIAL FUNCTION

The function $f(x) = b^x$, where $b > 0$, $b \neq 1$, and x is a real number, is called an **exponential function.** The number b is called the **base** of the function.

Now let's look at some graphs of exponential functions.

EXAMPLE 1   Graph $f(x) = 2^x$.

Solution   We make a table of values for x and $f(x)$.

$$f(-1) = 2^{-1} = \frac{1}{2}, \qquad f(0) = 2^0 = 1, \qquad f(1) = 2^1 = 2$$

Verify the other values in the table below. We then draw the graph.

x	f(x)
−2	$\frac{1}{4}$
−1	$\frac{1}{2}$
0	1
1	2
2	4
3	8

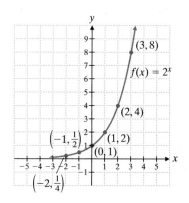

Notice how the curve of $f(x) = 2^x$ comes *very close to* the x-axis but *never touches* it. The x-axis is an **asymptote** for every exponential function. You should also notice that $f(x)$ is always positive, so the range of f is the set of all positive real numbers (whereas the domain is the set of all real numbers). When the base is greater than one, as x increases, $f(x)$ increases faster and faster (that is, the curve gets steeper).

Practice Problem 1   Graph $f(x) = 3^x$.

EXAMPLE 2   Graph $f(x) = \left(\frac{1}{2}\right)^x$.

Solution   We can write $f(x) = \left(\frac{1}{2}\right)^x$ as $f(x) = \left(\frac{1}{2}\right)^x = (2^{-1})^x = 2^{-x}$ and evaluate it for a few values of x. We then draw the graph.

x	f(x)
−3	8
−2	4
−1	2
0	1
1	$\frac{1}{2}$
2	$\frac{1}{4}$

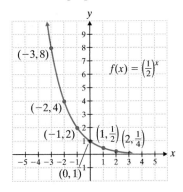

Note that as x increases, $f(x)$ decreases.

Practice Problem 2   Graph $f(x) = \left(\frac{1}{3}\right)^x$.

TO THINK ABOUT: Comparing Graphs   Look at the graph of $f(x) = 2^x$ in Example 1 and the graph of $f(x) = \left(\frac{1}{2}\right)^x = 2^{-x}$ in Example 2. How are the two graphs related?

EXAMPLE 3   Graph $f(x) = 3^{x-2}$.

Solution   We will make a table of values for a few values of x. Then we will graph the function.

$$f(0) = 3^{0-2} = 3^{-2} = \frac{1}{3^2} = \frac{1}{9}$$

$$f(1) = 3^{1-2} = 3^{-1} = \frac{1}{3}$$

$$f(2) = 3^{2-2} = 3^0 = 1$$

$$f(3) = 3^{3-2} = 3^1 = 3$$

$$f(4) = 3^{4-2} = 3^2 = 9$$

x	f(x)
0	$\frac{1}{9}$
1	$\frac{1}{3}$
2	1
3	3
4	9

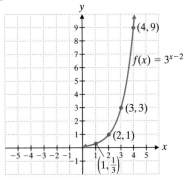

We observe that the curve is that of $f(x) = 3^x$ except that it has been shifted 2 units to the right.

Practice Problem 3   Graph $f(x) = 3^{x+2}$.

PRACTICE PROBLEM 2

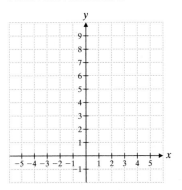

Graphing Calculator

Exploration

Using a graphing calculator, graph $f(x) = 2^x$. Then, on the same screen, graph $g(x) = 2^{x+3}$. Describe the shift that occurs. What will the graph of $g(x) = 2^{x+5}$ look like? Verify using the graphing calculator.

Using the graphing calculator, graph $f(x) = 2^{x-2}$. Describe the shift that occurs. What will the graph of $g(x) = 2^{x-3}$ look like? Verify using the graphing calculator.

Based on your experience with functions, what would the graph of $f(x) = 2^x + 3$ look like? How about $f(x) = 2^x - 4$? Verify using the graphing calculator.

PRACTICE PROBLEM 3

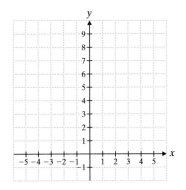

TO THINK ABOUT: Graph Shifts   How is the graph of $f(x) = 3^{x+2}$ related to the graph of $f(x) = 3^x$? Without making a table of values, draw the graph of $f(x) = 3^{x+3}$. Draw the graph of $f(x) = 3^{x-3}$.

For the next example we need to discuss a special number that is denoted by the letter e. The letter e is a number like π. It is an irrational number. It occurs in many formulas that describe real-world phenomena, such as the growth of cells and radioactive decay. We need an approximate value for e to use this number in calculations: $e \approx$ **2.7183.**

An extremely useful function is the exponential function $f(x) = e^x$. We usually obtain values for e^x by using a calculator or a computer. If you have a scientific calculator, use the $\boxed{e^x}$ key. (Many scientific calculators require you to press $\boxed{\text{SHIFT}}$ $\boxed{\ln}$ or $\boxed{\text{2nd F}}$ $\boxed{\ln}$ or $\boxed{\text{INV}}$ $\boxed{\ln}$ to obtain the operation e^x.) If you have a calculator that is not a scientific calculator, use $e \approx 2.7183$ as an approximate value. If you don't have any calculator, use Table A-2 in the appendix.

EXAMPLE 4   Graph $f(x) = e^x$.

Solution   We evaluate $f(x)$ for some negative and some positive values of x. We begin with $f(-2)$. Notice that the x-column in Table A-2 has only positive values. To find the value of $f(-2) = e^{-2}$, we must locate 2 in the x-column and then read across to the value in the column under e^{-x}. Thus, we see that $f(-2) \approx 0.1353$, or 0.14 rounded to the nearest hundredth.

To find $f(2) = e^2$ on a scientific calculator, we enter 2 $\boxed{e^x}$ and obtain 7.389056099 as an approximation. (On some scientific calculators you will need to use the keystrokes 2 $\boxed{\text{2nd F}}$ $\boxed{\ln}$ or 2 $\boxed{\text{SHIFT}}$ $\boxed{\ln}$ or 2 $\boxed{\text{INV}}$ $\boxed{\ln}$.) Thus, $f(2) = e^2 \approx 7.39$ to the nearest hundredth.

x	f(x)
−2	0.14
−1	0.37
0	1
1	2.72
2	7.39

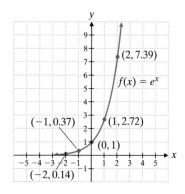

PRACTICE PROBLEM 4

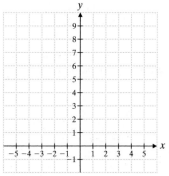

NOTE TO STUDENT: Fully worked-out solutions to all of the Practice Problems can be found at the back of the text starting at page SP-1

Practice Problem 4   Graph $f(x) = e^{x-2}$.

TO THINK ABOUT: Graph Shifts   Look at the graphs of $f(x) = e^x$ and $f(x) = e^{x-2}$. Describe the shift that occurs. Without making a table of values, draw the graph of $f(x) = e^{x+3}$.

2 Solving Elementary Exponential Equations

All the usual laws of exponents are true for exponential functions. We also have the following important property to help us solve exponential equations.

PROPERTY OF EXPONENTIAL EQUATIONS

If $b^x = b^y$, then $x = y$ for $b > 0$ and $b \neq 1$.

EXAMPLE 5   Solve $2^x = \dfrac{1}{16}$.

Solution   To use the property of exponential equations, we must have the same base on both sides of the equation.

$$2^x = \frac{1}{16}$$

$$2^x = \frac{1}{2^4} \qquad \text{Because } 2^4 = 16.$$

$$2^x = 2^{-4} \qquad \text{Because } \frac{1}{2^4} = 2^{-4}.$$

$$x = -4 \qquad \text{Property of exponential equations.}$$

Practice Problem 5   Solve $2^x = \dfrac{1}{32}$.

Solving Applications Requiring the Use of an Exponential Equation

An exponential function can be used to solve compound interest exercises. If a principal P is invested at an interest rate r compounded annually, the amount of money A accumulated after t years is $A = P(1 + r)^t$.

EXAMPLE 6   If a young married couple invests $5000 in a mutual fund that pays 16% interest compounded annually, how much will they have in 3 years?

Solution   Here $P = 5000$, $r = 0.16$, and $t = 3$.

$$\begin{aligned}
A &= P(1 + r)^t \\
&= 5000(1 + 0.16)^3 \\
&= 5000(1.16)^3 \\
&= 5000(1.560896) \\
&= 7804.48
\end{aligned}$$

The couple will have $7804.48.

If you have a scientific calculator, you can find the value of $5000(1.16)^3$ immediately by using the $\boxed{\times}$ key and the $\boxed{y^x}$ key. On most scientific calculators you can use the following keystrokes.

$$5000 \;\boxed{\times}\; 1.16 \;\boxed{y^x}\; 3 \;\boxed{=}\; 7804.48$$

Practice Problem 6   If Uncle Jose invests $4000 in a mutual fund that pays 11% interest compounded annually, how much will he have in 2 years?

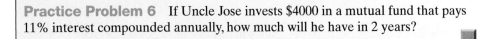

Interest is often compounded quarterly or monthly or even daily. Therefore, a more useful form of the interest formula that allows for variable compounding is needed. If a principal P is invested at an annual interest rate r that is compounded n times a year, then the amount of money A accumulated after t years is

$$A = P\left(1 + \frac{r}{n}\right)^{nt}.$$

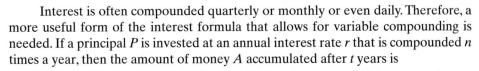

EXAMPLE 7 If we invest $8000 in a fund that pays 15% annual interest compounded monthly, how much will we have after 6 years?

Solution In this situation $P = 8000$, $r = 15\% = 0.15$, and $n = 12$. The interest is compounded monthly or twelve times per year. Finally, $t = 6$ since the interest will be compounded for 6 years.

$$A = 8000\left(1 + \frac{0.15}{12}\right)^{(12)(6)}$$
$$= 8000(1 + 0.0125)^{72}$$
$$= 8000(1.0125)^{72}$$
$$\approx 8000(2.445920268)$$
$$\approx 19{,}567.36215$$

Rounding to the nearest cent, we obtain the answer $19,567.36. Using a scientific calculator, we could have found the answer directly by using the following keystrokes.

8000 $\boxed{\times}$ 1.0125 $\boxed{y^x}$ 72 $\boxed{=}$ 19,567.36215

Depending on your calculator, your answer may contain fewer or more digits.

NOTE TO STUDENT: Fully worked-out solutions to all of the Practice Problems can be found at the back of the text starting at page SP-1

Practice Problem 7 How much money would Collette have if she invested $1500 for 8 years at 8% annual interest if the interest is compounded quarterly?

An exponential function is used to describe radioactive decay. The equation $A = Ce^{kt}$ tells us how much of a radioactive element is left in a sample after a specified time.

Graphing Calculator

Exploration

Graph the function $f(t) = 10e^{-0.0016008t}$ from Example 8 for $t = 0$ to $t = 100$ years. Now graph the function for $t = 0$ to $t = 2000$ years. What significant change is there in the two graphs? From the graphs, estimate a value of t for which $f(t) = 5.0$. (Round your value of t to the nearest hundredth.)

EXAMPLE 8 The radioactive decay of the chemical element americium 241 can be described by the equation

$$A = Ce^{-0.0016008t},$$

where C is the original amount of the element in the sample, A is the amount of the element remaining after t years, and $k = -0.0016008$, the decay constant for americium. If 10 milligrams (mg) of americium 241 is sealed in a laboratory container today, how much will theoretically be present in 2000 years? Round your answer to the nearest hundredth.

Solution Here $C = 10$ and $t = 2000$.
$$A = 10e^{-0.0016008(2000)} = 10e^{-3.2016}$$

Using a calculator or Table A-2, we have
$$A \approx 10(0.040697) = 0.40697 \approx 0.41 \text{ mg.}$$

The expression $10e^{-3.2016}$ can be found directly on some scientific calculators as follows:

10 $\boxed{\times}$ 3.2016 $\boxed{+/-}$ $\boxed{e^x}$ $\boxed{=}$ 0.406970366

(Scientific calculators with no $\boxed{e^x}$ key will require the keystrokes $\boxed{INV}$ $\boxed{\ln}$ or $\boxed{2nd F}$ $\boxed{\ln}$ or $\boxed{SHIFT}$ $\boxed{\ln}$ in place of the $\boxed{e^x}$.)

Thus, 0.41 milligrams of americium 241 would be present in 2000 years.

Practice Problem 8 If 20 milligrams of americium 241 is present in a sample now, how much will theoretically be present in 5000 years? Round your answer to the nearest thousandth.

Verbal and Writing Skills

1. The exponential function is an equation of the form _____ .

2. The irrational number *e* is a number that is approximately equal to _____ . (Give your answer with four decimal places.)

Graph each function.

3. $f(x) = 3^x$

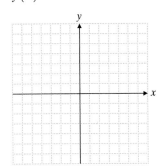

4. $f(x) = 2^x$

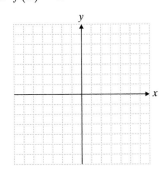

5. $f(x) = 2^{-x}$

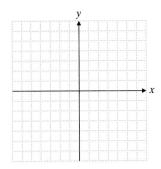

6. $f(x) = 5^{-x}$

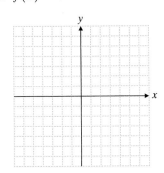

7. $f(x) = 3^{-x}$

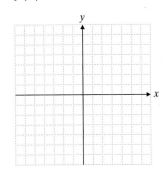

8. $f(x) = 4^{-x}$

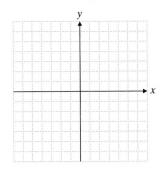

9. $f(x) = 2^{x+3}$

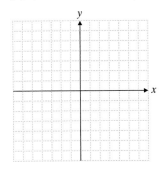

10. $f(x) = 2^{x+2}$

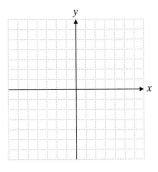

11. $f(x) = 3^{x-3}$

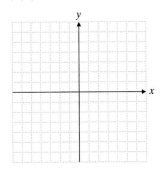

12. $f(x) = 3^{x-1}$

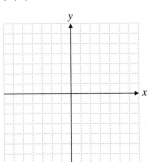

13. $f(x) = 2^x + 2$

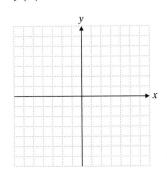

14. $f(x) = 2^x - 3$

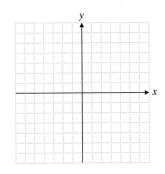

Graph each function. Use a calculator or Table A-2.

15. $f(x) = e^{x-1}$

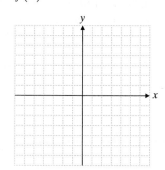

16. $f(x) = e^{x+1}$

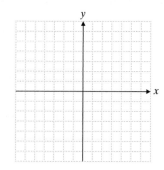

17. $f(x) = 2e^x$

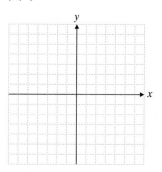

18. $f(x) = 3e^x$

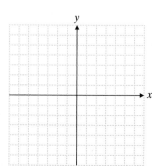

19. $f(x) = e^{1-x}$

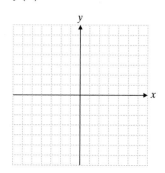

20. $f(x) = e^{2-x}$

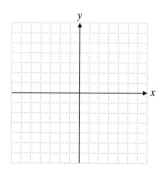

Solve for x.

21. $2^x = 4$

22. $2^x = 8$

23. $2^x = 1$

24. $2^x = 2$

25. $2^x = \dfrac{1}{2}$

26. $2^x = \dfrac{1}{64}$

27. $3^x = 81$

28. $3^x = 27$

29. $3^x = 1$

30. $3^x = 243$

31. $3^{-x} = \dfrac{1}{9}$

32. $3^{-x} = \dfrac{1}{3}$

33. $4^x = 256$

34. $4^{-x} = \dfrac{1}{16}$

35. $5^{x+1} = 125$

36. $6^{x-3} = 216$

37. $8^{3x-1} = 64$

38. $5^{5x-1} = 625$

Applications

To solve Exercises 39–42, use the interest formula $A = P\left(1 + \dfrac{r}{n}\right)^{nt}$. Round your answers to the nearest cent.

39. *Investments* Alicia is investing $2000 at an annual rate of 6.3% compounded annually. How much money will Alicia have after 3 years?

40. *Investments* Manza is investing $5000 at an annual rate of 7.1% compounded annually. How much money will Manza have after 4 years?

41. *Investments* How much money will Isabela have in 6 years if she invests $3000 at a 3.2% annual rate of interest compounded quarterly? How much will she have if it is compounded monthly?

42. *Investments* How much money will Waheed have in 3 years if he invests $5000 at a 3.85% annual rate of interest compounded quarterly? How much will he have if it is compounded monthly?

43. Bacteria Culture The number of bacteria in a culture is given by $B(t) = 4000(2^t)$, where t is the time in hours. How many bacteria will grow in the culture in the first 3 hours? In the first 9 hours?

44. College Tuition Suppose that the cost of a college education is increasing 4% per year. The equation $C(t) = P(1.04)^t$ forecasts the tuition cost t years from now and is based on the present cost P in dollars. How much will a college now charging $3000 for tuition charge in 10 years? How much will a college now charging $12,000 for tuition charge in 15 years?

45. Diving Depth U.S. Navy divers off the coast of Nantucket are searching for the wreckage of an old World War II–era submarine. They have found that if the water is relatively clear and the surface is calm, the ocean filters out 18% of the sunlight for each 4 feet they descend. How much sunlight is available at a depth of 20 feet? The divers need to use underwater spotlights when the amount of sunlight is less than 10%. Will they need spotlights when working at a depth of 48 feet?

46. Sewer Systems The city of Manchester just put in a municipal sewer to solve an underground water contamination problem, and many homeowners would like to have sewer lines connected to their homes. It is expected that each year the number of homeowners who use their own private septic tanks rather than the public sewer system will decrease by 8%. What percentage of people will still be using their private septic tanks in 5 years? The city feels that the underground water contamination problem will be solved when the number of homeowners still using septic tanks is less than 10%. Will that goal be achieved in the next 25 years?

Use an exponential equation to solve each problem. Round your answers to the nearest hundredth.

47. Radium Decay The radioactive decay of radium 226 can be described by the equation $A = Ce^{-0.0004279t}$, where C is the original amount of radium and A is the amount of radium remaining after t years. If 6 milligrams of radium are sealed in a container now, how much radium will be in the container after 1000 years?

48. Radon Decay The radioactive decay of radon 222 can be described by the equation $A = Ce^{-0.1813t}$, where C is the original amount of radon and A is the amount of radon after t days. If 1.5 milligrams are in a laboratory container today, how much was there in the container 10 days ago?

Atmospheric Pressure *Use the following information for Exercises 49 and 50. The atmospheric pressure measured in pounds per square inch is given by the equation $P = 14.7e^{-0.21d}$, where d is the distance in miles above sea level. Round your answers to the nearest hundredth.*

49. What is the pressure in pounds per square inch on an American Airlines jet plane flying 10 miles above sea level?

50. What is the pressure in pounds per square inch experienced by a man on a Colorado mountain that is 2 miles above sea level?

Stocks *The total number of shares N (in millions) that were traded on the New York Stock Exchange in any given year between 1940 and 2000 can be approximated by the equation $N = 0.00472e^{0.11596t}$, where t is the number of years since 1900. Use this information for Exercises 51 and 52. Source: New York Stock Exchange.*

51. Using the given equation, determine how many stocks were traded in 1980. In 1990. What was the percent of increase from 1980 to 1990?

52. Using the given equation, determine how many stocks were traded in 1985. In 2000. What was the percent of increase from 1985 to 2000?

World Population *The population of the world is growing exponentially. The following table and graph contain population data for selected years and show a pattern of significant increases.*

Year	AD1	1650	1850	1930	1975	1995	2000	2004
Approximate World Population in Billions	0.2	0.5	1	2	4	5.68	6.07	6.38

Source: Statistical Division of the United Nations.

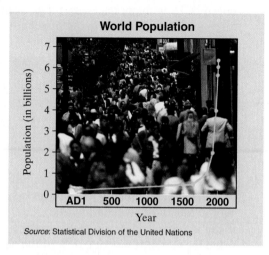

World Population

Source: Statistical Division of the United Nations

53. Based on the graph, in approximately what year did the world's population reach three billion people?

54. Based on the graph, what was the approximate world population in 1900?

55. The growth rate of the world's population during the period 1980–1990 was 1.7% per year. If that rate were to continue from 1995 to 2005, what would the world's population be in 2005?

56. The growth rate of the world's population during the period 1990–1997 was 1.4% per year. If that rate were to continue from 1995 to 2010, what would the world's population be in 2010?

Optional Graphing Calculator Problems

57. Let $f(x) = \dfrac{e^x + e^{-x}}{2}$. Evaluate $f(x)$ when $x = -1$, $-0.5, 0, 0.5, 1, 1.5,$ and 2. Now use these values to graph the function. (f defines a special function called *the hyperbolic cosine*. This function is used in advanced mathematics and science to study a variety of technical applications.)

58. Let $g(x) = \dfrac{e^x - e^{-x}}{2}$. Evaluate $g(x)$ when $x = -2$, $-1, -0.5, 0, 0.5, 1, 1.5,$ and 2. Now graph the function using these values. (g defines a special function called *the hyperbolic sine*.)

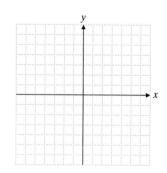

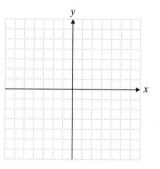

Cumulative Review

Solve for x.

59. $5 - 2(3 - x) = 2(2x + 5) + 1$

60. $\dfrac{7}{12} + \dfrac{3}{4}x + \dfrac{5}{4} = -\dfrac{1}{6}x$

Logarithms were invented about 400 years ago by the Scottish mathematician John Napier. Napier's amazing invention reduced complicated exercises to simple subtraction and addition. Astronomers quickly saw the immense value of logarithms and began using them. The work of Johannes Kepler, Isaac Newton, and others would have been much more difficult without logarithms.

The most important thing to know for this chapter is that a logarithm is an exponent. In Section 11.1 we solved the equation $2^x = 8$. We found that $x = 3$. The question we faced was, "To what power do we raise 2 to get 8?" The answer was 3. Mathematicians have to solve this type of problem so often that we have invented a short-hand notation for asking the question. Instead of asking, "To what power do we raise 2 to get 8?" we say instead, "What is $\log_2 8$?" Both questions mean the same thing.

Now suppose we had a general equation $x = b^y$ and someone asked, "To what power do we raise b to get x?" We would abbreviate this question by asking, "What is $\log_b x$?" Thus, we see that $y = \log_b x$ is an equivalent form of the equation $x = b^y$.

The key concept you must remember is that a logarithm is an exponent. We write $\log_b x = y$ to mean that the logarithm of x to the base b is equal to y. y is the exponent.

Student Learning Objectives

After studying this section, you will be able to:

1 Write exponential equations in logarithmic form.

2 Write logarithmic equations in exponential form.

3 Solve elementary logarithmic equations.

4 Graph a logarithmic function.

DEFINITION OF LOGARITHM

The **logarithm**, base b, of a *positive* number x is the power (exponent) to which the base b must be raised to produce x. That is, $y = \log_b x$ is the same as $x = b^y$, where $b > 0$ and $b \neq 1$.

Often you will need to convert logarithmic expressions to exponential expressions, and vice versa, to solve equations.

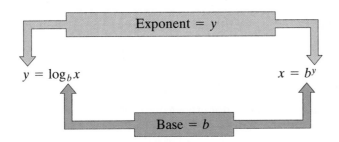

1 Writing Exponential Equations in Logarithmic Form

We begin by converting exponential expressions to logarithmic expressions.

> **EXAMPLE 1** Write in logarithmic form.
>
> **(a)** $81 = 3^4$ **(b)** $\dfrac{1}{100} = 10^{-2}$
>
> **Solution** We use the fact that $x = b^y$ is equivalent to $\log_b x = y$.
>
> **(a)** $81 = 3^4$
> Here $x = 81$, $b = 3$, and $y = 4$. So $4 = \log_3 81$.
>
> **(b)** $\dfrac{1}{100} = 10^{-2}$

Here $x = \dfrac{1}{100}$, $b = 10$, and $y = -2$. So $-2 = \log_{10}\left(\dfrac{1}{100}\right)$.

NOTE TO STUDENT: Fully worked-out solutions to all of the Practice Problems can be found at the back of the text starting at page SP-1

Practice Problem 1 Write in logarithmic form.

(a) $49 = 7^2$ **(b)** $\dfrac{1}{64} = 4^{-3}$

2 Writing Logarithmic Equations in Exponential Form

If we have an equation with a logarithm in it, we can write it in the form of an exponential equation. This is a very important skill. Carefully study the following example.

EXAMPLE 2 Write in exponential form.

(a) $2 = \log_5 25$ **(b)** $-4 = \log_{10}\left(\dfrac{1}{10{,}000}\right)$

Solution

(a) $2 = \log_5 25$

Here $y = 2$, $b = 5$, and $x = 25$. Thus, since $x = b^y$, $25 = 5^2$.

(b) $-4 = \log_{10}\left(\dfrac{1}{10{,}000}\right)$

Here $y = -4$, $b = 10$, and $x = \dfrac{1}{10{,}000}$. So $\dfrac{1}{10{,}000} = 10^{-4}$.

Practice Problem 2 Write in exponential form.

(a) $3 = \log_5 125$ **(b)** $-2 = \log_6\left(\dfrac{1}{36}\right)$

3 Solving Elementary Logarithmic Equations

Many logarithmic equations are fairly easy to solve if we first convert them to an equivalent exponential equation.

EXAMPLE 3 Solve for the variable.

(a) $\log_5 x = -3$ **(b)** $\log_a 16 = 4$

Solution

(a) $5^{-3} = x$ **(b)** $a^4 = 16$
$\dfrac{1}{5^3} = x$ $a^4 = 2^4$
$\dfrac{1}{125} = x$ $a = 2$

Practice Problem 3 Solve for the variable.

(a) $\log_b 125 = 3$ **(b)** $\log_{1/2} 32 = x$

With this knowledge we have the ability to solve an additional type of exercise.

EXAMPLE 4   Evaluate $\log_3 81$.

Solution Now, what exactly is the exercise asking for? It is asking, "To what power must we raise 3 to get 81?" Since we do not know the power, we call it x. We have

$$\log_3 81 = x$$
$$81 = 3^x \quad \text{Write an equivalent exponential equation.}$$
$$3^4 = 3^x \quad \text{Write 81 as } 3^4.$$
$$x = 4 \quad \text{If } b^x = b^y, \text{ then } x = y \text{ for } b > 0 \text{ and } b \neq 1.$$

Thus, $\log_3 81 = 4$.

Practice Problem 4   Evaluate $\log_{10} 0.1$.

4 Graphing a Logarithmic Function

We found in Chapter 10 that the graphs of a function and its inverse have an interesting property. They are symmetric to one another with respect to the line $y = x$. We also found in Chapter 10 that the procedure for finding the inverse of a function is to interchange the x and y variables. For example, $y = 2x + 3$ and $x = 2y + 3$ are inverse functions. In similar fashion, $y = 2^x$ and $x = 2^y$ are inverse functions. Another way to write $x = 2^y$ is the logarithmic equation $y = \log_2 x$. Thus, the logarithmic function $y = \log_2 x$ is the **inverse** of the exponential function $y = 2^x$. If we graph the function $y = 2^x$ and $y = \log_2 x$ on the same set of axes, the graph of one is the reflection of the other about the line $y = x$.

EXAMPLE 5   Graph $y = \log_2 x$.

Solution   If we write $y = \log_2 x$ in exponential form, we have $x = 2^y$. We make a table of values and graph the function $x = 2^y$.
    In each case, we pick a value of y as a first step.

If $y = -2$, $x = 2^y = 2^{-2} = \dfrac{1}{2^2} = \dfrac{1}{4}$.

If $y = -1$, $x = 2^{-1} = \dfrac{1}{2}$.

If $y = 0$, $x = 2^0 = 1$.

If $y = 1$, $x = 2^1 = 2$.

If $y = 2$, $x = 2^2 = 4$.

x	y
$\dfrac{1}{4}$	-2
$\dfrac{1}{2}$	-1
1	0
2	1
4	2

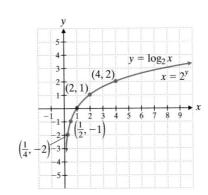

NOTE TO STUDENT: Fully worked-out solutions to all of the Practice Problems can be found at the back of the text starting at page SP-1

Practice Problem 5 Graph $y = \log_{1/2} x$.

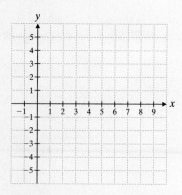

$f(x) = a^x$ and $f(x) = \log_a x$ are inverse functions. As such they have all the properties of inverse functions. We will review a few of these properties as we study the graphs of two inverse functions, $y = 2^x$ and $y = \log_2 x$.

EXAMPLE 6 Graph $y = \log_2 x$ and $y = 2^x$ on the same set of axes.

Solution Make a table of values (ordered pairs) for each equation. Then draw each graph.

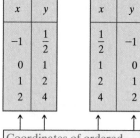

$y = 2^x$	
x	y
-1	$\frac{1}{2}$
0	1
1	2
2	4

$y = \log_2 x$	
x	y
$\frac{1}{2}$	-1
1	0
2	1
4	2

Coordinates of ordered pairs are reversed

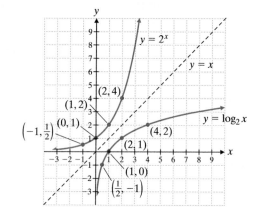

Note that $y = \log_2 x$ is the inverse of $y = 2^x$ because the ordered pairs (x, y) are reversed. The sketch of the two equations shows that they are inverses. If we reflect the graph of $y = 2^x$ about the line $y = x$, it will coincide with the graph of $y = \log_2 x$.

Recall that in function notation, f^{-1} means the inverse function of f. Thus, if we write $f(x) = \log_2 x$, then $f^{-1}(x) = 2^x$.

Practice Problem 6 Graph $y = \log_6 x$ and $y = 6^x$ on the same set of axes.

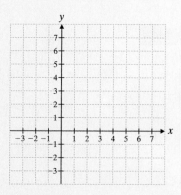

11.2 EXERCISES

| Student Solutions Manual | CD/ Video | PH Math Tutor Center | MathXL®Tutorials on CD | MathXL® | MyMathLab® | Interactmath.com |

Verbal and Writing Skills

1. A logarithm is an _____.

2. In the equation $y = \log_b x$, the value b is called the _____.

3. In the equation $y = \log_b x$, the domain (the set of permitted values of x) is _____.

4. In the equation $y = \log_b x$, the permitted values of b are _____.

Write in logarithmic form.

5. $49 = 7^2$

6. $512 = 8^3$

7. $36 = 6^2$

8. $100 = 10^2$

9. $0.001 = 10^{-3}$

10. $0.01 = 10^{-2}$

11. $\dfrac{1}{32} = 2^{-5}$

12. $\dfrac{1}{64} = 2^{-6}$

13. $y = e^5$

14. $y = e^{-8}$

Write in exponential form.

15. $2 = \log_3 9$

16. $2 = \log_2 4$

17. $0 = \log_{17} 1$

18. $0 = \log_{13} 1$

19. $\dfrac{1}{2} = \log_{16} 4$

20. $\dfrac{1}{2} = \log_{100} 10$

21. $-2 = \log_{10}(0.01)$

22. $-1 = \log_{10}(0.1)$

23. $-4 = \log_3 \dfrac{1}{81}$

24. $-7 = \log_2 \dfrac{1}{128}$

25. $-\dfrac{3}{2} = \log_e x$

26. $-\dfrac{5}{4} = \log_e x$

Solve.

27. $\log_2 x = 4$

28. $3 = \log_3 x$

29. $\log_{10} x = -3$

30. $\log_{10} x = -2$

31. $\log_4 64 = y$

32. $\log_7 343 = y$

33. $\log_8\left(\dfrac{1}{64}\right) = y$

34. $\log_3\left(\dfrac{1}{243}\right) = y$

35. $\log_a 121 = 2$

36. $\log_a 81 = 4$

37. $\log_a 1000 = 3$

38. $\log_a 100 = 2$

39. $\log_{25} 5 = w$

40. $\log_8 2 = w$

41. $\log_3\left(\dfrac{1}{3}\right) = w$

42. $\log_{37} 1 = w$

43. $\log_{15} w = 0$

44. $\log_{10} w = -4$

45. $\log_w 3 = \dfrac{1}{2}$

46. $\log_w 2 = \dfrac{1}{3}$

Evaluate.

47. $\log_{10}(0.001)$

48. $\log_{10}(0.0001)$

49. $\log_2 128$

50. $\log_5 125$

51. $\log_{23} 1$

52. $\log_{18} \dfrac{1}{18}$

53. $\log_6 \sqrt{6}$

54. $\log_7 \sqrt{7}$

55. $\log_{57} 1$

56. $\log_3 \dfrac{1}{27}$

Graph.

57. $\log_3 x = y$

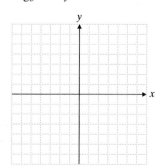

58. $\log_4 x = y$

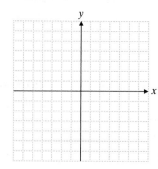

59. $\log_{1/4} x = y$

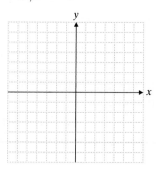

60. $\log_{1/3} x = y$

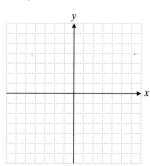

61. $\log_{10} x = y$

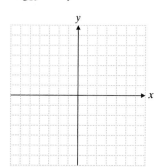

62. $\log_8 x = y$

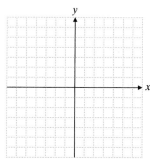

On one coordinate plane, graph the function f and the function f^{-1}. Then graph a dashed line for the equation $y = x$.

63. $f(x) = \log_3 x, f^{-1}(x) = 3^x$

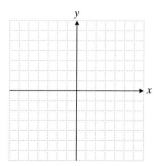

64. $f(x) = \log_4 x, f^{-1}(x) = 4^x$

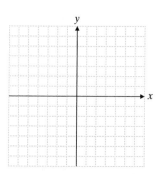

Applications

pH *Solutions* *To determine whether a solution is an acid or a base, chemists check the solution's* pH. *A solution is an acid if its* pH *is less than 7 and a base if its* pH *is greater than 7. The* pH *is defined by* $\text{pH} = -\log_{10}[\text{H}^+]$, *where* $[\text{H}^+]$ *is the concentration of hydrogen ions in the solution.*

65. The concentration of hydrogen ions in a lemon is approximately 10^{-2}. What is the pH of a lemon?

66. The concentration of hydrogen ions in cranberries is approximately $10^{-2.5}$. What is the pH of cranberries?

67. A scientist at the Center for Disease Control has mixed a solution with a pH of 8. Find the concentration of hydrogen ions $[\text{H}^+]$ in the solution.

68. A construction team repairing the Cape Cod Canal has created a cement mixture with a pH of 9. Find the concentration of hydrogen ions $[\text{H}^+]$ in the solution.

69. The chef of The Depot Restaurant has prepared a special balsamic vinaigrette salad dressing. What is the pH of the dressing if the concentration of hydrogen ions is 1.103×10^{-3}? The logarithm, base 10, of 0.001103 is approximately -2.957424488. Round your answer to three decimal places.

70. The EPA is testing a batch of experimental industrial solvent with a pH of 9.25. Find the concentration of hydrogen ions $[\text{H}^+]$ in the solution. Give your answer in scientific notation rounded to three decimal places.

Map Software   *Maptech produces map software for people who want to use their home computers to locate any place in the United States and print a detailed map. In the first 2 years of sales of the software, they have found that for N sets of software to be sold, they need to invest d dollars for advertising according to the equation*

$$N = 1200 + (2500)(\log_{10} d),$$

where d is always a positive number not less than 1.

71. How many sets of software were sold when they spent $10,000 on advertising?

72. How many sets of software were sold when they spent $100,000 on advertising?

73. They have a goal for 2 years from now of selling 18,700 sets of software. How much should they spend on advertising?

74. They have a goal for next year of selling 16,200 sets of software. How much should they spend on advertising?

75. Evaluate $5^{\log_5 4}$.

76. Evaluate $\log_2 \sqrt[4]{2}$.

Cumulative Review

77. Graph $y = -\dfrac{2}{3}x + 5$.

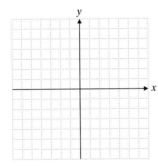

78. Graph $6x + 3y = -6$.

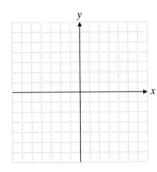

79. Find the equation of the line perpendicular to $y = -\dfrac{2}{3}x + 4$ that contains $(-4, 1)$.

80. Find the slope of a straight line containing $(-6, 3)$ and $(-1, 2)$.

81. *Viral Culture*   The number of viral cells in a laboratory culture in a biology research project is given by $A(t) = 9000(2^t)$, where t is measured in hours.
   (a) How many viral cells will grow in the culture in the first 2 hours?
   (b) How many viral cells will grow in the culture in the first 12 hours?

82. *College Tuition*   Assume the cost of a college education is increasing at 4% per year. The equation $C(t) = P(1.04)^t$ forecasts the tuition cost t years from now, based on the present cost P in dollars.
   (a) How much will a college now charging $4400 for tuition charge in 5 years?
   (b) How much will a college now charging $16,500 for tuition charge in 10 years?

83. *Commuting Costs*   John works outside the city of Philadelphia. If he takes the train to work, it costs him $98 for a monthly pass. If he drives to work, it takes about $2.50 per day to cover the cost of gas. Because the train station is so far from work, when he takes the train he must also take a bus. It costs him $0.75 per trip from the bus stop to work or from work to the bus stop. The cost of parking in a garage is $120 per month. If he drives to work, he pays $1 per day for the toll bridge. Which is the least-expensive way for John to commute between work and home each month? (Assume 23 working days per month.) How much would he save each month by using that method?

84. *Picnic Sandwiches*   There are 118 children at a picnic. Marcia Salzman and her daughter Lexi are in charge of the picnic. At this picnic, the parents are serving 30 peanut butter and jelly sandwiches, 34 turkey sandwiches, and 28 cheese sandwiches. If the sandwiches are cut into four pieces, how many children can eat one of each of the three types of sandwiches?

Student Learning Objectives

After studying this section, you will be able to:

1 Use the property $\log_b MN = \log_b M + \log_b N$.

2 Use the property $\log_b\left(\dfrac{M}{N}\right) = \log_b M - \log_b N$.

3 Use the property $\log_b M^p = p \log_b M$.

4 Solve simple logarithmic equations.

NOTE TO STUDENT: Fully worked-out solutions to all of the Practice Problems can be found at the back of the text starting at page SP-1

1 Using the Property $\log_b MN = \log_b M + \log_b N$

We have already said that logarithms reduce complex expressions to addition and subtraction. The following properties show us how to use logarithms in this way.

PROPERTY 1 THE LOGARITHM OF A PRODUCT

For any positive real numbers M and N and any positive base $b \neq 1$,

$$\log_b MN = \log_b M + \log_b N.$$

To see that this property is true, we let

$$\log_b M = x \quad \text{and} \quad \log_b N = y,$$

where x and y are any values. Now we write the expressions in exponential notation:

$$b^x = M \quad \text{and} \quad b^y = N.$$

Then

$$MN = b^x b^y = b^{x+y} \qquad \text{Laws of exponents.}$$

If we convert this equation to logarithmic form, then we have the following:

$$\log_b MN = x + y \qquad \text{Definition of logarithm.}$$
$$\log_b MN = \log_b M + \log_b N \qquad \text{By substitution.}$$

Note that the logarithms must have the same base.

EXAMPLE 1 Write $\log_3 XZ$ as a sum of logarithms.

Solution By property 1, $\log_3 XZ = \log_3 X + \log_3 Z$.

Practice Problem 1 Write as a sum of logarithms $\log_4 WXY$.

EXAMPLE 2 Write as a single logarithm $\log_3 16 + \log_3 x + \log_3 y$.

Solution If we extend our rule, we have $\log_b MNP = \log_b M + \log_b N + \log_b P$. Thus,

$$\log_3 16 + \log_3 x + \log_3 y = \log_3 16xy.$$

Practice Problem 2 Write as a single logarithm $\log_7 w + \log_7 8 + \log_7 x$.

2 Using the Property $\log_b\left(\dfrac{M}{N}\right) = \log_b M - \log_b N$

Property 2 is similar to property 1 except that it involves two expressions that are divided, not multiplied.

PROPERTY 2 THE LOGARITHM OF A QUOTIENT

For any positive real numbers M and N and any positive base $b \neq 1$,

$$\log_b\left(\frac{M}{N}\right) = \log_b M - \log_b N.$$

Property 2 can be proved using a similar approach to the one used to prove property 1. The proof will be left as an exercise for you to try.

EXAMPLE 3 Write as the difference of two logarithms $\log_3\left(\dfrac{29}{7}\right)$.

Solution $\log_3\left(\dfrac{29}{7}\right) = \log_3 29 - \log_3 7$

Practice Problem 3 Write as the difference of two logarithms $\log_3\left(\dfrac{17}{5}\right)$.

EXAMPLE 4 Express as a single logarithm $\log_b 36 - \log_b 9$.

Solution $\log_b 36 - \log_b 9 = \log_b\left(\dfrac{36}{9}\right) = \log_b 4$

Practice Problem 4 Express as a single logarithm $\log_b 132 - \log_b 4$.

CAUTION: Be sure you understand property 2!

$$\frac{\log_b M}{\log_b N} \neq \log_b M - \log_b N$$

Do you see why?

③ Using the Property $\log_b M^p = p\log_b N$

We now introduce the third property. We will not prove it now, but you will have a chance to verify this property as an exercise.

> **PROPERTY 3 THE LOGARITHM OF A NUMBER RAISED TO A POWER**
>
> For any positive real number M, any real number p, and any positive base $b \neq 1$,
> $$\log_b M^p = p\log_b M.$$

EXAMPLE 5 Write as a single logarithm $\dfrac{1}{3}\log_b x + 2\log_b w - 3\log_b z$.

Solution   First, we must eliminate the coefficients of the logarithm terms.

$$\log_b x^{1/3} + \log_b w^2 - \log_b z^3 \quad \text{By property 3.}$$

Now we can combine either the sum or difference of the logarithms. We'll do the sum.

$$\log_b x^{1/3}w^2 - \log_b z^3 \quad \text{By property 1.}$$

Now we combine the difference.

$$\log_b\left(\frac{x^{1/3}w^2}{z^3}\right) \quad \text{By property 2.}$$

Practice Problem 5 Write as one logarithm $\dfrac{1}{3}\log_7 x - 5\log_7 y$.

EXAMPLE 6 Write as a sum or difference of logarithms $\log_b\left(\dfrac{x^4 y^3}{z^2}\right)$.

Solution
$$\log_b\left(\frac{x^4 y^3}{z^2}\right) = \log_b x^4 y^3 - \log_b z^2 \qquad \text{By property 2.}$$
$$= \log_b x^4 + \log_b y^3 - \log_b z^2 \qquad \text{By property 1.}$$
$$= 4\log_b x + 3\log_b y - 2\log_b z \qquad \text{By property 3.}$$

NOTE TO STUDENT: Fully worked-out solutions to all of the Practice Problems can be found at the back of the text starting at page SP-1

Practice Problem 6 Write as a sum or difference of logarithms $\log_3\left(\dfrac{x^4 y^5}{z}\right)$.

4 Solving Simple Logarithmic Equations

A major goal in solving many logarithmic equations is to obtain a logarithm on one side of the equation and no logarithm on the other side. In Example 7 we will use property 1 to combine two separate logarithms that are added.

EXAMPLE 7 Find x if $\log_2 x + \log_2 5 = 3$.

Solution
$$\log_2 5x = 3 \qquad \text{Use property 1.}$$
$$5x = 2^3 \qquad \text{Convert to exponential form.}$$
$$5x = 8 \qquad \text{Simplify.}$$
$$x = \frac{8}{5} \qquad \text{Divide both sides by 5.}$$

Practice Problem 7 Find x if $\log_4 x + \log_4 5 = 2$.

In Example 8, two logarithms are subtracted on the left side of the equation. We can use property 2 to combine these two logarithms. This will allow us to obtain the form of one logarithm on one side of the equation and no logarithm on the other side.

EXAMPLE 8 Find x if $\log_3(x + 4) - \log_3(x - 4) = 2$.

Solution
$$\log_3\left(\frac{x + 4}{x - 4}\right) = 2 \qquad \text{Use property 2.}$$
$$\frac{x + 4}{x - 4} = 3^2 \qquad \text{Convert to exponential form.}$$
$$x + 4 = 9(x - 4) \qquad \text{Multiply each side by } (x - 4).$$
$$x + 4 = 9x - 36 \qquad \text{Simplify.}$$
$$40 = 8x$$
$$5 = x$$

Practice Problem 8 Find x if $\log_{10} x - \log_{10}(x + 3) = -1$.

To solve some logarithmic equations, we need a few additional properties of logarithms. We state these properties now. The proofs of some of them will be left as exercises for you.

The following properties are true for all positive values of $b \neq 1$ and all positive values of x and y.

Property 4 $\log_b b = 1$

Property 5 $\log_b 1 = 0$

Property 6   If $\log_b x = \log_b y$, then $x = y$.

We now illustrate each property in Example 9.

EXAMPLE 9

(a) Evaluate $\log_7 7$.

(b) Evaluate $\log_5 1$.

(c) Find x if $\log_3 x = \log_3 17$.

Solution

(a) $\log_7 7 = 1$ because $\log_b b = 1$.   Property 4.

(b) $\log_5 1 = 0$ because $\log_b 1 = 0$.   Property 5.

(c) If $\log_3 x = \log_3 17$, then $x = 17$.   Property 6.

Practice Problem 9   Evaluate.

(a) $\log_7 1$ **(b)** $\log_8 8$ **(c)** Find y if $\log_{12} 13 = \log_{12}(y + 2)$.

We now have the mathematical tools needed to solve a variety of logarithmic equations.

EXAMPLE 10   Find x if $2 \log_7 3 - 4 \log_7 2 = \log_7 x$.

Solution   We can use property 3 in two cases.

$$2 \log_7 3 = \log_7 3^2 = \log_7 9$$

$$4 \log_7 2 = \log_7 2^4 = \log_7 16$$

By substituting these results, we have the following.

$$\log_7 9 - \log_7 16 = \log_7 x$$

$$\log_7 \left(\frac{9}{16} \right) = \log_7 x \quad \text{Property 2.}$$

$$\frac{9}{16} = x \qquad \text{Property 6.}$$

Practice Problem 10   Find x if $\log_3 2 - \log_3 5 = \log_3 6 + \log_3 x$.

Student Solutions Manual | CD/Video | PH Math Tutor Center | MathXL®Tutorials on CD | MathXL® | MyMathLab® | Interactmath.com

Express as a sum of logarithms.

1. $\log_3 AB$

2. $\log_{12} CD$

3. $\log_5(7 \cdot 11)$

4. $\log_6(13 \cdot 5)$

5. $\log_b 9f$

6. $\log_b 5d$

Express as a difference of logarithms.

7. $\log_9\left(\dfrac{2}{7}\right)$

8. $\log_{11}\left(\dfrac{23}{17}\right)$

9. $\log_b\left(\dfrac{H}{10}\right)$

10. $\log_a\left(\dfrac{G}{7}\right)$

11. $\log_a\left(\dfrac{E}{F}\right)$

12. $\log_6\left(\dfrac{8}{M}\right)$

Express as a product.

13. $\log_8 a^7$

14. $\log_5 b^{10}$

15. $\log_b A^{-2}$

16. $\log_a B^{-5}$

17. $\log_5 \sqrt{w}$

18. $\log_6 \sqrt{z}$

Mixed Practice

Write each expression as a sum or difference of logarithms of x, y, and z.

19. $\log_8 x^2 y$

20. $\log_4 x y^3$

21. $\log_{11}\left(\dfrac{6M}{N}\right)$

22. $\log_5\left(\dfrac{2C}{7}\right)$

23. $\log_2\left(\dfrac{5xy^4}{\sqrt{z}}\right)$

24. $\log_5\left(\dfrac{3x^5\sqrt[3]{y}}{z^4}\right)$

25. $\log_a\left(\sqrt[3]{\dfrac{x^4}{y}}\right)$

26. $\log_a \sqrt[5]{\dfrac{y}{z^3}}$

Write as a single logarithm.

27. $\log_4 13 + \log_4 y + \log_4 3$

28. $\log_8 15 + \log_8 a + \log_8 b$

29. $5 \log_3 x - \log_3 7$

30. $3 \log_8 5 - \log_8 z$

31. $2 \log_b 7 + 3 \log_b y - \dfrac{1}{2}\log_b z$

32. $4 \log_b 2 + \dfrac{1}{3}\log_b z - 5 \log_b y$

Use the properties of logarithms to simplify each of the following.

33. $\log_3 3$

34. $\log_7 7$

35. $\log_e e$

36. $\log_{10} 10$

37. $\log_9 1$

38. $\log_e 1$

39. $3 \log_7 7 + 4 \log_7 1$

40. $\dfrac{1}{2}\log_5 5 - 8 \log_5 1$

Find x in each of the following.

41. $\log_8 x = \log_8 7$

42. $\log_9 x = \log_9 5$

43. $\log_5(2x + 7) = \log_5(29)$

44. $\log_{15}(26) = \log_{15}(3x - 1)$

45. $\log_3 1 = x$

46. $\log_8 1 = x$

47. $\log_7 7 = x$

48. $\log_5 5 = x$

49. $\log_{10} x + \log_{10} 25 = 2$

50. $\log_{10} x + \log_{10} 5 = 1$

51. $\log_2 7 = \log_2 x - \log_2 3$

52. $\log_5 1 = \log_5 x - \log_5 8$

53. $3 \log_5 x = \log_5 8$

54. $\dfrac{1}{2}\log_3 x = \log_3 4$

55. $\log_e x = \log_e 5 + 1$

56. $\log_e x + \log_e 7 = 2$

57. $\log_6(5x + 21) - \log_6(x + 3) = 1$

58. $\log_3(4x + 6) - \log_3(x - 1) = 2$

59. It can be shown that $y = b^{\log_b y}$. Use this property to evaluate $5^{\log_5 4} + 3^{\log_3 2}$.

60. It can be shown that $x = \log_b b^x$. Use this property to evaluate $\log_7 \sqrt[4]{7} + \log_6 \sqrt[12]{6}$.

To Think About

61. Prove that $\log_b\left(\dfrac{M}{N}\right) = \log_b M - \log_b N$ by using an argument similar to the proof of property 1.

62. Prove that $\log_b M^p = p \log_b M$ by using an argument similar to the proof of property 1.

Cumulative Review

▲ **63.** Find the volume of a cylinder with a radius of 2 meters and a height of 5 meters.

▲ **64.** Find the area of a circle whose radius is 4 meters.

65. Solve the system for (x, y).

$5x + 3y = 9$
$7x - 2y = 25$

66. Solve the system for (x, y, z).

$2x - y + z = 3$
$x + 2y + 2z = 1$
$4x + y + 2z = 0$

67. *Carbon Dioxide Emissions* Carbon dioxide emissions in China increased from 8.01×10^8 metric tons of carbon per year in 1996 to 9.30×10^8 metric tons in 2000. What is the percent of increase during this 4-year period? If an equal percent of increase took place from 2000 to 2004, what was the level of carbon dioxide emissions in China in 2004? *Source:* U.S. Energy Information Administration.

68. *Carbon Dioxide Emissions* Carbon dioxide emissions in Japan decreased from 3.04×10^8 metric tons of carbon per year in 1996 to 2.73×10^8 metric tons in 2000. What is the percent of decrease during this 4-year period? If an equal percent of decrease took place from 2000 to 2004, what was the level of carbon dioxide emissions in Japan in 2004? *Source:* U.S. Energy Information Administration.

69. *Braking* While traveling at 38 feet per second, on a back road in North Dakota, the driver of a Ford Explorer equipped with antilock brakes approaches a patch of ice. He immediately hits the brakes hard and stops the vehicle in 3 seconds. What is the vehicle's speed in miles per hour 2 seconds after he hits the brake? (Assume that he decelerates at a constant rate.) Was he exceeding the 35-mile-per-hour speed limit on this back road before he hit the brakes?

70. *Braking* The driver of a Corvette rapidly accelerates from a standstill to 45 miles per hour in 5 seconds. If he continues at a constant rate of acceleration, how many seconds will it take him to reach the speed limit of 65 miles per hour from the standstill? If he hits the brakes at 65 mph because of an object in the road and comes to a complete stop 14 seconds after starting from a standstill, how many feet did he travel while he was braking?

How are you doing with your homework assignments in Sections 11.1 to 11.3? Do you feel you have mastered the material so far? Do you understand the concepts you have covered? Before you go further in the textbook, take some time to do each of the following problems.

11.1

1. Sketch the graph of $f(x) = 2^{-x}$. Plot at least four points.

2. Solve for x: $3^{2x-1} = 27$

3. Solve for x: $2^x = \dfrac{1}{32}$

4. Solve for x: $125 = 5^{3x+4}$

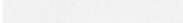

5. When a principal amount P is invested at interest rate r compounded annually, the amount of money A earned after t years is given by the equation $A = P(1 + r)^t$. How much money will Nancy have in 4 years if she invests \$10,000 in a mutual fund that pays 12% interest compounded annually?

11.2

6. Write in logarithmic form: $\dfrac{1}{49} = 7^{-2}$

7. Write in exponential form: $-3 = \log_{10}(0.001)$

8. Solve for x: $\log_5 x = 3$

9. Solve for x: $\log_x 81 = -2$

10. Evaluate $\log_{10}(10,000)$.

11.3

11. Write as a sum or difference of logarithms $\log_5\left(\dfrac{x^2 y^5}{z^3}\right)$.

12. Express as a single logarithm $\dfrac{1}{2}\log_4 x - 3 \log_4 w$.

13. Find x if $\log_3 x + \log_3 2 = 4$.　　**14.** Find x if $\log_7 x = \log_7 8$.

15. Find x if $\log_9 1 = x$.　　**16.** Find x if $\log_3 2x = 2$.

17. Find x if $1 = \log_4 3x$.　　**18.** Find x if $\log_e x + \log_e 3 = 1$.

Now turn to page SA-47 for the answers to each of these problems. Each answer also includes a reference to the objective in which the problem is first taught. If you missed any of these problems, you should stop and review the Examples and Practice Problems in the referenced objective. A little review now will help you master the material in the upcoming sections of the text.

1. _____
2. _____
3. _____
4. _____
5. _____
6. _____
7. _____
8. _____
9. _____
10. _____
11. _____
12. _____
13. _____
14. _____
15. _____
16. _____
17. _____
18. _____

Student Learning Objectives

After studying this section, you will be able to:

1 Find common logarithms.

2 Find the antilogarithm of a common logarithm.

3 Find natural logarithms.

4 Find the antilogarithm of a natural logarithm.

5 Evaluate a logarithm to a base other than 10 or e.

1 Finding Common Logarithms on a Scientific Calculator

Although we can find a logarithm of a number for any positive base except 1, the most frequently used bases are 10 and e. Base 10 logarithms are called *common logarithms* and are usually written with no subscript.

> **DEFINITION**
>
> For all real numbers $x > 0$, the **common logarithm** of x is
> $$\log x = \log_{10} x.$$

Before the advent of calculators and computers, people used tables of common logarithms. Now most work with logarithms is done with the aid of a scientific calculator or a graphing calculator. We will take that approach in this section of the text. To find the common logarithm of a number on a scientific calculator, enter the number and then press the $\boxed{\log x}$ or $\boxed{\log}$ key.

EXAMPLE 1 On a scientific calculator or a graphing calculator, find a decimal approximation for each of the following.

(a) log 7.32   **(b)** log 73.2   **(c)** log 0.314

Solution

(a) 7.32 $\boxed{\log}$ ≈ 0.864511081 ←

(b) 73.2 $\boxed{\log}$ ≈ 1.864511081 ←

> Note that the only difference in the two answers is the 1 before the decimal point.

(c) 0.314 $\boxed{\log}$ ≈ −0.503070352

Note: Your calculator may display fewer or more digits in the answer.

NOTE TO STUDENT: Fully worked-out solutions to all of the Practice Problems can be found at the back of the text starting at page SP-1

Practice Problem 1 On a scientific calculator or a graphing calculator, find a decimal approximation for each of the following.

(a) log 4.36   **(b)** log 436   **(c)** log 0.2418

TO THINK ABOUT: Decimal Point Placement Why is the difference in the answers to Example 1 **(a)** and **(b)** equal to 1.00? Consider the following.

$$\log 73.2 = \log(7.32 \times 10^1) \qquad \text{Use scientific notation.}$$
$$= \log 7.32 + \log 10^1 \qquad \text{By property 1.}$$
$$= \log 7.32 + 1 \qquad \text{Because } \log_b b = 1.$$
$$\approx 0.864511081 + 1 \qquad \text{Use a calculator.}$$
$$\approx 1.864511081 \qquad \text{Add the decimals.}$$

2 Finding the Antilogarithm of a Common Logarithm on a Scientific Calculator

We have previously discussed the function $f(x) = \log x$ and the corresponding inverse function $f^{-1}(x) = 10^x$. The inverse of a logarithmic function is an exponential function. There is another name for this function. It is called an **antilogarithm.**

If $f(x) = \log x$ (here the base is understood to be 10), then $f^{-1}(x) = $ antilog $x = 10^x$.

EXAMPLE 2   Find an approximate value for x if $\log x = 4.326$.

Solution   Here we are given the value of the logarithm, and we want to find the number that has that logarithm. In other words, we want the antilogarithm. We know that $\log_{10} x = 4.326$ is equivalent to $10^{4.326} = x$. So to solve this problem, we want to find the value of 10 raised to the 4.326 power. Using a calculator, we have the following.

$$4.326 \boxed{10^x} \approx 21183.61135$$

Thus, $x \approx 21,183.61135$. (If your scientific calculator does not have a $\boxed{10^x}$ key, you can usually use $\boxed{\text{2nd Fn}}$ $\boxed{\log}$ or $\boxed{\text{INV}}$ $\boxed{\log}$ or $\boxed{\text{SHIFT}}$ $\boxed{\log}$ to perform the operation.)

Practice Problem 2   Using a scientific calculator, find an approximate value for x if $\log x = 2.913$.

EXAMPLE 3   Evaluate antilog (-1.6784).

Solution   Asking what is antilog (-1.6784) is equivalent to asking what the value is of $10^{-1.6784}$. To determine this on most scientific calculators, it will be necessary to enter the numbers 1.6784 followed by the $\boxed{+/-}$ key. You may need slightly different steps on a graphing calculator.

$$1.6784 \boxed{+/-} \boxed{10^x} \approx 0.020970076$$

Thus, antilog$(-1.6784) \approx 0.020970076$.

Practice Problem 3   Evaluate antilog(-3.0705).

EXAMPLE 4   Using a scientific calculator, find an approximate value for x.

(a) $\log x = 0.07318$ **(b)** $\log x = -3.1621$

Solution

(a) $\log x = 0.07318$ is equivalent to $10^{0.07318} = x$.

$$0.07318 \boxed{10^x} \approx 1.183531987$$

Thus, $x \approx 1.183531987$.

(b) $\log x = -3.1621$ is equivalent to $10^{-3.1621} = x$.

$$3.1621 \boxed{+/-} \boxed{10^x} \approx 0.0006884937465.$$

Thus, $x \approx 0.0006884937465$.
(Some calculators may give the answer in scientific notation as $6.884937465 \times 10^{-4}$. This is often displayed on the calculator screen as $6.884937465 - 4$.)

Practice Problem 4   Using a scientific calculator, find an approximate value for x.

(a) $\log x = 0.06134$ **(b)** $\log x = -4.6218$

3 Finding Natural Logarithms on a Scientific Calculator

For most theoretical work in mathematics and other sciences, the most useful base for logarithms is e. Logarithms with base e are known as *natural logarithms* and are usually written $\ln x$.

> **DEFINITION**
> For all real numbers $x > 0$, the **natural logarithm** of x is
> $$\ln x = \log_e x.$$

On a scientific calculator we can usually approximate natural logarithms with the $\boxed{\ln x}$ or $\boxed{\ln}$ key.

EXAMPLE 5 On a scientific calculator, approximate the following values.

(a) $\ln 7.21$ **(b)** $\ln 72.1$ **(c)** $\ln 0.0356$

Solution

(a) $7.21 \boxed{\ln} \approx 1.975468951$

(b) $72.1 \boxed{\ln} \approx 4.278054044$

(c) $0.0356 \boxed{\ln} \approx -3.335409641$

Note that there is no simple relationship between the answers to parts **(a)** and **(b).** Do you see why these are different from common logarithms?

Practice Problem 5 On a scientific calculator, approximate the following values.

(a) $\ln 4.82$ **(b)** $\ln 48.2$ **(c)** $\ln 0.0793$

NOTE TO STUDENT: Fully worked-out solutions to all of the Practice Problems can be found at the back of the text starting at page SP-1

Finding the Antilogarithm of a Natural Logarithm on a Scientific Calculator

EXAMPLE 6 On a scientific calculator, find an approximate value of x for each equation.

(a) $\ln x = 2.9836$ **(b)** $\ln x = -1.5619$

Solution

(a) If $\ln x = 2.9836$, then $e^{2.9836} = x$.
$$2.9836 \boxed{e^x} \approx 19.75882051$$

(b) If $\ln x = -1.5619$, then $e^{-1.5619} = x$.
$$1.5619 \boxed{+/-} \boxed{e^x} \approx 0.209737192$$

Practice Problem 6 On a scientific calculator, find an approximate value of x for each equation.

(a) $\ln x = 3.1628$ **(b)** $\ln x = -2.0573$

An alternative notation is sometimes used. This is antilog$_e(x)$.

⑤ Evaluating a Logarithm to a Base Other Than 10 or *e*

Although a scientific calculator or a graphing calculator has specific keys for finding common logarithms (base 10) and natural logarithms (base *e*), there are no keys for finding logarithms with other bases. What do we do in such cases? The logarithm of a number for a base other than 10 or *e* can be found with the following formula.

CHANGE OF BASE FORMULA

$$\log_b x = \frac{\log_a x}{\log_a b},$$

where a, b, and $x > 0$, $a \neq 1$, and $b \neq 1$.

Let's see how this formula works. If we want to use common logarithms to find $\log_3 56$, we must first note that the value of b in the formula is 3. We then write

$$\log_3 56 = \frac{\log_{10} 56}{\log_{10} 3} = \frac{\log 56}{\log 3}.$$

Do you see why?

EXAMPLE 7 Evaluate using common logarithms. $\log_3 5.12$

Solution $\log_3 5.12 = \dfrac{\log 5.12}{\log 3}$

On a calculator, we find the following.

5.12 $\boxed{\log}$ $\boxed{\div}$ 3 $\boxed{\log}$ $\boxed{=}$ 1.486561234

Our answer is an approximate value with nine decimal places. Your answer may have more or fewer digits depending on your calculator.

Practice Problem 7 Evaluate using common logarithms. $\log_9 3.76$

If we desire to use base *e*, then the change of base formula is used with natural logarithms.

EXAMPLE 8 Obtain an approximate value for $\log_4 0.005739$ using natural logarithms.

Solution Using the change of base formula, with $a = e$, $b = 4$, and $x = 0.005739$, we have the following.

$$\log_4 0.005739 = \frac{\log_e 0.005739}{\log_e 4} = \frac{\ln 0.005739}{\ln 4}$$

This is done on some scientific calculators as follows.

0.005739 $\boxed{\ln}$ $\boxed{\div}$ 4 $\boxed{\ln}$ $\boxed{=}$ −3.722492455

Thus, we have $\log_4 0.005739 \approx -3.722492455$.

Check: To check our answer we want to know the following.

$$4^{-3.722492455} \overset{?}{=} 0.005739$$

Using a calculator, we can verify this with the $\boxed{y^x}$ key.

$4 \boxed{y^x} 3.722492455 \boxed{+/-} \boxed{=} 0.005739 \checkmark$

Practice Problem 8 Obtain an approximate value for $\log_8 0.009312$ using natural logarithms.

Graphing Calculator

Graphing Logarithmic Functions

You can use the change of base formula to graph logarithmic functions on a graphing calculator.

To graph $y = \log_2 x$ in Example 9 on a graphing calculator, enter the function $y = \dfrac{\log x}{\log 2}$ into the Y = editor of your calculator.

Display:

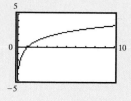

NOTE TO STUDENT: *Fully worked-out solutions to all of the Practice Problems can be found at the back of the text starting at page SP-1*

EXAMPLE 9 Using a scientific calculator, graph $y = \log_2 x$.

Solution If we use common logarithms ($\log_{10} x$), then for each value of x, we will need to calculate $\frac{\log x}{\log 2}$. Therefore, to find y when $x = 3$, we need to calculate $\frac{\log 3}{\log 2}$. On most scientific calculators, we would enter $3 \boxed{\log} \boxed{\div} 2 \boxed{\log} \boxed{=}$ and obtain 1.584962501. Rounded to the nearest tenth, we have $x = 3$ and $y = 1.6$. In a similar fashion we find other table values and then graph them.

x	y = log₂ x
0.5	−1
1	0
2	1
3	1.6
4	2
6	2.6
8	3

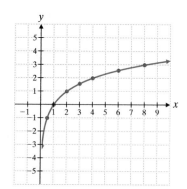

Practice Problem 9 Using a scientific calculator, graph $y = \log_5 x$.

11.4 EXERCISES

Student Solutions Manual   CD/Video   PH Math Tutor Center   MathXL®Tutorials on CD   MathXL®   MyMathLab®   Interactmath.com

Verbal and Writing Skills

1. Try to find log(−5.08) on a scientific calculator. What happens? Why?

2. Try to find log(−6.63) on a scientific calculator or a graphing calculator. What happens? Why?

Use a scientific calculator to approximate the following.

3. log 12.3

4. log 11.7

5. log 25.6

6. log 83.8

7. log 8

8. log 6

9. log 125,000

10. log 78,500

11. log 0.0123

12. log 0.567

Find an approximate value of x using a scientific calculator or a graphing calculator.

13. $\log x = 2.016$

14. $\log x = 2.754$

15. $\log x = -0.3562$

16. $\log x = -0.1398$

17. $\log x = 3.9304$

18. $\log x = 3.9576$

19. $\log x = 6.4683$

20. $\log x = 5.6274$

21. $\log x = -3.3893$

22. $\log x = -4.0458$

23. $\log x = -1.5672$

24. $\log x = -1.0075$

Approximate the following with a scientific calculator or a graphing calculator.

25. antilog (7.6215)

26. antilog (4.3894)

27. antilog(−1.0826)

28. antilog(−3.1145)

29. ln 5.62

30. ln 8.81

31. ln 1.53

32. ln 2.06

33. ln 136,000

34. ln 129,000

35. ln 0.00579

36. ln 0.00134

Find an approximate value of x using a scientific calculator or a graphing calculator.

37. $\ln x = 0.95$

38. $\ln x = 0.55$

39. $\ln x = 2.4$

40. $\ln x = 4.4$

41. $\ln x = -0.05$

42. $\ln x = -0.03$

43. $\ln x = -2.7$

44. $\ln x = -3.8$

Approximate the following with a scientific calculator or a graphing calculator.

45. antilog$_e$(6.1582)

46. antilog$_e$(1.9047)

47. antilog$_e$(−2.1298)

48. antilog$_e$(−3.3712)

613

Use a scientific calculator or a graphing calculator and **common logarithms** *to evaluate the following.*

49. $\log_3 9.2$

50. $\log_2 6.13$

51. $\log_7(7.35)$

52. $\log_9(9.85)$

53. $\log_6 0.127$

54. $\log_5 0.173$

55. $\log_{15} 12$

56. $\log_{17} 18$

Use a scientific calculator or a graphing calculator and **natural logarithms** *to evaluate the following.*

57. $\log_4 0.07733$

58. $\log_7 0.004462$

59. $\log_{21} 436$

60. $\log_{30} 913$

Mixed Practice

Use a scientific calculator or a graphing calculator to find an approximate value for each of the following.

61. $\ln 1537$

62. $\log 92.81$

63. $\text{antilog}_e(-1.874)$

64. $\log_6 0.5437$

Find an approximate value for x for each of the following.

65. $\log x = 8.5634$

66. $\ln x = 7.9631$

67. $\log_4 x = 0.8645$

68. $\log_3 x = 0.5649$

 Use a graphing calculator to graph the following.

69. $y = \log_6 x$

70. $y = \log_4 x$

71. $y = \log_{0.4} x$

72. $y = \log_{0.2} x$

Applications

Median Age The median age of people in the United States is slowly increasing as the population becomes older. In 1980 the median age of the population was 30.0 years. This means that approximately half the population of the country was under 30.0 years old and approximately half the population of the country was over 30.0 years old. By 1987 the median age had increased to 32.0 years. An equation that can be used to predict the median age N (in years) of the population of the United States is $N = 32.82 + 1.0249 \ln x$, where x is the number of years since 1990 and $x \geq 1$. Source: U.S. Census Bureau.

73. Use the equation to find the median age of the U.S. population in 2000 and in 2010. If this model is correct, by what percent will the median age increase from 2000 to 2010?

74. Use the equation to find the median age of the U.S. population in 1995 and in 2005. If this model is correct, by what percent will the median age increase from 1995 to 2005?

Earthquakes   *Suppose that we want to measure the magnitude of an earthquake. If an earthquake has a shock wave x times greater than the smallest shock wave that can be measured by a seismograph, then its magnitude R on the Richter scale is given by the equation $R = \log x$.*

An earthquake that has a shock wave 25,000 times greater than the smallest shock wave that can be detected will have a magnitude of $R = \log 25,000 \approx 4.40$. (Usually we round the magnitude of an earthquake to the nearest hundredth.)

75. What is the magnitude of an earthquake that has a shock wave that is 56,000 times greater than the smallest shock wave that can be detected?

76. What is the magnitude of an earthquake that has a shock wave that is 184,000 times greater than the smallest shock wave that can be detected?

77. If the magnitude of an earthquake is $R = 6.6$ on the Richter scale, what can you say about the size of the earthquake's shock wave?

78. If the magnitude of an earthquake is $R = 5.4$ on the Richter scale, what can you say about the size of the earthquake's shock wave?

Cumulative Review

Solve the quadratic equations. Simplify your answers.

79. $3x^2 - 11x - 5 = 0$

80. $2y^2 + 4y - 3 = 0$

Highway Exits   *On a specific portion of Interstate 91, there are six exits. There is a distance of 12 miles between odd-numbered exits. There is a distance of 15 miles between even-numbered exits. The total distance between Exit 1 and Exit 6 is 36 miles.*

81. Find the distance between Exit 1 and Exit 2. Find the distance between Exit 1 and Exit 3.

82. Find the distance between Exit 1 and Exit 4. Find the distance between Exit 1 and Exit 5.

1 Solving Logarithmic Equations

In general, when solving logarithmic equations we try to obtain all the logarithms on one side of the equation and all the numerical values on the other side. Then we seek to use the properties of logarithms to obtain a single logarithmic expression on one side.

We can describe a general procedure for solving logarithmic equations.

Step 1 If an equation contains some logarithms and some terms without logarithms, try to get one logarithm alone on one side and one numerical value on the other.

Step 2 Then convert to an exponential equation using the definition of a logarithm.

Step 3 Solve the equation.

Graphing Calculator

Solving Logarithmic Equations

Example 1 could be solved with a graphing calculator in the following way. First write the equation as

$$\log 5 + \log(x + 3) - 2 = 0$$

and then graph the function

$$y = \log 5 + \log(x + 3) - 2$$

to find an approximate value for x when $y = 0$. If you set the appropriate window and use your Zoom feature, you should be able to obtain $x = 17.0$, rounded to the nearest tenth. Now use your graphing calculator to solve the following.

1. $\log x + \log(x + 1) = 1$

2. $\log(x - 6) = 2 - \log(x + 15)$

3. $\ln(x + 2) = 12$

EXAMPLE 1 Solve $\log 5 = 2 - \log(x + 3)$.

Solution

$$\log 5 + \log(x + 3) = 2 \qquad \text{Add } \log(x + 3) \text{ to each side.}$$
$$\log[5(x + 3)] = 2 \qquad \text{Property 1.}$$
$$\log(5x + 15) = 2 \qquad \text{Simplify.}$$
$$5x + 15 = 10^2 \qquad \text{Write the equation in exponential form.}$$
$$5x + 15 = 100 \qquad \text{Simplify.}$$
$$5x = 85 \qquad \text{Subtract 15 from each side.}$$
$$x = 17 \qquad \text{Divide each side by 5.}$$

Check: $\log 5 \overset{?}{=} 2 - \log(17 + 3)$

$\log 5 \overset{?}{=} 2 - \log 20$

Since these are common logarithms (base 10), the easiest way to check the answer is to find decimal approximations for each logarithm on a calculator.

$$0.698970004 \overset{?}{=} 2 - 1.301029996$$
$$0.698970004 = 0.698970004 \quad \checkmark$$

Practice Problem 1 Solve $\log(x + 5) = 2 - \log 5$.

EXAMPLE 2 Solve $\log_3(x + 6) - \log_3(x - 2) = 2$.

Solution

$$\log_3\left(\frac{x + 6}{x - 2}\right) = 2 \qquad \text{Property 2.}$$

$$\frac{x + 6}{x - 2} = 3^2 \qquad \text{Write the equation in exponential form.}$$

$$\frac{x + 6}{x - 2} = 9 \qquad \text{Evaluate } 3^2.$$

$$x + 6 = 9(x - 2) \qquad \text{Multiply each side by } (x - 2).$$
$$x + 6 = 9x - 18 \qquad \text{Simplify.}$$
$$24 = 8x \qquad \text{Add } 18 - x \text{ to each side.}$$
$$3 = x \qquad \text{Divide each side by 8.}$$

Check: $\log_3(3 + 6) - \log_3(3 - 2) \overset{?}{=} 2$

$$\log_3 9 - \log_3 1 \overset{?}{=} 2$$

$$2 - 0 \overset{?}{=} 2$$

$$2 = 2 \quad \checkmark$$

Practice Problem 2 Solve $\log(x + 3) - \log x = 1$.

NOTE TO STUDENT: *Fully worked-out solutions to all of the Practice Problems can be found at the back of the text starting at page SP-1*

Some equations consist of logarithmic terms only. In such cases we may be able to use property 6 to solve them. Recall that this rule states that if $b > 0$, $b \neq 1$, $x > 0$, $y > 0$, and $\log_b x = \log_b y$, then $x = y$.

What if one of our possible solutions is the logarithm of a negative number? Can we evaluate the logarithm of a negative number? Look again at the graph of $y = \log_2 x$ on page 612. Note that the domain of this function is $x > 0$. (The curve is located on the positive side of the *x*-axis.) Therefore, the logarithm of a negative number is *not defined*.

You should be able to see this by using the definition of logarithms. If $\log(-2)$ were valid, we could write the following.

$$y = \log_{10}(-2)$$
$$10^y = -2$$

Obviously, no value of y can make this equation true. Thus, we see that **it is not possible to take the logarithm of a negative number.**

Sometimes when we attempt to solve a logarithmic equation, we obtain a possible solution that leads to the logarithm of a negative number. We can immediately discard such a solution.

EXAMPLE 3 Solve $\log(x + 6) + \log(x + 2) = \log(x + 20)$.

Solution

$$\log(x + 6)(x + 2) = \log(x + 20)$$
$$\log(x^2 + 8x + 12) = \log(x + 20)$$
$$x^2 + 8x + 12 = x + 20$$
$$x^2 + 7x - 8 = 0$$
$$(x + 8)(x - 1) = 0$$
$$x + 8 = 0 \qquad x - 1 = 0$$
$$x = -8 \qquad x = 1$$

Check: $\log(x + 6) + \log(x + 2) = \log(x + 20)$

$$x = 1: \quad \log(1 + 6) + \log(1 + 2) \overset{?}{=} \log(1 + 20)$$

$$\log(7) + \log(3) \overset{?}{=} \log(21)$$

$$\log(7 \cdot 3) \overset{?}{=} \log 21$$

$$\log 21 = \log 21 \quad \checkmark$$

$$x = -8: \quad \log(-8 + 6) + \log(-8 + 2) \overset{?}{=} \log(-8 + 20)$$

$$\log(-2) + \log(-6) \neq \log(12)$$

We can discard -8 because it leads to taking the logarithm of a negative number, which is not allowed. Only $x = 1$ is a solution. The only solution is 1.

Practice Problem 3 Solve $\log 5 - \log x = \log(6x - 7)$. Check your solution.

2 Solving Exponential Equations

You might expect that property 6 can be used in the reverse direction. It seems logical, for example, that if $x = 3$, we should be able to state that $\log_4 x = \log_4 3$. This is exactly the case, and we will formally state it as a property.

> **PROPERTY 7**
>
> If x and $y > 0$ and $x = y$, then $\log_b x = \log_b y$, where $b > 0$ and $b \neq 1$.

Property 7 is often referred to as "taking the logarithm of each side of the equation." Usually we will take the common logarithm of each side of the equation, but any base can be used.

EXAMPLE 4   Solve $2^x = 7$. Leave your answer in exact form.

Solution

$$\log 2^x = \log 7 \qquad \text{Take the logarithm of each side (property 7).}$$
$$x \log 2 = \log 7 \qquad \text{Property 3.}$$
$$x = \frac{\log 7}{\log 2} \qquad \text{Divide each side by log 2.}$$

NOTE TO STUDENT: *Fully worked-out solutions to all of the Practice Problems can be found at the back of the text starting at page SP-1*

Practice Problem 4   Solve $3^x = 5$. Leave your answer in exact form.

When we solve exponential equations, it will often be useful to find an approximate value for the answer.

EXAMPLE 5   Solve $3^x = 7^{x-1}$. Approximate your answer to the nearest thousandth.

Solution

$$\log 3^x = \log 7^{(x-1)}$$
$$x \log 3 = (x - 1) \log 7$$
$$x \log 3 = x \log 7 - \log 7$$
$$x \log 3 - x \log 7 = -\log 7$$
$$x(\log 3 - \log 7) = -\log 7$$
$$x = \frac{-\log 7}{\log 3 - \log 7}$$

We can approximate the value for x on most scientific calculators by using the following keystrokes.

7 log +/− ÷ (3 log − 7 log) = 2.296606943

Rounding to the nearest thousandth, we have $x \approx 2.297$.

Practice Problem 5   Solve $2^{3x+1} = 9^{x+1}$. Approximate your answer to the nearest thousandth.

If the exponential equation involves e raised to a power, it is best to take the natural logarithm of each side of the equation.

EXAMPLE 6   Solve $e^{2.5x} = 8.42$. Round your answer to the nearest ten-thousandth.

Solution

$$\ln e^{2.5x} = \ln 8.42 \qquad \text{Take the natural logarithm of each side.}$$
$$(2.5x)(\ln e) = \ln 8.42 \qquad \text{Property 3.}$$
$$2.5x = \ln 8.42 \qquad \ln e = 1.$$
$$x = \frac{\ln 8.42}{2.5} \qquad \text{Divide each side by 2.5.}$$

On most scientific calculators, the value of x can be approximated with the following keystrokes.

$$8.42 \boxed{\ln} \boxed{\div} 2.5 \boxed{=} 0.85224393$$

Rounding to the nearest ten-thousandth, we have $x \approx 0.8522$.

Practice Problem 6   Solve $20.98 = e^{3.6x}$. Round your answer to the nearest ten-thousandth.

③ Solving Applications Using Logarithmic or Exponential Equations

We now return to the compound interest formula and consider some other exercises that can be solved with it. For example, perhaps we would like to know how long it will take for a deposit to grow to a specified goal.

EXAMPLE 7   If P dollars are invested in an account that earns interest at 12% compounded annually, the amount available after t years is $A = P(1 + 0.12)^t$. How many years will it take for $300 in this account to grow to $1500? Round your answer to the nearest whole year.

Solution

$$1500 = 300(1 + 0.12)^t \qquad \text{Substitute } A = 1500 \text{ and } P = 300.$$
$$1500 = 300(1.12)^t \qquad \text{Simplify.}$$
$$\frac{1500}{300} = (1.12)^t \qquad \text{Divide each side by 300.}$$
$$5 = (1.12)^t \qquad \text{Simplify.}$$
$$\log 5 = \log(1.12)^t \qquad \text{Take the common logarithm of each side.}$$
$$\log 5 = t(\log 1.12) \qquad \text{Property 3.}$$
$$\frac{\log 5}{\log 1.12} = t \qquad \text{Divide each side by log 1.12.}$$

On a scientific calculator we have the following.

$$5 \boxed{\log} \boxed{\div} 1.12 \boxed{\log} \boxed{=} 14.20150519$$

Thus, it would take approximately 14 years. Look at the graph on the next page to get a visual image of how the investment increases.

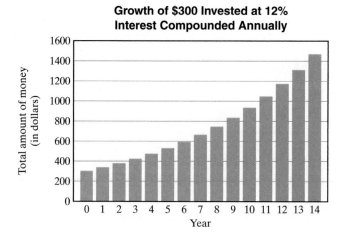

Growth of $300 Invested at 12% Interest Compounded Annually

NOTE TO STUDENT: Fully worked-out solutions to all of the Practice Problems can be found at the back of the text starting at page SP-1

Practice Problem 7 Mon Ling's father has an investment account that earns 8% interest compounded annually. How many years would it take for $4000 to grow to $10,000 in that account? Round your answer to the nearest whole year.

The growth equation for things that appear to be growing continuously is $A = A_0e^{rt}$, where A is the final amount, A_0 is the original amount, r is the rate at which things are growing in a unit of time, and t is the total number of units of time.

For example, if a laboratory starts with 5000 cells and they reproduce at a rate of 35% per hour, the number of cells in 18 hours can be described by the following equation.

$$A = 5000e^{(0.35)(18)} = 5000e^{6.3} \approx 5000(544.57) = 2{,}722{,}850 \text{ cells}$$

EXAMPLE 8 If world population is eight billion people and it continues to grow at the current rate of 2% per year, how many years will it take for the population to increase to fourteen billion people?

Solution

$$A = A_0e^{rt}$$

To make our calculation easier, we write the values for the population in terms of billions. This will allow us to avoid writing numbers like 14,000,000,000 and 8,000,000,000. Do you see why we can do this?

$14 = 8e^{0.02t}$ — Substitute known values.

$\dfrac{14}{8} = e^{0.02t}$ — Divide each side by 8.

$\ln\left(\dfrac{14}{8}\right) = \ln e^{0.02t}$ — Take the natural logarithm of each side.

$\ln(1.75) = (0.02t) \ln e$ — Property 3.

$\ln 1.75 = 0.02t$ — Since $\ln e = 1$.

$\dfrac{\ln 1.75}{0.02} = t$ — Divide each side by 0.02.

Using our calculator, we obtain the following.

$1.75 \boxed{\ln} \boxed{\div} 0.02 \boxed{=} 27.98078940$

Rounding to the nearest whole year, we find that the population will grow from eight billion to fourteen billion in about 28 years if the growth continues at the same rate.

Practice Problem 8   The wildlife management team in one region of Alaska has determined that the black bear population is growing at the rate of 4.3% per year. This region currently has approximately 1300 black bears. A food shortage will develop if the population reaches 5000. If the growth rate remains unchanged, how many years will it take for this food shortage problem to occur?

EXAMPLE 9   The magnitude of an earthquake is measured by the formula $R = \log\left(\dfrac{I}{I_0}\right)$, where I is the intensity of the earthquake and I_0 is the minimum measurable intensity. The 1964 earthquake in Anchorage, Alaska, had a magnitude of 8.4. The 1906 earthquake in Taiwan had a magnitude of 7.1. How much more energy was released from the Anchorage earthquake than from the Taiwan earthquake? *Source:* National Oceanic and Atmospheric Administration.

Solution

Let I_A = intensity of the Alaska earthquake. Then

$$8.4 = \log\left(\frac{I_A}{I_0}\right) = \log I_A - \log I_0.$$

Solving for $\log I_0$ gives

$$\log I_0 = \log I_A - 8.4.$$

Let I_T = intensity of the Taiwan earthquake. Then

$$7.1 = \log\left(\frac{I_T}{I_0}\right) = \log I_T - \log I_0.$$

Solving for $\log I_0$ gives

$$\log I_0 = \log I_T - 7.1.$$

Therefore,

$$\log I_A - 8.4 = \log I_T - 7.1.$$

$$\log I_A - \log I_T = 8.4 - 7.1$$

$$\log \frac{I_A}{I_T} = 1.3$$

$$10^{1.3} = \frac{I_A}{I_T}$$

$$19.95262315 = \frac{I_A}{I_T} \qquad \text{Use a calculator.}$$

$$20 \approx \frac{I_A}{I_T} \qquad \begin{array}{l}\text{Round to the}\\\text{nearest whole}\\\text{number.}\end{array}$$

$$20I_T \approx I_A$$

The Alaska earthquake had approximately twenty times the intensity of the Taiwan earthquake.

Practice Problem 9   The 1933 earthquake in Japan had a magnitude of 8.9. The 1989 earthquake in San Francisco had a magnitude of 7.1. How much more energy was released from the Japan earthquake than from the San Francisco earthquake? *Source:* National Oceanic and Atmospheric Administration.

Student Solutions Manual   CD/ Video   PH Math Tutor Center   MathXL®Tutorials on CD   MathXL®   MyMathLab®   Interactmath.com

Solve each logarithmic equation and check your solutions.

1. $\log_7\left(\dfrac{2}{3}x + 3\right) + \log_7 3 = 2$

2. $\log_5\left(\dfrac{1}{2}x - 2\right) + \log_5 2 = 1$

3. $\log_6(x + 3) + \log_6 4 = 2$

4. $\log_2 4 + \log_2(x - 1) = 5$

5. $\log_2\left(x + \dfrac{4}{3}\right) = 5 - \log_2 6$

6. $\log_5(3x + 1) = 1 - \log_5 2$

7. $\log(30x + 40) = 2 + \log(x - 1)$

8. $1 + \log x = \log(9x + 1)$

9. $2 + \log_6(x - 1) = \log_6(12x)$

10. $\log_2 x = \log_2(x + 5) - 1$

11. $\log(75x + 50) - \log x = 2$

12. $\log_{11}(x + 7) = \log_{11}(x - 3) + 1$

13. $\log_3(x + 6) + \log_3 x = 3$

14. $\log_8 x + \log_8(x - 2) = 1$

15. $1 + \log(x - 2) = \log(6x)$

16. $\log_5(2x) - \log_5(x - 3) = 3\log_5 2$

17. $\log_2(x + 5) - 2 = \log_2 x$

18. $\log_5(5x + 10) - 2 = \log_5(x - 2)$

19. $2\log_7 x = \log_7(x + 4) + \log_7 2$

20. $\log x + \log(x - 1) = \log 12$

21. $\ln(10) - \ln x = \ln(x - 3)$ **22.** $\ln(2 + 2x) = 2 \ln(x + 1)$

Solve each exponential equation. Leave your answers in exact form. Do not approximate.

23. $7^{x+3} = 12$ **24.** $4^{x-2} = 6$

25. $2^{3x+4} = 17$ **26.** $5^{2x-1} = 11$

Solve each exponential equation. Use your calculator to approximate your solutions to the nearest thousandth.

27. $8^{2x-1} = 90$ **28.** $15^{3x-2} = 230$ **29.** $5^x = 4^{x+1}$

30. $3^x = 2^{x+3}$ **31.** $28 = e^{x-2}$ **32.** $e^{x+2} = 88$

33. $88 = e^{2x+1}$ **34.** $3 = e^{1-x}$

Applications

Compound Interest *When a principal P earns an annual interest rate r compounded yearly, the amount A after t years is $A = P(1 + r)^t$. Use this information to solve Exercises 35–38. Round all answers to the nearest whole year.*

35. How long will it take $1500 to grow to $5000 at 8% compounded annually?

36. How long will it take $1000 to grow to $4500 at 7% compounded annually?

37. How long will it take for a principal to triple at 6% compounded annually

38. How long will it take for a principal to double at 5% compounded annually

39. What interest rate would be necessary to obtain $6500 in 6 years if $5000 is the amount of the original investment and the interest is compounded yearly? (Express the interest rate as a percent rounded to the nearest tenth.)

40. If $3000 is invested for 3 years with annual interest compounded yearly, what interest rate is needed to achieve an amount of $3600? (Express the interest rate as a percent rounded to the nearest tenth.)

World Population *The growth of the world's population can be described by the equation $A = A_0 e^{rt}$, where time t is measured in years, A_0 is the population of the world at time $t = 0$, r is the annual growth rate, and A is the population at time t. Assume that $r = 2\%$ per year. Use this information to solve Exercises 41–44. Round your answers to the nearest whole year.*

41. How long will it take a population of seven billion to increase to twelve billion?

42. How long will it take a population of six billion to increase to nine billion?

43. How long will it take for the world's population to double?

44. How long will it take for the world's population to quadruple (become four times as large)?

Cell Phone Industry *The number N of employees in the cellular telephone industry in the United States can be approximated by the equation $N = 20,800(1.264)^x$, where x is the number of years since 1990. Use this equation to answer the following questions. Source: Federal Communications Commission.*

45. Approximately how many employees were there in 2003? Round to the nearest thousand.

46. Approximately how many employees will there be in 2006? Round to the nearest thousand.

47. In what year did the number of employees reach 274,000?

48. In what year will the number of employees reach 699,000?

Use the equation $A = A_0 e^{rt}$ to solve Exercises 49–54. Round your answers to the nearest whole number.

49. Population The population of Bethel is 80,000 people, and it is growing at the rate of 1.5% per year. How many years will it take for the population to grow to 120,000 people?

50. Population The population of Melbourne, Australia, is approximately three million people. If the growth rate is 3% per year, in how many years will there be 3.5 million people?

51. Skin Grafts The number of new skin cells on a revolutionary skin graft is growing at a rate of 4% per hour. How many hours will it take for 200 cells to become 1800 cells?

52. Workforce The workforce in a state is increasing at the rate of 1.5% per year. During the last measured year, the workforce was 3.5 million. If this rate continues, how many years will it be before the workforce reaches 4.5 million?

53. Lyme Disease Unfortunately U.S. deer carry ticks that spread Lyme disease. The number of people who are infected by the virus is increasing by 5% every year. If 24,500 people were confirmed to have Lyme disease in 1997, how many will be infected by the end of the year 2010?

54. Video Rentals In the city of Scranton, the number of videotape rentals is increasing by 7.5% per year. For the last year that data are available, 1.3 million videos were rented. How many years will it be before 2.0 million videos are rented per year?

To Think About

Earthquakes *The magnitude of an earthquake (amount of energy released) is described by the formula $R = \log\left(\dfrac{I}{I_0}\right)$, where I is the intensity of the earthquake and I_0 is the minimum measurable intensity. Use this formula to solve Exercises 55–58. Round answers to the nearest tenth.*

55. October 17, 1989, brought tragedy to the San Francisco/Oakland area when an earthquake measuring 7.1 on the Richter scale centered in the Loma Prieta area (Santa Cruz Mountains) and collapsed huge sections of freeway, killing sixty-three people. Almost 6 years later, an earthquake measuring 8.2 on the Richter scale killed 190 people in the Kurile Islands of Japan and Russia. How much more energy was released from the Kurile earthquake than from the Loma Prieta earthquake? *Source:* National Oceanic and Atmospheric Administration.

56. On January 17, 1993, in Northridge, California, residents experienced an earthquake that measured 6.8 on the Richter scale, killed sixty-one people, and undermined supposedly earthquake-proof steel-framed buildings. Exactly 1 year later near Kobe, Japan, an earthquake measuring 7.2 on the Richter scale killed more than 5300 people, injured more than 35,000, and destroyed nearly 200,000 homes, in spite of construction codes reputed to be the best in the world. How much more energy was released from the Japan earthquake than from the Northridge earthquake? *Source:* National Oceanic and Atmospheric Administration.

57. The 1906 earthquake in San Francisco had a magnitude of 8.3. In 1971 an earthquake in Japan measured 6.8. How much more energy was released from the San Francisco earthquake than from the Japan earthquake? *Source:* National Oceanic and Atmospheric Administration.

58. The 1933 Japan earthquake had a magnitude of 8.9. In Turkey a 1975 earthquake had a magnitude of 6.7. How much more energy was released from the Japan earthquake than from the Turkey earthquake? *Source:* National Oceanic and Atmospheric Administration.

Optional Graphing Calculator Problems

59. *Fish Population* In Crystal Lake, north of Amherst, Nova Scotia, the fish population has been out of balance for several years because of an abundance of catfish. Environmentalists have taken a number of measures to increase the number of brook trout so that the populations of the two types of fish are at the same level. After several years of dealing with industrial pollution, the environmentalists have succeeded in cleaning the lake sufficiently enough so that the brook trout can reproduce more readily. The growth in the number of brook trout is now described by the equation $y = 300e^{0.12x}$, where x is the number of years from now. The growth in the number of catfish is given by the equation $y = 750 + 100x$, where x is the number of years from now. How many years will it take until the two fish populations are equal in number? Round your answer to the nearest tenth of a year.

60. *Wolf Population* Suppose that the population of wolves in one region of Alaska is growing according to the equation $y_1 = 34.572x + 850$, where x is the number of years from now. Suppose also that the food supply for wolves is growing according to the equation $y_2 = 1000e^{0.02x}$, where x is the number of years from now. In how many years (rounded to the nearest tenth of a year) will the food supply become inadequate for the number of wolves? (When will y_1 be greater than y_2?)

Cumulative Review

Simplify. Assume that x and y are positive real numbers.

61. $\left(\sqrt{3} + 2\sqrt{2}\right)\left(\sqrt{6} - \sqrt{2}\right)$

62. $\sqrt{98x^3y^2}$

63. *Spelling Bees* 248 students competed in the 2000 national spelling bee. A total of fifty-four students were 12 years old. Thirty-three more students were 13 years old than were 12 years old. Eighty students were 14 or 15 years old. All the remaining students were younger. Only one student, the youngest student in the competition, was 9 years old. Ten more students were 11 years old than were 10 years old. How many students were there in each age category from age 9 to 15?

64. *London Subway* The London subway system has a staff of 16,000 and provides 2.5 million passenger journeys each day. An expansion of the system begun in 1993 will extend the system 16 kilometers and will cost 2.85 billion British pounds. How many dollars per mile will this extension cost? (Use 1 kilometer = 0.62 miles and 1 U.S. dollar = 0.63 British pounds.) *Source: British Bureau of Tourism.*

Putting Your Skills to Work

The Growth of E-Commerce Retail Sales

If you were to buy airplane tickets in the coming year, would you purchase them online? Have you purchased any clothing over the Internet in the last year? When you want to buy CDs or theatre tickets or college textbooks, do you first look online? If so, you are among a growing number of people in this country who are shopping online. Just how rapidly are these online sales growing? Can you use your math skills to predict Internet sales for the coming years? You will have a chance to see how exponential functions can be used to help us make predictions as you answer the following questions.

   The table below shows total retail sales and e-commerce (Internet) retail sales for the years 2000–2002 and for the first 3 quarters of 2003. The data are given in millions of dollars.

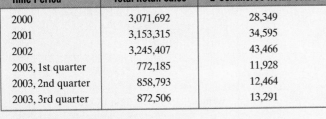

Time Period	Total Retail Sales	E-Commerce Retail Sales
2000	3,071,692	28,349
2001	3,153,315	34,595
2002	3,245,407	43,466
2003, 1st quarter	772,185	11,928
2003, 2nd quarter	858,793	12,464
2003, 3rd quarter	872,506	13,291

Source: U.S. Commerce Department

Problems for Individual Investigation and Analysis

1. What percent of total retail sales were e-commerce sales in the year 2000? Round to the nearst tenth of a percent.

2. What percent of total retail sales were e-commerce sales in the 3rd quarter of 2003?

3. In 2002, 4th quarter total retail sales were 36.3% of the sum of the first 3 quarters' sales. If that trend holds true in 2003, what will the 4th quarter sales be? Round to the nearest million dollars.

4. If the percent of e-commerce sales is the same in the 4th quarter of 2003 as it was in the 3rd quarter, what will the e-commerce sales be in the 4th quarter of 2003? Round to the nearest million dollars.

Problems for Group Investigation and Cooperative Study

U.S. online retail sales are expected to reach \$65 billion in 2004 and to continue to grow by a compound annual growth rate of 17% through 2008.

5. Find the estimated online retail sales in 2005 and in 2006. Use your work to write a formula for predicting future online retail sales. Use the variable n to represent years after 2004.

6. Use your formula to fill in the table below. Round the sales figures to the nearest hundredth of a billion.

Year	Estimated Online Retail Sales (in billions of dollars)
2005	
2006	
2007	
2008	

Topic	Procedure	Examples
Exponential function, p. 584.	$f(x) = b^x$, where $b > 0$, $b \neq 1$, and x is a real number.	Graph $f(x) = \left(\frac{2}{3}\right)^x$.
Property of exponential equations, p. 586.	When $b > 0$ and $b \neq 1$, if $b^x = b^y$, then $x = y$.	Solve for x, $2^x = \frac{1}{32}$. $$2^x = \frac{1}{2^5}$$ $$2^x = 2^{-5}$$ $$x = -5$$
Definition of logarithm, p. 593.	$y = \log_b x$ is the same as $x = b^y$, where $x > 0$, $b > 0$ and $b \neq 1$.	Write in exponential form $\log_3 17 = 2x$. $$3^{2x} = 17$$ Write in logarithmic form $18 = 3^x$ $$\log_3 18 = x$$ Solve for x $\log_6\left(\frac{1}{36}\right) = x$. $$6^x = \frac{1}{36}$$ $$6^x = 6^{-2}$$ $$x = -2$$
Properties of logarithms, pp. 600–603.	Suppose that $M > 0$, $N > 0$, $b > 0$, and $b \neq 1$. $\log_b MN = \log_b M + \log_b N$ $\log_b\left(\frac{M}{N}\right) = \log_b M - \log_b N$ $\log_b M^p = p \log_b M$ $\log_b b = 1$ $\log_b 1 = 0$ If $\log_b x = \log_b y$, then $x = y$. If $x = y$, then $\log_b x = \log_b y$,	Write as separate logarithms of x, y, and w. $$\log_3\left(\frac{x^2 \sqrt[3]{y}}{w}\right)$$ $$= 2 \log_3 x + \frac{1}{3}\log_3 y - \log_3 w$$ Write as one logarithm. $$5 \log_6 x - 2 \log_6 w - \frac{1}{4}\log_6 z$$ $$= \log_6\left(\frac{x^5}{w^2 \sqrt[4]{z}}\right)$$ Simplify. $$\log 10^5 + \log_3 3 + \log_5 1$$ $$= 5 \log 10 + \log_3 3 + \log_5 1$$ $$= 5 + 1 + 0$$ $$= 6$$
Finding logarithms, p. 608.	On a scientific calculator: $\log x = \log_{10} x$, for all $x > 0$ $\ln x = \log_e x$, for all $x > 0$	Find $\log 3.82$. 3.82 $\boxed{\log}$ $\log 3.82 \approx 0.5820634$ Find $\ln 52.8$. 52.8 $\boxed{\ln}$ $\ln 52.8 \approx 3.9665112$

Topic	Procedure	Examples
Finding antilogarithms, p. 609.	If $\log x = b$, then $10^b = x$. If $\ln x = b$, then $e^b = x$. Use a calculator or a table to solve.	Find x if $\log x = 2.1416$ $$10^{2.1416} = x$$ $$2.1416 \boxed{10^x} \approx 138.54792$$ Find x if $\ln x = 0.6218$. $$e^{0.6218} = x$$ $$0.6218 \boxed{e^x} \approx 1.8622771$$
Finding a logarithm to a different base, p. 611.	Change of base formula: $$\log_b x = \frac{\log_a x}{\log_a b},$$ where $a, b,$ and $x > 0$, $a \neq 1$, and $b \neq 1$.	Evaluate $\log_7 1.86$. $$\frac{\log 1.86}{\log 7}$$ $$1.86 \boxed{\log} \boxed{\div} 7 \boxed{\log} \boxed{=} 0.3189132$$
Solving logarithmic equations, p. 616.	**1.** If some but not all of the terms of an equation have logarithms, try to rewrite the equation with one single logarithm on one side and one numerical value on the other. Then convert the equation to exponential form. **2.** If an equation contains logarithmic terms only, try to get only one logarithm on each side of the equation. Then use the property that if $\log_b x = \log_b y$, $x = y$. *Note:* Always check your solutions when solving logarithmic equations.	Solve for x, $\log_5 3x - \log_5(x^2 - 1) = \log_5 2$. $$\log_5 3x = \log_5 2 + \log_5(x^2 - 1)$$ $$\log_5 3x = \log_5[2(x^2 - 1)]$$ $$3x = 2x^2 - 2$$ $$0 = 2x^2 - 3x - 2$$ $$0 = (2x + 1)(x - 2)$$ $$2x + 1 = 0 \qquad x - 2 = 0$$ $$x = -\frac{1}{2} \qquad x = 2$$ **Check:** $x = 2$: $\log_5 3(2) - \log_5(2^2 - 1) \stackrel{?}{=} \log_5 2$ $$\log_5 6 - \log_5 3 \stackrel{?}{=} \log_5 2$$ $$\log_5\left(\frac{6}{3}\right) \stackrel{?}{=} \log_5 2$$ $$\log_5 2 = \log_5 2 \ \checkmark$$ $x = -\frac{1}{2}$: For the expression $\log_5(3x)$, we would obtain $\log_5(-1.5)$. You cannot take the logarithm of a negative number. $x = -\frac{1}{2}$ is not a solution. The solution is 2.
Solving exponential equations, p. 618.	**1.** See whether each expression can be written so that only one base appears on one side of the equation and the same base appears on the other side. Then use the property that if $b^x = b^y$, $x = y$. **2.** If you can't do step 1, take the logarithm of each side of the equation and use the properties of logarithms to solve for the variable.	Solve for x, $2^{x-1} = 7$, $$\log 2^{x-1} = \log 7$$ $$(x - 1)\log 2 = \log 7$$ $$x \log 2 - \log 2 = \log 7$$ $$x \log 2 = \log 7 + \log 2$$ $$x = \frac{\log 7 + \log 2}{\log 2}$$ (We can approximate the answer as $x \approx 3.8073549$.)

Graph the functions in Exercises 1 and 2.

1. $f(x) = 4^{3+x}$

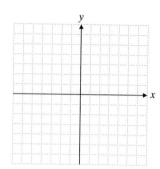

2. $f(x) = e^{x-3}$

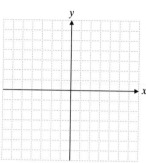

3. Solve: $3^{3x+1} = 81$

4. Write in exponential form: $-2 = \log(0.01)$

5. Change to logarithmic form: $8 = 4^{\frac{3}{2}}$

Solve.

6. $\log_w 16 = 4$ **7.** $\log_3 x = -2$ **8.** $\log_8 x = 0$ **9.** $\log_7 w = -1$ **10.** $\log_w 64 = 3$

11. $\log_{10} w = -1$ **12.** $\log_{10} 1000 = x$ **13.** $\log_2 64 = x$ **14.** $\log_2\left(\dfrac{1}{4}\right) = x$ **15.** $\log_3 243 = x$

16. Graph the equation $\log_3 x = y$.

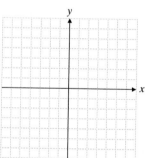

Write each expression as the sum or difference of $\log_2 x$, $\log_2 y$, *and* $\log_2 z$.

17. $\log_2\left(\dfrac{5x}{\sqrt{w}}\right)$

18. $\log_2 x^3\sqrt{y}$

Write as a single logarithm.

19. $\log_3 x + \log_3 w^{1/2} - \log_3 2$

20. $4\log_8 w - \dfrac{1}{3}\log_8 z$

21. Evaluate $\log_e e^6$.

Solve.

22. $\log_5 100 - \log_5 x = \log_5 4$

23. $\log_8 x + \log_8 3 = \log_8 75$

Find the value with a scientific calculator.

24. $\log 23.8$

25. $\log 0.0817$

26. $\ln 3.92$

27. $\ln 803$

28. Find n if $\log n = 1.1367$.

29. Find n if $\ln n = 1.7$.

30. $\log_8 2.81$

Solve each equation and check your solutions.

31. $\log_{11}\left(\frac{4}{3}x + 7\right) + \log_{11} 3 = 2$

32. $\log_8(x - 3) = \log_8 6x - 1$

33. $\log_5(x + 1) - \log_5 8 = \log_5 x$

34. $\log_{12}(x + 2) + \log_{12} 3 = 1$

35. $\log_2(x - 2) + \log_2(x + 5) = 3$

36. $\log_5(x + 1) + \log_5(x - 3) = 1$

37. $\log(2t + 3) + \log(4t - 1) = 2 \log 3$

38. $\log(2t + 4) - \log(3t + 1) = \log 6$

Solve each equation. Leave your answers in exact form. Do not approximate.

39. $3^x = 14$

40. $5^{x+3} = 130$

41. $e^{2x-1} = 100$

42. $e^{2x} = 30.6$

Solve each equation. Round your answers to the nearest ten-thousandth.

43. $2^{3x+1} = 5^x$

44. $3^{x+1} = 7$

45. $e^{3x-4} = 20$

46. $(1.03)^x = 20$

Compound Interest For Exercises 47–50, use $A = P(1 + r)^t$, the formula for exercises involving interest that is compounded annually.

47. How long will it take Frances to double the money in her account if the interest rate is 8% compounded annually? (Round your answer to nearest year.)

48. How much money would Chou Lou have after 4 years if he invested $5000 at 6% compounded annually?

49. Melinda invested $12,000 at 7% compounded annually. How many years will it take for it to amount to $20,000? (Round your answer to the nearest year.)

50. Robert invested $3500 at 5% compounded annually. His brother invested $3500 at 6% compounded annually. How many years will it take for Robert's amount to be $500 less than his brother's amount? (Round your answer to the nearest year.)

Populations The growth of the world's population can be described by the equation $A = A_0 e^{rt}$, where time t is measured in years, A_0 is the population of the world at time $t = 0$, r is the annual growth rate, and A is the population at time t. Use this information to solve Exercises 51–54. Round your answers to the nearest whole year.

51. How long will it take a population of seven billion to increase to sixteen billion if $r = 2\%$ per year?

52. How long will it take a population of six billion to increase to ten billion if $r = 2\%$ per year?

53. The number of moose in northern Maine is increasing at a rate of 3% per year. It is estimated in one county that there are now 2000 moose. If the growth rate remains unchanged, how many years will it be until there are 2600 moose in that county?

54. A town is growing at the rate of 8% per year. How long will it take the town to grow from 40,000 to 95,000 in population?

55. ***Gas Volume*** The work W done by a volume of gas expanding at a constant temperature from volume V_0 to volume V_1 is given by $W = p_0 V_0 \ln\left(\frac{V_1}{V_0}\right)$, where p_0 is the pressure at volume V_0.
(a) Find W when $p_0 = 40$ pounds per cubic inch, $V_0 = 15$ cubic inches, and $V_1 = 24$ cubic inches.
(b) If the amount of work is 100 pounds per cubic inch, $V_0 = 8$ cubic inches, and $V_1 = 40$ cubic inches, find p_0.

56. ***Earthquakes*** An earthquake's magnitude is given by $M = \log\left(\frac{I}{I_0}\right)$, where I is the intensity of the earthquake and I_0 is the minimum measurable intensity. The 1964 earthquake in Anchorage, Alaska, had a magnitude of 8.4. The 1975 earthquake in Turkey had a magnitude of 6.7. How much more energy was released from the Alaska earthquake than from the Turkey earthquake? *Source:* National Oceanic and Atmospheric Administration.

Remember to use your Chapter Test Prep Video CD to see the worked-out solutions to the test problems you want to review.

1. Graph $f(x) = 3^{x+1}$.

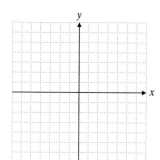

2. Graph $f(x) = \log_2 x$.

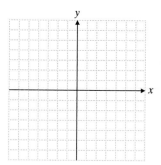

1. _____

2. _____

3. _____

4. _____

5. _____

6. _____

7. _____

8. _____

9. _____

10. _____

11. _____

12. _____

13. _____

14. _____

15. _____

16. _____

17. _____

3. Solve $4^{x+3} = 64$.

In Exercises 4 and 5, solve for the variable.

4. $\log_w 125 = 3$

5. $\log_8 x = -2$

6. Write as a single log: $2 \log_7 x + \log_7 y - \log_7 4$

Evaluate using a calculator. Round your answers to the nearest ten-thousandth.

7. $\ln 5.99$

8. $\log 23.6$

9. $\log_3 1.62$

Use a scientific calculator to approximate x.

10. $\log x = 3.7284$

11. $\ln x = 0.14$

Solve the equation and check your solutions for Exercises 12 and 13.

12. $\log_8(x + 3) - \log_8 2x = \log_8 4$

13. $\log_8 2x + \log_8 6 = 2$

14. Solve the equation. Leave your answer in exact form. Do not approximate. $e^{5x-3} = 57$

15. Solve $5^{3x+6} = 17$. Approximate your answer to the nearest ten-thousandth.

16. How much money will Henry have if he invests $2000 for 5 years at 8% annual interest compounded annually?

17. How long will it take for Barb to double her money if she invests it at 5% compounded annually? Round to the nearest whole year.

Approximately one-half of this test covers the content of Chapters 1–10. The remainder covers the content of Chapter 11.

1. Evaluate $2(-3) + 12 \div (-2) + 3\sqrt{36}$.

2. Solve for x: $3mx = 5(mx - y) + 1$ **3.** Graph $y = -\dfrac{2}{3}x + 4$.

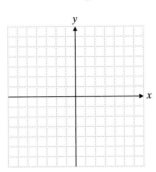

4. Factor $5ax + 5ay - 7wx - 7wy$.

5. Solve the system for (x, y, z).

$$3x - y + z = 6$$
$$2x - y + 2z = 7$$
$$x + y + z = 2$$

6. Simplify: $\left(3\sqrt{7} - \sqrt{3}\right)\left(\sqrt{7} + \sqrt{3}\right)$

7. Solve $x^4 - 5x^2 - 6 = 0$. Express imaginary solutions in i notation.

8. Solve for x and y.

$$2x - y = 4$$
$$4x - y^2 = 0$$

9. Solve $2x - 3 = \sqrt{7x - 3}$.

10. Rationalize the denominator: $\dfrac{3x}{\sqrt{6}}$

11. Graph $f(x) = 2^{3-2x}$.

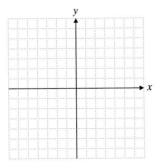

1. _____

2. _____

3. _____

4. _____

5. _____

6. _____

7. _____

8. _____

9. _____

10. _____

11. _____

Solve for the variable.

12. $\log_x\left(\dfrac{1}{64}\right) = 3$

12. _____

13. $5^{2x-1} = 25$

13. _____

Evaluate using a calculator or a table.

14. $\log 7.67$

14. _____

15. Find x if $\log x = 1.8209$.

15. _____

16. $\log_3 7$

16. _____

17. Find x if $\ln x = 1.9638$.

17. _____

Solve the equation.

18. $\log_9 x = 1 - \log_9(x - 8)$

18. _____

19. $\log_5 x = \log_5 2 + \log_5(x^2 - 3)$

19. _____

Solve for x.

20. $3^{x+2} = 5$
Approximate to the nearest thousandth.

20. _____

21. $33 = 66\, e^{2x}$
Leave your answer exact. Do not approximate.

21. _____

22. How much money will Frank and Linda have in 4 years if they invest $3000 at 9% compounded annually? Round to the nearest cent.

22. _____

Review the content areas of Chapters 1–11. Then try to solve the exercises in this Practice Final Examination.

Chapter 1

1. Evaluate
$(4 - 3)^2 + \sqrt{9} \div (-3) + 4.$

2. Write using scientific notation:
36,250,000

3. Simplify:
$3a + 6b - a + 5ab + 3a^2 + b$

4. Simplify.
$3[2x - 5(x + y)]$

5. $F = \dfrac{9}{5}C + 32.$ Find F when $C = -35.$

Chapter 2

6. Solve for y, $\dfrac{1}{3}y - 4 = \dfrac{1}{2}y + 1.$

7. Solve for b, $A = \dfrac{1}{2}a(b + c).$

8. Solve for x, $\left|\dfrac{2}{3}x - 4\right| = 2.$

9. Solve for x.
$2x - 3 < x - 2(3x - 2)$

▲ **10.** A piece of land is rectangular and has a perimeter of 1760 meters. The length is 200 meters less than twice the width. Find the dimensions of the land.

11. A man invested $4000, part at 12% interest and part at 14% interest. After 1 year he had earned $508 in interest. How much was invested at each interest rate?

12. Find the value of x that satisfies the given conditions.
$x + 5 \le -4$ or $2 - 7x \le 16$

13. Solve the inequality:
$|2x - 5| < 10$

Chapter 3

14. Find the intercepts and then graph the line $7x - 2y = -14.$

15. Graph the region $3x - 4y \le 6.$

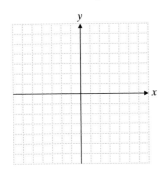

16. Find the slope of the line passing through $(1, 5)$ and $(-2, -3).$

17. Write the equation in standard form of the line that is parallel to $3x + 2y = 8$ and passes through $(-1, 4).$

Given the function defined by $f(x) = 3x^2 - 4x - 3$, find the following.

18. $f(3)$

19. $f(-2)$

1. _____

2. _____

3. _____

4. _____

5. _____

6. _____

7. _____

8. _____

9. _____

10. _____

11. _____

12. _____

13. _____

14. _____

15. _____

16. _____

17. _____

18. _____

19. _____

20. _____

21. _____

22. _____

23. _____

24. _____

25. _____

26. _____

27. _____

28. _____

29. _____

30. _____

31. _____

32. _____

33. _____

34. _____

35. _____

36. _____

20. Graph the function $f(x) = |2x - 4|$.

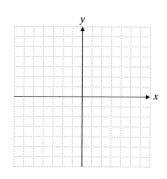

Chapter 4

21. Solve for x and y:

$$2x + y = 15$$
$$3x - 2y = 5$$

22. Solve for x and y.

$$4x - 3y = 12$$
$$3x - 4y = 2$$

23. Solve for x, y, and z.

$$2x + 3y - z = 16$$
$$x - y + 3z = -9$$
$$5x + 2y - z = 15$$

24. Solve for x and y:

$$3x - 2y = 7$$
$$-9x + 6y = 2$$

25. Reynoso sold 15 tickets for the drama club's play. Adult tickets cost $8 and children's tickets cost $3. Reynoso collected $100. How many of each kind of ticket did he sell?

26. Graph the region.

$$3y \geq 8x - 12$$
$$2x + 3y \leq -6$$

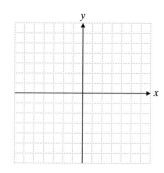

Chapter 5

27. Multiply and simplify your answer.
$$(3x - 2)(2x^2 - 4x + 3)$$

28. Divide.
$$(25x^3 + 9x + 2) \div (5x + 1)$$

Factor the following completely.

29. $8x^3 - 27$

30. $x^3 + 2x^2 - 4x - 8$

31. $2x^3 + 15x^2 - 8x$

32. Solve for x, $x^2 + 15x + 54 = 0$.

Chapter 6

Simplify the following.

33. $\dfrac{9x^3 - x}{3x^2 - 8x - 3}$

34. $\dfrac{x^2 - 9}{2x^2 + 7x + 3} \div \dfrac{x^2 - 3x}{2x^2 + 11x + 5}$

35. $\dfrac{3x}{x + 5} - \dfrac{2}{x^2 + 7x + 10}$

36. $\dfrac{\dfrac{3}{2x} - 1}{\dfrac{5}{2} + \dfrac{1}{x}}$

37. Solve for x, $\dfrac{x - 1}{x^2 - 4} = \dfrac{2}{x + 2} + \dfrac{4}{x - 2}$.

Chapter 7

38. Evaluate: $16^{\frac{3}{2}}$

39. Simplify: $\sqrt{44a^4b^7c}$

40. Combine like terms.
$5\sqrt{2} - 3\sqrt{50} + 4\sqrt{98}$

41. Rationalize the denominator:
$$\dfrac{5}{\sqrt{7} - 2}$$

42. Simplify and add together.
$i^3 + \sqrt{-25} + \sqrt{-16}$

43. Solve for x and check your solutions.
$$\sqrt{x + 7} = x + 5$$

44. If y varies directly with the square of x and $y = 15$ when $x = 2$, what will y be when $x = 3$?

Chapter 8

45. Solve for x, $5x(x + 1) = 1 + 6x$.

46. Solve for x, $5x^2 - 9x = -12x$.

47. Solve for x, $x^{2/3} + 5x^{1/3} - 14 = 0$.

48. Solve $3x^2 - 11x - 4 \geq 0$.

49. Graph the quadratic function $f(x) = -x^2 - 4x + 5$. Label the vertex and the intercepts.

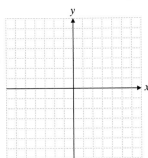

▲ **50.** The area of a rectangle is 52 square centimeters. The length of the rectangle is 1 centimeter longer than 3 times its width. Find the dimensions of the rectangle.

Chapter 9

51. Place the equation of the circle in standard form. Find its center and radius.
$$x^2 + y^2 + 6x - 4y = -9$$

Identify and graph.

52. $\dfrac{x^2}{16} + \dfrac{y^2}{25} = 1$

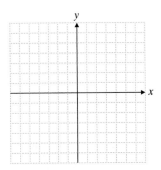

53. $\dfrac{x^2}{4} - \dfrac{y^2}{9} = 1$

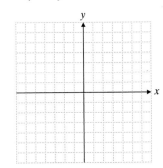

37. _____

38. _____

39. _____

40. _____

41. _____

42. _____

43. _____

44. _____

45. _____

46. _____

47. _____

48. _____

49. _____

50. _____

51. _____

52. _____

53. _____

54. _____

54. $x = (y - 3)^2 + 5$

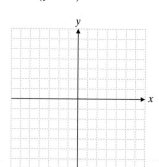

55. Solve the system of equations.
$$x^2 + y^2 = 16$$
$$x^2 - y = 4$$

55. _____

56. _____

Chapter 10

56. Let $f(x) = 3x^2 - 2x + 5$.
(a) Find $f(-1)$.
(b) Find $f(a)$.
(c) Find $f(a + 2)$.

57. If $f(x) = 5x^2 - 3$ and $g(x) = -4x - 2$, find $f[g(x)]$.

57. _____

58. _____

58. If $f(x) = \dfrac{1}{2}x - 7$, find $f^{-1}(x)$.

59. _____

59. Is M a one-to-one function? $M = \left\{ (3, 7), (2, 8), \left(7, \dfrac{1}{2} \right), (-3, 7) \right\}$

60. _____

Chapter 11

60. Graph $f(x) = 2^{1-x}$. Plot three points.

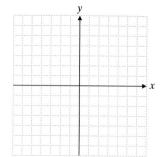

61. _____

62. _____

63. _____

Solve for the variable.

61. $\log_6 1 = x$

62. $\log_4 (3x + 1) = 3$

64. _____

63. $\log_{10} 0.01 = y$

64. $\log_2 6 + \log_2 x = 4 + \log_2(x - 5)$

Intermediate Algebra Glossary

Absolute value inequalities (2.8) Inequalities that contain at least one absolute value expression.

Absolute value of a number (1.2) The distance between the number and 0 on the number line. The absolute value of a number x is written as $|x|$. The absolute value can be determined by

$$|x| = \begin{cases} x, & \text{if } x \geq 0 \\ -x, & \text{if } x < 0 \end{cases}$$

Algebraic expression (1.5) A collection of numerical values, variables, and operation symbols. $\sqrt{5xy}$ and $3x - 6y + 3yz$ are algebraic expressions.

Algebraic fraction (6.1) An expression of the form $\dfrac{P}{Q}$, where P and Q are polynomials and Q is not zero. Algebraic fractions are also called *rational expressions*. For example, $\dfrac{x + 3}{x - 4}$ and $\dfrac{5x^2 + 1}{6x^3 - 5x}$ are algebraic fractions.

Approximate value (1.3) A value that is not exact. The approximate value of $\sqrt{3}$, correct to the nearest tenth, is 1.7. The symbol $\approx$ is used to indicate "is approximately equal to." We write $\sqrt{3} \approx 1.7$.

Associative property of addition (1.1) For all real numbers a, b, and c: $a + (b + c) = (a + b) + c$.

Associative property of multiplication (1.1) For all real numbers a, b, and c: $a(bc) = (ab)c$.

Asymptote (9.4) A line that a curve continues to approach but never actually touches. Often an asymptote is a helpful reference in making a sketch of a curve, such as a hyperbola.

Augmented matrix (Appendix C) A matrix derived from a linear system of equations. It consists of the coefficients of each variable in a linear system and the constants. The augmented matrix of the system $\begin{array}{l} -3x + 5y = -22 \\ 2x - y = 10 \end{array}$ is the matrix $\left[\begin{array}{cc|c} -3 & 5 & -22 \\ 2 & -1 & 10 \end{array}\right]$. Each row of the augmented matrix represents an equation of the system.

Axis of symmetry of a parabola (9.2) A line passing through the focus and the vertex of a parabola, about which the two sides of the parabola are symmetric. See the sketch.

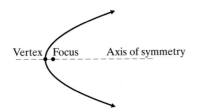

Vertex / Focus Axis of symmetry

Base (1.3) The number or variable that is raised to a power. In the expression 2^3, the number 2 is the base.

Base of an exponential function (11.1) The number b in the function $f(x) = b^x$.

Binomial (5.1) A polynomial of two terms. For example, $z^2 - 9$ is a binomial.

Cartesian coordinate system (3.1) Another name for the rectangular coordinate system named after its inventor, René Descartes.

Circle (1.6) A geometric figure that consists of a collection of points that are of equal distance from a fixed point called the *center*.

Center

Circumference of a circle (1.6) The distance around a circle. The circumference of a circle is given by the formulas $C = \pi d$ and $C = 2\pi r$, where d is the diameter of the circle and r is the radius of the circle.

Closure property of addition (1.1) For all real numbers a and b: the sum $a + b$ is a real number.

Closure property of multiplication (1.1) For all real numbers a and b: the product ab is a real number.

Coefficient (1.5) Any factor or group of factors in a term. In the term $8xy$, the coefficient of xy is 8. However, the coefficient of x is $8y$. In the term $abcd$, the coefficient of $abcd$ is 1.

Collect like terms (1.5) The process of adding and subtracting like terms. If we collect like terms in the expression $5x - 8y - 7x - 12y$, we obtain $-2x - 20y$.

Combined variation (7.7) When y varies directly with x and z and inversely with d^2, written $y = \dfrac{kxz}{d^2}$, where k is the constant of variation.

Common denominator (1.2) The same number or polynomial in the denominator of two fractions. The fractions $\frac{4}{13}$ and $\frac{7}{13}$ have a common denominator of 13.

Common logarithm (11.4) The common logarithm of a number x is given by $\log x = \log_{10} x$ for all $x > 0$. A common logarithm is a logarithm using base 10.

Commutative property of addition (1.1) For all real numbers a and b: $a + b = b + a$.

Commutative property of multiplication (1.1) For all real numbers a and b: $ab = ba$.

Complex fraction (also called a Complex rational expression) (6.3) A fraction made up of polynomials or numerical values in which the numerator or the denominator contains at least one fraction. Examples of complex fractions are

$$\frac{\frac{1}{3}+\frac{1}{5}}{\frac{2}{7}} \quad \text{and} \quad \frac{\frac{1}{x}+3}{2+\frac{5}{x}}.$$

Complex number (7.6) A number that can be written in the form $a + bi$, where a and b are real numbers and $i = \sqrt{-1}$.

Compound inequalities (2.7) Two inequality statements connected together by the word *and* or by the word *or*.

Conjugate of a binomial with radicals (7.4) The expressions $a\sqrt{x} + b\sqrt{y}$ and $a\sqrt{x} - b\sqrt{y}$. The conjugate of $2\sqrt{3} + 5\sqrt{2}$ is $2\sqrt{3} - 5\sqrt{2}$. The conjugate of $4 - \sqrt{x}$ is $4 + \sqrt{x}$.

Conjugate of a complex number (7.6) The expressions $a + bi$ and $a - bi$. The conjugate of $5 + 2i$ is $5 - 2i$. The conjugate of $7 - 3i$ is $7 + 3i$.

Coordinates of a point (3.1) An ordered pair of numbers (x, y) that specifies the location of a point on a rectangular coordinate system.

Counting numbers (1.1) The counting numbers are the natural numbers. They are the numbers in the infinite set

$$\{1, 2, 3, 4, 5, 6, 7, \dots\}.$$

Critical points of a quadratic inequality (8.6) In a quadratic inequality of the form $ax^2 + bx + c > 0$ or $ax^2 + bx + c < 0$, those points where $ax^2 + bx + c = 0$.

The degree of a polynomial (5.1) The degree of the highest-degree term in the polynomial. The polynomial $5x^3 + 4x^2 - 3x + 12$ is of degree 3.

The degree of a term (5.1) The sum of the exponents of the term's variables. The term $5x^2y^2$ is of degree 4.

Denominator (6.1) The bottom expression in a fraction. The denominator of $\frac{5}{11}$ is 11. The denominator of $\frac{x-7}{x+8}$ is $x + 8$.

Descending order for a polynomial (5.1) A polynomial is written in descending order if the term of the highest degree is first, the term of the next-to-highest degree is second, and so on, with each succeeding term of less degree. The polynomial $5y^4 - 3y^3 + 7y^2 + 8y - 12$ is in descending order.

Determinant (Appendix B) A square array of numbers written between vertical lines. For example $\begin{vmatrix} 1 & 5 \\ 2 & 4 \end{vmatrix}$ is a

2×2 determinant. It is also called a *second-order determinant*. $\begin{vmatrix} 1 & 7 & 8 \\ 2 & -5 & -1 \\ -3 & 6 & 9 \end{vmatrix}$ is a 3×3 determinant. It is also called a *third-order determinant*.

Different signs (1.2) When one number is positive and one number is negative, the two numbers are said to have different signs. The numbers 5 and -9 have different signs.

Direct variation (7.7) When a variable y varies directly with x, written $y = kx$, where k represents some real number that will stay the same over a range of exercises. This value k is called the *constant of variation*.

Discriminant of a quadratic equation (8.2) In the equation $ax^2 + bx + c = 0$, where $a \neq 0$, the expression $b^2 - 4ac$. It can be used to determine the nature of the roots of the quadratic equation. If the discriminant is *positive*, there are two rational or irrational roots. The two roots will be rational only if the discriminant is a perfect square. If the discriminant is *zero*, there is only one rational root. If the discriminant is *negative*, there are two complex roots.

Distance between two points (9.1) The distance between point (x_1, y_1) and point (x_2, y_2) is given by the formula $d = \sqrt{(x_2 - x_1)^2 + (y_2 - y_1)^2}$.

Distributive property of multiplication over addition (1.1) For any real numbers $a, b,$ and c: $a(b + c) = ab + ac$.

Dividend (5.2) The expression that is being divided by another. In $12 \div 4 = 3$, the dividend is 12. In $x - 5)\overline{5x^2 + 10x - 3}$, the dividend is $5x^2 + 10x - 3$.

Divisor (5.2) The expression that is divided into another. In $12 \div 4 = 3$, the divisor is 4. In $x + 3)\overline{2x^2 - 5x - 14}$, the divisor is $x + 3$.

Domain of a relation or a function (3.5) When the ordered pairs of a relation or a function are listed, all the different first items of each pair.

e (11.1) An irrational number that can be approximated by the value 2.7183.

Elements (1.1) The objects that are in a set.

Ellipse (9.3) The set of points in a plane such that for each point in the set, the sum of its distances to two fixed points is constant. Each of the fixed points is called a *focus*. Each of the following graphs is an ellipse.

Equation (2.1) A mathematical statement that two quantities are equal.

Equilateral hyperbola (9.4) A hyperbola for which $a = b$ in the equation of the hyperbola.

Equivalent equations (2.1) Equations that have the same solution(s).

Even integers (1.3) Integers that are exactly divisible by 2, such as $\ldots, -4, -2, 0, 2, 4, 6, \ldots$.

Exponent (1.3) The number that indicates the power of a base. If the number is a positive integer, it tells us how many factors of the base occur. In the expression 2^3, the exponent is 3. The number 3 tells us that there are 3 factors, each of which is 2 since $2^3 = 2 \cdot 2 \cdot 2$. If an exponent is negative, use the property that $x^{-n} = \dfrac{1}{x^n}$. If an exponent is zero, use the property that $x^0 = 1$, where $x \neq 0$.

Exponential function (11.1) $f(x) = b^x$, where $b > 0, b \neq 1$, and x is any real number.

Expression (1.3) Any combination of mathematical operation symbols with numbers or variables or both. Examples of mathematical expressions are $2x + 3y - 6z$ and $\sqrt{7xyz}$.

Extraneous solution to an equation (6.4) A correctly obtained potential solution to an equation that when substituted back into the original equation does not yield a true statement. For example, $x = 2$ is an extraneous solution to the equation

$$\frac{x}{x-2} - 4 = \frac{2}{x-2}$$

An extraneous solution is also called an *extraneous root*.

Factor (1.5 and 5.4) Each of the two or more numbers, variables, or algebraic expressions that is multiplied. In the expression $5st$, the factors are 5, s, and t. In the expression $(x - 6)(x + 2)$, the factors are $(x - 6)$ and $(x + 2)$.

First-degree equation (2.1) A mathematical equation such as $2x - 8 = 4y + 9$ or $7x = 21$ in which each variable has an exponent of 1. It is also called a *linear equation*.

First-degree equation in one unknown (2.1) An equation such as $x = 5 - 3x$ or $12x - 3(x + 5) = 22$ in which only one kind of variable appears and that variable has an exponent of 1. It is also called a *linear equation in one variable*.

Focus point of a parabola (9.2) The focus point of a parabola has many properties. For example, the focus point of a parabolic mirror is the point to which all incoming light rays that are parallel to the axis of symmetry will collect. A parabola is a set of points that is the same distance from a fixed line called the *directrix* and a fixed point. This fixed point is the focus.

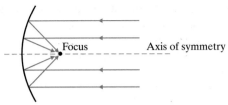

Formula (1.6) A rule for finding the value of a variable when the values of other variables in the expression are known. For example, the formula for finding the Fahrenheit temperature when the Celsius temperature is known is $F = 1.8C + 32$.

Fractional equation (6.4) An equation that contains a rational expression. Examples of fractional equations are $\dfrac{x}{3} + \dfrac{x}{4} = 7$ and $\dfrac{2}{3x-3} + \dfrac{1}{x-1} = \dfrac{-5}{12}$.

Function (3.5) A relation in which no different ordered pairs have the same first coordinate.

Graph of a function (3.5, 10.2) A graph in which a vertical line will never cross in more than one place. The following sketches represent the graphs of functions.

Graph of a linear inequality in two variables (3.4) A shaded region in two-dimensional space. It may or may not include the boundary line. If the line is included, the sketch shows a solid line. If it is not included, the sketch shows a dashed line. Two sketches follow.

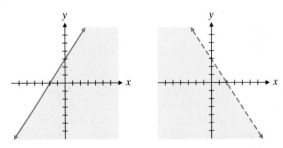

Graph of a one-to-one function (10.4) A graph of a function with the additional property that a horizontal line will never cross the graph in more than one place. The following sketches represent the graphs of one-to-one functions.

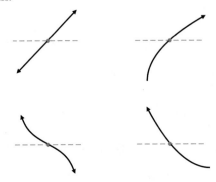

Greater than or equal to symbol (2.6) The $\geq$ symbol.

Greater than symbol (1.2, 2.6) The $>$ symbol. $5 > 3$ is read, "5 is greater than 3."

Greatest common factor of a polynomial (5.4) A common factor of each term of the polynomial that has the largest possible numerical coefficient and the largest possible exponent for each variable. For example, the greatest common factor of $50x^4y^5 - 25x^3y^4 + 75x^5y^6$ is $25x^3y^4$.

Higher-order equations (8.3) Equations of degree 3 or higher. Examples of higher-order equations are $x^4 - 29x^2 + 100 = 0$ and $x^3 + 3x^2 - 4x - 12 = 0$.

Higher-order roots (7.2) Cube roots, fourth roots, and roots with an index greater than 2.

Horizontal line (3.1) A straight line that is parallel to the x-axis. A horizontal line has a slope of zero. The equation of any horizontal line can be written in the form $y = b$, where b is a constant. A sketch is shown.

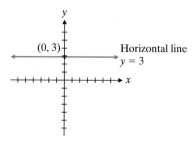

Horizontal parabolas (9.2) Parabolas that open to the right or to the left. The following graphs represent horizontal parabolas.

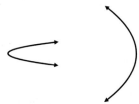

Hyperbola (9.4) The set of points in a plane such that for each point in the set, the absolute value of the difference of its distances to two fixed points is constant. Each of these fixed points is called a *focus*. The following sketches represent graphs of hyperbolas.

Hypotenuse of a right triangle (8.4) The side opposite the right angle in any right triangle. The hypotenuse is always the longest side of a right triangle. In the following sketch the hypotenuse is side c.

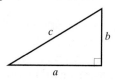

Identity property for addition (1.1) For any real number a, $a + 0 = a = 0 + a$.

Identity property for multiplication (1.1) For any real number a, $a(1) = a = 1(a)$.

Imaginary number (7.6) i, defined as $i = \sqrt{-1}$ and $i^2 = -1$.

Inconsistent system of equations (4.1) A system of equations for which no solution is possible.

Index of a radical (7.2) Indicates what type of a root is being taken. The index of a cube root is 3. In $\sqrt[3]{x}$, the 3 is the index of the radical. In $\sqrt[4]{y}$, the index is 4. The index of a square root is 2, but the index is not written in the square root symbol, as shown: $\sqrt{x}$.

Inequality (2.6) A mathematical statement expressing an order relationship. The following are inequalities:

$$x < 3, \qquad x \geq 4.5, \qquad 2x + 3 \leq 5x - 7,$$
$$x + 2y < 8, \qquad 2x^2 - 3x > 0$$

Infinite set (1.1) A set that has no end to the number of elements that are within it. An infinite set is often indicated by placing an ellipsis (...) after listing some of the elements of the set.

Integers (1.1) The numbers in the infinite set

$$\{ \ldots, -3, -2, -1, 0, 1, 2, 3, \ldots \}.$$

Interest (2.5) The charge made for borrowing money or the income received from investing money. Simple interest is calculated by the formula $I = prt$, where p is the principal that is borrowed, r is the rate of interest, and t is the amount of time the money is borrowed.

Inverse function of a one-to-one function (10.4) That function obtained by interchanging the first and second coordinates in each ordered pair of the function.

Inverse property of addition (1.1) For any real number a, $a + (-a) = 0 = (-a) + a$.

Inverse property of multiplication (1.1) For any real number $a \neq 0$, $a\left(\dfrac{1}{a}\right) = 1 = \left(\dfrac{1}{a}\right)a$.

Inverse variation (7.7) When a variable y varies inversely with x, written $y = \dfrac{k}{x}$, where k is the constant of variation.

Irrational numbers (1.1) Numbers whose decimal forms are nonterminating and nonrepeating. The numbers $\pi, e, \sqrt{2}$, and $1.56832574\ldots$ are irrational numbers.

Joint variation (7.7) When a variable y varies jointly with x and z, written $y = kxz$, where k is the constant of variation.

Least common denominator of algebraic fractions (6.2) A polynomial that is exactly divisible by each denominator. The LCD is the product of all the *different prime factors*. If a factor occurs more than once in a denominator, we must use the highest power of that factor.

For example, the LCD of

$$\frac{5}{2(x+2)(x-3)^2} \quad \text{and} \quad \frac{3}{(x-3)^4}$$

is $2(x+2)(x-3)^4$. The LCD of

$$\frac{5}{(x+2)(x-3)} \quad \text{and} \quad \frac{7}{(x-3)(x+4)}$$

is $(x+2)(x-3)(x+4)$.

Least common denominator of numerical fractions (2.1) The smallest whole number that is exactly divisible by all the denominators of a group of fractions. The least common denominator (LCD) of $\frac{1}{7}$, $\frac{9}{21}$, and $\frac{3}{14}$ is 42. The number 42 is the smallest number that can be exactly divided by 7, 21, and 14. The least common denominator is sometimes called the *lowest common denominator*.

Leg of a right triangle (8.4) One of the two shorter sides of a right triangle. In the following sketch, sides a and b are the legs of the right triangle.

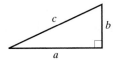

Less than or equal to symbol (2.6) The $\leq$ symbol.

Less than symbol (1.2) The $<$ symbol. $2 < 8$ is read, "2 is less than 8."

Like terms (1.5) Terms that have identical variables and identical exponents. In the mathematical expression $5x - 8syz + 7x + 15syz$, the terms $5x$ and $7x$ are like terms, and the terms $-8syz$ and $15syz$ are like terms.

Linear equation (2.1) A mathematical equation such as $3x + 7 = 5x - 2$ or $5x + 7y = 9$, in which each variable has an exponent of 1.

Linear inequality (2.6) An inequality statement in which each variable has an exponent of 1 and no variables are in the denominator. Some examples of linear inequalities are

$$2x + 3 > 5x - 6, \quad y < 2x + 1, \quad \text{and} \quad x < 8.$$

Literal equation (2.2) An equation that has other variables in it besides the variable for which we wish to solve. $I = prt$, $7x + 3y - 6z = 12$, and $P = 2w + 2l$ are examples of literal equations.

Logarithm (11.2) For a positive number x, the power to which the base b must be raised to produce x. That is, $y = \log_b x$ is the same as $x = b^y$, where $b > 0$ and $b \neq 1$. A logarithm is an exponent.

Logarithmic equation (11.2) An equation that contains at least one logarithm.

Magnitude of an earthquake (11.5) The magnitude of an earthquake is measured by the formula $M = \log\left(\frac{I}{I_0}\right)$, where I is the intensity of the earthquake and I_0 is the minimum measurable intensity.

Matrix (Appendix C) A rectangular array of numbers arranged in rows and columns. We use the symbol [ ] to indicate a matrix. The matrix $\begin{bmatrix} 3 & 4 & 5 \\ 6 & 7 & 8 \end{bmatrix}$ has two rows and three columns and is called a 2×3 *matrix*.

Minor of an element of a third-order determinant (Appendix B) The second-order determinant that remains after we delete the row and column in which the element appears. The minor of the element 6 in the determinant

$$\begin{vmatrix} 1 & 2 & 3 \\ 7 & 6 & 8 \\ -3 & 5 & 9 \end{vmatrix} \text{ is the second-order determinant } \begin{vmatrix} 1 & 3 \\ -3 & 9 \end{vmatrix}.$$

Monomial (5.1) A polynomial of one term. For example, $3a$ is a monomial.

Natural logarithm (11.4) For a number x, $\ln x = \log_e x$ for all $x > 0$. A natural logarithm is a logarithm using base e.

Negative integers (1.1) The numbers in the infinite set

$$\{-1, -2, -3, -4, -5, -6, -7, \dots \}.$$

Nonlinear system of equations (9.5) A system of equations in which at least one equation is not a linear equation.

Nonzero (1.4) A nonzero value is a value other than zero. If we say that the variable x is nonzero, we mean that x cannot have the value of zero.

Numerator (6.1) The top expression in a fraction. The numerator of $\frac{3}{19}$ is 3. The numerator of $\frac{x+5}{x^2+25}$ is $x + 5$.

Numerical coefficient (1.5) The numerical value multiplied by the variables in a term. The numerical coefficient of $-8xyw$ is -8. The numerical coefficient of abc is 1.

Odd integer (1.3) Integers that are not exactly divisible by 2, such as $\dots, -3, -1, 1, 3, 5, 7, \dots$.

One-to-one function (10.4) A function in which no two different ordered pairs have the same second coordinate.

Opposite of a number (1.2) That number with the same absolute value but a different sign. The opposite of -7 is 7. The opposite of 13 is -13.

Ordered pair (3.1) A pair of numbers represented in a specified order. An ordered pair is used to identify the location of a point. Every point on a rectangular coordinate system can be represented by an ordered pair (x, y).

Origin (3.1) The point determined by the intersection of the x-axis and the y-axis. It has the coordinates $(0,0)$.

Parabola (9.2) The set of points that is the same distance from some fixed line (called the *directrix*) and some fixed point (called the *focus*) that is not on the line. The graph of any equation of the form $y = ax^2 + bx + c$ or $x = ay^2 + by + c$, where a, b, and c are real numbers and $a \neq 0$, is a parabola. Some examples of the graphs of parabolas are shown.

Parallel lines (3.2) Two straight lines that never intersect. Parallel lines have the same slope.

Parallelogram (1.6) A four-sided geometric figure with opposite sides parallel. The opposite sides of a parallelogram are equal.

Percent (1.6) Hundredths or "per one hundred"; indicated by the % symbol. Thirty-seven hundredths means thirty-seven percent: $\frac{37}{100} = 37\%$.

Perfect square (1.3) If x is an integer and a is a positive real number such that $a = x^2$, then x is a square root of a and a is a perfect square. Some numbers that are perfect squares are $1, 4, 9, 16, 25, 36, 49, 64, 81,$ and 100.

Perfect square trinomials (5.6) Trinomials of the form $a^2 + 2ab + b^2$ or $a^2 - 2ab + b^2$.

Perpendicular lines (3.2) Two straight lines that meet at a 90-degree angle. If two nonvertical lines have slopes m_1 and m_2, and m_1 and $m_2 \neq 0$, then the lines are perpendicular if and only if $m_1 = -\dfrac{1}{m_2}$. pH of a solution (11.2) Defined by the equation pH $= -\log_{10}(\text{H}^+)$, where H^+ is the concentration of the hydrogen ion in the solution. The solution is an acid when the pH is less than 7 and a base when the pH is greater than 7.

Pi (1.6) An irrational number, denoted by the symbol π, which is approximately equal to 3.141592654. In most cases, 3.14 can be used as a sufficiently accurate approximation for π.

Point–slope form of the equation of a straight line (3.3) For a straight line passing through the point (x_1, y_1) and having slope m, $y - y_1 = m(x - x_1)$.

Polynomials (1.5 and 5.1) Variable expressions that contain terms with nonnegative integer exponents. A polynomial must contain no division by a variable. Some examples of polynomials are $5y^2 - 8y + 3, -12xy, 12a - 14b,$ and $7x$.

Positive integers (1.1) The numbers in the infinite set $\{1, 2, 3, 4, 5, 6, 7, \dots\}$. The positive integers are the natural numbers.

Power (1.3) When a number is raised to a power, the number's exponent is that power. Thus, two to the third power means 2^3. The power is the exponent, which is 3. In the expression x^5, we say, "x is raised to the fifth power."

Prime factors of a number (6.2) Those factors of a number that are prime. To write the number 40 as a product of prime factors, we would write $40 = 5 \times 2^3$. To write the number 462 as the product of prime factors, we would write $462 = 2 \times 3 \times 7 \times 11$.

Prime factors of a polynomial (6.2) Those factors of a polynomial that are prime. When a polynomial is completely factored, it is written as a product of prime factors. Thus, the prime factors of $x^4 - 81$ are written as $x^4 - 81 = (x^2 + 9)(x - 3)(x + 3)$.

Prime number (6.2) A positive integer that is greater than 1 and has no factors other than 1 and itself. The first ten prime numbers are $2, 3, 5, 7, 11, 13, 17, 19, 23,$ and 29.

Prime polynomial (5.7) A polynomial that cannot be factored. Examples of prime polynomials are $2x^2 + 100x - 19, 25x^2 + 9,$ and $x^2 - 3x + 5$.

Principal (1.6) In monetary exercises, the original amount of money invested or borrowed.

Principal square root (1.3 and 7.2) The positive square root of a number. The symbol indicating the principal square root is $\sqrt{}$. Thus, $\sqrt{4}$ means to find the principal square root of 4, which is 2.

Proportion (6.5) An equation stating that two ratios are equal. For example, $\dfrac{a}{b} = \dfrac{c}{d}$ is a proportion.

Pythagorean theorem (8.4) In any right triangle, if c is the length of the hypotenuse and a and b are the lengths of the two legs, then $c^2 = a^2 + b^2$.

Quadrants (3.1) The four regions into which the x-axis and the y-axis divide the rectangular coordinate system.

Quadratic equation in standard form (5.8, 8.1) An equation of the form $ax^2 + bx + c = 0$, where a, b, and c are real numbers and $a \neq 0$. A quadratic equation is classified as a second-degree equation.

Square root (1.3 and 7.2) If x is a real number and a is positive real number such that $a = x^2$, then x is a square root of a. One square root of 16 is 4 since $4^2 = 16$. Another square root of 16 is -4 since $(-4)^2 = 16$.

Standard form of the equation of a circle (9.1) For a circle with center at (h, k) and a radius of r,

$$(x - h)^2 + (y - k)^2 = r^2.$$

Standard form of the equation of an ellipse (9.3) For an ellipse with center at the origin,

$$\frac{x^2}{a^2} + \frac{y^2}{b^2} = 1, \quad \text{where } a \text{ and } b > 0.$$

This ellipse has intercepts at $(a, 0)$, $(-a, 0)$, $(0, b)$, and $(0, -b)$.

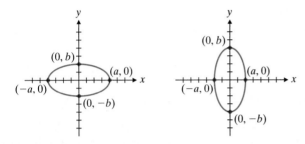

For an ellipse with center at (h, k),

$$\frac{(x - h)^2}{a^2} + \frac{(y - k)^2}{b^2} = 1, \quad \text{where } a \text{ and } b > 0.$$

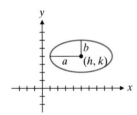

Standard form of the equation of a hyperbola with center at the origin (9.4) For a horizontal hyperbola with center at the origin,

$$\frac{x^2}{a^2} - \frac{y^2}{b^2} = 1, \quad \text{where } a \text{ and } b > 0.$$

The vertices are at $(-a, 0)$ and $(a, 0)$.

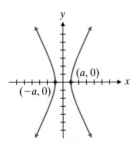

For a vertical hyperbola with center at the origin,

$$\frac{y^2}{b^2} - \frac{x^2}{a^2} = 1, \quad \text{where } a \text{ and } b > 0.$$

The vertices are at $(0, b)$ and $(0, -b)$.

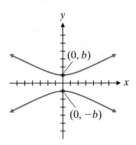

Standard form of the equation of a hyperbola with center at point (h, k) (9.4) For a horizontal hyperbola with center at (h, k),

$$\frac{(x - h)^2}{a^2} - \frac{(y - k)^2}{b^2} = 1, \quad \text{where } a \text{ and } b > 0.$$

The vertices are $(h - a, k)$ and $(h + a, k)$.

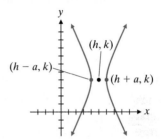

For a vertical hyperbola with center at (h, k),

$$\frac{(y - k)^2}{b^2} - \frac{(x - h)^2}{a^2} = 1, \quad \text{where } a \text{ and } b > 0.$$

The vertices are at $(h, k + b)$ and $(h, k - b)$.

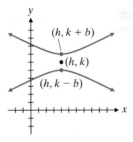

Standard form of the equation of a parabola (9.2) For a vertical parabola with vertex at (h, k),

$$y = a(x - h)^2 + k, \quad \text{where } a \neq 0.$$

For a horizontal parabola with vertex at (h, k),

$$x = a(y - k)^2 + h, \quad \text{where } a \neq 0.$$

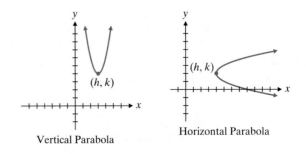

Vertical Parabola   Horizontal Parabola

Standard form of the equation of a straight line (3.3) $Ax + By = C$, where A, B, and C are real numbers.

Standard form of a quadratic equation (5.8) $ax^2 + bx + c = 0$, where a, b, and c are real numbers and $a \neq 0$. A quadratic equation is classified as a second-degree equation.

Subset (1.1) A set whose elements are members of another set. For example, the whole numbers are a subset of the integers.

System of dependent equations (4.1) A system of n linear equations in n variables in which some equations are dependent. It does not have a unique solution but an infinite number of solutions.

System of equations (4.1) A set of two or more equations that must be considered together. The solution is the value for each variable of the system that satisfies each equation.

$$x + 3y = -7$$
$$4x + 3y = -1$$

is a system of two equations in two unknowns. The solution is $(2, -3)$, or the values $x = 2$, $y = -3$.

System of inequalities (4.4) Two or more inequalities in two variables that are considered at one time. The solution is the region that satisfies every inequality at one time. An example of a system of inequalities is

$$y > 2x + 1$$
$$y < \frac{1}{2}x + 2.$$

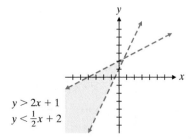

$$y > 2x + 1$$
$$y < \frac{1}{2}x + 2$$

Term (1.5) A real number, a variable, or a product or quotient of numbers and variables. The expression $5xyz$ is one term. The expression $7x + 5y + 6z$ has three terms.

Terminating decimal (1.1) A number in decimal form such as 0.18 or 0.3462, where the number of nonzero digits is finite.

Trapezoid (1.6) A four-sided geometric figure with two parallel sides. The parallel sides are called the *bases of the trapezoid*.

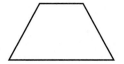

Triangle (1.6) A three-sided geometric figure.

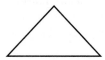

Trinomial (5.1) A polynomial of three terms. For example, $2x^2 + 3x - 4$ is a trinomial.

Unknown (2.1) A variable or constant whose value is not known.

Value of a second-order determinant (Appendix B) For a second-order determinant $\begin{vmatrix} a & b \\ c & d \end{vmatrix}$, $ad - cb$.

Value of a third-order determinant (Appendix B) For a third-order determinant $\begin{vmatrix} a_1 & b_1 & c_1 \\ a_2 & b_2 & c_2 \\ a_3 & b_3 & c_3 \end{vmatrix}$,

$$a_1b_2c_3 + b_1c_2a_3 + c_1a_2b_3 - a_3b_2c_1 - b_3c_2a_1 - c_3a_2b_1.$$

Variable (1.1) A letter used to represent a number.

Vertex of a parabola (9.2) In a vertical parabola, the lowest point on a parabola opening upward or the highest point on a parabola opening downward.

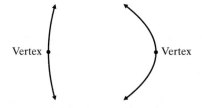

In a horizontal parabola, the leftmost point on a parabola opening to the right or the rightmost point on a parabola opening to the left.

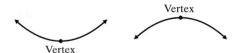

Vertical line (3.1) A straight line that is parallel to the y-axis. The slope of a vertical line is undefined. Therefore, a vertical line has no slope. The equation of a vertical line can be written in the form $x = a$, where a is a constant. A sketch of a vertical line is shown.

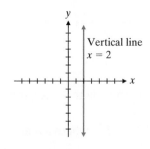

Vertical parabolas (9.2) Parabolas that open upward or downward. The following graphs represent vertical parabolas.

Whole numbers (1.1) The set of numbers containing the natural numbers as well as the number 0. The whole numbers can be written as the infinite set

$$\{0, 1, 2, 3, 4, 5, 6, 7, \ldots\}.$$

x-**intercept (3.1)** The ordered pair $(a, 0)$ in the line that crosses the *x*-axis. The *x*-intercept of the following line is $(5, 0)$.

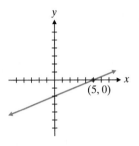

y-**intercept (3.1)** The ordered pair $(0, b)$ in the line that crosses the *y*-axis. The *y*-intercept of the following line is $(0, 4)$.

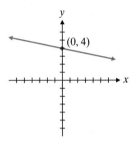

Appendix A Tables

Table A-1: Table of Square Roots

Square root values ending in 000 are exact. All other values are approximate and are rounded to the nearest thousandth.

x	$\sqrt{x}$	x	$\sqrt{x}$	x	$\sqrt{x}$	x	$\sqrt{x}$	x	$\sqrt{x}$
1	1.000	41	6.403	81	9.000	121	11.000	161	12.689
2	1.414	42	6.481	82	9.055	122	11.045	162	12.728
3	1.732	43	6.557	83	9.110	123	11.091	163	12.767
4	2.000	44	6.633	84	9.165	124	11.136	164	12.806
5	2.236	45	6.708	85	9.220	125	11.180	165	12.845
6	2.449	46	6.782	86	9.274	126	11.225	166	12.884
7	2.646	47	6.856	87	9.327	127	11.269	167	12.923
8	2.828	48	6.928	88	9.381	128	11.314	168	12.961
9	3.000	49	7.000	89	9.434	129	11.358	169	13.000
10	3.162	50	7.071	90	9.487	130	11.402	170	13.038
11	3.317	51	7.141	91	9.539	131	11.446	171	13.077
12	3.464	52	7.211	92	9.592	132	11.489	172	13.115
13	3.606	53	7.280	93	9.644	133	11.533	173	13.153
14	3.742	54	7.348	94	9.695	134	11.576	174	13.191
15	3.873	55	7.416	95	9.747	135	11.619	175	13.229
16	4.000	56	7.483	96	9.798	136	11.662	176	13.266
17	4.123	57	7.550	97	9.849	137	11.705	177	13.304
18	4.243	58	7.616	98	9.899	138	11.747	178	13.342
19	4.359	59	7.681	99	9.950	139	11.790	179	13.379
20	4.472	60	7.746	100	10.000	140	11.832	180	13.416
21	4.583	61	7.810	101	10.050	141	11.874	181	13.454
22	4.690	62	7.874	102	10.100	142	11.916	182	13.491
23	4.796	63	7.937	103	10.149	143	11.958	183	13.528
24	4.899	64	8.000	104	10.198	144	12.000	184	13.565
25	5.000	65	8.062	105	10.247	145	12.042	185	13.601
26	5.099	66	8.124	106	10.296	146	12.083	186	13.638
27	5.196	67	8.185	107	10.344	147	12.124	187	13.675
28	5.292	68	8.246	108	10.392	148	12.166	188	13.711
29	5.385	69	8.307	109	10.440	149	12.207	189	13.748
30	5.477	70	8.367	110	10.488	150	12.247	190	13.784
31	5.568	71	8.426	111	10.536	151	12.288	191	13.820
32	5.657	72	8.485	112	10.583	152	12.329	192	13.856
33	5.745	73	8.544	113	10.630	153	12.369	193	13.892
34	5.831	74	8.602	114	10.677	154	12.410	194	13.928
35	5.916	75	8.660	115	10.724	155	12.450	195	13.964
36	6.000	76	8.718	116	10.770	156	12.490	196	14.000
37	6.083	77	8.775	117	10.817	157	12.530	197	14.036
38	6.164	78	8.832	118	10.863	158	12.570	198	14.071
39	6.245	79	8.888	119	10.909	159	12.610	199	14.107
40	6.325	80	8.944	120	10.954	160	12.649	200	14.142

Table A-2: Exponential Values

x	e^x	e^{-x}	x	e^x	e^{-x}
0.00	1.0000	1.0000	1.6	4.9530	0.2019
0.01	1.0101	0.9900	1.7	5.4739	0.1827
0.02	1.0202	0.9802	1.8	6.0496	0.1653
0.03	1.0305	0.9704	1.9	6.6859	0.1496
0.04	1.0408	0.9608	2.0	7.3891	0.1353
0.05	1.0513	0.9512	2.1	8.1662	0.1225
0.06	1.0618	0.9418	2.2	9.0250	0.1108
0.07	1.0725	0.9324	2.3	9.9742	0.1003
0.08	1.0833	0.9231	2.4	11.023	0.0907
0.09	1.0942	0.9139	2.5	12.182	0.0821
0.10	1.1052	0.9048	2.6	13.464	0.0743
0.11	1.1163	0.8958	2.7	14.880	0.0672
0.12	1.1275	0.8869	2.8	16.445	0.0608
0.13	1.1388	0.8781	2.9	18.174	0.0550
0.14	1.1503	0.8694	3.0	20.086	0.0498
0.15	1.1618	0.8607	3.1	22.198	0.0450
0.16	1.1735	0.8521	3.2	24.533	0.0408
0.17	1.1853	0.8437	3.3	27.113	0.0369
0.18	1.1972	0.8353	3.4	29.964	0.0334
0.19	1.2092	0.8270	3.5	33.115	0.0302
0.20	1.2214	0.8187	3.6	36.598	0.0273
0.21	1.2337	0.8106	3.7	40.447	0.0247
0.22	1.2461	0.8025	3.8	44.701	0.0224
0.23	1.2586	0.7945	3.9	49.402	0.0202
0.24	1.2712	0.7866	4.0	54.598	0.0183
0.25	1.2840	0.7788	4.1	60.340	0.0166
0.26	1.2969	0.7711	4.2	66.686	0.0150
0.27	1.3100	0.7634	4.3	73.700	0.0136
0.28	1.3231	0.7558	4.4	81.451	0.0123
0.29	1.3364	0.7483	4.5	90.017	0.0111
0.30	1.3499	0.7408	4.6	99.484	0.0101
0.35	1.4191	0.7047	4.7	109.95	0.0091
0.40	1.4918	0.6703	4.8	121.51	0.0082
0.45	1.5683	0.6376	4.9	134.29	0.0074
0.50	1.6487	0.6065	5.0	148.41	0.0067
0.55	1.7333	0.5769	5.5	244.69	0.0041
0.60	1.8221	0.5488	6.0	403.43	0.0025
0.65	1.9155	0.5220	6.5	665.14	0.0015
0.70	2.0138	0.4966	7.0	1,096.6	0.00091
0.75	2.1170	0.4724	7.5	1,808.0	0.00055
0.80	2.2255	0.4493	8.0	2,981.0	0.00034
0.85	2.3396	0.4274	8.5	4,914.8	0.00020
0.90	2.4596	0.4066	9.0	8,103.1	0.00012
0.95	2.5857	0.3867	9.5	13,360	0.000075
1.0	2.7183	0.3679	10	22,026	0.000045
1.1	3.0042	0.3329	11	59,874	0.000017
1.2	3.3201	0.3012	12	162,755	0.0000061
1.3	3.6693	0.2725	13	442,413	0.0000023
1.4	4.0552	0.2466	14	1,202,604	0.0000008
1.5	4.4817	0.2231	15	3,269,017	0.0000003

Appendix B Determinants and Cramer's Rule

 Evaluating a Second-Order Determinant

Mathematicians have developed techniques to solve systems of linear equations by focusing on the coefficients of the variables and the constants in the equations. The computational techniques can be easily carried out by computers or calculators. We will learn to do them by hand so that you will have a better understanding of what is involved.

To begin, we need to define a matrix and a determinant. A **matrix** is any rectangular array of numbers that is arranged in rows and columns. We use the symbol [ ] to indicate a matrix.

$$\begin{bmatrix} 3 & 2 & 4 \\ -1 & 4 & 0 \end{bmatrix}, \quad \begin{bmatrix} 4 & -3 \\ 2 & \frac{1}{2} \\ 1 & 5 \end{bmatrix}, \quad [-4 \quad 1 \quad 6], \quad \text{and} \quad \begin{bmatrix} \frac{1}{4} \\ 3 \\ -2 \end{bmatrix}$$

are matrices. If you have a graphing calculator, you can enter the elements of a matrix and store them for future use. Let's examine two systems of equations.

$$\begin{aligned} 3x + 2y &= 16 \\ x + 4y &= 22 \end{aligned} \quad \text{and} \quad \begin{aligned} -6x &= 18 \\ x + 3y &= 9 \end{aligned}$$

We could write the coefficients of the variables in each of these systems as a matrix.

$$\begin{aligned} 3x + 2y \\ x + 4y \end{aligned} \Rightarrow \begin{bmatrix} 3 & 2 \\ 1 & 4 \end{bmatrix} \quad \text{and} \quad \begin{aligned} -6x \\ x + 3y \end{aligned} \Rightarrow \begin{bmatrix} -6 & 0 \\ 1 & 3 \end{bmatrix}$$

Now we define a determinant. A **determinant** is a *square* arrangement of numbers. We use the symbol | | to indicate a determinant.

$$\begin{vmatrix} 3 & 2 \\ 1 & 4 \end{vmatrix} \quad \text{and} \quad \begin{vmatrix} -6 & 0 \\ 1 & 3 \end{vmatrix}$$

are determinants. The value of a determinant is a *real number* and is defined as follows:

> **DEFINITION**
> The value of the second-order determinant $\begin{vmatrix} a & c \\ b & d \end{vmatrix}$ is $ad - bc$.

EXAMPLE 1   Find the value of each determinant.

(a) $\begin{vmatrix} -6 & 2 \\ -1 & 4 \end{vmatrix}$ **(b)** $\begin{vmatrix} 0 & -3 \\ -2 & 6 \end{vmatrix}$

Solution

(a) $\begin{vmatrix} -6 & 2 \\ -1 & 4 \end{vmatrix} = (-6)(4) - (-1)(2) = -24 - (-2) = -24 + 2 = -22$

(b) $\begin{vmatrix} 0 & -3 \\ -2 & 6 \end{vmatrix} = (0)(6) - (-2)(-3) = 0 - (+6) = -6$

Student Learning Objectives

After studying this section, you will be able to:

1 Evaluate a second-order determinant.

2 Evaluate a third-order determinant.

3 Solve a system of two linear equations with two unknowns using Cramer's rule.

4 Solve a system of three linear equations with three unknowns using Cramer's rule.

NOTE TO STUDENT: Fully worked-out solutions to all of the Practice Problems can be found at the back of the text starting at page SP-1

Practice Problem 1 Find the value of each determinant.

(a) $\begin{vmatrix} -7 & 3 \\ -4 & -2 \end{vmatrix}$ (b) $\begin{vmatrix} 5 & 6 \\ 0 & -5 \end{vmatrix}$

2 Evaluating a Third-Order Determinant

Third-order determinants have three rows and three columns. Again, each determinant has exactly one value.

DEFINITION

The value of the third-order determinant

$$\begin{vmatrix} a_1 & b_1 & c_1 \\ a_2 & b_2 & c_2 \\ a_3 & b_3 & c_3 \end{vmatrix}$$

is

$$a_1 b_2 c_3 + b_1 c_2 a_3 + c_1 a_2 b_3 - a_3 b_2 c_1 - b_3 c_2 a_1 - c_3 a_2 b_1.$$

Because this definition is difficult to memorize and cumbersome to use, we evaluate third-order determinants by a simpler method called **expansion by minors.** The **minor** of an element (number or variable) of a third-order determinant is the second-order determinant that remains after we delete the row and column in which the element appears.

EXAMPLE 2 Find **(a)** the minor of 6 and **(b)** the minor of -3 in the determinant.

$$\begin{vmatrix} 6 & 1 & 2 \\ -3 & 4 & 5 \\ -2 & 7 & 8 \end{vmatrix}$$

Solution

(a) Since the element 6 appears in the first row and the first column, we delete them.

$$\begin{vmatrix} \cancel{6} & \cancel{1} & \cancel{2} \\ \cancel{-3} & 4 & 5 \\ \cancel{-2} & 7 & 8 \end{vmatrix}$$

Therefore, the minor of 6 is $\begin{vmatrix} 4 & 5 \\ 7 & 8 \end{vmatrix}$

(b) Since -3 appears in the first column and the second row, we delete them.

$$\begin{vmatrix} \cancel{6} & 1 & 2 \\ \cancel{-3} & \cancel{4} & \cancel{5} \\ \cancel{-2} & 7 & 8 \end{vmatrix}$$

The minor of -3 is $\begin{vmatrix} 1 & 2 \\ 7 & 8 \end{vmatrix}.$

Practice Problem 2 Find **(a)** the minor of 3 and **(b)** the minor of -6 in the determinant.

$$\begin{vmatrix} 1 & 2 & 7 \\ -4 & -5 & -6 \\ 3 & 4 & -9 \end{vmatrix}$$

To evaluate a third-order determinant, we use expansion by minors of elements in the first column; for example, we have

$$\begin{vmatrix} a_1 & b_1 & c_1 \\ a_2 & b_2 & c_2 \\ a_3 & b_3 & c_3 \end{vmatrix} = a_1 \begin{vmatrix} b_2 & c_2 \\ b_3 & c_3 \end{vmatrix} - a_2 \begin{vmatrix} b_1 & c_1 \\ b_3 & c_3 \end{vmatrix} + a_3 \begin{vmatrix} b_1 & c_1 \\ b_2 & c_2 \end{vmatrix}$$

Note that the signs alternate. We then evaluate the second-order determinant according to our definition.

EXAMPLE 3 Evaluate the determinant $\begin{vmatrix} 2 & 3 & 6 \\ 4 & -2 & 0 \\ 1 & -5 & -3 \end{vmatrix}$ by expanding it by minors of elements in the first column.

Solution

$$\begin{vmatrix} 2 & 3 & 6 \\ 4 & -2 & 0 \\ 1 & -5 & -3 \end{vmatrix} = 2 \begin{vmatrix} -2 & 0 \\ -5 & -3 \end{vmatrix} - 4 \begin{vmatrix} 3 & 6 \\ -5 & -3 \end{vmatrix} + 1 \begin{vmatrix} 3 & 6 \\ -2 & 0 \end{vmatrix}$$

$$= 2[(-2)(-3) - (-5)(0)] - 4[(3)(-3) - (-5)(6)] + 1[(3)(0) - (-2)(6)]$$

$$= 2[6 - 0] - 4[-9 - (-30)] + 1[0 - (-12)]$$

$$= 2(6) - 4(21) + 1(12)$$

$$= 12 - 84 + 12$$

$$= -60$$

Practice Problem 3 Evaluate the determinant $\begin{vmatrix} 1 & 2 & -3 \\ 2 & -1 & 2 \\ 3 & 1 & 4 \end{vmatrix}$.

③ Solving a System of Two Linear Equations with Two Unknowns Using Cramer's Rule

We can solve a linear system of two equations with two unknowns by Cramer's rule. The rule is named for Gabriel Cramer, a Swiss mathematician who lived from 1704 to 1752. Cramer's rule expresses the solution to each variable of a linear system as the quotient of two determinants. Computer programs are available to solve a system of equations by Cramer's rule.

CRAMER'S RULE

The solution to

$$a_1 x + b_1 y = c_1$$
$$a_2 x + b_2 y = c_2$$

is

$$x = \frac{D_x}{D} \quad \text{and} \quad y = \frac{D_y}{D}, \qquad D \neq 0,$$

where

$$D_x = \begin{vmatrix} c_1 & b_1 \\ c_2 & b_2 \end{vmatrix}, \qquad D_y = \begin{vmatrix} a_1 & c_1 \\ a_2 & c_2 \end{vmatrix}, \qquad \text{and} \quad D = \begin{vmatrix} a_1 & b_1 \\ a_2 & b_2 \end{vmatrix}.$$

EXAMPLE 4   Solve by Cramer's rule.

$$-3x + y = 7$$
$$-4x - 3y = 5$$

Solution

$$D = \begin{vmatrix} -3 & 1 \\ -4 & -3 \end{vmatrix} \qquad D_x = \begin{vmatrix} 7 & 1 \\ 5 & -3 \end{vmatrix} \qquad D_y = \begin{vmatrix} -3 & 7 \\ -4 & 5 \end{vmatrix}$$

$$= (-3)(-3) - (-4)(1) \qquad = (7)(-3) - (5)(1) \qquad = (-3)(5) - (-4)(7)$$
$$= 9 - (-4) \qquad\qquad = -21 - 5 \qquad\qquad = -15 - (-28)$$
$$= 9 + 4 \qquad\qquad = -26 \qquad\qquad = -15 + 28$$
$$= 13 \qquad\qquad\qquad\qquad\qquad\qquad = 13$$

Hence,

$$x = \frac{D_x}{D} = \frac{-26}{13} = -2$$

$$y = \frac{D_y}{D} = \frac{13}{13} = 1.$$

The solution to the system is $x = -2$ and $y = 1$. Verify this.

NOTE TO STUDENT: Fully worked-out solutions to all of the Practice Problems can be found at the back of the text starting at page SP-1

Practice Problem 4   Solve by Cramer's rule.

$$5x + 3y = 17$$
$$2x - 5y = 13$$

4 Solving a System of Three Linear Equations with Three Unknowns Using Cramer's Rule

It is quite easy to extend Cramer's rule to three linear equations.

CRAMER'S RULE

The solution to the system

$$a_1x + b_1y + c_1z = d_1$$
$$a_2x + b_2y + c_2z = d_2$$
$$a_3x + b_3y + c_3z = d_3$$

is $\qquad x = \dfrac{D_x}{D}, \qquad y = \dfrac{D_y}{D}, \qquad$ and $\qquad z = \dfrac{D_z}{D}, \qquad D \neq 0,$

$$D = \begin{vmatrix} a_1 & b_1 & c_1 \\ a_2 & b_2 & c_2 \\ a_3 & b_3 & c_3 \end{vmatrix}, \qquad D_x = \begin{vmatrix} d_1 & b_1 & c_1 \\ d_2 & b_2 & c_2 \\ d_3 & b_3 & c_3 \end{vmatrix},$$

where

$$D_y = \begin{vmatrix} a_1 & d_1 & c_1 \\ a_2 & d_2 & c_2 \\ a_3 & d_3 & c_3 \end{vmatrix}, \qquad \text{and} \qquad D_z = \begin{vmatrix} a_1 & b_1 & d_1 \\ a_2 & b_2 & d_2 \\ a_3 & b_3 & d_3 \end{vmatrix}.$$

EXAMPLE 5 Use Cramer's rule to solve the system.

$$2x - y + z = 6$$
$$3x + 2y - z = 5$$
$$2x + 3y - 2z = 1$$

Solution

We will expand each determinant by the first column.

$$D = \begin{vmatrix} 2 & -1 & 1 \\ 3 & 2 & -1 \\ 2 & 3 & -2 \end{vmatrix}$$

$$= 2\begin{vmatrix} 2 & -1 \\ 3 & -2 \end{vmatrix} - 3\begin{vmatrix} -1 & 1 \\ 3 & -2 \end{vmatrix} + 2\begin{vmatrix} -1 & 1 \\ 2 & -1 \end{vmatrix}$$

$$= 2[-4 - (-3)] - 3[2 - 3] + 2[1 - 2]$$

$$= 2[-1] - 3[-1] + 2[-1]$$

$$= -2 + 3 - 2$$

$$= -1$$

$$D_x = \begin{vmatrix} 6 & -1 & 1 \\ 5 & 2 & -1 \\ 1 & 3 & -2 \end{vmatrix}$$

$$= 6\begin{vmatrix} 2 & -1 \\ 3 & -2 \end{vmatrix} - 5\begin{vmatrix} -1 & 1 \\ 3 & -2 \end{vmatrix} + 1\begin{vmatrix} -1 & 1 \\ 2 & -1 \end{vmatrix}$$

$$= 6[-4 - (-3)] - 5[2 - 3] + 1[1 - 2]$$

$$= 6[-1] - 5[-1] + 1[-1]$$

$$= -6 + 5 - 1$$

$$= -2$$

$$D_y = \begin{vmatrix} 2 & 6 & 1 \\ 3 & 5 & -1 \\ 2 & 1 & -2 \end{vmatrix}$$

$$= 2\begin{vmatrix} 5 & -1 \\ 1 & -2 \end{vmatrix} - 3\begin{vmatrix} 6 & 1 \\ 1 & -2 \end{vmatrix} + 2\begin{vmatrix} 6 & 1 \\ 5 & -1 \end{vmatrix}$$

$$= 2[-10 - (-1)] - 3[-12 - 1] + 2[-6 - 5]$$

$$= 2[-9] - 3[-13] + 2[-11]$$

$$= -18 + 39 - 22$$

$$= -1$$

$$D_z = \begin{vmatrix} 2 & -1 & 6 \\ 3 & 2 & 5 \\ 2 & 3 & 1 \end{vmatrix}$$

$$= 2\begin{vmatrix} 2 & 5 \\ 3 & 1 \end{vmatrix} - 3\begin{vmatrix} -1 & 6 \\ 3 & 1 \end{vmatrix} + 2\begin{vmatrix} -1 & 6 \\ 2 & 5 \end{vmatrix}$$

$$= 2[2 - 15] - 3[-1 - 18] + 2[-5 - 12]$$

$$= 2[-13] - 3[-19] + 2[-17]$$

$$= -26 + 57 - 34$$

$$= -3$$

$$x = \frac{D_x}{D} = \frac{-2}{-1} = 2; \qquad y = \frac{D_y}{D} = \frac{-1}{-1} = 1; \qquad z = \frac{D_z}{D} = \frac{-3}{-1} = 3$$

Graphing Calculator

Copying Matrices

If you are using a graphing calculator to evaluate the four determinants in Example 5 or similar exercises, first enter matrix D into the calculator. Then copy the matrix using the copy function to three additional locations. Usually we store matrix D as matrix A. Then store a copy of it as matrix B, C, and D. Finally, use the Edit function and modify one column of each of matrices B, C, and D so that they become D_x, D_y, and D_z. This allows you to evaluate all four determinants in a minimum amount of time.

NOTE TO STUDENT: Fully worked-out solutions to all of the Practice Problems can be found at the back of the text starting at page SP-1

Practice Problem 5   Find the solution to the system by Cramer's rule.

$$2x + 3y - z = -1$$
$$3x + 5y - 2z = -3$$
$$x + 2y + 3z = 2$$

Cramer's rule cannot be used for every system of linear equations. If the equations are dependent or if the system of equations is inconsistent, the determinant of coefficients will be zero. Division by zero is not defined. In such a situation the system will not have a unique answer.

If $D = 0$, then the following are true:

1. If $D_x = 0$ and $D_y = 0$ (and $D_z = 0$, if there are three equations), then the equations are *dependent*. Such a system will have an infinite number of solutions.

2. If at least one of D_x or D_y (or D_z if there are three equations) is nonzero, then the system of equations is *inconsistent*. Such a system will have no solution.

B EXERCISES

| Student Solutions Manual | CD/ Video | PH Math Tutor Center | MathXL®Tutorials on CD | MathXL® | MyMathLab® | Interactmath.com |

Evaluate each determinant.

1. $\begin{vmatrix} 5 & 6 \\ 2 & 1 \end{vmatrix}$

2. $\begin{vmatrix} 3 & 4 \\ 1 & 8 \end{vmatrix}$

3. $\begin{vmatrix} 2 & -1 \\ 3 & 6 \end{vmatrix}$

4. $\begin{vmatrix} -4 & 2 \\ 1 & 5 \end{vmatrix}$

5. $\begin{vmatrix} -\frac{1}{2} & -\frac{2}{3} \\ 9 & 8 \end{vmatrix}$

6. $\begin{vmatrix} 10 & 4 \\ -\frac{3}{2} & -\frac{2}{5} \end{vmatrix}$

7. $\begin{vmatrix} -5 & 3 \\ -4 & -7 \end{vmatrix}$

8. $\begin{vmatrix} 2 & -3 \\ -4 & -6 \end{vmatrix}$

9. $\begin{vmatrix} 0 & -6 \\ 3 & -4 \end{vmatrix}$

10. $\begin{vmatrix} -5 & 0 \\ 2 & -7 \end{vmatrix}$

11. $\begin{vmatrix} 2 & -5 \\ -4 & 10 \end{vmatrix}$

12. $\begin{vmatrix} -3 & 6 \\ 7 & -14 \end{vmatrix}$

13. $\begin{vmatrix} 0 & 0 \\ -2 & 6 \end{vmatrix}$

14. $\begin{vmatrix} -4 & 0 \\ -3 & 0 \end{vmatrix}$

15. $\begin{vmatrix} 0.3 & 0.6 \\ 1.2 & 0.4 \end{vmatrix}$

16. $\begin{vmatrix} 0.1 & 0.7 \\ 0.5 & 0.8 \end{vmatrix}$

17. $\begin{vmatrix} 7 & 4 \\ b & -a \end{vmatrix}$

18. $\begin{vmatrix} \frac{1}{4} & \frac{3}{5} \\ \frac{2}{3} & \frac{1}{5} \end{vmatrix}$

19. $\begin{vmatrix} \frac{3}{7} & -\frac{1}{3} \\ -\frac{1}{4} & \frac{1}{2} \end{vmatrix}$

20. $\begin{vmatrix} -3 & y \\ -2 & x \end{vmatrix}$

In the following determinant $\begin{vmatrix} 3 & -4 & 7 \\ -2 & 6 & 10 \\ 1 & -5 & 9 \end{vmatrix}$,

21. Find the minor of 3.

22. Find the minor of −2.

23. Find the minor of 10.

24. Find the minor of 9.

Evaluate each of the following determinants.

25. $\begin{vmatrix} 4 & 1 & 2 \\ 3 & -1 & 0 \\ 1 & 2 & 3 \end{vmatrix}$

26. $\begin{vmatrix} 2 & 3 & 1 \\ -3 & 1 & 0 \\ 2 & 1 & 4 \end{vmatrix}$

27. $\begin{vmatrix} -4 & 0 & -1 \\ 2 & 1 & -1 \\ 0 & 3 & 2 \end{vmatrix}$

28. $\begin{vmatrix} 3 & -4 & -1 \\ -2 & 1 & 3 \\ 0 & 1 & 4 \end{vmatrix}$

29. $\begin{vmatrix} \frac{1}{2} & 1 & -1 \\ \frac{3}{2} & 1 & 2 \\ 3 & 0 & -2 \end{vmatrix}$

30. $\begin{vmatrix} 1 & 2 & 3 \\ 4 & -2 & -1 \\ 5 & -3 & 2 \end{vmatrix}$

31. $\begin{vmatrix} 4 & 1 & 2 \\ -1 & -2 & -3 \\ 4 & -1 & 3 \end{vmatrix}$

32. $\begin{vmatrix} -\frac{1}{2} & 2 & 3 \\ \frac{5}{2} & -2 & -1 \\ \frac{3}{4} & -3 & 2 \end{vmatrix}$

33. $\begin{vmatrix} 2 & 0 & -2 \\ -1 & 0 & 2 \\ 3 & 4 & 3 \end{vmatrix}$

34. $\begin{vmatrix} 7 & 0 & 2 \\ 1 & 0 & -5 \\ 3 & 0 & 6 \end{vmatrix}$

35. $\begin{vmatrix} 6 & -4 & 3 \\ 1 & 2 & 4 \\ 0 & 0 & 0 \end{vmatrix}$

36. $\begin{vmatrix} 7 & 0 & 3 \\ 1 & 2 & 4 \\ 3 & 0 & -7 \end{vmatrix}$

Optional Graphing Calculator Problems

If you have a graphing calculator, use the determinant function to evaluate the following:

37. $\begin{vmatrix} 1.3 & 1.8 & 2.5 \\ 7.9 & 5.3 & 6.0 \\ 1.7 & 1.8 & 2.8 \end{vmatrix}$

38. $\begin{vmatrix} 0.7 & 5.3 & 0.4 \\ 1.6 & 0.3 & 3.7 \\ 0.8 & 6.7 & 4.2 \end{vmatrix}$

39. $\begin{vmatrix} -55 & 17 & 19 \\ -62 & 23 & 31 \\ 81 & 51 & 74 \end{vmatrix}$

40. $\begin{vmatrix} 82 & -20 & 56 \\ 93 & -18 & 39 \\ 65 & -27 & 72 \end{vmatrix}$

Solve each system by Cramer's rule.

41. $x + 2y = 8$
$\quad 2x + y = 7$

42. $x + 3y = 6$
$\quad 2x + y = 7$

43. $5x + 4y = 10$
$\quad -x + 2y = 12$

44. $3x + 5y = 11$
$\quad 2x + y = -2$

45. $x - 5y = 0$
$\quad x + 6y = 22$

46. $x - 3y = 4$
$\quad -3x + 4y = -12$

47. $0.3x + 0.5y = 0.2$
$\quad 0.1x + 0.2y = 0.0$

48. $0.5x + 0.3y = -0.7$
$\quad 0.4x + 0.5y = -0.3$

Solve by Cramer's rule. Round your answers to four decimal places.

49. $52.9634x - 27.3715y = 86.1239$
$\quad 31.9872x + 61.4598y = 44.9812$

50. $0.0076x + 0.0092y = 0.01237$
$\quad -0.5628x - 0.2374y = -0.7635$

Solve each system by Cramer's rule.

51. $2x + y + z = 4$
$\quad x - y - 2z = -2$
$\quad x + y - z = 1$

52. $x + 2y - z = -4$
$\quad x + 4y - 2z = -6$
$\quad 2x + 3y + z = 3$

53. $2x + 2y + 3z = 6$
$\quad x - y + z = 1$
$\quad 3x + y + z = 1$

54. $4x + y + 2z = 6$
$\quad x + y + z = 1$
$\quad -x + 3y - z = -5$

55. $x + 2y + z = 1$
$\quad 3x - 4z = 8$
$\quad 3y + 5z = -1$

56. $3x + y + z = 2$
$\quad 2y + 3z = -6$
$\quad 2x - y = -1$

Optional Graphing Calculator Problems

Round your answers to the nearest thousandth.

57. $10x + 20y + 10z = -2$
$\quad -24x - 31y - 11z = -12$
$\quad 61x + 39y + 28z = -45$

58. $121x + 134y + 101z = 146$
$\quad 315x - 112y - 108z = 426$
$\quad 148x + 503y + 516z = -127$

59. $28w + 35x - 18y + 40z = 60$
$\quad 60w + 32x + 28y = 400$
$\quad 30w + 15x + 18y + 66z = 720$
$\quad 26w - 18x - 15y + 75z = 125$

Appendix C Solving Systems of Linear Equations Using Matrices

1 Solving a System of Linear Equations Using Matrices

Student Learning Objective

After studying this section, you will be able to:

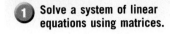 Solve a system of linear equations using matrices.

In Appendix B we defined a matrix as any rectangular array of numbers that is arranged in rows and columns.

$$\begin{bmatrix} 2 & 3 \\ 5 & 6 \end{bmatrix}$$ This is a 2 × 2 matrix with two rows and two columns.

$$\begin{bmatrix} 1 & -5 & -6 & 2 \\ 3 & 4 & -8 & -2 \\ 2 & 7 & 9 & -4 \end{bmatrix}$$ This is a 3 × 4 matrix with three rows and four columns.

A matrix that is derived from a linear system of equations is called the **augmented matrix** of the system. This augmented matrix is made up of two smaller matrices separated by a vertical line. The coefficients of each variable in a linear system are placed to the left of the vertical line. The constants are placed to the right of the vertical line.

The augmented matrix for the system of equations

$$-3x + 5y = -22$$
$$2x - y = 10$$

is the 2 × 3 matrix

$$\left[\begin{array}{rr|r} -3 & 5 & -22 \\ 2 & -1 & 10 \end{array}\right].$$

The augmented matrix for the system of equations

$$3x - 5y + 2z = 8$$
$$x + y + z = 3$$
$$3x - 2y + 4z = 10$$

is the 3 × 4 matrix

$$\left[\begin{array}{rrr|r} 3 & -5 & 2 & 8 \\ 1 & 1 & 1 & 3 \\ 3 & -2 & 4 & 10 \end{array}\right].$$

EXAMPLE 1
Write the solution to the system of linear equations represented by the following matrix.

$$\left[\begin{array}{rr|r} 1 & -3 & -7 \\ 0 & 1 & 4 \end{array}\right]$$

Solution This system is represented by the equations

$$x - 3y = -7 \quad \text{and}$$
$$0x + y = 4.$$

Since we know that $y = 4$, we can find x by substitution.

$$x - 3y = -7$$
$$x - 3(4) = -7$$
$$x - 12 = -7$$
$$x = 5$$

Thus, the solution to the system is $x = 5$; $y = 4$. We can also write the solution as $(5, 4)$.

NOTE TO STUDENT: Fully worked-out
solutions to all of the Practice Problems
can be found at the back of the text
starting at page SP-1

Practice Problem 1   Write the solution to the system of linear equations represented by the following matrix.

$$\begin{bmatrix} 1 & 9 & | & 33 \\ 0 & 1 & | & 3 \end{bmatrix}$$

To solve a system of linear equations in matrix form, we use three row operations of the matrix.

MATRIX ROW OPERATIONS

1. Any two rows of a matrix may be interchanged.

2. All the numbers in a row may be multiplied or divided by any nonzero number.

3. All the numbers in any row or any multiple of a row may be added to the corresponding numbers of any other row.

To obtain the values for x and y in a system of two linear equations, we use row operations to obtain an augmented matrix in a form similar to the form of the matrix in Example 1.

The desired form is

$$\begin{bmatrix} 1 & a & | & b \\ 0 & 1 & | & c \end{bmatrix} \quad \text{or} \quad \begin{bmatrix} 1 & a & b & | & d \\ 0 & 1 & c & | & e \\ 0 & 0 & 1 & | & f \end{bmatrix}.$$

The last row of the matrix will allow us to find the value of one of the variables. We can then use substitution to find the other variables.

EXAMPLE 2   Use matrices to solve the system.

$$4x - 3y = -13$$
$$x + 2y = 5$$

Solution   The augmented matrix for this system of linear equations is

$$\begin{bmatrix} 4 & -3 & | & -13 \\ 1 & 2 & | & 5 \end{bmatrix}.$$

First we want to obtain a 1 as the first element in the first row. We can obtain this by interchanging rows one and two.

$$\begin{bmatrix} 1 & 2 & | & 5 \\ 4 & -3 & | & -13 \end{bmatrix} \quad R_1 \longleftrightarrow R_2$$

Next we wish to obtain a 0 as the first element of the second row. To obtain this we multiply -4 by all the elements of row one and add this to row two.

$$\begin{bmatrix} 1 & 2 & | & 5 \\ 0 & -11 & | & -33 \end{bmatrix} \quad -4R_1 + R_2$$

Next, to obtain a 1 as the second element of the second row, we multiply each element of row two by $\left(-\frac{1}{11}\right)$.

$$\begin{bmatrix} 1 & 2 & | & 5 \\ 0 & 1 & | & 3 \end{bmatrix} \quad -\frac{1}{11}R_2$$

This final matrix is in the desired form. It represents the linear system

$$x + 2y = 5$$
$$y = 3.$$

Since we know that $y = 3$, we substitute this value into the first equation.

$$x + 2(3) = 5$$
$$x + 6 = 5$$
$$x = -1$$

Thus, the solution to the system is $(-1, 3)$.

Practice Problem 2   Use matrices to solve the system.

$$3x - 2y = -6$$
$$x - 3y = 5$$

Now we continue with a similar example involving three equations and three unknowns.

EXAMPLE 3   Use matrices to solve the system.

$$2x + 3y - z = 11$$
$$x + 2y + z = 12$$
$$3x - y + 2z = 5$$

Solution   The augmented matrix that represents this system of linear equations is

$$\left[\begin{array}{ccc|c} 2 & 3 & -1 & 11 \\ 1 & 2 & 1 & 12 \\ 3 & -1 & 2 & 5 \end{array}\right].$$

To obtain a one as the first element of the first row, we first need to interchange the first and second rows.

$$\left[\begin{array}{ccc|c} 1 & 2 & 1 & 12 \\ 2 & 3 & -1 & 11 \\ 3 & -1 & 2 & 5 \end{array}\right] \quad R_1 \longleftrightarrow R_2$$

Now, in order to obtain a 0 as the first element of the second row, we multiply row one by -2 and add the result to row two. In order to obtain a 0 as the first element of the third row, we multiply row one by -3 and add the result to row three.

$$\left[\begin{array}{ccc|c} 1 & 2 & 1 & 12 \\ 0 & -1 & -3 & -13 \\ 0 & -7 & -1 & -31 \end{array}\right] \quad \begin{array}{l} -2R_1 + R_2 \\ -3R_1 + R_3 \end{array}$$

To obtain a 1 as the second element of row two, we multiply all the elements of row two by -1.

$$\left[\begin{array}{ccc|c} 1 & 2 & 1 & 12 \\ 0 & 1 & 3 & 13 \\ 0 & -7 & -1 & -31 \end{array}\right] \quad -1R_2$$

Next, in order to obtain a 0 as the second element of row three, we add 7 times row two to row three.

$$\left[\begin{array}{ccc|c} 1 & 2 & 1 & 12 \\ 0 & 1 & 3 & 13 \\ 0 & 0 & 20 & 60 \end{array}\right] \quad 7R_2 + R_3$$

Finally, we multiply all the elements of row three by $\frac{1}{20}$. Thus, we have the following:

$$\left[\begin{array}{ccc|c} 1 & 2 & 1 & 12 \\ 0 & 1 & 3 & 13 \\ 0 & 0 & 1 & 3 \end{array}\right] \quad \frac{1}{20}R_3$$

From the final line of the matrix, we see that $z = 3$. If we substitute this value into the equation represented by the second line, we have

$$y + 3z = 13$$
$$y + 3(3) = 13$$
$$y + 9 = 13$$
$$y = 4.$$

Now we substitute the values obtained for y and for z into the equation represented by first line of the matrix.

$$x + 2y + z = 12$$
$$x + 2(4) + 3 = 12$$
$$x + 8 + 3 = 12$$
$$x + 11 = 12$$
$$x = 1$$

Thus, the solution to this linear system of three equations is $(1, 4, 3)$.

NOTE TO STUDENT: Fully worked-out solutions to all of the Practice Problems can be found at the back of the text starting at page SP-1

Practice Problem 3   Use matrices to solve the system.

$$2x + y - 2z = -15$$
$$4x - 2y + z = 15$$
$$x + 3y + 2z = -5$$

We could continue to use these row operations to obtain an augmented matrix of the form

$$\begin{bmatrix} 1 & 0 & | & a \\ 0 & 1 & | & b \end{bmatrix} \quad \text{or} \quad \begin{bmatrix} 1 & 0 & 0 & | & a \\ 0 & 1 & 0 & | & b \\ 0 & 0 & 1 & | & c \end{bmatrix}.$$

This form of the augmented matrix is given a special name. It is known as the **reduced row echelon form.** If the augmented matrix of a system of linear equations is placed in this form, we would immediately know the solution to the system. Thus, if a system of linear equations in the variables x, y, and z had an augmented matrix that could be placed in the form

$$\begin{bmatrix} 1 & 0 & 0 & | & 7 \\ 0 & 1 & 0 & | & 32 \\ 0 & 0 & 1 & | & 18 \end{bmatrix},$$

we could determine directly that $x = 7$, $y = 32$, and $z = 18$. A similar pattern is obtained for a system of four equations in four unknowns, and so on. Thus, if a system of linear equations in the variables w, x, y, and z had an augmented matrix that could be placed in the form

$$\begin{bmatrix} 1 & 0 & 0 & 0 & | & 23.4 \\ 0 & 1 & 0 & 0 & | & 48.6 \\ 0 & 0 & 1 & 0 & | & 0.73 \\ 0 & 0 & 0 & 1 & | & 5.97 \end{bmatrix},$$

we could directly conclude that $w = 23.4$, $x = 48.6$, $y = 0.73$, and $z = 5.97$. Reducing a matrix to reduced row echelon form is readily done on computers. Many mathematical software packages contain matrix operations that will obtain the reduced row echelon form of an augmented matrix. A number of the newer graphing calculators such as the TI-83 can be used to obtain the reduced row echelon form by using the **rref** command on a given matrix.

Graphing Calculator

Obtaining a Reduced Row Echelon Form of an Augmented Matrix

If your graphing calculator has a routine to obtain the **reduced row echelon form** of a matrix **(rref)**, then this routine will allow you to quickly obtain the solution of a system of linear equations if one exists. If your calculator has this capability, solve the following system.

$$5w + 2x + 3y + 4z = -8.3$$
$$-4w + 3x + 2y + 7z = -70.1$$
$$6w + x + 4y + 5z = -13.3$$
$$7w + 4x + y + 2z = 14.1$$

Answer:
$$w = 3.1, x = 2.2,$$
$$y = 4.6, z = -10.5$$

C EXERCISES

| Student Solutions Manual | CD/ Video | PH Math Tutor Center | MathXL®Tutorials on CD | MathXL® | MyMathLab® | Interactmath.com |

Solve each system of equations by the matrix method. Round your answers to the nearest tenth.

1. $2x + 3y = 5$
$5x + y = 19$

2. $3x + 5y = -15$
$2x + 7y = -10$

3. $2x + y = -3$
$5x - y = 24$

4. $x + 5y = -9$
$4x - 3y = -13$

5. $5x + 2y = 6$
$3x + 4y = 12$

6. $-5x + y = 24$
$x + 5y = 10$

7. $3x - 2y + 3 = 5$
$x + 4y - 1 = 9$

8. $3x + y - 4 = 12$
$-2x + 3y + 2 = -5$

9. $-7x + 3y = 2.7$
$6x + 5y = 25.7$

10. $x - 2y - 3z = 4$
$2x + 3y + z = 1$
$-3x + y - 2z = 5$

11. $x + y - z = -2$
$2x - y + 3z = 19$
$4x + 3y - z = 5$

12. $5x - y + 4z = 5$
$6x + y - 5z = 17$
$2x - 3y + z = -11$

13. $x + y - z = -3$
$x + y + z = 3$
$3x - y + z = 7$

14. $2x - y + z = 5$
$x + 2y - z = -2$
$x + y - 2z = -5$

15. $2x - 3y + z = 11$
$x + y + 2z = 8$
$x + 3y - z = -11$

16. $4x + 3y + 5z = 2$
$2y + 7z = 16$
$2x - y = 6$

17. $6x - y + z = 9$
$2x + 3z = 16$
$4x + 7y + 5z = 20$

18. $3x + 2y = 44$
$4y + 3z = 19$
$2x + 3z = -5$

▦ Optional Graphing Calculator Problems

If your graphing calculator has the necessary capability, solve the following exercises. Round your answers to the nearest tenth.

19. $5x + 6y + 7z = 45.6$
$1.4x - 3.2y + 1.6z = 3.12$
$9x - 8y + 22z = 70.8$

20. $2x + 12y + 9z = 37.9$
$1.6x + 1.8y - 2.5z = -20.53$
$7x + 8y + 4z = 39.6$

21. $6w + 5x + 3y + 1.5z = 41.7$
$2w + 6.7x - 5y + 7z = -21.92$
$12w + x + 5y - 6z = 58.4$
$3w + 8x - 15y + z = -142.8$

22. $2w + 3x + 11y - 14z = 6.7$
$5w + 8x + 7y + 3z = 25.3$
$-4w + x + 1.5y - 9z = -53.4$
$9w + 7x - 2.5y + 6z = 22.9$

Solutions to Practice Problems

Chapter 1

1.1 Practice Problems

1. (a) 1.26 is a rational number and a real number.

(b) 3 is a natural number, whole number, integer, rational number, and a real number.

(c) $\frac{3}{7}$ is a rational number and a real number.

(d) -2 is an integer, a rational number, and a real number.

(e) 5.182671 ... has no repeating pattern. Therefore, it is an irrational number and a real number.

2. (a) $9 + 8 = 8 + 9$ Commutative property of addition

(b) $17 + 0 = 17$ Identity property of addition

3. (a) $6 \cdot (2 \cdot w) = (6 \cdot 2) \cdot w$ Associative property of multiplication

(b) $4 \cdot \frac{1}{4} = 1$ Inverse property of multiplication

(c) $6(8 + 7) = 6 \cdot 8 + 6 \cdot 7$ Distributive property of multiplication over addition.

1.2 Practice Problems

1. (a) $|-4| = 4$

(b) $|3.16| = 3.16$

(c) $|8 - 8| = |0| = 0$

(d) $|12 - 7| = |5| = 5$

(e) $\left|2\frac{1}{3}\right| = 2\frac{1}{3}$

2. (a) $3.4 + 2.6 = 6.0$

(b) $-\frac{3}{4} + \left(-\frac{1}{6}\right) = -\frac{9}{12} + \left(-\frac{2}{12}\right)$
$$= -\frac{11}{12}$$

(c) $-5 + (-37) = -42$

3. (a) $24 + (-30) = -6$

(b) $-\frac{1}{5} + \frac{2}{3} = \frac{-3}{15} + \frac{10}{15} = \frac{7}{15}$

4. (a) $-8 - (-3) = -8 + 3 = -5$

(b) $\frac{1}{2} - \left(-\frac{1}{4}\right) = \frac{2}{4} + \frac{1}{4} = \frac{3}{4}$

(c) $-0.35 - 0.67 = -0.35 + (-0.67) = -1.02$

5. (a) $\left(-\frac{2}{5}\right)\left(\frac{3}{4}\right) = -\frac{6}{20} = -\frac{3}{10}$

(b) $\frac{150}{-30} = -5$

(c) $\frac{0.27}{-0.003} = -90$

(d) $\frac{-12}{-24} = \frac{1}{2}$

6. (a) $-\frac{2}{7} \cdot \frac{3}{5} = -\frac{6}{35}$

(b) $-60 \div (-5) = 12$

(c) $\left(\frac{3}{7}\right) \div \left(\frac{1}{5}\right) = \frac{3}{7} \cdot \frac{5}{1} = \frac{15}{7}$

7. $5 + 7(-2) - (-3) + 50 \div (-2)$
$= 5 + (-14) - (-3) + (-25)$
$= 5 + (-14) + 3 + (-25)$
$= -9 + 3 + (-25) = (-6) + (-25) = -31$

8. (a) $6(-2) + (-20) \div (2)(3) = -12 + (-10)(3)$
$$= -12 + (-30) = -42$$

(b) $\frac{7 + 2 - 12 - (-1)}{(-5)(-6) + 4(-8)} = \frac{-2}{-2} = 1$

1.3 Practice Problems

1. (a) $(-4)(-4)(-4)(-4) = (-4)^4$

(b) $z \cdot z \cdot z \cdot z \cdot z \cdot z \cdot z = z^7$

2. (a) $(-3)^5 = (-3)(-3)(-3)(-3)(-3) = -243$

(b) $(-3)^6 = (-3)(-3)(-3)(-3)(-3)(-3) = 729$

(c) $(-4)^4 = (-4)(-4)(-4)(-4) = 256$

(d) $-4^4 = -(4 \cdot 4 \cdot 4 \cdot 4) = -256$

(e) $\left(\frac{1}{5}\right)^2 = \left(\frac{1}{5}\right)\left(\frac{1}{5}\right) = \frac{1}{25}$

3. Since $(-7)^2 = 49$ and $7^2 = 49$ the square roots of 49 are -7 and 7. The principal square root is 7.

4. (a) $\sqrt{100} = 10$ because $10^2 = 100$.

(b) $\sqrt{1} = 1$ because $1^2 = 1$.

(c) $-\sqrt{36} = -6$ because $(6)^2 = 36$.

5. (a) $(0.3)^2 = (0.3)(0.3) = 0.09$, therefore, $\sqrt{0.09} = 0.3$.

(b) $\sqrt{\frac{4}{81}} = \frac{\sqrt{4}}{\sqrt{81}} = \frac{2}{9}$

(c) This is not a real number.

6. (a) $6(12 - 8) + 4 = 6(4) + 4$
$$= 24 + 4 = 28$$

(b) $5[6 - 3(7 - 9)] - 8 = 5[6 - 3(-2)] - 8$
$$= 5[6 + 6] - 8$$
$$= 5[12] - 8 = 60 - 8 = 52$$

7. $\frac{(-6)(3)(2)}{5 - 12 + 3} = \frac{-36}{-4} = 9$

8. (a) $\sqrt{(-5)^2 + 12^2} = \sqrt{25 + 144} = \sqrt{169} = 13$

(b) $|-3 - 7 + 2 - (-4)| = |-3 - 7 + 2 + 4|$
$$= |-10 + 6| = |-4| = 4$$

9. $-7 - 2(-3) + 4^3 = -7 + 6 + 64$
$$= -1 + 64 = 63$$

10. $5 + 6 \cdot 2 - 12 \div (-2) + 3\sqrt{4}$
$= 5 + 6 \cdot 2 - 12 \div (-2) + 3(2)$
$= 5 + 6 \cdot 2 + 6 + 3(2)$
$= 5 + 12 + 6 + 6$
$= 17 + 6 + 6 = 23 + 6 = 29$

11. $\frac{2(3) + 5(-2)}{1 + 2 \cdot 3^2 + 5(-3)} = \frac{6 - 10}{1 + 2 \cdot 9 + 5(-3)}$
$$= \frac{-4}{1 + 18 + (-15)} = \frac{-4}{4} = -1$$

1.4 Practice Problems

1. (a) $3^{-2} = \frac{1}{3^2} = \frac{1}{9}$

(b) $z^{-8} = \frac{1}{z^8}$

2. $\left(\frac{3}{4}\right)^{-2} = \frac{1}{\left(\frac{3}{4}\right)^2} = \frac{1}{\frac{9}{16}} = 1 \cdot \frac{16}{9} = \frac{16}{9}$

3. (a) $2^8 \cdot 2^{15} = 2^{23}$

(b) $x^2 \cdot x^8 \cdot x^6 = x^{16}$

(c) $(x + 2y)^4(x + 2y)^{10} = (x + 2y)^{14}$

4. (a) $(7w^3)(2w) = 14w^4$

(b) $(-5xy)(-2x^2y^3) = 10x^3y^4$

5. $(7xy^{-2})(2x^{-5}y^{-6}) = 14x^{-4}y^{-8} = \frac{14}{x^4y^8}$

6. (a) $\dfrac{w^8}{w^6} = w^{8-6} = w^2$

(b) $\dfrac{3^7}{3^3} = 3^{7-3} = 3^4$

(c) $\dfrac{x^5}{x^{16}} = x^{5-16} = x^{-11} = \dfrac{1}{x^{11}}$

(d) $\dfrac{4^5}{4^8} = 4^{5-8} = 4^{-3} = \dfrac{1}{4^3}$

7. (a) $6y^0 = 6(1) = 6$

(b) $(3xy)^0 = 1$. Note that the entire expression is raised to the zero power.

(c) $(5^{-3})(2a)^0 = (5^{-3})(1) = \dfrac{1}{5^3} = \dfrac{1}{125}$

8. (a) $\dfrac{30x^6y^5}{20x^3y^2} = \dfrac{30}{20}\cdot\dfrac{x^6}{x^3}\cdot\dfrac{y^5}{y^2} = \dfrac{3}{2}x^3y^3$ or $\dfrac{3x^3y^3}{2}$

(b) $\dfrac{-15a^3b^4c^4}{3a^5b^4c^2} = \dfrac{-15}{3}\cdot\dfrac{a^3}{a^5}\cdot\dfrac{b^4}{b^4}\cdot\dfrac{c^4}{c^2} = -5a^{-2}c^2 = -\dfrac{5c^2}{a^2}$

9. $\dfrac{2x^{-3}y}{4x^{-2}y^5} = \dfrac{1}{2}x^{-3-(-2)}y^{1-5} = \dfrac{1}{2}x^{-1}y^{-4} = \dfrac{1}{2xy^4}$

10. (a) $(w^3)^8 = w^{3\cdot8} = w^{24}$

(b) $(5^2)^5 = 5^{2\cdot5} = 5^{10}$

(c) $[(x-2y)^3]^3 = (x-2y)^{3\cdot3} = (x-2y)^9$

11. (a) $(4x^3y^4)^2 = 4^2x^6y^8 = 16x^6y^8$

(b) $\left(\dfrac{4xy}{3x^5y^6}\right)^3 = \dfrac{4^3x^3y^3}{3^3x^{15}y^{18}} = \dfrac{64x^3y^3}{27x^{15}y^{18}} = \dfrac{64}{27x^{12}y^{15}}$

(c) $(3xy^2)^{-2} = 3^{-2}x^{-2}y^{-4} = \dfrac{1}{9x^2y^4}$

12. (a) $\dfrac{7x^2y^{-4}z^{-3}}{8x^{-5}y^{-6}z^2} = \dfrac{7x^2\cdot x^5\cdot y^6}{8\cdot y^4\cdot z^3\cdot z^2} = \dfrac{7x^7y^2}{8z^5}$

(b) $\left(\dfrac{4x^2y^{-2}}{x^{-4}y^{-3}}\right)^{-3} = \dfrac{4^{-3}x^{-6}y^6}{x^{12}y^9} = \dfrac{y^6}{4^3x^{12}\cdot x^6y^9} = \dfrac{1}{64x^{18}y^3}$

13. $(2x^{-3})^2(-3xy^{-2})^{-3} = (2^2x^{-6})(-3^{-3}x^{-3}y^6)$
$= \dfrac{2^2y^6}{-3^3x^6x^3} = -\dfrac{4y^6}{27x^9}$

14. (a) $128{,}320 = 1.2832 \times 10^5$ **(b)** $476 = 4.76 \times 10^2$
(c) $0.0786 = 7.86 \times 10^{-2}$ **(d)** $0.007 = 7 \times 10^{-3}$
15. (a) $4.62 \times 10^6 = 4{,}620{,}000$ **(b)** $1.973 \times 10^{-3} = 0.001973$
(c) $4.931 \times 10^{-1} = 0.4931$

16. $\dfrac{(55{,}000)(3{,}000{,}000)}{5{,}500{,}000} = \dfrac{(5.5\times10^4)(3.0\times10^6)}{5.5\times10^6}$
$= \dfrac{3.0}{1}\times\dfrac{10^{10}}{10^6}$
$= 3.0\times10^{10-6} = 3.0\times10^4$

17. $6.0\times10^{24}\times3.4\times10^5$
$= 6.0\times3.4\times10^{24}\times10^5$
$= 20.4\times10^{29}$ kilograms
$= 2.04\times10^{30}$ kilograms

1.5 Practice Problems

1. (a) $7x - 2w^3$.
$7x$ is the product of a real number (7) and a variable (x), so $7x$ is a term.
$-2w^3$ is the product of a real number (-2) and a variable (w^3), so $-2w^3$ is a term.

(b) $5 + 6x + 2y$
5 is a real number, so it is a term. $6x$ is the product of a real number (6) and a variable (x), so $6x$ is a term. $2y$ is the product of a real number (2) and a variable (y), so $2y$ is a term.

2. (a) $5x^2y - 3.5w$
The numerical coefficient of the x^2y term is 5.
The numerical coefficient of the w term is -3.5.

(b) $\dfrac{3}{4}x^3 - \dfrac{5}{7}x^2y$
The numerical coefficient of the x^3 term is $\dfrac{3}{4}$.
The numerical coefficient of the x^2y term is $-\dfrac{5}{7}$.

(c) $-5.6abc - 0.34ab + 8.56bc$
The numerical coefficient of the abc term is -5.6, of the ab term is -0.34, and of the bc term is 8.56.

3. (a) $9x - 12x = (9-12)x = -3x$

(b) $4ab^2c + 15ab^2c = (4+15)ab^2c = 19ab^2c$

4. (a) $12x^3 - 5x^2 + 7x - 3x^3 - 8x^2 + x$
$= 12x^3 - 3x^3 - 8x^2 - 5x^2 + 7x + x$
$= 9x^3 - 13x^2 + 8x$

(b) $\dfrac{1}{3}a^2 - \dfrac{1}{5}a - \dfrac{4}{15}a^2 + \dfrac{1}{2}a + 5$
$= \dfrac{1}{3}a^2 - \dfrac{4}{15}a^2 + \dfrac{1}{2}a - \dfrac{1}{5}a + 5$
$= \dfrac{1}{15}a^2 + \dfrac{3}{10}a + 5$

(c) $4.5x^3 - 0.6x - 9.3x^3 + 0.8x$
$= 4.5x^3 - 9.3x^3 - 0.6x + 0.8x$
$= -4.8x^3 + 0.2x$

5. $-3x^2(2x-5) = -3x^2[2x + (-5)]$
$= (-3x^2)(2x) + (-3x^2)(-5)$
$= -6x^3 + 15x^2$

6. (a) $-5x(2x^2 - 3x - 1) = -10x^3 + 15x^2 + 5x$

(b) $3ab(4a^3 + 2b^2 - 6) = 12a^4b + 6ab^3 - 18ab$

7. (a) $(7x^2 - 8) = 1(7x^2 - 8) = 7x^2 - 8$

(b) $-(3x + 2y - 6) = -1(3x + 2y - 6) = -3x - 2y + 6$

(c) $-5x^2(x + 2xy) = -5x^3 - 10x^3y$

(d) $\dfrac{3}{4}(8x^2 + 12x - 3) = \dfrac{3}{4}(8x^2) + \dfrac{3}{4}(12x) - \dfrac{3}{4}(3)$
$= 6x^2 + 9x - \dfrac{9}{4}$

8. $-7(a + b) - 8a(2 - 3b) + 5a$
$= -7a - 7b - 16a + 24ab + 5a$
$= -18a + 24ab - 7b$

9. $-2\{4x - 3[x - 2x(1 + x)]\}$
$= -2\{4x - 3[x - 2x - 2x^2]\}$
$= -2\{4x - 3x + 6x + 6x^2\}$
$= -8x + 6x - 12x - 12x^2 = -14x - 12x^2$

1.6 Practice Problems

1. When $x = -3$,
$2x^2 + 3x - 8 = 2(-3)^2 + 3(-3) - 8$
$= 2(9) + 3(-3) - 8$
$= 18 - 9 - 8 = 9 - 8 = 1$.

2. When $x = -3$ and $y = 4$,
$(x - 3)^2 - 2xy$
$= [(-3) + (-3)]^2 - 2(-3)(4)$
$= (-6)^2 + 24 = 36 + 24 = 60$.

3. Evaluate when $x = -4$.
(a) $(-3x)^2 = [-3(-4)]^2$
$= (12)^2 = 144$
(b) $-3x^2 = -3(-4)^2$
$= -3(16) = -48$

4. When $C = 70°$,
$F = \dfrac{9}{5}C + 32$
$= \dfrac{9}{5}(70) + 32$
$= 9(14) + 32$
$= 158°$.

5. When $L = 128$ and $g = 32$,

$$T = 2\pi\sqrt{\frac{L}{g}}$$
$$= 2\pi\sqrt{\frac{128}{32}}$$
$$= 2\pi\sqrt{4}$$
$$= 2(3.14)(2)$$
$$= 12.56.$$

The period is about 12.6 seconds.

6. When $p = 600, r = 9\%$, and $t = 3$,

$$A = p(1 + rt)$$
$$= 600[1 + (0.09)(3)]$$
$$= 600(1 + 0.27)$$
$$= 600(1.27)$$
$$= 762.$$

The amount is \$762.

7. When $l = 0.76$ and $w = 0.38$,

$$p = 2l + 2w$$
$$= 2(0.76) + 2(0.38)$$
$$= 1.52 + 0.76$$
$$= 2.28.$$

The perimeter is 2.28 centimeters.

8. When $a = 12$ and $b = 14$,

$$A = \frac{1}{2}ab$$
$$= \frac{1}{2}(12)(14)$$
$$= 6(14)$$
$$= 84.$$

The area is 84 square meters.

9. When $r = 6$ and $h = 10$,

$$V = \pi r^2 h$$
$$= \pi(6)^2(10)$$
$$\approx 3.14(36)(10)$$
$$= 1130.4.$$

The volume is approximately 1130.4 cubic meters.

Chapter 2

2.1 Practice Problems

1. Replace a by the value $-\frac{3}{2}$ in the equation.

$$6\left(-\frac{3}{2}\right) - 5 \overset{?}{=} -4\left(-\frac{3}{2}\right) + 10$$
$$-9 - 5 \overset{?}{=} 6 + 10$$
$$-14 \neq 16$$

This last statement is not true. Thus $-\frac{3}{2}$ is not a solution of $6a - 5 = -4a + 10$.

2. Solve.

$$x + 5.2 = -2.8$$
$$x + 5.2 - 5.2 = -2.8 - 5.2$$
$$x = -8$$

Check.
$$x + 5.2 = -2.8$$
$$-8 + 5.2 \overset{?}{=} -2.8$$
$$-2.8 = -2.8 \checkmark$$

3. Solve.

$$\frac{1}{5}w = -6$$
$$5\left(\frac{1}{5}\right)w = -6(5)$$
$$w = -30$$

Check.
$$\frac{1}{5}w = -6$$
$$\frac{1}{5}(-30) \overset{?}{=} -6$$
$$-6 = -6 \checkmark$$

4. Solve.

$$8w - 3 = 2w - 7w + 4$$
$$8w - 3 = -5w + 4$$
$$8w + 5w - 3 = -5w + 5w + 4$$

$$13w - 3 = 4$$
$$13w - 3 + 3 = 4 + 3$$
$$13w = 7$$
$$\frac{13w}{13} = \frac{7}{13}$$
$$w = \frac{7}{13}$$

5. Solve.

$$a - 4(2a - 7) = 3(a + 6)$$
$$a - 8a + 28 = 3a + 18$$
$$-7a + 28 = 3a + 18$$
$$-7a - 3a + 28 = 3a - 3a + 18$$
$$-10a + 28 = 18$$
$$-10a + 28 - 28 = 18 - 28$$
$$-10a = -10$$
$$\frac{-10a}{-10} = \frac{-10}{-10}$$
$$a = 1$$

Check. $a - 4(2a - 7) = 3(a + 6)$
$$1 - 4[2(1) - 7] \overset{?}{=} 3(1 + 6)$$
$$1 - 4(2 - 7) \overset{?}{=} 3(1 + 6)$$
$$1 - 4(-5) \overset{?}{=} 3(1 + 6)$$
$$1 + 20 \overset{?}{=} 3(7)$$
$$21 = 21 \checkmark$$

6. Solve.

Multiply each term by the LCD = 12.
$$\frac{y}{3} + \frac{1}{2} = 5 + \frac{y - 9}{4}$$
$$12\left(\frac{y}{3}\right) + 12\left(\frac{1}{2}\right) = 12(5) + 12\left(\frac{y}{4}\right) - 12\left(\frac{9}{4}\right)$$
$$4y + 6 = 60 + 3y - 27$$
$$4y + 6 = 3y + 33$$
$$4y - 3y + 6 = 3y - 3y + 33$$
$$y + 6 = 33$$
$$y + 6 - 6 = 33 - 6$$
$$y = 27$$

Check. $\frac{y}{3} + \frac{1}{2} = 5 + \frac{y - 9}{4}$
$$\frac{27}{3} + \frac{1}{2} \overset{?}{=} 5 + \frac{27 - 9}{4}$$
$$9 + \frac{1}{2} \overset{?}{=} 5 + \frac{18}{4}$$
$$\frac{18}{2} + \frac{1}{2} \overset{?}{=} \frac{10}{2} + \frac{9}{2}$$
$$\frac{19}{2} = \frac{19}{2} \checkmark$$

7. Solve.

$$4(0.01x + 0.09) - 0.07(x - 8) = 0.83$$
$$0.04x + 0.36 - 0.07x + 0.56 = 0.83$$
Multiply each term by 100.
$$100(0.04x) + 100(0.36) - 100(0.07x) + 100(0.56) = 100(0.83)$$
$$4x + 36 - 7x + 56 = 83$$
$$-3x + 92 = 83$$
$$-3x + 92 - 92 = 83 - 92$$
$$-3x = -9$$
$$\frac{-3x}{-3} = \frac{-9}{-3}$$
$$x = 3$$

Check. $4(0.01x + 0.09) - 0.07(x - 8) = 0.83$
$$4[0.01(3) + 0.09] - 0.07(3 - 8) \overset{?}{=} 0.83$$
$$4(0.03 + 0.09) - 0.07(-5) \overset{?}{=} 0.83$$
$$4(0.12) + 0.35 \overset{?}{=} 0.83$$
$$0.48 + 0.35 \overset{?}{=} 0.83$$
$$0.83 = 0.83 \checkmark$$

8. $7 + 14x - 3 = 2(x - 4) + 12x$
$$7 + 14x - 3 = 2x - 8 + 12x$$
$$4 + 14x = 14x - 8$$
$$0 = -12$$

Clearly, no value of x will make this equation true. No solution.

9. $13x - 7(x + 5) = 4x - 35 + 2x$
$13x - 7x - 35 = 6x - 35$
$6x - 35 = 6x - 35$
$0 = 0$
Any real number is a solution.

2.2 Practice Problems

1. Solve for W.
$$P = 2L + 2W$$
$$P - 2L = 2W$$
$$\frac{P - 2L}{2} = \frac{2W}{2}$$
$$\frac{P - 2L}{2} = W$$

2. Solve for a.
$$H = \frac{3}{4}(a + 2b - 4)$$
$$H = \frac{3}{4}a + \frac{3}{2}b - 3$$
$$4(H) = 4\left(\frac{3}{4}a\right) + 4\left(\frac{3}{2}b\right) - 4(3)$$
$$4H = 3a + 6b - 12$$
$$4H - 6b + 12 = 3a + 6b - 6b - 12 + 12$$
$$4H - 6b + 12 = \frac{3a}{3}$$
$$\frac{4H - 6b + 12}{3} = a \quad \text{or} \quad \frac{4}{3}H - 2b + 4 = a$$

3. Solve for b.
$$-2(ab - 3x) + 2(8 - ab) = 5x + 4ab$$
$$-2ab + 6x + 16 - 2ab = 5x + 4ab$$
$$-4ab + 6x + 16 = 5x + 4ab$$
$$-4ab + 4ab + 6x + 16 = 5x + 4ab + 4ab$$
$$6x + 16 = 5x + 8ab$$
$$6x - 5x + 16 = 5x - 5x + 8ab$$
$$x + 16 = 8ab$$
$$\frac{x + 16}{8a} = b$$

4. Solve for x.
$$t = -0.4x + 81$$
$$10t = -4x + 810$$
$$4x = 810 - 10t$$
$$x = \frac{810 - 10t}{4}$$
$$x = \frac{405 - 5t}{2}$$
Solve for x when $t = 75$.
$$x = \frac{405 - 5(75)}{2}$$
$$x = \frac{405 - 375}{2}$$
$$x = \frac{30}{2}$$
$$x = 15$$
We estimate that this winning time will occur in the year $1990 + 15 = 2005$.

5. (a) Solve for h.
$$A = 2\pi rh + 2\pi r^2$$
$$A - 2\pi r^2 = 2\pi rh + 2\pi r^2 - 2\pi r^2$$
$$A - 2\pi r^2 = 2\pi rh$$
$$\frac{A - 2\pi r^2}{2\pi r} = \frac{2\pi rh}{2\pi r}$$
$$\frac{A - 2\pi r^2}{2\pi r} = h$$

(b) Solve for h when $A = 100$, $\pi \approx 3.14$, $r = 2.0$.
$$\frac{A - 2\pi r^2}{2\pi r} = h$$
$$\frac{100 - 2(3.14)(2.0)^2}{2(3.14)(2.0)} = h$$
$$\frac{100 - (6.28)(4)}{6.28(2.0)} = h$$
$$\frac{100 - 25.12}{12.56} = h$$
$$\frac{74.88}{12.56} = h$$
$$5.96 = h$$

2.3 Practice Problems

1. Solve $|3x - 4| = 23$ and check.
We have two equations.
$$3x - 4 = 23 \quad \text{or} \quad 3x - 4 = -23$$
$$3x = 27 \qquad\qquad 3x = -19$$
$$x = 9 \qquad\qquad x = -\frac{19}{3}$$

Check.   if $x = 9$
$$|3x - 4| = 23$$
$$|3(9) - 4| \overset{?}{=} 23$$
$$|27 - 4| \overset{?}{=} 23$$
$$|23| \overset{?}{=} 23$$
$$23 = 23 \checkmark$$

if $x = -\frac{19}{3}$
$$|3x - 4| = 23$$
$$\left|3\left(-\frac{19}{3}\right) - 4\right| \overset{?}{=} 23$$
$$|-19 - 4| \overset{?}{=} 23$$
$$|-23| \overset{?}{=} 23$$
$$23 = 23 \checkmark$$

2. Solve. $\left|\frac{2}{3}x + 4\right| = 2$
We have two equations.
$$\frac{2}{3}x + 4 = 2 \quad \text{or} \quad \frac{2}{3}x + 4 = -2$$
$$2x + 12 = 6 \qquad\quad 2x + 12 = -6$$
$$2x = -6 \qquad\qquad 2x = -18$$
$$x = -3 \qquad\qquad x = -9$$

Check.   if $x = -3$
$$\left|\frac{2}{3}(-3) + 4\right| \overset{?}{=} 2$$
$$|-2 + 4| \overset{?}{=} 2$$
$$|2| \overset{?}{=} 2$$
$$2 = 2 \checkmark$$

if $x = -9$
$$\left|\frac{2}{3}(-9) + 4\right| \overset{?}{=} 2$$
$$|-6 + 4| \overset{?}{=} 2$$
$$|-2| \overset{?}{=} 2$$
$$2 = 2 \checkmark$$

3. Solve $|2x + 1| + 3 = 8$.
First change the equation so that the absolute value expression is alone on one side of the equation.
$$|2x + 1| + 3 = 8$$
$$|2x + 1| + 3 - 3 = 8 - 3$$
$$|2x + 1| = 5$$
We have the two equations.
$$2x + 1 = 5 \quad \text{or} \quad 2x + 1 = -5$$
$$2x = 4 \qquad\qquad 2x = -6$$
$$x = 2 \qquad\qquad x = -3$$

Check.
if $x = 2$
$$|2(2) + 1| + 3 \overset{?}{=} 8$$
$$|5| + 3 \overset{?}{=} 8$$
$$8 = 8 \checkmark$$

if $x = -3$
$$|2(-3) + 1| + 3 \overset{?}{=} 8$$
$$|-5| + 3 \overset{?}{=} 8$$
$$8 = 8 \checkmark$$

4. Solve $|x - 6| = |5x + 8|$.

We write the two possible equations and solve each equation.

$$x - 6 = 5x + 8 \qquad \text{or} \qquad x - 6 = -(5x + 8)$$
$$-4x = 14 \qquad\qquad\qquad x - 6 = -5x - 8$$
$$x = \frac{14}{-4} = -\frac{7}{2} \qquad\qquad 6x = -2$$
$$x = \frac{-2}{6} = -\frac{1}{3}$$

Check. if $x = -\frac{7}{2}$

$$\left|-\frac{7}{2} - 6\right| \overset{?}{=} \left|5\left(-\frac{7}{2}\right) + 8\right|$$
$$\left|-\frac{7}{2} - 6\right| \overset{?}{=} \left|-\frac{35}{2} + 8\right|$$
$$\left|-\frac{7}{2} - \frac{12}{2}\right| \overset{?}{=} \left|-\frac{35}{2} + \frac{16}{2}\right|$$
$$\left|-\frac{19}{2}\right| \overset{?}{=} \left|-\frac{19}{2}\right|$$
$$\frac{19}{2} = \frac{19}{2} \checkmark$$

if $x = -\frac{1}{3}$

$$\left|-\frac{1}{3} - 6\right| \overset{?}{=} \left|5\left(-\frac{1}{3}\right) + 8\right|$$
$$\left|-\frac{1}{3} - 6\right| \overset{?}{=} \left|-\frac{5}{3} + 8\right|$$
$$\left|-\frac{19}{3}\right| \overset{?}{=} \left|\frac{19}{3}\right|$$
$$\frac{19}{3} = \frac{19}{3} \checkmark$$

2.4 Practice Problems

1. Let n = number of hours.

For 12 months per year, we multiply $400 for rent by 12.

Let $8n$ = the cost for using the computer for n hours.

$$400(12) + 8n = 7680$$
$$4800 + 8n = 7680$$
$$8n = 2880$$
$$n = 360$$

They used the computer for 360 hours.

2. Let n = the number of weeks on tour.

$$\text{weekly income} = 5(6000)(14) = 420{,}000$$
$$\text{weekly expenses} = 5(48{,}000) + 150{,}000 = 390{,}000$$
$$\text{earnings goal} = 5(60{,}000) = 300{,}000$$
$$420{,}000n - 390{,}000n = 300{,}000$$
$$30{,}000n = 300{,}000$$
$$n = 10$$

The group needs to tour for 10 weeks.

3. Let x = the length of the second side.

Then $2x$ = the length of the first side and $3x - 6$ = the length of the third side.

$$p = a + b + c$$
$$162 = x + 2x + 3x - 6$$
$$162 = 6x - 6$$
$$168 = 6x$$
$$x = 28$$

The 1st side is $2x = 2(28) = 56$ meters.

The 2nd side is $x = 28$ meters.

The 3rd side is $3x - 6 = 3(28) - 6 = 78$ meters.

2.5 Practice Problems

1. Let x = amount of sales last year,

and $0.15x$ = the increase in sales over last year.

$$x + 0.15x = 6900$$
$$1.15x = 6900$$
$$x = 6000$$

6000 computer workstations were sold last year.

2. Let a = the number of cars Alicia sold. Then $43 - a$ = the number of cars Heather sold. If Alicia doubles her sales next month, then she will sell $2a$ cars. If Heather triples hers, she will sell $3(43 - a)$ cars.

$$2a + 3(43 - a) = 108$$
$$2a + 129 - 3a = 108$$
$$-a + 129 = 108$$
$$-a = -21$$
$$a = 21$$

Therefore, Alicia sold 21 cars and Heather sold $43 - 21 = 22$ cars.

3. Let x = the amount invested at 8%, then $5500 - x$ = the amount invested at 12%.

$$0.08x + 0.12(5500 - x) = 540$$
$$0.08x + 660 - 0.12x = 540$$
$$-0.04x = -120$$
$$x = \$3000$$

Therefore, she invested $3000 at 8% and $5500 - 3000 = \$2500$ at 12%.

4.

	A	B	C
	Number of Grams of the Mixture	**% Pure Gold**	**Number of Grams of Pure Gold**
68% pure gold source	x	68%	$0.68x$
83% pure gold source	$200 - x$	83%	$0.83(200 - x)$
Final 80% mixture of pure gold	200	80%	$0.80(200)$

Now form an equation from the entries in Column C.

$$0.68x + 0.83(200 - x) = 0.80(200)$$
$$0.68x + 166 - 0.83x = 160$$
$$-0.15x = -6$$
$$x = 40$$

If $x = 40$, then $200 - x = 200 - 40 = 160$.

Thus, we have 40 grams of 68% pure gold and 160 grams of 83% pure gold.

5.

	r	t	d
Steady speed	x	4	$4x$
Slower speed	$x - 10$	2	$2(x - 10)$
Entire trip	Not appropriate	6	352

$$4x + 2(x - 10) = 352$$
$$4x + 2x - 20 = 352$$
$$6x - 20 = 352$$
$$6x = 372$$
$$x = 62$$

Thus, Wally drove 62 miles per hour for 4 hours and $x - 10 = 52$ miles per hour for 2 hours.

2.6 Practice Problems

1. (a) $-1 > -2$ **(b)** $\dfrac{2}{3} < \dfrac{3}{4}$ **(c)** $-0.561 < 0.5555$

2. (a) $(-8 - 2) \; ? \; (-3 - 12)$ **(b)** $|-15 + 8| \; ? \; |7 - 13|$
$(-10) > (-15)$ $|-7| \; ? \; |-6|$
    $7 > 6$

3. (a) $x > 3.5$

(b) $x \le -10$

(c) $x \geq -2$

(d) $-4 > x$

4. $x + 2 > -12$
$x > -14$

To check, we choose -13.
Is $-13 + 2 > -12$
$-11 > -12$ ✓ True

5. $8x - 8 \geq 5x + 1$
$3x \geq 9$
$x \geq 3$

6. $2 - 12x > 7(1 - x)$
$2 - 12x > 7 - 7x$
$-5x > 5$
$x < -1$

7. $0.5x - 0.4 \leq -0.8x + 0.9$
$5x - 4 \leq -8x + 9$
$13x \leq 13$
$x \leq 1$

8. $\dfrac{1}{5}(x - 6) < \dfrac{1}{3}(x - 2)$

$\dfrac{1}{5}x - \dfrac{6}{5} < \dfrac{1}{3}x - \dfrac{2}{3}$

$3x - 18 < 5x - 10$
$-2x < 8$
$x > -4$

9. Let x = the number of minutes after the first minute that they place the call. The cost must be less than or equal to $13.90

$3.50 + 0.65x \leq 13.90$
$0.65x \leq 10.40$ We subtract 3.50 from each side
$x \leq 16$ We divide each side by 0.65

Now we add the 16 minutes to the one minute that cost $3.50. This gives us 17 minutes. Thus the maximum amount of time they can talk is 17 minutes.

2.7 Practice Problems

1. $-8 < x$ and $x < -2$

2. $-1 \leq x \leq 5$

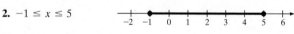

3. $-10 \leq x \leq -5.5$

4. $200 \leq x \leq 950$

5. $x < 8$ or $x > 12$

6. $x \leq -6$ or $x > 3$

7.

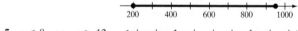

8. $3x - 4 < -1$ or $2x + 3 > 13$
$3x < 3$ $2x > 10$
$x < 1$ or $x > 5$

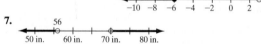

9. $3x + 6 > -6$ and $4x + 5 < 1$
$3x > -12$ $4x < -4$
$x > -4$ and $x < -1$
$-4 < x < -1$

10. $-2x + 3 < -7$ and $7x - 1 > -15$
$-2x < -10$ $7x > -14$
$x > 5$ and $x > -2$
$x > 5$ and at the same time $x > -2$. Thus $x > 5$ is the solution to the compound inequality.

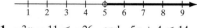

11. $-3x - 11 < -26$ and $5x + 4 < 14$
$-3x < -15$ $5x < 10$
$x > 5$ and $x < 2$
Now clearly it is impossible for one number to be greater than 5 and at the same time less than 2. There is no solution.

2.8 Practice Problems

1. $|x| < 2$
$-2 < x < 2$

$-2 < x < 2$

2. $|x - 6| < 15$
$-15 < x - 6 < 15$
$-15 + 6 < x - 6 + 6 < 15 + 6$
$-9 < x < 21$

$-9 < x < 21$

3. $\left| x + \dfrac{3}{4} \right| \leq \dfrac{7}{6}$

$-\dfrac{7}{6} \leq x + \dfrac{3}{4} \leq \dfrac{7}{6}$

$-14 \leq 12x + 9 \leq 14$

$-14 - 9 \leq 12x + 9 - 9 \leq 14 - 9$

$-23 \leq 12x \leq 5$

$\dfrac{-23}{12} \leq \dfrac{12x}{12} \leq \dfrac{5}{12}$

$-1\dfrac{11}{12} \leq x \leq \dfrac{5}{12}$

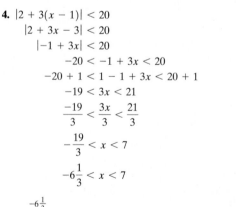

$-1\dfrac{11}{12} \leq x \leq \dfrac{5}{12}$

4. $|2 + 3(x - 1)| < 20$
$|2 + 3x - 3| < 20$
$|-1 + 3x| < 20$
$-20 < -1 + 3x < 20$
$-20 + 1 < 1 - 1 + 3x < 20 + 1$
$-19 < 3x < 21$
$\dfrac{-19}{3} < \dfrac{3x}{3} < \dfrac{21}{3}$
$-\dfrac{19}{3} < x < 7$
$-6\dfrac{1}{3} < x < 7$

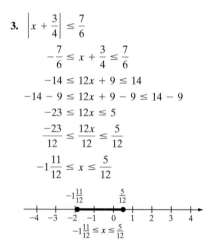

$-6\dfrac{1}{3} < x < 7$

5. $|x| > 2.5$
$x > 2.5$ or $x < -2.5$

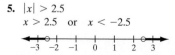

6. $|x + 6| > 2$

$\quad x + 6 > 2 \quad$ or $\quad x + 6 < -2$

$\qquad x > -4 \quad$ or $\qquad x < -8$

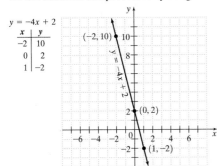

$\qquad\qquad x < -8 \text{ or } x > -4$

7. $|-5x - 2| > 13$

$\quad -5x - 2 > 13 \quad$ or $\quad -5x - 2 < -13$

$\quad\; -5x > 15 \quad$ or $\qquad -5x < -11$

$\qquad x < -3 \quad$ or $\qquad\quad x > \dfrac{11}{5}$

$\qquad\qquad\qquad\qquad\qquad\qquad x > 2\dfrac{1}{5}$

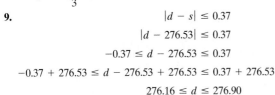

$\qquad x < -3 \quad$ or $\quad x > 2\dfrac{1}{5}$

8. $\left| 4 - \dfrac{3}{4}x \right| \geq 5$

$\quad 4 - \dfrac{3}{4}x \geq 5 \quad$ or $\quad 4 - \dfrac{3}{4}x \leq -5$

$\quad 16 - 3x \geq 20 \qquad\quad 16 - 3x \leq -20$

$\qquad\; -3x \geq 4 \qquad\qquad\; -3x \leq -36$

$\qquad\quad x \leq \dfrac{-4}{3} \qquad\qquad\quad x \geq 12$

$\qquad\quad x \leq -1\dfrac{1}{3}$

$\qquad x \leq -1\dfrac{1}{3} \quad$ or $\quad x \geq 12$

9.

$$|d - s| \leq 0.37$$

$$|d - 276.53| \leq 0.37$$

$$-0.37 \leq d - 276.53 \leq 0.37$$

$$-0.37 + 276.53 \leq d - 276.53 + 276.53 \leq 0.37 + 276.53$$

$$276.16 \leq d \leq 276.90$$

Thus the diameter of the transmission must be at least 276.16 millimeters, but not greater than 276.90 millimeters.

Chapter 3

3.1 Practice Problems

1. Let's choose $x = -2$, $x = 0$, and $x = 1$.

For $x = -2$, $y = -4(-2) + 2 = 10$.

For $x = 0$, $y = -4(0) + 2 = 2$.

We can condense this procedure by using a table.

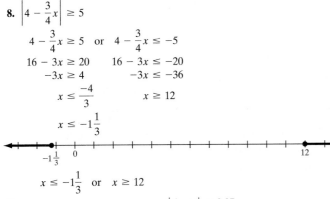

$y = -4x + 2$

x	y
-2	10
0	2
1	-2

2. Let $x = 0$.

$\quad 3(0) - 2y = -6$

$\qquad\quad -2y = -6$

$\qquad\qquad y = 3$

The y-intercept is $(0, 3)$.

Now let $y = 0$.

$3x - 2(0) = -6$

$\qquad 3x = -6$

$\qquad\; x = -2$

The x-intercept is $(-2, 0)$.

Let's pick $y = 9$.

$3x - 2(9) = -6$

$\quad 3x - 18 = -6$

$\qquad\quad 3x = -6 + 18$

$\qquad\quad 3x = 12$

$\qquad\quad x = \dfrac{12}{3} = 4$

The third point is $(4, 9)$.

x	y
-2	0
0	3
4	9

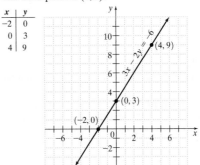

3. (a) The equation $x = 4$ means that for any value of y, x is 4. The graph of $x = 4$ is a vertical line 4 units to the right of the origin.

(b) The equation $3y + 12 = 0$ can be simplified.

$$3y + 12 = 0$$
$$3y = -12$$
$$y = -4$$

The equation $y = -4$ means that for any value of x, y is -4.

The graph of $y = -4$ is a horizontal line 4 units below the x-axis.

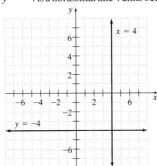

4. When $n = 0$, then $C = 300 + 0.15(0) = 300 + 0 = 300$.

When $n = 1000$, then $C = 300 + 0.15(1000) = 300 + 150 = 450$.

When $n = 2000$, then $C = 300 + 0.15(2000) = 300 + 300 = 600$.

We then have

n	C
0	300
1000	450
2000	600

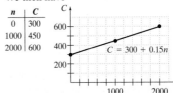

Since n varies from 0 to 2000 and C varies from 300 to 600, we let each square on the horizontal scale represent 200 products and each square on the vertical scale represent $100.

3.2 Practice Problems

1. Identify the y-coordinates and the x-coordinates for the points $(-6, 1)$ and $(-5, -2)$.

$$\overbrace{(-6, 1) \qquad\qquad (-5, -2)}^{y_2 - y_1}_{x_2 - x_1}$$

Use the formula.

$$\text{Slope} = m = \frac{y_2 - y_1}{x_2 - x_1} = \frac{-2 - 1}{-5 - (-6)} = \frac{-3}{1} = -3$$

2. (a) $m = \dfrac{-3.4 - (-6.2)}{-2.2 - 1.8} = \dfrac{2.8}{-4} = -0.7$

(b) $m = \dfrac{-\dfrac{3}{4} - \left(-\dfrac{1}{2}\right)}{\dfrac{4}{15} - \dfrac{1}{5}} = \dfrac{-\dfrac{1}{4}}{\dfrac{1}{15}} = -\dfrac{15}{4}$

3. slope $= \dfrac{\text{rise}}{\text{run}} = \dfrac{25.92}{1296} = 0.02$ or $\dfrac{1}{50}$

4. $m_l = \dfrac{-2 - 0}{6 - 5} = \dfrac{-2}{1} = -2$

$m_h = \dfrac{1}{2}$

5. $m_{AB} = \dfrac{1 - 5}{-1 - 1} = \dfrac{-4}{-2} = 2$

$m_{BC} = \dfrac{-1 - 1}{-2 - (-1)} = \dfrac{-2}{-2 + 1} = \dfrac{-2}{-1} = 2$

Since the slopes are the same and have a point in common, the points lie on the same line.

6. From the 4.5-mile point to the 3.5-mile point, the jet traveled 1200 feet downward over 1 horizontal mile. From the 3.5-mile point to the 1.3-mile point, the jet traveled 2640 feet downward over 2.2 horizontal miles. Thus, the first slope is 1200 feet per mile, and the second slope is 1200 feet per mile.

The slopes of the two portions of the flight are the same. The two portions of the flight have one point in common. Therefore the jet is descending in a straight line.

3.3 Practice Problems

1. $y = mx + b$

Substitute 4 for m and $-\dfrac{3}{2}$ for b.

$$y = (4)x + \left(-\dfrac{3}{2}\right)$$

$$y = 4x - \dfrac{3}{2}$$

2. y-intercept is $(0, -3)$, thus $b = -3$. Another point on the line is $(5, 0)$. Thus

$(x_2, y_2) = (0, -3)$

$(x_1, y_1) = (5, 0)$

$m = \dfrac{y_2 - y_1}{x_2 - x_1} = \dfrac{-3 - 0}{0 - 5} = \dfrac{3}{5}$

$y = mx + b$, substitute $\dfrac{3}{5}$ for m and -3 for b.

$$y = \dfrac{3}{5}x - 3$$

3. $3x - 4y = -8$

$-4y = -3x - 8$

$\dfrac{-4y}{-4} = \dfrac{-3x}{-4} + \dfrac{-8}{-4}$

$y = \dfrac{3}{4}x + 2$

The slope is $\dfrac{3}{4}$ and the y-intercept is $(0, 2)$.

Plot the point $(0, 2)$. From this point go up 3 units and to the right 4 units to locate a second point. Draw a straight line that contains these two points.

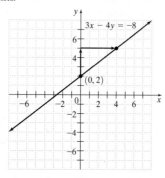

4. $y - y_1 = m(x - x_1)$

$y - (-2) = \dfrac{3}{4}(x - 5)$

$y + 2 = \dfrac{3}{4}x - \dfrac{15}{4}$

$4y + 4(2) = 4\left(\dfrac{3}{4}x\right) - 4\left(\dfrac{15}{4}\right)$

$4y + 8 = 3x - 15$

$-3x + 4y = -15 - 8$

$3x - 4y = 23$

5. $(-4, 1)$ and $(-2, -3)$

$m = \dfrac{y_2 - y_1}{x_2 - x_1} = \dfrac{-3 - 1}{-2 - (-4)} = \dfrac{-4}{-2 + 4} = \dfrac{-4}{2} = -2$

Substitute $m = -2$ and $(x_1, y_1) = (-4, 1)$ into the point–slope equation.

$y - y_1 = m(x - x_1)$

$y - 1 = -2[x - (-4)]$

$y - 1 = -2(x + 4)$

$y - 1 = -2x - 8$

$y = -2x - 7$

6. First, we need to find the slope of the line $5x - 3y = 10$. We do this by writing the equation in slope–intercept form

$5x - 3y = 10$

$-3y = -5x + 10$

$y = \dfrac{5}{3}x - \dfrac{10}{3}$

The slope is $\dfrac{5}{3}$. A line parallel to this passing through $(4, -5)$ would have an equation

$y - (-5) = \dfrac{5}{3}(x - 4)$

$y + 5 = \dfrac{5}{3}x - \dfrac{20}{3}$

$3y + 3(5) = 3\left(\dfrac{5}{3}x\right) - 3\left(\dfrac{20}{3}\right)$

$3y + 15 = 5x - 20$

$-5x + 3y = -35$

$5x - 3y = 35$

7. Find the slope of the line $6x + 3y = 7$ by rewriting it in slope–intercept form.

$6x + 3y = 7$

$3y = -6x + 7$

$y = -2x + \dfrac{7}{3}$

The slope is -2. A line perpendicular to this passing through $(-4, 3)$ would have a slope of $\dfrac{1}{2}$, and would have the equation

$y - 3 = \dfrac{1}{2}[x - (-4)]$

$y - 3 = \dfrac{1}{2}(x + 4)$

$y - 3 = \dfrac{1}{2}x + 2$

$2y - 2(3) = 2\left(\dfrac{1}{2}x\right) + 2(2)$

$2y - 6 = x + 4$

$-x + 2y = 10$

$x - 2y = -10$

3.4 Practice Problems

1. The boundary line is $y = 3x + 1$. Graph the boundary line with a dashed line since the inequality contains $>$. Substituting $(0, 0)$ into the inequality $y > 3x + 1$ gives $0 > 1$, which is false. Thus we shade on the side of the boundary opposite $(0, 0)$. The solution is the shaded region not including the dashed line.

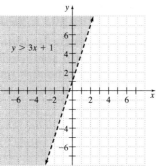

$y > 3x + 1$

2. The boundary line is $-4x + 5y = -10$. Graph the boundary line with a solid line because the inequality contains $\leq$. Substituting $(0,0)$ into the inequality $-4x + 5y \leq -10$ gives $0 \leq -10$ which is false. Therefore, shade the region on the side of the boundary opposite $(0,0)$. The solution is the shaded region including the solid line.

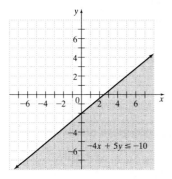

$-4x + 5y \leq -10$

3. The boundary line is $3y + x = 0$. We use a dashed line since the inequality contains $<$. We cannot use $(0,0)$ to test the inequality. let's pick $(-2, -3)$. Substituting $(-2, -3)$ into $3y + x < 0$ gives $-11 < 0$ which is true. Therefore, shade the side of the boundary line that contains $(-2, -3)$. The solution is the shaded region not including the dashed line.

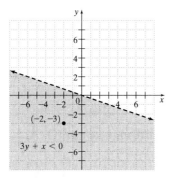

$(-2,-3)$

$3y + x < 0$

4. Simplify the original inequality by dividing each side by 6, which gives $y \leq 3$. We draw a solid horizontal line at $y = 3$. We use a solid line since the inequality contains $\leq$. The region we want to shade is the region below the horizontal line. The solution is the line $y = 3$ and the shaded region below that line.

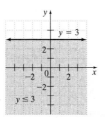

$y = 3$

$y \leq 3$

3.5 Practice Problems

1. Tables

Year	1936	1948	1960	1972	1984	1996	2000
Time	11.5	11.9	11.0	11.07	10.97	10.94	10.75

Ordered pairs

$\{(1936, 11.5), (1948, 11.9), (1960, 11.0), (1972, 11.07),$
$(1984, 10.97), (1996, 10.94), (2000, 10.75)\}$

Graph To save space, use the time values from 10 seconds to 12 seconds on the vertical axis. Plot the ordered pairs.

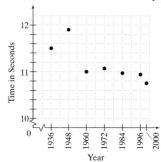

2. Recall the domain is the set of all the first items in each ordered pair. The range is the set of all the second items in each ordered pair. Thus,

Domain = $\{158, 160, 161, 163\}$
Range = $\{89.2, 94.9, 98.6, 101.4, 102.3\}$

$(161, 89.2)$ and $(161, 94.9)$ have the same first coordinate. This relation is not a function.

3. (a) This is a function. No vertical line will pass through more than one ordered pair on the curve.

(b) This is not a function. A vertical line could pass through $(2, 2)$ and $(2, -2)$.

4. (a) $f(-3) = 2(-3)^2 - 8 = 2(9) - 8 = 18 - 8 = 10$

(b) $f(4) = 2(4)^2 - 8 = 2(16) - 8 = 32 - 8 = 24$

(c) $f\left(\dfrac{1}{2}\right) = 2\left(\dfrac{1}{2}\right)^2 - 8 = 2\left(\dfrac{1}{4}\right) - 8 = \dfrac{1}{2} - \dfrac{16}{2} = -\dfrac{15}{2}$

3.6 Practice Problems

1. Income = $10,000 + 15\%$ of total sales
Let x = amount of total sales in dollars.

$$d(x) = 10,000 + 0.15(x)$$
$$d(0) = 10,000 + 0.15(0) = 10,000$$
$$d(40,000) = 10,000 + 0.15(40,000) = 16,000$$
$$d(80,000) = 10,000 + 0.15(80,000) = 22,000$$

Modify the table by recording x in thousands to make the task of graphing easier.

x	$d(x)$
0	10,000
40	16,000
80	22,000

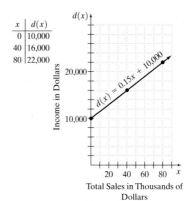

2. Find the function values for five values of x. The table of values and the resulting graph are

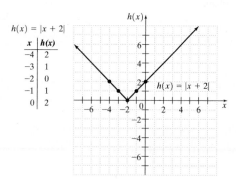

$h(x) = |x + 2|$

x	$h(x)$
−4	2
−3	1
−2	0
−1	1
0	2

3. For each function we will choose five values of x. Since these are not linear functions we use a curved line to connect the points. The tables of values and the resulting graphs are

(a)

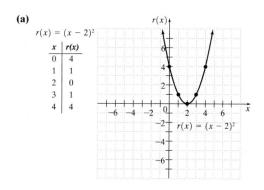

$r(x) = (x − 2)^2$

x	$r(x)$
0	4
1	1
2	0
3	1
4	4

(b)

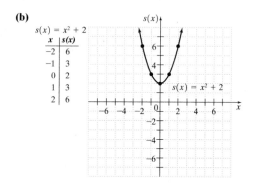

$s(x) = x^2 + 2$

x	$s(x)$
−2	6
−1	3
0	2
1	3
2	6

4. Pick five values of x, find the corresponding function values, and plot the four points to assist in sketching the graph.

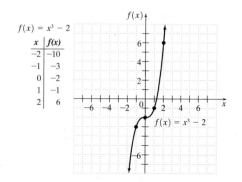

$f(x) = x^3 − 2$

x	$f(x)$
−2	−10
−1	−3
0	−2
1	−1
2	6

5. We cannot choose x to be zero. Therefore, choose five values for x greater than zero and five values for x less than zero.

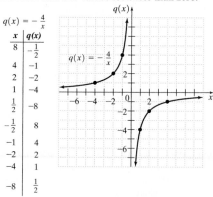

$q(x) = −\frac{4}{x}$

x	$q(x)$
8	$−\frac{1}{2}$
4	−1
2	−2
1	−4
$\frac{1}{2}$	−8
$−\frac{1}{2}$	8
−1	4
−2	2
−4	1
−8	$\frac{1}{2}$

6. (a)

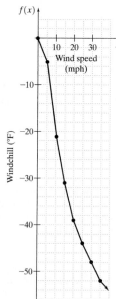

(b) The function is a linear function. As the number of patients per hour increases so does the weekly profit.

(c) We find that the value $x = 3$ on the graph corresponds to the values $y = 0$. Thus we would expect that if the doctor saw 3 patients per hour he would have 0 profit.

(d) We find that the value $x = 0$ on the graph corresponds to the value $y = −2000$. Thus we would expect that if the doctor saw 0 patients per hour he would have a loss of $2,000.00.

(e) No. In terms of mathematics, the function will increase indefinitely; however, is it reasonable to think that a doctor could see an infinite number of patients?

7. (a) Since windchill depends on wind speed, wind speed is the independent variable and windchill is the dependent variable. Label the vertical axis "Windchill" and the horizontal axis "Wind Speed". Plot the points. Use a smooth curve to connect the points.

(b) Find 32 along the horizontal axis. Move down to the curve, then move right until you intersect the vertical axis. The function value on the vertical axis for 32 mph is about −49°F.

(c) Since the last windchill number in the Table is -52, we need to extend the curve. At 40 mph the windchill is about $-54°F$.

(d) Find -24 along the vertical scale. Move to the right until you intersect the curve. Then move up until you interect the horizontal axis. This is slightly more than 11 mph.

(e) In theory based on our graph, the domain is all nonnegative real numbers. The range is all nonpositive real numbers.

8.

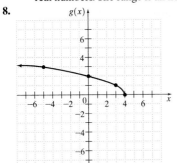

(a) Find $x = -3$ along the horizontal axis. Move up until you intersect the graph. Then move to the right until you intersect the vertical axis. Read the scale. Thus, $f(-3)$ is approximately 2.5.

(b) Find $f(x) = 1.5$ along the vertical axis. Move to the right until you intersect the graph. Then move down until you intersect the horizontal axis. Read the scale. Thus, x is approximately 1.6.

(c) The curve is moving upward more slowly. Thus, the slope is decreasing as x decreases.

Chapter 4

4.1 Practice Problems

1. Substitute $(-3, 4)$ into the first equation to see if the ordered pair is a solution.

$$2x + 3y = 6$$
$$2(-3) + 3(4) \overset{?}{=} 6$$
$$-6 + 12 \overset{?}{=} 6$$
$$6 = 6 \checkmark$$

Likewise, we will determine if $(-3, 4)$ is a solution to the second equation.

$$3x - 4y = 7$$
$$3(-3) - 4(4) \overset{?}{=} 7$$
$$-9 - 16 \overset{?}{=} 7$$
$$-25 \neq 7$$

Since $(-3, 4)$ is not a solution to each equation in the system, it is not a solution to the system itself.

2. You can use any method we developed in Chapter 3 to graph each line. We will change each equation to slope-intercept form to graph.

$$3x + 2y = 10$$
$$2y = -3x + 10$$
$$y = -\frac{3}{2}x + 5$$
$$x - y = 5$$
$$y = x - 5$$

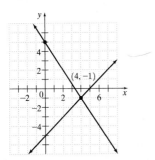

The lines intersect at the point $(4, -1)$. Thus $(4, -1)$ is a solution. We verify this by substituting $x = 4$ and $y = -1$ into the system of equations.

$$3x + 2y = 10 \qquad\qquad x - y = 5$$
$$3(4) + 2(-1) \overset{?}{=} 10 \qquad 4 - (-1) \overset{?}{=} 5$$
$$12 - 2 \overset{?}{=} 10 \qquad\qquad 5 = 5 \checkmark$$
$$10 = 10 \checkmark$$

3. $2x - y = 7 \quad [1]$
$3x + 4y = -6 \quad [2]$
Solve equation [1] for y.
$$-y = 7 - 2x$$
$$y = -7 + 2x \quad [3]$$
Substitute $-7 + 2x$ for y in equation [2].
$$3x + 4(-7 + 2x) = -6$$
$$3x - 28 + 8x = -6$$
$$11x - 28 = -6$$
$$11x = 22$$
$$x = 2$$
Substitute $x = 2$ into equation [3].
$$y = -7 + 2(2)$$
$$y = -7 + 4$$
$$y = -3$$
The solution is $(2, -3)$.

4. $\dfrac{x}{2} + \dfrac{2y}{3} = 1$

$\dfrac{x}{3} + y = -1$

First we clear both equations of fractions.
$$3x + 4y = 6 \quad [1]$$
$$x + 3y = -3 \quad [2]$$

Step 1 Solve for x in equation [2].
$$x + 3y = -3$$
$$x = -3y - 3$$

Steps 2 and 3 Substitute this expression for x into equation [1] and solve for y.
$$3(-3y - 3) + 4y = 6$$
$$-9y - 9 + 4y = 6$$
$$-5y = 15$$
$$y = -3$$

Step 4 Substitute $y = -3$ into equation [1] or [2].
$$x + 3y = -3$$
$$x + 3(-3) = -3$$
$$x - 9 = -3$$
$$x = 6$$
So our solution is $(6, -3)$.

5. $-3x + y = 5 \quad [1]$
$2x + 3y = 4 \quad [2]$
Multiply equation [1] by -3 and add to equation [2].
$$9x - 3y = -15$$
$$\underline{2x + 3y = 4}$$
$$11x = -11$$
$$x = -1$$
Now substitute $x = -1$ into equation [1].
$$-3(-1) + y = 5$$
$$3 + y = 5$$
$$y = 2$$
The solution is $(-1, 2)$.

6. $5x + 4y = 23 \quad [1]$
$7x - 3y = 15 \quad [2]$
Multiply equation [1] by 3 and equation [2] by 4.
$$15x + 12y = 69$$
$$\underline{28x - 12y = 60}$$
$$43x = 129$$
$$x = 3$$

Now substitute $x = 3$ into equation [1].
$$5(3) + 4y = 23$$
$$15 + 4y = 23$$
$$4y = 8$$
$$y = 2$$
The solution is $(3, 2)$.

7. $4x - 2y = 6$ [1]
$-6x + 3y = 9$ [2]
Multiply equation [1] by 3 and equation [2] by 2 and add together.
$$12x - 6y = 18$$
$$\underline{-12x + 6y = 18}$$
$$0 = 36$$
This statement is of course false. Thus, we conclude that this system of equations is inconsistent, so there is **no solution.**

8. $0.3x - 0.9y = 1.8$ [1]
$-0.4x + 1.2y = -2.4$ [2]
Multiply both equations by 10 to obtain a more convenient form.
$$3x - 9y = 18 \quad [3]$$
$$-4x + 12y = -24 \quad [4]$$
Multiply equation [3] by 4 and equation [4] by 3.
$$12x - 36y = 72$$
$$\underline{-12x + 36y = -72}$$
$$0 = 0$$
This statement is always true. Hence these are dependent equations. There are an infinite number of solutions.

9. (a) $3x + 5y = 1485$
$x + 2y = 564$

Solve for x in the second equation and solve using the substitution method.
$$x = -2y + 564$$
$$3(-2y + 564) + 5y = 1485$$
$$-6y + 1692 + 5y = 1485$$
$$-y = -207$$
$$y = 207$$
Substitute $y = 207$ into the second equation and solve for x.
$$x + 2(207) = 564$$
$$x + 414 = 564$$
$$x = 150 \qquad \text{The solution is } (150, 207).$$

(b) $7x + 6y = 45$
$6x - 5y = -2$

Using the addition method, multiply the first equation by -6 and the second equation by 7.
$$-6(7x) + (-6)(6y) = (-6)(45)$$
$$7(6x) - 7(5y) = 7(-2)$$
$$-42x - 36y = -270$$
$$\underline{42x - 35y = -14}$$
$$-71y = -284$$
$$y = 4$$
Substitute $y = 4$ into the first equation and solve for x.
$$7x + 6(4) = 45$$
$$7x + 24 = 45$$
$$7x = 21$$
$$x = 3 \qquad \text{The solution is } (3, 4).$$

4.2 Practice Problems

1. Substitute $x = 3$, $y = -2$, $z = 2$ into each equation.
$$2(3) + 4(-2) + (2) \stackrel{?}{=} 0$$
$$6 - 8 + 2 \stackrel{?}{=} 0$$
$$-2 + 2 \stackrel{?}{=} 0$$
$$0 = 0 \quad \checkmark$$
$$(3) - 2(-2) + 5(2) \stackrel{?}{=} 17$$
$$3 + 4 + 10 \stackrel{?}{=} 17$$
$$17 = 17 \quad \checkmark$$
$$3(3) - 4(-2) + 2 \stackrel{?}{=} 19$$
$$9 + 8 + 2 \stackrel{?}{=} 19$$
$$19 = 19 \quad \checkmark$$
Since we obtained three true statements, the ordered triple $(3, -2, 2)$ is a solution to the system.

2. $x + 2y + 3z = 4$ [1]
$2x + y - 2z = 3$ [2]
$3x + 3y + 4z = 10$ [3]
We eliminate x by multiplying equation [1] by -2 and adding it to equation [2].
$$-2x - 4y - 6z = -8 \quad [4]$$
$$\underline{2x + y - 2z = 3} \quad [2]$$
$$-3y - 8z = -5 \quad [5]$$
Now we eliminate x by multiplying equation [1] by -3 and adding it to equation [3].
$$-3x - 6y - 9z = -12 \quad [6]$$
$$\underline{3x + 3y + 4z = 10} \quad [3]$$
$$-3y - 5z = -2 \quad [7]$$
We now eliminate y and solve for z in the system formed by equation [5] and equation [7].
$$-3y - 8z = -5 \quad [5]$$
$$-3y - 5z = -2 \quad [7]$$
To do this we multiply equation [5] by -1 and add it to equation [7].
$$3y + 8z = 5 \quad [8]$$
$$\underline{-3y - 5z = -2} \quad [7]$$
$$3z = 3$$
$$z = 1$$
Substitute $z = 1$ into equation [8] and solve for y.
$$3y + 8(1) = 5$$
$$3y + 8 = 5$$
$$3y = -3$$
$$y = -1$$
Substitute $z = 1$, $y = -1$ into equation [1] and solve for x.
$$x + 2(-1) + 3(1) = 4$$
$$x + 1 = 4$$
$$x = 3$$
The solution is $(3, -1, 1)$.

3. $2x + y + z = 11$ [1]
$4y + 3z = -8$ [2]
$x - 5y = 2$ [3]
Multiply equation [1] by -3 and add the results to equation [2], thus eliminating the z terms.
$$-6x - 3y - 3z = -33 \quad [4]$$
$$\underline{4y + 3z = -8} \quad [2]$$
$$-6x + y = -41 \quad [5]$$
We can solve the system formed by equation [3] and equation [5].
$$x - 5y = 2 \quad [3]$$
$$-6x + y = -41 \quad [5]$$
Multiply equation [3] by 6 and add the results to equation [5].
$$6x - 30y = 12 \quad [6]$$
$$\underline{-6x + y = -41} \quad [5]$$
$$-29y = -29$$
$$y = 1$$
Now substitute $y = 1$ into equation [2] and solve for z.
$$4(1) + 3z = -8$$
$$4 + 3z = -8$$
$$3z = -12$$
$$z = -4$$
Now substitute $y = 1$, $z = -4$ into equation [1] and solve for x.
$$2x + 1 + (-4) = 11$$
$$2x - 3 = 11$$
$$2x = 14$$
$$x = 7$$
The solution is $(7, 1, -4)$.

4.3 Practice Problems

1. Let x = the number of baseballs purchased and y = the number of bats purchased.

Last week: $6x + 21y = 318$ [1]
This week: $5x + 17y = 259$ [2]

Multiply equation [1] by 5 and equation [2] by -6.

$$\begin{array}{rl}
30x + 105y = & 1590 \quad [3] \\
-30x - 102y = & -1554 \quad [4] \\
\hline
3y = & 36 \quad \text{Add equations [3] and [4].} \\
y = & 12
\end{array}$$

Substitute $y = 12$ into equation [2].

$$5x + 17(12) = 259$$
$$5x + 204 = 259$$
$$5x = 55$$
$$x = 11$$

Thus 11 baseballs and 12 bats were purchased.

2. Let $x =$ the number of small chairs and $y =$ the number of large chairs. (*Hint:* Change all hours to minutes)

$$30x + 40y = 1560 \quad [1]$$
$$75x + 80y = 3420 \quad [2]$$

Multiply equation [1] by -2 and add the results to equation [2].

$$\begin{array}{rl}
-60x - 80y = & -3120 \quad [3] \\
75x + 80y = & 3420 \quad [2] \\
\hline
15x = & 300
\end{array}$$
$$x = 20$$

Substitute $x = 20$ in either equation [1] or [2] and solve for y.

$$30(20) + 40y = 1560$$
$$600 + 40y = 1560$$
$$40y = 960$$
$$y = 24$$

Therefore, the company can make 20 small chairs and 24 large chairs each day.

3. Let $a =$ the speed of the airplane in still air in kilometers per hour and $w =$ the speed of the wind in kilometers per hour.

	R	$\cdot\ T$	$=\ D$
Against the wind	$a - w$	3	1950
With the wind	$a + w$	2	1600

We obtain a system of equations from the chart.

$$(a - w)3 = 1950$$
$$(a + w)2 = 1600$$

We remove the parentheses.

$$3a - 3w = 1950 \quad [1]$$
$$2a + 2w = 1600 \quad [2]$$

Multiply equation [1] by 2 and equation [2] by 3 and add the resulting equations.

$$\begin{array}{rl}
6a - 6w = & 3900 \\
6a + 6w = & 4800 \\
\hline
12a = & 8700
\end{array}$$
$$a = 725$$

Substituting $a = 725$ into equation [2] we have

$$2(725) + 2w = 1600$$
$$1450 + 2w = 1600$$
$$2w = 150$$
$$w = 75$$

Thus, the speed of the plane in still air is 725 kilometers per hour and the speed of the wind is 75 kilometers per hour.

4. $\begin{aligned}
A + B + C &= 260 \quad [1] \\
3A + 2B &= 390 \quad [2] \\
3B + 4C &= 655 \quad [3]
\end{aligned}$

Multiply equation [1] by -3 and add it to equation [2].

$$\begin{array}{rl}
-3A - 3B - 3C = & -780 \quad [4] \\
3A + 2B = & 390 \quad [2] \\
\hline
-B - 3C = & -390 \quad [5]
\end{array}$$

Now multiply equation [5] by 3 and add it to equation [3].

$$\begin{array}{rl}
-3B - 9C = & -1170 \quad [6] \\
3B + 4C = & 655 \quad [3] \\
\hline
-5C = & -515
\end{array}$$
$$C = 103$$

Substitute $C = 103$ into equation [3] and solve for B.

$$3B + 4(103) = 655$$
$$3B + 412 = 655$$
$$3B = 243$$
$$B = 81$$

Now substitute $B = 81$ into equation [2] and solve for A.

$$3A + 2(81) = 390$$
$$3A + 162 = 390$$
$$3A = 228$$
$$A = 76$$

Machine A wraps 76 boxes per hour, machine B wraps 81 boxes per hour, and machine C wraps 103 boxes per hour.

4.4 Practice Problems

1.

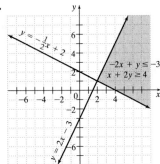

2.

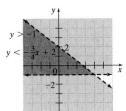

3.

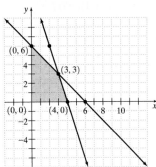

The vertices of the solution are $(3, 3)$, $(0, 6)$, $(4, 0)$, and $(0, 0)$.

Chapter 5

5.1 Practice Problems

1. (a) $3x^5 - 6x^4 + x^2$

    This is a trinomial of degree 5.

   (b) $5x^2 + 2$

    This is a binomial of degree 2.

   (c) $3ab + 5a^2b^2 - 6a^4b$

    This is a trinomial of degree 5.

   (d) $16x^4y^6$

    This is a monomial of degree 10.

2. $p(x) = 2x^4 - 3x^3 + 6x - 8$

   (a) $p(-2) = 2(-2)^4 - 3(-2)^3 + 6(-2) - 8$

    $= 2(16) - 3(-8) + 6(-2) - 8$

    $= 32 + 24 - 12 - 8$

    $= 36$

(b) $p(5) = 2(5)^4 - 3(5)^3 + 6(5) - 8$
$= 2(625) - 3(125) + 6(5) - 8$
$= 1250 - 375 + 30 - 8$
$= 897$

3. $(-7x^2 + 5x - 9) + (2x^2 - 3x + 5)$
We remove the parentheses and combine like terms.
$= -7x^2 + 2x^2 + 5x - 3x - 9 + 5$
$= -5x^2 + 2x - 4$

4. $(2x^2 - 14x + 9) - (-3x^2 + 10x + 7)$
We add the opposite of the second polynomial to the first polynomial.
$= (2x^2 - 14x + 9) + (3x^2 - 10x - 7)$
$= 2x^2 + 3x^2 - 14x - 10x + 9 - 7 = 5x^2 - 24x + 2$

5.
$(7x + 3)(2x - 5)$
$14x^2 - 35x + 6x - 15 = 14x^2 - 29x - 15$

6.
$(3x^2 - 2)(5x - 4)$
$15x^3 - 12x^2 - 10x + 8$

7. (a) $(7x - 2y)(7x + 2y) = (7x)^2 - (2y)^2$
$= 49x^2 - 4y^2$

8. (a) $(4u + 5v)^2 = (4u)^2 + 2(4u)(5v) + (5v)^2$
$= 16u^2 + 40uv + 25v^2$
(b) $(7x^2 - 3y^2)^2 = (7x^2)^2 - 2(7x^2)(3y^2) + (3y^2)^2$
$= 49x^4 - 42x^2y^2 + 9y^4$

9.
$$
\begin{array}{r}
2x^2 - 3x + 1 \\
\underline{x^2 - 5x} \\
-10x^3 + 15x^2 - 5x \\
\underline{2x^4 - 3x^3 + x^2} \\
2x^4 - 13x^3 + 16x^2 - 5x
\end{array}
$$

10. $(2x^2 - 3x + 1)(x^2 - 5x)$
$= (2x^2 - 3x + 1)(x^2) + (2x^2 - 3x + 1)(-5x)$
$= 2x^4 - 3x^3 + x^2 - 10x^3 + 15x^2 - 5x$
$= 2x^4 - 3x^3 - 10x^3 + x^2 + 15x^2 - 5x$
$= 2x^4 - 13x^3 + 16x^2 - 5x$

5.2 Practice Problems

1. $\dfrac{-16x^4 + 16x^3 + 8x^2 + 64x}{8x}$
$= \dfrac{-16x^4}{8x} + \dfrac{16x^3}{8x} + \dfrac{8x^2}{8x} + \dfrac{64x}{8x}$
$= -2x^3 + 2x^2 + x + 8$

2.
$$
\begin{array}{r}
2x^2 - x - 3 \\
4x - 3 \overline{)8x^3 - 10x^2 - 9x + 14} \\
\underline{8x^3 - 6x^2} \\
-4x^2 - 9x \\
\underline{-4x^2 + 3x} \\
-12x + 14 \\
\underline{-12x + 9} \\
5
\end{array}
$$
The answer is $2x^2 - x - 3$ remainder 5 or
$2x^2 - x - 3 + \dfrac{5}{4x - 3}$.

3.
$$
\begin{array}{r}
4x^2 - 6x + 9 \\
2x + 3 \overline{)8x^3 + 0x^2 + 0x + 27} \\
\underline{8x^3 + 12x^2} \\
-12x^2 + 0x \\
\underline{-12x^2 - 18x} \\
18x + 27 \\
\underline{18x + 27} \\
0
\end{array}
$$
The answer is $4x^2 - 6x + 9$.

4.
$$
\begin{array}{r}
x^2 - 3x + 1 \\
x^2 + 0x - 1 \overline{)x^4 - 3x^3 + 0x^2 + 3x + 4} \\
\underline{x^4 - 0x^3 - x^2} \\
-3x^3 + x^2 + 3x \\
\underline{-3x^3 - 0x^2 + 3x} \\
x^2 + 0x + 4 \\
\underline{x^2 + 0x - 1} \\
5
\end{array}
$$
The answer is $x^2 - 3x + 1 + \dfrac{5}{x^2 - 1}$.

5.3 Practice Problems

1.
$$
\begin{array}{r|rrrr}
-3 & 1 & -3 & 4 & -5 \\
& & -3 & +18 & -66 \\
\hline
& 1 & -6 & 22 & \boxed{-71}
\end{array}
$$
The quotient is $x^2 - 6x + 22 + \dfrac{-71}{x + 3}$.

2.
$$
\begin{array}{r|rrrrr}
3 & 2 & 0 & -1 & 5 & -12 \\
& & 6 & 18 & 51 & 168 \\
\hline
& 2 & 6 & 17 & 56 & \boxed{156}
\end{array}
$$
The quotient is $2x^3 + 6x^2 + 17x + 56 + \dfrac{156}{x - 3}$.

3.
$$
\begin{array}{r|rrrrr}
3 & 2 & -9 & 5 & 13 & -3 \\
& & 6 & -9 & -12 & 3 \\
\hline
& 2 & -3 & -4 & 1 & \boxed{0}
\end{array}
$$
The quotient is $2x^3 - 3x^2 - 4x + 1$.

5.4 Practice Problems

1. (a) $19x^3 - 38x^2 = 19x^2(x - 2)$
(b) $100a^4 - 50a^2 = 50a^2(2a^2 - 1)$
2. (a) $21x^3 - 18x^2y + 24xy^2 = 3x(7x^2 - 6xy + 8y^2)$
(b) $12xy^2 - 14x^2y + 20x^2y^2 + 36x^3y$
$= 2xy(6y - 7x + 10xy + 18x^2)$
3. $9a^3 - 12a^2b^2 - 15a^4 = 3a^2(3a - 4b^2 - 5a^2)$
To check, we multiply.
$$3a^2(3a - 4b^2 - 5a^2) = 9a^3 - 12a^2b^2 - 15a^4$$
This is the original polynomial. It checks.
4. $7x(x + 2y) - 8y(x + 2y) - (x + 2y)$
$= (x + 2y)(7x - 8y - 1)$
5. $bx + 5by + 2wx + 10wy = b(x + 5y) + 2w(x + 5y)$
$= (x + 5y)(b + 2w)$
6. To factor $5x^2 - 12y + 4xy - 15x$, rearrange the terms. Then factor.
$5x^2 - 15x + 4xy - 12y$
$= 5x(x - 3) + 4y(x - 3)$
$= (x - 3)(5x + 4y)$
7. To factor $xy - 12 - 4x + 3y$, rearrange the terms. Then factor.
$xy - 4x + 3y - 12$
$= x(y - 4) + 3(y - 4)$
$= (y - 4)(x + 3)$
8. To factor $2x^3 - 15 - 10x + 3x^2$, rearrange the terms. Then factor.
$2x^3 - 10x + 3x^2 - 15$
$= 2x(x^2 - 5) + 3(x^2 - 5)$
$= (x^2 - 5)(2x + 3)$

5.5 Practice Problems

1. $x^2 - 10x + 21 = (x - 7)(x - 3)$
2. $x^2 - 13x - 48 = (x - 16)(x + 3)$
3. $x^4 + 9x^2 + 8 = (x^2 + 8)(x^2 + 1)$
4. (a) $a^2 + 2a - 48 = (a + 8)(a - 6)$
(b) $x^4 + 2x^2 - 15 = (x^2 + 5)(x^2 - 3)$
5. (a) $x^2 - 16xy + 15y^2 = (x - 15y)(x - y)$
(b) $x^2 + xy - 42y^2 = (x + 7y)(x - 6y)$

6. $4x^2 - 44x + 72 = 4(x^2 - 11x + 18) = 4(x - 9)(x - 2)$

7. Factor $3x^2 + 2x - 8$.

The grouping number is -24. Two numbers whose product is -24 and whose sum is 2 are 6 and -4.

$$3x^2 + 6x - 4x - 8$$
$$= 3x(x + 2) - 4(x + 2)$$
$$= (x + 2)(3x - 4)$$

8. Factor $10x^2 - 9x + 2$.

The grouping number is 20. Two numbers whose product is 20 and whose sum is -9 are -5 and -4.

$$10x^2 - 5x - 4x + 2$$
$$= 5x(2x - 1) - 2(2x - 1)$$
$$= (2x - 1)(5x - 2)$$

9. $9x^3 - 15x^2 - 6x = 3x(3x^2 - 5x - 2)$
$$= 3x(3x^2 - 6x + x - 2)$$
$$= 3x[3x(x - 2) + 1(x - 2)]$$
$$= 3x(x - 2)(3x + 1)$$

10. $8x^2 - 6x - 5 = (4x - 5)(2x + 1)$

11. $6x^4 + 13x^2 - 5 = (2x^2 + 5)(3x^2 - 1)$

5.6 Practice Problems

1. $x^2 - 9 = (x + 3)(x - 3)$

2. $64x^2 - 121y^2 = (8x + 11y)(8x - 11y)$

3. $49x^2 - 25y^4 = (7x + 5y^2)(7x - 5y^2)$

4. $7x^2 - 28 = 7(x^2 - 4) = 7(x + 2)(x - 2)$

5. $9x^2 - 30x + 25 = (3x - 5)^2$

6. $25x^2 - 70x + 49 = (5x - 7)^2$

7. $242x^2 + 88x + 8 = 2(121x^2 + 44x + 4)$
$$= 2(11x + 2)^2$$

8. (a) $49x^4 + 28x^2 + 4 = (7x^2 + 2)^2$

(b) $36x^4 + 84x^2y^2 + 49y^4 = (6x^2 + 7y^2)^2$

9. $8x^3 + 125y^3 = (2x + 5y)(4x^2 - 10xy + 25y^2)$

10. $64x^3 - 125y^3 = (4x - 5y)(16x^2 + 20xy + 25y^2)$

11. $27w^3 - 125z^6 = (3w - 5z^2)(9w^2 + 15wz^2 + 25z^4)$

12. $54x^3 - 16 = 2(27x^3 - 8)$
$$= 2(3x - 2)(9x^2 + 6x + 4)$$

13. $64a^6 - 1$

Use the difference of two squares first.
$$(8a^3 + 1)(8a^3 - 1)$$
Now use the formula for the sum and difference of two cubes.
$$(2a + 1)(4a^2 - 2a + 1)(2a - 1)(4a^2 + 2a + 1)$$

5.7 Practice Problems

1. (a) $7x^5 + 56x^2 = 7x^2(x^3 + 8)$
$$= 7x^2(x + 2)(x^2 - 2x + 4)$$

(b) $125x^2 + 50xy + 5y^2$
$$= 5(25x^2 + 10xy + y^2)$$
$$= 5(5x + y)^2$$

(c) $12x^2 - 75 = 3(4x^2 - 25)$
$$= 3(2x + 5)(2x - 5)$$

(d) $3x^2 - 39x + 126$
$$= 3(x^2 - 13x + 42)$$
$$= 3(x - 7)(x - 6)$$

(e) $6ax + 6ay + 18bx + 18by$
$$= 6(ax + ay + 3bx + 3by)$$
$$= 6[a(x + y) + 3b(x + y)]$$
$$= 6(x + y)(a + 3b)$$

(f) $6x^3 - x^2 - 12x$
$$= x(6x^2 - x - 12)$$
$$= x(6x^2 - 9x + 8x - 12)$$
$$= x[3x(2x - 3) + 4(2x - 3)]$$
$$= x(2x - 3)(3x + 4)$$

2. $3x^2 - 10x + 4$

Prime. There are no factors of 12 whose sum is -10.

3. $16x^2 + 81$

Prime. Binomials of the form $a^2 + b^2$ cannot be factored.

5.8 Practice Problems

1.
$$x^2 + x = 56$$
$$x^2 + x - 56 = 0$$
$$(x + 8)(x - 7) = 0$$
$$x + 8 = 0 \qquad x - 7 = 0$$
$$x = -8 \qquad x = 7$$

2. $12x^2 - 11x + 2 = 0$
$$(4x - 1)(3x - 2) = 0$$
$$4x - 1 = 0 \qquad 3x - 2 = 0$$
$$x = \frac{1}{4} \qquad x = \frac{2}{3}$$

3. $7x^2 - 14x = 0$
$$7x(x - 2) = 0$$
$$7x = 0 \qquad x - 2 = 0$$
$$x = 0 \qquad x = 2$$

4.
$$16x(x - 2) = 8x - 25$$
$$16x^2 - 32x = 8x - 25$$
$$16x^2 - 32x - 8x + 25 = 0$$
$$16x^2 - 40x + 25 = 0$$
$$(4x - 5)^2 = 0$$
$$4x - 5 = 0 \qquad 4x - 5 = 0$$
$$x = \frac{5}{4} \text{ is a double root.}$$

5.
$$3x^3 + 6x^2 = 45x$$
$$3x^3 + 6x^2 - 45x = 0$$
$$3x(x^2 + 2x - 15) = 0$$
$$3x(x + 5)(x - 3) = 0$$
$$3x = 0 \qquad x + 5 = 0 \qquad x - 3 = 0$$
$$x = 0 \qquad x = -5 \qquad x = 3$$

6. $A = \frac{1}{2}ab$

Let the base $= x$, the altitude $= x + 5$.

$$52 = \frac{1}{2}(x + 5)(x)$$
$$104 = (x + 5)(x)$$
$$104 = x^2 + 5x$$
$$0 = x^2 + 5x - 104$$
$$0 = x^2 - 8x + 13x - 104$$
$$0 = (x - 8)(x + 13)$$
$$x - 8 = 0 \qquad x + 13 = 0$$
$$x = 8 \qquad x = -13$$

The base of a triangle must be a positive number, so we disregard -13. Thus,

base $= x = 8$ feet

altitude $= x + 5 = 13$ feet

7. Let $x = $ the length in square feet of last year's garden, then $3x + 2 = $ the length in square feet of this year's garden.

$$(3x + 2)^2 = 112 + x^2$$
$$9x^2 + 12x + 4 = 112 + x^2$$
$$8x^2 + 12x - 108 = 0$$
$$4(2x^2 + 3x - 27) = 0$$
$$4(x - 3)(2x + 9) = 0$$
$$x - 3 = 0 \qquad 2x + 9 = 0$$
$$x = 3 \qquad x = -\frac{9}{2}$$

Length cannot be negative, so we reject the negative answer. We use $x = 3$. Last year's garden is a square with each side measuring 3 feet. This year's garden measures $3x + 2 = 3(3) + 2 = 11$ feet on each side.

Chapter 6

6.1 Practice Problems

1. Solve the equation $x^2 - 9x - 22 = 0$.
$(x + 2)(x - 11) = 0$
$x + 2 = 0 \qquad x - 11 = 0$
$\qquad x = -2 \qquad\qquad x = 11$
The domain of $y = f(x)$ is all real numbers except -2 and 11.

2. $\dfrac{x^2 - 36y^2}{x^2 - 3xy - 18y^2} = \dfrac{(x + 6y)\cancel{(x - 6y)}}{(x + 3y)\cancel{(x - 6y)}} = \dfrac{x + 6y}{x + 3y}$

3. $\dfrac{9x^2y}{3xy^2 + 6x^2y} = \dfrac{\overset{3}{\cancel{9}}\,\overset{x}{\cancel{x^2}}y}{\cancel{3xy}(y + 2x)} = \dfrac{3x}{y + 2x}$

4. $\dfrac{2x^2 - 8x - 10}{2x^2 - 20x + 50} = \dfrac{2\cancel{(x-5)}(x + 1)}{2\cancel{(x-5)}(x - 5)} = \dfrac{x + 1}{x - 5}$

5. $\dfrac{-3x + 6y}{x^2 - 7xy + 10y^2} = \dfrac{-3\cancel{(x - 2y)}}{\cancel{(x - 2y)}(x - 5y)} = \dfrac{-3}{x - 5y}$

6. $\dfrac{7a^2 - 23ab + 6b^2}{4b^2 - 49a^2} = \dfrac{\cancel{(7a - 2b)}(a - 3b)}{\cancel{(2b - 7a)}(2b + 7a)} = -\dfrac{a - 3b}{7a + 2b}$

7. $\dfrac{2x^2 + 5x + 2}{4x^2 - 1} \cdot \dfrac{2x^2 + x - 1}{x^2 + x - 2}$

$= \dfrac{\cancel{(2x + 1)}\cancel{(x + 2)}}{\cancel{(2x - 1)}\cancel{(2x + 1)}} \cdot \dfrac{\cancel{(2x - 1)}(x + 1)}{\cancel{(x + 2)}(x - 1)} = \dfrac{x + 1}{x - 1}$

8. $\dfrac{9x + 9y}{5ax + 5ay} \cdot \dfrac{10ax^2 - 40x^2b^2}{27ax^2 - 54bx^2}$

$= \dfrac{9\cancel{(x + y)}}{\cancel{5}a\cancel{(x + y)}} \cdot \dfrac{\overset{2}{\cancel{10}}x^2(a - 4b^2)}{\underset{3}{\cancel{27}}x^2(a - 2b)}$

$= \dfrac{2(a - 4b^2)}{3(a - 2b)} \text{ or } \dfrac{2a - 8b^2}{3a - 6b}$

9. $\dfrac{8x^3 + 27y^3}{64x^3 - y^3} \div \dfrac{4x^2 - 9y^2}{16x^2 + 4xy + y^2}$

$= \dfrac{\cancel{(2x + 3y)}(4x^2 - 6xy + 9y^2)}{(4x - y)\cancel{(16x^2 + 4xy + y^2)}} \cdot \dfrac{\cancel{16x^2 + 4xy + y^2}}{(2x - 3y)\cancel{(2x + 3y)}}$

$= \dfrac{4x^2 - 6xy + 9y^2}{(4x - y)(2x - 3y)}$

10. $\dfrac{4x^2 - 9}{2x^2 + 11x + 12} \div (-6x + 9)$

$= \dfrac{4x^2 - 9}{2x^2 + 11x + 12} \cdot \dfrac{1}{-6x + 9}$

$= \dfrac{\cancel{(2x + 3)}\cancel{(2x - 3)}}{\cancel{(2x + 3)}(x + 4)} \cdot \dfrac{1}{-3\cancel{(2x - 3)}} = -\dfrac{1}{3(x + 4)}$

6.2 Practice Problems

1. Find the LCD. $\dfrac{8}{x^2 - x - 12}, \dfrac{3}{x - 4}$
Factor each denominator completely.
$x^2 - x - 12 = (x - 4)(x + 3)$
$x - 4$ cannot be factored.
The LCD is the product of all the different prime factors.
LCD $= (x - 4)(x + 3)$

2. Find the LCD. $\dfrac{2}{15x^3y^2}, \dfrac{13}{25xy^3}$
Factor each denominator.
$15x^3y^2 = 3 \cdot 5 \cdot \quad x \cdot x \cdot x \cdot y \cdot y$
$25xy^3 = \quad 5 \cdot 5 \cdot x \cdot y \cdot y \cdot y$
LCD $= 3 \cdot 5 \cdot 5 \cdot x \cdot x \cdot x \cdot y \cdot y \cdot y$
LCD $= 75x^3y^3$

3. $\dfrac{4x}{(x + 6)(2x - 1)} - \dfrac{3x + 1}{(x + 6)(2x - 1)} = \dfrac{x - 1}{(x + 6)(2x - 1)}$

4. $\dfrac{8}{(x - 4)(x + 3)} + \dfrac{3}{x - 4}$

LCD $= (x - 4)(x + 3)$

$= \dfrac{8}{(x - 4)(x + 3)} + \dfrac{3}{(x - 4)} \cdot \dfrac{(x + 3)}{(x + 3)}$

$= \dfrac{8 + 3x + 9}{(x - 4)(x + 3)} = \dfrac{3x + 17}{(x - 4)(x + 3)}$

5. $\dfrac{5}{x + 4} + \dfrac{3}{4x}$

LCD $= 4x(x + 4)$

$= \dfrac{5}{x + 4} \cdot \dfrac{4x}{4x} + \dfrac{3}{4x} \cdot \dfrac{(x + 4)}{(x + 4)}$

$= \dfrac{20x + 3x + 12}{4x(x + 4)} = \dfrac{23x + 12}{4x(x + 4)}$

6. $\dfrac{7}{4ab^3} + \dfrac{1}{3a^3b^2}$

LCD $= 12a^3b^3$

$= \dfrac{7}{4ab^3} \cdot \dfrac{(3a^2)}{(3a^2)} + \dfrac{1}{3a^3b^2} \cdot \dfrac{(4b)}{(4b)} = \dfrac{21a^2 + 4b}{12a^3b^3}$

7. $\dfrac{4x + 2}{x^2 + x - 12} - \dfrac{3x + 8}{x^2 + 6x + 8}$

$= \dfrac{4x + 2}{(x - 3)(x + 4)} - \dfrac{3x + 8}{(x + 2)(x + 4)}$

LCD $= (x + 2)(x - 3)(x + 4)$

$= \dfrac{4x + 2}{(x + 4)(x - 3)} \cdot \dfrac{(x + 2)}{(x + 2)} - \dfrac{3x + 8}{(x + 2)(x + 4)} \cdot \dfrac{(x - 3)}{(x - 3)}$

$= \dfrac{4x^2 + 10x + 4}{(x + 4)(x - 3)(x + 2)} - \dfrac{3x^2 - x - 24}{(x + 4)(x - 3)(x + 2)}$

$= \dfrac{x^2 + 11x + 28}{(x + 4)(x - 3)(x + 2)} = \dfrac{(x + 4)(x + 7)}{(x + 4)(x - 3)(x + 2)}$

$= \dfrac{x + 7}{(x - 3)(x + 2)}$

8. $\dfrac{7x - 3}{4x^2 + 20x + 25} - \dfrac{3x}{4x + 10}$

$= \dfrac{7x - 3}{(2x + 5)(2x + 5)} - \dfrac{3x}{2(2x + 5)}$

LCD $= 2(2x + 5)(2x + 5)$

$= \dfrac{7x - 3}{(2x + 5)(2x + 5)} \cdot \dfrac{2}{2} - \dfrac{3x}{2(2x + 5)} \cdot \dfrac{(2x + 5)}{(2x + 5)}$

$= \dfrac{14x - 6 - 6x^2 - 15x}{2(2x + 5)(2x + 5)} = \dfrac{-6x^2 - x - 6}{2(2x + 5)^2}$

6.3 Practice Problems

1. You can use either method 1 or method 2 to simplify a complex fraction.

$$\dfrac{y + \dfrac{3}{y}}{\dfrac{2}{y^2} + \dfrac{5}{y}}$$

METHOD 1
Simplify the numerator.
$$y + \dfrac{3}{y} = \dfrac{y^2}{y} + \dfrac{3}{y} = \dfrac{y^2 + 3}{y}$$
Simplify the denominator.
$$\dfrac{2}{y^2} + \dfrac{5}{y} = \dfrac{2}{y^2} + \dfrac{5y}{y^2} = \dfrac{2 + 5y}{y^2}$$

Divide the numerator by the denominator.

$$\frac{\dfrac{y^2 + 3}{y}}{\dfrac{2 + 5y}{y^2}}$$

$$= \frac{y^2 + 3}{y} \cdot \frac{y^2}{2 + 5y}$$

$$= \frac{y(y^2 + 3)}{2 + 5y}$$

METHOD 2

Find the LCD of all the fractions in the numerator and denominator.

LCD $= y^2$

Multiply the numerator and denominator by the LCD.

$$\frac{y + \dfrac{3}{y}}{\dfrac{2}{y^2} + \dfrac{5}{y}} \cdot \frac{y^2}{y^2}$$

$$= \frac{y(y^2) + \dfrac{3}{y}(y^2)}{\dfrac{2}{y^2}(y^2) + \dfrac{5}{y}(y^2)}$$

$$= \frac{y^3 + 3y}{2 + 5y}$$

$$= \frac{y(y^2 + 3)}{2 + 5y}$$

2. Simplify.

$$\frac{\dfrac{4}{16x^2 - 1} + \dfrac{3}{4x + 1}}{\dfrac{x}{4x - 1} + \dfrac{5}{4x + 1}}$$

METHOD 1

Simplify the numerator.

$$\frac{4}{(4x + 1)(4x - 1)} + \frac{3(4x - 1)}{(4x + 1)(4x - 1)} = \frac{12x + 1}{(4x + 1)(4x - 1)}$$

Simplify the denominator.

$$\frac{x(4x + 1)}{(4x - 1)(4x + 1)} + \frac{5(4x - 1)}{(4x + 1)(4x - 1)} = \frac{4x^2 + 21x - 5}{(4x - 1)(4x + 1)}$$

To divide the numerator by the denominator, we multiply the numerator by the reciprocal of the denominator.

$$\frac{12x + 1}{(4x + 1)(4x - 1)} \cdot \frac{(4x + 1)(4x - 1)}{4x^2 + 21x - 5} = \frac{12x + 1}{4x^2 + 21x - 5}$$

METHOD 2

Multiply the numerator and denominator by the LCD of all the rational expressions in the numerator and denominator.

LCD $= (4x + 1)(4x - 1)$. Notice we factored $16x^2 - 1$.

$$= \frac{\left[\dfrac{4}{(4x + 1)(4x - 1)} + \dfrac{3}{(4x + 1)}\right](4x + 1)(4x - 1)}{\left[\dfrac{x}{4x - 1} + \dfrac{5}{4x + 1}\right](4x + 1)(4x - 1)}$$

$$= \frac{\dfrac{4(4x + 1)(4x - 1)}{(4x + 1)(4x - 1)} + \dfrac{3(4x + 1)(4x - 1)}{4x + 1}}{\dfrac{x}{4x - 1} \cdot (4x + 1)(4x - 1) + \dfrac{5}{4x + 1} \cdot (4x + 1)(4x - 1)}$$

$$= \frac{4 + 3(4x - 1)}{x(4x + 1) + 5(4x - 1)} = \frac{4 + 12x - 3}{4x^2 + x + 20x - 5} = \frac{12x + 1}{4x^2 + 21x - 5}$$

3. Simplify by **METHOD 1.**

$$\frac{\dfrac{4 + x}{x - \dfrac{16}{x}}}{} = \frac{\dfrac{4 + x}{x}}{\dfrac{x}{1} \cdot \dfrac{x}{x} - \dfrac{16}{x}}$$

$$= \frac{\dfrac{4 + x}{x}}{\dfrac{x^2 - 16}{x}} = \frac{4 + x}{1} \div \frac{x^2 - 16}{x}$$

$$= \frac{4 + x}{1} \cdot \frac{x}{(x - 4)(x + 4)} = \frac{x}{x - 4}$$

4. Simplify by **METHOD 2.**

LCD $= y(y + 3)$

$$\frac{\dfrac{7}{y + 3} - \dfrac{3}{y}}{\dfrac{2}{y} + \dfrac{5}{y + 3}} \cdot \frac{y(y + 3)}{y(y + 3)}$$

$$= \frac{\dfrac{7}{y + 3} \cdot y(y + 3) - \dfrac{3}{y} \cdot y(y + 3)}{\dfrac{2}{y} \cdot y(y + 3) + \dfrac{5}{y + 3} \cdot y(y + 3)}$$

$$= \frac{7y - 3(y + 3)}{2(y + 3) + 5(y)} = \frac{7y - 3y - 9}{2y + 6 + 5y} = \frac{4y - 9}{7y + 6}$$

6.4 Practice Problems

1. $\dfrac{4}{3x} + \dfrac{x + 1}{x} = \dfrac{1}{2}$

$$6x\left[\frac{4}{3x}\right] + 6x\left[\frac{x + 1}{x}\right] = 6x\left[\frac{1}{2}\right]$$

$$8 + 6(x + 1) = 3x$$

$$8 + 6x + 6 = 3x$$

$$3x = -14$$

$$x = -\frac{14}{3}$$

Check.

$$\frac{4}{3\left(\dfrac{-14}{3}\right)} + \frac{\dfrac{-14}{3} + 1}{\dfrac{-14}{3}} \overset{?}{=} \frac{1}{2}$$

$$\frac{4}{-14} + \frac{\dfrac{-11}{3}}{\dfrac{-14}{3}} \overset{?}{=} \frac{1}{2}$$

$$-\frac{4}{14} + \frac{11}{14} \overset{?}{=} \frac{1}{2}$$

$$\frac{7}{14} \overset{?}{=} \frac{1}{2}$$

$$\frac{1}{2} = \frac{1}{2} \quad \checkmark$$

2. $\dfrac{1}{3x - 9} = \dfrac{1}{2x - 6} - \dfrac{5}{6}$

$$\frac{1}{3(x - 3)} = \frac{1}{2(x - 3)} - \frac{5}{6}$$

$$6(x - 3)\left[\frac{1}{3(x - 3)}\right] = 6(x - 3)\left[\frac{1}{2(x - 3)}\right] - 6(x - 3)\left(\frac{5}{6}\right)$$

$$2(1) = 3(1) - (x - 3)(5)$$

$$2 = 3 - 5x + 15$$

$$2 = 18 - 5x$$

$$-16 = -5x$$

$$\frac{16}{5} = x$$

Check.

$$\frac{1}{3\left(\dfrac{16}{5}\right) - 9} \overset{?}{=} \frac{1}{2\left(\dfrac{16}{5}\right) - 6} - \frac{5}{6}$$

$$\frac{1}{\dfrac{48}{5} - 9} \overset{?}{=} \frac{1}{\dfrac{32}{5} - 6} - \frac{5}{6}$$

$$\frac{\frac{1}{3}}{\frac{3}{5}} \overset{?}{=} \frac{\frac{1}{2} - \frac{5}{6}}{\frac{3}{5}}$$

$$\frac{5}{3} = \frac{5}{3} \checkmark$$

3. $\dfrac{y^2 + 4y - 2}{y^2 - 2y - 8} = 1 + \dfrac{4}{y - 4}$

$\dfrac{y^2 + 4y - 2}{(y + 2)(y - 4)} = 1 + \dfrac{4}{y - 4}$

$(y + 2)(y - 4)\left[\dfrac{y^2 + 4y - 2}{(y + 2)(y - 4)}\right]$

$\qquad = (y + 2)(y - 4)(1) + (y + 2)(y - 4)\left(\dfrac{4}{y - 4}\right)$

$y^2 + 4y - 2 = y^2 - 2y - 8 + 4y + 8$

$4y - 2 = 2y$

$2y = 2$

$y = 1$

4. $\dfrac{2x - 1}{x^2 - 7x + 10} + \dfrac{3}{x - 5} = \dfrac{5}{x - 2}$

$\dfrac{2x - 1}{(x - 2)(x - 5)} + \dfrac{3}{x - 5} = \dfrac{5}{x - 2}$

$(x - 2)(x - 5)\left[\dfrac{2x - 1}{(x - 2)(x - 5)}\right] + (x - 2)(x - 5)\left(\dfrac{3}{x - 5}\right)$

$\qquad = (x - 2)(x - 5)\left(\dfrac{5}{x - 2}\right)$

$2x - 1 + 3(x - 2) = 5(x - 5)$

$2x - 1 + 3x - 6 = 5x - 25$

$5x - 7 = 5x - 25$

$0 = -18$

Of course $0 \neq -18$. Therefore, no values of x makes the original equation true. Hence the equation has **no solution.**

5. $\dfrac{y}{y - 2} - 3 = 1 + \dfrac{2}{y - 2}$

$(y - 2)\left(\dfrac{y}{y - 2}\right) - (y - 2)(3) = (y - 2)(1) + (y - 2)\left(\dfrac{2}{y - 2}\right)$

$y - 3(y - 2) = 1(y - 2) + 2$

$y - 3y + 6 = y - 2 + 2$

$-2y + 6 = y$

$6 = 3y$

$2 = y$

Check. $\dfrac{2}{2 - 2} - 3 \overset{?}{=} 1 + \dfrac{2}{2 - 2}$

$\dfrac{2}{0} - 3 \overset{?}{=} 1 + \dfrac{2}{0}$

Division by zero is not defined. The value $y = 2$ is therefore not a solution to the original equation. There is **no solution.**

6.5 Practice Problems

1. Solve for t.

$\dfrac{1}{t} = \dfrac{1}{c} + \dfrac{1}{d}$

$cdt\left[\dfrac{1}{t}\right] = cdt\left[\dfrac{1}{c}\right] + cdt\left[\dfrac{1}{d}\right]$

$cd = dt + ct$

$cd = t(d + c)$

$\dfrac{cd}{d + c} = t$

2. Solve for p_1. $C = \dfrac{Bp_1p_2}{d^2}$

$Cd^2 = Bp_1p_2$

$\dfrac{Cd^2}{Bp_2} = p_1$

3. $\dfrac{21}{2} = \dfrac{x}{168}$

$2x = 3528$

$x = 1764$

The number of students enrolled should be 1764 to maintain that ratio.

4. We will draw the picture as two similar triangles.

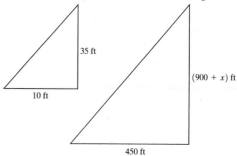

We write a proportion and solve.

$\dfrac{10}{35} = \dfrac{450}{900 + x}$

LCD $= 35(900 + x)$

$35(900 + x)\left(\dfrac{10}{35}\right) = 35(900 + x)\left(\dfrac{450}{900 + x}\right)$

$10(900 + x) = 35(450)$

$9000 + 10x = 15{,}750$

$10x = 6750$

$x = 675$

The helicopter is about 675 feet over the building.

5. We construct a table for the data.

	Rate of Work per Hour	Time Worked in Hours	Fraction of Task Done
Alfred	$\dfrac{1}{4}$	x	$\dfrac{x}{4}$
Son	$\dfrac{1}{5}$	x	$\dfrac{x}{5}$

The fraction of work done by Alfred plus the fraction of work done by his son equals 1 completed job.

Write an equation and solve.

$\dfrac{x}{4} + \dfrac{x}{5} = 1$

$5x + 4x = 20$

$9x = 20$

$x = 2.2\overline{2}$

Alfred and his son can mow the lawn working together in approximately 2.2 hours.

Chapter 7

7.1 Practice Problems

1. $\left(\dfrac{3x^{-2}y^4}{2x^{-5}y^2}\right)^{-3} = \dfrac{(3x^{-2}y^4)^{-3}}{(2x^{-5}y^2)^{-3}}$

$= \dfrac{3^{-3}(x^{-2})^{-3}(y^4)^{-3}}{2^{-3}(x^{-5})^{-3}(y^2)^{-3}}$

$= \dfrac{3^{-3}x^6y^{-12}}{2^{-3}x^{15}y^{-6}}$

$$= \frac{3^{-3}}{2^{-3}} \cdot \frac{x^6}{x^{15}} \cdot \frac{y^{-12}}{y^{-6}}$$

$$= \frac{2^3}{3^3} \cdot x^{6-15} \cdot y^{-12+6}$$

$$= \frac{8}{27} x^{-9} y^{-6} \text{ or } \frac{8}{27x^9 y^6}$$

2. (a) $(x^4)^{3/8} = x^{(4/1)(3/8)} = x^{3/2}$

(b) $\dfrac{x^{3/7}}{x^{2/7}} = x^{3/7 - 2/7} = x^{1/7}$

(c) $x^{-7/5} \cdot x^{4/5} = x^{-7/5 + 4/5} = x^{-3/5}$

3. (a) $(-3x^{1/4})(2x^{1/2}) = -6x^{1/4 + 1/2} = -6x^{1/4 + 2/4} = -6x^{3/4}$

(b) $\dfrac{13x^{1/12} y^{-1/4}}{26x^{-1/3} y^{1/2}} = \dfrac{x^{1/12 - (-1/3)} y^{-1/4 - 1/2}}{2}$

$$= \frac{x^{1/12 + 4/12} y^{-1/4 - 2/4}}{2}$$

$$= \frac{x^{5/12} y^{-3/4}}{2}$$

$$= \frac{x^{5/12}}{2y^{3/4}}$$

4. $-3x^{1/2}(2x^{1/4} + 3x^{-1/2}) = -6x^{1/2 + 1/4} - 9x^{1/2 - 1/2}$

$$= -6x^{2/4 + 1/4} - 9x^0$$

$$= -6x^{3/4} - 9$$

5. (a) $(4)^{5/2} = (2^2)^{5/2} = 2^{2/1 \cdot 5/2} = 2^5 = 32$

(b) $(27)^{4/3} = (3^3)^{4/3} = 3^{3/1 \cdot 4/3} = 3^4 = 81$

6. $3x^{1/3} + x^{-1/3} = 3x^{1/3} + \dfrac{1}{x^{1/3}}$

$$= \frac{x^{1/3}}{x^{1/3}} (3x^{1/3}) + \frac{1}{x^{1/3}}$$

$$= \frac{3x^{2/3} + 1}{x^{1/3}}$$

7. $4y^{3/2} - 8y^{5/2} = 4y^{2/2 + 1/2} - 8y^{2/2 + 3/2}$

$$= 4(y^{2/2})(y^{1/2}) - 8(y^{2/2})(y^{3/2})$$

$$= 4y(y^{1/2} - 2y^{3/2})$$

7.2 Practice Problems

1. (a) $\sqrt[3]{216} = \sqrt[3]{(6)^3} = 6$

(b) $\sqrt[5]{32} = \sqrt[5]{(2)^5} = 2$

(c) $\sqrt[3]{-8} = \sqrt[3]{(-2)^3} = -2$

(d) $\sqrt[4]{-81}$ is not a real number.

2. (a) $f(3) = \sqrt{4(3) - 3} = \sqrt{12 - 3} = \sqrt{9} = 3$

(b) $f(4) = \sqrt{4(4) - 3} = \sqrt{16 - 3} = \sqrt{13} \approx 3.6$

(c) $f(7) = \sqrt{4(7) - 3} = \sqrt{28 - 3} = \sqrt{25} = 5$

3. $0.5x + 2 \geq 0$

$$0.5x \geq -2$$

$$x \geq -4$$

The domain is all real numbers x where $x \geq -4$.

4.

x	$f(x)$
3	0
4	1.7
5	2.4
6	3
15	6

$f(x) = \sqrt{3x - 9}$

5. (a) $\sqrt[3]{x^3} = (x^3)^{1/3} = x^{3/3} = x^1 = x$

(b) $\sqrt[4]{y^4} = (y^4)^{1/4} = y^{4/4} = y^1 = y$

6. (a) $\sqrt[4]{x^3} = (x^3)^{1/4} = x^{3/4}$

(b) $\sqrt[5]{(xy)^7} = [(xy)^7]^{1/5} = (xy)^{7/5}$

7. (a) $\sqrt[4]{81x^{12}} = (3^4 x^{12})^{1/4} = 3x^3$

(b) $\sqrt[3]{27x^6} = [(3)^3 x^6]^{1/3} = 3x^2$

(c) $(32x^5)^{3/5} = (2^5 x^5)^{3/5} = 2^3 x^3 = 8x^3$

8. (a) $(xy)^{3/4} = \sqrt[4]{(xy)^3} = \sqrt[4]{x^3 y^3}$

(b) $y^{-1/3} = \dfrac{1}{y^{1/3}} = \dfrac{1}{\sqrt[3]{y}}$

(c) $(2x)^{4/5} = \sqrt[5]{(2x)^4} = \sqrt[5]{16x^4}$

(d) $2x^{4/5} = 2\sqrt[5]{x^4}$

9. (a) $8^{2/3} = \left(\sqrt[3]{8}\right)^2 = 2^2 = 4$

(b) $(-8)^{4/3} = \left(\sqrt[3]{-8}\right)^4 = (-2)^4 = 16$

(c) $100^{-3/2} = \dfrac{1}{100^{3/2}} = \dfrac{1}{\left(\sqrt{100}\right)^3} = \dfrac{1}{10^3} = \dfrac{1}{1000}$

10. (a) $\sqrt[5]{(-3)^5} = -3$

(b) $\sqrt[4]{(-5)^4} = |-5| = 5$

(c) $\sqrt[4]{w^4} = |w|$

(d) $\sqrt[7]{y^7} = y$

11. (a) $\sqrt{36x^2} = 6|x|$

(b) $\sqrt[4]{16y^8} = 2|y^2| = 2y^2$

(c) $\sqrt[3]{125x^3 y^6} = \sqrt[3]{(5)^3 (x)^3 (y^2)^3} = 5xy^2$

7.3 Practice Problems

1. $\sqrt{20} = \sqrt{4 \cdot 5} = \sqrt{4} \cdot \sqrt{5} = 2\sqrt{5}$

2. $\sqrt{27} = \sqrt{9 \cdot 3} = \sqrt{9} \cdot \sqrt{3} = 3\sqrt{3}$

3. (a) $\sqrt[3]{24} = \sqrt[3]{8} \cdot \sqrt[3]{3} = 2\sqrt[3]{3}$

(b) $\sqrt[3]{-108} = \sqrt[3]{-27} \cdot \sqrt[3]{4} = -3\sqrt[3]{4}$

4. $\sqrt[4]{64} = \sqrt[4]{16} \cdot \sqrt[4]{4} = 2\sqrt[4]{4}$

5. (a) $\sqrt{45x^6 y^7} = \sqrt{9 \cdot 5 \cdot x^6 \cdot y^6 \cdot y} = \sqrt{9x^6 y^6} \cdot \sqrt{5y}$

$$= 3x^3 y^3 \sqrt{5y}$$

(b) $\sqrt{27a^7 b^8 c^9} = \sqrt{9 \cdot 3a^6 \cdot a \cdot b^8 \cdot c^8 \cdot c}$

$$= \sqrt{9a^6 b^8 c^8} \cdot \sqrt{3ac}$$

$$= 3a^3 b^4 c^4 \sqrt{3ac}$$

6. $19\sqrt{xy} + 5\sqrt{xy} - 10\sqrt{xy} = (19 + 5 - 10)\sqrt{xy} = 14\sqrt{xy}$

7. $4\sqrt{2} - 5\sqrt{50} - 3\sqrt{98}$

$$= 4\sqrt{2} - 5\sqrt{25} \cdot \sqrt{2} - 3\sqrt{49} \cdot \sqrt{2}$$

$$= 4\sqrt{2} - 5(5)\sqrt{2} - 3(7)\sqrt{2}$$

$$= 4\sqrt{2} - 25\sqrt{2} - 21\sqrt{2}$$

$$= (4 - 25 - 21)\sqrt{2}$$

$$= -42\sqrt{2}$$

8. $4\sqrt{2x} + \sqrt{18x} - 2\sqrt{125x} - 6\sqrt{20x}$

$$= 4\sqrt{2x} + \sqrt{9} \cdot \sqrt{2x} - 2\sqrt{25} \cdot \sqrt{5x} - 6\sqrt{4} \cdot \sqrt{5x}$$

$$= 4\sqrt{2x} + 3\sqrt{2x} - 2(5)\sqrt{5x} - 6(2)\sqrt{5x}$$

$$= 4\sqrt{2x} + 3\sqrt{2x} - 10\sqrt{5x} - 12\sqrt{5x}$$

$$= 7\sqrt{2x} - 22\sqrt{5x}$$

9. $3x\sqrt[3]{54x^4} - 3\sqrt[3]{16x^7}$

$= 3x\sqrt[3]{27x^3} \cdot \sqrt[3]{2x} - 3\sqrt[3]{8x^6} \cdot \sqrt[3]{2x}$

$= 3x(3x)\sqrt[3]{2x} - 3(2x^2)\sqrt[3]{2x}$

$= 9x^2\sqrt[3]{2x} - 6x^2\sqrt[3]{2x}$

$= 3x^2\sqrt[3]{2x}$

7.4 Practice Problems

1. $(-4\sqrt{2})(-3\sqrt{13x}) = (-4)(-3)\sqrt{2 \cdot 13x} = 12\sqrt{26x}$

2. $\sqrt{2x}(\sqrt{5} + 2\sqrt{3x} + \sqrt{8})$

$= (\sqrt{2x})(\sqrt{5}) + (\sqrt{2x})(2\sqrt{3x}) + (\sqrt{2x})(\sqrt{8})$

$= \sqrt{10x} + 2\sqrt{6x^2} + \sqrt{16x}$

$= \sqrt{10x} + 2\sqrt{x^2}\sqrt{6} + \sqrt{16}\sqrt{x}$

$= \sqrt{10x} + 2x\sqrt{6} + 4\sqrt{x}$

3. $(\sqrt{7} + 4\sqrt{2})(2\sqrt{7} - 3\sqrt{2})$

$= 2\sqrt{49} - 3\sqrt{14} + 8\sqrt{14} - 12\sqrt{4}$

$= 2(7) + 5\sqrt{14} - 12(2)$

$= 14 + 5\sqrt{14} - 24$

$= -10 + 5\sqrt{14}$

4. $(2 - 5\sqrt{5})(3 - 2\sqrt{2}) = 6 - 4\sqrt{2} - 15\sqrt{5} + 10\sqrt{10}$

5. $(\sqrt{5x} + \sqrt{10})^2 = (\sqrt{5x} + \sqrt{10})(\sqrt{5x} + \sqrt{10})$

$= \sqrt{25x^2} + \sqrt{50x} + \sqrt{50x} + \sqrt{100}$

$= 5x + 2\sqrt{25}\sqrt{2x} + 10$

$= 5x + 2(5)\sqrt{2x} + 10$

$= 5x + 10\sqrt{2x} + 10$

6. (a) $\sqrt[3]{2x}(\sqrt[3]{4x^2} + 3\sqrt[3]{y})$

$= (\sqrt[3]{2x})(\sqrt[3]{4x^2}) + (\sqrt[3]{2x})(3\sqrt[3]{y})$

$= \sqrt[3]{8x^3} + 3\sqrt[3]{2xy}$

$= 2x + 3\sqrt[3]{2xy}$

(b) $(\sqrt[3]{7} + \sqrt[3]{x^2})(2\sqrt[3]{49} - \sqrt[3]{x})$

$= 2\sqrt[3]{343} - \sqrt[3]{7x} + 2\sqrt[3]{49x^2} - \sqrt[3]{x^3}$

$= 2\sqrt[3]{7^3} - \sqrt[3]{7x} + 2\sqrt[3]{49x^2} - x$

$= 2(7) - \sqrt[3]{7x} + 2\sqrt[3]{49x^2} - x$

$= 14 - \sqrt[3]{7x} + 2\sqrt[3]{49x^2} - x$

7. (a) $\dfrac{\sqrt{75}}{\sqrt{3}} = \sqrt{\dfrac{75}{3}} = \sqrt{25} = 5$

(b) $\sqrt[3]{\dfrac{27}{64}} = \dfrac{\sqrt[3]{27}}{\sqrt[3]{64}} = \dfrac{3}{4}$

(c) $\dfrac{\sqrt{54a^3b^7}}{\sqrt{6b^5}} = \sqrt{\dfrac{54a^3b^7}{6b^5}} = \sqrt{9a^3b^2} = 3ab\sqrt{a}$

8. $\dfrac{7}{\sqrt{3}} = \dfrac{7}{\sqrt{3}} \cdot \dfrac{\sqrt{3}}{\sqrt{3}} = \dfrac{7\sqrt{3}}{\sqrt{9}} = \dfrac{7\sqrt{3}}{3}$

9. $\dfrac{8}{\sqrt{20x}} = \dfrac{8}{\sqrt{4}\sqrt{5x}} = \dfrac{8}{2\sqrt{5x}} \cdot \dfrac{\sqrt{5x}}{\sqrt{5x}} = \dfrac{8\sqrt{5x}}{10x} = \dfrac{4\sqrt{5x}}{5x}$

10. $\sqrt[3]{\dfrac{6}{5x}} = \dfrac{\sqrt[3]{6}}{\sqrt[3]{5x}} = \dfrac{\sqrt[3]{6}}{\sqrt[3]{5x}} \cdot \dfrac{\sqrt[3]{25x^2}}{\sqrt[3]{25x^2}} = \dfrac{\sqrt[3]{150x^2}}{\sqrt[3]{125x^3}} = \dfrac{\sqrt[3]{150x^2}}{5x}$

11. $\dfrac{4}{2 + \sqrt{5}} = \dfrac{4}{2 + \sqrt{5}} \cdot \dfrac{2 - \sqrt{5}}{2 - \sqrt{5}}$

$= \dfrac{4(2 - \sqrt{5})}{2^2 - (\sqrt{5})^2}$

$= \dfrac{4(2 - \sqrt{5})}{4 - 5}$

$= \dfrac{4(2 - \sqrt{5})}{-1}$

$= -(8 - 4\sqrt{5})$

$= -8 + 4\sqrt{5}$

12. $\dfrac{\sqrt{11} + \sqrt{2}}{\sqrt{11} - \sqrt{2}} \cdot \dfrac{\sqrt{11} + \sqrt{2}}{\sqrt{11} + \sqrt{2}}$

$= \dfrac{\sqrt{121} + \sqrt{22} + \sqrt{22} + \sqrt{4}}{(\sqrt{11})^2 - (\sqrt{2})^2}$

$= \dfrac{11 + 2\sqrt{22} + 2}{11 - 2} = \dfrac{13 + 2\sqrt{22}}{9}$

7.5 Practice Problems

1. $\sqrt{3x - 8} = x - 2$

$(\sqrt{3x - 8})^2 = (x - 2)^2$

$3x - 8 = x^2 - 4x + 4$

$0 = x^2 - 7x + 12$

$0 = (x - 3)(x - 4)$

$x - 3 = 0 \quad \text{or} \quad x - 4 = 0$

$x = 3 \qquad\qquad x = 4$

Check:

For $x = 3$: $\quad \sqrt{3(3) - 8} \overset{?}{=} 3 - 2$

$\sqrt{1} \overset{?}{=} 1$

$1 = 1 \quad \checkmark$

For $x = 4$: $\quad \sqrt{3(4) - 8} \overset{?}{=} 4 - 2$

$\sqrt{4} \overset{?}{=} 2$

$2 = 2 \quad \checkmark$

The solutions are 3 and 4.

2. $\sqrt{x + 4} = x + 4$

$(\sqrt{x + 4})^2 = (x + 4)^2$

$x + 4 = x^2 + 8x + 16$

$0 = x^2 + 7x + 12$

$0 = (x + 3)(x + 4)$

$x + 3 = 0 \quad \text{or} \quad x + 4 = 0$

$x = -3 \qquad\qquad x = -4$

Check:

For $x = -3$: $\quad \sqrt{-3 + 4} \overset{?}{=} -3 + 4$

$\sqrt{1} \overset{?}{=} 1$

$1 = 1 \quad \checkmark$

For $x = -4$: $\quad \sqrt{-4 + 4} \overset{?}{=} -4 + 4$

$\sqrt{0} \overset{?}{=} 0$

$0 = 0 \quad \checkmark$

The solutions are -4 and -3.

3. $\sqrt{2x+5} - 2\sqrt{2x} = 1$

$$\sqrt{2x+5} = 2\sqrt{2x} + 1$$
$$\left(\sqrt{2x+5}\right)^2 = \left(2\sqrt{2x} + 1\right)^2$$
$$2x + 5 = \left(2\sqrt{2x} + 1\right)\left(2\sqrt{2x} + 1\right)$$
$$2x + 5 = 8x + 4\sqrt{2x} + 1$$
$$-6x + 4 = 4\sqrt{2x}$$
$$-3x + 2 = 2\sqrt{2x}$$
$$(-3x + 2)^2 = \left(2\sqrt{2x}\right)^2$$
$$9x^2 - 12x + 4 = 8x$$
$$9x^2 - 20x + 4 = 0$$
$$(9x - 2)(x - 2) = 0$$
$$9x - 2 = 0 \quad \text{or} \quad x - 2 = 0$$
$$x = \frac{2}{9} \qquad\qquad x = 2$$

Check: For $x = \frac{2}{9}$:

$$\sqrt{2\left(\frac{2}{9}\right) + 5} - 2\sqrt{2\left(\frac{2}{9}\right)} \stackrel{?}{=} 1$$
$$\sqrt{\frac{4}{9} + 5} - 2\sqrt{\frac{4}{9}} \stackrel{?}{=} 1$$
$$\sqrt{\frac{49}{9}} - 2\sqrt{\frac{4}{9}} \stackrel{?}{=} 1$$
$$\frac{7}{3} - \frac{4}{3} \stackrel{?}{=} 1$$
$$\frac{3}{3} \stackrel{?}{=} 1$$
$$1 = 1 \quad \checkmark$$

For $x = 2$:
$$\sqrt{2(2) + 5} - 2\sqrt{2(2)} \stackrel{?}{=} 1$$
$$\sqrt{9} - 2\sqrt{4} \stackrel{?}{=} 1$$
$$3 - 4 \stackrel{?}{=} 1$$
$$-1 \neq 1$$

The only solution is $\frac{2}{9}$.

4.
$$\sqrt{y-1} + \sqrt{y-4} = \sqrt{4y-11}$$
$$\left(\sqrt{y-1} + \sqrt{y-4}\right)^2 = \left(\sqrt{4y-11}\right)^2$$
$$\left(\sqrt{y-1} + \sqrt{y-4}\right)\left(\sqrt{y-1} + \sqrt{y-4}\right) = 4y - 11$$
$$y - 1 + 2\left(\sqrt{y-1}\right)\left(\sqrt{y-4}\right) + y - 4 = 4y - 11$$
$$2y - 5 + 2\left(\sqrt{y-1}\right)\left(\sqrt{y-4}\right) = 4y - 11$$
$$2\left(\sqrt{y-1}\right)\left(\sqrt{y-4}\right) = 2y - 6$$
$$\left(\sqrt{y-1}\right)\left(\sqrt{y-4}\right) = y - 3$$
$$\left(\sqrt{y^2 - 5y + 4}\right)^2 = (y-3)^2$$
$$y^2 - 5y + 4 = y^2 - 6y + 9$$
$$y - 5 = 0$$
$$y = 5$$

Check: $\sqrt{5-1} + \sqrt{5-4} \stackrel{?}{=} \sqrt{4(5) - 11}$
$$2 + 1 \stackrel{?}{=} 3$$
$$3 = 3 \qquad \checkmark$$

The solution is 5.

7.6 Practice Problems

1. (a) $\sqrt{-49} = \sqrt{-1}\sqrt{49} = (i)(7) = 7i$

(b) $\sqrt{-31} = \sqrt{-1}\sqrt{31} = i\sqrt{31}$

2. $\sqrt{-98} = \sqrt{-1}\sqrt{98} = i\sqrt{98} = i\sqrt{49}\sqrt{2} = 7i\sqrt{2}$

3. $\sqrt{-8} \cdot \sqrt{-2} = \sqrt{-1}\sqrt{8} \cdot \sqrt{-1}\sqrt{2}$
$$= i\sqrt{8} \cdot i\sqrt{2}$$
$$= i^2\sqrt{16}$$
$$= -1(4) = -4$$

4. $-7 + 2yi\sqrt{3} = x + 6i\sqrt{3}$
$$x = -7, \qquad 2y\sqrt{3} = 6\sqrt{3}$$
$$y = 3$$

5. $(3 - 4i) - (-2 - 18i)$
$$= [3 - (-2)] + [-4 - (-18)]i$$
$$= (3 + 2) + (-4 + 18)i$$
$$= 5 + 14i$$

6. $(4 - 2i)(3 - 7i)$
$$= (4)(3) + (4)(-7i) + (-2i)(3) + (-2i)(-7i)$$
$$= 12 - 28i - 6i + 14i^2$$
$$= 12 - 28i - 6i + 14(-1)$$
$$= 12 - 28i - 6i - 14 = -2 - 34i$$

7. $-2i(5 + 6i)$
$$= (-2)(5)i + (-2)(6)i^2$$
$$= -10i - 12i^2$$
$$= -10i - 12(-1) = 12 - 10i$$

8. (a) $i^{42} = (i^{40+2}) = (i^{40})(i^2) = (i^4)^{10}(i^2) = (1)^{10}(-1) = -1$

(b) $i^{53} = (i^{52+1}) = (i^{52})(i) = (i^4)^{13}(i) = (1)^{13}(i) = i$

9. $\dfrac{4 + 2i}{3 + 4i} \cdot \dfrac{3 - 4i}{3 - 4i} = \dfrac{12 - 16i + 6i - 8i^2}{9 - 16i^2}$

$$= \frac{12 - 10i - 8(-1)}{9 - 16(-1)}$$
$$= \frac{12 - 10i + 8}{9 + 16} = \frac{20 - 10i}{25}$$
$$= \frac{5(4 - 2i)}{25} = \frac{4 - 2i}{5}$$

10. $\dfrac{5 - 6i}{-2i} \cdot \dfrac{2i}{2i}$

$$= \frac{10i - 12i^2}{-4i^2} = \frac{10i - 12(-1)}{-4(-1)}$$
$$= \frac{10i + 12}{4} = \frac{2(5i + 6)}{4} = \frac{6 + 5i}{2}$$

7.7 Practice Problems

1. Let s = speed,
  h = horsepower.
  $s = k\sqrt{h}$
  Substitute $s = 128$ and $h = 256$.
  $128 = k\sqrt{256}$
  $128 = 16k$
  $8 = k$
  Now we know the value of k so
  $s = 8\sqrt{h}$.
  when $h = 225$
  $s = 8\left(\sqrt{225}\right)$
  $s = 8(15)$
  $s = 120$ miles per hour

2. $y = \dfrac{k}{x}$
  Substitute $y = 45$ and $x = 16$.
  $45 = \dfrac{k}{16}$
  $720 = k$
  We now write the equation $y = \dfrac{720}{x}$.
  Find the value of y when $x = 36$.
  $y = \dfrac{720}{36}$
  $y = 20$

3. Let r = resistance,

c = amount of current.

$$r = \frac{k}{c^2}$$

Find the value of k when r = 800 ohms and c = 0.01 amps.

$$800 = \frac{k}{(0.01)^2}$$

$$0.08 = k$$

We now write the equation $r = \frac{0.08}{c^2}$.

Now substitute c = 0.04 and solve for r.

$$r = \frac{0.08}{(0.04)^2}$$

$$r = \frac{0.08}{0.0016}$$

$$r = 50 \text{ ohms}$$

4. $y = \frac{kzw^2}{x}$

To find the value of k substitute $y = 20$, $z = 3$, $w = 5$ and $x = 4$. Solve for k.

$$20 = \frac{k(3)(5)^2}{4}$$

$$20 = \frac{75k}{4}$$

$$\frac{80}{75} = k$$

$$\frac{16}{15} = k$$

We now substitute $\frac{16}{15}$ for k.

$$y = \frac{16zw^2}{15x}$$

We use this equation to find y when $z = 4$, $w = 6$, and $x = 2$.

$$y = \frac{16(4)(6)^2}{15(2)} = \frac{2304}{30}$$

$$y = \frac{384}{5}$$

Chapter 8

8.1 Practice Problems

1. $x^2 - 121 = 0$

$$x^2 = 121$$

$$x = \pm 11$$

Check:

$(11)^2 - 121 \overset{?}{=} 0$ $(-11)^2 - 121 \overset{?}{=} 0$

$121 - 121 \overset{?}{=} 0$ $121 - 121 \overset{?}{=} 0$

$\qquad\qquad 0 = 0$ ✓ $0 = 0$ ✓

2. $x^2 = 18$

$$x = \pm\sqrt{18}$$

$$x = \pm 3\sqrt{2}$$

3. $5x^2 + 1 = 46$

$$5x^2 = 45$$

$$x = \pm\sqrt{9}$$

$$x = \pm 3$$

4. $3x^2 = -27$

$$x = \pm\sqrt{-9}$$

$$x = \pm 3i$$

Check:

$3(3i)^2 \overset{?}{=} -27$ $3(-3i)^2 \overset{?}{=} -27$

$3(9)(-1) \overset{?}{=} -27$ $3(9)(-1) \overset{?}{=} -27$

$\qquad -27 = -27$ ✓ $-27 = -27$ ✓

5. $(2x + 3)^2 = 7$

$$(2x + 3) = \pm\sqrt{7}$$

$$2x + 3 = \pm\sqrt{7}$$

$$2x = -3 \pm \sqrt{7}$$

$$x = \frac{-3 \pm \sqrt{7}}{2}$$

6. $x^2 + 8x + 3 = 0$

$$x^2 + 8x = -3$$

$$x^2 + 8x + (4)^2 = -3 + (4)^2$$

$$(x + 4)^2 = 13$$

$$x + 4 = \pm\sqrt{13}$$

$$x = -4 \pm \sqrt{13}$$

7. $2x^2 + 4x + 1 = 0$

$$x^2 + 2x = \frac{-1}{2}$$

$$x^2 + 2x + (1)^2 = \frac{-1}{2} + 1$$

$$(x + 1)^2 = \frac{1}{2}$$

$$(x + 1) = \pm\sqrt{\frac{1}{2}}$$

$$x + 1 = \pm\frac{1}{\sqrt{2}}$$

$$x = -1 \pm \frac{\sqrt{2}}{2} \quad \text{or} \quad \frac{-2 \pm \sqrt{2}}{2}$$

8.2 Practice Problems

1. $x^2 + 5x = -1 + 2x$

$$x^2 + 3x + 1 = 0$$

$$a = 1, b = 3, c = 1$$

$$x = \frac{-3 \pm \sqrt{3^2 - 4(1)(1)}}{2(1)}$$

$$x = \frac{-3 \pm \sqrt{5}}{2}$$

2. $2x^2 + 7x + 6 = 0$

$$a = 2, b = 7, c = 6$$

$$x = \frac{-7 \pm \sqrt{7^2 - 4(2)(6)}}{2(2)}$$

$$x = \frac{-7 \pm \sqrt{49 - 48}}{4}$$

$$x = \frac{-7 \pm \sqrt{1}}{4}$$

$$x = \frac{-7 + 1}{4} \quad \text{or} \quad x = \frac{-7 - 1}{4}$$

$$x = -\frac{6}{4} = -\frac{3}{2} \qquad x = -2$$

3. $2x^2 - 26 = 0$

$$a = 2, b = 0, c = -26$$

$$x = \frac{-0 \pm \sqrt{0^2 - 4(2)(-26)}}{2(2)}$$

$$x = \frac{\pm\sqrt{208}}{4} = \frac{\pm 4\sqrt{13}}{4} = \pm\sqrt{13}$$

4. $0 = -100x^2 + 4800x - 52{,}559$

$$a = -100, b = 4800, c = -52{,}559$$

$$x = \frac{-4800 \pm \sqrt{(4800)^2 - 4(-100)(-52{,}559)}}{2(-100)}$$

$$x = \frac{-4800 \pm \sqrt{23{,}040{,}000 - 21{,}023{,}600}}{-200}$$

$$x = \frac{-4800 \pm \sqrt{2{,}016{,}400}}{-200}$$

$$x = \frac{-4800 \pm 1420}{-200}$$

$$x = \frac{-4800 + 1420}{-200} = 16.9 \approx 17$$

or

$$x = \frac{-4800 - 1420}{-200} = 31.1 \approx 31$$

5. $\dfrac{1}{x} + \dfrac{1}{x-1} = \dfrac{5}{6}$ LCD is $6x(x-1)$.

$$6x(x-1)\left[\frac{1}{x}\right] + 6x(x-1)\left[\frac{1}{x-1}\right] = 6x(x-1)\left[\frac{5}{6}\right]$$

$$6(x-1) + 6x = 5(x^2 - x)$$

$$6x - 6 + 6x = 5x^2 - 5x$$

$$0 = 5x^2 - 17x + 6$$

$a = 5, b = -17, c = 6$

$$x = \frac{-(-17) \pm \sqrt{(-17)^2 - 4(5)(6)}}{2(5)}$$

$$x = \frac{17 \pm \sqrt{289 - 120}}{10}$$

$$x = \frac{17 \pm \sqrt{169}}{10}$$

$$x = \frac{17 \pm 13}{10}$$

$$x = \frac{17 + 13}{10} = \frac{30}{10} = 3, \qquad x = \frac{17 - 13}{10} = \frac{2}{5}$$

6. $2x^2 - 4x + 5 = 0$

$a = 2, b = -4, c = 5$

$$x = \frac{-(-4) \pm \sqrt{(-4)^2 - 4(2)(5)}}{2(2)}$$

$$x = \frac{4 \pm \sqrt{-24}}{4}$$

$$x = \frac{4 \pm 2i\sqrt{6}}{4} = \frac{2 \pm i\sqrt{6}}{2}$$

7. $9x^2 + 12x + 4 = 0$

$a = 9, b = 12, c = 4$

$b^2 - 4ac = 12^2 - 4(9)(4) = 144 - 144 = 0$

Since the discriminant is 0, there is one rational solution.

8. (a) $x^2 - 4x + 13 = 0$

  $a = 1, b = -4, c = 13$

  $b^2 - 4ac = (-4)^2 - 4(1)(13) = 16 - 52 = -36$

  Since the discriminant is negative, there are two complex solutions containing i.

(b) $9x^2 + 6x + 7 = 0$

  $a = 9, b = 6, c = 7$

  $b^2 - 4ac = 6^2 - 4(9)(7)$

  $= 36 - 252 = -216$

  Since the discriminant is negative, there are two complex solutions containing i.

9. $x = -10 \qquad x = -6$

  $x + 10 = 0 \qquad x + 6 = 0$

  $(x + 10)(x + 6) = 0$

  $x^2 + 6x + 10x + 60 = 0$

  $x^2 + 16x + 60 = 0$

10. $x = 2i\sqrt{3} \qquad x = -2i\sqrt{3}$

  $x - 2i\sqrt{3} = 0 \qquad x + 2i\sqrt{3} = 0$

  $\left(x - 2i\sqrt{3}\right)\left(x + 2i\sqrt{3}\right) = 0$

  $x^2 - 4i^2\left(\sqrt{9}\right) = 0$

  $x^2 - 4(-1)(3) = 0$

  $x^2 + 12 = 0$

8.3 Practice Problems

1. $x^4 - 5x^2 - 36 = 0$

Let $y = x^2$. Then $y^2 = x^4$.

Thus, our new equation is

$y^2 - 5y - 36 = 0.$

$(y - 9)(y + 4) = 0$

$y - 9 = 0$	$y + 4 = 0$
$y = 9$	$y = -4$
$x^2 = 9$	$x^2 = -4$
$x = \pm\sqrt{9}$	$x = \pm\sqrt{-4}$
$x = \pm 3$	$x = \pm 2i$

2. $x^6 - 5x^3 + 4 = 0$

Let $y = x^3$. Then $y^2 = x^6$.

$y^2 - 5y + 4 = 0$

$(y - 1)(y - 4) = 0$

$y - 1 = 0$	$y - 4 = 0$
$y = 1$	$y = 4$
$x^3 = 1$	$x^3 = 4$
$x = 1$	$x = \sqrt[3]{4}$

3. $3x^{4/3} - 5x^{2/3} + 2 = 0$

Let $y = x^{2/3}$ and $y^2 = x^{4/3}$.

$3y^2 - 5y + 2 = 0$

$(3y - 2)(y - 1) = 0$

$3y - 2 = 0$	$y - 1 = 0$
$y = \dfrac{2}{3}$	$y = 1$
$x^{2/3} = \dfrac{2}{3}$	$x^{2/3} = 1$
$(x^{2/3})^3 = \left(\dfrac{2}{3}\right)^3$	$(x^{2/3})^3 = 1^3$
$x^2 = \dfrac{8}{27}$	$x^2 = 1$
$x = \pm\sqrt{\dfrac{8}{27}}$	$x = \pm\sqrt{1}$
$x = \pm\dfrac{2\sqrt{2}}{3\sqrt{3}}$	$x = \pm 1$

$$x = \pm\frac{2\sqrt{2}}{3\sqrt{3}} \cdot \frac{\sqrt{3}}{\sqrt{3}}$$

$$x = \pm\frac{2\sqrt{6}}{9}$$

Check: for $x = \dfrac{2\sqrt{6}}{9}$

$$3\left(\frac{2\sqrt{6}}{9}\right)^{4/3} - 5\left(\frac{2\sqrt{6}}{9}\right)^{2/3} + 2 \overset{?}{=} 0$$

$$3\left(\frac{4}{9}\right) - 5\left(\frac{2}{3}\right) + 2 \overset{?}{=} 0$$

$$\frac{4}{3} - \frac{10}{3} + 2 \overset{?}{=} 0$$

$$0 = 0 \ \checkmark$$

for $x = -\dfrac{2\sqrt{6}}{9}$

$$3\left(-\frac{2\sqrt{6}}{9}\right)^{4/3} - 5\left(-\frac{2\sqrt{6}}{9}\right)^{2/3} + 2 \overset{?}{=} 0$$

$$3\left(\frac{4}{9}\right) - 5\left(\frac{2}{3}\right) + 2 \overset{?}{=} 0$$

$$\frac{4}{3} - \frac{10}{3} + 2 \overset{?}{=} 0$$

$$0 = 0 \ \checkmark$$

for $x = 1$

$$3(1)^{4/3} - 5(1)^{2/3} + 2 \overset{?}{=} 0$$

$$3(1) - 5(1) + 2 \overset{?}{=} 0$$

$$0 = 0 \ \checkmark$$

for $x = -1$
$$3(-1)^{4/3} - 5(-1)^{2/3} + 2 \overset{?}{=} 0$$
$$3(1) - 5(1) + 2 \overset{?}{=} 0$$
$$0 = 0 \quad \checkmark$$

4. $\qquad 3x^{1/2} = 8x^{1/4} - 4$
$3x^{1/2} - 8x^{1/4} + 4 = 0$
Let $y = x^{1/4}$ and $y^2 = x^{1/2}$.
$$3y^2 - 8y + 4 = 0$$
$$(3y - 2)(y - 2) = 0$$

$3y - 2 = 0$	$y - 2 = 0$
$y = \dfrac{2}{3}$	$y = 2$
$x^{1/4} = \dfrac{2}{3}$	$x^{1/4} = 2$
$(x^{1/4})^4 = \left(\dfrac{2}{3}\right)^4$	$(x^{1/4})^4 = (2)^4$
$x = \dfrac{16}{81}$	$x = 16$

Check: for $x = \dfrac{16}{81}$ $\qquad$ for $x = 16$

$$3\left(\frac{16}{81}\right)^{1/2} \overset{?}{=} 8\left(\frac{16}{81}\right)^{1/4} - 4 \qquad 3(16)^{1/2} \overset{?}{=} 8(16)^{1/4} - 4$$

$$3\left(\frac{4}{9}\right) \overset{?}{=} 8\left(\frac{2}{3}\right) - 4 \qquad\qquad 3(4) \overset{?}{=} 8(2) - 4$$

$$\frac{4}{3} = \frac{4}{3} \quad \checkmark \qquad\qquad\qquad 12 = 12 \quad \checkmark$$

8.4 Practice Problems

1. $V = \dfrac{1}{3}\pi r^2 h$ Solve for r.

$$\frac{3V}{\pi h} = r^2$$

$$\pm\sqrt{\frac{3V}{\pi h}} = r \quad \text{so} \quad r = \sqrt{\frac{3V}{\pi h}}$$

2. $2y^2 + 9wy + 7w^2 = 0$ Solve for y.
$$(2y + 7w)(y + w) = 0$$

$2y + 7w = 0$	$y + w = 0$
$2y = -7w$	$y = -w$

$$y = -\frac{7}{2}w$$

3. $3y^2 + 2fy - 7g = 0$ Solve for y.
Use the quadratic formula.
$a = 3, b = 2f, c = -7g$

$$y = \frac{-2f \pm \sqrt{(2f)^2 - 4(3)(-7g)}}{2(3)}$$

$$y = \frac{-2f \pm \sqrt{4f^2 + 84g}}{6}$$

$$y = \frac{-2f \pm \sqrt{4(f^2 + 21g)}}{6}$$

$$y = \frac{-2f + 2\sqrt{(f^2 + 21g)}}{6}$$

$$y = \frac{-f \pm \sqrt{f^2 + 21g}}{3}$$

4. $d = \dfrac{n^2 - 3n}{2}$ Solve for n.

Multiply each term by 2.
$2d = n^2 - 3n$
$0 = n^2 - 3n - 2d$

Use the quadratic formula.
$a = 1, b = -3, c = -2d$

$$n = \frac{-(-3) \pm \sqrt{(-3)^2 - 4(1)(-2d)}}{2}$$

$$n = \frac{3 \pm \sqrt{9 + 8d}}{2}$$

5. (a) $a^2 + b^2 = c^2$ Solve for b.
$$b^2 = c^2 - a^2$$
$$b = \sqrt{c^2 - a^2}$$

(b) $b = \sqrt{c^2 - a^2}$
$$b = \sqrt{(26)^2 - (24)^2}$$
$$b = \sqrt{676 - 576}$$
$$b = \sqrt{100}$$
$$b = 10$$

6. $x + x - 7 + c = 30$
$$2x - 7 + c = 30$$
$$c = -2x + 37$$
$a = x, b = x - 7, c = -2x + 37$
By the Pythagorean theorem,
$$x^2 + (x - 7)^2 = (-2x + 37)^2$$
$$x^2 + x^2 - 14x + 49 = 4x^2 - 148x + 1369$$
$$-2x^2 + 134x - 1320 = 0$$
$$x^2 - 67x + 660 = 0$$
By the quadratic formula,
$a = 1, b = -67, c = 660$

$$x = \frac{67 \pm \sqrt{(67)^2 - 4(1)(660)}}{2}$$

$$x = \frac{67 \pm \sqrt{4489 - 2640}}{2}$$

$$x = \frac{67 \pm \sqrt{1849}}{2}$$

$$x = \frac{67 \pm 43}{2}$$

$$x = \frac{67 + 43}{2} = 55 \quad \text{or} \quad x = \frac{67 - 43}{2} = 12$$

The only answer that makes sense is $x = 12$; therefore,
$$x = 12$$
$$x - 7 = 5$$
$$-2x + 37 = 13$$

The legs are 5 miles and 12 miles long. The hypotenuse of the triangle is 13 miles long.

7. $A = \pi r^2$
$A = \pi(6)^2$
$\quad = 36\pi$
Let $x = $ the radius of the new pipe.
(area of new pipe) minus (area of old pipe) $= 45\pi$
$$\pi x^2 - 36\pi = 45\pi$$
$$\pi x^2 = 45\pi + 36\pi$$
$$x^2 = 81$$
$$x = \pm 9$$

Since the radius must be positive, we select $x = 9$. The radius of the new pipe is 9 inches. The radius of the new pipe has been increased by 3 inches.

8. Let $x = $ width. Then $2x - 3 = $ the length.
$$x(2x - 3) = 54$$
$$2x^2 - 3x = 54$$
$$2x^2 - 3x - 54 = 0$$
$$(2x + 9)(x - 6) = 0$$

$2x + 9 = 0$	$x - 6 = 6$
$x = -\dfrac{9}{2}$	$x = 6$

We do not use the negative value.
Thus, width = 6 feet
length = $2x - 3 = 2(6) - 3 = 9$ feet

9.

	Distance	Rate	Time
Secondary Road	150	x	$\dfrac{150}{x}$
Better Road	240	$x + 10$	$\dfrac{240}{x + 10}$
TOTAL	390	(not used)	7

$$\frac{150}{x} + \frac{240}{x + 10} = 7$$

The LCD of this equation is $x(x + 10)$. Multiply each term by the LCD.

$$x(x + 10)\left[\frac{150}{x}\right] + x(x + 10)\left[\frac{240}{x + 10}\right] = x(x + 10)[7]$$

$$150(x + 10) + 240x = 7x(x + 10)$$
$$150x + 1500 + 240x = 7x^2 + 70x$$
$$7x^2 - 320x - 1500 = 0$$
$$(x - 50)(7x + 30) = 0$$
$$x - 50 = 0 \qquad 7x + 30 = 0$$
$$x = 50 \qquad\quad x = \frac{-30}{7}$$

We disregard the negative answer. Thus, $x = 50$ mph, so Carlos drove 50 mph on the secondary road and 60 mph on the better road.

8.5 Practice Problems

1. $f(x) = x^2 - 6x + 5$
$a = 1, b = -6, c = 15$

Step 1 The vertex occurs at $x = \dfrac{-b}{2a}$. Thus, $x = \dfrac{-(-6)}{2(1)} = 3$

The vertex has an x-coordinate of 3.
To find the y-coordinate, we evaluate $f(3)$.
$$f(3) = 3^2 - 6(3) + 5$$
$$= 9 - 18 + 5$$
$$= -4$$
Thus, the vertex is $(3, -4)$.

Step 2 The y-intercept is at $f(0)$.
$$f(0) = 0^2 - 6(0) + 5$$
$$= 5$$
The y-intercept is $(0, 5)$.

Step 3 The x-intercept is at $f(x) = 0$.
$$x^2 - 6x + 5 = 0$$
$$(x - 5)(x - 1) = 0$$
$$x - 5 = 0 \quad x - 1 = 0$$
$$x = 5 \qquad x = 1$$
Thus, the x-intercepts are $(5, 0)$ and $(1, 0)$.

2. $g(x) = x^2 - 2x - 2$
$a = 1, b = -2, c = -2$

Step 1 The vertex occurs at
$$x = \frac{-b}{2a}$$
$$x = \frac{-(-2)}{2(1)} = \frac{2}{2} = 1$$
The vertex has an x-coordinate of 1. To find the y-coordinate, we evaluate $f(1)$.
$$g(1) = 1^2 - 2(1) - 2$$
$$= 1 - 2 - 2$$
$$= -3$$
Thus, the vertex is $(1, -3)$.

Step 2 The y-intercept is at $g(0)$.
$$g(0) = 0^2 - 2(0) - 2$$
$$= -2$$
The y-intercept is $(0, -2)$.

Step 3 The x-intercepts occur when $g(x) = 0$. We set $x^2 - 2x - 2 = 0$ and solve for x. The equation does not factor, so we use the quadratic formula.

$$x = \frac{-(-2) \pm \sqrt{12}}{2} = \frac{2 \pm 2\sqrt{3}}{2} = 1 \pm \sqrt{3}$$

The x-intercepts are approximately $(2.7, 0)$ and $(-0.7, 0)$.

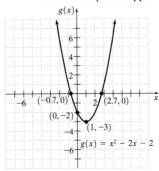

3. $g(x) = -2x^2 - 8x - 6$
$a = -2, b = -8, c = -6$
Since $a < 0$, the parabola opens downward.
The vertex occurs at
$$x = \frac{-b}{2a} = \frac{-(-8)}{2(-2)} = -2.$$
To find the y-coordinate, evaluate $g(-2)$.
$$g(-2) = -2(-2)^2 - 8(-2) - 6$$
$$= -8 + 16 - 6$$
$$= 2$$
Thus, the vertex is $(-2, 2)$.
The y-intercept is at $g(0)$.
$$g(0) = -2(0)^2 - 8(0) - 6$$
$$= -6$$
The y-intercept is $(0, -6)$.
The x-intercepts occur when $g(x) = 0$.
Using the quadratic formula.

$$x = \frac{-(-8) \pm \sqrt{64 - 4(-2)(-6)}}{2(-2)}$$
$$= \frac{8 \pm \sqrt{16}}{-4} = -2 \pm -1$$
$$x = -3, x = -1$$
The x-intercepts are $(-3, 0)$ and $(-1, 0)$.

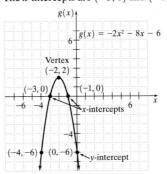

8.6 Practice Problems

1. $x^2 - 2x - 8 < 0$
Replace the inequality by an equal sign and solve.
$$(x - 4)(x + 2) = 0$$
$$x = 4 \qquad x = -2$$

Region I $x < -2$ $(-3)^2 - 2(-3) - 8$
$\quad\quad\quad$ $x = -3$ $\quad = 9 + 6 - 8 = 7 > 0$

Region II $-2 < x < 4$ $0^2 - 2(0) - 8 = -8 < 0$
$\quad\quad\quad$ $x = 0$

Region III $x > 4$ $(5)^2 - 2(5) - 8 = 7 > 0$
$\quad\quad\quad$ $x = 5$

Thus, $x^2 - 2x - 8 < 0$ when $-2 < x < 4$.

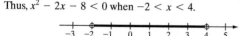

2. $3x^2 - x - 2 \geq 0$

$(3x + 2)(x - 1) = 0$

$3x + 2 = 0 \quad x - 1 = 0$

$x = -\frac{2}{3} \quad\quad x = 1$

Region I $x < -\frac{2}{3}$
$\quad\quad\quad$ $x = -1 \quad 3(-1)^2 + 1 - 2 = 2 > 0$

Region II $-\frac{2}{3} \leq x \leq 1$
$\quad\quad\quad$ $x = 0 \quad 3(0) - 0 - 2 = -2 < 0$

Region III $x > 1$
$\quad\quad\quad$ $x = 2 \quad 3(2)^2 - 2 - 2 = 8 > 0$

Thus, $3x^2 - x - 2 \geq 0$ when $x \leq -\frac{2}{3}$ or when $x \geq 1$.

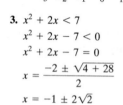

3. $x^2 + 2x < 7$

$x^2 + 2x - 7 < 0$

$x^2 + 2x - 7 = 0$

$x = \dfrac{-2 \pm \sqrt{4 + 28}}{2}$

$x = -1 \pm 2\sqrt{2}$

$x \approx 1.8 \quad$ and $\quad x \approx -3.8$

Region I $x < -3.8$
$\quad\quad\quad$ $x = -5 \quad\quad (-5)^2 + 2(-5) - 7 = 8 > 0$

Region II $-3.8 < x < 1.8$
$\quad\quad\quad$ $x = 0 \quad\quad (0)^2 + 2(0) - 7 = -7 < 0$

Region III $x > 1.8$
$\quad\quad\quad$ $x = 3 \quad\quad (3)^2 + 2(3) - 7 = 8 > 0$

Thus, $x^2 + 2x - 7 < 0$ when $-1 - 2\sqrt{2} < x < -1 + 2\sqrt{2}$. Approximately $-3.8 < x < 1.8$.

about −3.8   about 1.8
(number line from −4 to 4)

Chapter 9

9.1 Practice Problems

1. Let $(x_1, y_1) = (-6, -2)$ and $(x_2, y_2) = (3, 1)$

$d = \sqrt{(x_2 - x_1)^2 + (y_2 - y_1)^2}$

$\quad = \sqrt{[3 - (-6)]^2 + [1 - (-2)]^2}$

$\quad = \sqrt{(3 + 6)^2 + (1 + 2)^2}$

$\quad = \sqrt{(9)^2 + (3)^2}$

$\quad = \sqrt{81 + 9} = \sqrt{90} = 3\sqrt{10}$

2. $(x + 1)^2 + (y + 2)^2 = 9$
If we compare this to $(x - h)^2 + (y - k)^2 = r^2$, we can write it in the form

$$[x - (-1)]^2 + [y - (-2)]^2 = 3^2.$$

Thus, we see the center is $(h, k) = (-1, -2)$ and the radius is $r = 3$.

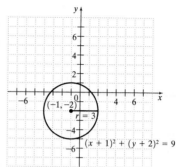

$(x + 1)^2 + (y + 2)^2 = 9$

3. We are given that $(h, k) = (-5, 0)$ and $r = \sqrt{3}$. Thus, $(x - h)^2 + (y - k)^2 = r^2$ becomes

$$[x - (-5)]^2 + (y - 0)^2 = \left(\sqrt{3}\right)^2$$
$$(x + 5)^2 + y^2 = 3$$

4. To write $x^2 + 4x + y^2 + 2y - 20 = 0$ in standard form, we complete the square.

$$x^2 + 4x + \underline{\quad} + y^2 + 2y + \underline{\quad} = 20$$
$$x^2 + 4x + 4 + y^2 + 2y + 1 = 20 + 4 + 1$$
$$x^2 + 4x + 4 + y^2 + 2y + 1 = 25$$
$$(x + 2)^2 + (y + 1)^2 = 25$$

The circle has its center at $(-2, -1)$ and the radius is 5.

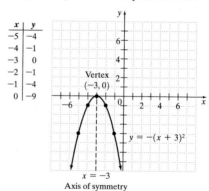

$(x + 2)^2 + (y + 1)^2 = 25$

9.2 Practice Problems

1. Make a table of values. Begin with $x = -3$ in the middle of the table because $(-3 + 3) = 0$. Plot the points and draw the graph.

x	y
−5	−4
−4	−1
−3	0
−2	−1
−1	−4
0	−9

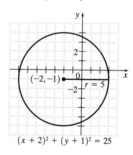

$y = -(x + 3)^2$

$x = -3$
Axis of symmetry

The vertex is $(-3, 0)$, and the axis of symmetry is the line $x = -3$. $y = -(x + 3)^2$

2. This graph looks like the graph of $y = x^2$, except that it is shifted 6 units to the right and 4 units up.
The vertex is $(6, 4)$. The axis of symmetry is $x = 6$.

If $x = 0$, $y = (0 - 6)^2 + 4 = 36 + 4 = 40$, so the y-intercept is $(0, 40)$.

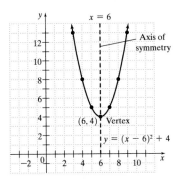

3. **Step 1**   The equation has the form $y = a(x - h)^2 + k$, where $a = \frac{1}{4}$, $h = 2$, and $k = 3$, so it is a vertical parabola.

   Step 2 $a > 0$; so the parabola opens upward.

   Step 3   We have $h = 2$ and $k = 3$. Therefore, the vertex is $(2, 3)$.

   Step 4   The axis of symmetry is the line $x = 2$. Plot a few points on either side of the axis of symmetry. At $x = 4$, $y = 4$. Thus, the point is $(4, 4)$. The image from symmetry is $(0, 4)$. At $x = 6$, $y = 7$. Thus the point is $(6, 7)$. The image from symmetry is $(-2, 7)$.

   Step 5   At $x = 0$,
   $$y = \frac{1}{4}(0 - 2)^2 + 3$$
   $$= \frac{1}{4}(4) + 3$$
   $$= 1 + 3 = 4.$$
   Thus, the y-intercept is $(0, 4)$.

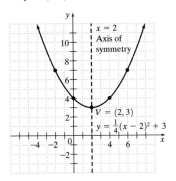

4. Make a table, plot points and draw the graph. Choose values of y and find x. Begin with $y = 0$. $x = -2y^2 + 4$

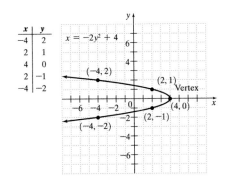

x	y
-4	2
2	1
4	0
2	-1
-4	-2

   The vertex is at $(4, 0)$. The axis of symmetry is the x-axis.

5. **Step 1**   The equation has the form $x = a(y - k)^2 + h$, so it is a horizontal parabola.

   Step 2 $a < 0$; so the parabola opens to the left.

   Step 3   We have $k = -1$ and $h = -3$. Therefore, the vertex is $(-3, -1)$.

Step 4   The line $y = -1$ is the axis of symmetry. At $y = 0$, $x = -4$. Thus, we have the point $(-4, 0)$ and $(-4, -2)$ from symmetry. At $y = 1$, $x = -7$. Thus, we have the point $(-7, 1)$ and $(-7, -3)$ from symmetry.

Step 5   At $y = 0$,
$$x = -(0 + 1)^2 - 3$$
$$= -(1) - 3$$
$$= -1 - 3 = -4$$
Thus, the x-intercept is $(-4, 0)$.

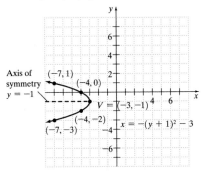

6. Since the y-term is squared, we have a horizontal parabola. The standard form is $x = a(y - k)^2 + h$.
$$x = y^2 - 6y + 13$$
$$= y^2 - 6y + \left(\frac{6}{2}\right)^2 - \left(\frac{6}{2}\right)^2 + 13 \quad \text{Complete the square.}$$
$$= (y^2 - 6y + 9) + 4$$
$$= (y - 3)^2 + 4$$
Therefore, we know that $a = 1$, $k = 3$, and $h = 4$. The vertex is at $(4, 3)$. The axis of symmetry is $y = 3$. If $y = 0$, $x = (-3)^2 + 4 = 9 + 4 = 13$. So the x-intercept is $(13, 0)$.

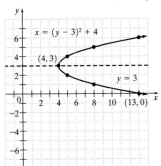

7. $y = 2x^2 + 8x + 9$

Since the x-term is squared, we have a vertical parabola. The standard form is
$$y = a(x - h)^2 + k$$
$$y = 2x^2 + 8x + 9$$
$$= 2(x^2 + 4x + \underline{\quad}) - \underline{\quad} + 9 \quad \text{Complete the square.}$$
$$= 2(x^2 + 4x + 4) - 2(4) + 9$$
$$= 2(x + 2)^2 + 1$$
The parabola opens upward. The vertex is $(-2, 1)$, and the y-intercept is $(0, 9)$. The axis of symmetry is $x = -2$

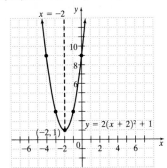

9.3 Practice Problems

1. Write the equation in standard form.

$$4x^2 + y^2 = 16$$

$$\frac{4x^2}{16} + \frac{y^2}{16} = \frac{16}{16}$$

$$\frac{x^2}{4} + \frac{y^2}{16} = 1$$

Thus, we have:

$$a^2 = 4 \quad \text{so} \quad a = 2$$
$$b^2 = 16 \quad \text{so} \quad b = 4$$

The x-intercepts are $(2, 0)$ and $(-2, 0)$ and the y-intercepts are $(0, 4)$ and $(0, -4)$.

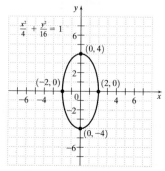

2. $\dfrac{(x-2)^2}{16} + \dfrac{(y+3)^2}{9} = 1$

The center is $(h, k) = (2, -3)$, $a = 4$, and $b = 3$. We start at $(2, -3)$ and measure to the right and to the left 4 units, and up and down 3 units.

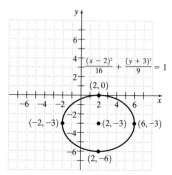

2. Write the equation in standard form.

$$y^2 - 4x^2 = 4$$

$$\frac{y^2}{4} - \frac{4x^2}{4} = \frac{4}{4}$$

$$\frac{y^2}{4} - \frac{x^2}{1} = 1$$

This is the equation of a vertical hyperbola with center $(0, 0)$, where $a = 1$ and $b = 2$. The vertices are $(0, 2)$ and $(0, -2)$.
The fundamental rectangle has corners at $(1, 2)$, $(1, -2)$, $(-1, 2)$, and $(-1, -2)$.

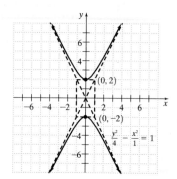

3. $\dfrac{(y+2)^2}{9} - \dfrac{(x-3)^2}{16} = 1$

This is a vertical hyperbola with center at $(3, -2)$, where $a = 4$ and $b = 3$. The vertices are $(3, 1)$ and $(3, -5)$. The fundamental rectangle has corners at $(7, 1)$, $(7, -5)$, $(-1, 1)$, and $(-1, -5)$.

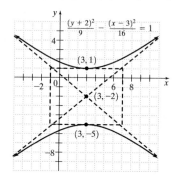

9.4 Practice Problems

1. $\dfrac{x^2}{16} - \dfrac{y^2}{25} = 1$

This is the equation of a horizontal hyperbola with center $(0, 0)$, where $a = 4$ and $b = 5$. The vertices are $(-4, 0)$ and $(4, 0)$.

Construct a fundamental rectangle with corners at $(4, 5)$, $(4, -5)$, $(-4, 5)$, and $(-4, -5)$. Draw extended diagonals as the asymptotes.

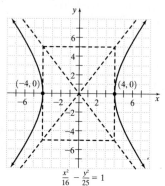

9.5 Practice Problems

1. $\dfrac{x^2}{4} - \dfrac{y^2}{4} = 1 \qquad x + y + 1 = 0$

$$x^2 - y^2 = 4 \quad (1) \qquad y = -x - 1 \quad (2)$$

Substitute (2) into (1).

$$x^2 - (-x - 1)^2 = 4$$
$$x^2 - (x^2 + 2x + 1) = 4$$
$$x^2 - x^2 - 2x - 1 = 4$$
$$-2x - 1 = 4$$
$$-2x = 5$$
$$x = \frac{5}{-2} = -2.5$$

Now substitute the value for x in the equation $y = -x - 1$.
For $x = -2.5$: $y = -(-2.5) - 1 = 2.5 - 1 = 1.5$.
The solution is $(-2.5, 1.5)$.

2. $2x - 9 = y \quad (1)$

$xy = -4 \quad (2)$ Solve equation (2) for y.

$$y = \frac{-4}{x} \quad (3)$$

Substitute equation (3) into (1) and solve for x.

$$2x - 9 = \frac{-4}{x}$$
$$2x^2 - 9x + 4 = 0$$
$$(2x - 1)(x - 4) = 0$$
$$2x - 1 = 0 \qquad x - 4 = 0$$
$$x = \frac{1}{2} \quad \text{or} \quad x = 4$$
$$x = \frac{1}{2} \text{ and } x = 4$$

For $x = \frac{1}{2}$:
$$y = \frac{-4}{\frac{1}{2}} = -8.$$

For $x = 4$:
$$y = \frac{-4}{4} = -1.$$

The solutions are $(4, -1)$ and $\left(\frac{1}{2}, -8\right)$.

The graph of $y = \frac{-4}{x}$ is the graph of $y = \frac{1}{x}$ reflected across the x-axis and stretched by a factor of 4. The graph of $y = 2x - 9$ is a line with slope 2 passing through the point $(0, -9)$. The sketch show that the points $(4, -1)$ and $\left(\frac{1}{2}, -8\right)$ seem reasonable.

3. (1) $x^2 + y^2 = 12$ $4x^2 + 4y^2 = 48$ Multiply (1) by 4.
   (2) $3x^2 - 4y^2 = 8$ $\underline{3x^2 - 4y^2 = 8}$ Add the equations.
$$7x^2 = 56$$
$$x^2 = 8$$
$$x = \pm\sqrt{8}$$
$$x = \pm 2\sqrt{2}$$

If $x = 2\sqrt{2}$, then $x^2 = 8$. Substituting this value into equation (1) gives
$$8 + y^2 = 12$$
$$y^2 = 4$$
$$y = \pm\sqrt{4}$$
$$y = \pm 2$$

Similarly, if $x = -2\sqrt{2}$, then $y = \pm 2$.
Thus, the four solutions are $\left(2\sqrt{2}, 2\right), \left(2\sqrt{2}, -2\right), \left(-2\sqrt{2}, 2\right), \left(-2\sqrt{2}, -2\right)$.

Chapter 10

10.1 Practice Problems

1. (a) $g(a) = \frac{1}{2}a - 3$

(b) $g(a + 4) = \frac{1}{2}(a + 4) - 3$
$$= \frac{1}{2}a + 2 - 3$$
$$= \frac{1}{2}a - 1$$

(c) $g(a) = \frac{1}{2}a - 3$
$$g(4) = \frac{1}{2}(4) - 3 = 2 - 3 = -1$$

Thus, $g(a) + g(4) = \left(\frac{1}{2}a - 3\right) + (-1)$
$$= \frac{1}{2}a - 3 - 1$$
$$= \frac{1}{2}a - 4$$

2. (a) $p(-3) = -3(-3)^2 + 2(-3) + 4$
$$= -3(9) + 2(-3) + 4$$
$$= -27 - 6 + 4$$
$$= -29$$

(b) $p(a) = -3(a)^2 + 2(a) + 4$
$$= -3a^2 + 2a + 4$$

(c) $p(2a) = -3(2a)^2 + 2(2a) + 4$
$$= -3(4a^2) + 2(2a) + 4$$
$$= -12a^2 + 4a + 4$$

(d) $p(a - 3) = -3(a - 3)^2 + 2(a - 3) + 4$
$$= -3(a - 3)(a - 3) + 2(a - 3) + 4$$
$$= -3(a^2 - 6a + 9) + 2(a - 3) + 4$$
$$= -3a^2 + 18a - 27 + 2a - 6 + 4$$
$$= -3a^2 + 20a - 29$$

3. (a) $r(a + 2) = \frac{-3}{(a + 2) + 1}$
$$= \frac{-3}{a + 3}$$

(b) $r(a) = \frac{-3}{a + 1}$

(c) $r(a + 2) - r(a) = \frac{-3}{a + 3} - \left(\frac{-3}{a + 1}\right)$
$$= \frac{-3}{a + 3} + \frac{3}{a + 1}$$
$$= \frac{(a + 1)(-3)}{(a + 1)(a + 3)} + \frac{3(a + 3)}{(a + 1)(a + 3)}$$
$$= \frac{-3a - 3}{(a + 1)(a + 3)} + \frac{3a + 9}{(a + 1)(a + 3)}$$
$$= \frac{-3a - 3 + 3a + 9}{(a + 1)(a + 3)}$$
$$= \frac{6}{(a + 1)(a + 3)}$$

4. $g(x + h) = 2 - 5(x + h) = 2 - 5x - 5h$
$g(x) = 2 - 5x$
$g(x + h) - g(x) = (2 - 5x - 5h) - (2 - 5x)$
$$= 2 - 5x - 5h - 2 + 5x$$
$$= -5h$$
Therefore, $\frac{g(x + h) - g(x)}{h} = \frac{-5h}{h} = -5.$

5. (a) $S(r) = 16(3.14)r + 2(3.14)r^2 = 50.24r + 6.28r^2$

(b) $S(2) = 50.24(2) + 6.28(2)^2$
$$= 50.24(2) + 6.28(4)$$
$$= 125.6 \text{ square meters}$$

(c) $S(2 + e) = 50.24(2 + e) + 6.28(2 + e)^2$
$$= 50.24(2 + e) + 6.28(2 + e)(2 + e)$$
$$= 50.24(2 + e) + 6.28(4 + 4e + e^2)$$
$$= 100.48 + 50.24e + 25.12 + 25.12e + 6.28e^2$$
$$= 125.6 + 75.36e + 6.28e^2$$

(d) $S = 125.6 + 75.36e + 6.28e^2$
$$= 125.6 + 75.36(0.3) + 6.28(0.3)^2$$
$$= 125.6 + 22.608 + 0.5652$$
$$= 148.77 \text{ square meters}$$

Thus, if the radius of 2 square meters was incorrectly measured as 2.3 meters, the surface area would be approximately 23.17 square meters too large.

10.2 Practice Problems

1. By the vertical line test, this relation is not a function.

2. $f(x) = x^2$ $\qquad\qquad\qquad\qquad$ $h(x) = x^2 - 5$

x	f(x)
−2	4
−1	1
0	0
1	1
2	4

x	h(x)
−2	−1
−1	−4
0	−5
1	−4
2	−1

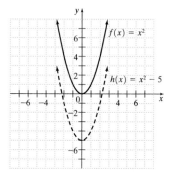

3. $f(x) = |x|$ $\qquad\qquad\qquad\qquad$ $p(x) = |x + 2|$

x	f(x)
−2	2
−1	1
0	0
1	1
2	2

x	p(x)
−4	2
−3	1
−2	0
−1	1
0	2
1	3
2	4

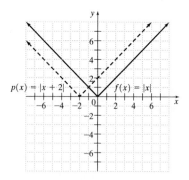

4. $f(x) = x^3$

x	f(x)
−2	−8
−1	−1
0	0
1	1
2	8

We recognize that $h(x)$ will have a similar shape, but the curve will be shifted 4 units to the left and 3 units upward.

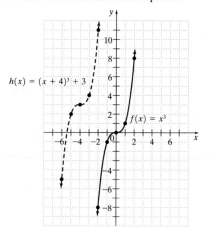

5. $f(x) = \dfrac{2}{x}$

x	f(x)
−4	$-\dfrac{1}{2}$
−2	−1
−1	−2
$-\dfrac{1}{2}$	−4
0	undefined
$\dfrac{1}{2}$	4
1	2
2	1
4	$\dfrac{1}{2}$

The graph of $g(x)$ is 1 unit to the left and 2 units below $f(x)$. We use each point on $f(x)$ to guide us in graphing $g(x)$. *Note:* $x = -1$ is undefined on $g(x)$.

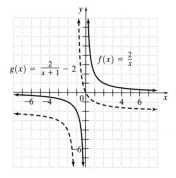

10.3 Practice Problems

1. (a) $(f + g)(x) = f(x) + g(x)$
$\qquad\qquad = (4x + 5) + (2x^2 + 7x - 8)$
$\qquad\qquad = 4x + 5 + 2x^2 + 7x - 8$
$\qquad\qquad = 2x^2 + 11x - 3$

(b) Using the formula obtained in **(a)**,
$\qquad (f + g)(x) = 2x^2 + 11x - 3$
$\qquad (f + g)(4) = 2(4)^2 + 11(4) - 3$
$\qquad\qquad\qquad = 2(16) + 11(4) - 3$
$\qquad\qquad\qquad = 32 + 44 - 3$
$\qquad\qquad\qquad = 73$

2. (a) $(fg)(x) = f(x) \cdot g(x)$
$= (3x + 2)(x^2 - 3x - 4)$
$= 3x^3 - 9x^2 - 12x + 2x^2 - 6x - 8$
$= 3x^3 - 7x^2 - 18x - 8$

(b) Using the formula obtained in **(a)**,
$(fg)(x) = 3x^3 - 7x^2 - 18x - 8$
$(fg)(2) = 3(2)^3 - 7(2)^2 - 18(2) - 8$
$= 3(8) - 7(4) - 18(2) - 8$
$= 24 - 28 - 36 - 8$
$= -48$

3. (a) $\left(\dfrac{g}{h}\right)(x) = \dfrac{5x + 1}{3x - 2}$,   where $x \neq \dfrac{2}{3}$

(b) $\left(\dfrac{g}{p}\right)(x) = \dfrac{5x + 1}{5x^2 + 6x + 1} = \dfrac{5x + 1}{(5x + 1)(x + 1)} = \dfrac{1}{(x + 1)}$,

where $x \neq -\dfrac{1}{5}, x \neq -1$

(c) $\left(\dfrac{g}{h}\right)(x) = \dfrac{5x + 1}{3x - 2}$

$\left(\dfrac{g}{h}\right)(3) = \dfrac{5(3) + 1}{3(3) - 2} = \dfrac{15 + 1}{9 - 2} = \dfrac{16}{7}$

4. $f[g(x)] = f(3x - 4)$
$= 2(3x - 4) - 1$
$= 6x - 8 - 1$
$= 6x - 9$

5. (a) $f[g(x)] = f(x + 2)$
$= 2(x + 2)^2 - 3(x + 2) + 1$
$= 2(x^2 + 4x + 4) - 3x - 6 + 1$
$= 2x^2 + 8x + 8 - 3x - 6 + 1$
$= 2x^2 + 5x + 3$

(b) $g[f(x)] = g(2x^2 - 3x + 1)$
$= (2x^2 - 3x + 1) + 2$
$= 2x^2 - 3x + 1 + 2$
$= 2x^2 - 3x + 3$

6. (a) $(g \circ f)(x) = g[f(x)]$
$= g(3x + 1)$
$= \dfrac{2}{(3x + 1) - 3}$
$= \dfrac{2}{3x + 1 - 3}$
$= \dfrac{2}{3x - 2}$

(b) $(g \circ f)(-3) = \dfrac{2}{3(-3) - 2} = \dfrac{2}{-9 - 2} = \dfrac{2}{-11} = -\dfrac{2}{11}$

10.4 Practice Problems

1. (a) A is a function. No two pairs have the same second coordinate. Thus, A is a one-to-one function.
(b) B is a function. The pair $(1, 1)$ and $(-1, 1)$ share a common second coordinate. Therefore, B is not a one-to-one function.
2. The graphs of **(a)** and **(b)** do not represent one-to-one functions. A horizontal line exists that intersects the graphs more than once.
3. The inverse of B is obtained by interchanging x- and y-values for each ordered pair.
$$B^{-1} = \{(2, 1), (8, 7), (7, 8), (12, 10)\}$$
4.
$y = 4 - 6x$
$x = 4 - 6y$
$x - 4 = -6y$
$-x + 4 = 6y$
$\dfrac{-x + 4}{6} = y$
$g^{-1}(x) = \dfrac{-x + 4}{6}$

5.
$y = 0.75 + 0.55(x - 1)$
$x = 0.75 + 0.55(y - 1)$
$x = 0.75 + 0.55y - 0.55$
$x = 0.55y + 0.2$
$x - 0.2 = 0.55y$
$\dfrac{x - 0.2}{0.55} = y$
$f^{-1}(x) = \dfrac{x - 0.2}{0.55}$

6.
$f(x) = -\dfrac{1}{4}x + 1$
$y = -\dfrac{1}{4}x + 1$
$x = -\dfrac{1}{4}y + 1$
$x - 1 = -\dfrac{1}{4}y$
$-4x + 4 = y$
$f^{-1}(x) = -4x + 4$
Now graph each line.

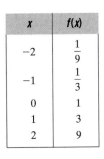

Chapter 11

11.1 Practice Problems

1. Graph $f(x) = 3^x$.
$f(x) = 3^x$
$f(-2) = 3^{-2} = \left(\dfrac{1}{3}\right)^2 = \dfrac{1}{9}$
$f(-1) = 3^{-1} = \left(\dfrac{1}{3}\right) = \dfrac{1}{3}$
$f(0) = 3^0 = 1$
$f(1) = 3^1 = 3$
$f(2) = 3^2 = 9$

x	f(x)
-2	$\dfrac{1}{9}$
-1	$\dfrac{1}{3}$
0	1
1	3
2	9

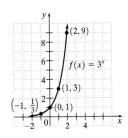

2. Graph $f(x) = \left(\frac{1}{3}\right)^x$.

$f(x) = \left(\frac{1}{3}\right)^x$

$f(-2) = \left(\frac{1}{3}\right)^{-2} = 3^2 = 9$

$f(-1) = \left(\frac{1}{3}\right)^{-1} = 3^1 = 3$

$f(0) = \left(\frac{1}{3}\right)^{0} = 1$

$f(1) = \left(\frac{1}{3}\right)^{1} = \frac{1}{3}$

$f(2) = \left(\frac{1}{3}\right)^{2} = \frac{1}{9}$

x	f(x)
-2	9
-1	3
0	1
1	$\frac{1}{3}$
2	$\frac{1}{9}$

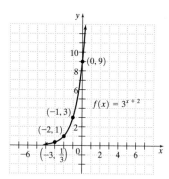

3. Graph $f(x) = 3^{x+2}$.

$f(-4) = 3^{-4+2} = 3^{-2} = \frac{1}{3^2} = \frac{1}{9}$

$f(-3) = 3^{-3+2} = 3^{-1} = \frac{1}{3}$

$f(-2) = 3^{-2+2} = 3^{0} = 1$

$f(-1) = 3^{-1+2} = 3^{1} = 3$

$f(0) = 3^{0+2} = 3^{2} = 9$

x	f(x)
-4	$\frac{1}{9}$
-3	$\frac{1}{3}$
-2	1
-1	3
0	9

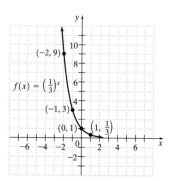

4. Graph $f(x) = e^{x-2}$.

$f(x) = e^{x-2}$ (values rounded to nearest hundredth)

$f(4) = e^{4-2} = e^{2} = 7.39$

$f(3) = e^{3-2} = e^{1} = 2.72$

$f(2) = e^{2-2} = e^{0} = 1$

$f(1) = e^{1-2} = e^{-1} = 0.37$

$f(0) = e^{0-2} = e^{-2} = 0.14$

x	f(x)
4	7.39
3	2.72
2	1
1	0.37
0	0.14

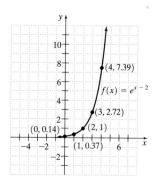

5. Solve $2^x = \frac{1}{32}$.

$2^x = \frac{1}{32}$

$2^x = \frac{1}{2^5}$ Because $2^5 = 32$.

$2^x = 2^{-5}$ Because $\frac{1}{2^5} = 2^{-5}$.

$x = -5$ Property of exponential equations.

6. Here $P = 4000, r = 0.11$, and $t = 2$.

$A = P(1 + r)^t$

$\quad = 4000(1 + 0.11)^2$

$\quad = 4000(1.11)^2$

$\quad = 4000(1.2321)$

$\quad = 4928.4$

Uncle Jose will have $4928.40.

7. How much money would Collette have if she invested $1500 for 8 years at 8% annual interest if the interest is compounded quarterly (four times a year)?

Here $P = 1500, r = 0.08, t = 8$, and $n = 4$.

$A = P\left(1 + \frac{r}{n}\right)^{nt}$

$\quad = 1500\left(1 + \frac{.08}{4}\right)^{[4(8)]}$

$\quad = 1500(1 + .02)^{32}$

$\quad = 1500(1.02)^{32}$

$\quad \approx 1500(1.884540592)$

$\quad \approx 2826.81089$

Collette will have approximately $2826.81.

8. Here $C = 20$ and $t = 5000$.

$A = 20e^{-0.0016008(5000)}$

$A = 20e^{-8.004}$

$A \approx 20(0.0003341) = 0.006682$

Thus, 0.007 milligrams of americium 241 would be present in 5000 years.

11.2 Practice Problems

1. Use the fact that $x = b^y$ is equivalent to $\log_b x = y$.

(a) Here $x = 49, b = 7$, and $y = 2$. So $2 = \log_7 49$.

(b) Here $x = \frac{1}{64}, b = 4$, and $y = -3$. So $-3 = \log_4\left(\frac{1}{64}\right)$.

2. (a) Here $y = 3, b = 5$ and $x = 125$. Thus, since $x = b^y$, $125 = 5^3$.

(b) Here $y = -2, b = 6$, and $x = \frac{1}{32}$. So $\frac{1}{36} = 6^{-2}$.

3. (a) $\log_b 125 = 3$; then $125 = b^3$
$$5^3 = b^3$$
$$b = 5$$

(b) $\log_{1/2} 32 = x$; then $32 = \left(\dfrac{1}{2}\right)^x$
$$2^5 = \left(\dfrac{1}{2}\right)^x$$
$$\dfrac{1}{2^{-5}} = \left(\dfrac{1}{2}\right)^x$$
$$\left(\dfrac{1}{2}\right)^{-5} = \left(\dfrac{1}{2}\right)^x$$
$$x = -5$$

4. $\log_{10} 0.1 = x$
$$0.1 = 10^x$$
$$10^{-1} = 10^x$$
$$-1 = x$$
Thus, $\log_{10} 0.1 = -1$.

5. Graph $y = \log_{1/2} x$.

To graph $y = \log_{1/2} x$, we first write $x = \left(\dfrac{1}{2}\right)^y$. We make a table of values.

x	y
$\dfrac{1}{4}$	2
$\dfrac{1}{2}$	1
1	0
2	-1
4	-2

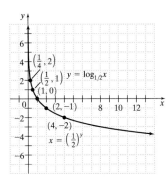

6. Make a table of values for each equation.

$y = \log_6 x$

x	y
$\dfrac{1}{6}$	-1
1	0
6	1
36	2

$y = 6^x$

x	y
-1	$\dfrac{1}{6}$
0	1
1	6
2	36

Ordered pairs reversed

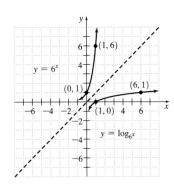

11.3 Practice Problems

1. Property 1 can be extended to three logarithms.
$$\log_b MNP = \log_b M + \log_b N + \log_b P$$
Thus,
$$\log_4 WXY = \log_4 W + \log_4 X + \log_4 Y.$$

2. $\log_b M + \log_b N + \log_b P = \log_b MNP$
Therefore,
$$\log_7 w + \log_7 8 + \log_7 x = \log_7(w \cdot 8 \cdot x) = \log_7 8wx.$$

3. $\log_3\left(\dfrac{17}{5}\right) = \log_3 17 - \log_3 5$

4. $\log_b 132 - \log_b 4 = \log_b\left(\dfrac{132}{4}\right) = \log_b 33$

5. $\dfrac{1}{3}\log_7 x - 5\log_7 y = \log_7 x^{1/3} - \log_7 y^5$ By property 3.
$$= \log_7\left(\dfrac{x^{1/3}}{y^5}\right) \quad \text{By property 2.}$$

6. $\log_3\left(\dfrac{x^4 y^5}{z}\right) = \log_3 x^4 y^5 - \log_3 z$ By property 2.
$$= \log_3 x^4 + \log_3 y^5 - \log_3 z \quad \text{By property 1.}$$
$$= 4\log_3 x + 5\log_3 y - \log_3 z \quad \text{By property 3.}$$

7. $\log_4 x + \log_4 5 = 2$
   $\log_4 5x = 2$ By property 1.
Converting to exponential form, we have
$$4^2 = 5x$$
$$16 = 5x$$
$$\dfrac{16}{5} = x$$
$$x = \dfrac{16}{5}$$

8. $\log_{10} x - \log_{10}(x + 3) = -1$
$$\log_{10}\left[\dfrac{x}{(x+3)}\right] = -1 \qquad \text{By property 2.}$$
$$\left[\dfrac{x}{x+3}\right] = 10^{-1} \qquad \text{Convert to exponential form.}$$
$$x = \dfrac{1}{10}(x + 3) \qquad \text{Multiply each side by } (x+3).$$
$$x = \dfrac{1}{10}x + \dfrac{3}{10} \qquad \text{Simplify.}$$
$$\dfrac{9}{10}x = \dfrac{3}{10}$$
$$x = \dfrac{1}{3}$$

9. (a) $\log_7 1 = 0$ By property 5.
(b) $\log_8 8 = 1$ By property 4.
(c) $\log_{12} 13 = \log_{12}(y + 2)$
$$13 = y + 2 \qquad \text{By property 6.}$$
$$y = 11$$

10. $\log_3 2 - \log_3 5 = \log_3 6 + \log_3 x$
$$\log_3\left(\dfrac{2}{5}\right) = \log_3 6x \qquad \text{By property 2 and property 3.}$$
$$\dfrac{2}{5} = 6x \qquad \text{By property 6.}$$
$$x = \dfrac{1}{15}$$

11.4 Practice Problems

1. (a) $4.36\ \boxed{\log} \approx 0.639486489$
(b) $436\ \boxed{\log} \approx 2.639486489$
(c) $0.2418\ \boxed{\log} \approx -0.616543703$

2. We know that $\log x = 2.913$ is equivalent to $10^{2.913} = x$. Using a calculator we have
$$2.913\ \boxed{10^x} \approx 818.46479.$$
Thus, $x \approx 818.46479$.

3. To evaluate antilog (-3.0705) using a scientific calculator, we have

$$3.0705 \boxed{+/-} \boxed{10^x} \approx 8.5015869 \times 10^{-4}.$$

Thus, antilog $(-3.0705) \approx 0.00085015869$.

4. (a) $\log x = 0.06134$ is equivalent to $10^{0.06134} = x$.

$$0.06134 \boxed{10^x} \approx 1.1517017$$
Thus, $x \approx 1.1517017$.

(b) $\log x = -4.6218$ is equivalent to $10^{-4.6218} = x$.

$$4.6218 \boxed{+/-} \boxed{10^x} \approx 2.3889112 \times 10^{-5}$$
Thus, $x \approx 0.000023889112$.

5. (a) $4.82 \boxed{\ln} \approx 1.572773928$

(b) $48.2 \boxed{\ln} \approx 3.875359021$

(c) $0.0793 \boxed{\ln} \approx -2.53451715$

6. (a) If $\ln x = 3.1628$, then $e^{3.1628} = x$.

$$3.1628 \boxed{e^x} \approx 23.636686$$
Thus, $x \approx 23.636686$.

(b) If $\ln x = -2.0573$, then $e^{-2.0573} = x$.

$$2.0573 \boxed{+/-} \boxed{e^x} \approx 0.1277986$$
Thus, $x \approx 0.1277986$.

7. To evaluate $\log_9 3.76$, we use the change of base formula.

$$\log_9 3.76 = \frac{\log 3.76}{\log 9}$$

On a calculator, find the following.

$$3.76 \boxed{\log} \boxed{\div} 9 \boxed{\log} \boxed{=} 0.602769044.$$

8. By the change of base formula,

$$\log_8 0.009312 = \frac{\log_e 0.009312}{\log_e 8} = \frac{\ln 0.009312}{\ln 8}.$$

On a calculator, find the following.

$$0.009312 \boxed{\ln} \boxed{\div} 8 \boxed{\ln} \boxed{=} -2.24889774$$
Thus, $\log_8 0.009312 \approx -2.24889774$.

9. Make a table of values.

x	$y = \log_5 x$
0.2	-1
1	0
2	0.4
3	0.7
5	1
8	1.3

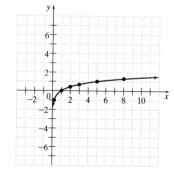

11.5 Practice Problems

1.
$$\log(x + 5) = 2 - \log 5$$
$$\log(x + 5) + \log 5 = 2$$
$$\log[5(x + 5)] = 2$$
$$\log(5x + 25) = 2$$
$$5x + 25 = 10^2$$
$$5x + 25 = 100$$
$$5x = 75$$
$$x = 15$$

Check: $\log(15 + 5) \stackrel{?}{=} 2 - \log 5$

$$\log 20 \stackrel{?}{=} 2 - \log 5$$
$$1.301029996 \stackrel{?}{=} 2 - 0.698970004$$
$$1.301029996 = 1.301029996 \ \checkmark$$

2. $\log(x + 3) - \log x = 1$

$$\log\left(\frac{x + 3}{x}\right) = 1$$
$$\frac{x + 3}{x} = 10^1$$
$$\frac{x + 3}{x} = 10$$
$$x + 3 = 10x$$
$$3 = 9x$$
$$\frac{1}{3} = x$$

Check: $\log\left(\frac{1}{3} + 3\right) - \log\frac{1}{3} \stackrel{?}{=} 1$

$$\log\frac{10}{3} - \log\frac{1}{3} \stackrel{?}{=} 1$$
$$\log\left(\frac{\frac{10}{3}}{\frac{1}{3}}\right) \stackrel{?}{=} 1$$
$$\log 10 \stackrel{?}{=} 1$$
$$1 = 1 \ \checkmark$$

3. $\log 5 - \log x = \log(6x - 7)$

$$\log\left(\frac{5}{x}\right) = \log(6x - 7)$$
$$\frac{5}{x} = 6x - 7$$
$$5 = 6x^2 - 7x$$
$$0 = 6x^2 - 7x - 5$$
$$0 = (3x - 5)(2x + 1)$$
$$3x - 5 = 0 \quad \text{or} \quad 2x + 1 = 0$$
$$3x = 5 \qquad\qquad 2x = -1$$
$$x = \frac{5}{3} \qquad\qquad x = -\frac{1}{2}$$

Check: $\log 5 - \log x = \log(6x - 7)$

$x = \frac{5}{3}$: $\log 5 - \log\left(\frac{5}{3}\right) \stackrel{?}{=} \log\left[6\left(\frac{5}{3}\right) - 7\right]$

$$\log 5 - \log\frac{5}{3} \stackrel{?}{=} \log(10 - 7)$$
$$\log 5 - \log\frac{5}{3} \stackrel{?}{=} \log 3$$
$$\log\left[\frac{5}{\frac{5}{3}}\right] \stackrel{?}{=} \log 3$$
$$\log 3 = \log 3 \ \checkmark$$

$x = -\frac{1}{2}$: $\log 5 - \log\left(-\frac{1}{2}\right) \stackrel{?}{=} \log\left[6\left(-\frac{1}{2}\right) - 7\right]$

Since logarithms of negative numbers do not exist, $x = -\frac{1}{2}$ is not a valid solution.

The only solution is $\frac{5}{3}$.

4. Take the logarithm of each side.

$$\log 3^x = \log 5$$
$$x \log 3 = \log 5$$
$$x = \frac{\log 5}{\log 3}$$

5. Take the logarithm of each side.

$$\log 2^{3x+1} = \log 9^{x+1}$$
$$(3x + 1)\log 2 = (x + 1)\log 9$$
$$3x \log 2 + \log 2 = x \log 9 + \log 9$$
$$3x \log 2 - x \log 9 = \log 9 - \log 2$$
$$x(3 \log 2 - \log 9) = \log 9 - \log 2$$
$$x = \frac{\log 9 - \log 2}{3 \log 2 - \log 9}$$

Use the following keystrokes.

$$(\boxed{9} \boxed{\log} \boxed{-} \boxed{2} \boxed{\log} \boxed{)} \boxed{\div} (\boxed{3} \boxed{\times} \boxed{2} \boxed{\log}$$

$$\boxed{-} \boxed{9} \boxed{\log} \boxed{)} \boxed{=} -12.76989838$$

Rounding to the nearest thousandth, we have $x \approx -12.770$.

6. Take the natural logarithm of each side.

$\ln 20.98 = \ln e^{3.6x}$

$\ln 20.98 = (3.6x)(\ln e)$

$\ln 20.98 = 3.6x$

$\dfrac{\ln 20.98}{3.6} = x$

On a scientific calculator, find the following.

$$20.98 \boxed{\ln} \boxed{\div} 3.6 \boxed{=} 0.845436001$$

Rounding to the nearest ten-thousandth, we have $x \approx 0.8454$.

7. We use the formula $A = P(1 + r)^t$, where
$A = \$10,000$, $P = \$4000$, and $r = 0.08$.

$10,000 = 4000(1 + 0.08)^t$

$10,000 = 4000(1.08)^t$

$\dfrac{10,000}{4000} = (1.08)^t$

$2.5 = (1.08)^t$

$\log(2.5) = \log(1.08)^t$

$\log 2.5 = t(\log 1.08)$

$\dfrac{\log 2.5}{\log 1.08} = t$

On a scientific calculator,

$$2.5 \boxed{\log} \boxed{\div} 1.08 \boxed{\log} \boxed{=} 11.905904.$$

Thus, it would take approximately 12 years.

8.
$A = A_0 e^{rt}$

$5000 = 1300 e^{0.043t}$ Substitute known values.

$\dfrac{5000}{1300} = e^{0.043t}$ Divide each side by 1300.

$\ln\left(\dfrac{5000}{1300}\right) = \ln e^{0.043t}$ Take the natural logarithm of each side.

$\ln\left(\dfrac{50}{13}\right) = 0.043t \ln e$

$\ln 50 - \ln 13 = 0.043t$

$\dfrac{\ln 50 - \ln 13}{0.043} = t$

Using a scientific calculator,

$$(\boxed{50} \boxed{\ln} \boxed{-} \boxed{13} \boxed{\ln} \boxed{)} \boxed{\div} 0.043 \boxed{=} 31.32729414$$

Rounding to the nearest whole years, the food shortage will develop in about 31 years.

9. Let I_J = intensity of the Japan earthquake.
Let I_S = intensity of the San Francisco earthquake.

$8.9 = \log\left(\dfrac{I_J}{I_{10}}\right) = \log I_J - \log I_0$

Solving for $\log I_0$ gives $\log I_0 = \log I_J - 8.9$

$7.1 = \log\left(\dfrac{I_S}{I_0}\right) = \log I_S - \log I_0$

Solving for $\log I_0$ gives $\log I_0 = \log I_S - 7.1$
Therefore, $\log I_J - 8.9 = \log I_S - 7.1$

$\log I_J - \log I_S = 1.8$

$\log\left(\dfrac{I_J}{I_S}\right) = 1.8$

$\dfrac{I_J}{I_S} = 10^{1.8}$

$\dfrac{I_J}{I_S} = 63.09573445$

$\dfrac{I_J}{I_S} \approx 63$

$I_J \approx 63 I_S$

The earthquake in Japan was about sixty-three times as intense as the San Francisco earthquake.

Appendix B

Practice Problems

1. (a) $(-7)(-2) - (-4)(3) = 14 - (-12) = 26$

(b) $(5)(-5) - (0)(6) = -25 - 0 = -25$

2. (a) Since 3 appears in the first column and third row, we delete them.

$$\begin{vmatrix} 1 & 2 & 7 \\ -4 & -5 & -6 \\ 3 & 4 & -9 \end{vmatrix}$$

The minor of 3 is $\begin{vmatrix} 2 & 7 \\ -5 & -6 \end{vmatrix}$.

(b) Since -6 appears in second row and third column, we delete them.

$$\begin{vmatrix} 1 & 2 & 7 \\ -4 & -5 & -6 \\ 3 & 4 & -9 \end{vmatrix}$$

The minor of -6 is $\begin{vmatrix} 1 & 2 \\ 3 & 4 \end{vmatrix}$.

3. We can find the determinant by expanding it by minors of elements in the first column.

$$\begin{vmatrix} 1 & 2 & -3 \\ 2 & -1 & 2 \\ 3 & 1 & 4 \end{vmatrix} = 1\begin{vmatrix} -1 & 2 \\ 1 & 4 \end{vmatrix} - 2\begin{vmatrix} 2 & -3 \\ 1 & 4 \end{vmatrix} + 3\begin{vmatrix} 2 & -3 \\ -1 & 2 \end{vmatrix}$$

$= 1[(-1)(4) - (1)(2)] - 2[(2)(4) - (1)(-3)]$
$\quad + 3[(2)(2) - (-1)(-3)]$

$= 1[-4 - 2] - 2[8 - (-3)] + 3[4 - 3]$

$= 1(-6) - 2(11) + 3(1)$

$= -6 - 22 + 3$

$= -25$

4. $D = \begin{vmatrix} 5 & 3 \\ 2 & -5 \end{vmatrix}$ $D_x = \begin{vmatrix} 17 & 3 \\ 13 & -5 \end{vmatrix}$

$\quad = (5)(-5) - (2)(3)$ $= (17)(-5) - (13)(3)$

$\quad = -25 - 6$ $= -85 - 39$

$\quad = -31$ $= -124$

$$D_y = \begin{vmatrix} 5 & 17 \\ 2 & 13 \end{vmatrix}$$

$= (5)(13) - (2)(17)$

$= 65 - 34$

$= 31$

$x = \dfrac{D_x}{D} = \dfrac{-124}{-31} = 4$, $y = \dfrac{D_y}{D} = \dfrac{31}{-31} = -1$

5. We will expand each determinant by the first column.

$$D = \begin{vmatrix} 2 & 3 & -1 \\ 3 & 5 & -2 \\ 1 & 2 & 3 \end{vmatrix} = 2\begin{vmatrix} 5 & -2 \\ 2 & 3 \end{vmatrix} - 3\begin{vmatrix} 3 & -1 \\ 2 & 3 \end{vmatrix} + 1\begin{vmatrix} 3 & -1 \\ 5 & -2 \end{vmatrix}$$

$= 2[15 - (-4)] - 3[9 - (-2)] + 1[-6 - (-5)]$

$= 2(19) - 3(11) + 1(-1)$

$= 4$

$$D_x = \begin{vmatrix} -1 & 3 & -1 \\ -3 & 5 & -2 \\ 2 & 2 & 3 \end{vmatrix} = -1\begin{vmatrix} 5 & -2 \\ 2 & 3 \end{vmatrix} - (-3)\begin{vmatrix} 3 & -1 \\ 2 & 3 \end{vmatrix} + 2\begin{vmatrix} 3 & -1 \\ 5 & -2 \end{vmatrix}$$

$= -1[15 - (-4)] + 3[9 - (-2)]$
$\quad + 2[-6 - (-5)]$

$= -1(19) + 3(11) + 2(-1)$

$= 12$

$$D_y = \begin{vmatrix} 2 & -1 & -1 \\ 3 & -3 & -2 \\ 1 & 2 & 3 \end{vmatrix} = 2\begin{vmatrix} -3 & -2 \\ 2 & 3 \end{vmatrix} - 3\begin{vmatrix} -1 & -1 \\ 2 & 3 \end{vmatrix} + 1\begin{vmatrix} -1 & -1 \\ -3 & -2 \end{vmatrix}$$

$$= 2[-9 - (-4)] - 3[-3 - (-2)] + 1[2 - 3]$$
$$= 2(-5) - 3(-1) + 1(-1)$$
$$= -8$$

$$D_z = \begin{vmatrix} 2 & 3 & -1 \\ 3 & 5 & -3 \\ 1 & 2 & 2 \end{vmatrix} = 2\begin{vmatrix} 5 & -3 \\ 2 & 2 \end{vmatrix} - 3\begin{vmatrix} 3 & -1 \\ 2 & 2 \end{vmatrix} + 1\begin{vmatrix} 3 & -1 \\ 5 & -3 \end{vmatrix}$$

$$= 2[10 - (-6)] - 3[6 - (-2)]$$
$$+ 1[-9 - (-5)]$$
$$= 2(16) - 3(8) + 1(-4)$$
$$= 4$$

$$x = \frac{D_x}{D} = \frac{12}{4} = 3; \ y = \frac{D_y}{D} = \frac{-8}{4} = -2; \ z = \frac{D_z}{D} = \frac{4}{4} = 1$$

Appendix C

Practice Problems

1. The system is represented by the equations $x + 9y = 33$ and $0x + y = 3$.

Since we know that $y = 3$, we can find x by substitution.

$$x + 9y = 33$$
$$x + 9(3) = 33$$
$$x + 27 = 33$$
$$x = 6$$

The solution to the system is $(6, 3)$.

2. The augmented matrix for this system of equations is

$$\begin{bmatrix} 3 & -2 & | & -6 \\ 1 & -3 & | & 5 \end{bmatrix}.$$

Interchange rows one and two so there is a 1 as the first element in the first row.

$$\begin{bmatrix} 1 & -3 & | & 5 \\ 3 & -2 & | & -6 \end{bmatrix} \quad R_1 \leftrightarrow R_2$$

We want a 0 as the first element in the second row. Multiply row one by -3 and add this to row two.

$$\begin{bmatrix} 1 & -3 & | & 5 \\ 0 & 7 & | & -21 \end{bmatrix} \quad -3R_1 + R_2$$

Now, to obtain a 1 as the second element of row two, multiply each element of row two by $\frac{1}{7}$.

$$\begin{bmatrix} 1 & -3 & | & 5 \\ 0 & 1 & | & -3 \end{bmatrix}. \quad \frac{1}{7}R_2$$

This represents the linear system $\begin{array}{l} x - 3y = 5 \\ \quad\quad\ y = -3. \end{array}$

We know $y = -3$. Substitute this value into the first equation.

$$x - 3(-3) = 5$$
$$x + 9 = 5$$
$$x = -4$$

The solution to the system is $(-4, -3)$.

3. The augmented matrix is

$$\begin{bmatrix} 2 & 1 & -2 & | & -15 \\ 4 & -2 & 1 & | & 15 \\ 1 & 3 & 2 & | & -5 \end{bmatrix}$$

$$\begin{bmatrix} 1 & 3 & 2 & | & -5 \\ 4 & -2 & 1 & | & 15 \\ 2 & 1 & -2 & | & -15 \end{bmatrix} \quad R_1 \leftrightarrow R_3$$

$$\begin{bmatrix} 1 & 3 & 2 & | & -5 \\ 0 & -14 & -7 & | & 35 \\ 0 & -5 & -6 & | & -5 \end{bmatrix} \quad \begin{array}{l} -4R_1 + R_2 \\ -2R_1 + R_3 \end{array}$$

$$\begin{bmatrix} 1 & 3 & 2 & | & -5 \\ 0 & 1 & \frac{1}{2} & | & -\frac{5}{2} \\ 0 & -5 & -6 & | & -5 \end{bmatrix} \quad -\frac{1}{14}R_2$$

$$\begin{bmatrix} 1 & 3 & 2 & | & -5 \\ 0 & 1 & \frac{1}{2} & | & -\frac{5}{2} \\ 0 & 0 & -\frac{7}{2} & | & -\frac{35}{2} \end{bmatrix} \quad 5R_2 + R_3$$

$$\begin{bmatrix} 1 & 3 & 2 & | & -5 \\ 0 & 1 & \frac{1}{2} & | & -\frac{5}{2} \\ 0 & 0 & 1 & | & 5 \end{bmatrix} \quad -\frac{2}{7}R_3$$

We now know that $z = 5$. Substitute this value into the second equation to find y.

$$y + \frac{1}{2}(5) = -\frac{5}{2}$$
$$y + \frac{5}{2} = -\frac{5}{2}$$
$$y = -5$$

Now substitute $y = -5$ and $z = 5$ into the first equation.

$$x + 3y + 2z = -5$$
$$x + 3(-5) + 2(5) = -5$$
$$x - 15 + 10 = -5$$
$$x - 5 = -5$$
$$x = 0$$

The solution to this system is $(0, -5, 5)$.

Use the following keystrokes.

$$\boxed{-}\ 9\ \boxed{\log}\ \boxed{)}\ \boxed{=}\ -12.76989838$$

Rounding to the nearest thousandth, we have $x \approx -12.770$.

6. Take the natural logarithm of each side.

$$\ln 20.98 = \ln e^{3.6x}$$
$$\ln 20.98 = (3.6x)(\ln e)$$
$$\ln 20.98 = 3.6x$$
$$\frac{\ln 20.98}{3.6} = x$$

On a scientific calculator, find the following.

$$20.98\ \boxed{\ln}\ \boxed{\div}\ 3.6\ \boxed{=}\ 0.845436001$$

Rounding to the nearest ten-thousandth, we have $x \approx 0.8454$.

7. We use the formula $A = P(1 + r)^t$, where
$A = \$10{,}000$, $P = \$4000$, and $r = 0.08$.

$$10{,}000 = 4000(1 + 0.08)^t$$
$$10{,}000 = 4000(1.08)^t$$
$$\frac{10{,}000}{4000} = (1.08)^t$$
$$2.5 = (1.08)^t$$
$$\log(2.5) = \log(1.08)^t$$
$$\log 2.5 = t(\log 1.08)$$
$$\frac{\log 2.5}{\log 1.08} = t$$

On a scientific calculator,

$$2.5\ \boxed{\log}\ \boxed{\div}\ 1.08\ \boxed{\log}\ \boxed{=}\ 11.905904.$$

Thus, it would take approximately 12 years.

8.
$$A = A_0 e^{rt}$$
$$5000 = 1300 e^{0.043t} \quad \text{Substitute known values.}$$
$$\frac{5000}{1300} = e^{0.043t} \quad \text{Divide each side by 1300.}$$
$$\ln\left(\frac{5000}{1300}\right) = \ln e^{0.043t} \quad \text{Take the natural logarithm of each side.}$$
$$\ln\left(\frac{50}{13}\right) = 0.043t \ln e$$
$$\ln 50 - \ln 13 = 0.043t$$
$$\frac{\ln 50 - \ln 13}{0.043} = t$$

Using a scientific calculator,

$$\boxed{(}\ 50\ \boxed{\ln}\ \boxed{-}\ 13\ \boxed{\ln}\ \boxed{)}\ \boxed{\div}\ 0.043\ \boxed{=}\ 31.32729414$$

Rounding to the nearest whole years, the food shortage will develop in about 31 years.

9. Let I_J = intensity of the Japan earthquake.
Let I_S = intensity of the San Francisco earthquake.

$$8.9 = \log\left(\frac{I_J}{I_{10}}\right) = \log I_J - \log I_0$$

Solving for $\log I_0$ gives $\log I_0 = \log I_J - 8.9$

$$7.1 = \log\left(\frac{I_S}{I_0}\right) = \log I_S - \log I_0$$

Solving for $\log I_0$ gives $\log I_0 = \log I_S - 7.1$
Therefore, $\log I_J - 8.9 = \log I_S - 7.1$

$$\log I_J - \log I_S = 1.8$$
$$\log\left(\frac{I_J}{I_S}\right) = 1.8$$
$$\frac{I_J}{I_S} = 10^{1.8}$$
$$\frac{I_J}{I_S} = 63.09573445$$
$$\frac{I_J}{I_S} \approx 63$$
$$I_J \approx 63 I_S$$

The earthquake in Japan was about sixty-three times as intense as the San Francisco earthquake.

Appendix B

Practice Problems

1. (a) $(-7)(-2) - (-4)(3) = 14 - (-12) = 26$
(b) $(5)(-5) - (0)(6) = -25 - 0 = -25$

2. (a) Since 3 appears in the first column and third row, we delete them.

$$\begin{vmatrix} 1 & 2 & 7 \\ -4 & -5 & -6 \\ 3 & 4 & -9 \end{vmatrix}$$

The minor of 3 is $\begin{vmatrix} 2 & 7 \\ -5 & -6 \end{vmatrix}$.

(b) Since -6 appears in second row and third column, we delete them.

$$\begin{vmatrix} 1 & 2 & 7 \\ -4 & -5 & -6 \\ 3 & 4 & -9 \end{vmatrix}$$

The minor of -6 is $\begin{vmatrix} 1 & 2 \\ 3 & 4 \end{vmatrix}$.

3. We can find the determinant by expanding it by minors of elements in the first column.

$$\begin{vmatrix} 1 & 2 & -3 \\ 2 & -1 & 2 \\ 3 & 1 & 4 \end{vmatrix} = 1\begin{vmatrix} -1 & 2 \\ 1 & 4 \end{vmatrix} - 2\begin{vmatrix} 2 & -3 \\ 1 & 4 \end{vmatrix} + 3\begin{vmatrix} 2 & -3 \\ -1 & 2 \end{vmatrix}$$

$$= 1[(-1)(4) - (1)(2)] - 2[(2)(4) - (1)(-3)]$$
$$\quad + 3[(2)(2) - (-1)(-3)]$$
$$= 1[-4 - 2] - 2[8 - (-3)] + 3[4 - 3]$$
$$= 1(-6) - 2(11) + 3(1)$$
$$= -6 - 22 + 3$$
$$= -25$$

4. $D = \begin{vmatrix} 5 & 3 \\ 2 & -5 \end{vmatrix}$ $D_x = \begin{vmatrix} 17 & 3 \\ 13 & -5 \end{vmatrix}$

$$\quad = (5)(-5) - (2)(3) \qquad = (17)(-5) - (13)(3)$$
$$\quad = -25 - 6 \qquad\qquad\quad = -85 - 39$$
$$\quad = -31 \qquad\qquad\qquad\ = -124$$

$$D_y = \begin{vmatrix} 5 & 17 \\ 2 & 13 \end{vmatrix}$$

$$= (5)(13) - (2)(17)$$
$$= 65 - 34$$
$$= 31$$

$$x = \frac{D_x}{D} = \frac{-124}{-31} = 4, \qquad y = \frac{D_y}{D} = \frac{31}{-31} = -1$$

5. We will expand each determinant by the first column.

$$D = \begin{vmatrix} 2 & 3 & -1 \\ 3 & 5 & -2 \\ 1 & 2 & 3 \end{vmatrix} = 2\begin{vmatrix} 5 & -2 \\ 2 & 3 \end{vmatrix} - 3\begin{vmatrix} 3 & -1 \\ 2 & 3 \end{vmatrix} + 1\begin{vmatrix} 3 & -1 \\ 5 & -2 \end{vmatrix}$$

$$= 2[15 - (-4)] - 3[9 - (-2)] + 1[-6 - (-5)]$$
$$= 2(19) - 3(11) + 1(-1)$$
$$= 4$$

$$D_x = \begin{vmatrix} -1 & 3 & -1 \\ -3 & 5 & -2 \\ 2 & 2 & 3 \end{vmatrix} = -1\begin{vmatrix} 5 & -2 \\ 2 & 3 \end{vmatrix} - (-3)\begin{vmatrix} 3 & -1 \\ 2 & 3 \end{vmatrix} + 2\begin{vmatrix} 3 & -1 \\ 5 & -2 \end{vmatrix}$$

$$= -1[15 - (-4)] + 3[9 - (-2)]$$
$$\quad + 2[-6 - (-5)]$$
$$= -1(19) + 3(11) + 2(-1)$$
$$= 12$$

$$D_y = \begin{vmatrix} 2 & -1 & -1 \\ 3 & -3 & -2 \\ 1 & 2 & 3 \end{vmatrix} = 2\begin{vmatrix} -3 & -2 \\ 2 & 3 \end{vmatrix} - 3\begin{vmatrix} -1 & -1 \\ 2 & 3 \end{vmatrix} + 1\begin{vmatrix} -1 & -1 \\ -3 & -2 \end{vmatrix}$$

$$= 2[-9 - (-4)] - 3[-3 - (-2)] + 1[2 - 3]$$
$$= 2(-5) - 3(-1) + 1(-1)$$
$$= -8$$

$$D_z = \begin{vmatrix} 2 & 3 & -1 \\ 3 & 5 & -3 \\ 1 & 2 & 2 \end{vmatrix} = 2\begin{vmatrix} 5 & -3 \\ 2 & 2 \end{vmatrix} - 3\begin{vmatrix} 3 & -1 \\ 2 & 2 \end{vmatrix} + 1\begin{vmatrix} 3 & -1 \\ 5 & -3 \end{vmatrix}$$

$$= 2[10 - (-6)] - 3[6 - (-2)]$$
$$+ 1[-9 - (-5)]$$
$$= 2(16) - 3(8) + 1(-4)$$
$$= 4$$

$$x = \frac{D_x}{D} = \frac{12}{4} = 3; \; y = \frac{D_y}{D} = \frac{-8}{4} = -2; \; z = \frac{D_z}{D} = \frac{4}{4} = 1$$

Appendix C

Practice Problems

1. The system is represented by the equations $x + 9y = 33$ and $0x + y = 3$.

Since we know that $y = 3$, we can find x by substitution.
$$x + 9y = 33$$
$$x + 9(3) = 33$$
$$x + 27 = 33$$
$$x = 6$$
The solution to the system is $(6, 3)$.

2. The augmented matrix for this system of equations is
$$\begin{bmatrix} 3 & -2 & | & -6 \\ 1 & -3 & | & 5 \end{bmatrix}.$$
Interchange rows one and two so there is a 1 as the first element in the first row.
$$\begin{bmatrix} 1 & -3 & | & 5 \\ 3 & -2 & | & -6 \end{bmatrix} \quad R_1 \leftrightarrow R_2$$
We want a 0 as the first element in the second row. Multiply row one by -3 and add this to row two.
$$\begin{bmatrix} 1 & -3 & | & 5 \\ 0 & 7 & | & -21 \end{bmatrix} \quad -3R_1 + R_2$$
Now, to obtain a l as the second element of row two, multiply each element of row two by $\frac{1}{7}$.
$$\begin{bmatrix} 1 & -3 & | & 5 \\ 0 & 1 & | & -3 \end{bmatrix}. \quad \frac{1}{7}R_2$$
This represents the linear system $\begin{aligned} x - 3y &= 5 \\ y &= -3. \end{aligned}$
We know $y = -3$. Substitute this value into the first equation.
$$x - 3(-3) = 5$$
$$x + 9 = 5$$
$$x = -4$$
The solution to the system is $(-4, -3)$.

3. The augmented matrix is
$$\begin{bmatrix} 2 & 1 & -2 & | & -15 \\ 4 & -2 & 1 & | & 15 \\ 1 & 3 & 2 & | & -5 \end{bmatrix}$$

$$\begin{bmatrix} 1 & 3 & 2 & | & -5 \\ 4 & -2 & 1 & | & 15 \\ 2 & 1 & -2 & | & -15 \end{bmatrix} \quad R_1 \leftrightarrow R_3$$

$$\begin{bmatrix} 1 & 3 & 2 & | & -5 \\ 0 & -14 & -7 & | & 35 \\ 0 & -5 & -6 & | & -5 \end{bmatrix} \quad \begin{aligned} -4R_1 + R_2 \\ -2R_1 + R_3 \end{aligned}$$

$$\begin{bmatrix} 1 & 3 & 2 & | & -5 \\ 0 & 1 & \frac{1}{2} & | & -\frac{5}{2} \\ 0 & -5 & -6 & | & -5 \end{bmatrix} \quad -\frac{1}{14}R_2$$

$$\begin{bmatrix} 1 & 3 & 2 & | & -5 \\ 0 & 1 & \frac{1}{2} & | & -\frac{5}{2} \\ 0 & 0 & -\frac{7}{2} & | & -\frac{35}{2} \end{bmatrix} \quad 5R_2 + R_3$$

$$\begin{bmatrix} 1 & 3 & 2 & | & -5 \\ 0 & 1 & \frac{1}{2} & | & -\frac{5}{2} \\ 0 & 0 & 1 & | & 5 \end{bmatrix} \quad -\frac{2}{7}R_3$$

We now know that $z = 5$. Substitute this value into the second equation to find y.
$$y + \tfrac{1}{2}(5) = -\tfrac{5}{2}$$
$$y + \tfrac{5}{2} = -\tfrac{5}{2}$$
$$y = -5$$
Now substitute $y = -5$ and $z = 5$ into the first equation.
$$x + 3y + 2z = -5$$
$$x + 3(-5) + 2(5) = -5$$
$$x - 15 + 10 = -5$$
$$x - 5 = -5$$
$$x = 0$$
The solution to this system is $(0, -5, 5)$.

Chapter 2

2.1 Exercises

1. No; when you replace x by -20 in the equation, you do not get a true statement.

3. Yes; when you replace x by $\frac{2}{7}$ in the equation, you get a true statement.

5. Multiply each term of the equation by the LCD, 12, to clear the fractions.   **7.** No; it would be easier to subtract 3.6 from both sides of the equation since the coefficient of x is 1.   **9.** 8   **11.** -5 **13.** 5

15. $\frac{3}{2}$ or $1\frac{1}{2}$ or 1.5   **17.** -3 **19.** 3   **21.** -2 **23.** $-\frac{1}{3}$ **25.** 5   **27.** 12   **29.** $-\frac{23}{3}$ or $-7\frac{2}{3}$ **31.** -2 **33.** 3   **35.** 0

37. 6   **39.** 1.5   **41.** 19   **43.** -2 **45.** 8   **47.** 16   **49.** $\frac{1}{4}$ **51.** 0   **53.** $\frac{29}{2}$ or $14\frac{1}{2}$ **55.** -1 **57.** No solution

59. Any real number is a solution.   **61.** No solution   **63.** Any real number is a solution.   **65.** -5 **66.** $\frac{27y^3}{8x^3}$ **67.** 16

68. $\frac{1}{4x^6y^2}$ **69.** (a) 6750   (b) 8895

2.2 Exercises

1. $x = \frac{3-5y}{6}$ **3.** $x = \frac{24-3y}{8}$ **5.** $x = \frac{3y+12}{2}$ **7.** $y = \frac{-12x+8}{9}$ **9.** $\frac{A}{w} = l$ **11.** $B = \frac{2A-hb}{h}$ **13.** $r = \frac{A}{2\pi h}$

15. $b = \frac{3H-2a}{4}$ **17.** $x = \frac{-6y}{a}$ **19.** (a) $a = \frac{2A}{b}$ (b) $a = 16$ **21.** (a) $n = \frac{A-a+d}{d}$ (b) $n = 2\frac{2}{3}$ or $\frac{8}{3}$

23. $x = \frac{160-t}{0.7}$ or $\frac{1600-10t}{7}$; the year 2005   **25.** (a) $m = 1.15k$ (b) 33.35 miles per hour   **27.** (a) $D = \frac{C-5.8263}{0.6547}$ (b) \$5.7 billion

29. $\frac{x^6}{4y^2}$ **30.** $\frac{y^{15}}{125x^{18}}$ **31.** 1   **32.** $9x - 3y - 15$ **33.** \$9610   **34.** 19.8 miles per gallon

2.3 Exercises

1. It will always have 2 solutions. One solution is when $x = b$ and one when $x = -b$. Since $b > 0$ the values of b and $-b$ are always different numbers.   **3.** You must first isolate the absolute value expression. To do this you add 2 to each side of the equation. The result will be $|x + 7| = 10$. Then you solve the two equations $x + 7 = 10$ and $x + 7 = -10$. The final answer is $x = -17$ and $x = 3$.   **5.** 30, -30

7. 22, -10 **9.** 9, -4 **11.** $-\frac{3}{2}, 4$ **13.** 10, 2   **15.** $\frac{13}{3}, 11$ **17.** 6, -10 **19.** $-2, \frac{10}{3}$ **21.** $-\frac{8}{3}, \frac{16}{3}$ **23.** $-\frac{1}{5}, \frac{13}{15}$ **25.** 9, -1

27. $-\frac{7}{3}, -1$ **29.** 3, 1   **31.** 0, 12   **33.** $-0.59, -3.29$ **35.** 0, -8 **37.** $\frac{3}{5}$ **39.** No solution   **41.** $\frac{7}{4}, -\frac{3}{4}$ **43.** 11, $-\frac{1}{5}$ or -0.2

45. $5xy^3$ **47.** Each beaker was \$30. Each Bunsen burner was \$75.   **48.** The Pennsylvania cost is a better deal. To buy 3 bags of chips in PA costs \$5.37. To buy 3 bags of chips in SC costs \$5.38.

2.4 Exercises

1. -90 **3.** \$48   **5.** 18 lbs   **7.** 72 weeks   **9.** 3 weeks   **11.** Melissa drives 38 miles, Marcia drives 19 miles, and John drives 55 miles.
13. The width is 44 yards. The length is 126 yards.   **15.** longest side = 23 feet; shortest side = 15 feet; third side = 21 feet   **17.** \$888 for the Saugus Shop; \$1166 for the Salem Shop   **19.** (a) Reliable = 550 minutes; Clear Call = 600 minutes; Nationwide = 500 minutes
(b) Reliable offers the lowest cost per minute.   **21.** Identity property of addition   **22.** Associative property of multiplication
23. 3   **24.** -94

How Am I Doing? Sections 2.1–2.4

1. -3.7 (obj. 2.1.1)   **2.** -18 (obj. 2.1.1)   **3.** $-\frac{10}{7}$ or $-1\frac{3}{7}$ (obj. 2.1.1)   **4.** -100 (obj. 2.1.1)   **5.** $\frac{5x-15}{8}$ (obj. 2.2.1)

6. $a = \frac{b+24}{11b}$ (obj. 2.2.1)   **7.** $r = \frac{A-P}{Pt}$ (obj. 2.2.1)   **8.** $\frac{3}{50}$ or 0.06 (obj. 2.2.1)   **9.** 3, $-\frac{5}{3}$ (obj. 2.3.1)   **10.** 4, 12 (obj. 2.3.2)
11. 2.5, -5.5 (obj. 2.3.1)   **12.** 5, 0.75 (obj. 2.3.3)   **13.** 9 cm $\times$ 23 cm (obj. 2.4.1)   **14.** 26 (obj. 2.4.1)   **15.** Cindi picked up 250 lb; Alan picked up 205 lb (obj. 2.4.1)   **16.** shortest side = 14.5 ft; longest side = 24 ft; third side = 23.5 ft (obj. 2.4.1)

2.5 Exercises

1. Approximately 203.8 million people   **3.** \$425   **5.** 1140 residents   **7.** 230 hemlocks, 460 spruces, 710 balsams   **9.** Walker earned \$260/week; Angela earned \$240/week   **11.** packet A = 5 mg; packet B = 3 mg   **13.** \$216   **15.** \$232.50   **17.** She invested \$3900 at 5% and \$2500 at 8%.   **19.** He invested \$12,000 in the CD and \$6000 in the fixed interest account.   **21.** She should mix 12 grams of cheese that contains 45% fat and 18 grams of cheese that contains 20% fat.   **23.** He should mix 75 lb of the hamburger with 30% fat with 25 lb of hamburger with 10% fat.   **25.** They need to mix 45 gal of 25% fertilizer with 105 gal of 15% fertilizer.   **27.** Her speed on secondary roads was 35 mph.   **29.** They each used the treadmill for $\frac{3}{4}$ hour.   **31.** The profit was \$18,000,000.   **33.** The area of the parallelogram is 32 cm^2.   **35.** 7   **36.** 15   **37.** 11   **38.** 1

2.6 Exercises

1. True   **3.** True   **5.** False   **7.** >   **9.** <   **11.** >   **13.** <   **15.** >   **17.** <   **19.** (number line: point at -2, shaded right; -4, -2, 0)

21. (number line: open circle at 15, shaded right; 13, 15, 17)   **23.** $x \le 1$ (number line: closed at 1; -1, 1, 3)   **25.** $x < -2$ (number line: open at -2, shaded left; -5 to 2)

27. $x > -1$ (number line: open at -1, shaded right; -3, -1, 1)   **29.** $x > 4$ **31.** $x \le -2$ **33.** $x > -\dfrac{5}{3}$ **35.** $x < -3$ **37.** $x \le -6$

39. $x \le 3$ **41.** $x \le 0$ **43.** $x \le 1$ **45.** $x \le -26$ **47.** More than 10 tables   **49.** A maximum of 18 minutes
51. A maximum of 13 computers   **53.** At least 100 time-shares per month   **55.** At least 4 times   **57.** $6xy - x^2y + 4x^2$
58. $4a^2b - \dfrac{4}{3}ab^2 + 6ab$ **59.** $\dfrac{27x^3w^{12}}{8y^6}$ **60.** $\dfrac{x^4z^{12}}{16y^8}$

2.7 Exercises

1. (number line: 3, 8)   **3.** (number line: -4, 2)   **5.** (number line: 7, 9)

7. (number line: -3 to 2)   **9.** (number line: 2, 8)   **11.** (number line: -3, -2, 4)

13. (number line: -10, 40)   **15.** (number line: -4 to 3)   **17.** (number line: -5, 1.5, 2)

19. No solution   **21.** $t < 10.9$ or $t > 11.2$ **23.** $5000 \le c \le 12{,}000$ **25.** $-4° \le F \le 51.8°$ **27.** $\$143.53 \le d \le \$259.81$
29. $-2 < x < 2$ **31.** $-2 \le x \le 1$ **33.** $x < -3$ or $x \ge 1$ **35.** $x \le 2$ **37.** $x \ge 2$ **39.** No solution   **41.** $x = -3$

43. $x < 5$ **45.** All real numbers   **47.** $\dfrac{2}{3} < x \le \dfrac{7}{3}$ **49.** $x = \dfrac{3y-8}{5}$ **50.** $y = \dfrac{-7x-12}{6}$ **51.** -1 **52.** 50

53. $\$240, \$310, \$380, \$420$ **54.** In order from the first to seventh branch, the plant has 68, 80, 102, 40, 40, 42, and 42 leaves.

2.8 Exercises

1. $-8 \le x \le 8$ (number line: -8, -4, 0, 4, 8)   **3.** $-9.5 < x < 0.5$ (number line)   **5.** $-2 \le x \le 8$

7. $-1 \le x \le 6$ **9.** $-\dfrac{2}{5} \le x \le \dfrac{6}{5}$ **11.** $-5 < x < 15$ **13.** $-32 < x < 16$ **15.** $-7 < x < 9$ **17.** $-\dfrac{10}{3} < x < \dfrac{14}{3}$

19. $x > 5$ or $x < -5$ **21.** $x > 3$ or $x < -7$ **23.** $x \ge 3$ or $x \le -1$ **25.** $x \ge 5$ or $x \le \dfrac{1}{3}$ **27.** $x < 10$ or $x > 110$

29. $x < -\dfrac{19}{2}$ or $x > \dfrac{21}{2}$ **31.** $-13 < x < 17$ **33.** $-\dfrac{22}{3} < x < 4$ **35.** $x < -2$ or $x > \dfrac{22}{9}$ **37.** $18.53 \le m \le 18.77$
39. $9.63 \le n \le 9.73$ **41.** The statement says $12 < -12$, which is a false statement.   **43.** 2   **44.** -11 **45.** 9.94 seconds
46. 7.85 seconds   **47.** First rack is 29 cents per CD; second rack is 26 cents; third rack is 32 cents. The second rack is the least expensive.

Putting Your Skills to Work

1. (a) 425.5 miles   **(b)** 606.9 miles   **2.** 1400 mi/month; 43.75 gal; 23.3 gal   **3.** $C = 0.065625x$ **4.** $C = 0.035x$
5. 15 months   **6.** 39 months

Chapter 2 Review Problems

1. $-\dfrac{5}{4}$ or -1.25 **2.** -28 **3.** $\dfrac{10}{3}$ or $3\dfrac{1}{3}$ **4.** $\dfrac{14}{5}$ or $2\dfrac{4}{5}$ **5.** -3 **6.** 15   **7.** -63 **8.** -2 **9.** $\dfrac{4x-5}{8}$ **10.** $m = \dfrac{3R}{n}$

11. $a = -\dfrac{2y}{3x}$ **12.** $b = \dfrac{-3a-4}{-10}$ **13. (a)** $F = \dfrac{9C+160}{5}$ **(b)** $50°$ **14. (a)** $W = \dfrac{P-2L}{2}$ **(b)** 29.5 m   **15.** $3, -\dfrac{1}{2}$ **16.** $6, -\dfrac{22}{3}$

17. $2, \dfrac{8}{3}$ **18.** $12, -\dfrac{4}{3}$ **19.** $44, -20$ **20.** $-3, \dfrac{29}{7}$ **21.** $\dfrac{13}{2}, \dfrac{3}{2}$ **22.** $\dfrac{21}{2}, -\dfrac{1}{2}$ **23.** Length = 15 ft, width = 6 ft

24. 120 men, 160 women   **25.** 350 miles   **26.** $7\dfrac{1}{5}$ miles   **27.** Retirement = \$10, state tax = \$23, federal tax = \$69

28. Nicholas sold 35 tickets, Emma sold 65 tickets, Jackson sold 80 tickets   **29.** 2650   **30.** 65,000 two-door sedans, 195,000 four-door sedans
31. \$4500 at 12%, \$2500 at 8%   **32.** 8 liters at 2%, 16 liters at 5%   **33.** 12 lb at \$4.25, 18 lb at \$4.50   **34.** 500 full-time students, 390 part-time students   **35.** $x < -4$ **36.** $x > 1$ **37.** $x < 7$ **38.** $x < -2$ **39.** $x > 1$ **40.** $x \le 6$ **41.** $x < 3$ **42.** $x > 4$

43. $x \le 1$ **44.** (number line: -3, 2)   **45.** (number line: -5 to 6)

46. (number line: -8, -4)   **47.** (number line: -9, -6)   **48.** (number line: -2, 5)

49. (number line: -3, 6)   **50.** (number line: -5, -1)   **51.** (number line: -8, -3)

52. (number line: 4, 5)   **53.** $x > 9$ or $x < -1$ **54.** No solution   **55.** $-6 < x < 3$ **56.** $-4 \le x \le \dfrac{5}{3}$

57. $\dfrac{9}{5} \le x < 5$ **58.** All real numbers   **59.** $-22 < x < 8$ **60.** $-27 < x < 9$ **61.** $-\dfrac{15}{2} < x < -\dfrac{1}{2}$ **62.** $-26 < x < -4$

63. $x \ge 5$ or $x \le -4$ **64.** $x \ge 1$ or $x \le -\dfrac{1}{3}$ **65.** $x \ge \dfrac{8}{3}$ or $x \le -\dfrac{2}{3}$ **66.** $x \ge 5$ or $x \le 1$ **67.** He may talk a maximum of 15 minutes.

68. A maximum of 18 packages can be carried.   **69.** He can order a maximum of 8 cubic yards.   **70.** It could weigh a maximum of 15 ounces.

71. 5 bolts   **72.** $2{,}404{,}480 \le x \le 3{,}025{,}240$ **73.** $-\dfrac{1}{2}$ **74.** $B = \dfrac{4H + 64}{3}$ or $\dfrac{4}{3}(H + 16)$ **75.** 80 grams of 77% pure copper, 20 grams of 92% pure copper   **76.** $x > 6$ ———○———→ (6)   **77.** $x \ge 3$ ———●———→ (3)

78. $-3 \le x \le 3$ —●———●— (−3, 0, 3)   **79.** $x < -4$ or $x > 3$ ←———○———○——→ (−4, 0, 3)

80. 4, 3   **81.** $-\dfrac{15}{4} \le x \le \dfrac{21}{4}$ **82.** $x < -3$ or $x > \dfrac{11}{5}$

How Am I Doing? Chapter 2 Test

1. $-\dfrac{2}{11}$ (obj. 2.1.1)   **2.** $\dfrac{1}{2}$ (obj. 2.1.1)   **3.** 1 (obj. 2.1.1)   **4.** $\dfrac{17}{14}$ or $1\dfrac{3}{14}$ (obj. 2.1.1)   **5.** $n = \dfrac{L - a + d}{d}$ (obj. 2.2.1)

6. $b = \dfrac{2A}{h}$ (obj. 2.2.1)   **7.** 3 cm (obj. 2.2.1)   **8.** $r = \dfrac{4H - 12b + 1}{2}$ (obj. 2.2.1)   **9.** $-7, \dfrac{39}{5}$ (obj. 2.3.1)   **10.** 6, −18 (obj. 2.3.2)

11. 1st side = 16 meters; 2nd side = 32 meters; 3rd side = 21 meters (obj. 2.4.1)   **12.** $2620 (obj. 2.5.1)
13. 2.5 gal at 90%, 7.5 gal at 50% (obj. 2.5.1)   **14.** $1800 at 6%, $3200 at 10% (obj. 2.5.1)
15. $x > -2$ ———○———→ (−4, −3, −2, −1, 0, 1, 2) (obj. 2.6.3)   **16.** $x \le -1$ ←———●——→ (−5, −4, −3, −2, −1, 0, 1, 2, 3) (obj. 2.6.3)

17. $-5 < x \le -1$ (obj. 2.7.3)   **18.** $x \le -2$ or $x \ge 1$ (obj. 2.7.3)   **19.** $-\dfrac{15}{7} \le x \le 3$ (obj. 2.8.1)   **20.** $x < -\dfrac{8}{3}$ or $x > 2$ (obj. 2.8.2)

Cumulative Test for Chapters 1–2

1. $-12, -3, 0, \dfrac{1}{4}, 2.16, 2.333\ldots, -\dfrac{5}{8}, 3$ **2.** Associative property of addition   **3.** 58   **4.** $-4x^3y^4z^2$ **5.** $-\dfrac{b^5}{2a^4}$ **6.** 1

7. 153.86 square inches   **8.** $-x + 15y$ **9.** 3   **10.** $b = \dfrac{3h - 2d}{2}$ **11.** 1st side = 25 meters; 2nd side = 35 meters; 3rd side = 45 meters

12. $80   **13.** 6 gallons at 80%, 3 gallons at 50%   **14.** $2000 at 12%, $4500 at 10%
15. $x > -10$ ———○———→ (−10)   **16.** $x \le 4$ ←———●——→ (4)   **17.** $-2 < x < 3$

18. $x \le -9$ or $x \ge -2$ **19.** $-20 \le x \le 12$ **20.** $x < -\dfrac{7}{3}$ or $x > 5$

Chapter 3

3.1 Exercises

1. Variables   **3.** To locate the point (a, b), assuming that $a, b > 0$, we move a units to the right and b units up. To locate the point (b, a), we move b units to the right and a units up. If $a \ne b$, the graphs of the points will be different. Thus, the order of the numbers in (a, b) matters. $(1, 3)$ is not the same as $(3, 1)$.   **5.** −13   **7.** −4

The coordinates that are used to plot the points to graph each equation in the selected answers to exercises 11–31 are suggested coordinates. You may choose other replacements for x and y and hence generate a different set of points. The graph of the equation, however, will be the same.

9.

x	y
0	−3
2	1
4	5

11.

x	y
−1	6
0	4
1	2

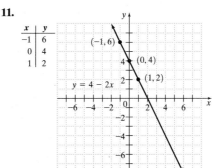

13.

x	y
−3	−6
0	−4
3	−2

15.

x	y
−2	0
0	3
2	6

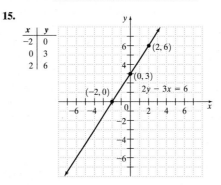

17.

x	y
0	−6
3	0
2	−2

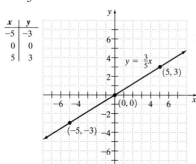

19.

x	y
0	−2
$-2\frac{2}{3}$	0
−4	1
4	−5

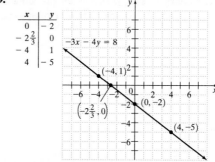

21. $y = \frac{3}{5}x$

x	y
−5	−3
0	0
5	3

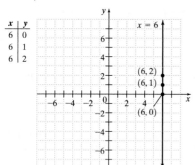

23. $x = -5$; vertical line

x	y
−5	0
−5	3
−5	2

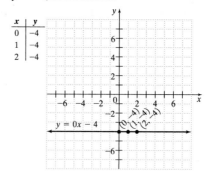

25. $x = 6$; vertical line

x	y
6	0
6	1
6	2

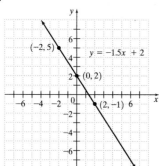

27. $y = -4$; horizontal line

x	y
0	−4
1	−4
2	−4

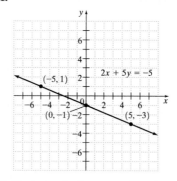

29.

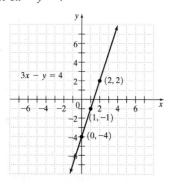

31.

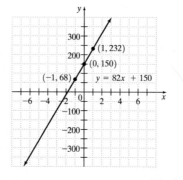

33. $3x - y = 4$

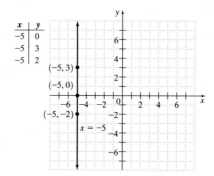

35. Vertical scale: 1 square = 50 units

37. Between 1995 and 2000   **39.** In 1990   **41.** 101.5%   **43. (a)** 120; 88; 56; 24; −8

(b)   **(c)** The baseball is moving downward instead of upward at $T = 4$ seconds.

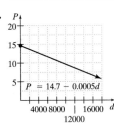

45. 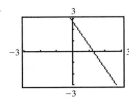   **47.** 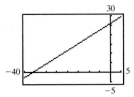   **49.**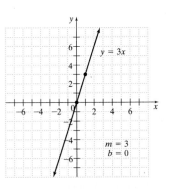

51. −3   **52.** $x \geq -3$ **53.** 520 red, 260 green, 1560 blue, 260 yellow, 130 white   **54.** $170,000

3.2 Exercises

1. vertical; horizontal   **3.** 0   **5.** No, division by zero is undefined.   **7.** $m = -2$ **9.** $m = \dfrac{8}{7}$ **11.** $m = \dfrac{2}{3}$

13. undefined slope   **15.** $m = 0$ **17.** $m = \dfrac{1}{8}$ **19.** $\dfrac{3}{5}$ or 0.6   **21.** $\dfrac{1}{4}$ or 0.25   **23.** 80 feet   **25.** $m_{\parallel} = -\dfrac{2}{9}$ **27.** $m_{\parallel} = -2$

29. $m_{\parallel} = \dfrac{3}{2}$ **31.** $m_{\perp} = -\dfrac{5}{3}$ **33.** $m_{\perp} = \dfrac{1}{3}$ **35.** $m_{\perp} = -2$ **37.** Yes. Each line has the same slope.

39. $m_{AD} = m_{BC} = -\dfrac{1}{6}$; $m_{AB} = m_{CD} = 1$ **41. (a)** $\dfrac{1}{12}$ **(b)** 2 feet   **(c)** 20.4 feet   **43.** 2   **44.** −56   **45.** $\dfrac{5x^{10}}{y^3}$ **46.** $8x^2 - 10x - 2y$

3.3 Exercises

1. First determine the slope from the coordinates of the points. Then substitute the slope and the coordinates of one point into the point-slope form of the equation of a line. You may rewrite the equation in standard form or in slope-intercept form.

3. $y = \dfrac{3}{4}x - 9$ **5.** $3x - 4y = -2$

7. 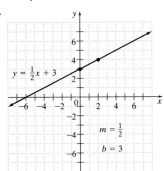   **9.** 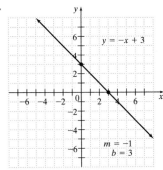   **11.**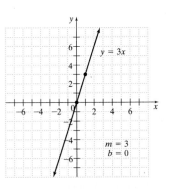

13. 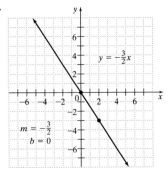   **15.**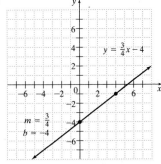

17. $y = x - 5; m = 1; (0, -5)$ **19.** $y = \frac{5}{4}x + 5; m = \frac{5}{4}; (0, 5)$ **21.** $y = -\frac{9}{2}x - 3; m = -\frac{9}{2}; (0, -3)$

23.

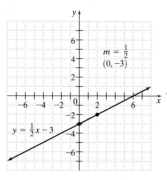

25.

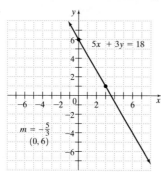

27. $y = -\frac{2}{3}x + 8$ **29.** $y = 5x + 33$ **31.** $y = -\frac{1}{5}x + \frac{6}{5}$ **33.** $y = \frac{5}{7}x + \frac{13}{7}$ **35.** $y = -\frac{2}{3}x - \frac{8}{3}$ **37.** $y = -3$

39. $5x - y = -10$ **41.** $x - 3y = 8$ **43.** $2x - 3y = 15$ **45.** $7x - y = -27$ **47.** neither   **49.** parallel   **51.** perpendicular
53. yes   **55.** \$267,060   **57.** answers may vary; approximately \$217,000   **59.** 146,740,000 housing units

61. answers may vary; approximately 131,980,000 housing units   **62.** $x = -\frac{3}{2}$ **63.** $x = -4$ **64.** $x = 13$ **65.** $x = 1$

How Am I Doing? Sections 3.1–3.3

1. $a = -\frac{24}{5}$ or -4.8 (obj. 3.1.1)

2. (obj. 3.1.1)

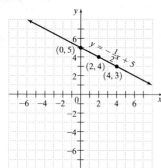

3. (obj. 3.1.2)

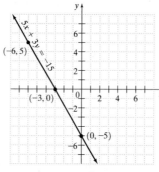

4. (obj. 3.1.1)

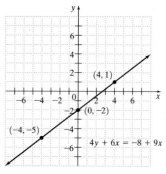

5. $m = -9$ (obj. 3.2.1)   **6.** $m_{\parallel} = -36$ (obj. 3.2.2)   **7.** $m_{\perp} = -\frac{9}{2}$ (obj. 3.2.2)   **8.** 65 feet (obj. 3.2.1)

9. $m = -\frac{6}{7}$; y-intercept $= (0, 2)$ (obj. 3.3.1)   **10.** $y = -2x + 11$ (obj. 3.3.2)   **11.** $y = -\frac{5}{3}x - \frac{11}{3}$ (obj. 3.3.3)

12. $y = 5x - 3$ (obj. 3.3.2)

3.4 Exercises

1. You need to use a dashed line when graphing a linear inequality that contains the $>$ symbol or the $<$ symbol.
3. To graph the region $x > 5$, you should shade the region to the right of the line $x = 5$.   **5.** The point $(0, 0)$ is on the line; false.
7.

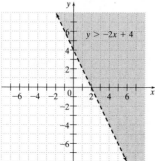

9.

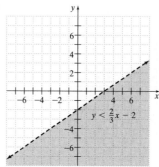

11.

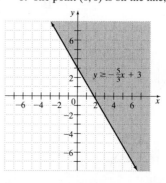

13.
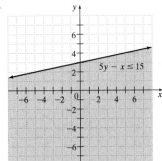

15.
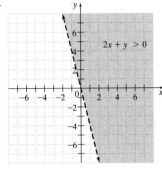

17.
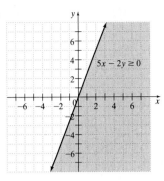

19.
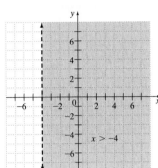

21.
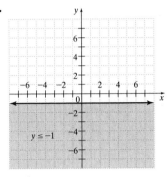

23.
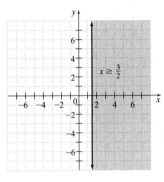

25.
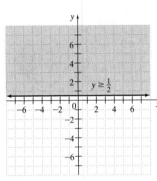

27.
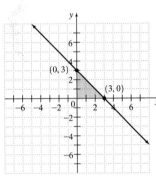

29.
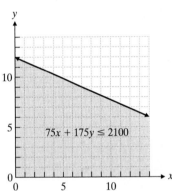

31. -2 **32.** $A = 24$ **33.** 9 packages

3.5 Exercises

1. A relation is any set of ordered pairs. A function is a set of ordered pairs in which no two different ordered pairs have the same first coordinate.   **3.** as a set of ordered pairs; as an equation; as a graph   **5.** domain $= \{0, 5, 7\}$; range $= \{0, 11, 13\}$; *not* a function
7. domain $= \{85, 16, -102, 62\}$; range $= \{-12, 4, 48\}$; function   **9.** domain $= \{6, 8, 10, 12, 14\}$; range $= \{38, 40, 42, 44, 46\}$; function
11. domain $= \{$Jan., Feb., Mar., Apr., May, June, July, Aug., Sept., Oct., Nov., Dec.$\}$; range $= \{79, 80, 81\}$; function
13. domain $= \{$Chicago, New York$\}$; range $= \{1046, 1127, 1136, 1250, 1350, 1454\}$; *not* a function
15. domain $= \{10, 20, 30, 40, 50\}$; range $= \{11.51, 23.02, 34.53, 46.04, 57.55\}$; function   **17.** function   **19.** not a function   **21.** function

23. not a function   **25.** function   **27.** -9.8 **29.** -4 **31.** -2 **33.** $\dfrac{10}{3}$ **35.** 7   **37.** 1.42   **39.** -1 **41.** -16 **43.** 21

45. -27 **47.** 2   **49.** 2   **51.** range $= \{3, 4, 7\}$ **53.** domain $= \{12, 11, 9, -4\}$ **54.** $x = 4, -3$ **55.** $-5 \le x \le -1$
56. \$433.93.   **57.** 3140 tons

3.6 Exercises

1.
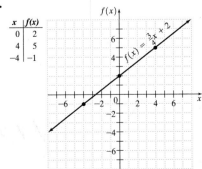

3.
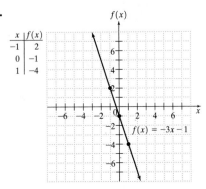

5. $C(x) = 0.15x + 25; C(0) = 25; C(100) = 40; C(200) = 55; C(300) = 70$

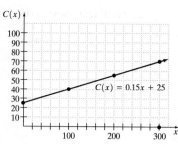

7.

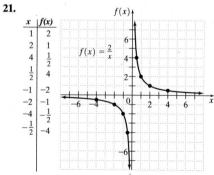

$P(x) = -1500x + 45,000; P(0) = 45,000;$
$P(10) = 30,000; P(30) = 0$

With 30 tons of pollutants, there will be no fish.

9.

x	f(x)
−1	2
0	1
1	0
2	1
3	2

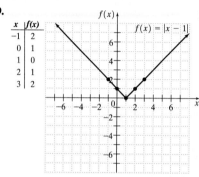

11.

x	g(x)
−2	−3
−1	−4
0	−5
1	−4
2	−3

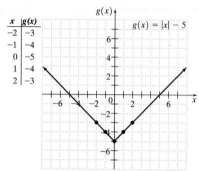

13.

x	g(x)
0	−4
1	−3
2	0
3	5
−1	−3
−2	0
−3	5

$g(x) = x^2 - 4$

15.

x	g(x)
−3	4
−2	1
−1	0
0	1
1	4

$g(x) = (x + 1)^2$

17.

x	g(x)
−1	−4
0	−3
1	−2
2	5

$g(x) = x^3 - 3$

19.

x	p(x)
−2	8
−1	1
0	0
1	−1
2	−8

$p(x) = -x^3$

21.

x	f(x)
1	2
2	1
4	$\frac{1}{2}$
$\frac{1}{2}$	4
−1	−2
−2	−1
−4	$-\frac{1}{2}$
$-\frac{1}{2}$	−4

$f(x) = \frac{2}{x}$

23.

x	h(x)
−6	1
−3	2
−2	3
−1	6
1	−6
2	−3
3	−2
6	−1

$h(x) = -\frac{6}{x}$

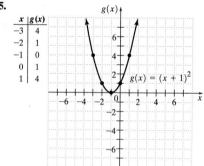

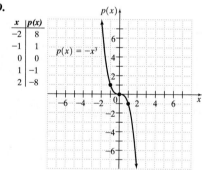

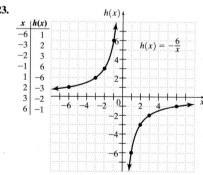

25. When the coefficient of x^2 is greater than 1, the curve is closer to the y-axis. When the coefficient is less than 1, the curve is farther away from the y-axis.

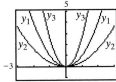

27.

x	$f(x)$
-5	-2
3	2
5	3

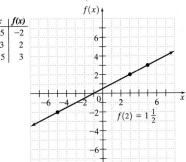

$f(2) = 1\frac{1}{2}$

29.

x	$f(x)$
0	3
1	2.5
3	2.2
-1	4
-2	6

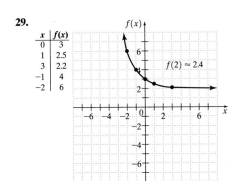

$f(2) \approx 2.4$

31.

x	$f(x)$
-2	-8
-1	-3
0	0
1	1
3	-3

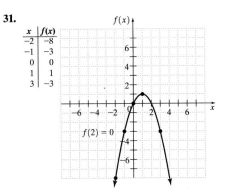

$f(2) = 0$

33. (a)

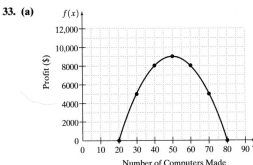

(b) 50   **(c)** between 40 and 60   **(d)** They will operate at a loss.   **(e)** $8700

35. $x = \dfrac{8y + 15}{a}$ **36.** $x = 7.25$ **37.** $x = 4$ **38.** 8.698×10^{13} square feet   **39.** It will breathe 21,840 times.

40. Charlene is only trying to buy speakers and both sets sound the same. The CD player and the value of the subwoofers is not relevant. The second set will cost $446.25, so it is cheaper. The second set is the better deal.

Putting Your Skills to Work

1. Y = yield = $\{50, 49, 48, 47, 46, 45, 40, 35\}$ **2.** 25 pounds   **3.** Yes, for each density, this formula gives the yield.   **4.** $Y = 80 - D$
5. $H = 80D - D^2$ **6.** H = yield = $\{0, 700, 1200, 1500, 1600, 1500, 1200, 700, 0\}$
7. 40   **8.** $752

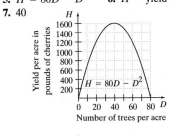

Chapter 3 Review Problems

1.
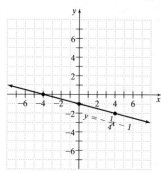

2.
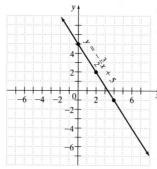

3.
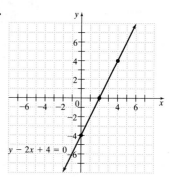

4.
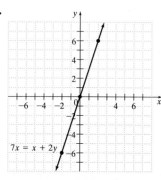

5.
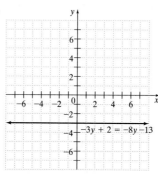

6.
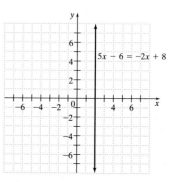

7. $m = -2$ **8.** $m = -\dfrac{1}{10}$ **9.** Slope is undefined.   **10.** $m = 0$ **11.** $m = \dfrac{1}{10}$ **12.** $m = -2$ **13.** $2x - 3y = 12$

14. $4x + y = 0$ **15.** $y = 1$ **16.** $x \geq 15$ **17.** $13x - 12y = -7$ **18.** $x = -6$ **19.** $8x - 7y = -51$ **20.** $3x - 2y = 13$

21. $y = \dfrac{3}{4}x - 4$

$m = \dfrac{3}{4}$

y-intercept $= (0, -4)$

22. $y = -3x + 5$

$m = -3$

y-intercept $= (0, 5)$

23. $y = 2$

$m = 0$

y-intercept $= (0, 2)$

24.
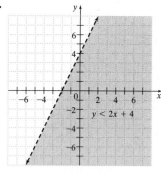

25.
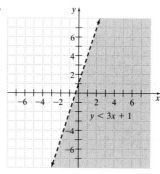

26.
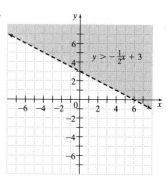

27.
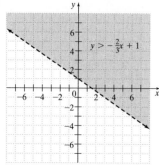

28.
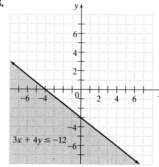

29.
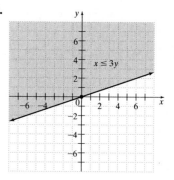

30.
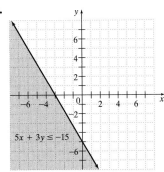

31.
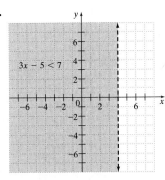

32.
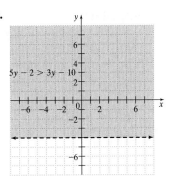

33. domain $= \{-20, -18, -16, -12\}$; range $= \{14, 16, 18\}$; function   **34.** domain $= \{0, 1, 2, 3\}$; range $= \{0, 1, 4, 9, 16\}$; *not* a function

35. function   **36.** function   **37.** not a function   **38.** $f(-2) = -14; f(-3) = -17$ **39.** $g(-3) = 22; g(2) = -3$

40. $h(-1) = 14$ **41.** $p(3) = 21$

42.
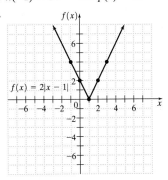

43.

x	g(x)
−3	4
−2	−1
−1	−4
0	−5
1	−4
2	−1
3	4

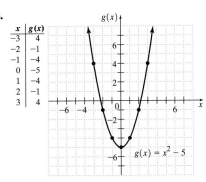

44.

x	h(x)
−2	−5
−1	2
0	3
1	4

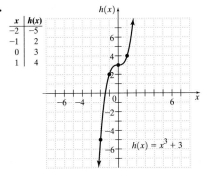

45.

x	f(x)
−1	5
−3	−3
−4	−4
−5	−3
−6	0
−7	5

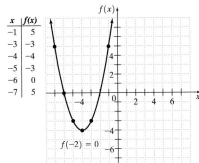

46.

x	f(x)
−3	4
−1	2
0	1
1	0
2	−1
3	0
4	1

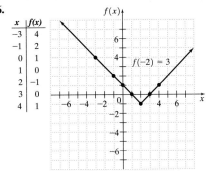

47.

x	f(x)
−5	7
0	3
10	−5

48.

x	f(x)
−3	31
0	4
4	24

49.

x	f(x)
−1	−7
0	−4
2	20

50.

x	f(x)
−2	−7
0	$\frac{7}{3}$
2	1

51. $m = \dfrac{1}{8}$ **52.** $m = -\dfrac{5}{8}$ **53.** $m = 7$ **54.** $5x - 6y = 30$ **55.** $y = 5x - 10$ **56.** $y = \dfrac{1}{4}x + \dfrac{19}{4}$ **57.** $2x + y = -5$

58. $x = 5$ **59.** $f(x) = 35 + 0.15x$ **60.** $f(x) = 15{,}000 + 500x$ **61.** $f(x) = 18{,}000 - 65x$

SA-14 **Answers to Selected Exercises**

How Am I Doing? Chapter 3 Test

1. (obj. 3.1.1)

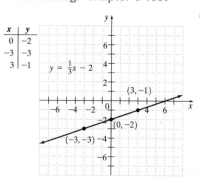

x	y
0	-2
-3	-3
3	-1

2. (obj. 3.1.3)

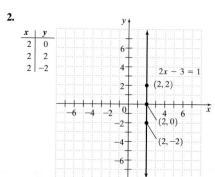

x	y
2	0
2	2
2	-2

3. (obj. 3.1.1)

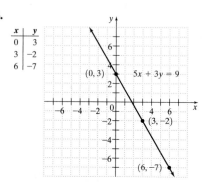

x	y
0	3
3	-2
6	-7

4. (obj. 3.1.1)

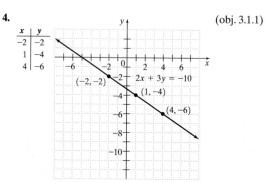

x	y
-2	-2
1	-4
4	-6

5. $m = 2$ (obj. 3.2.1)   **6.** $m = 0$ (obj. 3.2.1)   **7.** $m = -\dfrac{9}{7}$ (obj. 3.3.1)   **8.** $7x + 6y = -12$ (obj. 3.3.3)   **9.** $x + 8y = -11$ (obj. 3.3.2)

10. $y = 2$ (obj. 3.3.2)   **11.** $y = -5x - 8$ (obj. 3.3.1)

12. (obj. 3.4.1)

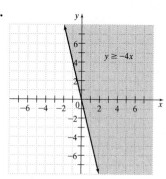

13. (obj. 3.4.1)

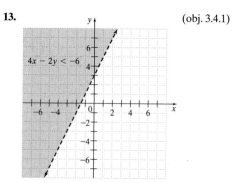

14. domain = $\{0, 1, 2\}$; range = $\{-4, -1, 0, 1, 4\}$ (obj. 3.5.1)   **15.** $f\left(\dfrac{3}{4}\right) = -\dfrac{3}{2}$ (obj. 3.5.3)   **16.** $g(-4) = 11$ (obj. 3.5.3)

17. $h(-9) = 10$ (obj. 3.5.3)   **18.** $p(-2) = 22$ (obj. 3.5.3)

19. (obj. 3.6.1)

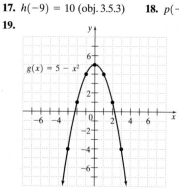

20. (obj. 3.6.1)

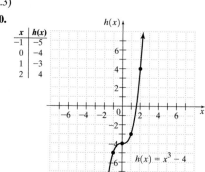

x	h(x)
-1	-5
0	-4
1	-3
2	4

21. 10 miles

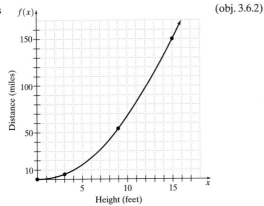

(obj. 3.6.2)

Cumulative Test for Chapters 1–3

1. Inverse property of addition   **2.** 12   **3.** $\dfrac{y^{12}}{81x^8}$ **4.** $7x^2 - 15xy - 12$ **5.** 4.37×10^{-4}

6. $x < -2$ or $x > 4$ **7.** $x = \dfrac{6a + y}{9}$

8. width = 15 centimeters; length = 31 centimeters   **9.** $1700 at 5%; $1300 at 8%   **10.** 14.13 square inches

11.

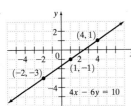

12. $m = \dfrac{1}{2}$ **13.** $2x + y = 11$ **14.** $3x + 2y = -12$

15. domain $= \left\{ \dfrac{1}{2}, 2, 3, 5 \right\}$; range $= \{-1, 2, 7, 8\}$; function   **16.** $f(-3) = -5$

17.

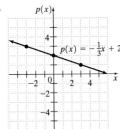

18.

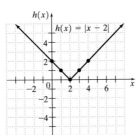

19.

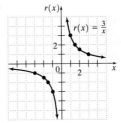

20.

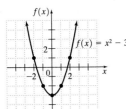

21.

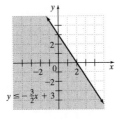

22.

x	$f(x)$
-2	20
0	4
3	-50

23. $f(x) = 32{,}500 - 1400x$ **24.** 11,500 cars per day

Chapter 4

4.1 Exercises

1. There is no solution. There is no point (x, y) that satisfies both equations. The graph of such a system yields two parallel lines.

3. It may have one solution, it may have no solution, or it may have an infinite number of solutions.

5. $\left(\dfrac{3}{2}, -1\right)$ is a solution to the system.

7. 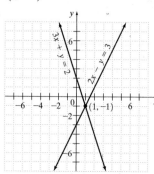   **9.**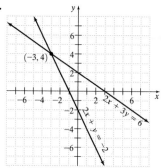

11. 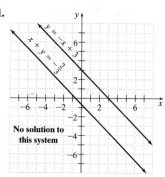   **13.**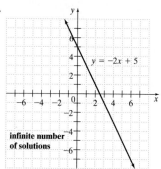

15. $(23, -43)$ **17.** $(2, -2)$ **19.** $(-1, 2)$ **21.** $(4, 0)$ **23.** $(0, 1)$ **25.** $\left(1, \dfrac{5}{3}\right)$ **27.** $(1, -3)$ **29.** $(3, -2)$ **31.** $(2, 8)$

33. $(-4, 5)$ **35.** $(6, -8)$ **37.** No solution; inconsistent system of equations   **39.** Infinite number of solutions; dependent equations   **41.** $(5, -3)$ **43.** No solution; inconsistent system of equations   **45.** $(16, 8)$ **47.** $(0, 2)$ **49.** Infinite number of solutions; dependent equations   **51. (a)** $y = 300 + 30x, \quad y = 200 + 50x$

(b)

x	$y = 300 + 30x$
0	300
4	420
8	540

x	$y = 200 + 50x$
0	200
4	400
8	600

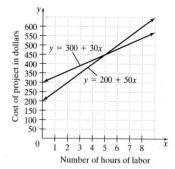

(c) The cost will be the same for 5 hours of installing new tile.   **(d)** The cost will be less for Modern Bathroom Headquarters.

53. $(2.46, -0.38)$ 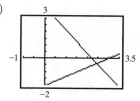   **55.** $(-2.45, 6.11)$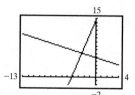

57. Approximately \$0.01 per pound   **58.** 341,889 cars

4.2 Exercises

1. $(2, 1, -4)$ is a solution.   **3.** No, $(-1, 5, 1)$ is not a solution.   **5.** $(1, 3, -2)$ **7.** $(3, -1, 4)$ **9.** $(0, -2, 5)$ **11.** $(1, -1, 2)$

13. $(2, 1, -4)$ **15.** $(4, 0, 2)$ **17.** $(3, -1, -2)$ **19.** $x = 1.10551, y = 2.93991, z = 1.73307$ **21.** $(3, -2, 1)$ **23.** $\left(\dfrac{1}{2}, \dfrac{2}{3}, \dfrac{5}{6}\right)$

25. $(1, 3, 5)$ **27.** $(-2, -5, 4)$ **29.** Infinite number of solutions; dependent equations   **31.** No solution; inconsistent system of equations

33. $x = -1, x = 4$ **34.** 7.63×10^7 **35.** $m = \dfrac{4 - 3}{1 + 2} = \dfrac{1}{3}, y - 4 = \dfrac{1}{3}(x - 1), x - 3y = -11$

36. $m(\perp \text{line}) = \dfrac{3}{2}, y - 2 = \dfrac{3}{2}(x + 4), 3x - 2y = -16$ **37.** He will buy 57 sheep and 22 cattle. After the purchase he will have 346 horses, 602 sheep, and 623 cattle.   **38.** 12.5 miles per hour

How Am I Doing? Sections 4.1–4.2

1. $(1, 5)$ (obj. 4.1.3)   **2.** $(5, -3)$ (obj. 4.1.4)   **3.** $(9, 9)$ (obj. 4.1.4)   **4.** $(3, -2)$ (obj. 4.1.4)   **5.** Infinite number of solutions; dependent equations (obj. 4.1.5)   **6.** $(2, -2)$ (obj. 4.1.4)   **7.** No solution; inconsistent system of equations (obj. 4.1.5)

8. No, $(-1, -2, 3)$ is not a solution. (obj. 4.2.1)   **9.** $(1, 3, 0)$ (obj. 4.2.2)   **10.** $\left(-1, 1, \dfrac{2}{3}\right)$ (obj. 4.2.2)   **11.** $(-2, 3, 4)$ (obj. 4.2.2)
12. $(-1, 3, 2)$ (obj. 4.2.2)

4.3 Exercises

1. 62 is the larger number; 25 is the smaller number   **3.** 16 heavy equipment operators; 19 general laborers   **5.** 51 tickets for regular coach seats; 47 tickets for sleeper car seats   **7.** 45 experienced managers; 10 newly hired managers   **9.** 30 packages of old fertilizer; 25 packages of new fertilizer   **11.** One doughnut costs $0.45; one large coffee costs $0.89   **13.** Speed of plane in still air 216 mph; speed of wind 36 mph   **15.** Speed of boat is 14 mph; speed of current is 2 mph   **17.** He scored 10 free throws and 11 2-point baskets.   **19.** 220 weekend minutes; 405 weekday minutes   **21.** The department pays $10,258 for a car and $17,300 for a truck.   **23.** She bought 5 pens, 4 notebooks, and 3 highlighters.   **25.** A total of 80 adults, 170 high school students, and 50 children not yet in high school attended.   **27.** A total of 800 senior citizens, 10,000 adults, and 1200 children under 12 ride during the rush hour.   **29.** 5 small pizzas, 8 medium pizzas, and 7 large pizzas were delivered.   **31.** She can prepare 2 of Box A, 3 of Box B, and 4 of Box C.   **33.** First angle measures 20°, second angle measures 40°, third angle measures 120°   **35.** The scientist should use 2 of packet A, 1 of packet B, 5 of packet C, and 3 of packet D.

36. $x = \dfrac{26}{7}$ or $3\dfrac{5}{7}$ **37.** $x = \dfrac{7}{18}$ **38.** $y = \dfrac{5}{3}$ or $1\dfrac{2}{3}$ **39.** $x = \dfrac{18ay - 3}{5a}$

4.4 Exercises

1. They would be dashed. The boundary lines are not included in the solution of a system of inequalities whenever the system contains only $<$ or $>$ symbols.   **3.** She could substitute $(3, -4)$ into each inequality. Since $(3, -4)$ does not satisfy either inequality, we know that the solution does not contain that ordered pair. Therefore, that point would not lie in the shaded region.

5.
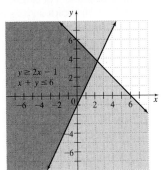

7.
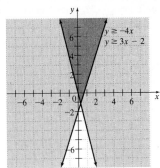

9.
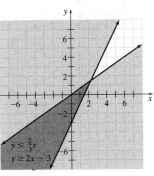

11.
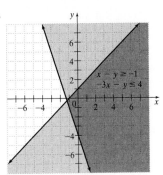

13.
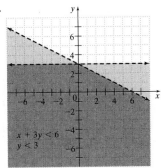

15.
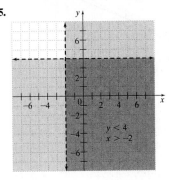

17.

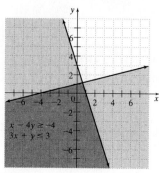

$x - 4y \geq -4$
$3x + y \leq 3$

19.

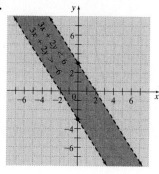

$3x + 2y \leq 6$
$3x + 2y > -6$

21.

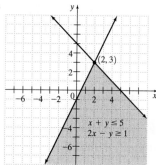

$(2, 3)$

$x + y \leq 5$
$2x - y \geq 1$

23.

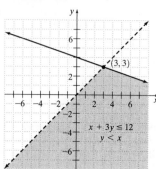

$(3, 3)$

$x + 3y \leq 12$
$y < x$

25.

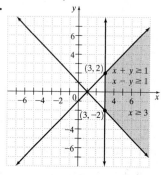

$(3, 2)$
$x + y \geq 1$
$x - y \geq 1$
$(3, -2)$
$x \geq 3$

27.

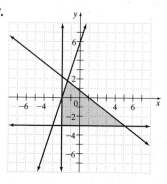

29. (a)

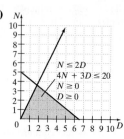

$N \leq 2D$
$4N + 3D \leq 20$
$N \geq 0$
$D \geq 0$

(b) Yes   **(c)** No   **31.** $-\dfrac{1}{2}$ **32.** Slope $= -\dfrac{3}{4}$; y-intercept $= (0, -2)$ **33.** 13

34. $-4x + 4y$ **35.** The Cinema takes in $4200 on a rainy day and $3000 on a sunny day.   **36.** The volunteers establish 43 ft of bicycle trails each day. The professionals establish 65 ft of bicycle trails each day.   **37.** They had each worked for 12 weeks and had each sold $100,000 worth of goods.   **38.** One roast beef sandwich costs $2.50. One order of french fries costs $1.75. One soda costs $0.95.

Putting Your Skills to Work

1. Plan A: $C = 20 + 0.52x$ Plan B: $C = 30 + 0.4x$

2.

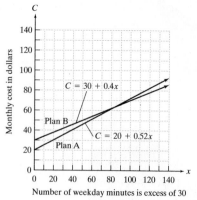

$C = 30 + 0.4x$
Plan B
$C = 20 + 0.52x$
Plan A
Monthly cost in dollars
Number of weekday minutes is excess of 30

The graphs seem to intersect when x is about 83. This represents 113 weekday minutes of cell phone use per month.
3. $x = 83.3$, yes   **4.** Plan A would be more economical for Gina. Plan B would be more economical for Aaron.
5. $258   **6.** Plan C   **7.** Plan D

Chapter 4 Review Problems

1.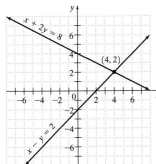

2.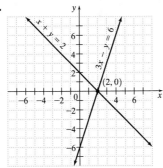

3.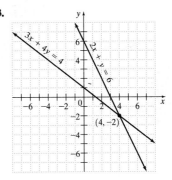

4. $(-1, 3)$ **5.** $(1, -7)$ **6.** $(0, -5)$ **7.** $(1, 3)$ **8.** $(1, -2)$ **9.** $(-3, 0)$ **10.** $(2, 3)$ **11.** $(5, -11)$

12. No solution; inconsistent system of equations **13.** Infinite number of solutions; dependent equations **14.** $(2, -4)$

15. $\left(-\dfrac{1}{3}, \dfrac{1}{2}\right)$ **16.** $(0, 3)$ **17.** $\left(\dfrac{1}{2}, \dfrac{2}{3}\right)$ **18.** $\left(\dfrac{4}{3}, -\dfrac{1}{2}\right)$ **19.** Infinite number of solutions, dependent equations **20.** $\left(0, \dfrac{2}{3}\right)$

21. No solution; inconsistent system of equations **22.** No solution; inconsistent system of equations **23.** $(5, 2)$ **24.** $(1, 1, -2)$

25. $(1, -2, 3)$ **26.** $(5, -3, 8)$ **27.** $\left(7, \dfrac{1}{2}, -3\right)$ **28.** $(3, 0, -2)$ **29.** $(1, -2, 3)$ **30.** $(1, 2, -4)$ **31.** $(-2, -4, -8)$

32. Speed of plane in still air = 264 mph; speed of wind = 24 mph **33.** 8 touchdowns; 3 field goals **34.** Laborers = 15; mechanics = 10
35. Children's tickets = 340; adult tickets = 250 **36.** Hats = $3; shirts = $15; pants = $12 **37.** Jess's score is 92; Nick's score is 85;
Chris's score is 72. **38.** One jar of jelly = $0.70; one jar of peanut butter = $1.00; one jar of honey = $0.80

39. Buses = 2; station wagons = 4; sedans = 3 **40.** $(0, 1)$ **41.** $\left(\dfrac{4}{3}, \dfrac{1}{3}\right)$ **42.** $(-3, 2)$ **43.** $(0, 2)$ **44.** $(10, 0)$ **45.** $(-1, -2)$

46. $(-3, -4)$ **47.** $(2, 5)$ **48.** $(11, 6)$ **49.** $(3, 2)$ **50.** $(-3, -2, 2)$ **51.** $(1, 0, -1)$ **52.** $(3, -1, -2)$ **53.** $(5, -5, -20)$

54.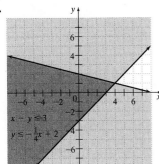

55.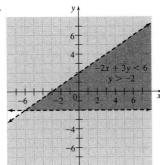

56.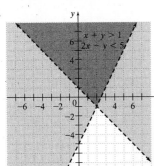

57.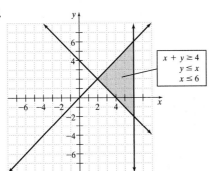

How Am I Doing? Chapter 4 Test

1. $(10, 7)$ (obj. 4.1.3) **2.** $(3, -4)$ (obj. 4.1.4) **3.** $(3, 4)$ (obj. 4.1.3) **4.** $\left(\dfrac{1}{2}, \dfrac{3}{2}\right)$ (obj. 4.1.4) **5.** $(1, 2)$ (obj. 4.1.4)

6. No solution; inconsistent system of equations (obj. 4.1.5) **7.** $(2, -1, 3)$ (obj. 4.2.2) **8.** $(-2, 3, 5)$ (obj. 4.2.3)
9. $(-4, 1, -1)$ (obj. 4.2.2) **10.** Speed of plane in still air is 450 mph; speed of wind is 50 mph (obj. 4.3.1) **11.** Each pen is $1.50,
each mug is $4.00, and each T-shirt is $10.00. (obj. 4.3.2) **12.** They charge $30 per day and $0.20 per mile. (obj. 4.3.1)

13.

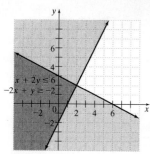

(obj. 4.4.1)

14.

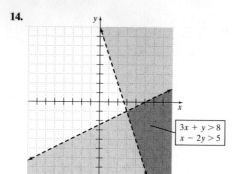

(obj.4.4.1)

Cumulative Test for Chapters 1–4

1. Identity property of addition   **2.** 2   **3.** $15x^{-6}y^2$ or $\dfrac{15y^2}{x^6}$ **4.** $22x + 12$ **5.** $P = \dfrac{A}{3 + 4rt}$ **6.** $x = 68$

7.

x	y
0	−1.25
4	0.75
7	2.25

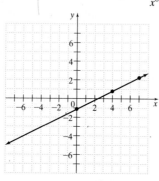

8. $m = \dfrac{1}{10}$ **9.** $x > -10$ **10.** $5 \le x \le 11$

11. $6x - 5y = 27$ **12.** 1st side = 17 m, 2nd side = 24 m, 3rd side = 28 m   **13.** \$1500 at 7%; \$4500 at 9%   **14.** $(2, -4)$
15. $(4, -1)$ **16.** Shirts = \$21, slacks = \$30   **17.** $(5, 3)$ **18.** $(1, 2, -2)$ **19.** Infinite number of solutions; dependent equations
20.

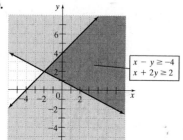

Chapter 5

5.1 Exercises

1. Trinomial, 2nd degree   **3.** Monomial, 8th degree   **5.** Binomial, 4th degree   **7.** 6   **9.** −18   **11.** 6   **13.** $-3x - 6$
15. $10m^3 + 4m^2 - 6m - 1.3$ **17.** $5a^3 - a^2 + 12$ **19.** $\dfrac{5}{6}x^2 - 6\dfrac{3}{4}x$ **21.** $-3.2x^3 + 1.8x^2 - 4$ **23.** $10x^2 + 61x + 72$
25. $15aw + 6ad - 20bw - 8bd$ **27.** $-12x^2 + 11xy - 2y^2$ **29.** $-28ar - 77rs^2 + 4as^2 + 11s^4$ **31.** $25x^2 - 64y^2$
33. $25a^2 - 20ab + 4b^2$ **35.** $49m^2 - 14m + 1$ **37.** $16 - 9x^4$ **39.** $9m^6 + 6m^3 + 1$ **41.** $6x^3 - 10x^2 + 2x$
43. $-\dfrac{2}{3}x^2y + 2xy^2 - 5xy$ **45.** $2x^3 - 5x^2 + 5x - 3$ **47.** $6x^3 - 7x^2y - 10xy^2 + 6y^3$ **49.** $\dfrac{3}{2}x^4 + 2x^3 - 10x^2 + 8x - 6$
51. $5a^4 - 18a^3 + 11a^2 - 10a + 12$ **53.** $2x^3 - 7x^2 - 7x + 30$ **55.** $3a^3 - a^2 - 22a + 24$ **57.** $6x^3 + 25x^2 + 49x + 40 \text{ cm}^3$
59. 77.73 parts per million   **61.** 3 parts per million   **63.** $x \ge 30$ **64.** $x = 0$ **65.** 9.2 minutes   **66.** 1,280,000 pages.

5.2 Exercises

1. $6x^2 - 2x - 11$ **3.** $3x^3 - x^2 + 7x$ **5.** $4b^2 - 3b - \dfrac{1}{2}$ **7.** $9a^2 + 6a - 2$ **9.** $5x - 2$ **11.** $3x + 4$ **13.** $7x - 2$

15. $x^2 - 2x + 13 - \dfrac{14}{x + 1}$ **17.** $2x^2 + 3x + 6 + \dfrac{5}{x - 2}$ **19.** $2x^2 - 4x + 2 - \dfrac{5}{2x + 1}$ **21.** $x^3 + 8x + 4$ **23.** $2t^2 - 3t + 2$

25. $3x - 1$ meters   **27.** The graphs of $y_1 = \dfrac{2x^2 - x - 10}{2x - 5}$ and $y_2 = x + 2$ should coincide.   **29.** $m = -\dfrac{6}{5}$ **30.** $m = -\dfrac{3}{2}$

31. $\dfrac{12 - 4y}{3} = x$ **32.** $\dfrac{9y + 2}{2} = x$ **33.** Curt   **34.** It could be Sylvia or Fritz.

5.3 Exercises

1. $2x + 1 + \dfrac{-2}{x - 6}$ **3.** $3x^2 - 2x + 1 + \dfrac{3}{x + 1}$ **5.** $x^2 + 4x + 5$ **7.** $7x^2 + 20x + \dfrac{-15}{x - 2}$ **9.** $x^2 - 4x + 8 + \dfrac{-8}{x + 2}$

11. $6x^3 - 5x^2 + 15x - 10 + \dfrac{6}{x + 3}$ **13.** $2x^3 + x^2 + 2 + \dfrac{3}{x + 1}$ **15.** $3x^4 - 3x^3 + 3x^2 - 3x + 4 + \dfrac{-5}{x + 1}$

17. $7x^4 - 7x^3 + 6x^2 - 3x + 3 + \dfrac{-1}{x + 1}$ **19.** $x^5 - x^4 + x^3 - 6x^2 + 7x - 7 + \dfrac{19}{x + 1}$ **21.** $x^2 + 3.7x + 0.84$ remainder 6.408   **23.** $a = 6$

25. $2x^2 - 6x + 9 + \dfrac{-21}{2x + 3}$ **27.** We are using the basic property of fractions that for any nonzero polynomials a, b, and c, $\dfrac{ac}{bc} = \dfrac{a}{b}$.

29. 268,000 cubic feet   **30.** 2.1 feet deep   **31.** 116

5.4 Exercises

1. $10(8 - y)$ **3.** $5a(a - 5)$ **5.** $3c(cx^3 - 3x - 2)$ **7.** $6y^2(5y^2 + 4y + 3)$ **9.** $5ab(3b + 1 - 2a^2)$
11. $12xy^2(y - 2x^2 + 3xy^2 - 5x^3y)$ **13.** $(x + y)(3x - 2)$ **15.** $(a - 3b)(5b + 8)$ **17.** $(a + 5b)(3x + 1)$ **19.** $(3x - y)(2a^2 - 5b^3)$
21. $(5x + y)(3x - 8y - 1)$ **23.** $(a - 6b)(2a - 3b - 2)$ **25.** $(x + 5)(x^2 + 3)$ **27.** $(x + 3)(2 - 3a)$ **29.** $(b - 4)(a - 3)$
31. $(x - 6)(5 - 2y)$ **33.** $(x - 3)(2 - 3y)$ **35.** $(z^2 + 5)(y - 3)$ **37.** $(sr - 1)(s^2 + t)$ **39.** $x\left(\dfrac{1}{3}x^2 + \dfrac{1}{2}x + \dfrac{1}{6}\right)$

41.

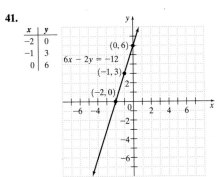

x	y
-2	0
-1	3
0	6

42.

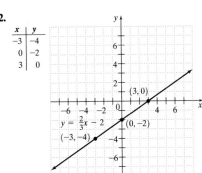

x	y
-3	-4
0	-2
3	0

43. -1 **44.** $m = -3, b = -\dfrac{3}{2}$

45. He answered 8 multiple-choice questions correctly.   **46.** Assuming he only wants to swim complete laps, he would need to swim 17 laps.

5.5 Exercises

1. $(x + 7)(x + 1)$ **3.** $(x - 7)(x - 2)$ **5.** $(x - 6)(x - 4)$ **7.** $(a + 9)(a - 5)$ **9.** $(x - 7y)(x + 6y)$ **11.** $(x - 14y)(x - y)$
13. $(x^2 - 8)(x^2 + 5)$ **15.** $(x^2 + 7y^2)(x^2 + 9y^2)$ **17.** $2(x + 11)(x + 2)$ **19.** $x(x + 5)(x - 4)$ **21.** $(2x + 1)(x - 1)$
23. $(3x - 5)(2x + 1)$ **25.** $(3a - 5)(a - 1)$ **27.** $(4a + 9)(2a - 1)$ **29.** $(2x + 3)(x + 5)$ **31.** $(3x^2 + 1)(x^2 - 3)$
33. $(3x + y)(2x + 11y)$ **35.** $(7x - 3y)(x + 2y)$ **37.** $x(2x + 5)(2x - 3)$ **39.** $5x^2(2x + 1)(x + 1)$ **41.** $(x - 9)(x + 7)$
43. $(3x + 2)(2x - 1)$ **45.** $(x - 17)(x - 3)$ **47.** $(5x + 2)(3x - 1)$ **49.** $2(x + 8)(x - 6)$ **51.** $3(3x + 2)(2x + 1)$
53. $9a(3x - 1)(x + 4)$ **55.** $2x(3x - 2)(x + 5)$ **57.** $(3x^2 - 5)(x^2 + 1)$ **59.** $(3a + b)(3a - 7b)$ **61.** $(x^3 - 13)(x^3 + 3)$
63. $2xy(2x - 1)(x + 1)$ **65.** One possibility is to have $6x + 5$ rows with $5x - 1$ trees in each row. Another possibility
is to have $5x - 1$ rows with $6x + 5$ trees in each row.   **67.** 28.26 in.^{2}

68. $\dfrac{3A - 4a}{3} = b$ **69. (a)** $\dfrac{4}{13}$ **(b)** Yes, because this hill has a slope of approximately 30.8%.

70.

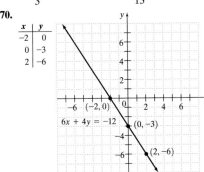

x	y
-2	0
0	-3
2	-6

71. They should stock at least 45 bike racks and 75 helmets.
72. 24 first class seats and 160 coach seats

How Am I Doing? Sections 5.1–5.5

1. $x^2 - 11x + 4$ (obj. 5.1.3)   **2.** $2x^3 - 9x^2 + x + 12$ (obj. 5.1.7)   **3.** $5a^2 - 43a + 56$ (obj. 5.1.4)   **4.** $4y^2 - 9$ (obj. 5.1.5)

5. $9x^4 + 24x^2 + 16$ (obj. 5.1.6)   **6.** $P(-3) = -80$ (obj. 5.1.2)   **7.** $5x - 6y - 10$ (obj. 5.2.1)   **8.** $3y^2 + y + 4 + \dfrac{7}{y - 2}$ (obj. 5.2.2)

9. $x^3 + 2x^2 - x - 2$ (obj. 5.2.2)   **10.** $2x^3 + 4x^2 - x - 3$ (obj. 5.3.1)   **11.** $12a^3b^2(2 + 3a - 5b)$ (obj. 5.4.1)
12. $(4x - 3y)(3x - 2)$ (obj. 5.4.1)   **13.** $(5w + 3z)(2x - 5y)$ (obj. 5.4.2)   **14.** $(5a - 4b)(2a - b)$ (obj. 5.4.2)
15. $(x - 5)(x - 2)$ (obj. 5.5.1)   **16.** $(2y - 5)(2y + 3)$ (obj. 5.5.2)   **17.** $(7x - 3y)(4x - y)$ (obj. 5.5.2)
18. $(x + 5)(2x + 7)$ (obj. 5.5.2)   **19.** $3(x + 4)(x - 6)$ (obj. 5.5.2)   **20.** $(4x - 3)(2x - 3)$ (obj. 5.5.2)

5.6 Exercises

1. The problem will have two terms. It will be in the form $a^2 - b^2$. One term is positive and one term is negative. The values and variables for the first and second terms are both perfect squares. So each one will be of the form 1, 4, 9, 16, 25, 36, and/or x^2, x^4, x^6, etc.
3. There will be two terms added together. It will be of the form $a^3 + b^3$. Each term will contain a number or variable cubed or both. They will be of the form 1, 8, 27, 64, 125, and/or x^3, x^6, x^9, etc.   **5.** $(a + 8)(a - 8)$ **7.** $(4x - 9)(4x + 9)$ **9.** $(8x - 1)(8x + 1)$
11. $(7m - 3n)(7m + 3n)$ **13.** $(10y - 9)(10y + 9)$ **15.** $(1 - 9xy)(1 + 9xy)$ **17.** $2(4x + 3)(4x - 3)$ **19.** $5x(1 + 2x)(1 - 2x)$
21. $(3x - 1)^2$ **23.** $(7x - 1)^2$ **25.** $(9w + 2t)^2$ **27.** $(6x+5y)^2$ **29.** $2(2x + 3)^2$ **31.** $3x(x - 4)^2$ **33.** $(x - 3)(x^2 + 3x + 9)$
35. $(x + 5)(x^2 - 5x + 25)$ **37.** $(4x - 1)(16x^2 + 4x + 1)$ **39.** $(5x - 2)(25x^2 + 10x + 4)$ **41.** $(1 - 3x)(1 + 3x + 9x^2)$
43. $(4x + 5)(16x^2 - 20x + 25)$ **45.** $(4s^2 + t^2)(16s^4 - 4s^2t^2 + t^4)$ **47.** $6(y - 1)(y^2 + y + 1)$ **49.** $3(x - 2)(x^2 + 2x + 4)$
51. $x^2(x - 2y)(x^2 + 2xy + 4y^2)$ **53.** $(5w^2 + 1)(5w^2 - 1)$ **55.** $(b^2 + 3)^2$ **57.** $(7m^3 - 9)(7m^3 + 9)$ **59.** $(6y^3 - 5)^2$
61. $2(a^4 - 5)(a^4 + 5)$ **63.** $(5m + 2n)(25m^2 - 10mn + 4n^2)$ **65.** $3(2a - b)(4a^2 + 2ab + b^2)$ **67.** $(2w - 5z)^2$
69. $9(2a - 3b)(2a + 3b)$ **71.** $(4x^2 + 9y^2)(2x - 3y)(2x + 3y)$ **73.** $(5m^2 + 2)(25m^4 - 10m^2 + 4)$ **75.** $(5x + 4)(5x + 1)$
77. $(4x - 3)(x - 3)$ **79.** $A = (4x + y)(4x - y)$ square feet   **80.** $A = \pi(2y + x)(2y - x)$ square inches
81. The year 2005   **82.** She invested $1400 at 8% and $2600 at 11%.   **83.** 1st side is 20 cm; 2nd side is 30 cm; 3rd side is 16 cm
84. Melinda paid $388; Hector paid $278; Alice paid $192

5.7 Exercises

1. To remove a common factor if possible   **3.** You cannot factor polynomials of the form $a^2 + b^2$. All such polynomials that are the sum of two squares are prime. You can factor polynomials of the form $a^2 - b^2$, but you do NOT have that form in this problem.   **5.** $3y(x - 2z)$
7. $(y - 2)(y + 9)$ **9.** $(3x - 5)(x - 1)$ **11.** $(a - 2y)(x + 3w)$ **13.** $(2x - 5y)(4x^2 + 10xy + 25y^2)$ **15.** $x(x + 2y - z)$
17. Prime   **19.** $(8y - 5z)(8y + 5z)$ **21.** $(6x + 1)(x - 4)$ **23.** Prime   **25.** $x(x - 5)(x - 6)$ **27.** $(5x - 4)^2$
29. $6(a - 3)(a + 2)$ **31.** $(3x - y)(x - 1)$ **33.** $(9a^2 + 1)(3a + 1)(3a - 1)$ **35.** $2x(x - 3)(x + 3)(x^2 + 1)$
37. $2ab(2a - 5b)(2a + 5b)$ **39.** $2(2x^2 - 4x - 3)$ **41.** $(x - 10)(5x + 8y)$ square feet $= 10x^2 - 100x + 16xy - 160y$ square feet
43. $x \leq -9$ **44.** $-\dfrac{1}{5} < x < \dfrac{3}{5}$ **45.** $x < -\dfrac{7}{4}$ or $x > \dfrac{17}{4}$ **46.** $x \geq 11$ or $x \leq 4$ **47.** $295 million
48. Approximately 49.1%   **49.** Approximately $435.4 million   **50.** Approximately $791.7 million

5.8 Exercises

1. $x = 3, x = -2$ **3.** $x = 0, x = \dfrac{6}{5}$ **5.** $x = -\dfrac{6}{5}, x = \dfrac{6}{5}$ **7.** $x = -\dfrac{4}{3}, x = 2$ **9.** $x = \dfrac{3}{4}, x = -\dfrac{1}{2}$ **11.** $x = \dfrac{3}{8}, x = 1$
13. $x = 0, x = -1$ **15.** $x = -\dfrac{1}{5}$, double root   **17.** $x = 0, x = -3, x = -2$ **19.** $x = 0, x = 8, x = -6$ **21.** $x = 0, x = -3, x = 3$
23. $x = -7, x = 2, x = 0$ **25.** $x = -\dfrac{3}{7}, x = 1$ **27.** $x = -\dfrac{3}{2}, x = 4$ **29.** $x = 0, x = \dfrac{8}{5}$ **31.** $c = -2, x = 2$
33. Altitude is 18 in.; base is 20 in.   **35. (a)** Altitude is 8 ft; base is 26 ft   **(b)** Altitude is $2\frac{2}{3}$ yards; base is $8\frac{2}{3}$ yards
37. (a) Width is 28 cm; length is 32 cm.   **(b)** Width is 280 mm; length is 320 mm.   **39.** 12 feet   **41.** Length is 9 inches; height is 11 inches
43. Width is 6 mi; length is 9 mi.   **45.** Old side is 7 cm; new side is 15 cm.   **47.** 20 units   **49.** 6 units   **51.** 13,360   **53.** 1981
55. $200x^{11}y^{10}$ **56.** $\dfrac{a^4}{2b^2}$ **57.** $(2, -3)$ **58.** $y = 2x + 2$

Putting Your Skills to Work

1. 200,000 in 1900 and 156,200 in 1930   **2.** 138,800 in 1940 and 99,800 in 1960   **3.** There were 30,800 in 1990, and this was a decrease of 84.6%.   **4.** There were 5000 in 2000, and this was a decrease of 96.4%.   **5.** 38,000; 40,000; 38,000; 36,000; 40,000; 55,000
6.   **7.** 87,000   **8.** Changes in technology, the number of jobs may reach its maximum level, etc.

$f(x) = 0.11x^3 - 1.9x^2 + 9.4x + 26$

Number of jobs in thousands

Number of years since 1998

Chapter 5 Review Problems

1. $-x^2 - 10x + 13$ **2.** $x^2y - 5xy - 8y$ **3.** $-11x^2 + 10xy + 6y^2$ **4.** $-11x^2 + 15x - 15$ **5.** $12x + 4$ **6.** $-x^2 + 4x - 1$
7. -199 **8.** 2   **9.** 46   **10.** -36 **11.** -176 **12.** -2 **13.** -49 **14.** -44 **15.** -4 **16.** $3x^3y - 3x^2y^2 + 3xy^3$

17. $6x^3 - 3x^2 + 2x - 1$ **18.** $25x^4 + 30x^2 + 9$ **19.** $2x^3 - 7x^2 - 7x + 30$ **20.** $-2x^4 + 7x^3 - 7x^2 + 7x - 2$

21. $9x^3 - 9x^2 - 22x + 20$ **22.** $36x^2y^2 - 49$ **23.** $15a^2 - 26ab^2 + 8b^4$ **24.** $-5x^2 + 3x + 20$ **25.** $4x - 8 + \dfrac{12}{3x + 2}$

26. $x^2 - x + 1 - \dfrac{2}{2x + 3}$ **27.** $3y^2 + 9y + 25 + \dfrac{80}{y - 3}$ **28.** $5a^2 - a + 3 + \dfrac{-a + 7}{3a^2 - 1}$ **29.** $2x^3 + 2x^2 + x + 7 + \dfrac{10}{x - 1}$

30. $2x^3 - 3x^2 + x - 4$ **31.** $3x^3 - x^2 + x - 1$ **32.** $3ab(2a - b - 1)$ **33.** $x^2(x^3 - 3x^2 + 2)$ **34.** $4m(3n - 2)$

35. $(x + 3)(2 - y)$ **36.** $(x^2 + 1)(8y + b)$ **37.** $(3a - 2)(b - 5)$ **38.** $(x - 11)(x + 2)$ **39.** $(4x + 3)(x - 2)$

40. $(3x + 7)(2x - 3)$ **41.** $(10x + 7)(10x - 7)$ **42.** $(2x - 7)^2$ **43.** $(2a - 3)(4a^2 + 6d + 9)$ **44.** $(3x - 11)(3x + 11)$

45. $(5x - 1)(x - 2)$ **46.** $x(x + 6)(x + 2)$ **47.** $(x + 4w)(x - 2y)$ **48.** Prime   **49.** Prime   **50.** $(x + 9y)(x - 3y)$

51. $x(3x - 1)(9x^2 + 3x + 1)$ **52.** $(7a + 2b)(3a + 2b)$ **53.** $-a^2b^3(3a - 2b + 1)$ **54.** $a^2b^4(a + 3)(a - 2)$ **55.** $(3x^2 + 1)(x^2 - 2)$

56. $b(3a + 7)(3a - 2)$ **57.** Prime   **58.** Prime   **59.** $y^2(4y - 9)(y - 1)$ **60.** $y^2(y + 7)(y - 5)$ **61.** $4x^2(y^2 - 3y - 2)$

62. $(3x^2 + 2)(x^2 - 3)$ **63.** $(a + b^3)(a + 4b^3)$ **64.** $(2x + 3)(x + 2)(x - 2)$ **65.** $2(x - 3)(x + 3)(x^2 + 3)$ **66.** $4(2 - x)(a + b)$

67. $(4x^2 + 3y^2)(2x^2 + 7y^2)$ **68.** $2x(2x - 1)(x + 3)$ **69.** $x(2a - 1)(a - 7)$ **70.** $(4x^2y - 7)^2$ **71.** $2xy(8x - 1)(8x + 1)$

72. $(5x + 4y)(b - 7)$ **73.** $3ab(3c - 2)(3c + 2)$ **74.** $x = -\dfrac{1}{5}, x = 2$ **75.** $x = \dfrac{3}{2}, x = 4$ **76.** $x = -\dfrac{5}{6}, x = 3$ **77.** $x = 0, x = 3$

78. $x = 0, x = -\dfrac{10}{3}$ **79.** $x = 0, x = -3, x = -4$ **80.** Base is 11 meters; altitude is 14 meters.   **81.** Width is 4 miles; length is 10 miles.

82. 5 calculators   **83.** Old side is 1 yard; new side is 5 yards.

How Am I Doing? Chapter 5 Test

1. $-4x^2y - 1$ (obj. 5.1.3)   **2.** $5a^2 - 9a - 2$ (obj. 5.1.3)   **3.** $-2x^2 - 6xy + 8x$ (obj. 5.1.7)   **4.** $4x^2 - 12xy^2 + 9y^4$ (obj. 5.1.6)
5. $2x^3 - 3x^2 - 3x + 2$ (obj. 5.1.7)   **6.** $5x^2 + 4x - 7$ (obj. 5.2.1)   **7.** $x^3 - x^2 + x - 2$ (obj. 5.2.2)   **8.** $x^2 - 3x + 1$ (obj. 5.2.2)

9. $x^3 - 1 - \dfrac{2}{x + 1}$ (obj. 5.3.1)   **10.** $(11x - 5y)(11x + 5y)$ (obj. 5.6.1)   **11.** $(3x + 5y)^2$ (obj. 5.6.2)   **12.** $x(x - 2)(x - 24)$ (obj. 5.7.1)

13. $4x^2y(x + 2y + 1)$ (obj. 5.4.1)   **14.** $(x + 3y)(x - 2w)$ (obj. 5.4.2)   **15.** Prime (obj. 5.7.2)   **16.** $3(6x - 5)(x + 1)$ (obj. 5.5.2)
17. $2a(3a - 2)(9a^2 + 6a + 4)$ (obj. 5.7.1)   **18.** $x(3x^2 - y)^2$ (obj. 5.7.1)   **19.** $(3x^2 + 2)(x^2 + 5)$ (obj. 5.7.1)
20. $(x + 2y)(3 - 5a)$ (obj. 5.7.1)   **21.** -18 (obj. 5.1.2)   **22.** 17 (obj. 5.1.2)   **23.** $x = -2, x = 7$ (obj. 5.8.1)

24. $x = -\dfrac{1}{3}, x = 4$ (obj. 5.8.1)   **25.** $x = 0, x = \dfrac{2}{7}$ (obj. 5.8.1)   **26.** Base is 14 in.; altitude is 10 in. (obj. 5.8.2)

Cumulative Test for Chapters 1–5

1. Associative property of multiplication   **2.** 2   **3.** $x - 3 + 6y$ **4.** $x = \dfrac{2 - 7y}{5}$ **5.** $x = -\dfrac{5}{3}$ **6.** $m = \dfrac{8}{3}$

7.
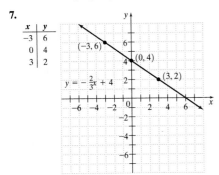

x	y
-3	6
0	4
3	2

$y = -\dfrac{2}{3}x + 4$

8.
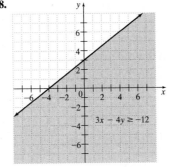

$3x - 4y \geq -12$

9. $x > \dfrac{1}{3}$ **10.** Width is 6 m; length is 17 m.   **11.** 7   **12.** $-a^2b - 8ab$ **13.** $10x^3 - 19x^2 - 14x + 8$ **14.** $-3x^2 + 2x - 4$

15. $2x^2 + x + 5 + \dfrac{6}{x - 2}$ **16.** $2x^2(x - 5)$ **17.** $(8x + 7)(8x - 7)$ **18.** $(3x + 4)(x - 2)$ **19.** $(5x + 6)^2$ **20.** $3(x - 7)(x + 2)$

21. $2(x + 2)(x + 10)$ **22.** Prime   **23.** $x(3x + 1)(2x + 3)$ **24.** $x(3x + 4)(9x^2 - 12x + 16)$ **25.** $(x - 3)(2 - 5y)$

26. $x = -\dfrac{2}{3}, x = 2$ **27.** $x = 11, x = -3$ **28.** Base is 8 meters; altitude is 17 meters.

Chapter 6

6.1 Exercises

1. All real numbers except 3   **3.** All real numbers except -4 and 9   **5.** $-\dfrac{3x^2}{2y^5}$ **7.** $\dfrac{x^2}{2}$ **9.** $\dfrac{3x}{4x - 5}$ **11.** $\dfrac{y(x - 3)}{2x^2(1 - 2y)}$ **13.** $\dfrac{1}{x - 5}$

15. $y - 2$ **17.** $-\dfrac{x + 2}{x}$ **19.** $-\dfrac{2y + 5}{2 + y}$ **21.** $-\dfrac{4n^4}{m}$ **23.** $\dfrac{a(a - 2)}{a + 2}$ **25.** $\dfrac{3}{x - 3}$ **27.** $(x - 8y)(x + 7y)$ **29.** $\dfrac{y + 2}{2(y - 1)}$

31. $\dfrac{y(x - 5)}{x^2}$ **33.** $\dfrac{1}{5}$ **35.** $\dfrac{b - 3}{2b - 1}$ **37.** $\dfrac{x - 3y}{x(x^2 + 2)}$ **39.** $\dfrac{1}{3x^2y^2}$ **41.** $\dfrac{x}{2}$ **43.** $\dfrac{3(x + 7)}{xy}$ **45.** $\dfrac{x^2(x - 3)}{y(x - 2)}$

47. Cannot be simplified   **49.** All real numbers except $x \approx -1.4$ and $x \approx 0.9$ **51.** 150   **53.** 225

55.

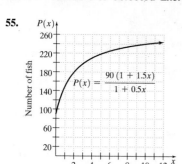

$$P(x) = \frac{90(1 + 1.5x)}{1 + 0.5x}$$

Number of fish (y-axis); Number of months since fish placed in aquarium (x-axis)

57.

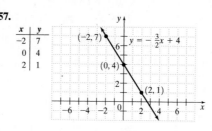

x	y
-2	7
0	4
2	1

$(-2, 7)$, $(0, 4)$, $(2, 1)$; $y = -\frac{3}{2}x + 4$

58.

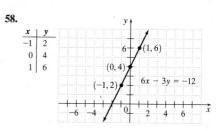

x	y
-1	2
0	4
1	6

$(1, 6)$, $(0, 4)$, $(-1, 2)$; $6x - 3y = -12$

59. $(4, 0)$ **60.** $x > -3$ **61.** 3960 inquiries   **62.** No   **63.** 72,103 people died.

6.2 Exercises

1. The factors are $5, x$, and y. The factor y is repeated in one fraction three times. Since the highest power of y is 3, the LCD is $5xy^3$.

3. $(x - 1)^2$ **5.** $2m^3n^2$ **7.** $(x + 2)(3x + 4)^3$ **9.** $x(6x - 1)(3x + 2)$ **11.** $\dfrac{3x - 10}{(x + 4)(x - 4)}$ **13.** $\dfrac{12y + 2x}{5x^2y}$

15. $\dfrac{8x - 15}{x(x - 4)(x - 3)}$ **17.** $\dfrac{14x - 20}{2x - 5}$ **19.** $\dfrac{-5y^2 + 11y + 6}{(y + 1)(y - 1)^2}$ **21.** $\dfrac{a^2 + a + 4}{2(a + 2)(a - 2)}$ **23.** $\dfrac{2x + 17}{(x - 4)(x + 1)}$ **25.** $\dfrac{-2x + 7}{(x + 1)^2(x - 2)}$

27. $\dfrac{-y^2 + 6y + 3}{(y + 1)(y + 2)}$ **29.** $\dfrac{3a^2 + 4a - 13}{3a - 5}$ **31.** $P(x) = \dfrac{12x^2 - 68x + 120}{x^2 - 5x + 6}$ **33.** $11,429   **35.** $\dfrac{27x^3 - 37x^2 + 26x - 8}{(3x - 2)^2}$

37. $-\dfrac{1}{2}$ **38.** 3.51×10^{-4} **39.** Assuming people were working 24 hours per day, it would take 1460 hours, or approximately 60.83 days.

40. (a) 63.96%   **(b)** approximately 69 actual medications   **41.** Tony's car cost $6000, Melissa's car cost $13,000, and Alreda's car cost $7500.
42. Number of liters at 15% is 40 liters. Number of liters at 30% is 20 liters.   **43.** Speed of boat in still water is 20 km/hr; speed of the current is 5 km/hr.   **44.** 5000 restaurants

6.3 Exercises

1. $\dfrac{7y}{3}$ **3.** $\dfrac{2(x - 1)}{x(x + 5)}$ **5.** $\dfrac{5y - 6}{5(y + 3)}$ **7.** $\dfrac{y^2 - 3}{3}$ **9.** $\dfrac{2}{(y - 3)(y + 6)}$ **11.** $\dfrac{(x + 5)(x - 2)}{2(x^2 - 2x + 2)}$ **13.** $-\dfrac{4(x - 1)}{x + 2}$ **15.** $\dfrac{x(2x + 5)}{5(2x + 3)}$

17. $-\dfrac{6 + x - y}{2y}$ **19.** $\dfrac{1}{x(x - a)}$ **21.** $-\dfrac{1}{y - 3}$ **23.** $\dfrac{2x(x + 1)}{3(2x + 1)}$ or $\dfrac{2x^2 + 2x}{6x + 3}$ **25.** $x = -\dfrac{2}{3}$ or $x = 2$ **26.** $x = -5$ or $x = 15$

27. $-\dfrac{3}{5} < x < \dfrac{9}{5}$ **28.** $x \geq 1.5$ or $x \leq -2.5$ **29. (a)** $253,440,000 was spent per mile in the 1970s.   **(b)** Starting in the year 2002, a total of 6.5 miles can be built with that budget limit.   **30.** She should ship 1750 kilograms by airfreight and 3850 kilograms by ocean freighter.

How Am I Doing? Sections 6.1–6.3

1. $\dfrac{7x + 3}{x + 1}$ (obj. 6.1.1)   **2.** $\dfrac{x + 3}{x + 8}$ (obj. 6.1.1)   **3.** $\dfrac{2x + 1}{x - 5}$ (obj. 6.1.1)   **4.** $\dfrac{2(9a - 8)}{2a - 5}$ (obj. 6.1.2)   **5.** $\dfrac{15(x - 5y)}{2y^4(x + 5y)}$ (obj. 6.1.3)

6. 3 (obj. 6.1.2)   **7.** $\dfrac{x^2 - 4x + 8}{3x(x - 2)}$ (obj. 6.2.2)   **8.** $\dfrac{12x + 5}{(x + 5)(x - 5)}$ (obj. 6.2.2)   **9.** $\dfrac{10y - 6}{(y + 3)(y - 3)(y + 4)}$ (obj. 6.2.2)

10. $\dfrac{-3x}{(x + 4)(x - 4)}$ (obj. 6.2.2)   **11.** $\dfrac{21x}{8}$ (obj. 6.3.1)   **12.** $\dfrac{x}{(2x - 1)(6x + 1)}$ (obj. 6.3.1)   **13.** $\dfrac{5 + 3x}{6 - 2x}$ (obj. 6.3.1)

14. $\dfrac{x^2 + 5x + 10}{x^2 + 7x + 4}$ (obj. 6.3.1)

6.4 Exercises

1. $3 = x$ **3.** $x = \dfrac{3}{4}$ **5.** $x = 6$ **7.** $y = -2$ **9.** No solution   **11.** $x = 2$ **13.** $x = 3$ **15.** $y = -2$ **17.** $x = 0$

19. $y = 1$ **21.** $z = \dfrac{34}{3}$ **23.** $x = 0$ **25.** No solution   **27.** $x = -\dfrac{5}{3}$ **29.** When the solved value of the variable causes the denominator of any fraction to equal 0, or when the variable drops out.   **31.** $x \approx -1.9$ **33.** $7(x + 3)(x - 3)$ **34.** $2(x + 5)^2$

35. $(4x - 3y)(16x^2 + 12xy + 9y^2)$ **36.** $(3x - 7)(x - 2)$ **37.** 80,000 will attend counseling. 20,000 will likely remain married.
38. 52,800

6.5 Exercises

1. $m = \dfrac{y - b}{x}$ **3.** $b = \dfrac{af}{a - f}$ **5.** $h = \dfrac{V}{lw}$ **7.** $\dfrac{2F}{y + z} = x$ **9.** $V = \dfrac{4\pi r^3}{3}$ **11.** $\dfrac{Er}{R + r} = e$ **13.** $T_1 = \dfrac{P_1 V_1 T_2}{P_2 V_2}$

15. $w = \dfrac{S - 2lh}{2h + 2l}$ **17.** $T_1 = \dfrac{ET_2}{T_2 - 1}$ **19.** $x_1 = \dfrac{mx_2 - y_2 + y_1}{m}$ **21.** $D = \dfrac{2Vt + at^2}{2}$ **23.** $t_2 = \dfrac{kAt_1 - QL}{kA}$ **25.** $T_2 = \dfrac{-qT_1}{W - q}$

27. $v_0 = \dfrac{s - s_0 - gt^2}{t}$ **29.** $T \approx T_0 + 0.1702\left(\dfrac{V}{V_0}\right) - 0.1656$ **31.** Approximately 119.2 kilometers   **33.** Approximately 54.55 mph

35. 80 grizzly bears   **37.** 26 officers, 91 seamen   **39.** 44 people in marketing, 143 people in sales   **41.** width is 12 inches, length is 15 inches
43. 60 people prefer the new software.   **45.** 24 power boats, 54 sailboats   **47.** 6.4 feet high   **49.** 196 times   **51.** 3.6 hours

53. 6 hours   **55.** 282 feet   **57.** $m = \dfrac{7}{3}; b = -\dfrac{8}{3}$ **58.** $m = -\dfrac{1}{5}$ **59.** $y = -\dfrac{1}{3}x + 3$ **60.** $(5, -1)$

61. Partial scholarships are $4444.44 per student. Full scholarships are $8888.89 per student.

Putting Your Skills to Work

1. $247.27 per month   **2.** $142.60   **3.** $8901.72; $10,267.20   **4.** $513.72; $1879   **5.** $230.53 per month at 5.9%; $328.13 per month at 0%
6. $18,148.16 at 5.9%; $17,500 at 0%   **7.** The 5.9% plan has a lower down payment and lower monthly payments. The 0% plan has a lower overall cost. Other answers may be given.   **8.** advantages: lower monthly payments and no down payment   disadvantages: you don't own the car at the end of the lease.

Chapter 6 Review Problems

1. $\dfrac{x}{2}$ **2.** $\dfrac{3x^3}{x-4}$ **3.** $\dfrac{2x^2}{3y^2}$ **4.** $\dfrac{7c^3}{4a^3}$ **5.** $\dfrac{2x-3}{3x+5}$ **6.** $\dfrac{a-b}{3(x-2)}$ **7.** $\dfrac{x(2x-1)}{x-2}$ **8.** $\dfrac{1}{8xy}$ **9.** $\dfrac{y-5}{y-4}$ **10.** $\dfrac{x^2(x+5)}{2y(x+4)}$

11. $\dfrac{x+6}{3(x+2)}$ **12.** $\dfrac{b^3(x+2)}{5a(2x+1)}$ **13.** $\dfrac{2y+5}{2y-5}$ **14.** $\dfrac{2a+5}{2a(2a^2-7a-13)}$ **15.** $-\dfrac{x^2+10x-9}{(2x+1)(x-2)}$ **16.** $\dfrac{-7x+20}{4x(x+4)}$ **17.** $\dfrac{-6y-11}{36y}$ or

$-\dfrac{6y+11}{36y}$ **18.** $\dfrac{7y-18}{(y+5)(y-5)}$ **19.** $\dfrac{4y^2-y+3}{(y+1)^2(y-1)}$ **20.** $-\dfrac{y+2}{y+3}$ **21.** $\dfrac{a-2}{a+3}$ **22.** $\dfrac{-x^2+2}{(x+4)^2(2x+1)}$ **23.** $\dfrac{5b^2+13b-5}{b+3}$

24. $\dfrac{4x^2+6x+5}{2x}$ **25.** $\dfrac{x}{x-5}$ **26.** $\dfrac{4(x-2)}{2x+5}$ **27.** $\dfrac{y^2+y+1}{y^2-y-1}$ **28.** $\dfrac{5}{2}$ **29.** $\dfrac{2(2x-1)}{3x}$ **30.** $-\dfrac{1}{x+y}$ **31.** $\dfrac{x+1}{x-1}$

32. $-\dfrac{(x+3)(x+4)}{x(x-4)}$ **33.** $x = -1$ **34.** $x = 6$ **35.** $x = 1$ **36.** $x = 1$ **37.** $a = -6$ **38.** $a = 5$ **39.** $y = \dfrac{3}{4}$ **40.** $y = \dfrac{1}{7}$

41. $a = -\dfrac{1}{2}$ **42.** $a = -\dfrac{1}{2}$ **43.** No solution   **44.** $x = 0$ **45.** $M = \dfrac{mV}{N} - N$ **46.** $x = \dfrac{y - y_0 + mx_0}{m}$ **47.** $a = \dfrac{bf}{b-f}$

48. $t = \dfrac{2S}{V_1 + V_2}$ **49.** $R_2 = \dfrac{dR_1}{L-d}$ **50.** $r = \dfrac{S-P}{Pt}$ **51.** $\dfrac{x+6}{x+5}$ **52.** $\dfrac{2x-1}{x+2}$ **53.** $\dfrac{4x^2+17x+11}{x+4}$ **54.** $\dfrac{1}{x+1}$ **55.** $x = \dfrac{5}{7}$
56. 161 scientific calculators, 92 graphing calculators   **57.** 21 one-story homes, 91 two-story homes   **58.** 35 inches wide, 49 inches long
59. 5.6 hours   **60.** 500 rabbits   **61.** 28 officers   **62.** 12.25 nautical miles   **63.** 182 feet tall   **64.** 7.2 hours   **65.** 6 minutes
66. 973 million messages   **67.** 1185 million messages   **68.** 1093 million messages   **69.** 598 million messages

How am I doing? Chapter 6 Test

1. $\dfrac{x+2}{x-3}$ (obj. 6.1.1)   **2.** $-\dfrac{5p^3}{9r^3}$ (obj. 6.1.1)   **3.** $\dfrac{2(2y-1)}{3y+5}$ (obj. 6.1.2)   **4.** $-\dfrac{6}{(2x-1)(x+1)}$ (obj. 6.1.3)   **5.** $\dfrac{x+3}{x(x+1)}$ (obj. 6.2.2)

6. $\dfrac{3x^2+8x+6}{(x+3)^2(x+2)}$ (obj. 6.2.2)   **7.** $-\dfrac{2}{5}$ (obj. 6.3.1)   **8.** $-x+1$ (obj. 6.3.1)   **9.** $x = \dfrac{16}{3}$ (obj. 6.4.1)   **10.** $x = -2$ (obj. 6.4.1)

11. $y = 5$ (obj. 6.4.1)   **12.** $x = 3$ (obj. 6.4.1)   **13.** $W = \dfrac{S-2Lh}{2h+2L}$ (obj. 6.5.1)   **14.** $h = \dfrac{3V}{\pi r^2}$ (obj. 6.5.1)

15. 39 employees got the bonus; 247 did not. (obj. 6.5.2)   **16.** 1500 feet wide, 2550 feet long (obj. 6.5.2)

Cumulative Test for Chapters 1–6

1. $\dfrac{9}{y}$ **2.** $x = \dfrac{55}{24}$ **3.**

x	y
0	-6
1	-3
2	0

$-6x + 2y = -12$

$(2, 0)$
$(1, -3)$
$(0, -6)$

4. $5x - 6y = 13$ **5.** $700 at 5% interest; $6300 at 8% interest

6. $x < -1$ **7.** -22 **8.** $-2 \le x \le \dfrac{14}{3}$ **9.** $(2x - 5y)(4x^2 + 10xy + 25y^2)$ **10.** $x(9x - 5y)^2$

11. $x = -18, x = -2$ **12.** $x = 4, x = -\dfrac{1}{3}$ **13.** $\dfrac{7(x-2)}{x+4}$ **14.** $\dfrac{x(x+1)}{2x+5}$ **15.** $\dfrac{x}{3(2x+1)}$ **16.** $\dfrac{4x+1}{3(x+1)(x-1)}$

17. $\dfrac{2(x+1)(2x-1)}{16x^2-7}$ **18.** $\dfrac{7x-12}{(x-6)(x+4)}$ **19.** $x = -4$ **20.** $x = -2$ **21.** $b = \dfrac{5H-2x}{3+4H}$ **22.** 693 on foot, 2541 in squad cars

Chapter 7

7.1 Exercises

1. $\dfrac{81x^4}{y^4z^8}$ **3.** $-\dfrac{8b^3}{27a^3}$ **5.** $\dfrac{y^3}{8x^6}$ **7.** $\dfrac{y^{10}}{9x^2}$ **9.** $x^{3/2}$ **11.** y^8 **13.** $x^{2/5}$ **15.** $x^{1/2}$ **17.** $x^{5/2}$ **19.** $x^{4/7}$ **21.** $a^{7/8}$ **23.** $y^{1/2}$

25. $\dfrac{1}{x^{3/4}}$ **27.** $\dfrac{b^{1/3}}{a^{5/6}}$ **29.** $\dfrac{1}{6^{1/2}}$ **31.** $\dfrac{2}{a^{1/4}}$ **33.** 9 **35.** 8 **37.** -32 **39.** 9 **41.** $x^{5/6}y$ **43.** $-14x^{7/12}y^{1/12}$ **45.** $6^{4/3}$

47. $2x^{7/10}$ **49.** $-\dfrac{4x^{5/2}}{y^{6/5}}$ **51.** $2ab$ **53.** $9x^{4/5}y^3z^{2/3}$ **55.** $x^2 - x^{13/15}$ **57.** $m^{3/8} + 2m^{15/8}$ **59.** $\dfrac{1}{2}$ **61.** $\dfrac{1}{343}$ **63.** 32

65. $\dfrac{3y+1}{y^{1/2}}$ **67.** $\dfrac{1+6^{4/3}x^{1/3}}{x^{1/3}}$ **69.** $2a(5a^{1/4} - 2a^{3/5})$ **71.** $3x(2x^{3/4} - 5x^{1/2})$ **73.** $a = -\dfrac{3}{8}$ **75.** radius $= 1.86$ meters

77. radius $= 5$ feet **79.** $x = -\dfrac{3}{2}$ **80.** $b = \dfrac{2A - ah}{h}$ or $b = \dfrac{2A}{h} - a$ **81.** 147 milligrams **82.** 5 years old

7.2 Exercises

1. A square root of a number is a value that when multiplied by itself is equal to the original number.

3. $\sqrt[3]{-8} = -2$ because $(-2)(-2)(-2) = -8$ **5.** 10 **7.** 13 **9.** $-\dfrac{1}{3}$ **11.** Not a real number **13.** 0.2 **15.** $4.6, 4.9, 6, 3; x \geq -7$

17. $0, 1, 2, 2.2; x \geq 6$ **19.** 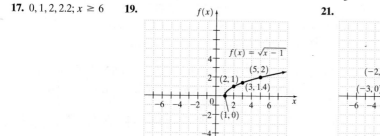 **21.** 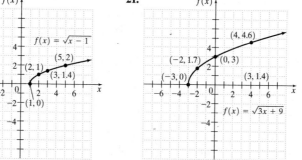 **23.** 4 **25.** -10 **27.** 3

29. 3 **31.** 8 **33.** 5 **35.** $-\dfrac{1}{4}$ **37.** $y^{1/3}$ **39.** $m^{3/5}$ **41.** $(2x)^{1/5}$ **43.** $(a+b)^{3/7}$ **45.** $x^{1/6}$ **47.** $(3x)^{5/6}$ **49.** 12

51. xy^2 **53.** $6x^4y^2$ **55.** $2a^2b$ **57.** $-5x^{10}$ **59.** $\left(\sqrt[7]{y}\right)^4$ **61.** $\dfrac{1}{\sqrt[3]{49}}$ **63.** $\left(\sqrt[7]{2a+b}\right)^5$ **65.** $\left(\sqrt[5]{-x}\right)^3$ **67.** $\sqrt[5]{8x^3y^3}$ **69.** 8

71. $\dfrac{2}{5}$ **73.** 2 **75.** $\dfrac{1}{5x^2}$ **77.** $11x^2$ **79.** $12a^3b^{12}$ **81.** $6x^3y^4z^5$ **83.** $6ab^3c^4$ **85.** $5|x|$ **87.** $-2x^2$ **89.** x^2y^4 **91.** $|a^3b|$

93. $2x^4y^2$ **95.** \$1215 **97.** 6.3938×10^{16} Btu **98.** 20.39%

7.3 Exercises

1. $2\sqrt{2}$ **3.** $3\sqrt{2}$ **5.** $2\sqrt{7}$ **7.** $2\sqrt{11}$ **9.** $3x\sqrt{x}$ **11.** $2a^3b^3\sqrt{10b}$ **13.** $3xz^2\sqrt{10xy}$ **15.** 2 **17.** $2\sqrt[3]{5}$ **19.** $3\sqrt[3]{2a^2}$

21. $2ab^2\sqrt[3]{b^2}$ **23.** $2x^2y^3\sqrt[3]{3y^2}$ **25.** $3p^5\sqrt[4]{kp^3}$ **27.** $-2xy\sqrt[5]{y}$ **29.** $a = 4$ **31.** $12\sqrt{5}$ **33.** $4\sqrt{3} - 4\sqrt{7}$ **35.** $8\sqrt{2}$

37. $11\sqrt{3}$ **39.** $-5\sqrt{2}$ **41.** $-12\sqrt{2}$ **43.** 0 **45.** $8\sqrt{3x}$ **47.** $29\sqrt{2x}$ **49.** $2\sqrt{11} - \sqrt{7x}$ **51.** $6x\sqrt{2x}$ **53.** $11\sqrt[3]{2}$

55. $-10xy\sqrt[3]{y} + 6xy^2$ **57.** $20.78460969 = 20.78460969$ **59.** 7.071 amps **61.** 3.14 seconds **63.** $x(4x - 7y)^2$

64. $y(9x+5)(9x-5)$ **65.** 2.5 servings of scallops and 2 servings of skim milk **66.** 5 servings of scallops; 4 servings of skim milk

67. approximately 18.6% **68.** approximately 0.37% per year **69.** approximately 8.8%

7.4 Exercises

1. $\sqrt{35}$ **3.** $-30\sqrt{10}$ **5.** $-12\sqrt{3}$ **7.** $-3\sqrt{5xy}$ **9.** $-12x^2\sqrt{5y}$ **11.** $15\sqrt{ab} - 25\sqrt{a}$ **13.** $-2\sqrt{2xy} + 6\sqrt{5y}$

15. $-a + 2\sqrt{ab}$ **17.** $14\sqrt{3x} - 35x$ **19.** $22 - 5\sqrt{2}$ **21.** $4 - 6\sqrt{6}$ **23.** $14 + 11\sqrt{35x} + 60x$ **25.** $\sqrt{15} + 3 + 2\sqrt{10} + 2\sqrt{6}$

27. $29 - 4\sqrt{30}$ **29.** $36 - 60\sqrt{a} + 25a$ **31.** $3x + 13 + 6\sqrt{3x+4}$

33. $3x\sqrt[3]{4} - 4x^2\sqrt[3]{x}$ **35.** $1 + \sqrt[3]{18} - \sqrt[3]{12}$ **37.** $\dfrac{7}{5}$ **39.** $\dfrac{2\sqrt{3x}}{7y^3}$ **41.** $\dfrac{2xy^2\sqrt[3]{x^2}}{3}$ **43.** $\dfrac{y^2\sqrt[3]{5y^2}}{3x}$ **45.** $\dfrac{3\sqrt{2}}{2}$ **47.** $\dfrac{2\sqrt{3}}{3}$

49. $\dfrac{\sqrt{5y}}{5y}$ **51.** $\dfrac{\sqrt{7ay}}{y}$ **53.** $\dfrac{\sqrt{3x}}{3x}$ **55.** $\dfrac{x\left(\sqrt{5}+\sqrt{2}\right)}{3}$ **57.** $2y\left(\sqrt{6}-\sqrt{5}\right)$ **59.** $\dfrac{x\sqrt{3} - \sqrt{2x}}{3x - 2}$ **61.** $4 + \sqrt{15}$

63. $\dfrac{3x - 3\sqrt{3xy} + 2y}{3x - y}$ **65.** $-12\sqrt{2}$ **67.** $-24 + \sqrt{6}$ **69.** $\dfrac{9\sqrt{2x}}{4x}$ **71.** $3 - \sqrt{5}$ **73.** $1.194938299, 1.194938299$; yes; yes

75. $-\dfrac{25}{8\left(\sqrt{3} - 2\sqrt{7}\right)}$ **77.** \$2.92 **79.** $\left(x + 8\sqrt{x} + 15\right)$ square millimeters **81.** $x = 2, y = 3$ **82.** $x = 1, y = -5, z = 3$

83. January 11 **84.** 5 cups of coffee and 6 cups of tea; on January 22 **85.** 72% **86.** 72,250 rings

How Am I Doing? Sections 7.1–7.4

1. $\dfrac{6y^{5/6}}{x^{1/4}}$ (obj. 7.1.1)   **2.** $-\dfrac{64y}{x^{3/4}}$ (obj. 7.1.1)   **3.** $6x^3y^{5/3}$ (obj. 7.1.1)   **4.** $\dfrac{9x^4}{y^6}$ (obj. 7.1.1)   **5.** $\dfrac{1}{81}$ (obj. 7.2.3)   **6.** -3 (obj. 7.2.1)

7. 9 (obj. 7.2.1)   **8.** $7x^3y^{10}$ (obj. 7.2.4)   **9.** $3a^4b^2c^5$ (obj. 7.2.4)   **10.** $(4x)^{5/6}$ (obj. 7.2.2)   **11.** $2x^5y^7$ (obj. 7.3.1)

12. $2x^2y^5\sqrt[3]{4x^2}$ (obj. 7.3.1)   **13.** $3\sqrt{11}$ (obj. 7.3.2)   **14.** $2y\sqrt{3y}+5\sqrt[3]{2}$ (obj. 7.3.2)   **15.** $108-57\sqrt{2}$ (obj. 7.4.1)

16. $\dfrac{3\sqrt{5x}}{5x}$ (obj. 7.4.3)   **17.** $-5-2\sqrt{6}$ (obj. 7.4.3)

7.5 Exercises

1. Isolate one of the radicals on one side of the equation.   **3.** $x=3$ **5.** $x=1$ **7.** $y=2;y=1$ **9.** $x=1$ **11.** No solution

13. $y=7$ **15.** $y=0,y=-1$ **17.** $x=3,x=7$ **19.** $x=0,x=\dfrac{1}{2}$ **21.** $x=\dfrac{5}{2}$ **23.** $x=7$ **25.** $x=12$

27. $x=3,x=0$ **29.** $x=\dfrac{1}{4}$ **31.** $x=5$ **33.** $x=0,x=8$ **35.** No solution   **37.** $x=9$ **39.** $x=4.9232,x=0.4028$

41. (a) $S=\dfrac{V^2}{12}$ **(b)** 27 feet   **43.** $x=0.055y^2+1.25y-10$ **45.** $c=9$ **47.** $16x^4$ **48.** $\dfrac{1}{2x^2}$ **49.** $-6x^2y^3$ **50.** $-2x^3y$

51. $(8x^3+16x^2+24x+27)$ cubic centimeters   **52.** $(4r^3+18r^2+26r+12)$ boxes   **53.** 3 miles per hour   **54.** 15 miles per hour

7.6 Exercises

1. No. There is no real number that, when squared, will equal -9.
3. No. To be equal, the real number parts must be equal, and the imaginary parts must be equal. $2\neq3$ and $3i\neq2i$ **5.** $5i$ **7.** $5i\sqrt{2}$

9. $\dfrac{2}{7}i$ **11.** $-9i$ **13.** $2+i\sqrt{3}$ **15.** $-1.5+9i$ **17.** $-3+2i\sqrt{6}$ **19.** $-\sqrt{6}$ **21.** -12 **23.** $x=5;y=-3$

25. $x=1.3;y=2$ **27.** $x=-6,y=3$ **29.** $-5+11i$ **31.** $1-i$ **33.** $1.2+2.1i$ **35.** -12 **37.** 42   **39.** $7+4i$

41. $8+3i$ **43.** $-10-12i$ **45.** $-\dfrac{3}{4}+i$ **47.** $-\sqrt{21}$ **49.** $12-\sqrt{10}+4i\sqrt{2}+3i\sqrt{5}$ **51.** i **53.** 1   **55.** -1 **57.** i

59. 0   **61.** $1+i$ **63.** $\dfrac{1+i}{2}$ **65.** $\dfrac{3+6i}{10}$ **67.** $-\dfrac{2+5i}{6}$ **69.** $-2i$ **71.** $\dfrac{35+42i}{61}$ **73.** $\dfrac{11-6i}{13}$ **75.** $7i\sqrt{2}$ **77.** $6-12i$

79. $-22-34i$ **81.** $\dfrac{1-8i}{5}$ **83.** $-2299.95+3293.32i$ **85.** $Z=\dfrac{2-3i}{3}$

87. 18 hours producing juice in glass bottles, 25 hours producing juice in cans, 62 hours producing juice in plastic bottles   **88.** $96,030

7.7 Exercises

1. Answers will vary. A person's weekly paycheck varies as the number of hours worked. $y=kx$, y is the weekly salary, k is the hourly salary, x is the number of hours.   **3.** $y=\dfrac{k}{x}$ **5.** $y=24$ **7.** 71.4 pounds per square inch   **9.** 160 feet   **11.** $y=160$ **13.** 3316 gallons

15. 30 miles per hour   **17.** 400 pounds   **19.** 980 pounds   **21.** 62.7 miles per hour   **23.** $x=\dfrac{2}{3},x=2$ **24.** $x=1,x=-8$

25. $460   **26.** 55 gallons   **27.** 65 gold leaf frames, 45 silver frames
28. first side is 16 centimeters; second side is 20 centimeters; third side is 14 centimeters

Putting Your Skills to Work

1. 1.059463094   **2.** 466.1637615 hertz   **3.** 415.3046976 hertz   **4.** 523.26 hertz   **5.** The frequency of the higher note is twice the frequency of the lower note. Yes.   **6.** 130.82 hertz   **7.** 220 hertz

Chapter 7 Review Problems

1. $\dfrac{15x^3}{y^{5/2}}$ **2.** $\dfrac{1}{2}x^{1/2}$ or $\dfrac{x^{1/2}}{2}$ **3.** $5a^{3/2}b^2$ **4.** $5^{3/4}$ **5.** $-6a^{5/6}b^{3/4}$ **6.** $\dfrac{x^{1/2}y^{3/10}}{2}$ **7.** $\dfrac{x}{32y^{1/2}z^4}$ **8.** $7a^5b$ **9.** $x^{1/2}y^{1/5}$

10. $3x^{n+1}$ **11.** $5^{12/7}$ **12.** $\dfrac{2x+1}{x^{2/3}}$ **13.** $3x(2x^{1/2}-3x^{-1/2})$ **14.** -4 **15.** -2 **16.** Not a real number   **17.** $-\dfrac{1}{5}$ **18.** 0.2

19. Not a real number   **20.** $-\dfrac{1}{2}$ **21.** $\dfrac{3}{4}$ **22.** 16   **23.** 625   **24.** $9xy^3z^5$ **25.** $5a^3b^{20}$ **26.** $-2a^4b^5c^7$ **27.** $7x^{11}y$ **28.** $a^{2/5}$

29. $y^{3/4}$ **30.** $(2b)^{1/2}$ **31.** $(6c)^{1/3}$ **32.** $(ab)^{5/6}$ **33.** $\sqrt{m}$ **34.** $\sqrt[4]{n}$ **35.** $\sqrt[5]{y^3}$ **36.** $\sqrt[3]{9z^2}$ **37.** $\sqrt[7]{8x^3}$ **38.** 8   **39.** 32

40. 9   **41.** -2 **42.** $\dfrac{1}{3}$ **43.** 0.7   **44.** 2   **45.** 6   **46.** $125a^3b^6$ **47.** $32a^{15}b^5$ **48.** $11\sqrt{2}$ **49.** $13\sqrt{7}$ **50.** $15\sqrt{2}$

51. $5x\sqrt{10x}$ **52.** $11\sqrt{2x}-5x\sqrt{2}$ **53.** $-6\sqrt[3]{2}$ **54.** $90\sqrt{2}$ **55.** $-24x\sqrt{5}$ **56.** $12x\sqrt{2}-36\sqrt{3x}$ **57.** $2\sqrt{5a}-5a\sqrt{3}$

58. $-x\sqrt{6y}+3y\sqrt{2x}$ **59.** $2b\sqrt{7a}-2b^2\sqrt{21c}$ **60.** $4-9\sqrt{6}$ **61.** $34-14\sqrt{3}$ **62.** $74-12\sqrt{30}$

63. $2x+2\sqrt[3]{3x^2}-\sqrt[3]{2xy}-\sqrt[3]{6y}$ **64. (a)** $f(16)=10$ **(b)** all real numbers x where $x\geq-4$

65. (a) $f(5)=4$ **(b)** all real numbers x where $x\leq9$ **66. (a)** $f(1)=\dfrac{1}{2}$ **(b)** all real numbers x where $x\geq\dfrac{2}{3}$ **67.** $\dfrac{x\sqrt{3y}}{y}$

68. $\dfrac{2\sqrt{3y}}{3y}$ **69.** $\sqrt{3}$ **70.** $2\sqrt{6} + 2\sqrt{5}$ **71.** $\dfrac{3x - \sqrt{xy}}{9x - y}$ **72.** $-\dfrac{\sqrt{35} + 3\sqrt{5}}{2}$ **73.** $\dfrac{2 + 3\sqrt{2}}{7}$

74. $\dfrac{10\sqrt{3} - 3\sqrt{2} + 5\sqrt{6} - 3}{3}$ **75.** $\dfrac{3x + 4\sqrt{xy} + y}{x - y}$ **76.** $\dfrac{\sqrt[3]{4x^2y}}{2y}$ **77.** $4i + 3i\sqrt{5}$ **78.** $x = \dfrac{-7 + \sqrt{6}}{2}; y = -3$

79. $-9 - 11i$ **80.** $-10 + 2i$ **81.** $29 - 29i$ **82.** $48 - 64i$ **83.** $-8 + 6i$ **84.** $-5 - 4i$ **85.** -1 **86.** i **87.** $\dfrac{13 - 34i}{25}$

88. $\dfrac{11 + 13i}{10}$ **89.** $-\dfrac{3 + 4i}{5}$ **90.** $\dfrac{18 + 30i}{17}$ **91.** $-2i$ **92.** $x = 9$ **93.** $x = 3$ **94.** $x = 4$ **95.** $x = 5$ **96.** $x = 5, x = 1$

97. $x = 1, x = \dfrac{3}{2}$ **98.** $y = 9.6$ **99.** $y = 12.5$ **100.** 168.1 feet   **101.** 3.5 seconds   **102.** $y = 0.5$

103. 16.8 pounds per square inch   **104.** $y = 1.3$ **105.** 160 cubic centimeters

How Am I Doing? Chapter 7 Test

1. $-6x^{5/6}y^{1/2}$ (obj. 7.1.1)   **2.** $\dfrac{7x^{9/4}}{4}$ (obj. 7.1.1)   **3.** $8^{3/2}x^{1/2}$ or $16(2x)^{1/2}$ (obj. 7.1.1)   **4.** $\dfrac{8}{27}$ (obj. 7.1.1)   **5.** -2 (obj. 7.2.1)

6. $\dfrac{1}{4}$ (obj. 7.1.1)   **7.** 32 (obj. 7.1.1)   **8.** $5a^2b^4\sqrt{3b}$ (obj. 7.3.1)   **9.** $7a^2b^5$ (obj. 7.3.1)   **10.** $3mn\sqrt[3]{2n^2}$ (obj. 7.3.1)

11. $6\sqrt{6} + 2\sqrt{2}$ (obj. 7.3.2)   **12.** $2\sqrt{10x} + \sqrt{3x}$ (obj. 7.3.2)   **13.** $-30y\sqrt{5x}$ (obj. 7.4.1)   **14.** $18\sqrt{2} - 10\sqrt{6}$ (obj. 7.4.1)

15. $12 + 39\sqrt{2}$ (obj. 7.4.1)   **16.** $\dfrac{6\sqrt{5x}}{x}$ (obj. 7.4.3)   **17.** $\dfrac{\sqrt{3xy}}{3}$ (obj. 7.4.3)   **18.** $\dfrac{9 + 7\sqrt{3}}{6}$ (obj. 7.4.3)   **19.** $x = 2, x = 1$ (obj. 7.5.1)

20. $x = 10$ (obj. 7.5.1)   **21.** $x = 6$ (obj. 7.5.2)   **22.** $2 + 14i$ (obj. 7.6.2)   **23.** $-1 + 4i$ (obj. 7.6.2)   **24.** $18 + i$ (obj. 7.6.3)

25. $\dfrac{-13 + 11i}{10}$ (obj. 7.6.5)   **26.** $27 + 36i$ (obj. 7.6.3)   **27.** $-i$ (obj. 7.6.4)   **28.** $y = 3$ (obj. 7.7.2)   **29.** $y = \dfrac{5}{6}$ (obj. 7.7.3)

30. about 83.3 feet (obj. 7.7.1)

Cumulative Test for Chapters 1–7

1. Associative property of addition   **2.** $-2a^2b + 3a^4 + ab^3$ **3.** -64 **4.** $x = \dfrac{-4y + 8}{3}$

5.

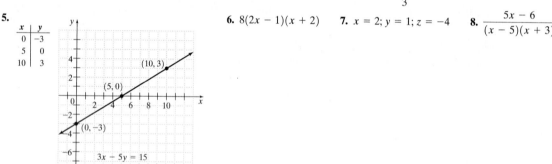

x	y
0	−3
5	0
10	3

$3x - 5y = 15$

6. $8(2x - 1)(x + 2)$ **7.** $x = 2; y = 1; z = -4$ **8.** $\dfrac{5x - 6}{(x - 5)(x + 3)}$

9. width = 7 meters; length = 17 meters   **10.** $a = \dfrac{y + 3b}{7x}$ **11.** $\dfrac{1}{2x^{1/2}y^{15/2}}$ **12.** $\dfrac{x^{1/6}}{3^{1/3}y^{2/3}}$ **13.** $\dfrac{1}{4}$ **14.** $2xy^3\sqrt[3]{5x^2}$

15. $4\sqrt{5x}$ **16.** $-34 + 3\sqrt{6}$ **17.** $-\dfrac{16 + 9\sqrt{3}}{13}$ **18.** $12i$ **19.** $-7 - 24i$ **20.** $\dfrac{13 + i}{10}$ **21.** $x = 8$ **22.** $x = -1$

23. $y = 75$ **24.** about 53.3 lumens

Chapter 8

8.1 Exercises

1. $x = \pm 10$ **3.** $x = \pm\sqrt{15}$ **5.** $x = \pm 2\sqrt{10}$ **7.** $x = \pm 9i$ **9.** $x = \pm 9i$ **11.** $x = 3 \pm 2\sqrt{3}$ **13.** $x = -9 \pm \sqrt{21}$

15. $x = \dfrac{-1 \pm \sqrt{7}}{2}$ **17.** $x = \dfrac{9}{4}, x = -\dfrac{3}{4}$ **19.** $x = 1, x = -6$ **21.** $x = \pm\dfrac{\sqrt{15}}{3}$ **23.** $x = -5 \pm 2\sqrt{5}$ **25.** $x = 4 \pm \sqrt{33}$

27. $x = 6, x = 8$ **29.** $x = \dfrac{-5 \pm \sqrt{41}}{2}$ **31.** $y = \dfrac{-5 \pm \sqrt{3}}{2}$ **33.** $x = \dfrac{-5 \pm \sqrt{31}}{3}$ **35.** $y = 2, y = -\dfrac{3}{2}$ **37.** $x = -1 \pm \sqrt{6}$

39. $x = 4, x = -2$ **41.** $x = \dfrac{1 \pm i\sqrt{11}}{6}$ **43.** $x = \dfrac{1 \pm i\sqrt{3}}{2}$ **45.** $x = \dfrac{3 \pm i\sqrt{7}}{4}$

47. $(-1 + \sqrt{6})^2 + 2(-1 + \sqrt{6}) - 5 \overset{?}{=} 0; 1 - 2\sqrt{6} + 6 - 2 + 2\sqrt{6} - 5 \overset{?}{=} 0; 0 = 0$ ✓   **49.** 16 feet   **51.** Approximately 0.88 second

53. 15 seconds   **55.** 8   **56.** 7   **57.** 40   **58.** 4

8.2 Exercises

1. Place the quadratic equation in standard form. Find a, b, and c. Substitute these values into the quadratic formula.   **3.** One real

5. $x = \dfrac{-1 \pm \sqrt{21}}{2}$ **7.** $x = \dfrac{-1 \pm \sqrt{33}}{4}$ **9.** $x = 0, x = \dfrac{2}{3}$ **11.** $x = 1, x = -\dfrac{2}{3}$ **13.** $x = \dfrac{-3 \pm \sqrt{41}}{8}$ **15.** $x = \dfrac{\pm\sqrt{21}}{3}$

17. $x = \dfrac{-1 \pm \sqrt{3}}{2}$ **19.** $x = \pm 3$ **21.** $x = \dfrac{2 \pm 3\sqrt{2}}{2}$ **23.** $x = 2 \pm \sqrt{10}$ **25.** $y = 9, y = 5$ **27.** $x = -2 \pm 2i\sqrt{2}$

29. $x = \dfrac{\pm i\sqrt{30}}{2}$ **31.** $x = \dfrac{4 \pm i\sqrt{5}}{3}$ **33.** Two irrational roots   **35.** Two rational roots   **37.** One rational root

39. Two nonreal complex roots   **41.** $x^2 - 11x - 26 = 0$ **43.** $x^2 + 17x + 60 = 0$ **45.** $x^2 + 16 = 0$ **47.** $2x^2 - x - 15 = 0$
49. $x = -2.7554, 1.0888$ **51.** $x = 1.4643, -2.0445$ **53.** 18 mountain bikes or 30 mountain bikes per day
55. The profit is \$3249 per day. 24 is the average of 18 and 30.   **57.** $-3x^2 - 10x + 11$ **58.** $-y^2 + 3y$
59. The width is 9 feet and length is 16 feet.   **60.** The suits cost \$95 and the goggles cost \$29 last year.

8.3 Exercises

1. $x = \pm\sqrt{5}, x = \pm 2$ **3.** $x = \pm\sqrt{3}, x = \pm 2i$ **5.** $x = \pm\dfrac{i\sqrt{6}}{3}, x = \pm 2$ **7.** $x = 2, x = -1$ **9.** $x = 0, x = \sqrt[3]{3}$

11. $x = \pm 2, x = \pm 1$ **13.** $x = \pm\dfrac{\sqrt[4]{54}}{3}$; these are the only real roots.   **15.** $x = -64, x = 27$ **17.** $x = \dfrac{1}{64}, x = -\dfrac{8}{27}$ **19.** $x = 81$

21. $x = 16$ **23.** $x = 1; x = -32$ **25.** $x = \sqrt[3]{7}, x = \sqrt[3]{-2}$ **27.** $x = -2, x = 1, x = \dfrac{-1 \pm \sqrt{13}}{2}$ **29.** $x = 9, x = 4$

31. $x = -\dfrac{1}{3}$ **33.** $x = \dfrac{5}{6}, x = \dfrac{3}{2}$ **35.** $x = -2, y = 3$ **36.** $\dfrac{15x + 6}{7 - 3x}$ **37.** $3\sqrt{10} - 12\sqrt{3}$ **38.** $6 + 6\sqrt{3} - 2\sqrt{10} - 2\sqrt{30}$

39. High school graduate 67.5%; Associate's degree 51.7%; Bachelor's degree 66.4%; Doctorate 71.0%   **40.** An additional 15.5 years
41. \$42,456.31 per performance   **42.** Approximately \$46.92 per person   **43.** \$316,341,965.80

How Am I Doing? Sections 8.1–8.3

1. $x = \pm 3\sqrt{2}$ (obj. 8.1.1)   **2.** $x = \dfrac{-4 \pm 2\sqrt{5}}{3}$ (obj. 8.1.1)   **3.** $x = 6, x = 2$ (obj. 8.1.2)   **4.** $x = \dfrac{2 \pm \sqrt{10}}{2}$ (obj. 8.1.2)

5. $x = \dfrac{1 \pm \sqrt{57}}{8}$ (obj. 8.2.1)   **6.** $x = -2 \pm \sqrt{11}$ (obj. 8.2.1)   **7.** $x = \dfrac{-3 \pm 2i\sqrt{2}}{2}$ (obj. 8.2.1)   **8.** $x = -2, x = \dfrac{6}{5}$ (obj. 8.2.1)

9. $x = 0, x = \dfrac{12}{7}$ (obj. 8.2.1)   **10.** $x = \dfrac{3 \pm i\sqrt{87}}{8}$ (obj. 8.2.1)   **11.** $x = -\dfrac{2}{3}, x = 3$ (obj. 8.2.1)   **12.** $x = 2, x = -1$ (obj. 8.3.1)

13. $w = \pm 8, w = \pm 2\sqrt{2}$ (obj. 8.3.2)   **14.** $x = \pm\sqrt[4]{3}, x = \pm\sqrt{2}$ (obj. 8.3.1)   **15.** $x = \dfrac{1}{32}, x = 243$ (obj. 8.3.2)

8.4 Exercises

1. $t = \pm\dfrac{\sqrt{S}}{4}$ **3.** $r = \pm\sqrt{\dfrac{S}{4\pi}}$ **5.** $x = \pm\sqrt{\dfrac{6H}{a}}$ **7.** $y = \pm\dfrac{\sqrt{7R - 4w + 5}}{2}$ **9.** $M = \pm\sqrt{\dfrac{2cQ}{3mw}}$ **11.** $r = \pm\sqrt{\dfrac{V - \pi R^2 h}{\pi h}}$

13. $x = 0, \dfrac{3a}{7b}$ **15.** $I = \dfrac{E \pm \sqrt{E^2 - 4RP}}{2R}$ **17.** $w = \dfrac{3q \pm \sqrt{9q^2 + 160}}{20}$ **19.** $r = \dfrac{-\pi h \pm \sqrt{\pi^2 h^2 + S\pi}}{\pi}$

21. $x = \dfrac{-5 \pm \sqrt{25 - 8aw - 8w}}{2a + 2}$ **23.** $b = 2\sqrt{5}$ **25.** $a = \sqrt{15}$ **27.** $b = \dfrac{24\sqrt{5}}{5}, a = \dfrac{12\sqrt{5}}{5}$ **29.** The legs are $5\sqrt{2}$ inches long.

31. The width is 0.13 miles and the length is 0.2 miles.   **33.** Width is 7 feet; length is 18 feet.   **35.** Base is 8 centimeters; altitude is
18 centimeters.   **37.** The rate of the car in rain was 45 mph; the rate of the car without rain was 50 mph.   **39.** 30 miles   **41.** 1,659,250

43. In the year 2004   **45.** $w = \dfrac{-7b - 21 \pm \sqrt{529b^2 + 294b + 441}}{10}$ **47.** $\dfrac{4\sqrt{3x}}{3x}$ **48.** $\dfrac{\sqrt{30}}{2}$ **49.** $\dfrac{3(\sqrt{x} - \sqrt{y})}{x - y}$

50. $-2 - 2\sqrt{2}$ **51.** $\dfrac{3\sqrt[3]{a^2 b}}{2}$

8.5 Exercises

1. $V(1, -9); I(0, -8); (4, 0); (-2, 0)$ **3.** $V(-2, 16); I(0, 12); (-6, 0); (2, 0)$ **5.** $V(-2, -9); I(0, 3); (-0.3, 0); (-3.7, 0)$

7. $V\left(-\dfrac{1}{3}, -\dfrac{17}{3}\right)$; no x-intercepts; y-intercept $(0, -6)$ **9.** $V\left(-\dfrac{1}{2}, -\dfrac{9}{2}\right); I(0, -4); (1, 0); (-2, 0)$

11.

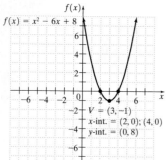

$f(x) = x^2 - 6x + 8$
$V = (3, -1)$
x-int. = (2, 0); (4, 0)
y-int. = (0, 8)

13.

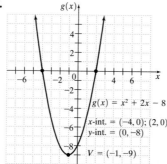

$g(x) = x^2 + 2x - 8$
x-int. = (-4, 0); (2, 0)
y-int. = (0, -8)
$V = (-1, -9)$

15.

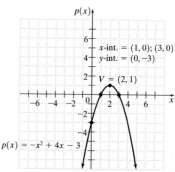

x-int. = (1, 0); (3, 0)
y-int. = (0, -3)
$V = (2, 1)$
$p(x) = -x^2 + 4x - 3$

17.

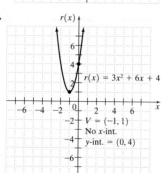

$r(x) = 3x^2 + 6x + 4$
$V = (-1, 1)$
No x-int.
y-int. = (0, 4)

19.

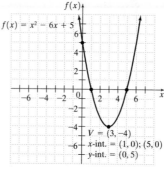

$f(x) = x^2 - 6x + 5$
$V = (3, -4)$
x-int. = (1, 0); (5, 0)
y-int. = (0, 5)

21.

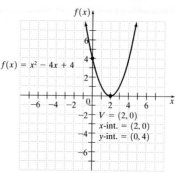

$f(x) = x^2 - 4x + 4$
$V = (2, 0)$
x-int. = (2, 0)
y-int. = (0, 4)

23.

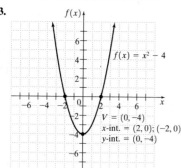

$f(x) = x^2 - 4$
$V = (0, -4)$
x-int. = (2, 0); (-2, 0)
y-int. = (0, -4)

25. $N(20) = 110{,}650; N(40) = 263{,}050; N(60) = 559{,}450; N(80) = 999{,}850; N(100) = 1{,}584{,}250;$
27. $N(70) \approx 750{,}000.$ Approximately 750,000 people scuba dive and have a mean income of \$70,000.   **29.** Approximately 50. This means that 390,000 people who scuba dive have a mean income of \$50,000.
31. $P(16) = -216; P(20) = 168; P(24) = 360; P(30) = 288; P(35) = -102$
33. 26 tables per day will give the maximum profit of \$384 per day.
35. 18 tables per day or 34 tables per day.
37. The maximum height is 56 feet. It will take about 2.9 seconds.
39. Vertex $(2.2, 2.75)$; y-intercept $(0, 7.59)$; no x-intercepts

41. x-intercepts $(-0.3, 0)$ and $(2.6, 0)$

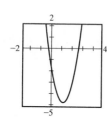

43. $a = 5, b = -6, c = 2$ **45.** $(-1, 3)$ **46.** $(11, 5)$
47. $(3, 1, 2)$ **48.** $(3, -5, 8)$

8.6 Exercises

1. The critical points divide the number line into regions. All values of x in a given region produce results that are greater than zero or else all the values of x in a given region produce results that are less than zero.
3. $-4 < x < 3$    **5.** $x \leq -2$ or $x \geq 2$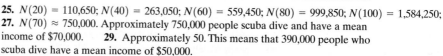

7. $-\dfrac{3}{2} < x < 1$    **9.** $x < -5$ or $x > 4$ **11.** $-\dfrac{1}{2} \leq x \leq \dfrac{3}{4}$ **13.** $x < -\dfrac{2}{3}$ or $x > \dfrac{3}{2}$

15. $-10 \leq x \leq 3$ **17.** All real numbers   **19.** $x = 2$ **21.** Approximately $x < -1.2$ or $x > 3.2$ **23.** Approximately $1.6 < x < 4.4$
25. All real numbers   **27.** No real number   **29.** For time t greater than 15 seconds but less than 25 seconds
31. (a) Approximately $11.5 < x < 208.5$ **(b)** \$122,000 **(c)** \$144,000   **33.** She must score a combined total of 167 points on the two tests. Any two test scores that total 167 will be sufficient to participate in synchronized swimming.   **34.** 52 ounces of potato chips, 122 ounces of peanuts, 62 ounces of popcorn, and 124 ounces of pretzels.   **35.** \$400 for the 2-hour trip, \$483 for the 3-hour trip.
36. 9 adults, 12 children, 2 elderly

Putting Your Skills to Work

1. $P(0) = 150.4$
$P(15) = 184.45$

2. $P(10) = 170.8$
$P(20) = 200.4$

3.

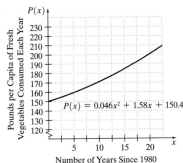

$P(x) = 0.046x^2 + 1.58x + 150.4$

Pounds per Capita of Fresh Vegetables Consumed Each Year

Number of Years Since 1980

4. $P(15) \approx 184$ **5.** 36.62 pounds
6. 43.34 pounds   **7.** 2008   **8.** 2010

Chapter 8 Review Problems

1. $x = \pm 2$ **2.** $x = 1, -17$ **3.** $x = -4 \pm \sqrt{3}$ **4.** $x = \dfrac{2 \pm \sqrt{3}}{2}$ or $1 \pm \dfrac{\sqrt{3}}{2}$ **5.** $x = \dfrac{5 \pm \sqrt{7}}{3}$ **6.** $x = 3 \pm \sqrt{13}$ **7.** $x = \dfrac{3}{2}$

8. $x = 7, -2$ **9.** $x = 0, \dfrac{9}{2}$ **10.** $x = \dfrac{5 \pm \sqrt{17}}{4}$ **11.** $x = 7, -4$ **12.** $x = \pm\sqrt{2}$ **13.** $x = \dfrac{2}{3}, 5$ **14.** $x = -1 \pm \sqrt{5}$

15. $x = \pm 2i\sqrt{3}$ **16.** $x = \dfrac{-5 \pm \sqrt{13}}{6}$ **17.** $x = -\dfrac{2}{3}, \dfrac{1}{3}$ **18.** $x = \dfrac{11 \pm \sqrt{21}}{10}$ **19.** $x = \pm\dfrac{\sqrt{5}}{3}$ **20.** $x = -\dfrac{1}{4}, -1$

21. $y = -\dfrac{5}{6}, -2$ **22.** $y = -\dfrac{5}{2}, \dfrac{3}{5}$ **23.** $y = -5, 3$ **24.** $y = 9$ **25.** $y = -3, 1$ **26.** $y = 0, -\dfrac{5}{2}$ **27.** $x = -2, 3$

28. $x = -3, 2$ **29.** Two rational solutions   **30.** Two irrational solutions   **31.** Two nonreal complex solutions
32. One rational solution   **33.** $x^2 - 25 = 0$ **34.** $x^2 + 9 = 0$ **35.** $x^2 - 32 = 0$ **36.** $8x^2 + 10x + 3 = 0$ **37.** $x = \pm 2, \pm\sqrt{2}$

38. $x = -\dfrac{\sqrt[3]{4}}{2}, \sqrt[3]{3}$ **39.** $x = 27, -1$ **40.** $x = 1$ **41.** $x = 1, 2$ **42.** $x = \pm 1, \pm\sqrt{2}$ **43.** $A = \pm\sqrt{\dfrac{3MN}{2}}$ **44.** $t = \pm\sqrt{3ay - 2b}$

45. $x = \dfrac{3 \pm \sqrt{9 + 28y}}{2y}$ **46.** $d = \dfrac{x}{4}, -\dfrac{x}{5}$ **47.** $y = \dfrac{2a \pm \sqrt{4a^2 - 6a}}{3}$ **48.** $x = \dfrac{-1 \pm \sqrt{1 - 15y^2 + 5PV}}{5}$ **49.** $c = \sqrt{22}$

50. $a = 4\sqrt{15}$ **51.** The car is approximately 3.3 miles from the observer.   **52.** Base is 7 cm; altitude is 20 cm
53. Width is 7 m; length is 29 m   **54.** 20 mph cruising; 5 mph trolling   **55.** 50 mph during first part, 45 mph during rain
56. The walkway should be approximately 2.4 feet wide.   **57.** The walkway should be 2 feet wide.   **58.** Vertex $(3, -2)$;
y-intercept $(0, -11)$; no x-intercepts   **59.** Vertex $(-5, 0)$; one x-intercept $(-5, 0)$; y-intercept $(0, 25)$

60.

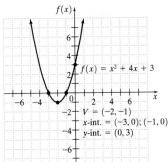

$f(x) = x^2 + 4x + 3$
$V = (-2, -1)$
x-int. $= (-3, 0); (-1, 0)$
y-int. $= (0, 3)$

61.

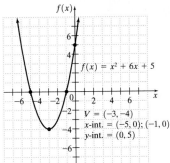

$f(x) = x^2 + 6x + 5$
$V = (-3, -4)$
x-int. $= (-5, 0); (-1, 0)$
y-int. $= (0, 5)$

62.

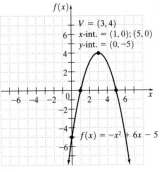

$V = (3, 4)$
x-int. $= (1, 0); (5, 0)$
y-int. $= (0, -5)$
$f(x) = -x^2 + 6x - 5$

63. The maximum height is 2540 feet. The amount of time for the complete flight is 25.1 seconds.
64. $R(x) = x(1200 - x)$; the maximum revenue will occur if the price is \$600 for each unit.
65. $-9 < x < 2$
66. $-7 < x < 3$
67. $x < 4$ or $x > 5$ **68.** $x < 4$ or $x > 7$
69. $-\dfrac{1}{3} \le x \le 2$ **70.** $-\dfrac{1}{2} \le x \le 3$ **71.** $x < -\dfrac{2}{3}$ or $x > \dfrac{2}{3}$ **72.** $x < -\dfrac{5}{4}$ or $x > \dfrac{5}{4}$ **73.** $x \le -6$ or $x \ge -2$
74. $x < \left(1 - \sqrt{5}\right)$ or $x > \left(1 + \sqrt{5}\right)$; approximately $x < -1.2$ or $x > 3.2$ **75.** $x < -8$ or $x > 2$ **76.** $x < 1.4$ or $x > 2.6$
77. No real solution   **78.** No real solution   **79.** $x < -4$ or $2 < x < 3$ **80.** $-4 < x < -1$ or $x > 2$

How Am I Doing? Chapter 8 Test

1. $x = 0, -\dfrac{9}{8}$ (obj. 8.1.2)   **2.** $x = \dfrac{3 \pm \sqrt{33}}{12}$ (obj. 8.1.2)   **3.** $x = 2, -\dfrac{2}{9}$ (obj. 8.2.1)   **4.** $x = -2, 10$ (obj. 8.2.1)

5. $x = \pm 2\sqrt{2}$ (obj. 8.1.1)   **6.** $x = \dfrac{7}{2}, -1$ (obj. 8.2.1)   **7.** $x = \dfrac{3 \pm i}{2}$ (obj. 8.2.1)   **8.** $x = \dfrac{3 \pm \sqrt{3}}{2}$ (obj. 8.2.1)

9. $x = \pm 3, \pm \sqrt{2}$ (obj. 8.3.1)   **10.** $x = \dfrac{1}{5}, -\dfrac{3}{4}$ (obj. 8.3.1)   **11.** $x = 64, -1$ (obj. 8.3.2)   **12.** $z = \pm\sqrt{\dfrac{xyw}{B}}$ (obj. 8.4.1)

13. $y = \dfrac{-b \pm \sqrt{b^2 - 30w}}{5}$ (obj. 8.4.1)   **14.** Width is 5 miles; length is 16 miles. (obj. 8.4.3)   **15.** $c = 4\sqrt{3}$ (obj. 8.4.2)

16. 2 mph during first part; 3 mph after lunch (obj. 8.4.3)

17. $V = (-3, 4)$; y-int. $(0, -5)$; x-int. $(-5, 0)$; $(-1, 0)$ (obj. 8.5.2)

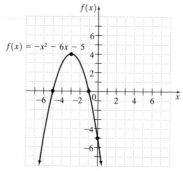

$f(x) = -x^2 - 6x - 5$

18. $x \le -\dfrac{9}{2}$ or $x \ge 3$ (obj. 8.6.1)   **19.** $-2 < x < 7$ (obj. 8.6.1)   **20.** $x < -4.5$ or $x > 1.5$ (obj. 8.6.2)

Cumulative Test for Chapters 1–8

1. $\dfrac{81y^{12}}{x^8}$ **2.** $\dfrac{1}{4}a^3 - a^2 - 3a$ **3.** $x = 12$ **4.**

x	y
0	4
-2	0

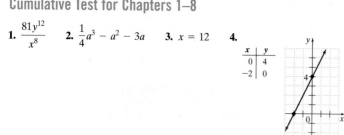

5. $x + 2y = 4$ **6.** $x = 1, y = 2$

7. $(5x - 3y)(25x^2 + 15xy + 9y^2)$ **8.** $6xy^3\sqrt{2x}$ **9.** $4\sqrt{6} + 5\sqrt{3}$ **10.** $\dfrac{x\sqrt{6}}{2}$ **11.** $x = 0, \dfrac{14}{3}$ **12.** $x = \dfrac{2}{3}, \dfrac{1}{4}$

13. $x = \dfrac{3 \pm 2\sqrt{3}}{2}$ **14.** $x = \dfrac{2 \pm i\sqrt{11}}{3}$ **15.** $x = 3$ **16.** $x = -27, -216$ **17.** $y = \dfrac{-5w \pm \sqrt{25w^2 + 56z}}{4}$

18. $y = \dfrac{\pm\sqrt{15w - 48z^2}}{3}$ **19.** $\sqrt{15}$ **20.** Base is 5 meters; altitude is 18 meters.

21. Vertex $(4, 4)$; y-intercept $(0, -12)$; x-intercepts $(2, 0)$; $(6, 0)$ **22.**

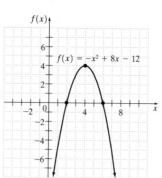

$f(x) = -x^2 + 8x - 12$

23. $-\dfrac{1}{2} \le x \le \dfrac{2}{3}$ **24.** $x < -5$ or $x > 3$

Chapter 9

9.1 Exercises

1. Subtract the value of the points and use the absolute value: $|-2 - 4| = 6$ **3.** Since the equation is given in standard form, we determine the values of h, k and r to find the center and radius. $h = 1$, $k = -2$, and $r = 3$. Thus, the center is $(1, -2)$ and the radius is 3.

5. $\sqrt{5}$ **7.** $4\sqrt{2}$ **9.** 10   **11.** $\dfrac{\sqrt{10}}{3}$ **13.** $\dfrac{2\sqrt{26}}{5}$ **15.** $5\sqrt{2}$ **17.** $y = 10, y = -6$ **19.** $y = 0, y = 4$ **21.** $x = 6, x = 8$

23. 4.4 miles   **25.** $(x + 3)^2 + (y - 7)^2 = 36$ **27.** $(x + 1.8)^2 + y^2 = \dfrac{4}{25}$ **29.** $\left(x - \dfrac{3}{8}\right)^2 + y^2 = 3$

31. 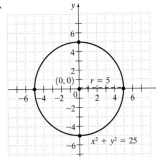   **33.** 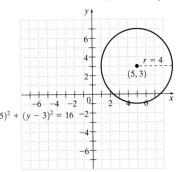   **35.**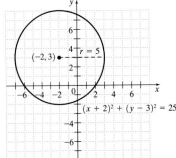

37. $(x + 4)^2 + (y - 3)^2 = 49$; center $(-4, 3)$, $r = 7$ **39.** $(x - 6)^2 + (y + 1)^2 = 49$; center $(6, -1)$, $r = 7$

41. $\left(x + \dfrac{3}{2}\right)^2 + y^2 = \dfrac{17}{4}$; center $\left(-\dfrac{3}{2}, 0\right)$, $r = \dfrac{\sqrt{17}}{2}$ **43.** $(x - 44.8)^2 + (y - 31.8)^2 = 640.09$ **45.**

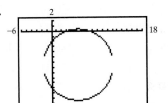

47. $x = -\dfrac{2}{3}, x = \dfrac{1}{3}$ **48.** $x = \dfrac{2}{3}, x = 1$ **49.** $x = \dfrac{-1 \pm \sqrt{5}}{4}$ **50.** $x = \dfrac{3 \pm 2\sqrt{11}}{5}$ **51.** 8.364×10^{10} cubic feet

52. approximately 81 seconds

9.2 Exercises

1. y-axis, x-axis   **3.** If it is in the standard form $y = a(x - h)^2 + k$, the vertex is (h, k). So in this case the vertex is $(3, 4)$.

5. 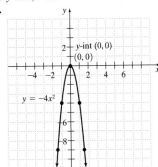   **7.** 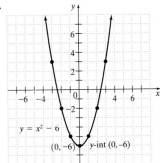   **9.**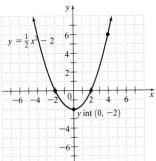

11. 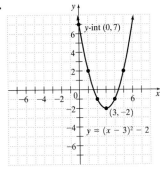   **13.** 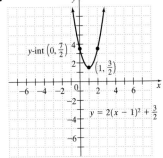   **15.**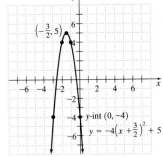

17.

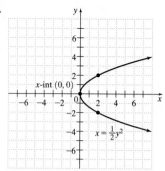

19.

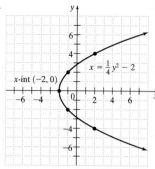

21.

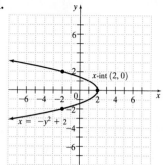

23.

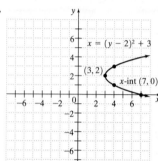

25.

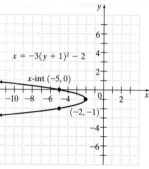

27. $y = (x - 2)^2 - 5$ **(a)** vertical   **(b)** opens upward   **(c)** vertex $(2, -5)$ **29.** $y = -2(x - 1)^2 + 7$ **(a)** vertical   **(b)** opens downward   **(c)** vertex $(1, 7)$ **31.** $x = (y + 4)^2 - 7$ **(a)** horizontal   **(b)** opens right   **(c)** vertex $(-7, -4)$

33. $y = \dfrac{1}{32}x^2$ **35.** 8 inches   **37.** vertex $(-1.62, -5.38)$; y-intercept $(0, -0.1312)$; x-intercepts $(0.020121947, 0)$ and $(-3.260121947, 0)$

39. maximum profit $= \$36,000$; number of items produced $= 40$ **41.** maximum yield $= 202,500$; number of trees planted per acre $= 450$

43. $5x\sqrt{2x}$ **44.** $2xy\sqrt[3]{5y}$ **45.** $2x\sqrt{2} - 8\sqrt{2x}$ **46.** $2x\sqrt[3]{2x} - 20x\sqrt[3]{2}$ **47.** approximately 42.2 miles   **48.** $27\dfrac{1}{3}$ miles

49. 7392 blooms   **50.** approximately 1098 buds

9.3 Exercises

1. In the ellipse $\dfrac{(x - h)^2}{a^2} + \dfrac{(y - k)^2}{b^2} = 1$ the center is (h, k). In this case the center of the ellipse is $(-2, 3)$.

3.

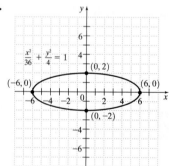

5.

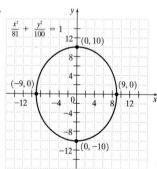

7.

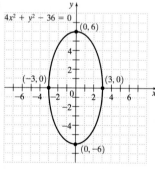

9.

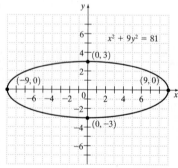

11.

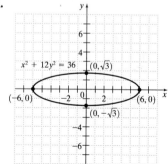

13.

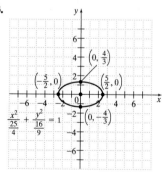

15.

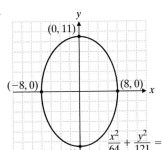

(0, 11)

(−8, 0)   (8, 0)

x

(0, −11) $\dfrac{x^2}{64} + \dfrac{y^2}{121} = 1$

Scale: Each unit = 2

17. $\dfrac{x^2}{169} + \dfrac{y^2}{144} = 1$ **19.** $\dfrac{x^2}{81} + \dfrac{y^2}{18} = 1$ **21.** 142 million miles

23.

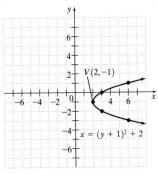

$\dfrac{(x-5)^2}{9} + \dfrac{(y-2)^2}{1} = 1$

(2, 2)   (5, 3)   C (5, 2)   (8, 2)   (5, 1)

25.

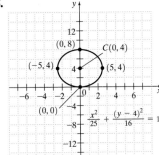

(0, 8)   C(0, 4)

(−5, 4)   (5, 4)

(0, 0) $\dfrac{x^2}{25} + \dfrac{(y-4)^2}{16} = 1$

27.

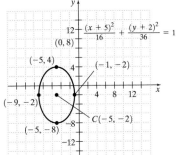

$\dfrac{(x+5)^2}{16} + \dfrac{(y+2)^2}{36} = 1$

(0, 8)   (−5, 4)   (−1, −2)   (−9, −2)   C(−5, −2)   (−5, −8)

29. $\dfrac{(x-4)^2}{4} + \dfrac{(y-3)^2}{16} = 1$ **31.** $\dfrac{(x-30)^2}{900} + \dfrac{(y-20)^2}{400} = 1$ **33.** (0, 7.2768), (0, 3.3232), (4.2783, 0), (2.9217, 0)

35. 22,376.0 square meters   **36.** $\dfrac{5\left(\sqrt{2x} + \sqrt{y}\right)}{2x - y}$ **37.** $30\sqrt{2} + 40\sqrt{3} - 2\sqrt{6} - 8$

38. $22\frac{2}{3}$ weeks   **39.** 7.9 inches

How Am I Doing? Sections 9.1–9.3

1. $(x - 8)^2 + (y + 2)^2 = 7$ (obj. 9.1.3)   **2.** $3\sqrt{5}$ (obj. 9.1.1)   **3.**

(obj. 9.1.4)

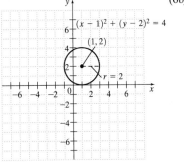

$(x - 1)^2 + (y - 2)^2 = 4$

(1, 2)

$r = 2$

4. The line $x = 3$. (obj. 9.2.1)   **5.** vertex $(-4, 6)$ (obj. 9.2.1)

6.

(obj. 9.2.2)   **7.**

(obj. 9.2.3)

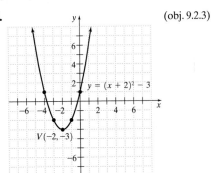

V(2, −1)

$x = (y + 1)^2 + 2$

$y = (x + 2)^2 - 3$

V(−2, −3)

8. $\dfrac{x^2}{100} + \dfrac{y^2}{49} = 1$ (obj. 9.3.1)

9. (obj. 9.3.1)

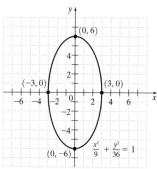

10. (obj. 9.3.2)

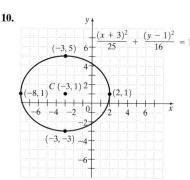

9.4 Exercises

1. The standard form of a horizontal hyperbola centered at the origin is $\dfrac{x^2}{a^2} - \dfrac{y^2}{b^2} = 1$ with a and b being positive real numbers.

3. This is a horizontal hyperbola, centered at the origin, with vertices at $(4,0)$ and $(-4,0)$. Draw a fundamental rectangle with corners at $(4,2)$, $(4,-2)$, $(-4,2)$, and $(-4,-2)$. Extend the diagonals through the rectangle as asymptotes of the hyperbola. Construct each branch of the hyperbola passing through the vertex and approaching the asymptotes.

5.

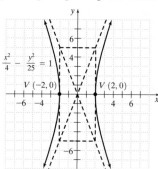

7.

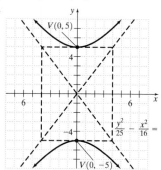

9.

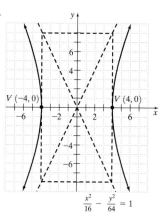

11.

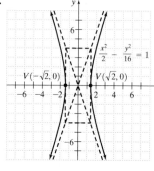

13.

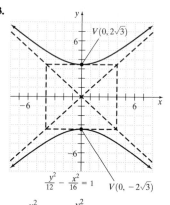

15. $\dfrac{x^2}{9} - \dfrac{y^2}{16} = 1$ **17.** $\dfrac{y^2}{121} - \dfrac{x^2}{169} = 1$ **19.** $\dfrac{x^2}{14,400} - \dfrac{y^2}{129,600} = 1$, where x and y are measured in millions of miles

21.

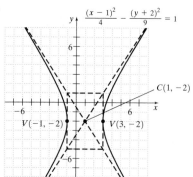

23.

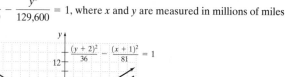
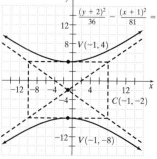

25. center $(-6,0)$ vertices $\left(-6 + \sqrt{7},0\right), \left(-6 - \sqrt{7},0\right)$ **27.** $\dfrac{(y+7)^2}{49} - \dfrac{(x-4)^2}{16} = 1$ **29.** $y = \pm 9.055385138$

31. $(4x + 3)(3x - 2)$ **32.** $2(x - 3)(x^2 + 3x + 9)$ **33.** $\dfrac{5x}{(x - 3)(x - 2)(x + 2)}$ **34.** $\dfrac{-x - 6}{(5x - 1)(x + 2)}$ or $-\dfrac{x + 6}{(5x - 1)(x + 2)}$

35. 20 people; 24 people   **36. (a)** 287 songs per day   **(b)** 292 minutes per day   **(c)** approximately 79.7%
37. 640,000,000 pencils   **38.** approximately 38.7%

9.5 Exercises

1.
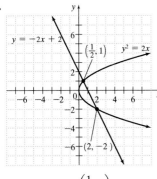

$(2, -2), \left(\dfrac{1}{2}, 1\right)$

3.
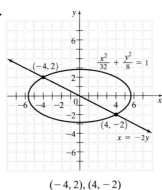

$(-4, 2), (4, -2)$

5. $(-2, 3), (1, 0)$ **7.** $(-5, 0), (4, 3)$

9. $\left(\dfrac{2}{3}, \dfrac{4}{3}\right), (2, 0)$ **11.** $\left(\dfrac{5}{2}, \dfrac{3}{2}\right)$ **13.** $(3, 2), (-3, 2), (3, -2), (-3, -2)$

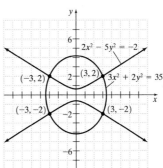

15. $\left(2, \sqrt{5}\right), \left(2, -\sqrt{5}\right), \left(-2, \sqrt{5}\right), \left(-2, -\sqrt{5}\right)$ **17.** $\left(\dfrac{\sqrt{30}}{3}, \dfrac{\sqrt{21}}{3}\right), \left(-\dfrac{\sqrt{30}}{3}, \dfrac{\sqrt{21}}{3}\right), \left(\dfrac{\sqrt{30}}{3}, -\dfrac{\sqrt{21}}{3}\right), \left(-\dfrac{\sqrt{30}}{3}, -\dfrac{\sqrt{21}}{3}\right)$

19. $\left(2, \sqrt{3}\right), \left(-2, \sqrt{3}\right), \left(2, -\sqrt{3}\right), \left(-2, -\sqrt{3}\right)$ **21.** $\left(5, \dfrac{1}{2}\right), \left(-2, -\dfrac{5}{4}\right)$ **23.** $(-3, 2), (1, -6)$ **25.** no real solution

27. $(0, -6), \left(\dfrac{147}{29}, \dfrac{120}{29}\right)$ **29.** Yes, the hyperbola intersects the circle; $(3290, 2270)$ **31.** $x^2 - 3x - 10$ **32.** $\dfrac{2x(x + 1)}{x - 2}$

33. 128,500 CD-ROMs   **34.** 25 miles per hour

Putting Your Skills to Work

1. $\dfrac{x^2}{900} + \dfrac{y^2}{36} = 1$ **2.** 9 hours   **3.** 2.25 days   **4.** 1:36 P.M.   **5.** 36 days   **6.** 61.3 hours   **7.** 2.6 days

Chapter 9 Review Problems

1. $\sqrt{73}$ **2.** $\sqrt{41}$ **3.** $(x + 6)^2 + (y - 3)^2 = 15$ **4.** $x^2 + (y + 7)^2 = 25$ **5.** $(x + 1)^2 + (y - 3)^2 = 5$; center $(-1, 3), r = \sqrt{5}$
6. $(x - 5)^2 + (y + 6)^2 = 9$; center $(5, -6), r = 3$
7.
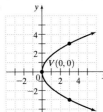
8.
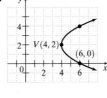
9.

10. $y = (x + 3)^2 - 5$; vertex at $(-3, -5)$; opens upward
11. $x = (y - 4)^2 - 6$; vertex at $(-6, 4)$; opens to the right

12.

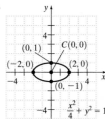

13.

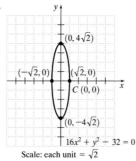

$16x^2 + y^2 - 32 = 0$

Scale: each unit $= \sqrt{2}$

14. center $(-5, -3)$; vertices are $(-3, -3)$, $(-7, -3)$, $(-5, 2)$, $(-5, -8)$

15. center $(-1, 2)$; vertices are $(2, 2)$, $(-4, 2)$, $(-1, 6)$, $(-1, -2)$

16.

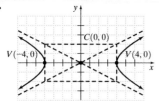

17.

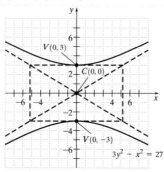

$3y^2 - x^2 = 27$

18. vertices are $(0, -3)$, $(4, -3)$; center $(2, -3)$ **19.** vertices are $(-5, 3)$, $(-5, 1)$; center $(-5, 2)$ **20.** $(-3, 0)$, $(2, 5)$

21. $(0, 2)$, $(2, 0)$ **22.** $(2, 3)$, $(-2, 3)$, $(2, -3)$, $(-2, -3)$ **23.** $(2, -1)$, $(-2, 1)$, $(1, -2)$, $(-1, 2)$ **24.** no real solution

25. $(0, 1)$, $\left(\sqrt{5}, 6\right)$, $\left(-\sqrt{5}, 6\right)$ **26.** $(1, 4)$, $(-1, -4)$, $\left(2\sqrt{2}, \sqrt{2}\right)$, $\left(-2\sqrt{2}, -\sqrt{2}\right)$ **27.** $(1, 2)$, $(-1, 2)$, $(1, -2)$, $(-1, -2)$

28. $(2, 2)$ **29.** $\left(\frac{1}{2}, -\frac{1}{2}\right)$, $(2, 1)$ **30.** 0.78 feet   **31.** 1.56 feet

How Am I Doing? Chapter 9 Test

1. $\sqrt{185}$ (obj. 9.1.1)

2. parabola; vertex $(4, 3)$; $x = (y - 3)^2 + 4$

(obj. 9.2.3)

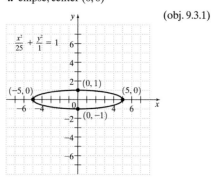

3. circle; center $(-3, 2)$; $(x + 3)^2 + (y - 2)^2 = 4$

(obj. 9.1.4)

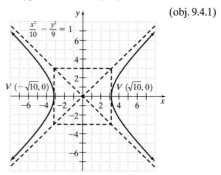

4. ellipse; center $(0, 0)$

(obj. 9.3.1)

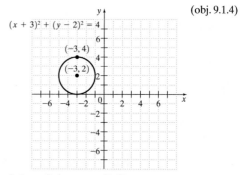

5. hyperbola; center $(0, 0)$

(obj. 9.4.1)

6. parabola; vertex $(-3, 4)$

(obj. 9.2.1)

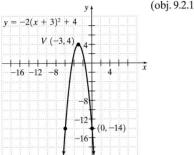

7. ellipse; center $(-2, 5)$

(obj. 9.3.2)

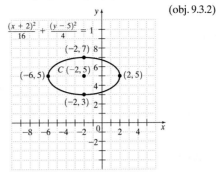

8. hyperbola; center $(0, 0)$

(obj. 9.4.1)

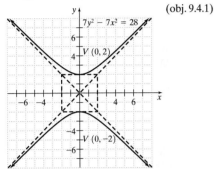

9. $(x - 3)^2 + (y + 5)^2 = 8$

(obj. 9.1.3)

10. $\dfrac{x^2}{9} + \dfrac{y^2}{25} = 1$ (obj. 9.3.1)   **11.** $x = (y - 3)^2 - 7$ (obj. 9.2.3)   **12.** $\dfrac{x^2}{9} - \dfrac{y^2}{25} = 1$ (obj. 9.4.1)   **13.** $(0, 5), (-4, -3)$ (obj. 9.5.1)

14. $(0, -3), (3, 0)$ (obj. 9.5.1)   **15.** $(1, 0), (-1, 0)$ (obj. 9.5.2)   **16.** $\left(3, -\sqrt{3}\right), \left(3, \sqrt{3}\right), \left(-3, \sqrt{3}\right), \left(-3, -\sqrt{3}\right)$ (obj. 9.5.2)

Cumulative Test for Chapters 1–9

1. Commutative property of multiplication   **2.** $8x + 12$ **3.** -19 **4.** $p = \dfrac{A - 3bt}{rt}$ **5.** $4x(x + 2)(x - 2)$ **6.** $\dfrac{3x^2 + 2x - 10}{(x - 2)(x - 1)}$

7. $x = \dfrac{7}{2}$ **8.** $x = 4, y = -3, z = 1$ **9.** $4\sqrt{3} + 6\sqrt{2} - \sqrt{6} - 3$ **10.** $8x\sqrt{3} - 3\sqrt{2x}$ **11.** $x < 0$ **12.** $x \geq 14$ **13.** $3\sqrt{10}$

14. parabola

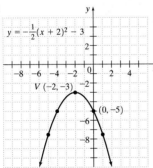

15. circle

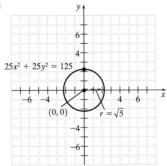

16. hyperbola

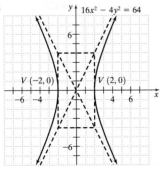

17. ellipse

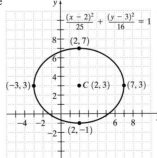

18. $(2, 8), (-1, 2)$ **19.** $(0, 1)$ **20.** $(-5, 0)(3, 4)$ **21.** $\left(\dfrac{15}{4}, -4\right)(-3, 5)$

Chapter 10

10.1 Exercises

1. -7 **3.** $3a - 17$ **5.** $\dfrac{1}{2}a - 4$ **7.** $\dfrac{3}{2}a$ **9.** $a - 5$ **11.** $\dfrac{1}{2}a^2 - \dfrac{1}{5}$ **13.** 2   **15.** $\dfrac{3}{4}$ **17.** $3a^2 + 10a + 5$ **19.** $\dfrac{4a^2}{3} - \dfrac{8a}{3} - 2$

21. 3   **23.** $2\sqrt{3}$ **25.** $\sqrt{a^2 + 4}$ **27.** $\sqrt{-2b + 5}$ **29.** $2\sqrt{a + 1}$ **31.** $\sqrt{b^2 + b + 5}$ **33.** $\dfrac{7}{4}$ **35.** 14   **37.** $\dfrac{7}{a^2 - 3}$

39. $\dfrac{7}{a - 1}$ **41.** $-\dfrac{7}{5}$ **43.** 2   **45.** $2x + h - 1$ **47. (a)** $P(w) = 2.5w^2$ **(b)** 1000 kilowatts   **(c)** $P(e) = 2.5e^2 + 100e + 1000$

(d) 1210 kilowatts   **49.** The percent would decrease by 13; 26   **51.** $0.002a^2 - 0.120a + 1.23$ **53.** $3a^2 + 8.78a + 5.891$

55. $A(x) = \dfrac{x^2 - 20x + 200}{8}$; $A(2) = 20.5$; $A(5) = 15.625$; $A(8) = 13$ **57.** $x = -3$ **58.** $x = 5$

59. approximately 17.9 times greater   **60.** approximately 26.3 times greater

10.2 Exercises

1. No, $f(x + 2)$ means to substitute $x + 2$ for x in the function $f(x)$. $f(x) + f(2)$ means to evaluate $f(x)$ and to evaluate $f(2)$ and then to add the two solutions.   **3.** up   **5.** not a function   **7.** function   **9.** function   **11.** not a function   **13.** function

15.

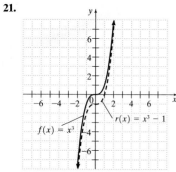

17.

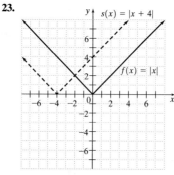

19.

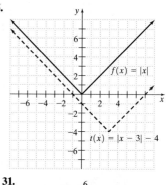

21.

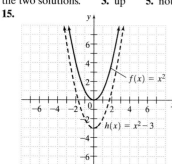

23.

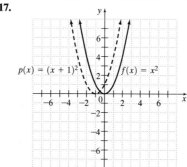

25.

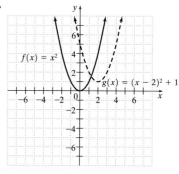

27.

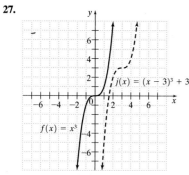

29.

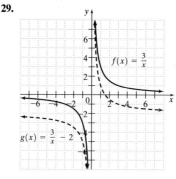

31.

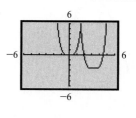

33. $15\sqrt{2} - 10\sqrt{3}$ **34.** $5x + 2\sqrt{10x} + 2$ **35.** $\dfrac{7 - 3\sqrt{5}}{4}$ **36.** 18 French fries or less   **37.** $227 per student

How Am I Doing? Sections 10.1–10.2

1. -12 (obj. 10.1.1)   **2.** $2a - 6$ (obj. 10.1.1)   **3.** $4a - 6$ (obj. 10.1.1)   **4.** $2a - 2$ (obj. 10.1.1)   **5.** 13 (obj. 10.1.1)

6. $5a^2 + 2a - 3$ (obj. 10.1.1)   **7.** $5a^2 + 12a + 4$ (obj. 10.1.1)   **8.** $45a^2 + 6a - 3$ (obj. 10.1.1)   **9.** $\dfrac{6(a^2 - 2)}{a(a + 2)}$ (obj. 10.1.1)

10. $\dfrac{18(a-1)}{5(3a+2)}$ (obj. 10.1.1)   **11.** function (obj. 10.2.1)   **12.** not a function (obj. 10.2.1)

13. (obj. 10.2.2)   **14.** (obj. 10.2.2)

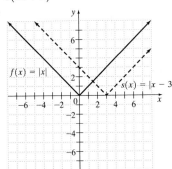

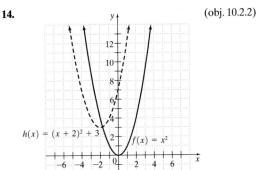

15. (obj. 10.2.2)   **16.** The curve would be the same shape exactly 2 units lower at each location. (obj. 10.2.2)

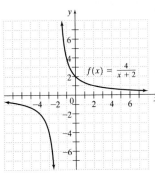

10.3 Exercises

1. (a) $2x+5$ (b) $-6x+1$ (c) 9 (d) 7 **3.** (a) $3x^2+4x+2$ (b) $3x^2-6x-2$ (c) 22 (d) 7 **5.** (a) $x^3+\dfrac{1}{2}x^2+\dfrac{3}{4}x-5$

(b) $x^3-\dfrac{3}{2}x^2+\dfrac{5}{4}x+5$ (c) $\dfrac{13}{2}$ (d) $\dfrac{5}{4}$ **7.** (a) $3\sqrt{x+6}$ (b) $-13\sqrt{x+6}$ (c) $6\sqrt{2}$ (d) $-13\sqrt{5}$

9. (a) $-4x^3 \pm 11x-3$ (b) 72 **11.** (a) $\dfrac{2(x-1)}{x}$ (b) $\dfrac{8}{3}$ **13.** (a) $-3x\sqrt{-2x+1}$ (b) $9\sqrt{7}$ **15.** (a) $\dfrac{x-6}{3x}, x \neq 0$ (b) $-\dfrac{2}{3}$

17. (a) $x+1, x \neq 1$ (b) 3 **19.** (a) $x+5, x \neq -5$ (b) 7 **21.** (a) $\dfrac{1}{x+2}, x \neq -2, x \neq \dfrac{1}{4}$ (b) $\dfrac{1}{4}$ **23.** $-x^2+5x+2$

25. $3x, x \neq 2$ **27.** -3 **29.** -3 **31.** $-6x-13$ **33.** $2x^2-4x+7$ **35.** $-5x^2-7$ **37.** $\dfrac{7}{2x+1}, x \neq -\dfrac{1}{2}$ **39.** $|2x+2|$

41. $9x^2+30x+27$ **43.** $3x^2+11$ **45.** 11 **47.** $\sqrt{x^2+1}$ **49.** $\dfrac{3\sqrt{2}}{2}+5$ **51.** $\sqrt{26}$ **53.** $K[C(F)]=\dfrac{5F+2297}{9}$

55. $v[r(h)]=384.65h^2$; 24,617.6 cubic feet   **57.** $(6x-1)^2$ **58.** $(5x^2+1)(5x^2-1)$ **59.** $(x+3)(x-3)(x+1)(x-1)$
60. $(3x-1)(x-2)$ **61.** 8 commercials that are 60 seconds long and 12 commercials that are 30 seconds long
62. 115 children; 335 adults

10.4 Exercises

1. have the same second coordinate.   **3.** $y=x$ **5.** Yes, it passes the vertical line test. No, it does not pass the horizontal line test.
7. not one-to-one   **9.** one-to-one   **11.** one-to-one   **13.** one-to-one   **15.** not one-to-one   **17.** not one-to-one

19. $J^{-1}=\{(2,8),(1,1),(0,0),(-2,-8)\}$ **21.** $f^{-1}(x)=\dfrac{x+5}{4}$ **23.** $f^{-1}(x)=\sqrt[3]{x-7}$

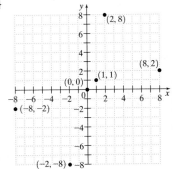

25. $f^{-1}(x)=-\dfrac{4}{x}$ **27.** $f^{-1}(x)=\dfrac{4}{x}+5$ or $\dfrac{4+5x}{x}$

29. $g^{-1}(x) = \dfrac{x-5}{2}$

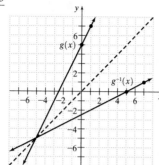

31. $h^{-1}(x) = 2x + 4$

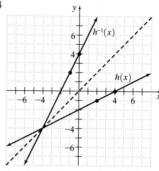

33. $r^{-1}(x) = -\dfrac{x+1}{3}$

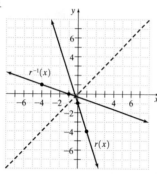

35. $f^{-1}(x) = \dfrac{x+4}{1.437}$; The inverse function would tell how many Irish pounds were given by the bank for x dollars. No. Because of the bank fee, it would not work for Sean's transaction.

37. No. $f(x)$ is not one-to-one.   **39.** $f[f^{-1}(x)] = 2\left[\dfrac{1}{2}x - \dfrac{3}{4}\right] + \dfrac{3}{2} = x$; $f^{-1}[f(x)] = \dfrac{1}{2}\left[2x + \dfrac{3}{2}\right] - \dfrac{3}{4} = x$ **41.** $x = 3$

42. $x = -64$; $x = -27$ **43.** $23:21$ **44.** 13 overtime hours   **45.** 800,000 people   **46.** Approximately 19.6%

Putting Your Skills to Work

1. 309 pounds   **2.** 886 pounds   **3.** 1238 pounds   **4.** 2320 pounds   **5.** 2675 pounds   **6.** 3757 pounds   **7.** 2726 pounds
8. 3416 pounds

Chapter 10 Review Problems

1. $\dfrac{1}{2}a + \dfrac{5}{2}$ **2.** $\dfrac{1}{2}a + 4$ **3.** $-\dfrac{1}{2}$ **4.** 1   **5.** $\dfrac{1}{2}b^2 + \dfrac{3}{2}$ **6.** $2b^2 + \dfrac{7}{2}$ **7.** -28 **8.** -6 **9.** $-8a^2 + 6a - 16$

10. $-18a^2 - 9a - 1$ **11.** $-2a^2 - 5a - 3$ **12.** $-2a^2 + 15a - 28$ **13.** 1   **14.** 11   **15.** $\left|\dfrac{1}{2}a - 1\right|$ **16.** $|3a - 1|$

17. $|2a^2 + 2a - 1|$ **18.** $|4a^2 - 6a - 1|$ **19.** $\dfrac{5}{3}$ **20.** 9   **21.** $\dfrac{6a - 15}{2a - 1}$ **22.** $\dfrac{3 - 3a}{5 - a}$ **23.** $\dfrac{30a + 36}{7a + 28}$ **24.** $-\dfrac{12}{a + 4}$ **25.** 7

26. 6   **27.** $4x + 2h - 5$ **28.** $-6x - 3h + 2$ **29.** (a) function (b) one-to-one   **30.** (a) not a function (b) not one-to-one
31. (a) function (b) not one-to-one   **32.** (a) function (b) not one-to-one   **33.** (a) not a function (b) not one-to-one
34. (a) function (b) one-to-one   **35.** $g(x) = (x + 2)^2 + 4$ **36.**

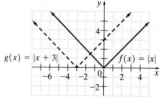

37.

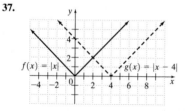

38.

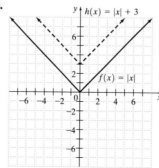

39.

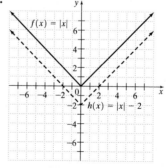

40.

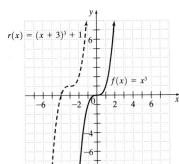

41.

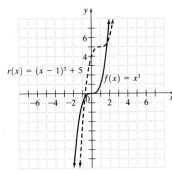

42.

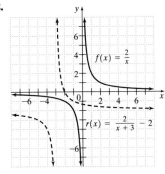

43.

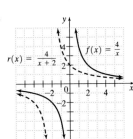

44. $\frac{5}{2}x + 2$ **45.** $2x^2 + 9$ **46.** $2x^2 - 6x - 1$ **47.** $-\frac{7}{2}x - 8$ **48.** -5 **49.** $\frac{5}{2}$

50. $\frac{6x + 10}{x}$, $x \neq 0$ **51.** $-x^3 - \frac{9}{2}x^2 + 7x - 12$ **52.** $\frac{2x - 8}{x^2 + x}$, $x \neq 0$, $x \neq -1$, $x \neq 4$ **53.** $\frac{2}{3x^2 + 5x}$, $x \neq 0$, $x \neq -\frac{5}{3}$ **54.** -6

55. $\frac{1}{6}$ **56.** $18x^2 + 51x + 39$ **57.** $-\frac{1}{2}\sqrt{x - 2} - 3$ **58.** $\sqrt{2x^2 - 3x + 2}$ **59.** $\sqrt{-\frac{1}{2}x - 5}$, $x \leq -10$ **60.** 2 **61.** 2

62. $f[g(x)] = \frac{6}{x} + 5 = \frac{6 + 5x}{x}$; $g[f(x)] = \frac{2}{3x + 5}$; $f[g(x)] \neq g[f(x)]$

63. $p[g(x)] = \frac{8}{x^2} - \frac{6}{x} + 4 = \frac{8 - 6x + 4x^2}{x^2}$; $g[p(x)] = \frac{2}{2x^2 - 3x + 4}$; $p[g(x)] \neq g[p(x)]$

64. (a) domain = $\{0, 3, 7\}$ (b) range = $\{-8, 3, 7, 8\}$ (c) not a function   (d) not one-to-one

65. (a) domain = $\{100, 200, 300, 400\}$ (b) range = $\{10, 20, 30\}$ (c) function   (d) not one-to-one

66. (a) domain = $\left\{-\frac{1}{3}, \frac{1}{4}, \frac{1}{2}, 4\right\}$ (b) range = $\left\{-3, \frac{1}{4}, 2, 4\right\}$ (c) function   (d) one-to-one

67. (a) domain = $\{-6, 0, 12\}$ (b) range = $\{-12, -1, 6\}$ (c) not a function   (d) not one-to-one

68. (a) domain = $\{0, 1, 2, 3\}$ (b) range = $\{-3, 1, 7\}$ (c) function   (d) not one-to-one

69. (a) domain = $\{-1, 0, 1, 2\}$ (b) range = $\{-2, 1, 2, 9\}$ (c) function   (d) one-to-one

70. $A^{-1} = \left\{\left(\frac{1}{3}, 3\right), \left(-\frac{1}{2}, -2\right), \left(-\frac{1}{4}, -4\right), \left(\frac{1}{5}, 5\right)\right\}$ **71.** $B^{-1} = \{(10, 1), (7, 3), (15, 12), (1, 10)\}$ **72.** $f^{-1}(x) = -\frac{4}{3}x + \frac{8}{3}$

73. $g^{-1}(x) = -\frac{1}{4}x - 2$ **74.** $h^{-1}(x) = \frac{6}{x} - 5$ or $\frac{6 - 5x}{x}$ **75.** $j^{-1}(x) = \frac{7}{x} + 2$ or $\frac{2x + 7}{x}$ **76.** $p^{-1}(x) = x^3 - 1$ **77.** $r^{-1}(x) = \sqrt[3]{x - 2}$

78. $f^{-1}(x) = -3x - 2$

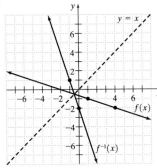

79. $f^{-1}(x) = -\frac{4}{3}x + \frac{4}{3}$

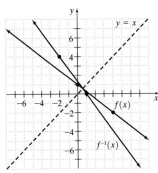

How Am I Doing? Chapter 10 Test

1. -8 (obj. 10.1.1)   **2.** $\frac{9}{4}a - 2$ (obj. 10.1.1)   **3.** $\frac{3}{4}a - \frac{3}{2}$ (obj. 10.1.1)   **4.** 124 (obj. 10.1.1)   **5.** $3a^2 + 4a + 5$ (obj. 10.1.1)

6. $3a^2 - 2a + 9$ (obj. 10.1.1)   **7.** $12a^2 + 4a + 2$ (obj. 10.1.1)   **8.** (a) function (obj. 10.2.1)   (b) not one-to-one (obj. 10.4.1)

9. (a) function (obj. 10.2.1)   **(b)** one-to-one (obj. 10.4.1)

10. (obj. 10.2.2)   **11.** (obj. 10.2.2)

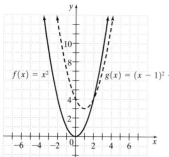

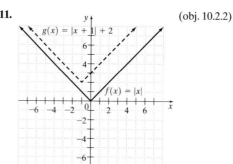

12. (a) $x^2 + 4x + 1$ (obj. 10.3.1)   **(b)** $5x^2 - 6x - 13$ (obj. 10.3.1)   **(c)** 19 (obj. 10.3.1)

13. (a) $\dfrac{6x - 3}{x}, x \neq 0$ (obj. 10.3.1)   **(b)** $\dfrac{3}{2x^2 - x}, x \neq 0, x \neq \dfrac{1}{2}$ (obj. 10.3.1)   **(c)** $\dfrac{3}{2x - 1}, x \neq \dfrac{1}{2}$ (obj. 10.3.2)

14. (a) $2x - \dfrac{1}{2}$ (obj. 10.3.2)   **(b)** $2x - 7$ (obj. 10.3.2)   **(c)** 0 (obj. 10.3.2)

15. (a) one-to-one (obj. 10.4.1)   **(b)** $B^{-1} = \{(8, 1), (1, 8), (10, 9), (9, -10)\}$ (obj. 10.4.2)

16. (a) not one-to-one (obj. 10.4.1)   **(b)** inverse cannot be found (obj. 10.4.2)

17. $f^{-1}(x) = \dfrac{x^3 + 1}{2}$ (obj. 10.4.2)   **18.** $f^{-1}(x) = -\dfrac{1}{3}x + \dfrac{2}{3}$ (obj. 10.4.3)

19. $f^{-1}[f(x)] = x$ (obj. 10.3.2)   **20.** -8 (obj. 10.1.1)

Cumulative Test for Chapters 1–10

1. $-27x^2 - 30xy$ **2.** 14   **3.** $x = 2$ **4.** $(4x^2 + 1)(2x + 1)(2x - 1)$ **5.** $2x^3 - x^2 - 6x + 5$ **6.** $x = -\dfrac{12}{5}$

7. $y = -3x + 5$ **8.** $(3, -2)$ **9.** $3x^2y^3z\sqrt{2xz}$ **10.** $-2\sqrt{6} - 8$ **11.** $3\sqrt{10}$ **12.** $(4x - 1)(3x - 2)$

13. $x(x - 7)(x + 2)$ **14.** $x^2 + (y + 5)^2 = 8$ **15. (a)** 17   **(b)** $3a^2 - 14a + 17$ **(c)** $3a^2 - 2a + 18$

16.   **17. (a)** $10x^3 - 19x^2 - 45x - 18$ **(b)** $\dfrac{2x^2 - 5x - 6}{5x + 3}, x \neq -\dfrac{3}{5}$ **(c)** $50x^2 + 35x - 3$

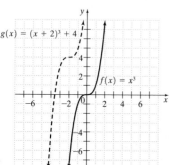

18. (a) function   **(b)** one-to-one   **(c)** $A^{-1} = \{(6, 3), (8, 1), (7, 2), (4, 4)\}$ **19.** $f^{-1}(x) = \dfrac{x^3 + 3}{7}$

20. (a) 544   **(b)** -168 **(c)** $40a^3 - 12a^2 - 6$ **21. (a)** $f^{-1}(x) = -\dfrac{3}{2}x + 3$ **(b)**   **22.** $f[f^{-1}(x)] = x$

Chapter 11

11.1 Exercises

1. $f(x) = b^x$, where $b > 0, b \neq 1$, and x is a real number.   **3.** $f(x) = 3^x$

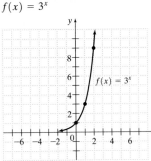

5. $f(x) = 2^{-x}$

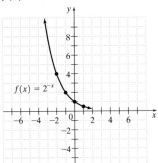

7. $f(x) = 3^{-x}$

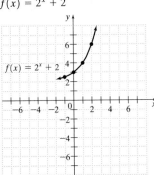

9.

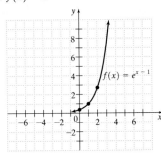

11.

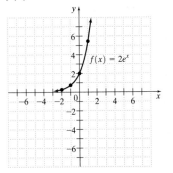

13. $f(x) = 2^x + 2$

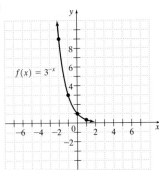

15. $f(x) = e^{x-1}$

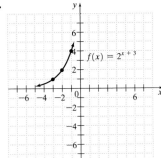

17. $f(x) = 2e^x$

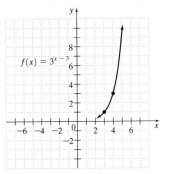

19. $f(x) = e^{1-x}$

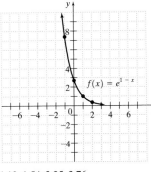

21. $x = 2$ **23.** $x = 0$ **25.** $x = -1$ **27.** $x = 4$ **29.** $x = 0$ **31.** $x = 2$
33. $x = 4$ **35.** $x = 2$ **37.** $x = 1$ **39.** \$2402.31   **41.** \$3632.24; \$3634.08
43. 32,000; 2,048,000   **45.** 37%; yes   **47.** 3.91 milligrams
49. 1.80 pounds per square inch   **51.** about 50.4 million; about 160.8 million; about
219%.   **53.** 1955   **55.** 6.7 billion people

57. 1.54; 1.13; 1; 1.13; 1.54; 2.35; 3.76

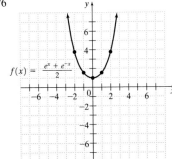

59. $x = -6$ **60.** $x = -2$

11.2 Exercises

1. exponent   **3.** $x > 0$ **5.** $\log_7 49 = 2$ **7.** $\log_6 36 = 2$ **9.** $\log_{10} 0.001 = -3$ **11.** $\log_2\left(\dfrac{1}{32}\right) = -5$ **13.** $\log_e y = 5$

15. $3^2 = 9$ **17.** $17^0 = 1$ **19.** $16^{1/2} = 4$ **21.** $10^{-2} = 0.01$ **23.** $3^{-4} = \dfrac{1}{81}$ **25.** $e^{-\frac{3}{2}} = x$ **27.** $x = 16$ **29.** $x = \dfrac{1}{1000}$

31. $y = 3$ **33.** $y = -2$ **35.** $a = 11$ **37.** $a = 10$ **39.** $w = \dfrac{1}{2}$ **41.** $w = -1$ **43.** $w = 1$ **45.** $w = 9$ **47.** -3 **49.** 7

51. 0 **53.** $\dfrac{1}{2}$ **55.** 0 **57.**

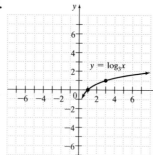

59.

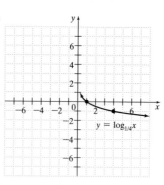

61.

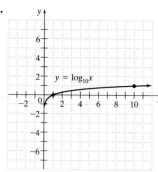

63.

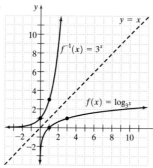

65. 2 **67.** 10^{-8} **69.** 2.957 **71.** $11,200$ sets
73. $\$10,000,000$ **75.** 4

77.

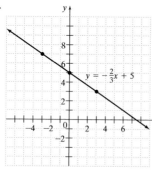

78.

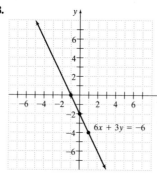

79. $y = \dfrac{3}{2}x + 7$ **80.** $m = -\dfrac{1}{5}$

81. **(a)** $36,000$ cells   **(b)** $36,864,000$ cells
82. **(a)** $\$5353.27$ **(b)** $\$24,424.03$
83. the train; $\$68$ per month   **84.** 112 children

11.3 Exercises

1. $\log_3 A + \log_3 B$ **3.** $\log_5 7 + \log_5 11$ **5.** $\log_b 9 + \log_b f$ **7.** $\log_9 2 - \log_9 7$ **9.** $\log_b H - \log_b 10$ **11.** $\log_a E - \log_a F$

13. $7\log_8 a$ **15.** $-2\log_b A$ **17.** $\dfrac{1}{2}\log_5 w$ **19.** $2\log_8 x + \log_8 y$ **21.** $\log_{11} 6 + \log_{11} M - \log_{11} N$

23. $\log_2 5 + \log_2 x + 4\log_2 y - \dfrac{1}{2}\log_2 z$ **25.** $\dfrac{4}{3}\log_a x - \dfrac{1}{3}\log_a y$ **27.** $\log_4 39y$ **29.** $\log_3\left(\dfrac{x^5}{7}\right)$ **31.** $\log_b\left(\dfrac{49y^3}{\sqrt{z}}\right)$ **33.** 1 **35.** 1

37. 0 **39.** 3 **41.** $x = 7$ **43.** $x = 11$ **45.** $x = 0$ **47.** $x = 1$ **49.** $x = 4$ **51.** $x = 21$ **53.** $x = 2$ **55.** $x = 5e$
57. $x = 3$ **59.** 6 **63.** approximately 62.83 cubic meters   **64.** approximately 50.27 square meters   **65.** $(3, -2)$ **66.** $(-1, -2, 3)$
67. about 16.1%; about 1.08×10^9 metric tons   **68.** about 10.2%; about 2.45×10^8 metric tons   **69.** approximately 8.64 miles per hour; no
70. approximately 7.22 seconds; approximately 323.18 feet

How Am I Doing? Sections 11.1–11.3

1.

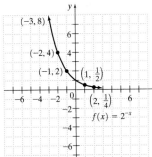

(obj. 11.1.1)   **2.** $x = 2$ (obj. 11.1.2)   **3.** $x = -5$ (obj. 11.1.2)   **4.** $x = -\dfrac{1}{3}$ (obj. 11.1.2)

5. \$15,735.19 (obj. 11.1.3)   **6.** $\log_7\left(\dfrac{1}{49}\right) = -2$ (obj. 11.2.1)   **7.** $10^{-3} = 0.001$ (obj. 11.2.2)

8. $x = 125$ (obj. 11.2.3)   **9.** $x = \dfrac{1}{9}$ (obj. 11.2.3)   **10.** 4 (obj. 11.2.3)

11. $2\log_5 x + 5\log_5 y - 3\log_5 z$ (obj. 11.3.3)   **12.** $\log_4\left(\dfrac{\sqrt{x}}{w^3}\right)$ (obj. 11.3.2)

13. $x = \dfrac{81}{2}$ (obj. 11.3.4)   **14.** $x = 8$ (obj. 11.3.4)   **15.** $x = 0$ (obj. 11.3.4)   **16.** $x = \dfrac{9}{2}$ (obj. 11.3.4)

17. $x = \dfrac{4}{3}$ (obj. 11.3.4)   **18.** $x = \dfrac{e}{3}$ (obj. 11.3.4)

11.4 Exercises

1. Error. You cannot take the log of a negative number.   **3.** 1.089905111   **5.** 1.408239965   **7.** 0.903089987   **9.** 5.096910013
11. −1.910094889   **13.** 103.752842   **15.** 0.440352027   **17.** 8519.223264   **19.** 2,939,679.609   **21.** 0.000408037   **23.** 0.027089438
25. 41,831,168.87   **27.** 0.082679911   **29.** 1.726331664   **31.** 0.425297735   **33.** 11.82041016   **35.** −5.15162299   **37.** 2.585709659
39. 11.02317638   **41.** 0.951229425   **43.** 0.067205513   **45.** 472.576671   **47.** 0.1188610637   **49.** 2.020006063   **51.** 1.02507318
53. −1.151699337   **55.** 0.91759992   **57.** −1.846414   **59.** 1.996254706   **61.** 7.3375877   **63.** 0.1535084   **65.** 3.6593167×10^8
67. 3.3149796   **69.** 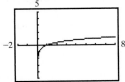   **71.**   **73.** 35.18; 35.89; approximately 2.0%

75. $R \approx 4.75$ **77.** The shock wave is about 3,981,000 times greater than the smallest detectable shock wave.   **79.** $x = \dfrac{11 \pm \sqrt{181}}{6}$

80. $y = \dfrac{-2 \pm \sqrt{10}}{2}$ **81.** 6 miles; 12 miles   **82.** 21 miles; 24 miles

11.5 Exercises

1. $x = 20$ **3.** $x = 6$ **5.** $x = 4$ **7.** $x = 2$ **9.** $x = \dfrac{3}{2}$ **11.** $x = 2$ **13.** $x = 3$ **15.** $x = 5$ **17.** $x = \dfrac{5}{3}$ **19.** $x = 4$

21. $x = 5$ **23.** $x = \dfrac{\log 12 - 3\log 7}{\log 7}$ **25.** $x = \dfrac{\log 17 - 4\log 2}{3\log 2}$ **27.** $x \approx 1.582$ **29.** $x \approx 6.213$ **31.** $x \approx 5.332$ **33.** $x \approx 1.739$

35. $t \approx 16$ years   **37.** $t \approx 19$ years   **39.** 4.5%   **41.** 27 years   **43.** 35 years   **45.** 437,000 employees   **47.** 2001   **49.** 27 years
51. 55 hours   **53.** 46,931 people   **55.** about 12.6 times greater   **57.** about 31.6 times greater   **59.** 17.8 years
61. $3\sqrt{2} + 4\sqrt{3} - \sqrt{6} - 4$ **62.** $7xy\sqrt{2x}$ **63.** 9 years old, 1 student; 10 years old, 8 students; 11 years old, 18 students; 12 years old,
54 students; 13 years old, 87 students; 14 or 15 years old, 80 students   **64.** approximately \$456 million per mile

Putting Your Skills to Work

1. 0.9%   **2.** 1.5%   **3.** 908,765 million dollars   **4.** 13,843 million dollars   **5.** onlines sales (in billions) $= 65(1.17)^n$
6.

Year	Estimated Online Retail Sales (in billions of dollars)
2005	76.05
2006	88.98
2007	104.10
2008	121.80

Chapter 11 Review Problems

1.

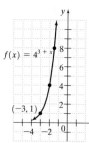

$f(x) = 4^{3+x}$

$(-3, 1)$

2.

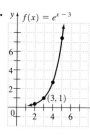

$f(x) = e^{x-3}$

$(3, 1)$

3. $x = 1$ **4.** $10^{-2} = 0.01$ **5.** $\log_4 8 = \dfrac{3}{2}$ **6.** $w = 2$ **7.** $x = \dfrac{1}{9}$ **8.** $x = 1$

9. $w = \dfrac{1}{7}$ **10.** $w = 4$ **11.** $w = 0.1$ or $\dfrac{1}{10}$ **12.** $x = 3$ **13.** $x = 6$ **14.** $x = -2$ **15.** $x = 5$

16.

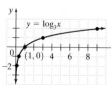

$y = \log_3 x$

$(1, 0)$

17. $\log_2 5 + \log_2 x - \dfrac{1}{2}\log_2 w$ **18.** $3\log_2 x + \dfrac{1}{2}\log_2 y$ **19.** $\log_3\left(\dfrac{x\sqrt{w}}{2}\right)$ **20.** $\log_8\left(\dfrac{w^4}{\sqrt[3]{z}}\right)$

21. 6   **22.** $x = 25$ **23.** $x = 25$ **24.** 1.376576957   **25.** -1.087777943 **26.** 1.366091654   **27.** 6.688354714

28. $n = 13.69935122$ **29.** $n = 5.473947392$ **30.** 0.49685671   **31.** $x = 25$ **32.** $x = 12$ **33.** $x = \dfrac{1}{7}$ **34.** $x = 2$ **35.** $x = 3$

36. $x = 4$ **37.** $t = \dfrac{3}{4}$ **38.** $t = -\dfrac{1}{8}$ **39.** $x = \dfrac{\log 14}{\log 3}$ **40.** $x = \dfrac{\log 130 - 3\log 5}{\log 5}$ **41.** $x = \dfrac{\ln 100 + 1}{2}$ **42.** $x = \dfrac{\ln 30.6}{2}$

43. $x \approx -1.4748$ **44.** $x \approx 0.7712$ **45.** $x \approx 2.3319$ **46.** $x \approx 101.3482$ **47.** 9 years   **48.** \$6312.38   **49.** 8 years   **50.** 9 years
51. 41 years   **52.** 26 years   **53.** 9 years   **54.** 11 years   **55. (a)** approximately 282 pounds   **(b)** 7.77 pounds per cubic inch
56. 50.1 times more

How Am I Doing? Chapter 11 Test

1.

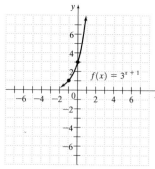

$f(x) = 3^{x+1}$

(obj. 11.1.1)

2.

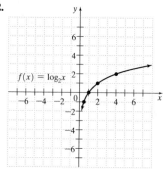

$f(x) = \log_2 x$

(obj. 11.2.4)

3. $x = 0$ (obj. 11.1.2)

4. $w = 5$ (obj. 11.2.2)   **5.** $x = \dfrac{1}{64}$ (obj. 11.2.2)   **6.** $\log_7\left(\dfrac{x^2 y}{4}\right)$ (obj. 11.3.3)   **7.** 1.7901 (obj. 11.4.3)   **8.** 1.3729 (obj. 11.4.1)

9. 0.4391 (obj. 11.4.5)   **10.** $x \approx 5350.569382$ (obj. 11.4.2)   **11.** $x \approx 1.150273799$ (obj. 11.4.4)   **12.** $x = \dfrac{3}{7}$ (obj. 11.5.1)

13. $x = \dfrac{16}{3}$ (obj. 11.5.1)   **14.** $x = \dfrac{\ln 57 + 3}{5}$ (obj. 11.5.2)   **15.** $x \approx -1.4132$ (obj. 11.5.2)   **16.** \$2938.66 (obj. 11.5.3)

17. 14 years (obj. 11.5.3)

Cumulative Test for Chapters 1–11

1. 6   **2.** $x = \dfrac{5y - 1}{2m}$ **3.**

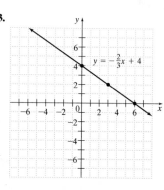

$y = -\dfrac{2}{3}x + 4$

4. $(x + y)(5a - 7w)$ **5.** $(1, -1, 2)$ **6.** $18 + 2\sqrt{21}$

7. $x = \pm\sqrt{6}; x = \pm i$ **8.** $x = 4, y = 4; x = 1, y = -2$ **9.** $x = 4$ **10.** $\dfrac{x\sqrt{6}}{2}$

11.
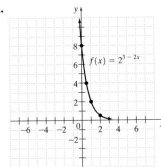

12. $x = \dfrac{1}{4}$ **13.** $x = \dfrac{3}{2}$ **14.** 0.884795364 **15.** $x \approx 66.20640403$ **16.** 1.771243749

17. $x \approx 7.1263558$ **18.** $x = 9$ **19.** $x = 2$ **20.** $x = -0.535$ **21.** $x = \dfrac{\ln 0.5}{2}$

22. $4234.74

Practice Final Examination

1. 4   **2.** 3.625×10^7 **3.** $2a + 7b + 5ab + 3a^2$ **4.** $-9x - 15y$ **5.** $F = -31$ **6.** $y = -30$ **7.** $b = \dfrac{2A - ac}{a}$

8. $x = 3, x = 9$ **9.** $x < 1$ **10.** Length $= 520$ m; width $= 360$ m   **11.** $2600 at 12%; $1400 at 14%   **12.** $x \le -9$ or $x \ge -2$

13. $-\dfrac{5}{2} < x < \dfrac{15}{2}$

14. x-intercept $(-2, 0)$; y-intercept $(0, 7)$

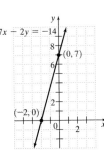

15.

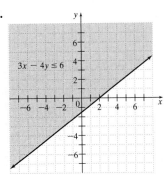

16. $m = \dfrac{8}{3}$

17. $3x + 2y = 5$ **18.** $f(3) = 12$ **19.** $f(-2) = 17$ **20.**

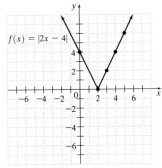

21. $(5, 5)$ **22.** $(6, 4)$

23. $(1, 4, -2)$ **24.** No solution   **25.** 11 adult, 4 children   **26.**

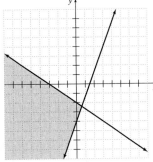

27. $6x^3 - 16x^2 + 17x - 6$

28. $5x^2 - x + 2$ **29.** $(2x - 3)(4x^2 + 6x + 9)$ **30.** $(x + 2)(x + 2)(x - 2)$ **31.** $x(2x - 1)(x + 8)$ **32.** $x = -6, x = -9$

33. $\dfrac{x(3x - 1)}{x - 3}$ **34.** $\dfrac{x + 5}{x}$ **35.** $\dfrac{3x^2 + 6x - 2}{(x + 5)(x + 2)}$ **36.** $\dfrac{3 - 2x}{5x + 2}$ **37.** $x = -1$ **38.** 64   **39.** $2a^2 b^3\sqrt{11bc}$

40. $18\sqrt{2}$ **41.** $\dfrac{5(\sqrt{7} + 2)}{3}$ **42.** $8i$ **43.** $x = -3$ **44.** $y = 33.75$ **45.** $x = \dfrac{1 \pm \sqrt{21}}{10}$ **46.** $x = 0, x = -\dfrac{3}{5}$

47. $x = 8, x = -343$ **48.** $x \le -\dfrac{1}{3}$ or $x \ge 4$ **49.**

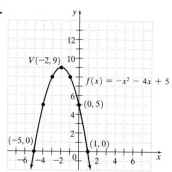

50. Width $= 4$ cm; length $= 13$ cm   **51.** $(x + 3)^2 + (y - 2)^2 = 4$; center at $(-3, 2)$; radius $= 2$

52. $\dfrac{x^2}{16} + \dfrac{y^2}{25} = 1$; ellipse

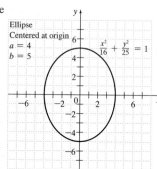

53. $\dfrac{x^2}{4} - \dfrac{y^2}{9} = 1$; hyperbola

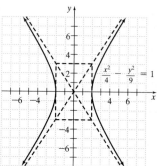

54. Parabola opening right

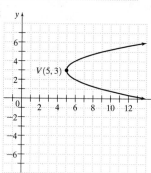

55. $(0, -4), \left(\sqrt{7}, 3\right), \left(-\sqrt{7}, 3\right)$

56. $f(-1) = 10; f(a) = 3a^2 - 2a + 5; \quad f(a + 2) = 3a^2 + 10a + 13$

57. $f[g(x)] = 80x^2 + 80x + 17$

58. $f^{-1}(x) = 2x + 14$

59. No   **60.** $f(x) = 2^{1-x}$

x	0	1	−2	4	−1
y	2	1	8	$\frac{1}{8}$	4

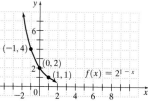

61. $x = 0$ **62.** $21 = x$ **63.** $y = -2$ **64.** $x = 8$

Appendix B

Exercises

1. -7 **3.** 15 **5.** 2 **7.** 47 **9.** 18 **11.** 0 **13.** 0 **15.** -0.6 **17.** $-7a - 4b$ **19.** $\dfrac{11}{84}$ **21.** $\begin{vmatrix} 6 & 10 \\ -5 & 9 \end{vmatrix}$ **23.** $\begin{vmatrix} 3 & -4 \\ 1 & -5 \end{vmatrix}$

25. -7 **27.** -26 **29.** 11 **31.** -27 **33.** -8 **35.** 0 **37.** -3.179 **39.** $18{,}553$ **41.** $x = 2; y = 3$ **43.** $x = -2; y = 5$

45. $x = 10; y = 2$ **47.** $x = 4; y = -2$ **49.** $x = 1.5795; y = -0.0902$ **51.** $x = 1; y = 1; z = 1$ **53.** $x = -\dfrac{1}{2}; y = \dfrac{1}{2}; z = 2$

55. $x = 4; y = -2; z = 1$ **57.** $x = -0.219; y = 1.893; z = -3.768$ **59.** $w = -3.105; x = 4.402; y = 15.909; z = 6.981$

Appendix C

Exercises

1. $(4, -1)$ **3.** $(3, -9)$ **5.** $(0, 3)$ **7.** $(2, 2)$ **9.** $(1.2, 3.7)$ **11.** $(3, -1, 4)$ **13.** $(1, -1, 3)$ **15.** $(0, -2, 5)$ **17.** $(0.5, -1, 5)$

19. $(3.6, 1.8, 2.4)$ **21.** $(4.2, -3.6, 8.8, 5.4)$

Applications Index

Subject Index

Photo Credits